QUARTZ CEMENTATION
IN SANDSTONES

SPECIAL PUBLICATION NUMBER 29 OF THE
INTERNATIONAL ASSOCIATION OF SEDIMENTOLOGISTS

# Quartz Cementation in Sandstones

EDITED BY

RICHARD H. WORDEN AND SADOON MORAD

Blackwell
Science

and published for them by
Blackwell Science Ltd
Editorial Offices:
Osney Mead, Oxford OX2 0EL
25 John Street, London WC1N 2BL
23 Ainslie Place, Edinburgh EH3 6AJ
350 Main Street, Malden
MA 02148 5018, USA
54 University Street, Carlton
Victoria 3053, Australia
10, rue Casimir Delavigne
75006 Paris, France

Other Editorial Offices:
Blackwell Wissenschafts-Verlag GmbH
Kurfürstendamm 57
10707 Berlin, Germany

Blackwell Science KK
MG Kodenmacho Building
7–10 Kodenmacho Nihombashi
Chuo-ku, Tokyo 104, Japan



First published 2000

Set by Graphicraft Ltd, Hong Kong
Printed and bound in Great Britain at the Alden Press, Oxford and Northampton



DISTRIBUTORS
Marston Book Services Ltd
PO Box 269
Abingdon, Oxon OX14 4YN
(*Orders*: Tel: 01235 465500
Fax: 01235 465555)

USA
Blackwell Science, Inc.
Commerce Place
350 Main Street
Malden, MA 02148 5018
(*Orders*: Tel: 800 759 6102
781 388 8250
Fax: 781 388 8255)

Canada
Login Brothers Book Company
324 Saulteaux Crescent
Winnipeg, Manitoba R3J 3T2
(*Orders*: Tel: 204 837 2987)

Australia
Blackwell Science Pty Ltd
54 University Street
Carlton, Victoria 3053
(*Orders*: Tel: 3 9347 0300
Fax: 3 9347 5001)

A catalogue record for this title is available from the British Library

ISBN 0-632-05482-4

Library of Congress
Cataloging-in-publication Data

Quartz cementation in sandstones/edited by R. Worden and S. Morad.
p. cm.—(Special publication number 29 of the International Association of Sedimentologists.)
Includes bibliographical references.
ISBN 0-632-05482-4
1. Sandstone. 2. Cementation (Petrology) 3. Quartz.
I. Worden, R. (Richard H.) II. Morad, S. (Sadoon)
III. Series: Special publication . . . of the International Association of Sedimentologists; no. 29.
QE471.15.S25Q32 2000
552′.5—dc21 99-32636
CIP

For further information on Blackwell Science, visit our website:
www.blackwell-science.com

# Contents

*Please note: Colour plates facing p. 98, p. 218, p. 254, p. 290 and p. 326.*

# Introduction

Quartz is a major porosity-destroying cement in many sandstones. Despite its simple chemistry and crystallography, it is the source of many conflicts and disagreements within the petrographic community about practically every aspect of its genesis. The ultimate goal of research into quartz cement is the ability to make robust predictions of the amount and distribution of quartz cement in the full spectrum of sedimentary-basin settings for the major lithofacies, and for the wide variety of burial, pressure and thermal histories. It was the primary objective of this book to focus attention on this debate, thus allowing future research to tackle the most critical elements.

The idea for this book was conceived originally at an international workshop on quartz cement held at the Queen's University of Belfast in the UK and convened by Richard Worden in May 1996. There were a total of 26 presentations at the workshop representing 12 countries from four continents. The IAS Special Publication on *Carbonate Cementation in Sandstones* (Number 26), edited by Sadoon Morad, was at an advanced stage of preparation at that time and an equivalent book on one of the other main diagenetic minerals in sandstones seemed to be a natural corollary. Many papers presented at the workshop are included in this volume. Other papers were specifically invited and solicited to add balance to the content of the volume.

The readership of the book will include: (i) sedimentologists and petrographers who are dealing with the occurrence and importance of quartz cement in sandstones; and (ii) geochemists attempting to unravel the factors controlling quartz cementation, taking into consideration an assessment of the rates and mechanisms of a wide variety of geochemical processes. The book will also be of immediate relevance to the wider petroleum geoscience community since quartz cement typically reduces the storage capacity for petroleum in deeply buried sandstone reservoirs and the rate at which petroleum can be produced.

RICHARD H. WORDEN
SADOON MORAD

# Acknowledgements

A silent minority involved in all books and publications of this type are the reviewers. Without their freely given time and effort, the book could not have been completed. We would therefore like to acknowledge the following reviewers of the 21 chapters in this book:

Stuart Barclay
Morten Bergan
Per Arne Bjørkum
Carsten Büker
Christelle Demars
Bob Ehrlich
Jon Evans
Jean-Pierre Girard
Colin Graham
Jon Gluyas
Jim Hendry
James Hickey
Ian Hutcheon
Jens Jahren
Attila Juhász
Simon Kelley
Kevin Knauss
Ralph Kugler
Lynton Land
Dave Manning
János Mátyás
Mark Osborne
Norman Oxtoby
John Parnell
John Popek
Tim Primmer
Mogens Ramm
Mohammed Rezaee
Alain Saada
Harry Shaw
Craig Smalley
Christoph Spötl
Fred Walgenwitz
Jeff Warner

but it generally decreases with increasing overall salinity. Note also that reaction 4 (albitization) will be enhanced by an increase in the dissolved-halite content of a formation water.

It is likely that 'pressure dissolution' will be inhibited slightly by an influx of high salinity brines because there is a direct, if minor, effect of density (and thus water salinity) on the solubility of silica (Fournier & Potter, 1982; Porter & James, 1986). Silica solubility is proportional to the density of formation water. Silica is more soluble in low density than in high density formation waters (all other factors being equal). The relative difference in solubility as a function of salinity variation is up to 10% (Porter & James, 1986). The practical consequence of this effect is that processes that depend on relative silica solubility differences (e.g. pressure dissolution) will be influenced by water salinity.

## Rock composition

The mineralogical composition of a sandstone has a major effect on the likelihood of finding quartz cement. It is controlled by a great many factors including: the geological and geomorphological nature of the terrain supplying the original sediment, the extent of coeval volcanism, the extent of chemical weathering (itself controlled by climate, rate of uplift, rate and distance of transportation), relative sea level (influencing the distance from source to site of deposition), etc. The primary lithofacies (depositional mineralogy) has been compared to the overall diagenetic style (Primmer *et al.*, 1997). Elevated initial quartz contents of the primary sand leads to a statistically greater chance of finding quartz cement in the sandstone (Fig. 6). There are a great many factors that lead to this correlation. Some involve the relative absence of other types of cement in clean quartz arenites (less early diagenetic, pore-filling cements). Others involve the relative absence of clay mineral and microquartz grain surfaces. Sandstones rich in micas, volcaniclastic and argillaceous rock fragments appear to have the greatest potential for minimizing quartz cementation (all other factors being equal) because the lithic fragments undergo ductile deformation during burial resulting in 'pseudo-matrix' that fills pores and covers detrital quartz grains (Worden *et al.*, 1997).

## Rock fabric

The fabric of a rock is defined by the shapes and sizes of detrital and diagenetic components. It has been widely noted that fine-grained sandstones undergo more pressure dissolution than coarse-grained sandstones (e.g. James *et al.*, 1986; Porter & James, 1986). On this basis, interbedded fine- and coarse-grained sandstones should exhibit dissolution of the fine strata and growth of quartz in the coarse strata. Such facies-selective transfer of silica has been reported (e.g. Trewin & Fallick, this volume, pp. 219–229).

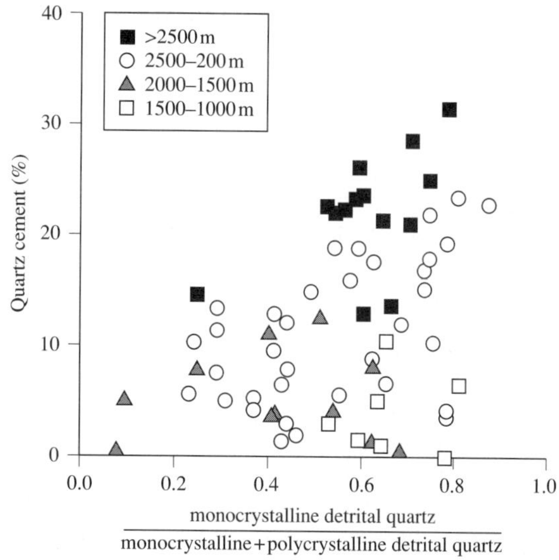


**Fig. 7.** Quartz cement content versus the ratio of polycrystalline to total detrital quartz in the Triassic sandstones of the Paris Basin, France (data from Worden *et al.*, 1994). Quartz cement is most abundant in sandstones most enriched in monocrystalline quartz grains.

It has also been noted that polycrystalline detrital quartz grains tend to have thinner quartz overgrowths than monocrystalline quartz grains within the same rock (Fig. 7). This is likely to be due to competitive growth between separate and differently orientated quartz nuclei on the surface of polycrystalline quartz grains. Such a phenomenon was noted for growth of feldspar cement on multiply twinned detrital feldspar grains where the overgrowth was composed of a multitude of subcrystals (Worden & Rushton 1992).

An important side-effect of microcrystalline quartz growth coatings is that the activity of silica in dissolution is likely to be elevated (Aase *et al.*, 1996). This will affect the feldspar-related sources of quartz cement by keeping the water in the K-feldspar stability-field for longer. It will also probably inhibit 'pressure dissolution' by reducing the concentration gradient between the grain contact and the pore, thus reducing the chemical potential gradient: the ultimate driving force for cementation.

## EFFECT OF OIL EMPLACEMENT ON QUARTZ CEMENTATION

It has widely been assumed that oil emplacement will quell all inorganic diagenetic phenomenon. However, recent data have been used to suggest that cementation may continue after oil has entered a reservoir. The debate about the consequences of oil emplacement on quartz cementation has been strongly and emotively argued from both angles. The hypothesis that oil emplacement inhibits cementation is based on the apparent preservation of porosity in some oil-bearing sandstone reservoirs (e.g. Dixon *et al.*, 1989; Emery *et al.*, 1993; Gluyas *et al.*, 1993). Conversely, the view that diagenesis is not affected by oil emplacement in sandstones is supported by: (i) primary oil inclusions being widespread in quartz cements (Walderhaug, 1990); (ii) according to kinetic calculations, precipitation (assumed to be unaffected by oil emplacement) is the rate-controlling process for quartz cementation below 130°C (Oelkers *et al.*, 1992); and (iii) some reservoirs do not show differences in quartz cement volumes between water and oil legs (Ramm, 1992).

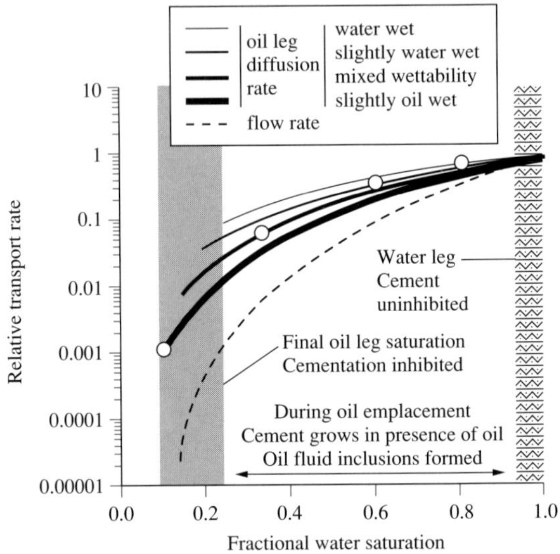


**Fig. 8.** Relative effects of oil emplacement on the rates of silica diffusion and advection of silica-bearing fluids (modified from Worden *et al.*, 1998). The rates of diffusion of silica in water have been calculated for (i) evolving water/oil ratios; and (ii) different oil–water wettability states. The large open circles represent the likely evolution path of a rock–oil–water system during oil filling (i.e. becoming more oil wet during filling). The rate of advection (flow) is affected very strongly by decreasing water/oil ratios as the relative permeability scaling factor decreases markedly as the amount ratio decreases.

### Empirical data on the effect of oil-filling on quartz cementation

#### *Significance of oil-filled inclusions in quartz cement*

The presence of primary oil inclusions completely enclosed within quartz cement shows that quartz cement can grow in the presence of petroleum. Oil-filled inclusions within diagenetic mineral cements have been recognized for a long time (e.g. Murray, 1957). The presence of primary oil inclusions within quartz cements is proof that *some* oil was present at the time that quartz cement was growing. This has been used to infer that quartz cementation may continue in the presence of elevated oil-contents within a sandstone, even at irreducible water saturations ($S_{wi}$) (Walderhaug, 1990; Saigal *et al.*, 1992). However, the presence of primary oil inclusions in quartz cement is not conclusive evidence that cement can grow at irreducible water saturations.

The key question is what oil saturation level is necessary during quartz cementation in order to give rise to oil inclusions. Unfortunately, a critical element in answering this question, the mechanism of oil inclusion formation within quartz cement, is not yet understood. There are two major issues to be resolved. First, quartz is normally water-wet and does not usually come into direct contact with oil. Second, most (though not all) oil inclusions do not contain visible aqueous fluids, even though the enclosing cement must have precipitated from an aqueous fluid (Kvenvolden & Roedder, 1971; Nedkvitne *et al.* 1993).

At high water saturation ($S_w$), during the earlier stages of oil emplacement, neither flow rates nor diffusion rates are significantly reduced, and quartz can continue to grow unhindered. During oil charging $S_w$ will decrease with time and may be highly heterogeneous within a developing oil column (e.g. Hirsch & Thompson, 1995) at all scales down to pore level. It may well be the case that petroleum inclusions are only trapped in that part of the pore network experiencing a particular range of $S_w$. The appropriate $S_w$ window must be $< 1$ (i.e. some oil is present), but significantly higher than $S_{wi}$, such that quartz cement could still grow at a sufficient rate to trap inclusions. A possible oil inclusion trapping window is illustrated in Fig. 8.

#### *Cement-volume trends in oil fields and their aquifers*

Differences in quartz cement volumes between the oil and water legs have long been reported in the literature (e.g. Lowry, 1956; Füchtbauer, 1967) but these studies do not unequivocally demonstrate that the differences occur within the same lithotype having the same potential

| Wetting state | Supply of $SiO_2$ | | Transport of dissolved $SiO_2$ | | Precipitation of $SiO_2$ as quartz |
|---|---|---|---|---|---|
| | internal | external | advection | diffusion | |
| Oil wet sandstone | STOP | NO EFFECT | STOP | STOP | STOP |
| Water wet sandstone | NO EFFECT | NO EFFECT | STOP | SLOW | NO EFFECT |

**Fig. 9.** Summary of effect of oil emplacement on the three component steps involved in quartz cementation (see Fig. 1) relative to the underlying aquifer. Source has been divided between internal and external sources of silica. The effect of oil emplacement describes the fundamental effect of oil emplacement and not the overall result (noting that the overall rate is limited to the rate of the slowest step). Transport has been divided between advection and diffusion. The wettability of the local oil–sandstone system is clearly critical. Note that many shades between fully water and fully oil wet exist in nature.

sources of cement and sites for precipitation and therefore do not prove that oil has influenced quartz cementation.

Conversely, the absence of a difference between quartz cement volumes in the oil and water legs in no way proves that oil does not inhibit cementation (e.g. data from North Sea Brent Group sandstones in Giles *et al.*, 1992): the timing, duration and rates of cementation and oil emplacement are paramount. For example, if oil was emplaced following quartz cementation then there is no reason why rocks in the oil and water zones should have different quantities of cement.

Before an assessment can be made of the controls on the distribution of quartz cement in sandstone oil reservoirs, it is crucial that (i) the paragenetic sequence is defined in relationship to the timing of oil emplacement; (ii) the sources of cement are defined (whether they are internally or externally derived); (iii) other factors that may locally inhibit quartz precipitation are accounted for (e.g. local coats of clay impurities or stratigraphically controlled differences in detrital quartz type); and (iv) factors that may influence transport rates are accounted for (e.g. grain size, tortuosity, porosity) throughout the reservoir. If any of these are not known then the origin of cement distribution patterns must remain ambiguous.

## Oil in sandstones

The fractional amount of water in a sandstone is known as the residual water saturation ($S_w$). The remaining water at maximum oil saturation is known as irreducible water and the amount of this water is known as the irreducible water saturation ($S_{wi}$). The $S_w$ value of a rock reflects the height above the oil–water transition zone and the rock fabric. $S_w$ values are also affected by the wettability of the oil–water–rock system. In qualitative terms, wettability describes the ability of a rock to allow oil to come into contact with the grain surfaces. The irreducible water saturation is generally lower in oil-wet reservoirs (0.1–0.15) than in water-wet reservoirs (0.2–0.25; Hearst & Nelson, 1985; Barclay & Worden, 1997; and this volume, pp. 103–117).

Most sandstone reservoirs have mixed wettability, i.e. the oil is in contact with mineral surfaces to a limited degree (Dullien *et al.*, 1990). Sandstones often contain a mixture of minerals: quartz, feldspar and illite are susceptible to water-wet behaviour (Fassi-Fihri *et al.*, 1991), while kaolinite and chlorite are susceptible to being oil-wet (Sincock & Black, 1988; Fassi-Fihri *et al.*, 1991; Barclay & Worden, this volume, pp. 103–117).

## The effect of oil emplacement on the original supply and precipitation of silica

This can be divided into two end-members: internally sourced and externally sourced silica. It is unlikely that a distant external supply of silica to a reservoir (typically requiring advection) can be affected by having oil in the potential host reservoir sandstone (Fig. 9). It is often assumed that (i) silicate minerals are water-wet; and (ii) that the water is present as a semicontinuous film on grain surfaces. If these are indeed the ways that water exists in oil fields then the ability of minerals to dissolve (and supply silica) will, most likely, *not* be impeded by the presence of oil. However, wettability is a complex function of oil chemistry, mineralogy, mineral chemistry, water chemistry, water pH, gas–oil ratio, pressure and temperature. It is unlikely that all silicate minerals are coated with water in oil fields; even quartz grains can be oil-wet under certain conditions of water pH and oil chemistry (Barclay & Worden, this volume, pp. 103–117). Internal supply from whatever source may be unaffected in water-wet reservoirs with a reasonably high residual water content. In contrast, internal supply of silica from mixed-wettability or oil-wet reservoirs is likely to be hindered or even stopped (Fig. 9).

In water-wet sandstone reservoirs with a typical residual water saturation, the presence of oil is also unlikely to affect significantly the rate of precipitation of quartz cement because the surface of grains will be coated with water. In comparison, in oil-wet systems, the surface of detrital quartz grains will be coated with oil and silica, in aqueous solution will be unable to gain access to allow precipitation and growth of quartz cement (Fig. 9).

### The effect of oil emplacement on the transport of silica

The transportation of silica can be divided into two end-members: (i) diffusion and (ii) advection (Fig. 9). The flow of water in a two-phase fluid system is controlled by principles of relative permeability. The actual permeability of a fluid phase in a two- (or more) phase fluid system is a product of the intrinsic permeability (as measured in routine core analysis) and the dimensionless relative permeability factor. The product is known as the effective permeability. In water-wet systems, the effective permeability tends towards zero as water saturations decrease towards 30%. In oil-wet systems, the effective permeability tends to zero at higher water saturations. The consequence of this is that silica-loaded water will be unable to flow into oil-filled sandstones because the effective permeability to water will be zero. Supply of silica by advection will be severely limited even at low oil saturations as the relative permeability decreases rapidly with increasing oil content. Oil emplacement will practically halt the advective supply of silica into sandstones (Fig. 9).

The rate of silica diffusion in sandstones will also be hampered by oil emplacement due to the increased tortuosity of the residual water network in the pore spaces (Worden *et al.*, 1998). The degree of hampering of diffusion is a direct function of the residual water saturation and the wettability. Diffusive transport is impeded most in oil-wet systems as the residual water fails to make a continuous circuit through which silica diffusion occurs. Rates of silica diffusion in oil fields will be two or more orders of magnitude slower than in the underlying aquifer (Figs 8 & 9 and Worden *et al.*, 1998).

### Rate controlling step in the presence of oil

The slowest step in a chain of reactions (the rate-controlling step) determines the overall rate of reaction. In water-filled sandstones, modelling work has concluded that the precipitation rate of quartz is the rate-controlling step for quartz cementation under most diagenetic conditions (e.g. Murphy *et al.*, 1989). However, the rate of diffusion in oil-filled sandstones is slower than the rate in the underlying aquifer. The degree of retardation was assessed quantitatively by Worden *et al.* (1998) and was shown to be a function of the water/oil ratio and wettability. At typical residual water saturations for water-wet sandstones, diffusion was slowed by several orders of magnitude relative to the aquifer. Thus even in water-wet sandstones in which surface processes (solution and precipitation) control the rate of cementation in the aquifer, diffusion will be retarded and will be become rate-limiting in the overlying oil leg. Consequently, although the fundamental rates of precipitation and dissolution of quartz are not necessarily affected by oil emplacement in water-wet systems, the rate of cementation will be slower in the oil leg than in the water leg because the rate of diffusion is slowed down and because the process becomes diffusion-controlled. Oil emplacement will thus at least retard quartz cementation in sandstones.

A significantly oil-wet sandstone case is more difficult to appraise quantitatively. The rate of diffusion in the oil leg relative to the water leg will be retarded greatly (e.g. Worden *et al.*, 1998) and $S_{wi}$ is lower than in water-wet cases leading to less of the aqueous medium in which silica can diffuse. However, the rates of precipitation and dissolution may also be retarded in this case because water will not be able to gain access to mineral surfaces. The rate of quartz cementation is bound to be much slower in the oil leg than in the water leg in oil-wet sandstones. The more oil-wet the system is, the greater will be the inhibiting effect of oil on quartz cementation. Oil-wetting increases with increasing capillary pressure, suggesting that the tendency to inhibit precipitation and supply rates will increase upwards through the petroleum column. Quartz cementation will thus be more inhibited towards the crests of oil fields.

## CONCLUSIONS

**1** The silica in quartz cement has no single source that can be universally predicted in sandstones. It can be sourced on a sandbody-scale by feldspar alteration reactions, pressure dissolution at grain contacts and in stylolites, dissolution of biogenic silica and volcanic fragments, and from the illitization of smectite. Redistribution of silica may also occur within different facies within sand sequences. In principle, silica may also be sourced externally to the sandstone, including mudrock-related sources, silica transported from deep in the basin (from down-dip equivalents of the sandstone in question; or via faults), or other external sources.

**2** Silica is transported largely by diffusional processes because advection-related transport requires unreasonably large amounts of fluids. Additionally, quartz-cemented fractures in sandstones are a rarity rather than the rule as would be expected if fault-related fluid flow was the main transport process.

**3** The debate over the duration and rate of quartz cementation is unresolved. Cementation may be a slow and continuous process or rapid and episodic process. There is a major discrepancy between experimental rates and empirical rates for cases assumed to have had slow and continuous growth. Cathodoluminscence zonation in quartz cement seems to suggest that cementation is not a continuous process but occurs in bursts.

**4** Quartz cementation tends to occur at elevated temperatures because temperature controls the solubility of quartz and the most important $SiO_2$-liberating reactions. The role of pressure is uncertain but the relative lack of quartz cement in strongly overpressured, yet hot, reservoirs suggests that it may be important. In terms of water chemistry, quartz cementation is slightly favoured in high pH, low potassium concentration and high salinity (NaCl-content) waters. High salinity formation waters, though, may inhibit the rate of diffusion as the solubility of silica is inversely proportional to water density (i.e. salinity). Quartz cementation is most likely, and most abundant, in clean quartz arenites that do not have early diagenetic pore-filling carbonate cements or grain-coating clay minerals or microquartz.

**5** Quartz cementation that is synchronous with, or after, oil emplacement in sandstones is probably strongly inhibited relative to the underlying aquifer. The exact degree of inhibition depends on wettability (itself dependent on oil, water and rock chemistry, pressure, pore geometry and time), the source of the silica in the quartz cement and the dominant transport mechanism. Quartz cementation will be least inhibited in water-wet oil fields with an internal source of quartz and a diffusion-dominated transport regime. Under these circumstances, the overall rate may be two or more orders of magnitude retarded relative to the aquifer. Available empirical data are too poorly constrained to help resolve the degree of quartz cement inhibition in oil-wet fields.

## ACKNOWLEDGEMENTS

The Swedish Natural Science Research Council (NFR) is thanked (SM) for supporting related research activities. Queen's University are most gratefully thanked (RHW) for providing such a deeply nourishing research environment. Jim Hendry (BG Technology) and Mark Osborne (BP) are thanked for providing careful and thoughtful reviews of this paper.

## REFERENCES

AASE, N.E., BJØRKUM, P.A. & NADEAU, P.H. (1996) The effect of grain-coating microquartz on preservation of reservoir porosity. *American Association of Petroleum Geologists Bulletin* **80**, 1654–1673.

ABERCROMBIE, H.J., HUTCHEON, I.E., BLOCH, J.D. & DE CARITAT, P. (1994) Silica activity and the smectite–illite reaction. *Geology* **22**, 539–542.

BARCLAY, S.A. & WORDEN, R.H. (1997) Reservoir wettability and its effect upon cementation in oil fields. In: *Extended Abstract, Volume of the Geofluids II, 1997 Conference.* (eds HENDRY, J., CAREY, P., PARNELL, J., RUFFELL, A. & WORDEN, R.H.) pp. 264–267. The Geofluids Research Group, Belfast.

BARCLAY, S.A. & WORDEN, R.H. (2000) (This volume) Effects of reservoir wettability on quartz cementation in oil fields. In: *Quartz Cementation in Sandstones*. (eds WORDEN, R.H. & MORAD, S.) pp. 103–117. Special Publication of the International Association of Sedimentologists 29. Blackwell Science, Oxford.

BENNET, P. & SIEGEL, D.I. (1987) Increased solubility of quartz in water due to complexing by organic compounds. *Nature* **326**, 684–686.

BJØRKUM, P.A. (1996) How important is pressure in causing dissolution of quartz in sandstones? *Journal of Sedimentary Research A* **66** (6), 147–154.

BJØRLYKKE, K. (1994) Pore water flow and mass transfer of solids in solution in sedimentary basins. In: *Quantitative Diagenesis: Recent Developments and Applications to Reservoir Geology.* (eds PARKER, A. & SELLWOOD, B.W.) pp. 189–221, Kluwer Dordrecht.

BOLES, J.R. (1982) Active albitization of plagioclase, Gulf Coast Tertiary. *American Journal of Science* **282**, 165–180.

BOLES, J.R. & FRANKS, S.G. (1979) Clay diagenesis in Wilcox sandstones of southwest Texas: implications of smectite diagenesis on sandstone cementation. *Journal of Sedimentary Petrology* **49**, 55–70.

BRADY, P.V. & WALTHER, J.V. (1990) Kinetics of quartz dissolution at low temperatures. *Chemical Geology* **82**, 253–264.

BURLEY, S.D. (1993) Models of burial diagenesis for deep exploration plays in Jurassic fault traps of the Central and Northern North Sea. In:. *Petroleum Geology of Northwest Europe: Proceedings of the 4th Conference of the Geological Society.* (ed. PARKER, J.R.) pp. 1353–1375. The Geological Society of London.

BURLEY, S.D., MULLIS, J. & MATTER, A. (1989) Timing diagenesis in the Tartan Reservoir (UK, North Sea): constraints from combined cathodoluminescence microscopy and fluid inclusion studies. *Marine and Petroleum Geology* **6**, 98–120.

CHANG, H.K., MACKENZIE, F.T. & SCHOMAKERS, J. (1986) Comparison between diagenesis of dioctahedral and trioctahedral smectite, Brazilian offshore basins. *Clays and Clay Minerals* **34**, 407–423.

CRERAR, D., HELLMANN, R. & DOVE, P. (1988) Dissolution kinetics of albite and quartz in hydrothermal solutions. *Chemical Geology* **70**, 77–87.

DIXON, S.A., SUMMERS, D.M. & SURDAM, R.C. (1989) Diagenesis and preservation of porosity in Norphlet Formation (Upper Jurassic), Southern Alabama. *American Association of Petroleum Geologists Bulletin* **73** (6), 707–728.

DULLIEN, F.A.L., ALLSOP, H.A., MACDONALD, I.F. & CHATZIS, I. (1990) Wettability and miscible displacement in Pembina Cardium Sandstone. *Journal of Canadian Petroleum Technology* **29** (4), 63–74.

EGEBERG, P.K. & AAGAARD, P. (1989) Origin and evolution of formation waters from oilfield on the Norwegian Shelf. *Applied Geochemistry* **4**, 131–142.

EHRENBERG, S.N. (1990) Relationship between diagenesis and reservoir quality in sandstones of the Garn Formation, Haltenbanken, Mid-Norwegian continental shelf. *American Association of Petroleum Geologists Bulletin* **74**, 1538–1558.

EMERY, D., SMALLEY, P.C. & OXTOBY, N.H. (1993) Synchronous oil migration and cementation in sandstone reservoirs demonstrated by quantitative description of diagenesis. *Philosophical Transactions of the Royal Society of London A* **344**, 115–125.

FASSI-FIHRI, O., ROBIN, M. & ROSENBERG, E. (1991) Wettability studies at the pore level: a new approach by the use of cryo-scanning electron microscopy. *Proceedings of the 66th Annual Technical Conference and Exhibition of the Society of Petroleum Engineers (Dallas)* **G. SPE 22596**, 97–110.

FEIN, J.B. (2000) (This volume) Experimental and field constraints on the role of Si-organic complexation and Si-microbial interaction during sediment diagenesis. In: *Quartz Cementation in Sandstones.* (eds WORDEN, R.H. & MORAD, S.) pp. 119–129. Special Publication of the International Association of Sedimentologists 29. Blackwell Science, Oxford.

FISHER, Q., KNIPE, R. & WORDEN, R.H. (2000) (This volume) The relationship between faulting, fractures, transport of silica and quartz cementation in North Sea oil fields. In: *Quartz Cementation in Sandstones.* (eds WORDEN, R.H. & MORAD, S.) pp. 129–146. Special Publication of the International Association of Sedimentologists 29. Blackwell Science, Oxford.

FOURNIER, R.O. & POTTER, R.W., I.I. (1982) An equation correlating the solubility of quartz in water from 25°C to 900°C at pressures up to 10 000 bars. *Geochimica et Cosmochimica Acta* **46**, 1969–73.

FÜCHTBAUER, H. (1967) Influence of different types of diagenesis on sandstone porosity. *7th World Petroleum Congress Proceedings* **2**, 353–368.

GILES, M.R., STEVENSON, S., MARTIN, S.V. *et al.* (1992) The reservoir properties and diagenesis of the Brent Group: a regional perspective. In: *Geology of the Brent Group.* (eds MORTON, A.C., HASZELDINE, R.S., GILES, M.R. & BROWN, S.) pp. 289–327. Special Publications of the Geological Society of London 61.

GLASSMAN, J.R. (1992) The fate of feldspar in the Brent Group reservoirs, North Sea: a regional synthesis of diagenesis in shallow, intermediate and deep burial environments. In: *Geology of the Brent Group.* (eds MORTON, A.C., HASZELDINE, R.S., GILES, M.R. & BROWN, S.) pp. 329–350. Special Publications of the Geological Society of London 61.

GLASMANN, J.R., CLARK, R.A., LARTER, S., BRIEDS, N.A. & LUNDEGARD, P.D. (1989) Diagenesis and hydrocarbon accumulation, Brent Sandstones (Jurassic), Bergen High, North Sea. *American Association of Petroleum Geologists Bulletin* **73**, 1341–1360.

GLUYAS, J.G. & COLEMAN, M.L. (1992) Material flux and porosity changes during sediment diagenesis. *Nature* **356**, 52–54.

GLUYAS, J.G., GRANT, S.M. & ROBINSON, A. (1993) Geochemical evidence for a temporal link control on sandstone cementation. In: *Diagenesis and Basin Development.* (eds HORBURY, A.D. & ROBINSON, A.G.) pp. 23–33. Association of Petroleum Geologists Studies in Geology 36.

GLUYAS, J., GARLAND, C., OXTOBY, N.H. & HOGG, A.J.C. (2000) (This volume) Quartz cement; the Miller's Tale. In: *Quartz cementation in sandstones.* (eds WORDEN, R.H. & MORAD, S.) pp. 199–217. Special Publication of the International Association of Sedimentologists 29. Blackwell Science, Oxford.

GRANT, S.M. & OXTOBY, N.H. (1992) The timing of quartz cementation in Mesozoic sandstones from Haltenbanken, offshore mid-Norway: fluid inclusion evidence. *Journal of the Geological Society London* **149**, 479–482.

HARRIS, N.B. (1989) Diagenetic quartzarenite and destruction of secondary porosity: an example from the Middle Jurassic Brent sandstone of northwest Europe. *Geology* **17**, 361–364.

HASZELDINE, R.S., SAMSON, I.M. & CORNFORD, C. (1984) Dating diagenesis in a petroleum basin, a new fluid inclusion method. *Nature* **307**, 354–357.

HEARST, J.R. & NELSON, P.H. (1985) *Well Logging for Physical Properties*. McGraw-Hill, New York, 571pp.

HENDRY, J.P. & TREWIN, N.H. (1995) Authigenic quartz microfabrics in Cretaceous turbidites: evidence for silica transformation processes in sandstones. *Journal of Sedimentary Research* **65**, 380–392.

HIRSCH, L.M. & THOMPSON, A.H. (1995) Minimum saturations and buoyancy in secondary migration. *American Association of Petroleum Geologists Bulletin* **79** (5), 696–710.

HOGG, A.J.C., PEARSON, M.J., FALLICK, A.E. & HAMILTON, P.J. (1995) An integrated thermal and isotopic study of the diagenesis of the Brent Group, Alwyn-South, UK North Sea. *Applied Geochemistry* **10**, 531–540.

HOUSEKNECHT, D.W. (1984) Influence of grain size and temperature on intergranular pressure solution, quartz cementation and porosity in a quartzose sandstone. *Journal of Sedimentary Petrology* **54**, 348–361.

HOWER, J., ESLINGER, E., HOWER, M.E. & PERRY, E.A. (1976) Mechanism of burial metamorphism of argilliceous sediment: 1. Mineralogical and chemical evidence. *Geological Society of America Bulletin* **87**, 725–737.

JAMES, W.C., WILMAR, G.C. & DAVIDSON, B.G. (1986) Role of quartz type and grain size in silica diagenesis, Nugget sandstone, South-Central Wyoming. *Journal of Sedimentary Petrology* **56**, 657–662.

KRAISHAN, G.M., REZAEE, M.R. & WORDEN, R.H. (2000) (This volume) Trace element composition of quartz cement: a key to reveal the origin of quartz cement In: *Quartz Cementation in Sandstones.* (eds WORDEN, R.H. & MORAD, S.) pp. 317–331. Special Publication of the International Association of Sedimentologists 29. Blackwell Science, Oxford.

KVENVOLDEN, K.A. & ROEDDER, E. (1971) Fluid inclusions in quartz crystals from South-West Africa. *Geochimica et Cosmochimica Acta* **35**, 1209–1229.

LASAGA, A. & KIRKPATRICK, R.J. (1981) Kinetics of geochemical processes. Reviews in Mineralogy, 8. Mineral Society of America, 398pp.

LOWRY, W.D. (1956) Factors in loss of porosity by quartzose sandstones of Virginia. *American Association of Petroleum Geologists Bulletin* **40**, 489–500.

MCBRIDE, E.F. (1989) Quartz cement in sandstones—a review. *Earth Science Reviews* **26**, 69–112.

MORAD, S. (1986) Albitization of K-feldspar grains in Proterozoic arkoses and greywackes from southern Sweden. *Neues Jahrbuch für Mineralogie Monatshefte* **1986**, 145–156.

MORAD, S., BERGAN, M., KNARUD, R. & NYSTUEN, J.P. (1990) Albitization of detrital plagioclase in Triassic reservoir sandstones from the Snorre Field, Norwegian North Sea. *Journal of Sedimentary Petrology* **60**, 411–425.

MORAD, S., BEN ISMAIL, H., AL-AASM, I.S. & DE ROS, L.F. (1994) Diagenesis and formation-water chemistry of Triassic reservoir sandstones from southern Tunisia. *Sedimentology* **41**, 1253–1272.

MURPHY, W.M., OELKERS, E.H. & LICHTNER, P.C. (1989) Surface reaction versus diffusion control of mineral dissolution and growth rates in geochemical processes. *Chemical Geology* **78**, 357–380.

MURRAY, R.C. (1957) Hydrocarbon fluid inclusions in quartz. *American Association of Petroleum Geologists Bulletin* **41**, 950–956.

NEDKVITINE, T., KARLSEN, D.A., BJØRLYKKE, K. & LARTER, S.R. (1993) Relationship between reservoir diagenetic evolution and petroleum emplacement in the Ula Field, North Sea. *Marine and Petroleum Geology* **10**, 255–270.

OELKERS, E.H., BJØRKUM, P.A. & MURPHY, W.M. (1992) The mechanism of porosity reduction, stylolite development and quartz cementation in North Sea sandstones. In: *Proceedings of the 7th International Symposium on Water–Rock Interaction*. (eds KHARAKA, Y.K. & MAEST, A.S.) pp. 1183–1186, Balkema, Rotterdam.

OELKERS, E.H., BJØRKUM, P.A. & MURPHY, W.M. (1996) A petrographic and computational investigation of quartz cementation and porosity reduction in North Sea sandstones. *American Journal of Science* **296**, 420–452.

OSBORNE, M. & HASZELDINE, R.S. (1993) Evidence for resetting fluid inclusion temperatures from quartz cements in oilfields. *Marine and Petroleum Geology* **10**, 271–278.

OSBORNE, M. & SWARBRICK, R.E. (1999) Diagenesis in North Sea HPHT reservoirs—consequences for porosity and overpressure prediction. *Marine and Petroleum Geology* **16**, 337–353.

PAGEL, M. (1975) Détermenation de conditions physicochimiques de la silicification diagénetique des grés Athabasca (Canada) au moyen de inclusions fluides. Academie de Sciences (Paris). *Comptes Rendus, Serie II* **280**, 2301–2304.

PASSCHIER, C.W. & TROUW, R.A.J. (1996) *Microtectonics*. Springer, Berlin, 289pp.

PORTER, E.W. & JAMES, W.C. (1986) Influence of pressure, salinity, temperature and grain size on silica diagenesis in quartzose sandstones. *Chemical Geology* **57**, 359–369.

PRIMMER, T.J., CADE, C.A., EVANS, I.J. *et al.* (1997) Global patterns in sandstone diagenesis: application to reservoir quality prediction for petroleum exploration. In: (eds KUPEZC, J., GLUYAS, J.G. & BLOCH, S.) pp. 61–78. American Association of Petroleum Geologists Memoir 69. *Reservoir quality prediction in sandstones and carbonates*.

RAMM, M. (1992) Porosity–depth trends in reservoir sandstones: theoretical models related to Jurassic sandstones, offshore Norway. *Marine and Petroleum Geology* **9**, 553–567.

RAMM, M. & BJØRLYKKE, K. (1994) Porosity/depth trends in reservoir sandstones: assessing the quantitative effects of varying pore pressure, temperature history and mineralogy, Norwegian Shelf data. *Clay Minerals* **29**, 475–490.

RENARD, F., BROSSE, E. & SOMMER, F. (2000) (This volume) The different processes involved in the mechanism of pressure solution in quartz-rich rocks and their interactions. In: *Quartz Cementation in Sandstones*. (eds WORDEN, R.H. & MORAD, S.) pp. 67–78. Special Publication of the International Association of Sedimentologists 29. Blackwell Science, Oxford.

ROBINSON, A. & GLUYAS, J. (1992) Duration of quartz cementation in sandstones, North Sea and Haltenbank Basins. *Marine and Petroleum Geology* **9**, 324–327.

RUTTER, E.H. (1976) The kinetics of rock deformation by pressure solution. *Philosophical Transactions of the Royal Society of London A* **283**, 203–220.

SAIGAL, G.C., BJØRLYKKE, K. & LARTER, S. (1992) The effects of oil emplacement in diagenetic processes – example from the Fulmar Reservoir sandstones, Central North Sea. *American Association of Petroleum Geologists Bulletin* **76**, 1024–1033.

SINCOCK, K.J. & BLACK, C.J.J. (1988) Validation of water/oil displacement scaling criteria using microvisualisation techniques. *Proceedings of the 64th Annual Technical Conference and Exhibition of the Society of Petroleum Engineers (San Antonio)*. **SPE 18294**, 339–347.

STEEFEL, C.I. & VAN CAPPELLEN, P. (1990) A. new kinetic approach to modelling water–rock interaction: the role of nucleation, precursors and Ostwald ripening. *Geochimica et Cosmochimica Acta* **54**, 2657–2677.

SULLIVAN, K.B. & MCBRIDE, E.F. (1991) Diagenesis of sandstone at shale contacts and diagenetic heterogeneity, Frio Formation, Texas. *American Association of Petroleum Geologists Bulletin* **75**, 121–138.

SWARBRICK, R.E. (1994) Reservoir diagenesis and hydrocarbon migration under hydrostatic palaeopressure conditions. *Clay Minerals* **29**, 463–473.

TADA, R. & SIEVER, R. (1989) Pressure solution during diagenesis. *Annual Review Earth Planetary Science Letters* **17**, 89–118.

THOMPSON, A. (1971) Preservation of porosity in the deep Woodbine/Tuscaloosa Trend, Louisiana. *Transactions of the Gulf Coast Association for Petroleum Geology*, **vol 30**, pp. 396–408.

TREWIN, N.H. & FALLICK, A.E. (2000). (This volume) Quartz cement origins and budget in the Tumblagooda Sandstone, W Australia. In: *Quartz Cementation in Sandstones*. (eds WORDEN, R.H. & MORAD, S.) pp. 219–229. Special Publication of the International Association of Sedimentologists 29. Blackwell Science, Oxford.

VAGLE, G.B., HURST, A. & DYPVIK, H. (1994) Origin of quartz cements in some sandstones from the Jurassic of the Inner Moray Firth (U.K.). *Sedimentology* **41**, 363–377.

WALDERHAUG, O. (1990) A. fluid inclusion study of quartz cemented sandstones from offshore mid-Norway—possible evidence for continued quartz cementation during oil emplacement. *Journal of Sedimentary Petrology* **60**, 203–210.

WALDERHAUG, O. (1994) Precipitation rates for quartz cement in sandstones determined by fluid inclusion microthermometry and temperature history modelling. *Journal of Sedimentary Petrology* **60**, 203–210.

WALDERHAUG, O. (1995) Discussion of 'Evidence for resetting of fluid inclusions temperatures from quartz cement in oilfields' by Osborne and Haszeldine (1993). *Marine and Petroleum Geology* **12**, 559–561.

WORDEN, R.H. & RUSHTON, J. (1992) Diagenetic K-feldspar textures: a TEM study and model for diagenetic K-feldspar growth. *Journal of Sedimentary Petrology* **62**, 779–789.

WORDEN, R.H., BRACH, M., COLEMAN, M.L. *et al.* (1994) *Geochemical studies of rocks and fluids to give predictive modelling of the permeability distribution in sedimentary basins.* Final Report, EC Contract, JOUF-0016C.

WORDEN, R.H., WARREN, E.A., SMALLEY, P.C., PRIMMER, T.J. & OXTOBY, N.H. (1995) Discussion of 'Evidence for resetting of fluid inclusions temperatures from quartz cement in oilfields' by Osborne and Haszeldine (1993). *Marine and Petroleum Geology* **12**, 566–570.

WORDEN, R.H., MAYALL, M.J. & EVANS, I.J. (1997) Predicting reservoir quality during exploration: lithic grains, porosity and permeability in Tertiary clastics of the South China Sea basin. In: *Petroleum Geology of S.E. Asia.* (eds FRASER. A.J., MATTHEWS, S.J. & MURPHY, R.W.) pp. 107–115. Special Publications of the Geological Society of London 126.

WORDEN, R.H., OXTOBY, N.H. & SMALLEY, P.C. (1998) Can oil emplacement stop quartz cementation in sandstones? *Petroleum Geoscience* **4**, 129–137.

*Spec. Publs int. Ass. Sediment.* (2000) **29**, 21–38

# The origin of large-scale quartz cementation: evidence from large data sets and coupled heat–fluid mass transport modelling

M. R. GILES, S. L. INDRELID, G. V. BEYNON *and* J. AMTHOR*

*Shell International Exploration and Production B.V. Research and Technological Services, Postbus 60, 2280AB Rijswijk, The Netherlands*

## ABSTRACT

Quartz is the most important porosity destroying cement in sandstone reservoirs. The large petrographic data sets, from 11 intervals from around the world, presented and referenced here, indicate that massive quartz cementation follows a predictable pattern. Cementation occurs in two distinct phases: an early minor phase during which a few percentage of authigenic quartz are precipitated, followed by the relatively abrupt start of massive deep burial cementation which can result in up to 20% reduction in porosity. Massive quartz cementation typically begins at 90 ± 10°C. The relative temperature dependence indicates some form of kinetic control.

Computer modelling indicates that it is not possible to import large quantities of silica into the reservoir. Thus the source of silica for quartz cementation must be derived from within the sandstone or directly adjacent to it. The pattern of cementation strongly suggest sources which have a temperature and, in some cases, a stress dependence. Temperature-dependent silica contributing reactions probably include the dissolution of feldspars, the transformation of smectitic clay to illite and the dissolution of heavy minerals (although only a minor contributor in many circumstances). Pressure dissolution within sandstones and adjacent siltstones also becomes important over the same stress/temperature range, partly because the dissolution of framework grains will increase the stress on the remaining framework grains. Finally, fluid movement within a non-isothermal reservoir promotes redistribution of quartz.

## INTRODUCTION

It has been recognized for many years that authigenic quartz cementation is the single most important chemical process responsible for the destruction of sandstone reservoir properties (e.g. Blatt, 1979). However, progress toward understanding the origins and controls on quartz cementation has been slow. Hypotheses for the origin of quartz cementation (McBride, 1989 for an extensive review) include:

**1** Circulation of meteoric fluids (Friedman, 1954; Riezebos, 1974).

**2** Pressure dissolution and reprecipitation of quartz, either by intergranular pressure dissolution or at stylolites (Waldschmidt, 1941; Heald, 1955, 1956, 1959; Füchtbauer, 1974; de Boer, 1977; Dewers & Ortoleva, 1990; Bjørlykke, 1994; Oelkers *et al.*, 1996; Walderhaug *et al.*, this volume, pp. 39–49).

**3** Dissolution of feldspars as a silica source (Sorby, 1890; Fothergill, 1955; Hawkins, 1978).

**4** As a result of illitization of feldspars (Bjørlykke, 1980).

**5** Oversaturated fluids derived from mudrocks (Towe, 1962; Burley, 1993; Scotchman *et al.*, 1989).

**6** Convection (Wood *et al.*, 1984).

**7** Movement of quartz cementation fronts (Gluyas & Coleman, 1992; Canals & Meunier, 1995).

**8** Up-dip cooling of fluids in equilibrium with dissolved silica (Leder & Park, 1986; Fang *et al.*, 1987; Robinson & Gluyas, 1992).

**9** Complexation of silica by organic anions (Coleman, personal communication).

**10** Movement of hot fluids up fault planes (Haszeldine *et al.*, 1984; Jourdan *et al.*, 1987; Scotchman *et al.*, 1989; Burley, 1993; Gaupp *et al.*, 1993).

The purpose of this paper is to examine the origin of quartz cementation based on: (i) large regional

* Present address: Petroleum Development Oman LLC, Muscat, Oman

petrographic data sets; and (ii) coupled heat–fluid mass transport modelling. By synthesizing the two approaches we hope to discriminate between the possible hypotheses listed above and to identify the origins of quartz cement.

## OBSERVATIONS OF THE DISTRIBUTION OF AUTHIGENIC QUARTZ IN THE SUBSURFACE

The data presented below come from a variety of basinal settings ranging from failed rift (Brent Group), passive margin (Gulf Coast) to complex mixed settings such as samples from the Haushi Group of Oman, which were originally deposited in a rift setting and are today in a foreland basin. Subsidence rates range from high, for instance the Gulf Coast samples of Land (1984) reach 150 m $Myr^{-1}$ (in terms of compacted thickness), to low for the Haima Group samples down to 9 m $Myr^{-1}$. Current geothermal gradients also show a considerable range (reaching 40°C $km^{-1}$ for the samples from offshore Sarawak); however, all are believed to be at their maximum burial depth and temperature today.

The authigenic quartz content of the Shell data sets quoted (Brent Group, North Sea, Haushi and Haima Groups of Oman, and the Oligo-Miocene sands of offshore Sarawak) have been carefully point-counted using both ordinary petrographic techniques and cathodoluminescence methods. Any trends referred to below have been subject to careful statistical analysis and have been shown to be highly significant (for instance, for the Etive samples see Giles *et al.*, 1992). The corresponding porosity depth curves for these intervals can be found in Giles (1997).

### Middle Jurassic Rannoch, Etive, Ness and Tarbert Formations of the Brent Group, North Sea

One of the most comprehensive data sets available to study quartz cementation is from the deltaic to shallow marine, Brent Group of the North Sea. Giles *et al.* (1992) presented the results of point-counting 633 thin sections from 35 wells across the Brent Province. Within the Brent Group, quartz is quantitatively the most important cement forming up to 28% of the bulk rock volume (% BV). In all cases it forms syntaxial overgrowths on detrital quartz grains. The pattern of quartz cementation observed in each of the four Brent Group Formations (Tarbert, Ness, Etive and Rannoch) is similar (Fig. 1a–d) with a steady increase in the volume of cement below depths of approximately 9000 ft (2700 m). Above this depth, only a few percentage of quartz cement are present as indicated from core samples, cuttings and logs. The pattern of quartz cementation of the Brent Group described above is also consistent with that described by Bjørlykke *et al.* (1992) and Harris (1992).

Comparison of the pattern of cementation observed in these sandstones with the porosity-loss curves (see Giles *et al.*, 1992; their fig. 6a,b) indicates that the porosities of the sandstones were reduced to about half their depositional values, largely due to mechanical compaction, prior to significant cementation.

The scatter in the data shown in Fig. 1(a–d), and in subsequent plots arises from two major causes:

**1** Facies variations reflecting the amount of detrital clay and textural variations within any one facies. Samples from facies rich in detrital clay tend to have lower quantities of authigenic quartz. For instance, in Fig. 4, only samples with less than 10% clay have been plotted and consequently the quartz cementation depth trend is enhanced.

**2** Measurement errors due to the inherent uncertainties in point-count results and/or the lack of representatives of thin section data for the reservoir as a whole. Large uncertainties exist in the figures derived from point-counting. The uncertainty falls with increasing number of points counted (van der Plas & Tobi, 1965). In addition, microscale variations in texture and clay content can influence the quantity of quartz cement visible in a single thin section, thus anyone thin section may not be representative of the average volume of quartz present in a given volume of rock (see Giles, 1997, p. 21).

### Permo-Carboniferous Ghaba Formation of the Haushi Group, Central Oman

The Haushi Group sandstones of Central Oman comprise fluvial and fluvio-glacial sediments of the Late Carboniferous to the Permian. The data presented are from a regional study undertaken by Giles, Gaarenstroom, Omatsola, and Links during the early 1980s. Authigenic quartz is again the volumetrically most important cement observed in most of these rocks. The quartz forms syntaxial overgrowths on the detrital quartz grains and displays 3–5 zones when viewed using cathodoluminescence (CL) microscopy. These zones may be traced over a large part of Central Oman. The pattern of quartz cementation observed is shown in Fig. 2. The figure shows that major quartz cementation is again prevalent below depths of 10 000 ft (3050 m), reaching values up to 27% BV. At shallower depths the volume of authigenic quartz rarely exceeds 4% BV. In the more deeply buried siltier sediments, or where the detrital quartz grains are coated with

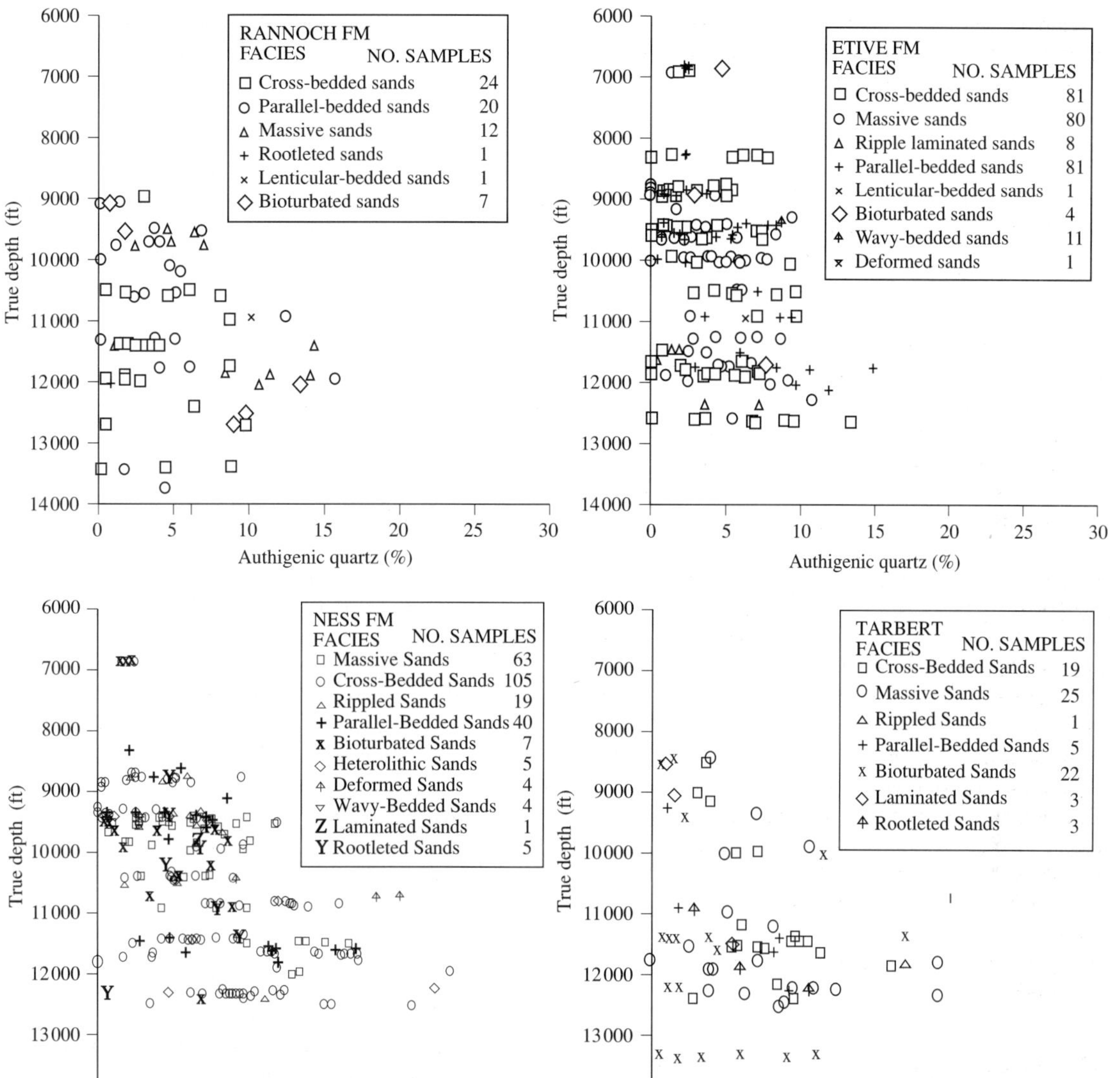


**Fig. 1.** Quartz cementation in the Brent Group of the North Sea (Tarbert, Ness, Etive and Rannoch Formations). Note the increase in quartz cementation below approximately 9000 ft (2700 m). From Giles *et al.* (1992).

illite, intense pressure dissolution was observed using CL microscopy. In the coarser grained sediments from comparable depths, little or no pressure dissolution was visible.

## Cambro-Ordovician Haima Group, Central Oman

Haima Group sediments are unconformably overlain by the Misfar and Haushi Groups. They form a suite of sandstones and mudstones whose origin ranges from continental through lacustrine to shallow marine. The data shown here were acquired in the mid-1990s as part of a regional review of the diagenesis of the Haima Group by Borgomano, Amthor, Terken, Wiemer and van der Zwan. The pattern of quartz cementation visible in Fig. 3 demonstrates a shallow zone (< 3.5 km, 11 482 ft) with less than 10% BV authigenic quartz, and a deeper zone in which many samples exceed this value.

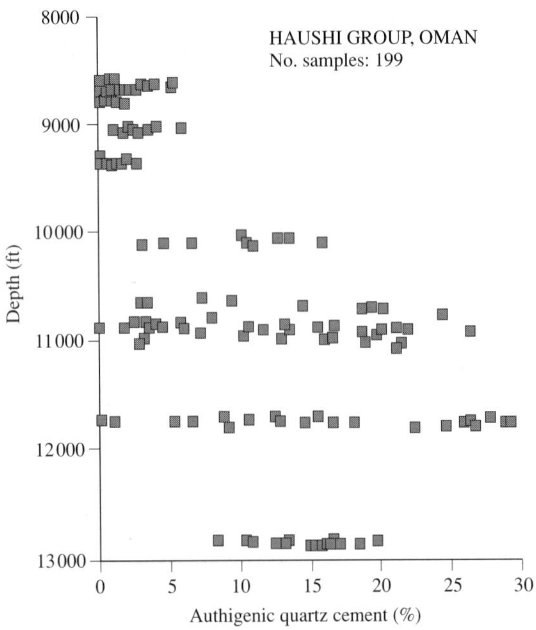


**Fig. 2.** Quartz cement in the Permo-Carboniferous Haushi Group of Central Oman.

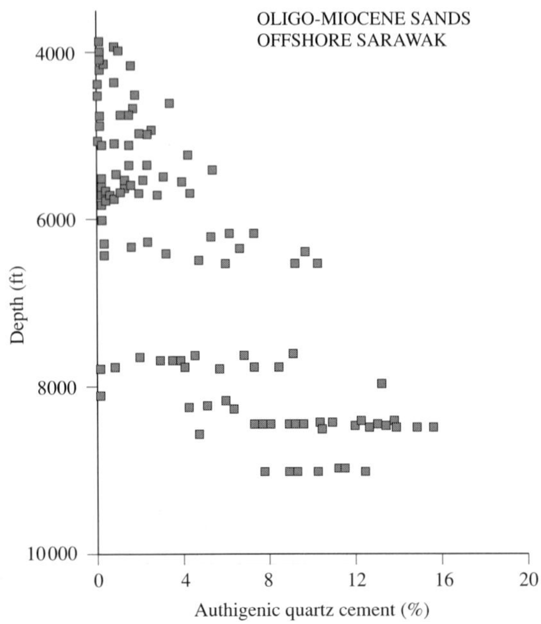


**Fig. 4.** Quartz cement in Oligo-Miocene sandstones, offshore Sarawak. The onset of significant quartz cementation seems to occur at shallower depths than in the other examples.

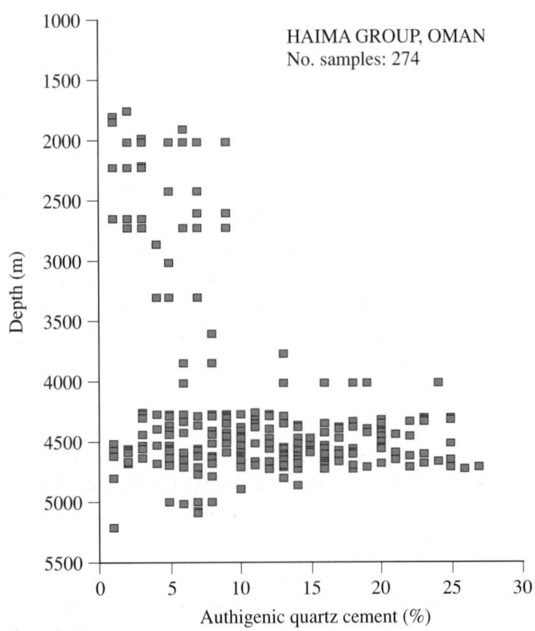


**Fig. 3.** Quartz cement in the Haima Group, Oman.

## Oligo-Miocene Sandstones, offshore Sarawak, Malaysia

The sediments investigated from the Balingian Province offshore Sarawak range in depositional environment from fluvial to shallow marine. The data set was acquired in the mid-1980s as part of a large regional review of the province by Gaarenstroom, Giles, Teyssen, and Links. Figure 4 shows the pattern of cementation as a function of depth. As in the previous examples only a small amount of quartz cement is present in the shallow subsurface, but in contrast to the other data sets major quartz cementation starts at much shallower depths of around 6000 ft (1828 m).

For the samples shown in Fig. 4, the present day burial depth represents the maximum that these sediments have encountered. However, time and depositionally-equivalent sediments closer to the present day coast of Sarawak have been significantly uplifted. It is possible to reconstruct the burial history of these sediments from sonic velocity (see Giles, 1997, fig. 10.9) and other data. In the uplifted sediments, the observed amount of quartz cement correlates with the maximum burial depth. In other words, the trend of maximum burial depth versus quartz cementation is comparable with Fig. 4.

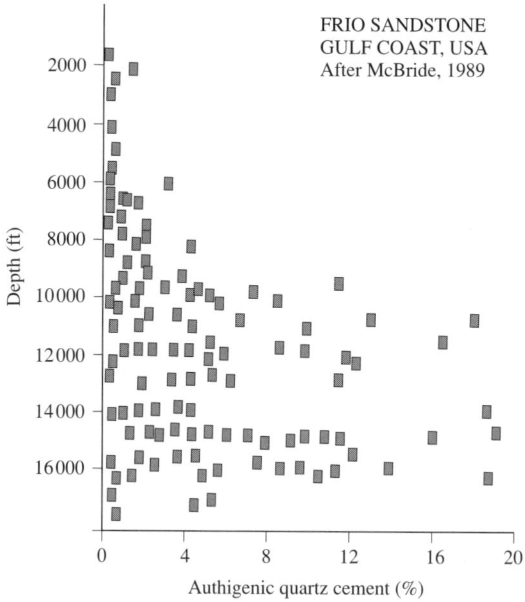


**Fig. 5.** Quartz cement in the Frio sandstones of the US Gulf Coast. From Land (1984).

### Tertiary Sandstones of the Gulf Coast

A number of authors have published data on the abundance of quartz cement in Tertiary Sandstones of the Gulf Coast. For instance, Land (1984; his fig. 5) showed a quartz-cement–depth plot based on 576 samples for the Frio Formation which is very similar to those shown here. Land (1984) noted that 'Quartz cementation is quite uncommon and not volumetrically significant above about 8000 ft (2440 m). Only some sands below about 10 000 ft (3050 m) contain significant amounts of quartz cement. The change in the degree of cementation corresponds to the type of geopressure'. In the previous examples shown only some of the Brent Group samples come from a significantly overpressured terrane, and overpressures are absent from the Haushi and Haima Group samples from Oman. Thus the conclusions of Land (1984) are unlikely to be correct.

McBride (1989; his fig. 8a–d) presented data from the Miocene sandstones of onshore Louisiana, Frio (Oligocene) of Texas, Vicksburg (Oligocene) of South Texas and the Wilcox (Eocene) of Central Texas. For comparison Fig. 5 shows McBride's data for the Frio Sandstones. The pattern of cementation observed in the Frio, and in the other formations reported by McBride (1989) are wholly consistent with the data presented from the North Sea, Oman, and Sarawak, i.e. little quartz cementation at shallow depths, followed by a rapid depth-dependent increase in the volume of quartz cement.

### Other areas

Similar patterns of quartz cementation have also been reported from the Jurassic Dogger Beta Sandstones of Western Germany (Füchtbauer, 1974), Tertiary sandstones of the Pattani Basin, Gulf of Thailand (Trevena & Clark, 1986), Lower Cretaceous of the Green River Basin of Wyoming (Dutton, unpublished data) and the Jurassic of the N.W. Shelf of Australia (unpublished data from Dutton).

One potential reason why these trends have not been observed more commonly is that the amount of quartz cementation reflects the maximum burial depth. Thus, in situations where there has been widespread differential uplift there will be no clear correlation between the volume of cement and the present day burial depth.

## QUARTZ CEMENTATION AND COMPACTION

Comparison of the porosity loss curve (Fig. 6) for the Sarawak data with the quartz cement depth data demonstrates that the onset of large-scale quartz cementation has no obvious impact on the gradient of the porosity loss curve. It should be noted that the transition does not conform to the changing patterns of cementation, e.g. carbonate giving way to quartz cementation. There appears to be a smooth transition from compactionally-dominated porosity reduction to cementation-dominated porosity reduction. As noted in the discussion of the Brent Group data, this occurs at about 50% of the depositional porosity. Similar patterns of behaviour can be shown for Brent Group and Haushi Group sediments. The continuity in the porosity loss curve suggests that the driving force for porosity reduction remains the same throughout the depth range of the data. Compactional porosity loss is driven by increasing vertical effective stress (e.g. Terzaghi & Peck, 1948; Giles, 1997) with a secondary temperature effect as discussed below.

## OTHER DIAGENETIC PHENOMENA OCCURRING OVER SIMILAR DEPTH RANGES

Data from the Brent Group study (Giles *et al.*, 1992) may be used to illustrate that a number of other processes also occur over the same depth range that quartz cementation starts to dominate porosity reduction. Of these processes

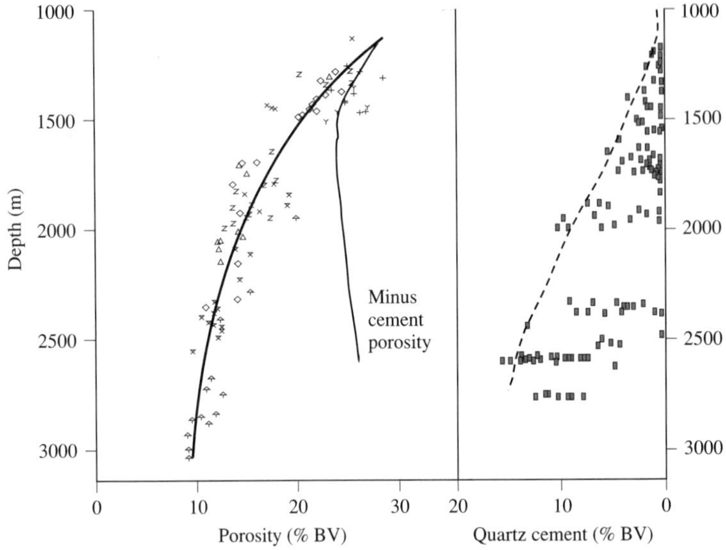


**Fig. 6.** Comparison of porosity loss and quartz cement abundance, offshore Sarawak. The onset of significant quartz cement does not seem to affect the smooth loss of porosity with depth. This indicates a smooth transition from mechanical processes of compaction to cementation. The minus cement porosity curve illustrates the total amount of pore space, including that filled by quartz cement. The increase in minus cement porosity with depth indicates that some material must be lost over the same depth range as quartz cementation occurs.

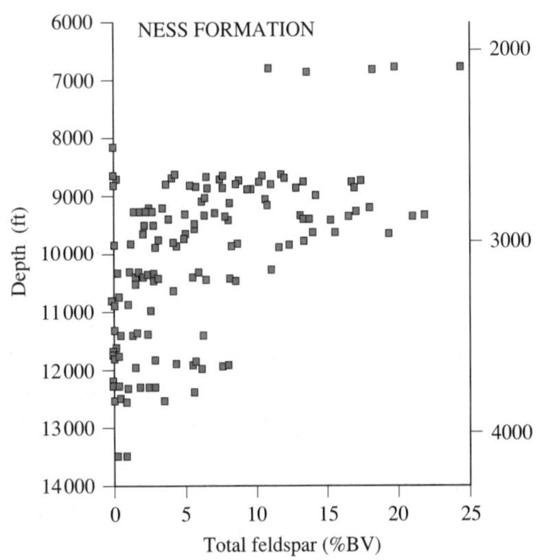


**Fig. 7.** Reduction in the volume of feldspars with depth in the Ness Formation of the Brent Group, North Sea. From Giles *et al.* (1992).

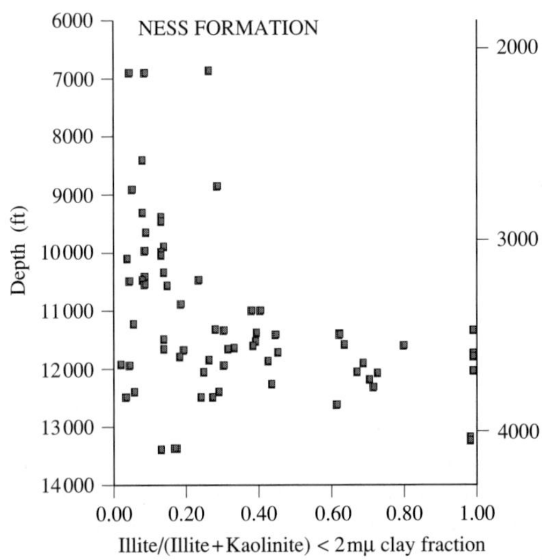


**Fig. 8.** Increase in the illite volume with depth in the Ness Formation of the Brent Group, North Sea. From Giles *et al.* (1992).

the rapid reduction in the volume of feldspars (Fig. 7), and the increase in the volume of illite (Fig. 8) are the most significant. Similar patterns of diagenetic modification are evident from the Oman and Sarawak data. In the latter case there is also a marked reduction in the volume of andesitic/basic volcanic rock fragments present.

## DISCUSSION OF THE OBSERVATIONAL EVIDENCE

The relatively abrupt onset of quartz cementation observed in Figs 1–5 suggests that quartz cementation may occur in two distinct phases: a minor phase as a result of shallow-burial processes, causing the precipitation of a few percentage of authigenic quartz, followed by deep-burial processes which contribute up to 20% of the total reduction in porosity. However, while the authigenic quartz abundance increases abruptly, porosity continues to fall gradually with burial depth (Fig. 6) in all the studied data sets.

Many of the models proposed for quartz cementation (see Introduction, or McBride, 1989; Bjørlykke & Egeberg, 1993) focus on individual formations or basins and attempt to explain local phenomena. Yet, the occurrence of similar patterns of authigenic quartz cementation world-wide suggests that quartz cementation with depth must reflect fundamental processes common to all basins. The differences in the depth at which significant quartz cementation starts between basins suggest that something more than depth or effective stress is the primary driving force.

The relationship between cementation and depth suggests that fundamental state variables which correlate with depth, such as temperature or stress, are responsible. Temperature history reconstructions (see for instance Giles *et al.*, 1992) demonstrate that although the depths at which significant quartz cementation varies, the temperature does not. In the North Sea, Oman, Sarawak and the Gulf Coast, the depth below which significant cementation occurs is approximately equivalent to the 90–100°C isotherm. Differences in depths of large-scale quartz cementation reflect differences in gross geothermal gradient between the areas. Fluid inclusion data summarized by Bjørlykke & Egeberg (1993) indicate that the onset of precipitation occurs at temperatures in the range of 70–100°C in the North Sea. At these temperatures pore fluids in general approach equilibrium between dissolved silica and alpha quartz (Giles & de Boer, 1990, their fig. 6).

Increasing temperature and stress may trigger reactions releasing silica (e.g. illitization of smectite) and the onset of pressure dissolution. The relatively modest increase in the solubility of silica in solution due to temperature increase alone, may be outstripped by the rise in the mass of silica produced by temperature dependent reactions, resulting in an oversaturated solution and hence in quartz cementation. The relationship of quartz cementation to other temperature controlled reactions, and the relatively rapid onset of this cementation, suggests that it is controlled by kinetics via a number of linked reactions.

Temperature, however, cannot be the only control on cementation. In the US Gulf Coast, Pliocene reservoir sandstones are significantly quartz cemented only below 4 km, whereas Miocene sandstones are cemented below 3 km (100°C). Mesozoic sands are cemented at still shallower depths. This suggests a kinetic control with, in addition to temperature, the length of time available for cementation playing a significant role.

In addition to both temperature and time, the vertical effective stress (see Giles, 1997, for definitions and further explanation) must also play a significant role. The grain to grain stresses which drive pressure dissolution are related to the vertical effective stress acting on the rock. Thus in undercompacted settings where significant overpressures have been present throughout the burial history, resulting in low vertical effective stress, compaction and pressure dissolution will be inhibited. This may explain why in some highly overpressured Fulmar reservoirs in the North Sea uncemented sands are present at depths in excess of 12 000 ft (3600 m).

Alternative explanations of the patterns of cementation are difficult to find given the wide range of geological settings and the consistency of the data. For instance, in situations such as the Brent Group in the North Sea, where data are available from adjacent fault blocks, with considerable offsets, the amount of quartz cement always follows that expected from the data shown in Fig. 1.

## INTRODUCTION TO MODELLING

In the following section we employ a computer-based physico-chemical model of diagenetic processes to analyse what, if any, are the fundamental controls on quartz cementation. The approach taken is to model porosity, permeability and mineralogical changes accompanying reactions between a pore fluid and a host rock, based on a continuum mechanical approach to the solution of the partial differential equations (PDEs) describing coupled heat–fluid mass transport. Details of the model are to be found in Giles (1997) and Beynon and Giles (in preparation). The equations are not dealt with here in detail, but the basic concepts are explained.

Cementation and dissolution occur as a result of reactions between the rock matrix and the pore fluid. To model the effect of all these processes on reservoir properties requires a number of critical components:

**1** Formulations of crystal growth and dissolution for each mineral.

**2** Chemical speciation model for the fluid.

**3** Mass transport model which includes diffusive, dispersive and advective flux terms.

**4** Numerical description of fluid movement, so that fluxes of reactants and the advective perturbation of the temperature field can be calculated.

**5** Calculation of the temperature field: most diagenetic reactions are surface controlled and therefore highly temperature sensitive.

These elements form the basis of a predictive diagenetic model capable of enhancing cur understanding of diagenetic processes. The general numerical framework to solve this problem is supplied by PDEs for solute mass, fluid mass and energy conservation, supplemented by equations describing the speciation and changes in the speciation of the pore fluid, and the dependence of permeability on grain size, specific surface and porosity. In the model presented here, the Kozeny–Carmen equation (Scheidegger, 1974) has been used for the calculation of permeability. The system of equations is described in detail by Giles (1997, chapters 11–14). These equations need to be solved as a fully coupled system in order to adequately allow for the feedback between all the processes.

Mineral–fluid reactions can be modelled either as an equilibrium thermodynamic process in which the fluid is in local equilibrium with the pore fluid, or as a kinetic process in which the reaction rates depend on the departure from equilibrium and on the temperature. Many low temperature systems, such as the Triassic sandstone aquifers of SW England (Walton, 1982) display considerable disequilibrium between the rock matrix and the pore fluid, suggesting that an equilibrium thermodynamic formulation is inappropriate, at least at low temperatures. This conclusion is supported by kinetic data for minerals such as quartz (Rimstidt & Barnes, 1980), which demonstrate that at low degrees of supersaturation and low to moderate temperatures, substantial lengths of time (> 1 Myr) are required to precipitate small quantities of a given mineral, in this case quartz. Thus, for at least some common minerals such as quartz, feldspars and carbonates, a kinetic rather than an equilibrium thermodynamic approach is required.

As a result of the large variation of groundwater velocities found in nature (Giles, 1987, 1997), it is necessary to model the flux of dissolved species, taking into account both advective water movement and diffusion. Where groundwater flow is significant, an additional process known as dispersion is important (Bear, 1972). Dispersion causes the solute front to spread out as a result of the inhomogeneities introduced by the pore structure. Larger-scale geological structures introduce further inhomogeneities which result in so-called macrodispersion. Like diffusion, the dispersive flux may be modelled as the product of the mechanical dispersion coefficient and the concentration gradient of the dissolved species. However, unlike diffusion, the process is driven by fluid movement. The magnitude of the dispersion coefficient, or dispersivity, is proportional to the flow rate and dependent on the scale over which the modelling is carried out (Kinzelbach, 1986). The length-scale dependence results from the increasing scale of inhomogeneities encountered along the flow path.

The model discussed here fully accommodates solute transport by moving pore fluids, diffusion and dispersion, and takes into account the feedback on changes in reservoir properties due to cementation and dissolution, both on the fluid velocity and the temperature field. The distribution of reactive components may be specified by the user. The chemical speciation model used is capable of accommodating complexation, but complexation has not been used in these calculations for several reasons:

**1** because of the lack of data on silica-organic complexes;

**2** calculations on metal complexation in real pore fluid compositions have shown it is insignificant in that context, due to the competing effects of other components (Giles *et al.*, 1994);

**3** complexation implies highly oversaturated solutions, but these have not been observed at temperatures above 50°C;

**4** the mechanisms by which the complex would transport silica along the migration path, and then suddenly become unstable allowing silica to precipitate, are unknown, and the process would seem to be an unlikely possibility.

In order to solve the PDEs describing the conservation equations it is necessary to specify a set of boundary and initial conditions. To specify the initial conditions the initial aquifer mineralogy, porosity and grain size along the length of the aquifer must be supplied by the user. The boundary conditions for mass transport are a specification of the fluid composition at the input point ($X = 0$) on the aquifer and a flux boundary condition at the far end of the model. A constant fluid pressure differential across the aquifer is used to drive fluid movement. Similarly a specified temperature (and therefore temperature difference) in the aquicludes (seals) above and below the aquifer supply the boundary conditions for the thermal model. The 1-D model presented is a steady state model (i.e. time derivatives of concentration, temperature, etc., are set to zero), whereas the 2-D model is fully transient. Every effort was used to remove any spurious numerical errors from the results.

For quartz and amorphous silica, the solubility and kinetic data of Rimstidt & Barnes (1980) have been employed. Note that silica dissolution/precipitation is a surface controlled reaction and therefore temperature sensitive. The solubility of silica increases with temperature, rising rapidly above 80°C. For instance, the solubility of dissolved silica roughly doubles between 80 and 110°C. Thus cooling of higher temperature fluids will produce a greater amount of cement than the same amount of cooling of a fluid below 80°C.

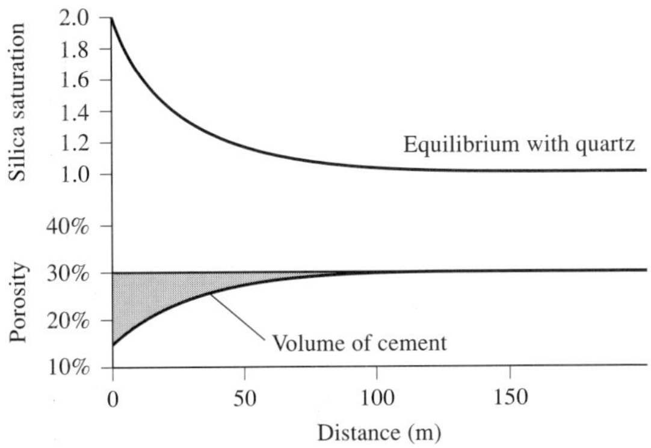


**Fig. 9.** Quartz cementation from an initially oversaturated fluid. Reservoir porosity and concentration of silica (relative to equilibrium with quartz) along the length of an aquifer, 0.1 Myr after a pore fluid, two times oversaturated with respect to quartz, started flowing into the aquifer. As quartz precipitates, the pore fluid becomes less saturated, and by the time it has travelled 70 m it is very close to equilibrium. This limits quartz cementation to the area close to the fluid entry point. This zone becomes more and more tightly cemented with time, until, eventually, the entrance to the aquifer becomes completely blocked. Results such as this indicate that oversaturated fluids cannot be transported over long distances in the subsurface and cannot cause cementation of large volumes of rock. Grain size = 0.23 mm; temperature = 50°C; initial permeability = 2000 mD; initial fluid flow velocity = 5.7 m yr$^{-1}$; initial porosity = 30%; initial quartz fraction = 70%.

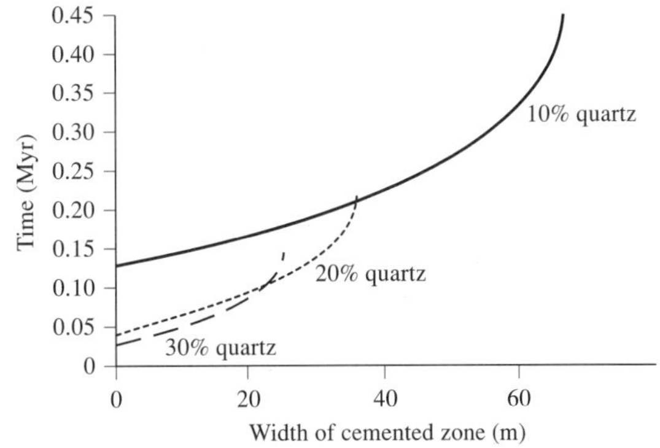


**Fig. 10.** Quartz cementation from an initially oversaturated fluid. The zone in which porosity is reduced by more than 5% expands through time until the entrance to the aquifer is completely blocked and no more silica can be transported in. A reservoir with a lower fraction of quartz grains (e.g. 10% quartz) has less sites on which quartz cement can nucleate and grow, hence the pore fluid is able to travel a greater distance before attaining equilibrium, and cementation occurs over a wider area. Grain size = 0.4 mm; initial permeability = 2000 mD; temperature = 50°C; flow rate = 5.7 m yr$^{-1}$; initial oversaturation with respect to quartz = 2; initial porosity = 30%.

## 1-D Modelling

In this section a 1-D version of the coupled heat–fluid mass transport model will be used to assess the feasibility of transporting silica from external sources compared with remobilizing internal sources. Given the applied boundary conditions, the model calculates the evolution of aquifer mineralogy, porosity, temperature, fluid velocity and fluid chemistry with time and distance along the aquifer.

The model will be used to investigate a number of possible scenarios for quartz cementation, these are firstly, the pore fluid is oversaturated with respect to alpha quartz; secondly, the pore fluid is in equilibrium with alpha quartz, but temperature differences drive changes in solubility of silica and hence cause cementation/dissolution; and thirdly, the silica is sourced from within the reservoir.

### *The pore fluid is oversaturated with respect to alpha quartz*

Figure 9 shows a simulation in which a fluid initially oversaturated (dissolved silica/silica required for equilibrium with alpha quartz = 2) with respect to alpha quartz flows through an isothermal aquifer at a temperature of 50°C (other details of the aquifer are given in the figure caption). The simulation shows that waters which are initially oversaturated will precipitate quartz as they travel through the reservoir and rapidly approach equilibrium downstream. Cementation occurs mainly where the oversaturated solution enters the aquifer. The area of the aquifer affected by this cementation is very narrow, and is determined primarily by the temperature (the higher the temperature, the narrower the cemented zone), the rate of fluid flow (the faster the flow rate, the wider the cemented zone) and the specific surface area. The specific surface area depends on the grain size and shape, the fraction of the minerals in the rock on which quartz can nucleate and grow, and the presence or absence of inhibitors (such as clay coatings) at the grain surface. A decrease in specific surface area results in a wider zone of cementation (Fig. 10). At typical reservoir temperatures (60–100°C) high flow rates (> 100 m yr$^{-1}$) are required for the cemented area to attain a width of more than 100 m. However, flow rates in the subsurface are generally much smaller (Giles, 1987, 1997). These simulations clearly indicate that cementation by the movement of oversaturated fluids in the subsurface is an untenable method of producing quartz cementation over wide areas.

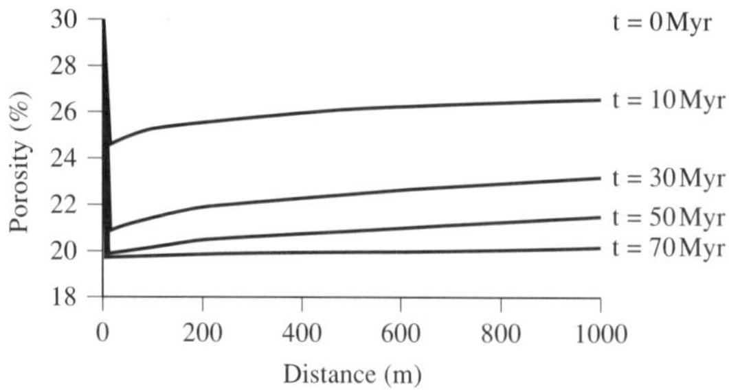


**Fig. 11.** Cooling of fluids initially in equilibrium with quartz. Pore fluids initially in equilibrium with the reservoir are cooled from 60°C to 50°C along the length of the aquifer, resulting in slow quartz precipitation. At higher temperatures, cementation occurs much more rapidly. Such models illustrate that up-dip cooling is a feasible method for quartz cementing large areas of a reservoir. Grain size = 0.4 mm; initial permeability = 2000 mD; initial porosity = 30%.

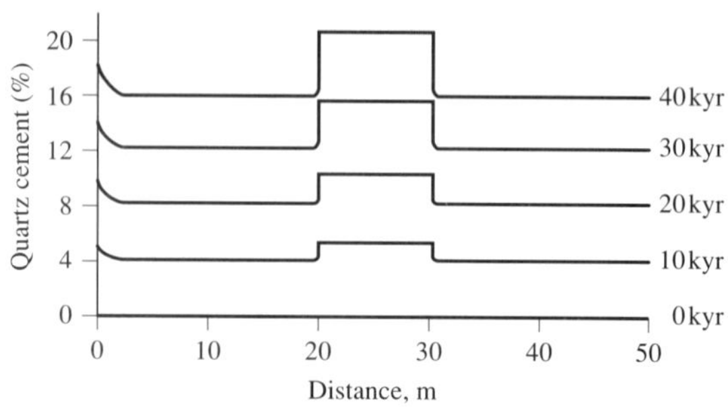


**Fig. 12.** Precipitation of quartz cement due to dissolution of amorphous silica. Quartz (initially 35%) and amorphous silica (35%) are distributed throughout the aquifer, but the amorphous silica has a smaller grain size in the central area (20–30 m). Dissolution of amorphous silica releases silica into the pore water. This silica is reprecipitated as quartz along the whole length of the aquifer. This redistribution occurs most rapidly where the grain size of the amorphous silica is smaller, because smaller grains are less stable due to their greater surface-area: volume ratio. Grain size = 0.4 mm; except amorphous silica (= 0.1 mm) at 20–30 m, initial porosity = 30%; temperature = 80°C; initial permeability = 2000 mD; fluid flow rate = 0.12 m $yr^{-1}$.

*The pore fluid is in equilibrium with the reservoir mineralogy*

Leder & Park (1986), Fang *et al.* (1987) and Robinson & Gluyas (1992) have proposed, on the basis of highly simplified models, that cooling pore fluids everywhere in equilibrium with the reservoir mineralogy are responsible for quartz cementation. Such a possibility follows as a result of the increasing solubility of silica with increasing temperature. Figure 11 indicates that this concept represents a feasible method of cementing large volumes of the reservoir over geological time scales. However, there are some problems with this hypothesis. Firstly, large volumes of fluid moving up-dip will, at velocities above 1 m $yr^{-1}$, affect the temperature field of the reservoir, reducing the temperature drop along the aquifer and therefore the amount of cementation (Giles, 1997). Secondly, in subsiding sedimentary basins the pore fluid is becoming progressively buried and hence it is heated through time (Berner, 1980). Given that in most geological settings significant convection is extremely unlikely (Giles, 1987, 1997; Bjørlykke *et al.*, 1988), then there are only two general settings where cooling of the pore fluid might occur: either the pore fluid is being pushed up-dip by a fluid potential gradient, or the temperature of the reservoir is dropping, e.g. due to uplift and erosion. Both of these conditions generally occur over limited areas, and although up-dip cooling may be able to explain or predict locally high abundances of authigenic quartz cement, it cannot explain widespread cementation over whole provinces.

*Silica supplied from within the reservoir*

Potential sources of silica within the reservoir include the dissolution of feldspars and heavy minerals, illitization of smectite, pressure dissolution of quartz, remobilization of metastable opaline or amorphous silica, and the remobilization of silica from small grains (silt and clay sized quartz grains, via Oswald ripening) to overgrowths on large quartz grains. The modelling techniques discussed above can also be used to evaluate this potential mechanism of quartz cementation.

For simplicity consider a reservoir composed of quartz and amorphous silica through which a pore fluid is moving which is initially in equilibrium with alpha quartz. As the solubility of amorphous silica is higher than that of alpha quartz, dissolution of the amorphous silica can provide the source for quartz cement. Precipitation (Fig. 12) occurs throughout the reservoir. The amount of cement is limited by the amount of silica which can be supplied by dissolution of the amorphous silica. The distribution of quartz cement is not significantly modified by downstream transport of silica (Fig. 13), except at unrealistically high flow rates. Thus the distribution of cement is determined by the distribution of the dissolving (or transforming) minerals and their grain size. Evidence from field studies indicates that porosity loss due to quartz cementation is dependent on grain size as well as on temperature (Houseknecht, 1986).

*Summary of results*

The results of the modelling exercises indicate that at

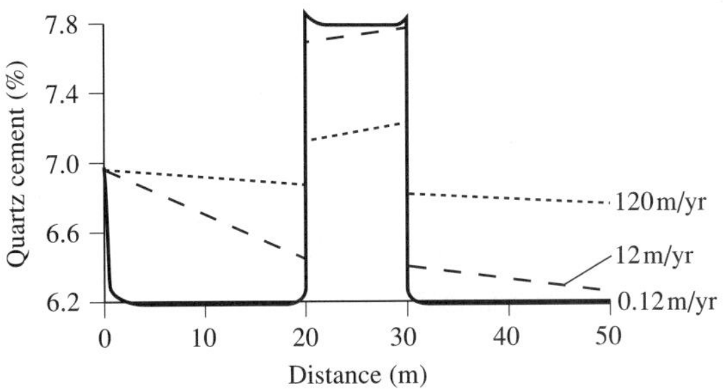


**Fig. 13.** Total quartz cement precipitated after 30 kyr at different flow rates. Even at flow rates as high as 120 m $yr^{-1}$, a more cemented area which mimics the initial grain size distribution can still be recognized. Flow rates in natural aquifers rarely exceed 5 m $yr^{-1}$, suggesting that most quartz cements precipitate near the silica source. All parameters as in Fig. 4, except temperature = 60°C.

flow velocities common in the subsurface (< 0.1 m $yr^{-1}$, see Giles, 1987, 1997) it is extremely difficult to cause widespread quartz cementation as a result of any form of oversaturated solution moving through the subsurface, be it due to the effects of advection (i.e. fluid movement), dispersion or diffusion. This arises because of the high reaction rates between the pore fluid and the detrital quartz which forms the largest single component of the sandstone. Thus, if mudstones export silica, then exported silica should be trapped close (within metres or centimetres) of the boundaries of the mudstone. The pattern of cementation around the mudstone should be characteristic of the process (see for instance the cementation pattern observed in Fig. 9). To import the volumes of quartz cement seen in the clastic reservoirs described by advection or diffusion from external sources would only be possible if:

**1** Fluid velocities were orders of magnitude higher than those used in the numerical experiments discussed in this paper. This is highly unlikely given the mass of subsurface fluid velocity data available (Giles, 1987, 1997).

**2** Silica solubilities were much higher than known from formation waters (see Giles, 1987) at temperatures greater than 50°C.

**3** Reaction rates were very much lower than known from experiments. This in turn would make the commonly observed control of dissolved silica concentration in natural waters by alpha quartz and other forms of silica difficult to explain.

Up-dip cooling of formation waters everywhere in equilibrium with quartz appears to have the potential to produce quartz cementation. However, relatively high fluid velocities are required, and additionally, in most subsiding sedimentary basins fluids are generally moving downwards, i.e. heating up, rather than cooling (see Berner, 1980), except locally, for instance around anticlines.

Any form of redistribution of silica from within the reservoir or close by is a viable mechanism for widespread quartz cementation providing the sources are evenly distributed throughout the reservoir. This topic will be investigated in the following 2-D model.

## 2-D Modelling

The example which will be discussed here is a simple anticline composed of a quartz sandstone of uniform grain size (0.025 mm) and a porosity of 30% embedded within shale aquicludes. A grid of 1071 nodes (51 by 21) has been used for the computation of the fully coupled PDEs representing the heat–solute-mass–fluid conservation equations, including a kinetic description of the quartz–fluid interaction and a full solute speciation model. Flow across the anticline is driven by a fluid pressure difference across the model. The dimensions of the model are 1.6 km in depth and 10 km in width.

The imposed pressure drop induced a horizontal component of fluid velocity ranging from 1 to 6 m $yr^{-1}$ (average around 3.5 m $yr^{-1}$) with the fluid moving from the left to the right in the model. The vertical component of velocity varies with position by ±2 m $yr^{-1}$. A large temperature gradient has been imposed across the model, with the base of the aquifer at a temperature of 130°C and the apex of the anticline at around 35°C. The reason for employing such a large temperature gradient was two fold; firstly to minimize rounding errors associated with small concentration differences, and secondly to enable the model to run in a reasonable time. Models with smaller temperature gradients produce similar results but the run times to produce the results seen in this model are prohibitive. Figure 14 shows the initial temperature distribution within the model. The position of the aquifer is shown by the streamlines. Note the elevated geotherm within the dome, arising from the combined effects of the advective heat transfer and the higher thermal conductivity of the aquifer compared with the surrounding shales. Pore fluid entering the model has a silica concentration in equilibrium with alpha quartz. Running the model enables us to observe the redistribution of silica which is driven by changes in temperature alone. Note that the model is constructed so that there is no net upward movement of the pore water.

### *Discussion of the modelling results*

Intuitively one would expect from the increased solubility

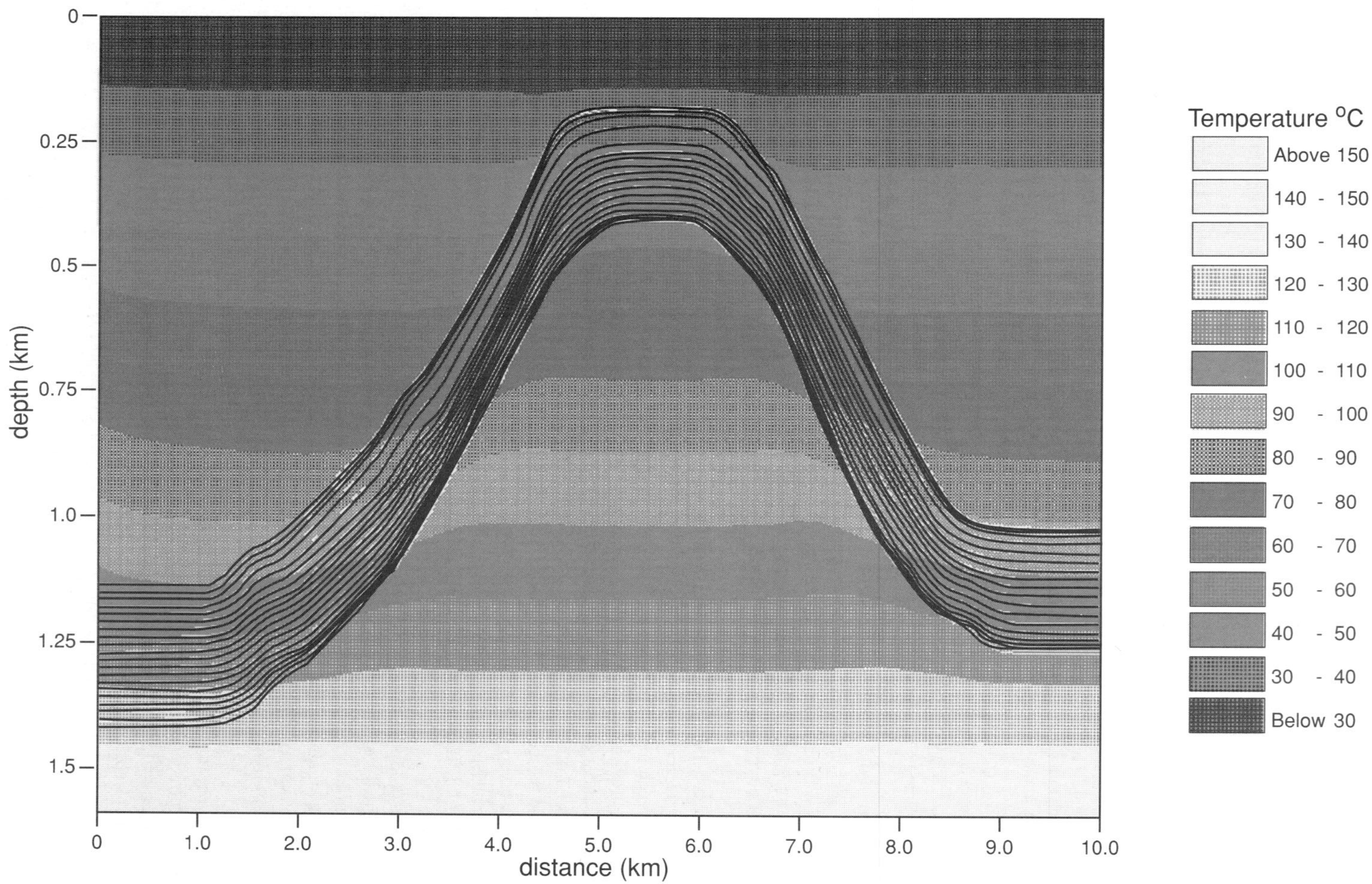


**Fig. 14.** Temperature distribution in the anticline model. In this model the anticline contains a bed of pure quartz sandstone with a grain size of 0.025 mm and an initial porosity of 30%, encased between shale aquicludes. The large temperature gradient (around 70°C $km^{-1}$) was imposed to decrease the model run-time and minimize errors. Note that the geotherm is elevated in the region of the crest, due to the combined effects of heat advection by the fluids in the aquifer, and conductivity contrasts between sandstone and shale. Solid lines are flow streamlines.

of alpha quartz with increased temperatures, that as the fluid cools on the up-going (left hand) limb of the anticline, cementation will occur whereas on the down-going limb dissolution will be prevalent. In gross terms this is indeed what is seen. However, in detail a much more complex pattern evolves.

Figure 15 shows the relative change in quartz grain size within the aquifer after 550 000 years. As expected, cementation is prevalent on the up-going limb, but interestingly, it is dominantly occurring toward the top of the bed, while dissolution is occurring toward the bottom of the bed. The reasons for this pattern involve at least three factors. Firstly, the tilt of the isotherms results in a maximum geothermal gradient oblique to the direction of fluid movement and inclined toward the top of the bed, so favouring cementation in that direction. Secondly, convective fluid movements resulting from the inclined isotherms (see Wood & Hewett, 1984; Bjørlykke *et al.*, 1988; Giles, 1997) will occur even though the fluid is in motion. Such movement will cause dissolution on the down-going limb and cementation on the up-going limb of the cell, resulting in cementation at the top of the bed and dissolution at the base. Finally fluid dispersion plays a major role. A fluid flowing in an infinitely thick reservoir will have dispersive fluxes which are equal upwards and downwards. However, in a confined aquifer (where the aquicludes supply no fluid), clearly some of the dispersive flux moving outwards from the centre of the aquifer will move up the temperature gradient, therefore heating, promoting dissolution toward the base of the bed. On the down-going limb, dissolution of quartz is the dominant feature. However, some cementation still occurs toward the top of the bed, for the same reasons as discussed above.

Figure 16 shows the porosity change after 1.35 Myr. As demonstrated by Fig. 16, the porosity structure is no longer the uniform 30% that it was initially, and, furthermore, the relatively simple pattern described above has also changed. The porosity difference section (Fig. 15) shows cementation on the up-going limb of the anticline. However, there are now two heavily cemented bands, one toward the top of the bed and the other approximately a third of the way from the base of the aquifer. This latter zone of cementation has developed because the higher porosity zone which formed toward the base of the bed captured a growing proportion of the flow as its permeability increased. Dispersion and convection out of this zone promoted cementation in the adjacent sandstone. Although the down-going limb of the fluid pathway is again dominated by dissolution, the two zones of cementation are again represented.

## VIABLE SOURCES OF SILICA AND MECHANISMS FOR CEMENTATION

The modelling so far has indicated that:

**1** It is not feasible to introduce silica in large volumes by flow of an oversaturated pore fluid through a sandstone aquifer. This mechanism can only produce localized cementation.

**2** Up-dip cooling of formation waters can account for quartz cementation in some areas. However, its general applicability is limited by the fact that in subsiding sedimentary basins, formation waters are in general moving downwards, i.e. being heated.

**3** The movement of formation waters within a non-isothermal aquifer system results in quartz redistribution, and leads to increased cementation toward the tops of beds and to the occurrence of more heavily cemented and less heavily cemented zones within an initially homogeneous aquifer.

**4** Redistribution of silica within the reservoir has none of the drawbacks encountered by other potential mechanisms.

The analysis suggests that the pattern of quartz cementation must either arise from redistribution of material already present within the reservoir or from a completely new process that we do not understand. Silica supply from adjacent mudstones could be important close to the mudstone beds, but as demonstrated by the 1-D modelling any free silica is unlikely to move very far into the sandstone before it reacts with detrital quartz. Assuming that silica is derived from within the reservoir, then the following are likely contributory sources:

Dissolution of feldspars and heavy minerals.

Illitization of dispersed smectite clays, or interbedded mudstones.

Dissolution of silt or clay sized quartz grains and reprecipitation on to larger quartz grains (Oswald ripening).

Dissolution of unstable silica polymorphs, e.g. amorphous silica.

Pressure dissolution both within the sandstone matrix and along stylolites.

Up-dip cooling of formation waters in structurally favourable positions.

Redistribution of silica within the reservoir due to flow within a non-isothermal temperature field.

Contribution from reactions in interbedded mudstones.

All of these are temperature/depth dependent processes and none of them alone can account for the volumes of authigenic quartz found in some reservoirs. However, in combination they almost certainly can.

In an earlier section it has been shown that the disappearance of feldspars with depth occurs over a similar

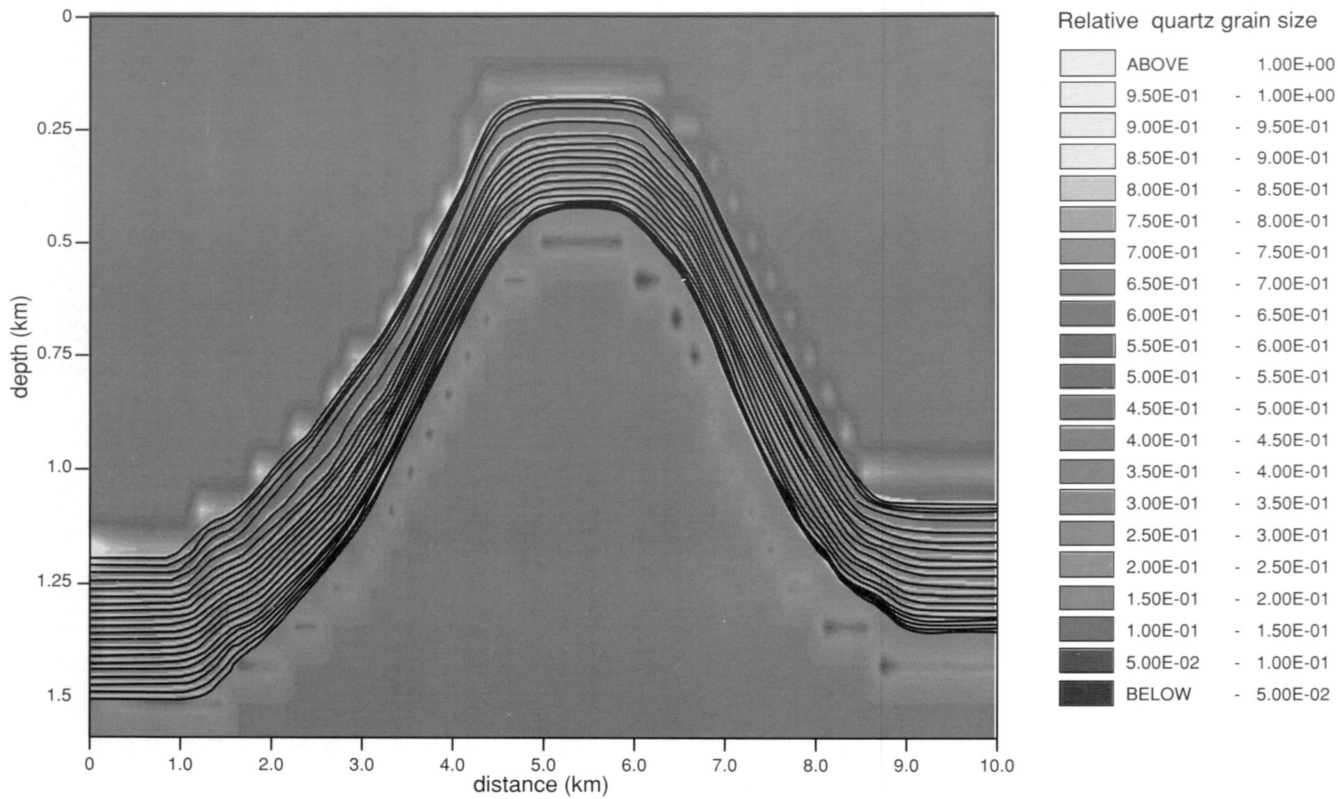


**Fig. 15.** Change in quartz grain size after 0.5 million years. The initial pattern of cementation is shown by the increase in quartz grain size occurring toward the top of the aquifer and a reduction in grain size toward the base.

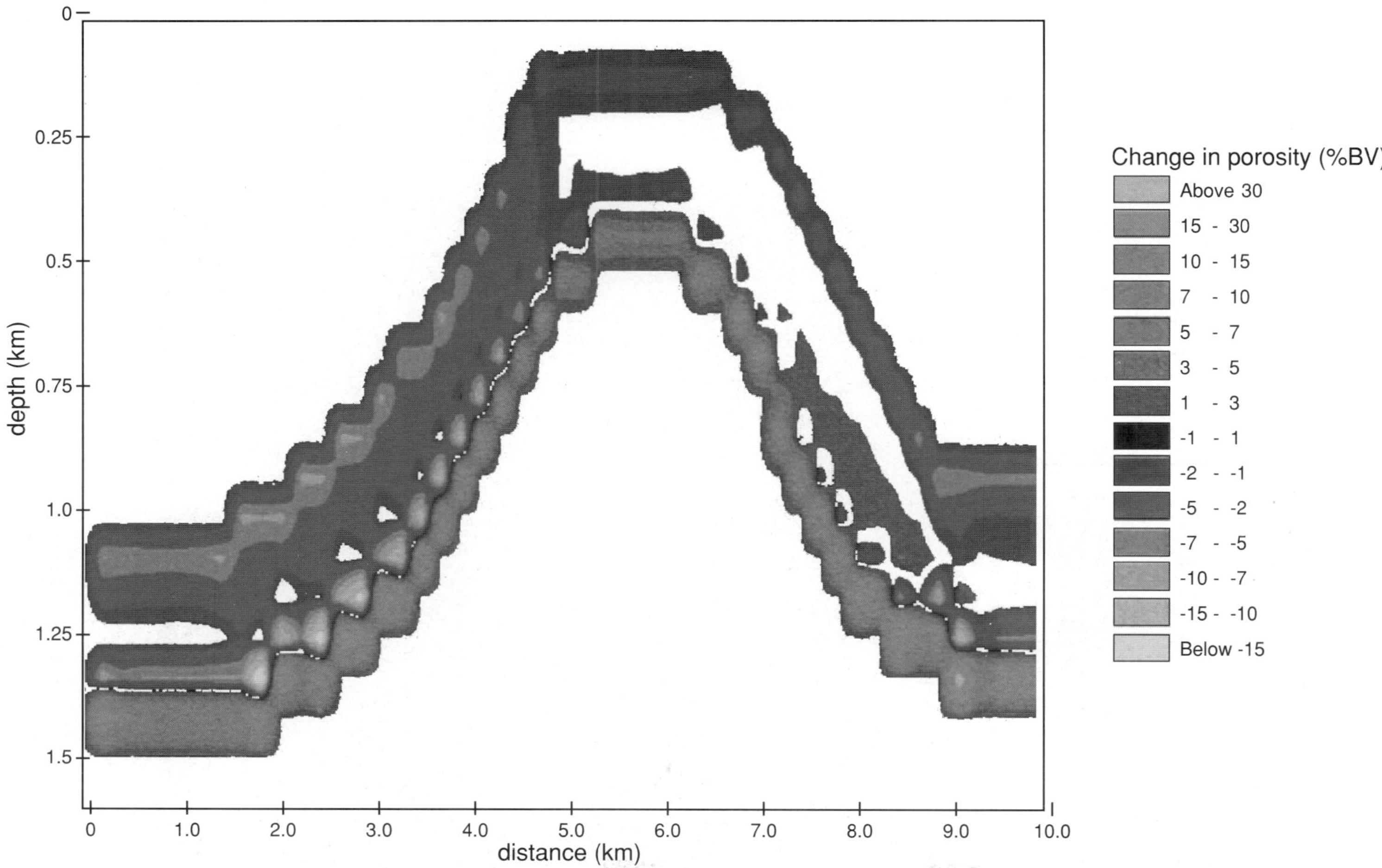


**Fig. 16.** Change in porosity after 1.35 million years. In general there is a decrease in porosity (i.e. cementation) on the up-going limb of the anticline. There are two heavily cemented bands—one at the top of the bed and one about a third of the way up the bed. The zone at the top of the bed has developed as a result of the temperature difference across the aquifer allowing fluid to cool. The second band is due to the high porosity zone developed at the base of the aquifer capturing the majority of the flow, combined with the effects of dispersion of silica out of the high velocity zone in the direction of falling temperature.

**Fig. 17.** Intense pressure dissolution within deeply buried (3600 m) Permo-Carboniferous Haushi Group sandstones, Oman. Cathodoluminescence techniques reveal the presence of authigenic quartz (AuQz) and pressure dissolution at grain contacts (arrows). Field of view approximately 0.5 mm.

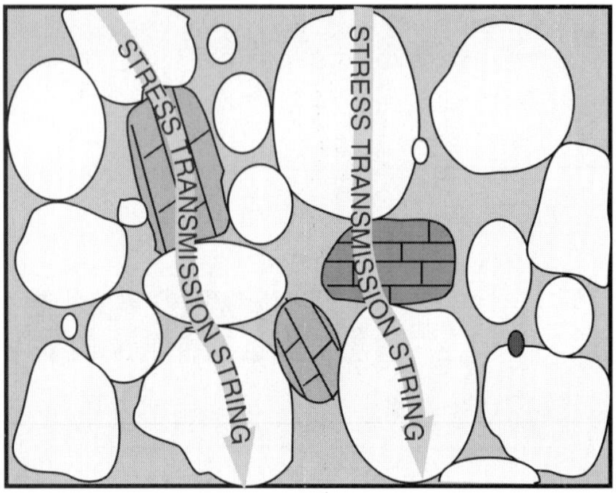


**Fig. 18.** Conceptual model of stress transmission strings. Not all grains within the sandstone will be load bearing. Hence, some grains are subjected to greater stresses than others.

depth/temperature range as massive quartz cementation. The decomposition of feldspar to authigenic quartz and kaolinite can yield between 41 and 45% of the original volume of the feldspar back as authigenic quartz. Thus, for instance in the Ness Formation of the Brent Group, the disappearance of up to 20% feldspar could yield up to 9% authigenic quartz.

Illitization of smectite clays could also provide silica for re-precipitation as authigenic quartz, as follows:

$$1.78 \text{ smectite} \rightarrow \text{illite} + 7.63 \text{ quartz} + \text{fluid}$$

This reaction (Leder & Park, 1986) produces 0.28 g of authigenic quartz for every gram of smectite consumed. Dispersed clays commonly form a few to 10 volume percentage of sandstones and consequently are a viable source of silica, as shown by the Brent Group data (Fig. 8). Similar reactions within the surrounding mudrocks might also contribute to the silica available for cementation. For cementation to occur the pore fluid must have access to the free surface of quartz grains and within mudrocks the dense distribution of clays may inhibit growth of quartz cement enabling the mudrocks to act as net exporters of silica to the surrounding sandstones. As shown by the 1-D modelling the exported silica will not penetrate far into the sandstones before reacting with detrital quartz. Numerous, thin, interbedded mudstones would thus be more efficient exporters of silica into sandstones than thick mudstone sequences.

Reactions such as these are strongly temperature dependent and can explain the temperature dependence of massive quartz cementation discussed above.

It is likely that all the reactions listed above provide a source of silica for quartz cementation. An additional source is provided by pressure dissolution along stylolites or within the sandstone matrix. However, the connection between pressure dissolution (e.g. Fig. 17) and the starting depth for massive quartz cementation is somewhat obscure. However, in the next section a possible link between the two will be examined.

## QUARTZ CEMENTATION AND COMPACTION

Earlier, a smooth transition from mechanical compaction to quartz cementation was demonstrated, this is suggestive of a continuum of processes operating. Rocks typically consist of a wide range of grain sizes. Some grains, generally clays, are sufficiently small enough to fit within pores. Thus, a sediment can be considered to be made up of load bearing and non-load bearing grains and porosity (Giles, 1997). The non-load bearing grains will be at a stress equivalent to the pore pressure. Hence, only part of the sandstone framework is actually transmitting the vertical load. Stress is probably transmitted through strings of larger grains, what will be termed stress transmission strings (Fig. 18). The existence of such strings was shown by the numerical simulations of Cunndle & Strack (1979) to determine how stresses were transmitted through grain aggregates. If such stress transmission strings did not exist, the presence of delicate fabrics often visible in thin sections (e.g. secondary porosity with preservation of the original grain rims) would not be possible.

The equations describing pressure-dissolution-driven redistribution of silica (de Boer, 1975, 1997a,b) are dependent both on the grain to grain stress within the stress transmission strings and the temperature, as the latter controls the rate of quartz dissolution and precipitation. The grain to grain stresses are related to the volume averaged stress carried by the stress transmission strings. Should the stress transmission strings become broken because of selective dissolution of some of the load bearing grains then the grain to grain stresses in the remaining strings will increase proportionally, promoting pressure dissolution.

Feldspars are typically large enough that they may form part of the load bearing framework of the sandstone. They can be preferentially dissolved, thereby breaking the stress transmission strings. Consequently, the disappearance of feldspars may also promote pressure dissolution.

Unfortunately, although this hypothesis goes some way to explain why pressure dissolution may become important in sandstones at depths in excess of 10 000 ft, it does not explain the common observation of enhanced pressure dissolution in the siltier sandstones and where the grains are coated with illite.

## CONCLUSIONS

**1** Quartz cementation follows similar patterns worldwide. This suggests that quartz cementation is controlled by fundamental processes that are common to all basins. It follows that it should be possible to predictively model quartz cementation.

**2** The depth below which abundant quartz cementation occurs decreases with increasing thermal gradient and age. This indicates that both the temperature (typically 90–100°C) and the residence time at temperatures above this figure control the onset of significant quartz cementation.

**3** Modelling indicates that the silica for quartz cementation must be derived from within the sandstone, though the directly adjacent mudstones may also contribute.

**4** The pattern of cementation strongly suggests sources which have a temperature and in some cases, a stress dependence. Temperature dependent contributory reactions include the dissolution of feldspars, heavy minerals and the transformation of smectite and kaolinite to illite. Pressure dissolution within sandstones and adjacent siltstones also becomes important over the same stress/temperature range, partly because the dissolution of framework grains will theoretically increase the stress on the remaining framework grains enhancing pressure dissolution. In addition, should fluid movement occur within a non-isothermal reservoir then quartz redistribution will be promoted.

## REFERENCES


BEAR, J. (1972) *Dynamics of Fluids in Porous Media.* Elsevier, New York.

BERNER, R.A. (1980) *Early diagenesis. A Theoretical Approach.* Princeton University Press, Princeton, NJ, 241pp.

BJØRLYKKE, K. (1980) Clastic diagenesis and basin evolution: Revista del Instituto de Investigaciones Geológicas. *Diputacion Provincial, Universidad de Barcelona* **34**, 21–44.

BJØRLYKKE, K. (1994) Fluid-flow processes and diagenesis in sedimentary basins. In: *Geofluids: Origin, Migration and Evolution of Fluids in Sedimentary Basins.* (ed. PARNELL, J.) pp. 127–140. Special Publication of the Geological Society of London 78.

BJØRLYKKE, K. & EGEBERG, P.K. (1993) Quartz cementation in sedimentary basins. *American Association of Petroleum Geologists Bulletin* **77**, 1538–1548.

BJØRLYKKE, K., MO, A. & PALM, E. (1988) Modelling of thermal convection in sedimentary basins and its relevance to diagenetic reactions. *Marine and Petroleum Geology* **5**, 338–353.

BJØRLYKKE, K., NEDKVITNE, T., RAMM, M. & SAIGAL, G.C. (1992) Diagenetic processes in the Brent Group (Middle Jurassic) reservoirs of the North Sea: an overview. In: *Geology of the Brent Group.* (eds MORTON, A.C., HASZELDINE, R.S., GILES, M.R. & BROWN, S.) pp. 263–287. Special Publication of the Geological Society of London 61.

BLATT, H. (1979) Diagenetic Processes. In: *Aspects of Diagenesis.* (eds SCHOLLE, P.A. & SCHLUGER, P.R.) pp. 141–157. Society of Economic Paleantologists and Mineralogists Special Publication 26.

DE BOER, R.B. (1975) *Influence of pore solutions on rock strength.* PhD Thesis, University of Utrecht.

DE BOER, R.B. (1977a) Pressure solution. Theory and Experiments, *Tectonophysics* **39** (1–3), 287–301

DE BOER, R.B. (1977b) On thermodynamics of pressure solution-interaction between chemical and mechanical forces. *Geochimica et Cosmochimica Acta* **41**, 249–256.

BURLEY, S.D. (1993) Models of burial diagenesis for deep exploration plays in Jurassic fault traps of the Central and Northern North Sea. In: *Petroleum Geology of Northwest Europe: Proceedings of the 4th Conference.* (ed. PARKER, J.R.) pp. 1353–1375. Geological Society of London.

CANALS, M. & MEUNIER, J.D. (1995) A model for porosity reduction in quartzite reservoirs by quartz cementation. *Geochimica et Cosmochimica Acta* **59** (4), 699–709.

CUNNDLE, P.A. & STRACK, O.D.L. (1979) The development of constitutive laws for soil using the distinct element method. In: *Proceedings of the 3rd International Conference on Numerical Methods in Geomechanics*, pp. 289–298.

DEWERS, T. & ORTOLEVA, P.J. (1990) Interaction of reaction, mass transport, and rock deformation during diagenesis: mathematical modelling of intergranular pressure solution, stylolites, and differential compaction/cementation. In: *Prediction of Reservoir Quality Through Chemical Modelling.* (eds MESHRI, I.D. & ORTOLEVA, P.J.) American Association of Petroleum Geologists Memoir 49, Tulsa, Oklahoma.

FANG, J.H., VISSCHER, P.B. & DAVIS, A.M.J. (1987) Porosity reduction in sandstone by quartz overgrowth: Discussion. *American Association of Petroleum Geologists Bulletin* **72** (12), 1515–1517.

FOTHERGILL, C.A. (1955) The cementation of oil reservoir sands and its origin. 4th World Petroleum Conference, Rome. *Proceedings of the Section* **1**, 301–314.

FRIEDMAN, M. (1954) Miocene orthoquartzite from New Jersey. *Journal of Sedimentary Petrology* **24**, 235–241.

FÜCHTBAUER, H. (1974) *Sediments and Sedimentary Rocks*, 1st edn. Wiley, New York.

GAUPP, R., MATTER, A., PLATT, J., RAMSEYER, K. & WALZEBUCK, J. (1993) Diagenesis and fluid evolution of deeply buried Permian (Rotliegende) gas reservoirs, northwest Germany. *American Association of Petroleum Geologists Bulletin* **77**, 1111–1128.

GILES, M.R. (1987) Mass transfer and the problems of secondary porosity creation in deeply buried hydrocarbon reservoirs. *Marine and Petroleum Geology* **4**, 188–204.

GILES, M.R. (1997) *Diagenesis: a Quantitative Perspective*. Kluwer, Amsterdam.

GILES, M.R. & DE BOER, R.B. (1990) Origin and significance of redistributional secondary porosity. *Marine and Petroleum Geology* **7**, 378–397.

GILES, M.R., DE BOER, R. & MARSHALL, J. (1994) How important are organic acids in generating secondary porosity in the subsurface? In: *The Role of Organic Acids in Diagenesis*. (eds PITTMAN, E. & LEWAN, M.) pp. 447–470. Springer Verlag, Berlin.

GILES, M.R., STEVENSON, S., MARTIN, S.V. *et al.* (1992) The reservoir properties of the Brent Group: A regional perspective. In: *Geology of the Brent Group*. (eds MORTON, A.C., HASZELDINE, R.S., GILES, M.R. & BROWN, S.) pp. 289–327. Special Publications of the Geological Society of London 61.

GLUYAS, J.G. & COLEMAN, M. (1992) Material flux and porosity changes during sediment diagenesis. *Nature* **356**, 52–54.

HARRIS, N.B. (1992) Burial diagenesis of the Brent sandstones: a study of the Statfjord, Hutton and Lyell fields. In: *Geology of the Brent Group*. (eds MORTON, A.C., HASZELDINE, R.S., GILES, M.R. & BROWN, S.) pp. 351–375. Special Publications of the Geological Society of London 61.

HASZELDINE, R.S., SAMSON, I.M. & CORNFORD, C. (1984) Quartz diagenesis and convective fluid movement; Beatrice Oilfield, UK North Sea. *Clay Minerals* **19**, 391–402.

HAWKINS, P.J. (1978) Relationship between diagenesis, porosity reduction, and oil emplacement in late Carboniferous sandstone reservoirs, Bothamsall oilfield, east Midlands. *Journal of Geological Society* **135**, 7–24.

HEALD, M.T. (1955) Stylolites in sandstones. *Journal of Geology* **63**, 101–114.

HEALD, M.T. (1956) Cementation of Simpson and St. Peter sandstones in parts of Oklahoma, Arkansas, and Missouri. *Journal of Geology* **64**, 16–30.

HEALD, M.T. (1959) Significance of stylolites in permeable sandstones. *Journal of Sedimentary Petrology* **29**, 251–253.

HOUSEKNECHT, D.W. (1986) Influence of grain size and temperature on intergranular pressure dissolution, quartz cementation, and porosity in quartz sandstones. *Journal of Sedimentary Petrology* **54**, 348–361.

JOURDAN, A., THOMAS, M., BREVART, O. *et al.* (1987) Diagenesis as the control of the Brent sandstone reservoir properties in the Greater Alwyn area (East Shetland Basin). In: *Petroleum Geology of North West Europe* (eds BROOKS, J. & GLENNIE, K.W.) pp. 951–961. Graham & Trotman, London.

KINZELBACH, W. (1986) *Groundwater Modeling: An Introduction with Sample Programs in Basic*. Elsevier, Amsterdam.

LAND, L.S. (1984) Frio Sandstone diagenesis, Texas Gulf Coast: A regional isotopic study. In: *Clastic Diagenesis*. (eds MCDONALD, D.A. & SURDAM, R.C.) pp. 47–62. American Association of Petroleum Geologists Memoir 37.

LEDER, F. & PARK, W.C. (1986) Porosity reduction in sandstone by quartz overgrowth. *American Association of Petroleum Geologists Bulletin* **70** (11), 1713–1728.

MCBRIDE, E.F. (1989) Quartz cementation in sandstones: a review. *Earth Science Reviews* **26**, 69–112.

OELKERS, E.H., BJØRKUM, P.A. & MURPHY, W.M. (1996) A petrographic and computational investigation of quartz cementation and porosity reduction in North Sea sandstones. *American Journal of Science* **296**, 420–452.

VAN DER PLAS, L. & TOBI, A.C. (1965) A chart for judging the reliability of point-count results. *American Journal of Science* **263**, 87–90.

RIEZEBOS, P.A. (1974) Scanning electron microscopical observations of weakly cemented Miocene sands. *Geologie en Mijnouw* **53**, 109–122.

RIMSTIDT, J.D. & BARNES, H.L. (1980) The kinetics of silica–water reactions. *Geochimica et Cosmochimica Acta* **44**, 1683–1699.

ROBINSON, A. & GLUYAS, J. (1992) Model calculations of loss of porosity in sandstones as a result of compaction and quartz cementation. *Marine and Petroleum Geology* **9**, 319–323.

SCHEIDEGGER, A.E. (1974) *The Physics of Flow through Porous Media*, 3rd edn, 353pp. University of Toronto Press, Toronto.

SCOTCHMAN, I., JOHNNES, L.H. & MILLER, R.S. (1989) Clay diagenesis and oil migration in Brent Group sandstones of NW Hutton field, UK North Sea. *Clay Minerals* **24**, 339–374.

SORBY, H.C. (1890) On the structure and origin of non-calcareous stratified rocks. *Quarterly Journal of the Geological Society London* **37**, 49–92.

TERZAGHI, K. & PECK, R.B. (1948) *Soil Mechanics in Engineering Practice*. Wiley, Chichester.

TOWE, K.M. (1962) Clay mineral diagenesis as a possible source of silica cement in sedimentary rocks. *Journal of Sedimentary Petrology* **32**, 26–28.

TREVENA, A.S. & CLARK, R.A. (1986) Diagenesis of sandstones reservoirs of Pattani Basin, Gult of Thailand. *American Association of Petroleum Geologists Bulletin* **70** (3), 299–308.

WALDERHAUG, O., LANDER, R.H., BJØRKUM, P.A. *et al.* (2000) (This volume) Modelling quartz cementation and porosity in reservoirs sandstones—examples from the Norwegian continental shelf. In: *Quartz Cementation in Sandstones*. (eds Worden, R.H. & Morad, S.) pp. 39–49. Special Publication of the International Association of Sedimentologists 29. Blackwell Science, Oxford.

WALDSCHMIDT, W.A. (1941) Cementing materials in sandstones and their influence on the migration of oil. American Association of Petroleum Geologists Bulletin **25**, 1839–1879.

WALTON, N.R.G. (1982) *A detailed hydrochemical study of groundwaters from the Triassic sandstone aquifer of south-west Britain*. Institute of Geological Sciences, Report no. 81/5.

WOOD, J.R. & HEWETT, T.A. (1984) Reservoir diagenesis and convective fluid flow. In: *Clastic Diagenesis*. (eds MCDONALD, D.A. & SURDAM, R.C.) pp. 99–111. American Association of Petroleum Geologists Memoir 37.

*Spec. Publs int. Ass. Sediment.* (2000) **29**, 39–49

# Modelling quartz cementation and porosity in reservoir sandstones: examples from the Norwegian continental shelf

O. WALDERHAUG[1]*, R. H. LANDER[2], P. A. BJØRKUM[3], E. H. OELKERS[4], K. BJØRLYKKE[5] *and* P. H. NADEAU[3]

[1]*Rogaland Research, PO Box 2503 Ullandhaug*, 4004 Stavanger, Norway;*
[2]*Geologica a.s., PO Box 2503 Ullandhaug, 4004 Stavanger, Norway;*
[3]*Statoil a.s., 4035 Stavanger, Norway;*
[4]*Université Paul Sabatier-CNRS, 38 rue de Trente-six Ponts, 31400 Toulouse, France; and*
[5]*University of Oslo, PO Box 1047 Blindern, 0316 Oslo 3, Norway*

## ABSTRACT

The volume of precipitated quartz cement and the resulting porosity loss in a quartzose sandstone can be calculated from the temperature history of the sandstone based on an equation relating the quartz precipitation rate per unit surface area and per unit time to temperature. In addition to temperature and time, the quartz surface area available for quartz cement precipitation will control the progress of quartz cementation within a given sandstone. Grain size, detrital grain mineralogy and abundance of grain coatings, factors which are controlled by provenance and depositional environment, are therefore also essential input parameters for modelling of quartz cementation.

Computed quartz cement volumes and porosities were compared with measured values for Brent Group sandstone samples from two wells in the northern North Sea. Porosities and quartz cement volumes in these sandstones currently vary from 8 to 19% and from 6 to 28%, respectively, due to large variations in grain size, grain coating abundance and quartz clast content. Despite these compositional and textural variations, modelled and measured values for both quartz cement and porosity in most cases differ by less than a few percent. Mean measured porosity and quartz cement volume differ from mean modelled porosity and quartz cement volume by less than one percent in both wells.

## INTRODUCTION

Extensive study of sandstones from the North Sea and from offshore mid-Norway during the last 15 years (e.g. Bjørlykke *et al.*, 1979, 1986; Ehrenberg, 1990; Walderhaug, 1990; Giles *et al.*, 1992; Bjørlykke & Egeberg, 1993, 1994a, 1994b) combined with recent chemical modelling of the quartz cementation process (Oelkers *et al.*, 1992, 1993, 1996; Walderhaug, 1996; Bjørkum *et al.*, 1998) suggests that quartz cementation of these sandstones has taken place as a three-step coupled process involving: quartz grain dissolution at the contacts between the quartz grains and clay or mica, short range diffusional transport of the dissolved silica, and precipitation as syntaxial quartz overgrowths on the surfaces of detrital quartz grains. This mechanism of quartz cementation thus has much in common with the process suggested 40 years ago by Heald (1955). In this paper a quantitative model for quartz cementation in sandstones via diffusional redistribution of silica within a closed system is presented and shown to accurately predict amounts of precipitated quartz cement and the resulting porosity-loss as a function of temperature history and mineralogical and textural parameters. As an example of the capabilities of the model, measured quartz cement volumes and porosities for sandstone samples from the Brent Group in the northern North Sea are compared with modelled values.

* Present address: Statoil a.s., 4035 Stavanger, Norway.

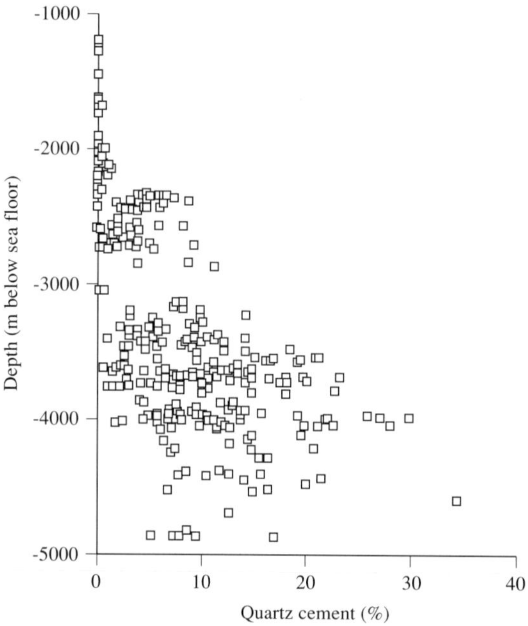


**Fig. 1.** Quartz cement content versus depth for 376 Jurassic and Palaeogene sandstone samples from the Norwegian shelf. Samples with more than 10% carbonate cement, detrital clay, mica or pyrite cement were not included on the plot, and samples with well-developed clay grain coatings and samples from uplifted areas were also excluded.

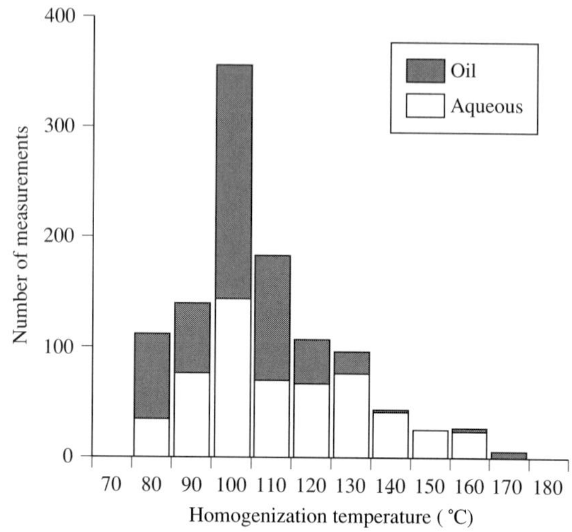


**Fig. 2.** Homogenization temperatures for 1080 fluid inclusions in quartz overgrowths in Jurassic, Triassic and Permian sandstones from the Norwegian shelf. Data from Walderhaug (1990, 1994a, 1995) plus 223 previously unpublished measurements.

## QUARTZ CEMENTATION IN SANDSTONES FROM THE NORWEGIAN SHELF

In Norwegian shelf sandstones that have not been subjected to major uplift, significant amounts of quartz cement are typically not encountered at depths of less than approximately 2 km. Below this depth the amount of quartz cement tends to increase as a function of depth (Fig. 1), and quartz cementation is the main cause of porosity loss at depths exceeding 3 km (Bjørlykke *et al.*, 1986, 1992; Ehrenberg, 1990). There is, however, considerable variation in the amount of quartz cement present at any specific depth (Fig. 1). This variation is in our opinion largely due to differences in grain size, quartz clast content and degree of grain coatings which in combination, cause large differences in the amount of quartz surface area available for quartz overgrowth formation. Also, a depth versus quartz cement plot does not adequately represent the effect of temperature history on quartz cementation because sandstones presently at the same depth may have reached this depth by a variety of burial paths, and geothermal gradients may vary both in space and time. However, with the exception of uplifted areas, the depth where significant quartz cementation is first encountered seems to normally correspond to a temperature of approximately 80°C in all parts of the Norwegian shelf (Bjørlykke *et al.*, 1989; Ehrenberg, 1990; Walderhaug, 1994a). Measurement of homogenization temperatures for fluid inclusions trapped in quartz overgrowths also suggests that significant quartz cementation starts at around 80°C (Fig. 2).

The abundant evidence for quartz grain dissolution at stylolites in deeply buried sandstones on the Norwegian shelf (e.g. Olaussen *et al.*, 1984; Bjørlykke *et al.*, 1986, 1992; Sverdrup & Prestholm, 1990; Walderhaug, 1994a; Bjørkum, 1996) combined with calculations strongly suggesting that fluid flow or long range diffusion (> 100 m) from shales cannot transport sufficient silica into thick sandstones (Bjørlykke, 1980; Bjørlykke & Egeberg, 1993), clearly points to stylolites as the dominant source of quartz cement. Note that our definition of a stylolite is a functional rather than a morphological definition, and thus encompasses all clay-rich or micaceous laminae where quartz is being dissolved at the interface and/or within the laminae due to the catalytic effect of clay and mica on quartz dissolution. Functional stylolites thus range in size from a single mica grain or a clay wisp between two quartz grains to laminae with tens of metres lateral extent. Primary clay drapes, clay rims around bioturbation structures, matrix-supported sandstone laminae and any other features where clay or mica is in contact with quartz can be included within this definition of a stylolite. Some quartz cement may also be supplied from illitization of kaolinite and K-feldspar within the sandstones (Hower *et al.*, 1976; Bjørlykke, 1980), although the volume of quartz cement produced from this reaction is less than half of the volume of the kaolinite consumed (Bjørkum *et al.*, 1993). Producing 10% quartz cement from illitization of kaolinite would therefore require consumption of more than 20% kaolinite plus dissolution of a similar volume of K-feldspar. Illitization of kaolinite therefore cannot be the main source of quartz cement in most strongly quartz cemented sandstones.

Dissolution of quartz at stylolites is thought not to be a pressure dissolution process, but rather a pressure-insensitive dissolution process induced by the presence of clay and mica (Bjørkum, 1996; Oelkers *et al.*, 1996). Within stylolites, the charge on mica and clay surfaces causes a different pore water composition close to the mica and clay. Quartz may be more soluble within these water layers than in the neighbouring pores, and the silica dissolved at stylolites may diffuse down the concentration gradient outwards from the stylolites and precipitate as syntaxial overgrowths on detrital quartz grains. The precipitation step appears to be the slowest step in the overall quartz cementation process, except when stylolite spacing is unusually large and diffusion distances correspondingly great (> 0.2–0.5 m depending upon temperature) (Oelkers *et al.*, 1992, 1993, 1996; Walderhaug, 1996; Bjørkum *et al.*, 1998). This is supported by the lack of obvious gradients in the amount of quartz cementation outwards from stylolites, for if diffusion were the rate-controlling step, then most quartz cement should precipitate close to the stylolites. Precipitation rate control makes the quartz cementation process strongly dependent upon the available quartz surface area, as indicated by the absence of significant quartz cementation in deeply buried sandstones with abundant grain coatings (e.g. Heald & Larese, 1974; Ehrenberg, 1993). Grain size and quartz clast content also affect quartz surface area and the amount of precipitated quartz cement, but temperature history is probably the most important factor for determining the abundance of quartz cement in deeply buried sandstones because quartz precipitation rates per unit surface area and time increase exponentially as a function of temperature (Rimstidt & Barnes, 1980; Walderhaug, 1994b).

## THE QUARTZ CEMENTATION MODEL

The present model simulates quartz cementation in settings where silica is supplied from clay- and mica-induced dissolution of quartz grains at stylolites, and where the silica diffuses short distances into the inter-stylolite regions where it precipitates on quartz grains as syntaxial overgrowths. This process has been mathematically represented in two ways, through a combined dissolution/diffusion/precipitation algorithm (Oelkers *et al.*, 1992, 1993, 1996), and by assuming precipitation rate control of the overall quartz cementation process (Walderhaug, 1994b, 1996). This latter precipitation model is now combined with a compaction algorithm in the Exemplar™ diagenetic modelling program (Lander *et al.*, 1995), and the modelling results presented in this paper were generated using this program.

The quartz cementation model implemented in the Exemplar™ program assumes that quartz cementation takes place uniformly within a 1 $cm^3$ reference volume located between stylolites. At constant temperature, the volume of quartz cement $Vq$ (in $cm^3$) precipitated in 1 $cm^3$ of sandstone containing a total quartz surface area A (in $cm^2$) during time t (in s) can be expressed as (Walderhaug, 1996):

$$Vq = MrAt/\rho \tag{1}$$

where M is the molar mass of quartz (60.09 g/mole), r is the quartz precipitation rate in moles/$cm^2$s, and ρ is the density of quartz (2.65 g/$cm^3$). Combined study of petrographic composition, fluid inclusions and temperature history (Walderhaug, 1994b) indicate that at diagenetic temperatures the temperature dependence of quartz precipitation rate can be approximated by a logarithmic function of temperature:

$$r = \mathrm{a}10^{\mathrm{b}T} \tag{2}$$

where T is temperature (°C) and a and b are constants with units of moles/$cm^2$s and per °C, respectively. For temperature histories consisting of time–temperature points linked by linear functions, Equation 2 can be replaced by an equation with time as the variable instead of temperature (Walderhaug, 1996):

$$r = \mathrm{a}10^{\mathrm{b}(c_n t + d_n)} \tag{3}$$

where $c_n$ is heating rate (°C/s), $d_n$ is the initial temperature (°C), and the index *n* refers to the relevant segment of the temperature history curve. By combining Equations 1 and 3, the volume of quartz cement precipitated in 1 $cm^3$ of sandstone from time $t_0$ to $t_m$ can be found as the sum of a series of integrals where each integral gives the volume of quartz cement precipitated during a time step:

$$Vq = \frac{M}{\rho}A_0\mathrm{a}\int_{t_0}^{t_1}10^{\mathrm{b}(c_1t+d_1)}\mathrm{d}t + \frac{M}{\rho}A_1\mathrm{a}\int_{t_1}^{t_2}10^{\mathrm{b}(c_2t+d_2)}\mathrm{d}t + \ldots\ldots \frac{M}{\rho}A_{m-1}\mathrm{a}\int_{t_{m-1}}^{t_m}10^{\mathrm{b}(c_mt+d_m)}\mathrm{d}t \tag{4}$$

Quartz surface area A is adjusted after each time step, and is calculated from the equation:

$$A = (1 - C)6fV\Phi/D\Phi_0 \qquad (5)$$

where C is the fraction of quartz grain surface coated by clay, other cements, etc., f is the volume fraction of detrital quartz in the volume V ($cm^3$) of sandstone under consideration, $\Phi$ is present porosity, $\Phi_0$ is depositional porosity and D is the grain diameter of the quartz clasts (cm). Equation 5 was originally derived for spherical grains, but in practice grain size is defined as mean quartz grain long axis as measured in thin-section. This is permissible because for most quartz grain populations the mean true nominal diameter (diameter of a sphere with the same volume as the particle under consideration) is within approximately 5% of the mean apparent long axis (Johnson, 1994). By calculating average long axis values, effects of sorting on surface area are also taken into account. Further details concerning the derivation of Equation 5 and the integration of Equation 4 are given in Walderhaug (1996).

The compaction algorithm in the Exemplar™ program models compaction as an exponential function of effective stress using lithology specific exponential constants (Smith, 1971; Ungerer *et al.*, 1990). Effective stress is calculated from the pressure history of the sandstone if such is available, otherwise hydrostatic conditions are assumed, and effective stress is calculated from the burial history. Rather than calculate porosity directly, the compaction function calculates intergranular volume (IGV) defined as the sum of porosity, cements and intergranular matrix in percent, and is given by:

$$\mathrm{IGV} = \mathrm{IGV}_{min} + (\Phi_0 + \mathrm{IM} - \mathrm{IGV}_{min})e^{-\beta\sigma} \qquad (6)$$

where $\mathrm{IGV}_{min}$ is the minimum intergranular volume (%), $\Phi_0$ is the initial porosity of the sediment (%), IM is the intergranular matrix (%), $\beta$ is the exponential rate of IGV decline ($MPa^{-1}$), and $\sigma$ is effective stress (MPa). If the user-defined minimum intergranular volume is reached, compaction is terminated. Optionally, compaction can also be turned off when a user-defined amount of quartz cement has precipitated in the modelled sandstone and thus strengthened the rock framework. Compaction is modelled as an irreversible process, and the modelled sandstone will not decompact if it is uplifted or if effective stress is reduced due to other factors.

## MODELLING QUARTZ CEMENTATION IN THE BRENT GROUP

To test the applicability of the proposed computational model, results from a detailed petrographic study of deeply buried Jurassic shallow marine Tarbert Formation quartzose sandstones from two wells in the northern part of the Norwegian sector of the North Sea were compared with quartz cement volumes and porosities calculated with the Exemplar™ program.

### Petrographic data

Forty-four and 48 thin-sections stained for carbonates and K-feldspar were available from the Tarbert Formation in two Norsk Hydro-operated wells A and B, respectively, and seven thin-sections were selected for point counting (500 points) from each of the two sample sets (Tables 1 and 2). Grain size was determined by measuring the long axis of 50–100 quartz grains per thin-section and calculating the average value for each sample. The percentage of the perimeter of each of these 50–100 quartz grains covered by detrital or authigenic clay, etc. was also estimated visually, and an average coating factor calculated for every sample. The point counted thin-sections were chosen to reflect the variations in grain size, degree of grain coating and abundance of quartz cement found in the Tarbert Formation in the two wells. Modelling of quartz cementation was not performed prior to point counting in order to avoid biasing sample selection. Samples with well-developed dust rims were selected where possible in order to facilitate discrimination between quartz cement and quartz overgrowths. Based on the absence of quartz overgrowths within matrix-supported areas and within areas filled by early carbonate cement, recycled quartz overgrowths from older sandstones are not thought to be present in the examined samples.

The point count data show that the Tarbert Formation in wells A and B comprises very fine-, fine- and medium-grained subarkosic and quartz arenitic sandstones with low to moderate detrital clay contents and present porosities in the range 8–19%. Quartz overgrowths form the only volumetrically important diagenetic phase, and account for 6–28% of sample volumes. Although detrital and authigenic clay is not abundant, thin clay coatings are common in some samples, and determined grain coating factors vary between 8% and 67% (Tables 1 and 2).

### Temperature and burial histories

A 1-D temperature and burial history generated with Norsk Hydro's internally developed program 1D Hydrobas was available for well A (Fig. 3), and a 2-D temperature and burial history constructed with 2D PetroMod (developed by IES Gmbh) for well B (Fig. 3). The temperature and burial histories were constructed for a single reference depth within the Tarbert Formation. Corrections

**Table 1.** Petrographic composition of Tarbert Formation samples, well A

| *Depth (mRKB)* | 4075.42 | 4077.74 | 4137.00 | 4142.45 | 4148.19 | 4153.31 | 4154.07 |
|---|---|---|---|---|---|---|---|
| *Depth (m below sea floor)* | 3947.42 | 3949.74 | 4009.00 | 4014.45 | 4020.19 | 4025.31 | 4026.07 |
| *Grain size (mm)* | 0.22 | 0.37 | 0.39 | 0.38 | 0.17 | 0.11 | 0.11 |
| *Coating factor (%)* | 49 | 45 | 45 | 67 | 53 | 27 | 22 |
| *Intergranular volume (%)* | 38.6 | 36.0 | 31.0 | 24.8 | 33.0 | 39.8 | 44.8 |
| *Clasts* | | | | | | | |
| quartz | 57.6 | 62.6 | 68.4 | 74.2 | 64.0 | 58.8 | 51.4 |
| chert | 0.2 | 0.2 | trace | 0.4 | — | — | 0.2 |
| K-feldspar | — | — | — | — | — | — | — |
| plagioclase | — | 0.2 | — | — | — | — | — |
| muscovite | 1.2 | 0.2 | — | trace | 1.2 | 0.4 | 1.4 |
| biotite | — | — | — | — | — | — | — |
| chlorite | — | — | — | — | — | — | — |
| heavy minerals | 0.4 | 0.2 | 0.2 | trace | 1.0 | 0.6 | 0.8 |
| clay clasts | 2.0 | 0.6 | 0.4 | 0.6 | 0.8 | 0.4 | 1.4 |
| detrital clay matrix | 8.8 | 6.6 | 3.8 | 7.0 | 4.6 | 6.4 | 8.4 |
| *Cements* | | | | | | | |
| quartz overgrowths | 13.2 | 10.6 | 11.8 | 6.2 | 9.6 | 22.8 | 28.2 |
| kaolinite/dickite | trace | 0.2 | 0.8 | 1.8 | — | — | trace |
| illite | 0.2 | 0.4 | — | trace | — | — | 0.2 |
| siderite | — | — | — | — | — | — | — |
| ferroan calcite | trace | 0.4 | — | — | — | — | — |
| dolomite | — | — | — | — | — | — | — |
| barite | 0.4 | trace | — | — | — | — | — |
| pyrite | — | — | — | 0.2 | 0.2 | 0.6 | — |
| *Porosity* | | | | | | | |
| primary | 10.0 | 13.4 | 13.0 | 5.8 | 14.6 | 6.4 | 3.8 |
| dissolution | 6.0 | 4.4 | 1.6 | 3.8 | 4.0 | 3.6 | 4.2 |

**Table 2.** Petrographic composition of Tarbert Formation samples, well B

| *Depth (mRKB)* | 3583.00 | 3598.00 | 3603.00 | 3613.00 | 3653.05 | 3669.00 | 3734.00 |
|---|---|---|---|---|---|---|---|
| *Depth (m below sea floor)* | 3461.00 | 3476.00 | 3481.00 | 3491.00 | 3531.05 | 3547.00 | 3612.00 |
| *Grain size (mm)* | 0.25 | 0.28 | 0.21 | 0.31 | 0.29 | 0.28 | 0.42 |
| *Coating factor (%)* | 11 | 24 | 32 | 29 | 8 | 18 | 16 |
| *Intergranular volume (%)* | 39.2 | 36.0 | 33.6 | 32.4 | 39.4 | 38.0 | 28.4 |
| *Clasts* | | | | | | | |
| quartz | 50.2 | 48.2 | 46.8 | 53.4 | 53.6 | 46.8 | 71.2 |
| chert | — | 0.2 | — | 0.6 | 0.2 | 0.2 | 0.2 |
| K-feldspar | 8.2 | 9.2 | 11.2 | 7.2 | 3.6 | 8.8 | ? |
| plagioclase | 1.2 | 1.0 | 0.6 | 1.0 | 0.6 | 1.2 | trace |
| muscovite | 0.8 | 2.4 | 3.6 | 3.0 | 1.8 | 1.8 | 0.2 |
| biotite | — | — | 0.2 | — | — | 0.2 | — |
| chlorite | — | — | — | 0.2 | — | — | — |
| heavy minerals | trace | 0.4 | 0.4 | 1.0 | 0.2 | 0.2 | trace |
| clay clasts | 0.4 | 2.6 | 3.6 | 1.2 | 0.6 | 2.8 | trace |
| detrital clay matrix | 0.8 | 1.6 | 0.4 | 1.4 | 1.0 | 2.2 | trace |
| *Cements* | | | | | | | |
| quartz overgrowths | 18.6 | 14.4 | 12.4 | 9.6 | 21.2 | 19.4 | 13.2 |
| kaolinite/dickite | trace | trace | trace | 0.6 | 1.0 | 0.2 | 0.4 |
| illite | 0.4 | 0.6 | 1.6 | 1.8 | 0.2 | 1.0 | — |
| siderite | — | 0.2 | trace | — | — | — | — |
| ferroan calcite | — | — | — | trace | — | — | — |
| dolomite | trace | — | — | — | trace | — | — |
| barite | — | — | — | — | — | — | — |
| pyrite | 0.6 | 0.2 | 0.2 | 0.8 | 0.6 | 0.2 | — |
| *Porosity* | | | | | | | |
| primary | 17.2 | 17.4 | 14.4 | 15.6 | 13.8 | 12.2 | 14.6 |
| dissolution | 1.6 | 1.6 | 4.6 | 2.6 | 1.6 | 2.8 | 0.2 |

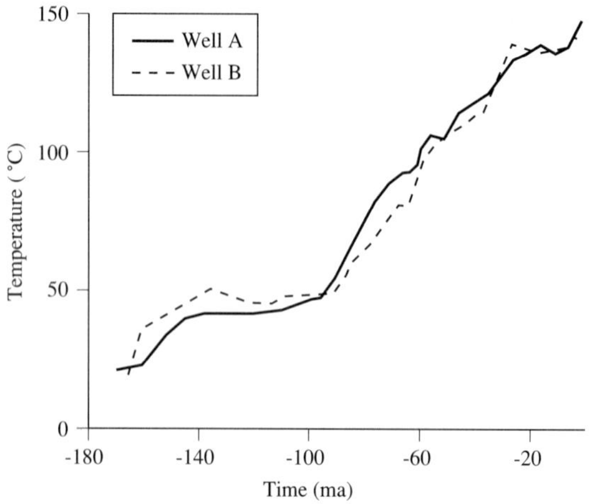


**Fig. 3.** Temperature histories for the top of the Tarbert Formation in wells A and B.

for slightly different sample depths were made by adding the present difference in depth between sample depths and the reference depth to the burial history. Temperatures were corrected by assuming a 5°C present surface temperature, calculating present mean geothermal gradient from the surface to the reference depth, and adding the product of the mean geothermal gradient multiplied with the depth difference between the modelled samples and the reference depth.

## Modelling procedures

The depositional composition of each sample was calculated from the point count data and an assumed surface porosity of 45%. Mechanical compaction was modelled assuming hydrostatic pressures calculated from the burial history using $\beta = 0.06$ MPa$^{-1}$, and minimum intergranular volume was for each sample set equal to the point counted IGV listed in Tables 1 and 2. Grain sizes and grain coating abundances were also taken from Tables 1 and 2. Grain coatings appear to be dominantly detrital and possibly also very early diagenetic, and were therefore modelled as being in place from one million years after deposition, i.e. long before quartz cementation was initiated. Based on typical observations from the northern North Sea (Bjørlykke *et al.*, 1989; Giles *et al.*, 1992), quartz cementation was assumed to start at 80°C, and modelling of quartz cement precipitation was therefore started at this temperature. Due to the very low rates of quartz cement precipitation at low temperatures, starting quartz cementation at, for instance 70°C, instead of at 80°C does not significantly change modelled results. Based on previous calibration of the quartz cementation model (Walderhaug, 1994b), the pre-exponential factor, a, in the equations controlling quartz precipitation rate was set to $2.0 \times 10^{-22}$ moles/cm$^2$s in all simulations. The exponential factor, b, was systematically varied between 0.018°C$^{-1}$ and 0.023°C$^{-1}$, and the optimal value for minimizing the differences between observed and modelled porosities and quartz cement volumes was determined. The calibration procedure was focused on the exponential factor because the uncertainty in this parameter has a far greater influence on precipitation rates than the uncertainty in the pre-exponential factor.

## Results

Modelled and measured quartz cement volumes and porosities are compared in Tables 3 and 4 and in Figs 4 and 5. Using $b = 0.0195$°C$^{-1}$, modelled quartz cement volumes in well A are within 0.2–3.9% of measured values except for sample 4148.19 mRKB where quartz cement volume is overestimated by 5.3%. Modelled porosities are within 0.2–4.0% except in the sample 4148.19 mRKB where overestimation of the quartz cement volume leads to a modelled porosity 4.5% below the measured value. Mean modelled porosity for well A samples is only 0.8% above measured mean porosity, and

**Table 3.** Measured and modelled quartz cement volumes and porosities, well A

| Sample (mRKB) | Measured quartz cement (%) | Modelled quartz cement (%) | Measured porosity (%) | Modelled porosity (%) |
|---|---|---|---|---|
| 4075.42 | 13.2 | 11.4 | 16.0 | 18.5 |
| 4077.74 | 10.6 | 8.3 | 17.8 | 20.8 |
| 4137.00 | 11.8 | 8.7 | 14.6 | 18.6 |
| 4142.45 | 6.2 | 4.0 | 9.6 | 13.2 |
| 4148.19 | 9.6 | 14.9 | 18.6 | 14.1 |
| 4153.31 | 22.8 | 26.7 | 10.0 | 6.7 |
| 4154.07 | 28.2 | 28.4 | 8.0 | 8.2 |
| *Mean* | 14.6 | 14.6 | 13.5 | 14.3 |

**Table 4.** Measured and modelled quartz cement volumes and porosities, well B

| Sample (mRKB) | Measured quartz cement (%) | Modelled quartz cement (%) | Measured porosity (%) | Modelled porosity (%) |
|---|---|---|---|---|
| 3583.00 | 18.6 | 19.0 | 18.8 | 18.9 |
| 3598.00 | 14.4 | 14.0 | 19.0 | 20.2 |
| 3603.00 | 12.4 | 14.8 | 19.0 | 17.5 |
| 3613.00 | 9.6 | 11.5 | 18.2 | 17.4 |
| 3653.05 | 21.2 | 19.7 | 15.4 | 17.4 |
| 3669.00 | 19.4 | 16.5 | 15.0 | 18.5 |
| 3734.00 | 13.2 | 15.4 | 14.8 | 13.7 |
| *Mean* | 15.5 | 15.8 | 17.2 | 17.7 |

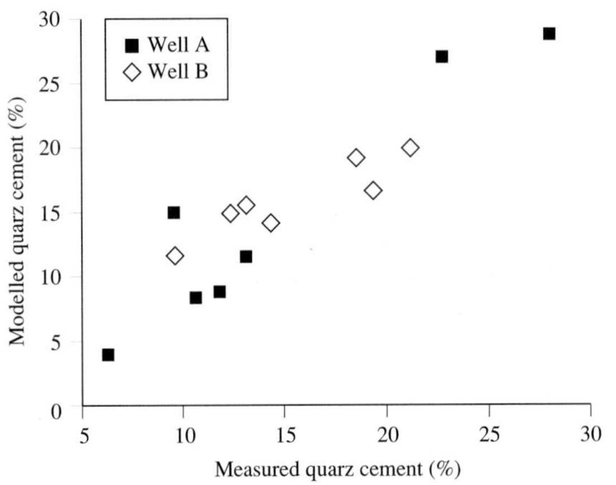


**Fig. 4.** Modelled quartz cement volumes versus measured quartz cement volumes.

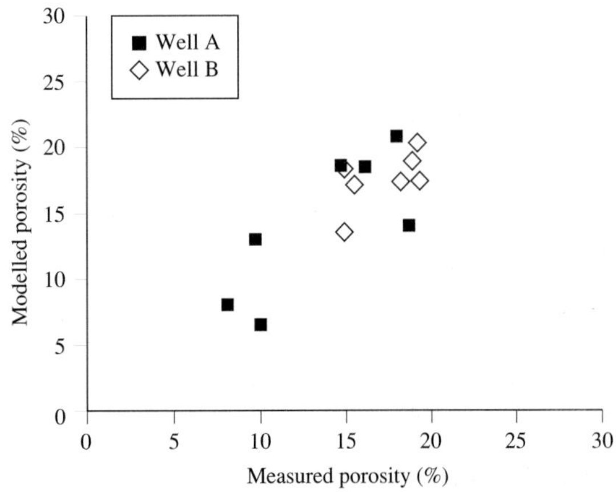


**Fig. 5.** Modelled porosities versus measured porosities.

mean modelled and measured quartz cement volumes are identical (Table 3).

In well B, optimal results were obtained with b = 0.0200°C$^{-1}$. Computed quartz cement volumes are then within 0.4–2.9% of measured values, and modelled porosities are within 0.1–3.5% of measured porosities. Mean modelled quartz cement content and mean modelled porosity deviate from their corresponding measured values by only 0.3% and 0.5%, respectively (Table 4).

## DISCUSSION

Comparison of computed and measured quartz cement volumes and porosities for wells A and B indicate that the computational model in most cases is able to calculate present porosities and quartz cement volumes to within a few percent of measured values, and that mean modelled and measured porosities for a set of samples can be estimated to within one percent of their measured counterparts (Tables 3 and 4). Since values determined by point counting probably have an uncertainty of a few percent, this implies that the modelled values are mostly within the uncertainty of the point count data. The modelled samples have a large range of grain sizes (0.11–0.42 mm) and of grain coating (8–67%) and also show a considerable variation in the percentage of detrital quartz (46.8–74.2%). The close correspondence of measured and modelled porosities and quartz cement volumes therefore indicates that the computational model adequately handles these textural and mineralogical variations.

Theoretically, the coefficients a and b controlling the rate of quartz cement precipitation per unit time and per unit quartz surface area should be essentially constant for all quartzose sandstones. However, this is only the case if temperature histories are accurate, and considerable uncertainty is indicated by the fact that the temperature history for well B, modelled with the 1D Hydrobas system, differs significantly from results obtained with the 2D PetroMod (Fig. 6). In practice therefore it will typically

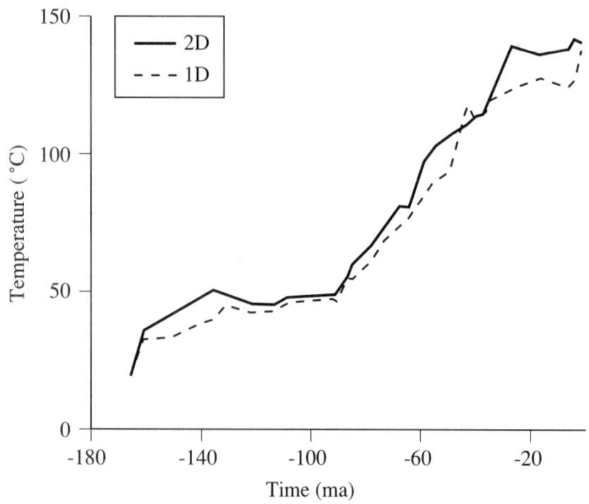


**Fig. 6.** Temperature histories for the top of the Tarbert Formation in well B constructed with two different basin modelling programs.

be necessary to adjust the b-parameter in order to obtain an optimal fit between modelled and measured porosities and quartz cement volumes. This does *not*, however, mean that any value of b can be used in order to obtain a fit between modelling and observations for each individual sample. All the samples from a given well must be modelled with the same b-value as long as they are being modelled with the same temperature history.

Although the presented model can compensate for inaccurate temperature histories when modelling results are calibrated against petrographic data, inaccurate temperature histories may cause some problems during pre-drill evaluation of reservoir quality in new prospects where no petrographic data are available. When determining the porosity of a prospect, a choice will have to be made concerning which quartz cementation kinetics should be used with the assumed temperature history for the prospect. The preferred option is to use the same methodology and the same basin modelling system for generating temperature histories of the prospect and for the calibration wells where optimal kinetics can be determined. The kinetics obtained by modelling of samples from the calibration wells will have the lowest degree of uncertainty relative to the prospect.

To illustrate the effect of uncertain temperature histories on model predictions, porosities and quartz cement volumes were calculated for well B samples using optimal kinetic parameters for the 2-D temperature history ($b = 0.0200°C^{-1}$) with both the 2-D and the colder 1-D temperature history. Modelled quartz cement volumes using the 1-D temperature history are then typically 4–5% lower than for the 2-D temperature history thus giving 4–5% higher porosities for the 1-D temperature history compared with the 2-D temperature history (Table 5).

Grain size, quantity of quartz clasts and amount of grain coatings also have a considerable effect on quartz cementation through their influence on quartz surface area. Of these three factors, quantity of quartz clasts is typically the least important because this factor rarely varies by more than a factor of two, and is also easier to predict for a prospect than the grain coating abundance and grain size. In contrast, grain size and coating abundance may vary by a factor of 10. This variation is related to sandstone provenance and depositional setting, and an understanding of how provenance and sedimentary facies control sandstone composition and texture will therefore be very important when modelling reservoir quality prior to drilling. The effect of variations in grain size, amount of quartz clasts and grain coating abundance can be illustrated by varying one of these three parameters at a time and modelling quartz cementation and porosity loss given the same temperature history (Tables 6–8).

In addition to uncertainties in input and calibration data, the model itself contains various approximations. The model does not take into account variations in silica supersaturations, and therefore in quartz precipitation

**Table 5.** Predicted porosities and quartz cement for well B using optimal kinetics for 2-D temperature history for both 2-D (warm) and 1-D (cold) temperature histories

| Sample (mRKB) | Quartz cement (%) 1-D | Quartz cement (%) 2-D | Porosity (%) 1-D | Porosity (%) 2-D |
|---|---|---|---|---|
| 3583.00 | 13.5 | 19.0 | 24.4 | 18.9 |
| 3598.00 | 9.9 | 14.0 | 24.5 | 20.2 |
| 3603.00 | 10.4 | 14.8 | 21.9 | 17.5 |
| 3613.00 | 8.0 | 11.5 | 20.9 | 17.4 |
| 3653.05 | 14.1 | 19.7 | 23.0 | 17.4 |
| 3669.00 | 11.5 | 16.5 | 23.6 | 18.5 |
| 3734.00 | 11.0 | 15.4 | 18.1 | 13.7 |
| *Mean* | 11.2 | 15.8 | 22.3 | 17.7 |

**Table 6.** Effect of grain coatings on modelled quartz cement and porosity

| Coating (%) | Modelled porosity (%) | Modelled quartz cement (%) |
|---|---|---|
| 0.0 | 13.5 | 13.8 |
| 25.0 | 16.1 | 11.2 |
| 50.0 | 19.2 | 8.1 |
| 75.0 | 22.9 | 4.4 |
| 90.0 | 25.5 | 1.8 |
| 100.0 | 27.3 | 0.0 |

Modelled with temperature history, composition and texture of well A sample 4137.00 mRKB and b = 0.0195.

**Table 7.** Effect of grain size on modelled quartz cement and porosity

| Grain size (mm) | Modelled porosity (%) | Modelled quartz cement (%) |
|---|---|---|
| 0.1 | 5.9 | 21.4 |
| 0.2 | 12.8 | 14.5 |
| 0.3 | 16.5 | 10.8 |
| 0.5 | 20.2 | 7.1 |
| 1.0 | 23.5 | 3.4 |
| 2.0 | 25.3 | 2.0 |

Modelled with temperature history, composition and clay coatings of well A sample 4137.00 mRKB and b = 0.0195.

**Table 8.** Effect of quartz clast abundance on modelled quartz cement and porosity

| Quartz clasts (%) at present time | Modelled porosity (%) | Modelled quartz cement (%) |
|---|---|---|
| 20.0 | 24.4 | 2.9 |
| 30.0 | 23.1 | 4.2 |
| 40.0 | 21.8 | 5.5 |
| 50.0 | 20.6 | 6.7 |
| 60.0 | 19.5 | 7.8 |
| 68.4 | 18.6 | 8.7 |

Modelled with temperature history, textural and compositional data for well A sample 4137.00 mRKB (except amount of quartz clasts) and b = 0.0195.

rate per unit surface and time, caused by (i) illitization of kaolinite within the sandstones; (ii) variations of stylolite spacing; (iii) distance from stylolites; and (iv) ratios of surface area available for quartz precipitation to surface area of stylolites (Oelkers *et al.*, 1992, 1993, 1996, Walderhaug, 1996). Similarly, pore pressure is not included as a variable in the quartz cementation algorithm because it has a negligible effect on quartz precipitation rates (Oelkers *et al.*, 1996), although it is taken into account in the compaction algorithm through the effective stress term. Furthermore, it is difficult to define a surface area function that can fully take into account the effects of variable sorting, grain shape and grain roughness, and slight porosity loss due to dissolution at individual grain contacts may have occurred in some of the samples. Nevertheless, the close correspondence between modelled and measured quartz cement volumes and porosities suggests that including the effects of the mentioned factors in the quartz cementation model would not greatly improve results, except in extreme cases with unusually large stylolite spacings or grain size contrasts.

The presence of hydrocarbons was assumed to have a negligible influence on quartz cementation rates because silica diffusion and quartz precipitation can still take place as long as the sandstones remain water-wet (Lowry, 1956; Rittenhouse, 1971), and comparison of water-filled and hydrocarbon-filled Brent Group reservoirs indicates that the hydrocarbon-filled sandstones are not more porous and less quartz cemented than the sandstones lacking hydrocarbons (Giles *et al.*, 1992). However, at high temperatures and large stylolite spacings, the quartz cementation process may tend to become more transport-controlled, and the presence of hydrocarbons may therefore in some cases significantly reduce the rate of quartz cementation (cf. Oelkers *et al.*, 1996; Bjørkum *et al.*, 1998).

## CONCLUSIONS

Quartz cementation of sandstones and the resulting porosity loss can be successfully predicted with a relatively simple mathematical model using temperature history, grain size, abundance of grain coatings and detrital composition as input.

Despite large variations in grain size, grain coating abundance and quantity of detrital quartz, computed quartz cement volumes and porosities are in most cases within less than a few percent of measured values for the studied Brent Group sandstone samples. Mean predicted porosity and quartz cement volume differ by less than a percent from mean measured porosity and quartz cement volume for both sample sets. The differences between computed and measured values are of approximately the same magnitude as the uncertainty of the point counts, which suggests that the model successfully simulates quartz cementation and the resulting porosity loss.

The accuracy of temperature histories may be a practical problem when modelling porosities in new prospects, and grain size and amount of grain coatings will also significantly affect results. However, the proposed computational model probably offers the best presently available technique for rapidly and quantitatively evaluating the effect of these parameters on sandstone porosity.

## ACKNOWLEDGEMENTS

The study of the Tarbert Formation was performed for Norsk Hydro a.s., and their permission to publish results from this study is gratefully acknowledged. Temperature histories for wells A and B were supplied by Dag Hermansen. The original manuscript was improved by the comments of referees Jon Gluyas, Lynton Land and Richard Worden.

## REFERENCES

BJØRKUM, P.A. (1996) How important is pressure in causing dissolution of quartz in sandstones? *Journal of Sedimentary Research A* **66**, 147–154.

BJØRKUM, P.A., WALDERHAUG, O. & AASE, N.E. (1993) A model for the effect of illitization on porosity and quartz cementation in sandstones. *Journal of Sedimentary Petrology* **63**, 1089–1091.

BJØRKUM, P.A., OELKERS, E.H., NADEAU, P.H., WALDERHAUG, O. & MURPHY, W.M. (1998) Porosity prediction in quartzose sandstones as a function of time, temperature, depth, stylolite frequency and hydrocarbon saturation. *American Association of Petroleum Geologists Bulletin* **82**, 637–648.

BJØRLYKKE, K. (1980) Clastic diagenesis and basin evolution. Revista del Instituto de Investigaciones Geológicas. *Diputacion Provincial Universidad de Barcelona* **34**, 21–44.

BJØRLYKKE, K. & EGEBERG, P.K. (1993) Quartz cementation in sedimentary basins. *American Association of Petroleum Geologists Bulletin* **77**, 1538–1548.

BJØRLYKKE, K., ELVERHØI, A. & MALM, A.O. (1979) Diagenesis in Mesozoic sandstones from Spitsbergen and the North Sea—a comparison. *Geologische Rundschau* **68**, 1152–1171.

BJØRLYKKE, K., AAGAARD, P., DYPVIK, H., HASTINGS, D.S. & HARPER, A.S. (1986) Diagenesis and reservoir properties of Jurassic sandstones from the Haltenbanken area, offshore mid-Norway. In: *Habitat of Hydrocarbons on the Norwegian Continental Shelf.* (ed. SPENCER, A.M.) pp. 275–286. Graham & Trotman, London.

BJØRLYKKE, K., RAMM, M. & SAIGAL, G.C. (1989) Sandstone diagenesis and porosity modification during basin evolution. *Geologische Rundschau* **78**, 243–268.

BJØRLYKKE, K., NEDKVITNE, T., RAMM, M., SAIGAL, G.C. (1992) Diagenetic processes in the Brent Group (Middle Jurassic) reservoirs of the North Sea: an overview. In: *Geology of the Brent Group.* (eds MORTON, A.C. *et al.*) pp. 263–287. Special Publications of the Geological Society of London 61.

EHRENBERG, S.N. (1990) Relationship between diagenesis and reservoir quality in sandstones of the Garn Formation, Haltenbanken, mid-Norwegian continental shelf. *American Association of Petroleum Geologists Bulletin* **74**, 1538–1558.

EHRENBERG, S.N. (1993) Preservation of anomalously high porosity in deeply buried sandstones by grain-coating chlorite: examples from the Norwegian continental shelf. *American Association of Petroleum Geologists Bulletin* **77**, 1260–1286.

GILES, M.R., STEVENSON, S., MARTIN, S.V. *et al.* (1992) The reservoir properties and diagenesis of the Brent Group: a regional perspective. In: *Geology of the Brent Group.* (ed. MORTON, A.C. *et al.*) Special Publications of the Geological Society of London 61, pp. 289–327.

HEALD, M.T. (1955) Stylolites in sandstones. *Journal of Geology* **63**, 101–114.

HEALD, M.T. & LARESE, R.E. (1974) Influence of coatings on quartz cementation. *Journal of Sedimentary Petrology* **44**, 1269–1274.

HOWER, J., ESLINGER, E.V., HOWER, M.E. & PERRY, E.A. (1976) Mechanism of burial metamorphism of argillaceous sediment, I. Mineralogical and chemical evidence. *Geological Society of America Bulletin* **87**, 725–737.

JOHNSON, M.R. (1994) Thin section grain size analysis revisited. *Sedimentology* **41**, 985–999.

LANDER, R.H., WALDERHAUG, O., LYON, A. & ANDERSEN, Å. (1995) Reservoir quality prediction through simulation of compaction and quartz cementation (abstract). *American Association of Petroleum Geologists 1995 Annual Convention Official Program*, 53App.

LOWRY, W.D. (1956) Factors in loss of porosity by quartzose sandstones of Virginia. *American Association of Petroleum Geologists Bulletin* **40**, 489–500.

OELKERS, E.H., BJØRKUM, P.A. & MURPHY, W.M. (1992) The mechanism of porosity reduction, stylolite development and quartz cementation in North Sea sandstones. In: *Water–Rock Interaction* (eds KHARAKA, Y.K. & MAEST, A.S.) pp. 1183–1186. Balkema, Rotterdam.

OELKERS, E.H., BJØRKUM, P.A. & MURPHY, W.M. (1993) Calculation of the rate and distribution of chemically driven quartz cementation in North Sea sandstones. In: *Proceedings from the Fourth International Symposium on Hydrothermal Reactions.* (eds CUNEY, M. & CATHELINEAN, M.) pp. 169–172. Institut Lorain des Geosciences, Nancy, France.

OELKERS, E.H., BJØRKUM, P.A. & MURPHY, W.M. (1996) A petrographic and computational investigation of quartz cement and porosity reduction in North Sea sandstones. *American Journal of Science* **296**, 420–452.

OLAUSSEN, S., DALLAND, A., GLOPPEN, T.G. & JOHANNESSEN, E. (1984) Depositional environment and diagenesis of Jurassic reservoir sandstones in the eastern part of Troms I area. In: *Petroleum Geology of the North European Margin* (eds SPENCER, A.M. *et al.*) pp. 61–79. Graham & Trotman, London.

RIMSTIDT, J.D. & BARNES, H.L. (1980) The kinetics of silica–water reactions. *Geochimica et Cosmochimica Acta* **44**, 1683–1699.

RITTENHOUSE, G. (1971) Pore space-reduction by solution and cementation. *American Association of Petroleum Geologists Bulletin* **55**, 80–91.

SMITH, J.E. (1971) The dynamics of shale compaction and evolution of fluid pressure. *Mathematical Geology* **3**, 239–269.

SVERDRUP, E. & PRESTHOLM, E. (1990) Synsedimentary deformation structures and their implications for stylolitization during deeper burial. *Sedimentary Geology* **68**, 201–210.

UNGERER, P., BURRUS, J., DOLIGEZ, B., CHÉNET, P.Y. & BESSIS, F. (1990) Basin evaluation by integrated two-dimensional modeling of heat transfer, fluid flow, hydrocarbon generation, and migration. *American Association of Petroleum Geologists Bulletin* **74**, 309–335.

WALDERHAUG, O. (1990) A fluid inclusion study of quartz-cemented sandstones from offshore mid-Norway—possible evidence for continued quartz cementation during oil emplacement. *Journal of Sedimentary Petrology* **60**, 203–210.

WALDERHAUG, O. (1994a) Temperatures of quartz cementation in Jurassic sandstones from the Norwegian continental shelf—evidence from fluid inclusions. *Journal of Sedimentary Research A* **64**, 311–323.

WALDERHAUG, O. (1994b) Precipitation rates for quartz cement in sandstones determined by fluid-inclusion microthermometry and temperature-history modelling. *Journal of Sedimentary Research A* **64**, 324–333.

WALDERHAUG, O. (1995) Discussion of 'Evidence for resetting of fluid inclusion temperatures from quartz cements in oilfields' by Osborne and Haszeldine (1993). *Marine and Petroleum Geology* **12**, 559–561.

WALDERHAUG, O. (1996) Kinetic modelling of quartz cementation and porosity loss in deeply buried sandstone reservoirs. *American Association of Petroleum Geologists Bulletin* **80**, 731–745.

*Spec. Publs int. Ass. Sediment.* (2000) **29**, 51–66

# Related quartz and illite cementation in the Brent sandstones: a modelling approach

É. BROSSE[1], J. MATTHEWS[1], B. BAZIN[1], Y. LE GALLO[1] *and* F. SOMMER[2]

[1]*Institut Français du Pétrole, Rueil-Malmaison, France;*
[2]*Total–CST, Saint-Rémi-lès-Chevreuse, France*

## ABSTRACT

DIAPHORE is a water–rock interaction code that simulates short-term diagenetic episodes (few millions of years). It couples rock–water interaction and the transport of dissolved chemical species. It explicitly takes kinetics into account. DIAPHORE is applied here to the diagenesis of subarkosic sandstone reservoirs. It shows the extent to which silica for quartz cementation can be derived from the dissolution of kaolinite and feldspars, and the role of advective transport. Simulation results are controlled by core and fluid data from the Brent Group in the Greater Alwyn area (North Sea) where diagenetic illite and quartz can be considered, at least in part, as being cogenetic and of Eocene age (≈ 105°C and 270 bars). Two different types of diagenetic transformations are depicted by modelling: (i) The first transformation is governed by the instability of minerals present in the system. Approximately half of the quartz formed during modelling corresponds to the rapid, isochemical transformation of kaolinite and K-feldspar to illite and quartz. This transformation can be modelled using a closed-system approach. (ii) The second transformation is governed by disequilibrium between inflowing water and the minerals. It can be addressed only with an open-system model. Once the less abundant phase in the kaolinite and K-feldspar pair has disappeared, additional quartz can precipitate from the residual silicates present in the system (e.g. albite). Reaction fronts form and propagate throughout the reservoir at a slower velocity than the fluid flow. The reactions are no longer isochemical. Moreover, charging the reservoir with hydrocarbons, or any other process that would stop the water flow during the progress of the reactions, can help to explain heterogeneities in the mineral composition observed at the field scale.

## INTRODUCTION

Water–rock interaction modelling can help to identify key parameters and provide a quantitative measure of the mass transfers occurring during diagenesis of reservoirs. This approach is applied here to the diagenetic evolution of arkosic or subarkosic sandstones, at relatively deep reservoir conditions. As temperature and pressure increase with burial, and interstitial water circulation may cause mineral dissolution and precipitation, silica is released and partly available for the formation of quartz cement. This is one of the possible sources of silica to explain deep-burial quartz cementation in sandstones (e.g. Milliken *et al.*, 1981; Bjørkum & Gjelswik, 1988; Ehrenberg, 1990; Harrison & Tempel, 1993; Bjørlykke *et al.*, 1995; Land & Milliken, this volume, pp. 183–198). Diagenetic illite and quartz are common in the Brent Group sediments of the North Sea. Core data acquired on the Dunbar field of the Greater Alwyn area provide quantitative controls for discussing and validating simulation results. Numerical modelling yields constraints on the proportion of total quartz cement that can be considered as cogenetic with illite. It points out the respective influence of mineral composition and of interstitial water flow in the cementation process.

A number of increasingly sophisticated water–rock interaction software applicable to diagenetic problems have recently been developed, first using the 'reaction-path' concept and closed-system conditions (e.g. Helgeson *et al.*, 1970; Parkhurst *et al.*, 1980; Fritz, 1981; Perkins *et al.*, 1990; Wolery *et al.*, 1990; Madé *et al.*, 1994), then coupling reaction and transport (e.g. Ortoleva *et al.*, 1987a; 1987b; Yeh & Tripathi, 1991; Lichtner, 1992; Steefel & Lasaga, 1994; Wang & Van Cappellon, 1996; Giles, 1997; Le Gallo *et al.*, 1998). The software used hereafter is a multidimensional water–rock interaction code called DIAPHORE and developed at Institut Français du Pétrole (IFP). For coupling reaction and transport, it solves the mass-balance equations of the chemical elements involved in the mineral system under consideration. The model is therefore well adapted to decipher the role of water in the redistribution of dissolved mineral matter at the reservoir scale.

At present, our approach does not take into account mechanical stress or compaction. In the Brent sandstones, the observed abundance of quartz overgrowths exceeds the proportion of quartz that, according to a geochemical model, can be drawn from illitization of silicates. The implication is that intergranular pressure-solution between quartz grains should also act as a source of silica (Houseknecht, 1988). Various other mechanisms of silicification have been proposed and discussed in the literature (Leder & Park, 1986; Gluyas & Coleman, 1992; Oelkers *et al.*, 1996; Aplin *et al.*, 1993; Bjørkum, 1996; Giles, 1997).

## PRINCIPLES OF THE MODELLING APPROACH USED IN DIAPHORE

### Key parameters and time scale

Diagenetic transformations occur only if the sediment is not at thermodynamic equilibrium. Increase of temperature and pressure with burial, and influx of allochthonous water undersaturated or oversaturated with respect to a given mineral are two main causes of disequilibrium. Parameters that are useful for describing a diagenetic transformation include: (i) factors related to the geological evolution of the reservoir, i.e. temperature, pressure, mechanical stress and composition and flow rate of water; these are referred to as 'regional' factors; (ii) reservoir characteristics, i.e. sedimentary architecture and connectivity, mineral composition and texture; these are referred to as 'local' factors.

The duration and timing of diagenetic processes can be highly variable, and are often difficult to assess. At moderate temperatures, some reactions are so slow that metastable assemblages can survive over geological time periods. K–Ar ages of illite in the Brent Province indicate formation within a few million years (Clauer *et al.*, 1992; Glasmann, 1992; Hamilton *et al.*, 1992). In other cases indirect dating of diagenetic phases can be achieved from stable isotopic compositions or fluid inclusions (Emery & Robinson, 1993). These data often indicate that diagenetic reactions occur at a relatively narrow range of temperatures and fluid compositions (e.g. Robinson & Gluyas, 1992; Fallick *et al.*, 1993; Gluyas *et al.*, 1993; Longstaffe, 1994), but not always (Williams *et al.*, 1997). As a first approximation for the purpose of numerical simplification DIAPHORE was designed to simulate the effect of relatively short-term diagenetic episodes (few millions of years) during which the 'regional' factors mentioned above may be fixed at constant values. Only 'local' characters of the reservoir are varying throughout the simulations. Long-term diagenetic processes, such as intergranular pressure-solution (Dewers & Ortoleva, 1990), are not taken into account.

### Geochemical system and kinetics

In the version of DIAPHORE presented here, the geochemical system consists of a set of mineral phases and the water phase. 'Primary' minerals are those present in the initial step of the simulation and 'secondary' minerals will eventually be formed during the simulation. Thermodynamic equilibrium is assumed between dissolved components. For heterogeneous reactions, in contrast, kinetics are considered. The thermodynamic database used in DIAPHORE was developed and maintained by Strasbourg University (Fritz, 1981). Correction of the pressure effect is made using SUPCRT 92 (Johnson *et al.*, 1992).

In a range of pH around neutrality, valid for most applications in reservoirs the rate of reaction of a mineral M is expressed in $mol \cdot kg_{water}^{-1} \cdot s^{-1}$ as the product of three terms (Lasaga, 1981; Lasaga *et al.*, 1994; Madé *et al.*, 1994):

$$\frac{d[M]}{dt} = k_{d/p,M} \cdot S_M \cdot \left(1 - \frac{Q_M}{K_M}\right) \quad (1)$$

where $k_{d,M}$ and $k_{p,M}$ are coefficients exponentially dependent on temperature according to Arrhenius' law (they have opposite signs), $S_M$ is the reactive surface area which is variable depending on the mineral type and texture and $(1 - Q_M/K_M)$ is the departure from equilibrium; $K_M$ is the equilibrium constant of the hydrolysis reaction, which depends on T and P, and $Q_M$ is the actual activity product. In the following application to the Brent sandstones, a unique value is considered for $k_{d,M}$ and $k_{p,M}$ (coefficient k in Table 1).

The physical concept of 'reactive surface area' is particularly difficult to assess (e.g. Anbeek, 1992). In water–rock interaction modelling, a convenient way to model variation of the reactive surface area is to link it to variation of the geometric surface area of grains with a simple shape. In the version of DIAPHORE used for the simulations presented below, a feed-back of the mineral reactions on the surface areas $S_M$ is obtained using a 'floating-spheres' model (Bildstein, 1998), where each type of mineral is represented by a sphere of a specific

**Table 1.** Data used for the minerals in the simulations

| | | Quartz | K-feldspar | Low-albite | Kaolinite | K-muscovite (illite) | Calcite |
|---|---|---|---|---|---|---|---|
| A | Initial proportion (vol.%) | 55.0 | 10.0 | 4.5 | 7.0 | 1.5 | 2.0 |
| | Grain-size (diameter, mm) | 0.1 | 0.1 | 0.1 | 0.01 | 0.01 | 0.02 |
| | Coefficient for reactive surface area | 0.1 | 0.1 | 0.1 | 0.2 | 0.2 | 0.0001 |
| B | Molar volume ($cm^3/mol$) | 22.69 | 108.72 | 100.07 | 99.52 | 140.71 | 36.93 |
| C | THERMODYNAMICS | | | | | | |
| 105°C | –log K | 3.041 | 19.027 | 16.931 | 33.605 | 47.614 | 9.062 |
| 270 bars | for aqueous species | $H_4SiO_4$ | $(Al(OH)_4)^-$ | $(Al(OH)_4)^-$ | $(Al(OH)_4)^-$, $H^+$ | $(Al(OH)_4)^-$, $H^+$ | $CO_3^=$ |
| | KINETICS | | | | | | |
| | coeff. k ($mol/m^2.yr$) | $1.24\ 10^{-3}$ | $4.34\ 10^{-3}$ | $1.09\ 10^{-3}$ | $1.88\ 10^{-5}$ | $1.47\ 10^{-5}$ | $1.1\ 10^{+3}$ |
| D | THERMODYNAMICS | | | | | | |
| 80°C | –log K | 3.23 | 20.15 | 17.85 | 35.316 | 50.037 | 8.842 |
| 220 bars | for aqueous species | $H_4SiO_4$ | $(Al(OH)_4)^-$ | $(Al(OH)_4)^-$ | $(Al(OH)_4)^-$, $H^+$ | $(Al(OH)_4)^-$, $H^+$ | $CO_3^=$ |
| | KINETICS | | | | | | |
| | coeff. k ($mol/m^2.yr$) | $2.29\ 10^{-4}$ | $1.52\ 10^{-3}$ | $3.83\ 10^{-4}$ | $1.06\ 10^{-5}$ | $8.39\ 10^{-6}$ | $7.1\ 10^{+2}$ |
| | Reference for coeff. k | [1] | [2] | [2] | [3], [4] | [3], [5] | [6] |

(A) Mineralogical composition assumed for Upper Massive Sands before illitization. Grain-size and ratio between reactive surface area and geometric surface area calculated from grain diameter. The very low ratio for calcite is to avoid numerical problems linked to the relatively high kinetic coefficient of this mineral. (B) Molar volumes. (C) and (D) Thermodynamic data from SUPCRT.92 (Johnson *et al.*, 1992) and kinetic coefficients in the range of neutral pH: (C) at 105°C, 270 bars; (D) at 80°C, 220 bars. From: [1] Brady & Walther (1990); [2] Helgeson *et al.* (1984); [3] Stumm & Wieland (1990); [4] Carroll-Webb & Walther (1988); [5] Knauss & Wolery (1988); [6] Sjoberg & Rickard (1984).

size that changes as the reactions proceed. An empirical corrective coefficient, characteristic of the mineral, can reduce the reactive efficiency of the resulting geometric surface area.

## Geometry, transport and reaction–transport coupling

The numerical model can be applied to a single cell, closed or open to the flow of water. In this box-type approach, there is no geometry involved, and simulations have either local or average meaning. Such applications are shown in Bildstein (1998). Nevertheless by design, DIAPHORE is applicable to a variety of reservoir geometries and heterogeneities, from simple linear one-dimensional (1-D) systems exhibiting homogeneous composition in the initial conditions, to complex three-dimensional (3-D) layerings dealing with several heterogeneous reservoir units (or facies). This paper presents only 1-D applications. Le Gallo *et al.* (1998) show 3-D results and discuss numerical aspects of the code. Assuming the pore system is saturated with water, DIAPHORE solves the conservation equations for water and for chemical elements. In a representative elementary volume, the coupling is expressed in the conservation equation of any element $E_i$:

$$\frac{\partial(\Phi\rho_w[E_i])}{\partial t} + \mathrm{div}(\Phi\rho_w[E_i]\vec{u}_w + \vec{J}_i^w) - \rho_r(1-\Phi)R_i + [E_i]^{inj}Q_w = 0 \quad (2)$$

where:
the subscript w represents water;
$\rho_w$ and $\rho_r$ are, respectively, the densities of water and matrix;
$Q_w$ is an eventual exchange term between the system and outside (injection term at the inlet, production term at the outlet);
$\vec{u}_w$ is the filtration velocity of water given by Darcy's law;
$\vec{J}_i^w$ is the diffusive–dispersive flux given by an extended Fick's law;
$\Phi$ is the porosity;
$[E_i]$ and $[E_i]^{inj}$ are, respectively, the molarities of $E_i$ in the interstitial water and in the injected water;
and $R_i$ is the rate of change of $E_i$ due to the kinetically controlled mineral reactions.

Various 2-step numerical architectures were implemented (Le Gallo *et al.*, 1998). It is beyond the scope of the present paper to examine their respective advantages and drawbacks, but choice is made partly on the role played by the basic mechanisms involved, i.e. advection, diffusion, and kinetics, and partly on the numerical efficiency that must be reached to perform the simulation in a reasonable time. For the theoretical and mathematical background to reaction–transport modelling, the reader is referred to Domenico & Palciauskas (1979); Rubin (1983); Bahr and Rubin (1987); Kirkner & Reeves (1988); Lichtner (1992, 1993); Steefel & Lasaga (1994); Boudreau (1997); and Giles (1997), among others.

The effect of mineral reactions on permeability is modelled using a relation linking permeability to porosity, mineral surface area and mineral proportion. In this respect, a model of diagenesis that is able to address the complete mineral composition has a considerable advantage because a particular importance can be assigned to a particular mineral (e.g. diagenetic illite) for determining the permeability evolution. An appropriate approach is to empirically calibrate the relation to be used, on the basis of petrophysical data available for the studied case. The choice of such a relation, and its calibration, are illustrated in Le Gallo *et al.* (1998).

### Input, output and boundary conditions

Input data for a typical DIAPHORE simulation are the following: (i) geometry of the reservoir, represented by a grid system; (ii) distribution of temperature versus space, assumed to be constant with time; (iii) imposed and fixed pressures at the boundaries, or impermeable limits; (iv) composition and velocity of the solution at the inlet (both are constant with time); (v) in each grid element, 'initial' values (i.e. values at the beginning of the simulated episode) of mineral and interstitial water compositions, of porosity, permeabilities and reactive surface area (linked to grain size); (vi) calibrated relation for feedback effect on permeability.

DIAPHORE outputs are descriptions of the temporal and spatial evolution of mineral and water compositions and of petrophysical parameters during the simulated diagenetic episode. They can be compared with available petrophysical and mineralogical data obtained from cores. Once a set of input parameters is able to generate outcomes matching observations made in drilled areas, the corresponding simulation can be considered as holding some predictive value for parts of the field which have not been drilled yet. However, two points must be kept in mind for using such modelling results: the uncertainty of the control data can be large; and the set of input parameters that allows one to reproduce field observations may not be a unique solution.

## CONDITIONS OF PRECIPITATION, AND DISTRIBUTION OF DIAGENETIC ILLITE AND QUARTZ IN THE BRENT SANDSTONES OF THE DUNBAR FIELD

During burial subarkosic sandstones commonly undergo potassium transfer from K-feldspar to authigenic illite that precipitates. This is notably the case in the Brent Group reservoirs (Middle Jurassic) of the North Sea, where this transformation is observed at depth of 3.6 km and more (Giles *et al.*, 1992). In order to have quantitative controls on conditions and mass balance of this diagenetic transformation, we used a data set collected from exploration and development wells drilled by Total in the Greater Alwyn area (Johnson & Eyssautier, 1987; Jourdan *et al.*, 1987; Inglis & Gérard, 1991), and more specifically in the Dunbar field (Hogg *et al.*, 1992; Baillie *et al.*, 1996) (Fig. 1).

### Geological and sedimentary framework

The Brent sandstones were deposited from the Bajocian to late Bathonian–early Callovian times, in the context of incipient rifting of the North Viking graben. Synsedimentary faults separate tilted blocks. Rifting ended in late Oxfordian–Berriasian times during deposition of the Kimmeridge Clay mudstones, the main source rock for oil in the basin. The Cimmerian unconformity marks a hiatus in the early Cretaceous sedimentation. From the Cenomanian onwards, the sedimentation rate was high (Lepercq & Gaulier, 1996).

The Dunbar field is structurally more complex than Alwyn North (Fig. 1). Two main compartments, separated by a major regional fault, are distinguished (Jourdan *et al.*, 1987; Baillie *et al.*, 1996): (i) the 'Western Flank' with relatively thin Jurassic sediments and severe erosional features at the top of the tilted block; (ii) the 'Frontal Panel' with much thicker Jurassic deposits, particularly from the Ness Formation upwards (Didier *et al.*, 1995). The simulations presented hereafter concern more specifically the northern part of the Frontal Panel, where reservoir sandstones exhibit their most complete development and where illitization has been maximal.

Most of the Brent Group deposits, in the northern North Sea, can be interpreted as the result of progradation/retrogradation of a deltaic complex. In the Greater Alwyn area, the transgressive part, i.e. Ness A2 to Tarbert, contains the main reservoir units. The Tarbert Formation is usually composed of shoreface sands. In Dunbar, however, syn-sedimentary faulting resulted in an unusual succession: the upper part of the Tarbert Formation, called the 'Upper Massive Sands' (UMS), is a thick unit

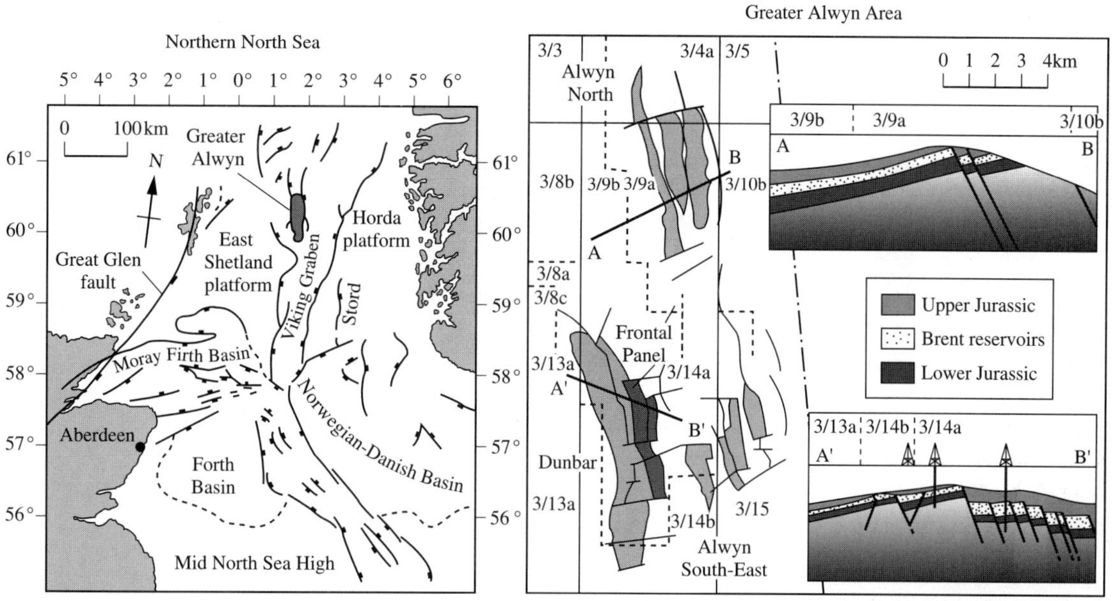


**Fig. 1.** Left. Location of the Greater Alwyn area in the northern North Sea. Right. Location of the oil fields in the Greater Alwyn area: Alwyn North, Dunbar, Alwyn South-East (Grant and Ellon). Cross-sections illustrate the geological structure of the Brent Group reservoirs (Middle Jurassic) at Alwyn North (depth *c.* 3200–3500 m) and at Dunbar (depth *c.* 3600–3900 m).

(commonly ≈ 150 m) of relatively coarse-grained sandstones deposited as upper shoreface or in a storm-influenced fan delta (Didier *et al.*, 1995). The simulations discussed below concern only the UMS unit.

## Distribution of illite and quartz at the reservoir scale and initial mineralogy

Five wells of the Frontal Panel were selected for this study. Figure 2 illustrates average mineralogical compositions of the UMS unit that were derived from the combination of the four following techniques (Durand *et al.*, unpublished results). Detrital quartz (dominant) and muscovite (always present in small amount) are not represented. The absolute precision of the mineral amounts given in Fig. 2 is estimated ± 1%, based on:

**1** Point counting of 124 thin sections, with poor precision (100 points only in most cases, see Van der Plas & Tobi, 1965), was judged appropriate for estimates of quartz, rock fragments, muscovite and kaolinite contents, and a rough estimate of quartz overgrowth.

**2** X-ray diffractometry (XRD) of the same 124 bulk samples and < 2 μm size fractions permitted quantification of K-feldspar, albite, calcite and illite contents.

**3** An appraisal of the abundance of quartz overgrowth for a limited set of samples from wells 3/14a-7 and 3/14a-8 was derived from SEM/cathodoluminescence microscopy (Hogg *et al.*, 1992).

**4** Bulk elemental composition of seven samples from well 3/14a-D03, six samples from well 3/14a-D05 and one sample from well 3/14a-15. This method proved to yield much more accurate results than routine point counting and XRD analysis, especially for the estimation of clay content (Nadeau & Hurst, 1991; Gluyas & Coleman, 1992; Ehrenberg & Boassen, 1993; Potdevin & Hassouta, 1997).

Except for quartz and illite, most minerals do not exceed a few volume per cent. Kaolinite is generally absent from the studied reservoirs, except in the structural crest of the UMS. A salient north-to-south trend is observed. The southern wells show relatively more pronounced quartz and illite cementation, and, correlatively, more intense dissolution of feldspars. Illite is also relatively more abundant in 3/14a-D05, which is close to the main fault limiting the Frontal Panel to the west. An important feature of the studied area is that the UMS unit is in the oil zone in all five wells. A challenge for modelling is to link the observed trend in mineralogy with the diagenetic evolution of the reservoir.

The reconstruction of the 'initial' amounts of minerals to be used as input in the simulations benefited from a data set acquired on much less illitized shoreface

Dunbar, Frontal Panel

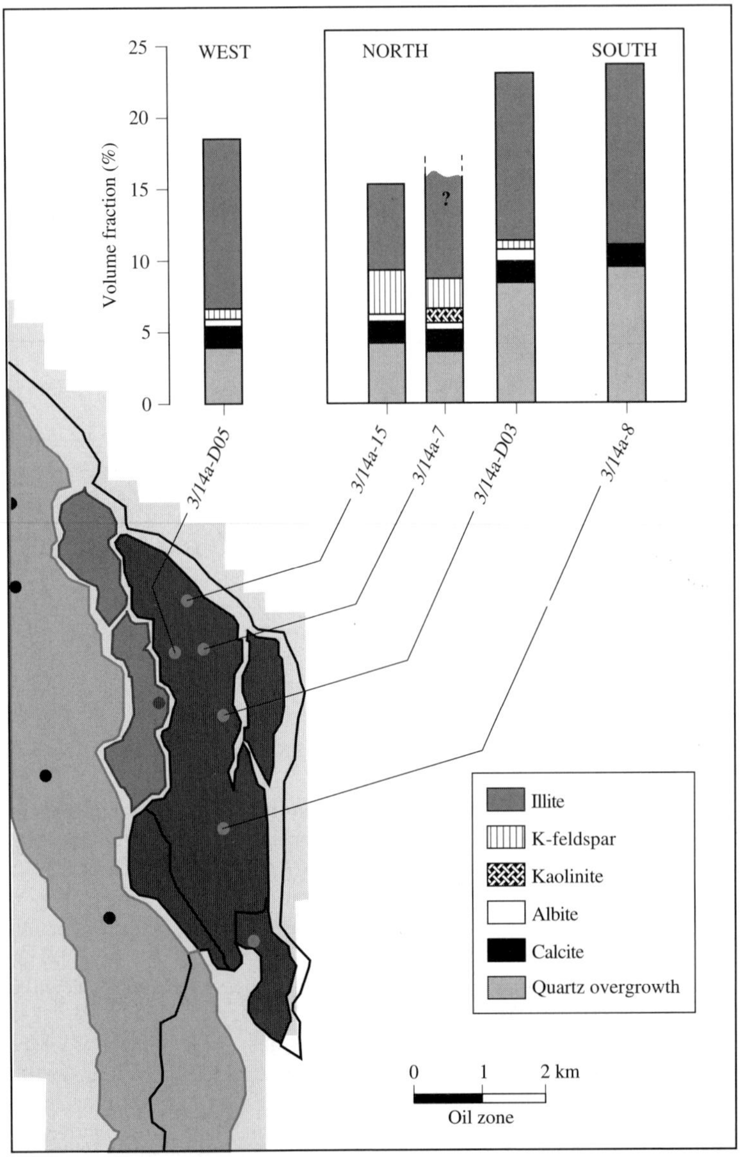


**Fig. 2.** Mineral composition of the UMS-type facies in the Dunbar Frontal Panel. The present study is focused on the Frontal Panel of Dunbar, i.e. the eastern border. The map shows the location of the five wells studied in the Dunbar field. Average mineral composition, derived from a combination of data sources (see text), are shown except for detrital quartz and porosity. The studied facies is representative of the UMS unit, totally included in the oil zone in the Frontal Panel.

reservoirs of the Tarbert Formation, that were sampled from Alwyn North and South-East fields (Fig. 1). Their petrographic composition was studied in the same way as explained above. In spite of a slight difference of lithofacies between the upper shoreface of the UMS unit at Dunbar and the shoreface of the Tarbert Formation at Alwyn North or South-East, a similar initial composition is assumed in the area for the two sediments deposited in the same deltaic complex. Unfortunately literature is not abundant on the question, nevertheless data published by Kairo *et al.* (1993) support this important assumption. With a single source for detrital material sedimentary processes can cause variations in mineral composition between distinct lithofacies, e.g. relative concentration of feldspars in fine-grained sands (Odom *et al.*, 1976). However, compositional variation within a single facies is probably much less than that between facies. Finally the average mineralogy considered for the UMS sand-

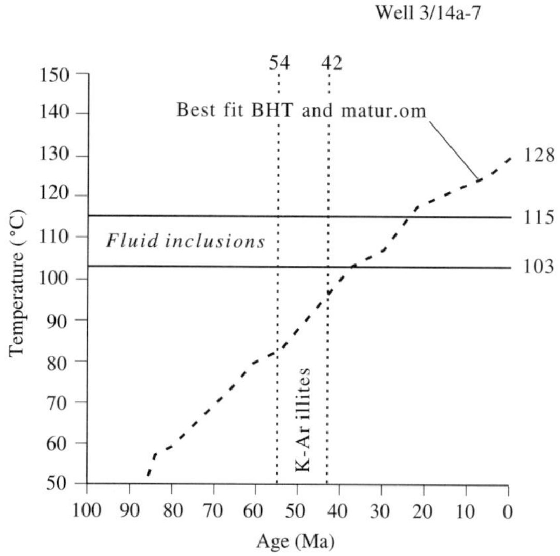


**Fig. 3.** The thermal history of the Brent reservoirs in well 3/14a-7 was calculated with the ©GENEX software. The simulation is validated using present-day temperatures in wildcats, and the degree of organic matter maturity constrained by Rock-Eval pyrolyses and vitrinite reflectance. The K–Ar dates obtained from illitic clay fractions, and the temperatures obtained from fluid inclusions, are reported.

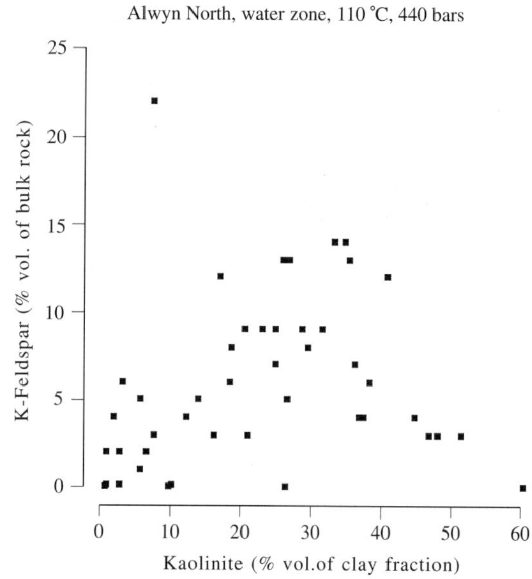


**Fig. 4.** Constraint from Alwyn North on illitization conditions at Dunbar. Proportions of kaolinite and K-feldspar measured in the water zone of the Alwyn North field. Temperature and pressure conditions are ≈ 110°C and 440 bars. Diagenetic illite is only present as traces.

stone, before illite and quartz precipitation, is reported in Table 1.

In addition, microprobe analyses revealed that diagenetic illites are close to muscovite in composition (Durand, unpublished results). Accordingly, the model uses thermodynamic data from K-muscovite for illite.

## Conditions of illite and quartz precipitation: temperature, pressure and interstitial water

Textural relations between secondary mineral phases observed from thin sections and from SEM images showed that at Dunbar, diagenetic illite and quartz could be assumed to be cogenetic. In addition, their isotopic composition is consistent with coeval precipitation from the same fluid (Hogg *et al.*, 1995). For the Frontal Panel, most K–Ar dates of the finest separated illitic fractions range from 57 to 42 Ma (Hogg *et al.*, 1993). Homogenization temperatures of aqueous fluid inclusions trapped in quartz overgrowths (always near the core–overgrowth boundary) range between 100°C and 115°C in 90% of the available data (Hogg *et al.*, 1992; Cordon & Guilhaumou, 1995). The consistency of the preceding measurements with the reconstructed thermal history of the Brent sandstones was examined using the basin modelling software ©GENEX from BEICIP-FRANLAB. The example of well 3/14a-7 is presented in Fig. 3. The heat-flow value, assumed constant since Cretaceous times, was chosen in order that the corrected temperatures obtained from well tests (DST) and the maturation state of organic matter measured by both Rock-Eval and vitrinite reflectance are correctly adjusted. Figure 3 shows that the closure of fluid inclusions slightly postdates the 'youngest' illitic fractions. Nevertheless the presence of undetected contaminants can never be totally ruled out in an illitic fraction (Ehrenberg & Nadeau, 1989), and illitization could occur at a moderately higher temperature than those deduced from K–Ar dating. In addition, the temperature of fluid inclusion closure in many instances is known to be higher than the temperature at which quartz cementation started for these rocks (e.g. Walderhaug, 1994a, 1994b; Williams *et al.*, 1997). Another important constraint on the conditions of illitization at Dunbar comes from the present-day water zone in the Alwyn North field. Core samples from this zone, where temperature is 110°C and pressure 440 bars, still contain large amounts of K-feldspar and kaolinite whereas diagenetic illite is only present as traces (Fig. 4). Despite some differences between present-day waters in Dunbar and Alwyn North, and between burial histories, we assumed that the Brent reservoirs at Dunbar also experienced a situation where at 100–110°C illitization was

**Table 2.** Water compositions used in the simulations

| | | Al | K | Na | Ca | Mg | Si | C | Cl | $pCO_2$ | pH |
|---|---|---|---|---|---|---|---|---|---|---|---|
| Brackish | 105°C | $2.0(10^{-7})$ | $4.5(10^{-3})$ | 0.400 | $5.0(10^{-3})$ | $1.5(10^{-3})$ | $1.0(10^{-3})$ | 0.010 | 0.420 | 2.86 | 6.4 |
| Brine | 105°C | $1.4(10^{-7})$ | $7.0(10^{-3})$ | 0.750 | $5.0(10^{-3})$ | $1.5(10^{-3})$ | $9.0(10^{-4})$ | 0.010 | 0.770 | 2.86 | 6.4 |
| Brine | 80°C | $1.0(10^{-9})$ | $7.0(10^{-3})$ | 0.760 | $1.32(10^{-2})$ | $1.5(10^{-3})$ | $1.8(10^{-3})$ | 0.017 | 0.788 | 0.8 | 6.12 |

Brackish water and brines were used for the simulations. Complete compositions were reconstituted from a regional compilation of water analyses (Bazin *et al.*, 1997). The concentrations are expressed in mol/l.

still very limited. Moreover, from a discussion involving several data sets of K–Ar dates available for the Brent group, Giles (1997) concluded that 100–110°C is the most probable range of temperatures for precipitation of such diagenetic illite. Considering all these aspects, DIAPHORE simulations presented in the following section were run at 105°C, with one additional simulation run at 80°C. The pressure value, 270 bars, was evaluated from Buhrig's review (1989). The kinetic coefficients considered for quartz, K-muscovite (*alias* illite), K-feldspar, low-albite and calcite at 105°C, 270 bars are reported in Table 1.

A numerical simulation of reservoir diagenesis requires some knowledge of the composition of the interstitial solution that invaded the reservoir during the simulated episode. This was discussed by Bazin *et al.* (1997). Input data comprise silica and aluminium which, along with pH, play an important role in controlling the saturation state of silicates. Unfortunately, Si and Al concentrations are rarely measured. However, in the Brent reservoirs of the Greater Alwyn area salinities measured in the fluid inclusions of quartz overgrowths are in the same range of values as the salinity range of the present-day waters: from 2.5 to 4.5% expressed in sodium chloride equivalent (chloride concentration ranges from 0.42 to 0.77 mol/l). An overall trend of decreasing salinity from the south-east to the north is observed in the present-day water compositions, and also in the palaeo-salinities. This feature probably is related to regional geology, as suggested by Egeberg & Aagaard (1989) and Warren & Smalley (1993, 1994). Accordingly, the compositions of the present-day Brent waters in the Greater Alwyn area have been used to reconstruct a possible composition range for the palaeo-waters. Missing Si, Al and pH values were calculated using partial equilibrium with silicates at 105°C, 270 bars, as explained by Bazin *et al.* (1997). The partial pressure of $CO_2$, 2.86 bars, is derived from Smith & Ehrenberg (1989). Finally a choice of two water compositions was made for the simulations: a brackish water typical of the present-day Alwyn North situation, and a brine closer to the present-day Dunbar water (Table 2).

## SIMULATION RESULTS

First a reference simulation is presented. Other simulations, designed to test the respective influence of water-flow composition, T and P conditions, and water-flow velocity, are then compared with the reference one.

### Reference simulation

The reservoir considered is one-dimensional, 1 km long, with a homogeneous mineralogical composition at the start of the simulation (Table 1 and Fig. 5). The system is infiltrated by the brackish water (Table 2) at an average velocity of 1 cm·yr$^{-1}$. Results as a function of distance are shown in Fig. 5. During a first, rapid evolution (≈ 20 000 years) the assemblage kaolinite/K-feldspar dissolves until the least abundant of the two minerals disappears, here kaolinite. Illite and quartz are produced. The overall reaction:

$$\text{K-feldspar} + \text{kaolinite} \rightarrow \text{illite} + 2\ \text{quartz} + H_2O \qquad (3)$$

is isochemical and occurs at the same rate in all parts of the system. During this stage, the infiltrating water has no influence on the evolution of the system.

In a second stage a very different pattern develops. Successive reaction fronts form and cross the system, from its inlet to its outlet, which correspond to the complete dissolution of K-feldspar and of albite, respectively. The movement of reaction fronts is ≈ 300 times slower than water flow velocity. The diagenetic evolution is controlled by the magnitude of disequilibrium between allochthonous water and reservoir. The overall reaction is no longer isochemical. The sodium yielded by albite dissolution is completely exported from the system. Additional illite and quartz form. Illite is able to use up aluminium released by the dissolution of either albite or K-feldspar, and takes a part of its potassium content from the solute charge imported by the inflowing water. With the initial mineral proportions considered here, approximately one-third (4.5%) of the total diagenetic illite (14%) and half of the quartz cement modelled (3% of

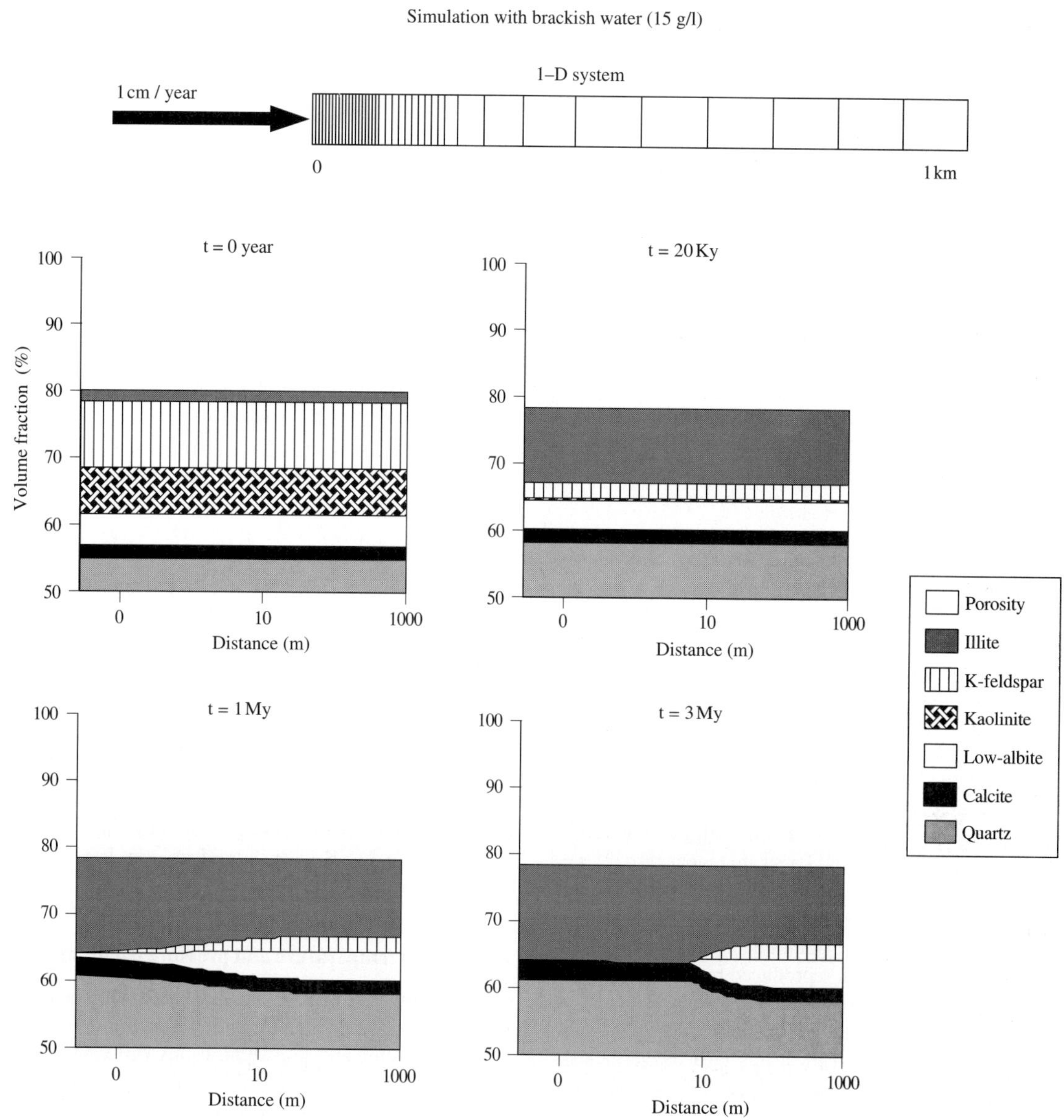


**Fig. 5.** Reference DIAPHORE simulation, brackish water. An irregular grid of 40 elements was designed to take into account the effects of disequilibrium which are more pronounced at the water inlet, on the left side. In examples illustrated here, the water-flow velocity is 1 cm·yr$^{-1}$ (average velocity, i.e. Darcy's velocity divided by effective porosity). The figure depicts the evolution of a UMS-type facies (K-feldspar initially more abundant than kaolinite), invaded at 105°C and 270 bars by brackish water (Table 2).

6.5%) are formed during the second stage where formation fluid composition and flow rate control the geochemical evolution of the sandstone. If a closed-system hypothesis had been used instead of the open-system one, the diagenetic evolution would have stopped shortly after the end of the first stage, because the residual mineral system would have been at thermodynamic equilibrium. In contrast, the simulations shown in Fig. 5 demonstrate that the open-system hypothesis is able to generate additional volumes of illite and quartz.

A consequence of the simulated mineral transformation is a slight increase, by 2%, in reservoir porosity. It results from the molar volumes of minerals involved in the diagenetic evolution, and of the lack of compaction in

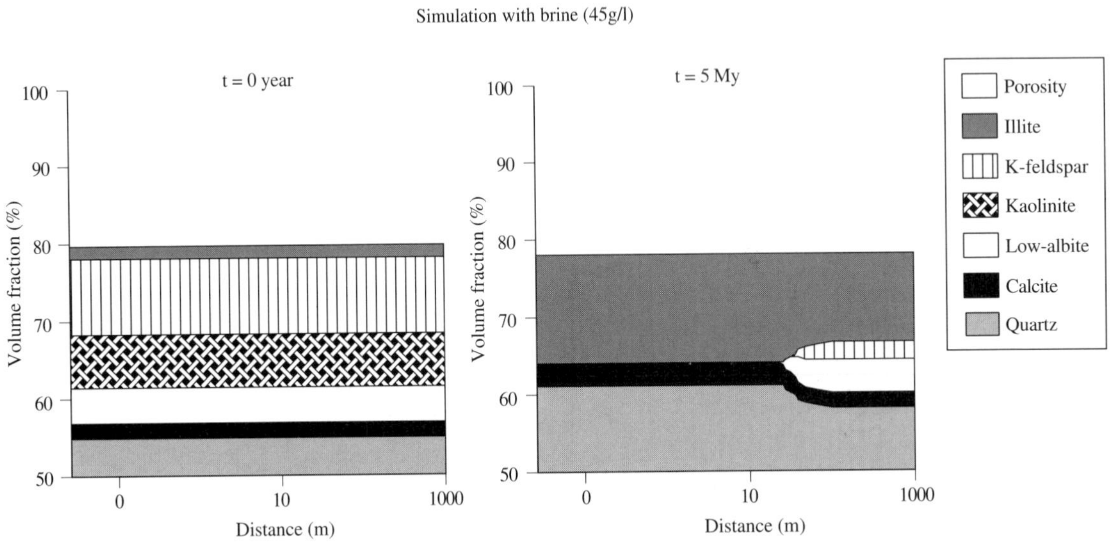


**Fig. 6.** DIAPHORE simulation, brine. Mineralogical evolution of the UMS-type facies invaded at 105°C and 270 bars by a brine. Composition of the brine is defined in Table 2.

the model. Influence of the illite proportion on the permeability distribution can be calculated in a way that is developed by Le Gallo *et al.* (1998).

The pattern obtained after 3 Myr in the model (Fig. 5) shows numerous similarities with the south-to-north trend of the average present-day mineralogy observed in the UMS unit of the Frontal Panel at Dunbar (Fig. 2), i.e. increasing diagenetic illite and quartz proportions, and disappearance of K-feldspar and albite towards the south. In contrast, the presence of kaolinite in the oil zone at the top of the structure is not reproduced by modelling. The circulation of water, from south to north at Dunbar, is consistent with the structure of the field and with the subsidence history of the basin in this area. Nevertheless, inlet of basinal water directly from the graben to the east is also likely. Water flow rates are discussed in a subsequent section.

### Influence of the water composition and effect of oversaturation with respect to quartz

Figure 6 shows that in the studied range of salinities and compositions the effect of water composition on diagenetic pattern is not perceptible. At 105°C in particular, the activity ratio between sodium and potassium required for albite stability is higher than those considered here.

Oversaturation of water with respect to quartz can be allowed in the model before quartz starts to precipitate. At 105°C, oversaturation values of 0.4 (log units) with respect to quartz are commonly observed in formation waters (Fig. 7, water A). Assuming this value, the diagenetic pattern remains similar to the reference one. Oversaturation would have to reach 0.6 (log unit) to block quartz precipitation (Fig. 7, water B). This behaviour can be understood when examining the activity diagram in the $K_2O$–$SiO_2$–$Al_2O_3$–$H_2O$ system (Bjørkum & Gjelswik, 1988; Ben Baccar *et al.*, 1993; Bazin *et al.*, 1997).

### Influence of temperature and pressure conditions

A simulation done at 80°C and 220 bars (Table 1) results in a much slower transformation for the water-flow driven reactions, and in a significantly different pattern (Fig. 8). In the example considered, transient transformation of albite to K-feldspar occurs. A pattern comparable to the observed distribution at Dunbar (Fig. 2) seems not to be reached in a reasonable time, considering that the water velocity of the presented simulation is 1 cm·yr$^{-1}$. Such a modelling result adds an argument in favour of diagenetic illite slightly younger than measured by K–Ar dating from separated size fractions.

### Influence of the water-flow velocity

A simulation at 105°C and 270 bars assuming a 1 mm·yr$^{-1}$ average velocity results in a water-flow driven transformation approximately 10 times slower than the reference one, without any significant difference in the

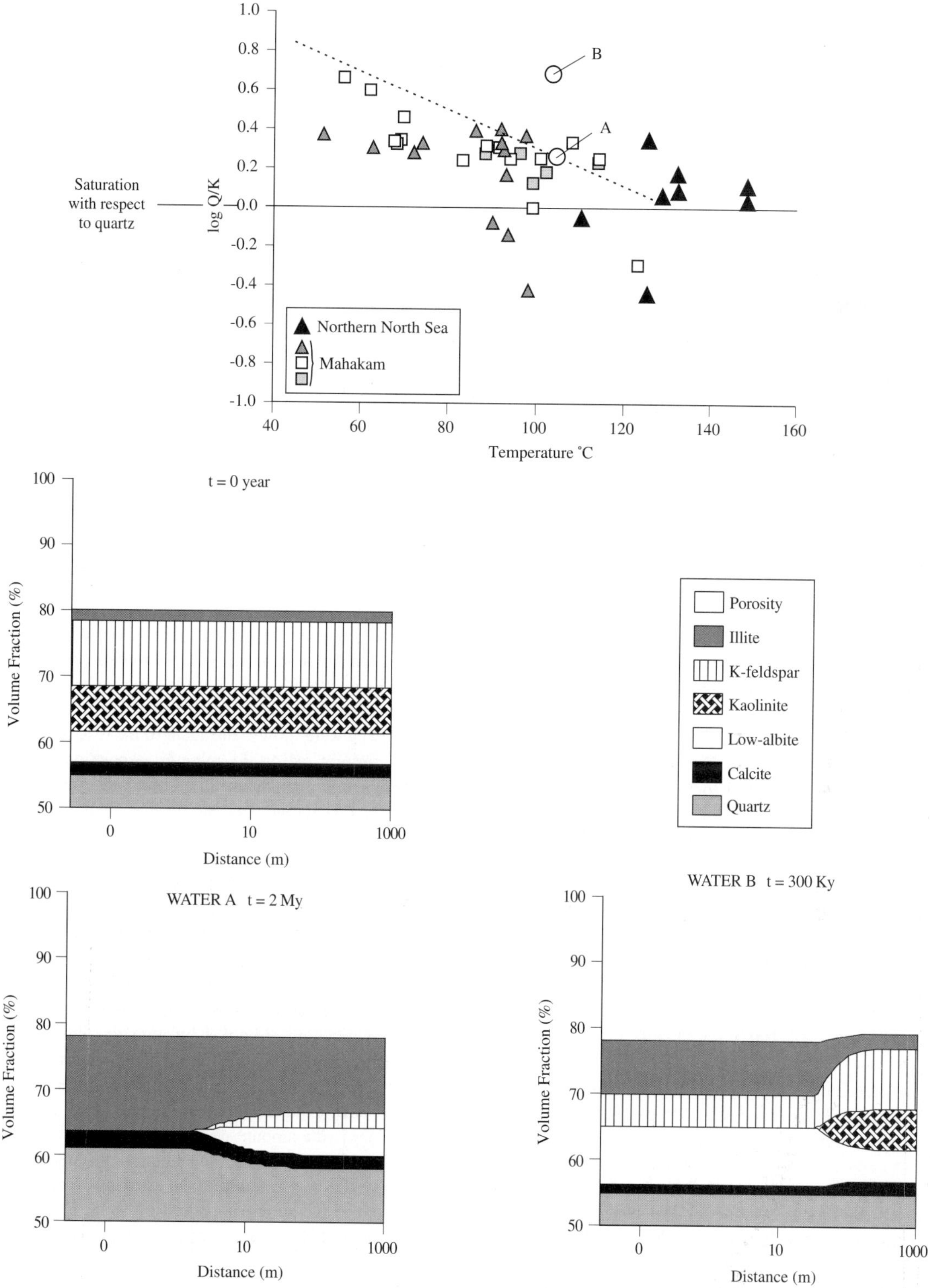


**Fig. 7.** DIAPHORE simulation, oversaturation with respect to quartz. In order to stabilize feldspars the oversaturation with respect to quartz of water entering the reservoir would have to reach values (water B) which are not observed. Observed oversaturation values (water A) lead to a diagenetic evolution similar to the reference one, depicted in Fig. 5.

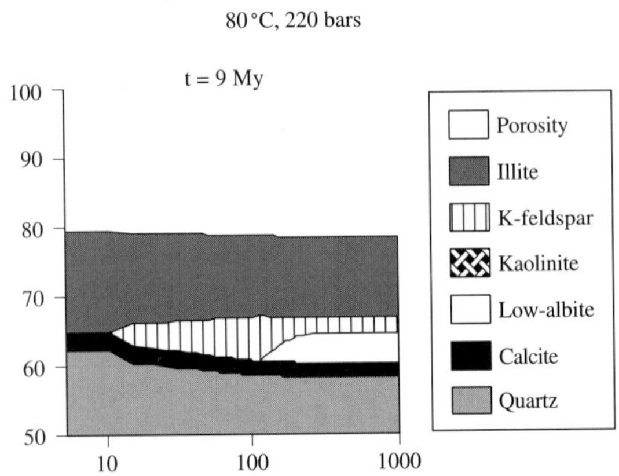


**Fig. 8.** DIAPHORE simulation, 80°C and 220 bars. Mineralogical evolution of the UMS-type facies invaded at 80°C and 220 bars by a brine. Composition of the water is defined in Table 2.

chemical budget. The same reaction fronts as shown in Fig. 5 would need 20 Myr to reach a 100-m distance from the inlet. If realized at Dunbar, such a delay would have resulted in an observable difference between ages of illite in wells 3/14a-7 and 3/14a-8, which is not the case (Hogg *et al.*, 1993). Alternatively this argues in favour of more rapid water-flow velocities (1 $cm{\cdot}yr^{-1}$ instead of 1 $mm{\cdot}yr^{-1}$).

In a section crossing the North Viking graben and bordering sub-basins the Temispack software calculates average water-flow velocities up to 0.5 $mm{\cdot}yr^{-1}$ in the Brent Group during Eocene and Oligocene times (Wendebourg, unpublished results). Nevertheless these values are average values for the whole Brent Group, which is not represented by a 100% proportion of sandstone such as the UMS unit in DIAPHORE. The possibility of flow focusing in the sandstones is worth consideration (Giles, 1997).

## DISCUSSION

The preceding simulations illustrate how numerical modelling of diagenetic reactions can contribute to the interpretation of petrographical and geochemical data. First, numerical modelling points out two distinct regimes of diagenetic evolution: (i) the reactions governed by the breakdown of a metastable mineral system, probably in relation to some temperature threshold; (ii) the reactions governed by a continuous but moderate disequilibrium between minerals and mobile pore water. Although the same kinetic coefficients are used throughout the simulations, the rates of the overall mineral transformation in the two preceding regimes are very different. The reason why a metastable assemblage can be maintained during geological periods, however, is still poorly understood. During the first regime, the mineral assemblage reacted quickly because it accumulated a strong internal potential of reaction expressed by a large departure from equilibrium. The phase rule indicates that the mineral assemblage cannot reach equilibrium at the considered T and P. At least one of the minerals must dissolve. The example of the kaolinite/K–feldspar association is typical. From a thermodynamic viewpoint, it should begin to react at low T and P conditions. Nevertheless it is observed up to 110°C and 440 bars (Fig. 4). When the metastable state is broken, the higher the T and P values, the faster the reaction. The solution composition has no influence on the reaction results. In addition, the transformation does not require abundant water volume since it is isochemical, achieved quickly and homogeneously everywhere. Such diagenetic effects should be able to take place in the oil zone, until the water saturation reaches such a low value that the continuity of the aqueous film between solids and oil is disrupted and the local transfer of elements between the reactive minerals cannot be achieved. In contrast, during the diagenetic transformations governed by water flow the mineral assemblage is close to equilibrium, except at the inlet of the system if the solution entering the reservoir comes from a very different geochemical environment. The water flow is the agent of the reaction and if it stops for any reason the mineral association reaches thermodynamic equilibrium in a short time and the reaction stops. Heterogeneity can develop as the result of propagating reaction fronts.

The question of elemental budgets can be easily addressed with modelling. In this study, constraint on the concentration of dissolved aluminium in water is lacking. Data on dissolved silica are rare in North Sea reservoir waters (Fig. 6). We assumed rather low Al and Si solubilities (Table 2). Al and Si budgets, and consequently the overall transformation, are controlled by these low values. The simulations do not show any significant import or export of Al or Si (Si is very moderately imported). The reverse is true for Na and K. When albite dissolves, all the produced Na is exported out of the system. At a 1 $mm{\cdot}yr^{-1}$ flow velocity, the amount of K removed from water by illitization is much less than the amount of K introduced into the system by advection. If the 1 km long reservoir is considered to have a cross-section of 100 × 100 m, the total amount of K brought in by a 1 $mm{\cdot}yr^{-1}$ water flow is of the order of $10^7$ moles in 1 million years, whereas the total quantity of K present in the diagenetic illite (the major part of this potassium comes from dissolved K-feldspar) is of the order of $10^4$ moles. Integrated over millions of years mineral transfers can be significant for the most soluble elements. In a subarkosic sandstone, both diagenetic regimes distinguished above, i.e. isochemical and allochemical, produce quartz from the silica released by dissolution of silicates (kaolinite and feldspars). In the particular case of the considered sandstones from Dunbar, the total amount of quartz precipitated in this way is ≈ 6 volume percentage, in the range of values observed by Hogg *et al.* (1992). Additional pressure solution cannot be excluded, but would be only required here as a moderate topping-up contribution.

The data set from Dunbar does not provide unambiguous information on the actual role played by water flow in the diagenetic patterns observed at the field scale (Fig. 2). The density of information is too low and the precision of analyses too poor. In addition, local variations in the respective amounts of detrital phases cannot be ruled out, even in a single lithofacies on a kilometric scale. Nevertheless, this data set provides a relatively well-constrained case study. Numerical simulations demonstrated that the observed heterogeneity of mineralogy can be created by a water-flow driven diagenesis involving flow rates between 0.1 and 1 $cm{\cdot}yr^{-1}$ at the burial depth considered (3 km). In any case, the rates of reaction front advance for silicate reactions are very slow at this depth and the diagenetic heterogeneity can only have a local extension.

Finally, modelling results also suggest that the present-day mineral composition observed in the UMS unit should still be locally reactive. The illitization seems incomplete, at least in the northern wells. We attribute this phenomenon to the oil filling, which could stop the diagenetic process governed by the water circulation. Unfortunately, at Dunbar, there is no equivalent Tarbert facies in the water zone that could be used for comparison.

## CONCLUSIONS

Numerical modelling allows one to explore the effects of water composition and velocity on the evolving diagenetic patterns, to calculate mass-transfer budgets and to examine what are the water-flow velocities required to generate significant amounts of cement, particularly of quartz, in reasonable periods of time. In a given diagenetic situation, it can help to point out the limiting mechanism of mineral transformations and, for instance, of quartz cementation. In revealing the formation of diagenetic fronts, it gives new insight into the problem of mineralogical and petrophysical heterogeneity at the facies scale. It can also give indirect arguments on the efficiency of the mineral reactions in the oil zone. However, outcomes of numerical simulations must be examined with great care and systematically compared with the available observations and measurements. In some cases, diagenetic modelling will bring to light the need for additional data, in order to address correctly a particular diagenetic feature. In that respect, a precise quantification of the mineral composition from core samples is an important prerequisite for validating simulation results in cored areas and to giving the model a predictive value in uncored areas.

Two very different types of diagenetic transformations are clearly depicted by the application of the DIAPHORE code: (i) the transformations governed by the instability of minerals present in the system; (ii) the transformations governed by disequilibrium between inflowing water and the minerals. The former reactions tend to be isochemical (no influence of the solution) and relatively rapid, and it does not generate any heterogeneity because it affects all the system in the same way at the same time. The latter ones result in the formation and in the slow propagation of reaction fronts, likely to create heterogeneity over geological periods of time, and to produce allochemical budgets (possible import/export of the most soluble elements). Only transformations of the second type are stopped by hydrocarbon filling. Conversely, transformations of the first type can take place as long as a low residual water saturation exists allowing the aqueous transfer of elements between reacting minerals. In the case of the Brent sediments from the Dunbar field, today located at ≈ 3.8 km burial depth, the simulations showed that the majority of the observed quartz cement can be explained by the evolution of silicates present in the reservoir, and demonstrates that open system alteration is essential.

## ACKNOWLEDGEMENTS

The DIAPHORE code was developed by several other persons at IFP, particularly by Olivier Bildstein. Some geochemical aspects of the Dunbar case have been studied by Claudine Durand (IFP). The mineralogical database has been created by Isabelle Pelletier (IFP). The application of DIAPHORE to the Dunbar case has been partly funded by the Thermie Programme of the EEC (Contract N. OG/233/94-FR/UK, 1995–1996). The manuscript benefited greatly from the detailed and useful corrections made by C. Büker, J.-P. Girard and S. Morad.

## REFERENCES

Anbeek, C. (1992) Surface roughness of minerals and implications for dissolution studies. *Geochimica et Cosmochimica Acta* **56**, 1461–1469.

APLIN, A.C., WARREN, E.J., GRANT, S.M. & ROBINSON, A.G. (1993) Mechanisms of quartz cementation in North Sea reservoir sands: constraints from fluid compositions. In: *Diagenesis and Basin Development.* (eds HORBURY, A.D. & ROBINSON, A.G.) pp. 5–22. Association of American Petroleum Geologists Studies in Geology 36.

BAHR, J.M. & RUBIN, J. (1987) Direct comparison of kinetic and local equilibrium formulations for solute transport affected by surface reactions. *Wat. Resource Research* **23**, 438–452.

BAILLIE, J., COOMBES, T. & RAE, S. (1996) Dunbar Reservoir Model, a multidisciplinary approach to update Brent Reservoir description and modelling. *Society of Petroleum Engineers* **35528**, 329–336.

BAZIN, B., BROSSE, E. & SOMMER, F. (1997) Chemistry of oil-field brines in relation to diagenesis of reservoirs. 2. Reconstruction of palaeo-water composition for modelling illite diagenesis in the Greater Alwyn area (North Sea). *Marine and Petroleum Geology* **14** (5), 497–511.

BEN BACCAR, M., FRITZ, B. & BRÉVART, O. (1993) Geochemical modelling of late diagenetic processes in the Brent Sandstone, Alwyn South area (East Shetland Basin, North Sea). 1. Estimation of the circulated fluids composition. *Chemical Geology* **109**, 135–147.

BILDSTEIN, O. (1998) *Modélisation des interactions géochimiques eau-gaz-roche. Application à la diagenèse minérale dans les réservoirs géologiques.* Thèse, University of Louis Pasteur, Strasbourg.

BJØRKUM, P.A. (1996) How important is pressure in causing dissolution of quartz in sandstones? *Journal of Sedimentary Research* **66** (1), 147–154.

BJØRKUM, P.A. & GJELSWIK, N. (1988) An isochemical model for formation of authigenic kaolinite, K-feldspar and illite in sediments. *Journal of Sedimentary Petrology* **58** (3), 506–511.

BJØRLYKKE, K., AAGAARD, P., EGEBERG, P.K. & SIMMONS, S.P. (1995) Geochemical constraints from formation water analyses from the North Sea and the Gulf Coast Basins on quartz, feldspar and illite precipitation in reservoir rocks. In: *The Geochemistry of Reservoirs.* (eds CUBITT, J.M. & ENGLAND, W.A.) pp. 33–50. Special Publications of the Geological Society of London 86.

BOUDREAU, B. (1997). *Diagenetic Models and their Implementation.* Springer-Verlag, Berlin.

BRADY, P.V. & WALTHER, J.V. (1990) Kinetics of quartz dissolution at low temperatures. *Chemical Geology* **82**, 253–264.

BUHRIG, C. (1989) Geopressured Jurassic reservoirs in the Viking graben: modelling and geological significance. *Marine and Petroleum Geology* **6**, 31–48.

CARROLL-WEBB, S.A. & WALTHER, J.V. (1988) A surface complex reaction model for pH-dependence of corundum and kaolinite dissolution rates. *Geochimica et Cosmochimica Acta* **52**, 2609–2623.

CLAUER, N., SAVIN, S.M. & CHAUDHURI, S. (1992) Isotopic compositions of clay minerals as indicators of the timing and conditions of sedimentation and burial diagenesis. In: *Isotopic Signatures and Sedimentary Records.* (eds CLAUER, N. & CHAUDHURI, S.) pp. 239–286. Lecture Notes in Earth Sciences 43.

CORDON, S. & GUILHAUMOU, N. (1995) Température de silicification des grès du Brent, Jurassique moyen, du champ de Dunbar, Mer du Nord, secteur britannique. *Comptes Rendue Academie Scientifique Paris*, **320**, sér. IIa, 563–569.

DEWERS, T. & ORTOLEVA, P.J. (1990) Interaction of reaction, mass transport, and rock deformation during diagenesis: mathematical modeling of intergranular pressure solution, stylolites, and differential compaction/cementation. In: *Prediction of Reservoir Quality Through Chemical Modeling.* (eds MESHRI, I.D. & ORTOLEVA, P.J.) pp. 147–160. American Association of Petroleum Geologists Memoir 49.

DIDIER, B., ESCHARD, R. & FÉROUL, J.M. (1995) Mode of transgression of a wave- and tide-dominated deltaic system in a tectonically active setting: the Brent Group on Alwyn South fields, North Sea, UK. *International Association of Sedimentologists 16th Regional Meeting of Sedimentology, April 24–26* 1995, *Aix-les-Bains, France.* (Abstract).

DOMENICO, P.A. & PALCIAUSKAS, V.V. (1979) The-averaged mass-transport equation for chemical diagenetic models. *Journal of Hydrology* **43**, 427–438.

EGEBERG, P.K. & AAGAARD, P. (1989) Origin and evolution of formation waters from oil fields on the Norwegian shelf. *Applied Geochemistry* **4**, 131–142.

EHRENBERG, S.N. (1990) Relationship between diagenesis and reservoir quality in sandstones of the Garn Formation, Haltenbanken, Mid-Norwegian Continental Shelf. *American Association. of Petroleum Geologists Bulletin* **74** (10), 1538–1558.

EHRENBERG, S.N. & BOASSEN, T. (1993) Factors controlling permeability variation in sandstones of the Garn Formation in Trestakk field, Norwegian continental shelf. *Journal of Sedimentary Petrology* **63** (5), 929–944.

EHRENBERG, S.N. & NADEAU, P. (1989) Formation of diagenetic illite in sandstones of the Garn Formation, Haltenbanken, mid-Norwegian continental shelf. *Clay Minerals* **24**, 233–253.

EMERY, D. & ROBINSON, A. (1993). *Inorganic Geochemistry: Applications to Petroleum Geology.* Blackwell Scientific Publications, Oxford.

FALLICK, A.E., MACAULAY, C.I. & HASZELDINE, R.S. (1993) Implication of linearly correlated oxygen and hydrogen isotopic compositions for kaolinite and illite in the Magnus Sandstone, North Sea. *Clays and Clay Minerals* **41**, 184–190.

FRITZ, B. (1981). *Etude Thermodynamique et Modélisation Des Réactions Hydrothermales et Diagénétiques. Sciences Géologiques Memoir N.* 65, Strasbourg University.

GILES, M.R. (1997). *Diagenesis: a Quantitative Perspective.* Kluwer, Dordrecht.

GILES, M.R., STEVENSON, S., MARTIN, S.V. *ET AL.* (1992) The reservoir properties and diagenesis of the Brent Group: a regional perspective. In: *Geology of the Brent Group.* (eds MORTON, A.C., HASZELDINE, R.S., GILES, M.R. & BROWN, S.) pp. 289–328. Special Publications of the Geological Society of London 61.

GLASMANN, J.R. (1992) The fate of feldspars in Brent group reservoirs, North Sea: a regional synthesis of diagenesis in shallow, intermediate and deep burial environments. In: *Geology of the Brent Group.* (eds MORTON, A.C., HASZELDINE, R.S., GILES, M.R. & BROWN, S.) pp. 329–350. Special Publications of the Geological Society of London 61.

GLUYAS, J.G. & COLEMAN, M. (1992) Material flux and porosity changes during sediment diagenesis. *Nature* **356**, 52–53.

GLUYAS, J.G., ROBINSON, A.G. & GRANT, S.M. (1993) Geochemical evidence for a temporal control on sandstone cementation. In: *Diagenesis and Basin Development.* (eds HORBURY, A.D. & ROBINSON, A.G.) pp. 23–34. American Association of Petroleum Geologists Studies in Geology 36.

HAMILTON, P.J., GILES, M.R. & AINSWORTH, P. (1992) K–Ar dating of illites in Brent Group reservoirs: a regional perspective. In: *Geology of the Brent Group.* (eds MORTON, A.C.,

HASZELDINE, R.S., GILES, M.R. & BROWN, S.) pp. 377–400. Special Publications of the Geological Society of London 61.

HARRISON, W.J. & TEMPEL, R.N. (1993) Diagenetic pathways in sedimentary basins. In: *Diagenesis and Basin Development.* (eds HORBURY, A.D. & ROBINSON, A.G.) pp. 69–86. American Association of Petroleum Geologists Studies in Geology 36.

HELGESON, H.C., BROWN, T., NIGRINI, A. & JONES, T. (1970) Calculation of mass transfer in geochemical processes involving aqueous solutions. *Geochimica et Cosmochimica Acta* **34**, 569–592.

HELGESON, H.C., MURPHY, W. & AAGAARD, P. (1984) Thermodynamic and kinetic constraints on reaction rates among minerals and aqueous solutions: II. Rate constants, effective surface area and hydrolysis of feldspars. *Geochimica et Cosmochimica Acta* **48**, 2405–2432.

HOGG, A.J.C., SELLIER, E. & JOURDAN, A.J. (1992) Cathodoluminescence of quartz cements in Brent Group sandstones, Alwyn South, UK North Sea. In: *Geology of the Brent Group.* (eds MORTON, A.C., HASZELDINE, R.S., GILES, M.R. & BROWN, S.) pp. 421–440. Special Publications of the Geological Society of London 61.

HOGG, A.J.C., HAMILTON, P.J. & MACINTYRE, R.M. (1993) Mapping diagenetic fluid flow within a reservoir: K-Ar dating in the Alwyn area (UK North Sea). *Marine and Petroleum Geology* **10**, 279–294.

HOGG, A.J.C., PEARSON, M.J., FALLICK, A.E. & HAMILTON, P.J. (1995) An integrated thermal and isotopic study of the diagenesis of the Brent Group, Alwyn South, U.K. North Sea. *Applied Geochemistry* **11**, 531–546.

HOUSEKNECHT, D.W. (1988) Intergranular pressure solution in four quartzose sandstones. *Journal of Sedimentary Petrology* **58** (2), 228–246.

INGLIS, I. & GÉRARD, J. (1991) The Alwyn North Field, Blocks 3/9a, 3/4a, UK North Sea. In: *United Kingdom Oil and Gas Fields, 25 Years Commemorative Volume.* (ed. ABBOTTS, I.L.) pp. 21–32. Geological Society Memoirs 14.

JOHNSON, A. & EYSSAUTIER, M. (1987) Alwyn North Field and its regional geologic context. In: *Petroleum Geology of North West Europe.* (eds BROOKS, J. & GLENNIE, K.W.) pp. 963–977. Graham & Trotman, London.

JOHNSON, J.W., OELKERS, E.H. & HELGESON, H.C. (1992) SUPCRT92: software package for calculating the standard molal thermodynamic properties of minerals, gases, aqueous species, and relations among them as functions of temperature and pressure. *Computers and Geoscience* **18**, 899–947.

JOURDAN, A., THOMAS, M., BREVART, O., ROBSON, P., SOMMER, F. & SULLIVAN, M. (1987) Diagenesis as the control of the Brent sandstones reservoir properties in the Greater Alwyn area (East Shetland Basin). In: *Petroleum Geology of North West Europe.* (eds BROOKS, J. & GLENNIE, K.W.) pp. 951–961. Graham & Trotman, London.

KAIRO, S., SUTTNER, L.J. & DUTTA, P.K. (1993) Variability in sandstone composition as a function of depositional environment in coarse-grained delta systems. *Geological Society of America Bulletin, Special Paper* **284**, 263–282.

KIRKNER, D.J. & REEVES, H. (1988) Multicomponent mass transport with homogeneous and heterogeneous chemical reactions: effect of the chemistry on the choice of numerical algorithm. *Journal of Theoretical Water Resources Research* **24** (10), 1719–1729.

KNAUSS, K.G. & WOLERY, T.J. (1988) The dissolution kinetics of quartz as a function of pH and time at 70°C. *Geochimica et Cosmochimica Acta* **52**, 43–53.

LASAGA, A.C. (1981) Transition state theory. In: *Kinetics of Geochemical Processes.* (eds LASAGA, A.C. & KIRKPATRICK, R.J.) pp. 135–169. Reviews in Mineralogy 4.

LASAGA, A.C., SOLER, J.M., GANOR, J., BURCH, T.E. & NAGY, K.L. (1994) Chemical weathering rate laws and global geochemical cycles. *Geochimica et Cosmochimica Acta* **58** (10), 2361–2386.

LE GALLO, Y., BILDSTEIN, O. & BROSSE, E. (1998). Coupled reaction-flow modeling of diagenetic changes in reservoir permeability, porosity and mineral compositions. *Journal of Hydrology* **209**, 366–388.

LEDER, F. & PARK, W.C. (1986) Porosity reduction in sandstone by quartz overgrowth. *American Association of Petroleum Geologists Bulletin* **70** (11), 1713–1728.

LEPERCQ, J.Y. & GAULIER, J.M. (1996) Two-stage rifting in the North Viking Graben area (North Sea): inferences from a new three-dimensional subsidence analysis. *Marine and Petroleum Geology* **13** (2), 129–148.

LICHTNER, P. (1992) Time–space continuum description of fluid/rock interaction in permeable media. *Water Resources Research* **28** (12), 3135–3155.

LICHTNER, P. (1993) Scaling properties of time–space kinetic mass transport equations and the local equilibrium limit. *American Journal of Science* **293**, 257–296.

LONGSTAFFE, F.J. (1994) Stable isotopic constraints on sandstone diagenesis in the Western Canada sedimentary basin. In: *Quantitative Diagenesis: Recent Developments and Applications to Reservoir Geology.* (eds PARKER, A. & SELLWOOD, B.W.) pp. 223–274. NATO ASI Series C, 453.

MADÉ, B., CLÉMENT, A. & FRITZ, B. (1994) Modélisation thermodynamique et cinétique des réactions diagénétiques dans les bassins sédimentaires. Présentation du modèle géochimique KINDISP. *Revue de l'Institut Français Du Pétrole* **49** (6), 569–602.

MILLIKEN, K.L., LAND, L.S. & LOUCKS, R.G. (1981) History of burial diagenesis determined from isotopic geochemistry, Frio Formation, Brazoria County, Texas. *American Association of Petroleum Geologists Bulletin* **65** (8), 1397–1413.

NADEAU, P.H. & HURST, A. (1991) Application of back-scattered electron microscopy to the quantification of clay mineral microporosity in sandstones. *Journal of Sedimentary Petrology* **61**, 921–925.

ODOM, I.E., DOE, T.W. & DOTT, R.H. JR (1976) Nature of feldspar-grain size relations in some quartz-rich sandstones. *Journal of Sedimentary Petrology* **46** (4), 862–870.

OELKERS, E.H., BJØRKUM, P.A. & MURPHY, W.M. (1996) A petrographic and computational investigation of quartz cementation and porosity reduction in North Sea sandstones. *American Journal of Science* **296**, 420–452.

ORTOLEVA, P. (1987a) Geochemical self-organization: I. Reaction–transport feedbacks and modeling approach. *American Journal of Science* **287**, 979–1007.

ORTOLEVA, P. (1987b) Geochemical self-organization: II. The reactive-infiltration instability. *American Journal of Science* **287**, 1008–1040.

PARKHURST, D.L., THORSTENTON, D. & PLUMMER, L.N. (1980) *PHREEQE: a computer program for geochemical calculations*. US Geological Survey, Water Resources Investigations Report, pp. 80–96.

PERKINS, E.H., KHARAKA, Y.K., GUNTER, W.D. & DEBRAAL, J.D. (1990) Geochemical modeling of water–rock interactions using SOLMINEQ.88. In: *Chemical Modeling of Aqueous Systems II.* (eds MELCHIOR, D.C. & BASSETT, R.L.) pp. 117–127. ACS Symposium Series 416.

POTDEVIN, J.L. & HASSOUTA, L. (1997) Bilan de matière des processus d'illitisation et de surcroissance de quartz dans un réservoir pétrolier du champ d'Ellon. *Bulletin of the Society of Geology of France* **168** (2), 219–230.

ROBINSON, A.G. & GLUYAS, J.G. (1992) Duration of quartz cementation in sandstones, North Sea and Haltenbanken basins. *Marine and Petroleum Geology* **9**, 324–327.

RUBIN, J. (1983) Transport of reacting solutes in porous media: relation between mathematical nature of problem formulation and chemical nature of reactions. *Water Resources Research* **19**, 1231–1252.

SJOBERG, E.L. & RICKARD, D.J. (1984) The influence of experimental design on the rate of calcite dissolution. *Geochimica et Cosmochimica Acta* **47**, 2281–2285.

SMITH, J.T. & EHRENBERG, S.N. (1989) Correlation of carbon dioxide abundance with temperature in clastic hydrocarbon reservoirs: relationship to inorganic chemical equilibrium. *Marine and Petroleum Geology* **6**, 129–135.

STEEFEL, C.I. & LASAGA, A.C. (1994) A coupled model for transport of multiple chemical species and kinetic precipitation/dissolution reactions with application to reactive flow in single phase hydrothermal systems. *American Journal of Science* **294**, 529–592.

STUMM, W. & WIELAND, E. (1990) Dissolution of oxides and silicate minerals: rates depend on surface speciation. In: *Aquatic Surface Chemistry.* (ed. STUMM, W.) pp. 367–400. Wiley, Chichester.

VAN DER PLAS, L. & TOBI, A.C. (1965) A chart for judging the reliability of point counting results. *American Journal of Science* **263**, 87–90.

WALDERHAUG, O. (1994a) Temperatures of quartz cementation in Jurassic sandstones from the Norwegian continental shelf—evidence from fluid inclusions. *Journal of Sedimentary Research A* **64**, 311–323.

WALDERHAUG, O. (1994b) Precipitation rates for quartz cement in sandstones determined by fluid-inclusion microthermometry and temperature-history modelling. *Journal of Sedimentary Research A* **64**, 324–333.

WANG, Y. & VAN CAPPELLEN, P. (1996) A multicomponent reactive transport model of early diagenesis: application to redox cycling in coastal marine sediments. *Geochimica et Cosmochimica Acta* **60** (16), 2993–3014.

WARREN, E.A. & SMALLEY, P.C. (1993) The chemical composition of North Sea formation waters: a review of their heterogeneity and potential applications. In: *Petroleum Geology of the Northwestern Europe: Proceedings of the 4th Conference* (ed. PARKER, J.R.) pp. 1347–1352. The Geological Society of London.

WARREN, E.A. & SMALLEY, P.C. (1994). *North Sea Formation Waters Atlas.* Geological Society Memoir 15.

WILLIAMS, L.B., HERVIG, R.L. & BJØRLYKKE, K. (1997) New evidence for the origin of quartz cements in hydrocarbon reservoirs revealed by oxygen isotope microanalyses. *Geochimica et Cosmochimica Acta* **61**, 2529–2538.

WOLERY, T.J., JACKSON, K.J., BOURCIER, W.L. *ET AL.* (1990) Current status of the EQ3/6 software package for geochemical modeling. In: *Chemical Modeling of Aqueous Systems II.* (eds MELCHIOR, D.C. & BASSETT, R.L.) pp. 104–116. American Chemical Society Symposium Series 416.

YEH, G.T. & TRIPATHI, V.S. (1991) A model for simulating transport of reactive multispecies components: model development and demonstration. *Water Resources Research* **2** (12), 3075–3094.

*Spec. Publs int. Ass. Sediment.* (2000) **29**, 67–78

# The different processes involved in the mechanism of pressure solution in quartz-rich rocks and their interactions

F. RENARD,[1,3] É. BROSSE[2] *and* J. P. GRATIER[1]

[1]*LGIT, CNRS-Observatoire, BP53X 38041 Grenoble, France;*
[2]*Institut Français du Pétrole, 1–4 avenue de Bois Préau, 92506 Rueil-Malmaison, France; and*
[3]*University of Oslo, Department of Geology, Box 1047, Blindern, 0316 Oslo, Norway*

## ABSTRACT

Pressure solution is a very efficient mechanism for rock deformation and compaction in the upper crust. It is controlled by stress, temperature, and grain size. The mechanism commonly used to describe deformation by pressure solution at a grain scale, the 'water film diffusion' mechanism, can be divided into three successive steps: (i) dissolution at the grain interface; (ii) diffusion of solutes along an adsorbed water film inside the contact between two grains; (iii) precipitation on the surface of the grains adjacent to the open pore. The slowest step controls the rate of the overall process.

Important geometrical variables controlling the pressure solution rate are the grain size and the geometry of the grain–grain interface. Knowledge of the geometry permits estimation of the path length for diffusion of the solutes from the contact between the grains to the pore. Physicochemical variables are important also, including: temperature, stress, chemistry of the pore water, mineralogy of the rock, and thickness of the water film trapped between the grains.

In pressure solution, different coupled variables are involved on several spatial scales: the water film thickness at a nanometre scale, the geometry of the surface of contact at a micrometre scale, and the grain size and texture of the rock at a millimetre to centimetre scale. To understand the mechanism of pressure solution, the values of all these variables must be evaluated to estimate the kinetics of the various coupled processes in which they are involved. These variables can be modelled to determine: (i) the conditions under which pressure solution is efficient in sedimentary basins; (ii) the parameters that control the rate of pressure solution and quartz cementation. The result is a model that can estimate the porosity variations and quartz cementation due to pressure solution, as a function of pressure, temperature, and rock texture.

## INTRODUCTION

The deformation of rocks by a dissolution–transport–precipitation mechanism, called pressure-solution by most authors, has been observed for many years, particularly in sandstones and limestones (Heald, 1955; Weyl, 1959; see McBride, 1989 for a complete review). A common mechanistic model of pressure solution is 'water film diffusion' (Weyl, 1959; Rutter, 1976). In this model it is assumed that mineral dissolution takes place at the grain–grain contact area with the solutes diffusing along an adsorbed water film. The precipitation of the solutes occurs on the mineral surface facing the pore. This simple model of deformation can be described by a three-step mechanism (Fig. 1a). First, a mineral dissolves at the contact between two grains because of a concentration of stress. Then solutes diffuse to the pore where, lastly, precipitation occurs. Thus, the whole mechanism, modifying the grain size and the volume of the pores, is a function of rock deformation. If one of the three steps in Fig. 1a is slower than the two others, it will limit the rate of deformation. When the kinetics of dissolution at the grain contacts or precipitation in the pore are the limiting step for deformation, the strain rate depends on the rate constant for dissolution (Table 1). When the slowest step is diffusion along the interface, the strain rate depends on the coefficient of diffusion inside the interface (Table 1). In these two cases, the

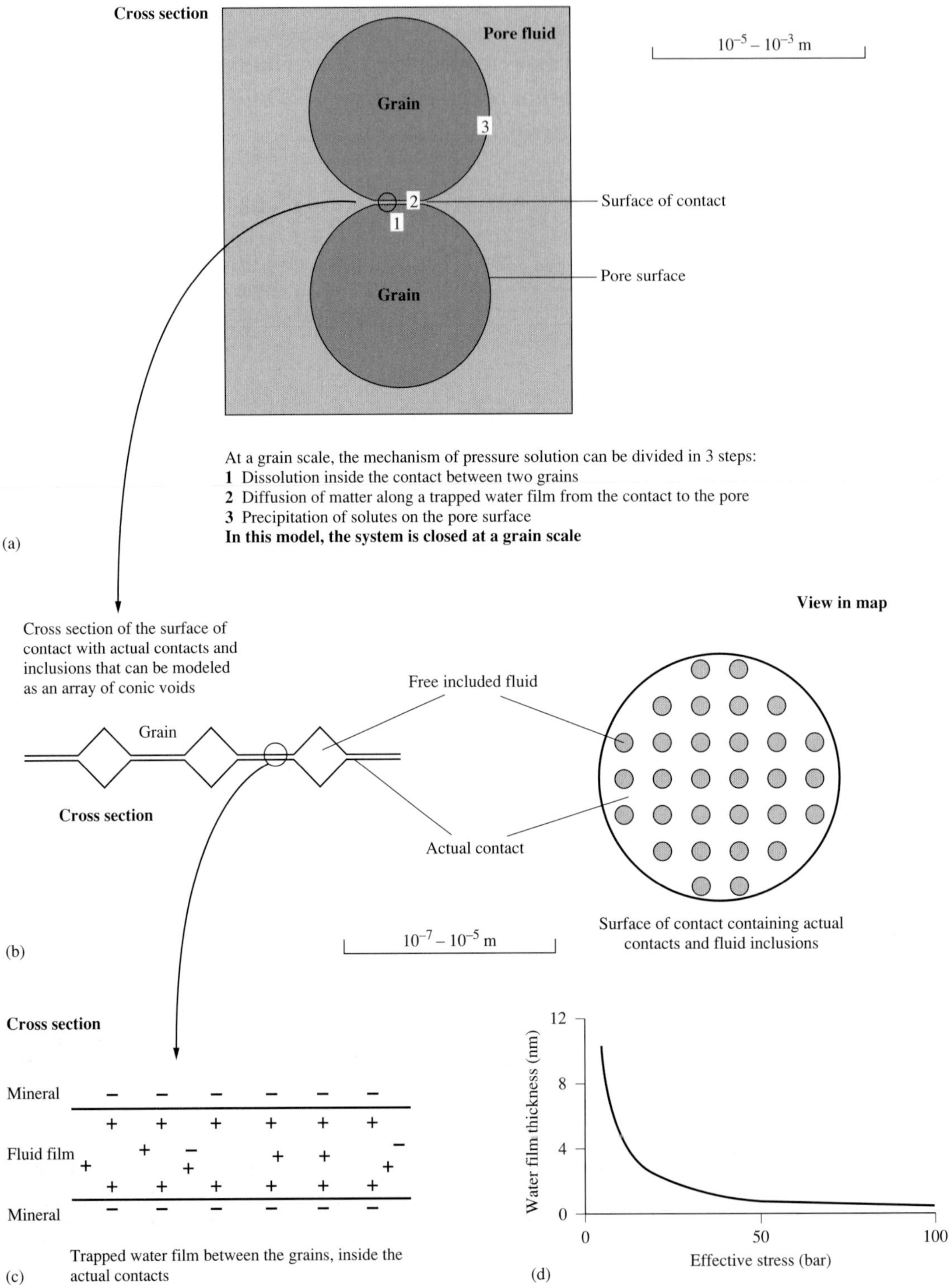
Cross section
Pore fluid
Grain
3
10⁻⁵ – 10⁻³ m
2
1
Surface of contact
Grain
Pore surface
At a grain scale, the mechanism of pressure solution can be divided in 3 steps:
1 Dissolution inside the contact between two grains
2 Diffusion of matter along a trapped water film from the contact to the pore
3 Precipitation of solutes on the pore surface
In this model, the system is closed at a grain scale
(a)
View in map
Cross section of the surface of contact with actual contacts and inclusions that can be modeled as an array of conic voids
Free included fluid
Grain
Cross section
Actual contact
Surface of contact containing actual contacts and fluid inclusions
10⁻⁷ – 10⁻⁵ m
(b)
Cross section
Mineral
Fluid film
Mineral
Trapped water film between the grains, inside the actual contacts
(c)
Water film thickness (nm)
12
8
4
0
0
50
100
Effective stress (bar)
(d)

**Table 1.** The different variables involved in the estimation of the strain rate (Poirier, 1985). If the mechanism of pressure solution depends on diffusion or on the kinetics of dissolution at the interface (Fig. 1a), the dependence on the grain size is different. In both cases, the driving force for deformation is a difference of normal stress along the grain surface

| | Geometric factors | Chemical factor | Kinetic factors | Driving force |
|---|---|---|---|---|
| Diffusion | $\frac{w}{d^3}$ | $\underline{c}$ | D | $\Delta\sigma_n$ |
| Dissolution | $\frac{1}{d^2}$ | $\underline{c}$ | k | $\Delta\sigma_n$ |

d: grain size; k: kinetic constant of dissolution; w: width of grain interface; D: coefficient of diffusion along the interface; $\underline{c}$: concentration of solutes; $\Delta\sigma_n$: difference of normal stress.

dependence of the strain rate on grain size is different (Poirier, 1985).

The presence of water in the pore is necessary to obtain significant deformation by pressure solution because water is the medium of dissolution, diffusion, and precipitation of solutes (Gratier & Guiguet, 1986). A high content of oil should lower the rate of deformation since quartz solubility in oil is negligible. When oil is present in the porosity, a thin water film can be still trapped into the contact between the grains. A small amount of dissolution into the water film will saturate it, but the presence of oil in the pores will inhibit precipitation on the pore walls. Oil should decrease the rate of the third step (the precipitation on the pore surface) in Fig. 1a if the rock is oil-wet (Worden *et al.*, 1998), and therefore decrease the overall rate of deformation by pressure solution.

## PHYSICAL AND CHEMICAL VARIABLES INVOLVED IN PRESSURE SOLUTION

### Stability and properties of a trapped water film

A key assumption in the theory of pressure solution is the nature of the thin water film that is believed to provide the reaction and transport medium in the 'water film diffusion' mechanism (Weyl, 1959; Rutter, 1976). To facilitate pressure solution, such a film must exist for large differences between the stress applied across the grain contact and the fluid pressure in the pore. This thin water film has a stress-dependent thickness (Horn *et al.*, 1988, 1989) and the coefficient of diffusion of solutes along this film to the pore fluid is probably higher than the coefficient of diffusion in solids.

Many authors have demonstrated that a water film can be trapped between the sheets of some minerals such as clays and micas (Low, 1992; Sposito, 1992) or sapphires (Horn *et al.*, 1988). The thickness of this film has been measured or calculated for different minerals and varies from a few tenths to several nanometres (Peschel & Aldfinger, 1971; Heidug, 1995). These authors show that under most sedimentary basin conditions, its thickness decreases exponentially with stress until a lower limit of 0.5 nm is reached, which represents the thickness of two layers of water molecules (Fig. 1d).

The thin water film acts as a semipermeable membrane: the ionic concentration inside the film is higher than the concentration in the pore fluid. This difference of concentration in charged species (such as $Na^+$ or $Cl^-$) creates an osmotic pressure effect that stabilizes the water film relative to the stress on the contact. The main origin of this physical dependency is the electric surface charge, and the stability of the film is caused by an osmotic/Debye–Hückel model (Renard & Ortoleva, 1997). Most minerals have an electric charge on their surface and these charges are strongly pH-dependent. Electric charges spread on the surfaces of a mineral attract oppositely-charged ions from the pore fluid (Fig. 1c). This local

**Fig. 1.** (*Opposite.*) The mechanism of pressure solution at different scales. (a) At a grain scale, dissolution occurs at the contact area between grains modelled as truncated spheres (step 1), then solutes diffuse along a water film to the pore (step 2), and silica precipitates on the pore surface (step 3). (b) At a micrometre scale, the contact surface between the grains is subdivided into points of actual contact, and fluid inclusions, modelled as an array of conic inclusions. If the inclusions are connected, they can form small channels. (c) At a nanometre scale, the actual contacts contain a trapped water film. Electric charges spread on the surface of quartz (negative in our example) attract positive ions from the pore. The water film acts as a semipermeable membrane and its stability is through an electric Debye–Hückel osmotic effect. (d) The water film thickness exponentially decreases as the effective stress (pressure normal to the grain contact minus pore fluid pressure) increases. The curve is calculated for a mineral electric surface charge of −0.1 Coulomb $m^{-2}$, like for quartz in a fluid at pH 8 (Renard & Ortoleva, 1997).

effect creates a difference of free energy between the pore fluid and the water film where ionic concentration is higher. In Fig. 1d, the thickness of the water film is given as a function of the effective pressure, i.e. the amount by which the fluid pressure acting on the fluid–solid interface exceeds the hydrostatic pressure in the bulk fluid (Renard & Ortoleva, 1997).

The diffusion coefficient within the water film controls the rate at which solutes are expelled from the stressed contact to the pore and is poorly defined. However, it may be not constant and should vary with the water film thickness such that when the film is thick enough the rate of diffusion should be similar to that in free water, and when the water film is only a few tenths of nanometre thick, the rate of diffusion is expected to be like that in solid grain boundaries.

The coefficient of diffusion of a particle in a large volume of liquid is described by the Stokes–Einstein equation:

$$D = \frac{kT}{6\pi\delta\eta} \tag{1}$$

where $D$ is the diffusion coefficient ($m^2\ s^{-1}$), $k$ the Boltzmann constant ($1.38 \times 10^{-23}\ m^3\ Pa.s^{-1}$), $\delta$ the diameter of the diffusing particle, and $\eta$ the viscosity of the liquid (Pa.s). Horn *et al.* (1989) have measured the viscosity of a 2 nm water film trapped between silica sheets and they have found a value in the order of magnitude of $10^{-3}$ Pa.s, an order of magnitude lower than for a free fluid. Taking the diameter of a molecule of hydrated silica ($\delta$) to be 0.5 nm, the diffusion coefficient of aqueous silica inside a 2 nm water film at 25°C is therefore $3.5 \times 10^{-10}\ m^2.s^{-1}$; one order of magnitude less than diffusion in free water (Mullis, 1993).

Other authors have deduced the product of the coefficient of diffusion and water film thickness from pressure solution experiments and found values from $10^{-18}$ to $10^{-21}\ m^3.s^{-1}$ (Rutter, 1976; Gratier & Guiguet, 1986; Gratier, 1993a). Their experiments were performed at high effective stress, so the water film thickness should not exceed 0.5 nm, corresponding to two molecular layers of water. Their experiments were made at different temperatures, allowing an activation energy for diffusion to be estimated. The coefficient of diffusion of silica in a 0.5 nm thick water film was estimated and ranged from $8 \times 10^{-12}$ to $8 \times 10^{-15}\ m^2.s^{-1}$ at 25°C, with an energy of activation of 10 to 20 $kJ.mole^{-1}$. Despite the range in these values, they are several orders of magnitude greater than that for diffusion in solids (Freer, 1981). The activation energy is similar to that for diffusion in free liquids (Applin, 1987), but the coefficient of diffusion is 1–3 orders of magnitude smaller.

### The driving force for pressure solution

The stress state of a solid has strong implications for its free energy; the effect of stress being to increase the molar free energy compared with a zero-stress state. Following the Gibbs (1961) and Paterson (1973) theory of heterogeneously stressed solids, it is possible to write the chemical potential of the solid as

$$\mu = \mu^0 + P\overline{V}_s + F\overline{V}_s^0 \tag{2}$$

where $\mu^0$ is the chemical potential at reference state for a stress free solid at temperature $T$ and pressure $P_0 = 1$ bar, $P$ is the actual normal pressure exerted by the fluid in contact with the solid, and $\overline{V}_s$ and $\overline{V}_s^0$ are the molar volumes of the solid under stress and with zero stress respectively. In Equation 2, the Helmholz free energy $F$ contains different contributions: elastic energy induced by stress, plastic energy created by the dislocations inside the mineral, and surface free energy for particles with a very small radius of curvature. Under the conditions found in sedimentary basins, the Helmholz free energy is one or two orders of magnitude less than the $P\overline{V}_s$ term in Equation 2 (Paterson, 1973); and to a first approximation, it can be ignored.

During pressure solution, the normal stress $P$ is greater inside the contact between two grains, where it is approximately lithostatic-like, than in the pore where it is the fluid pressure. Thus, by reference to Equation 2, the chemical potential of a solid can be seen to vary along the surface of a grain, allowing for the chemical diffusive flux of silica from the contact surface to the free pore. In an overpressured system, the fluid pressure can reach the lithostatic pressure. In this case the gradient of chemical potential between the contact and the pore disappears and pressure solution will be inhibited.

If the Helmholz free energy is neglected, the chemical potential of quartz is

$$\mu_{qz} = \mu_0 + P\overline{V}_s \tag{3}$$

By definition $\mu_{qz} = RT\ln(K_{eq})$, where $K_{eq}$ is the equilibrium constant for the reaction of quartz dissolution, thus

$$K_{eq} = K_0(T) \cdot \exp\left(\frac{P\overline{V}_s}{RT}\right) \tag{4}$$

where $K_0(T)$ is the equilibrium constant for an unstressed mineral of quartz, whose value depends only on temperature (Rimstidt, 1997).

The chemical reaction of dissolution or precipitation of quartz can be described by

$$SiO_{2(s)} + 2H_2O \leftrightarrow H_4SiO_{4(aq)}$$

and the equilibrium constant $K_{eq}$ for this reaction is

$$K_{eq} = \frac{a_{H_4SiO_4}}{a_{SiO_2} a^2_{H_2O}} \tag{5}$$

At equilibrium, if one assumes that the activities of solid quartz and water are 1 (Anderson & Crerar, 1993), the equilibrium constant is related to silica solubility $c_{H_4SiO_4}$ through

$$K_{eq} = \gamma_{H_4SiO_4} c_{H_4SiO_4} \tag{6}$$

where $\gamma_{H_4SiO_4}$ is the activity coefficient, assumed to be equal to 1 because the concentration of aqueous silica is low at the low temperature of the upper crust (Rimstidt, 1997). Using Equations 4 and 6, it is possible to estimate the quartz solubility in the contact and in the pore and the corresponding gradient of concentration that facilitates deformation.

### Effects on the kinetics of reactions

The rate constant for dissolution/precipitation of quartz increases with temperature and varies with pH and ionic concentrations (Dove, 1994). Some authors have evaluated the kinetics of quartz precipitation through laboratory measurements (Brady & Walther, 1990; Dove, 1994; Rimstidt, 1997) and found rates faster than in geological conditions (Oelkers *et al.*, 1996). Walderhaug (1994) has estimated precipitation rates in sandstones from the Norwegian margin and found a value much lower than derived by experiments. The kinetics of quartz dissolution can vary on one or two orders of magnitude for slight variations of the fluid composition, at constant temperature. For example, the presence of ions such as $Na^+$ increases the kinetics of dissolution and silica solubility (Dove, 1994).

## IMPORTANCE OF ROCK GEOMETRY

### Rock texture and grain size

During deformation by pressure solution, grain shape is modified while silica is dissolved at grain contacts and precipitation occurs on a grain surface facing a pore if a closed system is considered at a grain scale. A model of pressure solution should take into account dynamic variations of grain geometry and therefore the induced modifications of rock porosity. Dewers & Ortoleva (1990) proposed a model where a whole rock is described as a cubic array of grains represented as truncated spheres (Fig. 2). Four length variables facilitate the modelling of deformation by pressure solution: $L_x$, $L_y$, $L_z$, the truncations in the three directions in space, are characteristic of the dissolution at the grain contact; whereas $L_f$, the radius of the sphere, takes into account the overgrowth on the pore free surface. If the stress is not isotropic on the grain, the normal pressure perpendicular to the contacts is different in the three directions of space. In this case, the rate of dissolution of the contacts is different in these three directions and compaction is not isotropic.

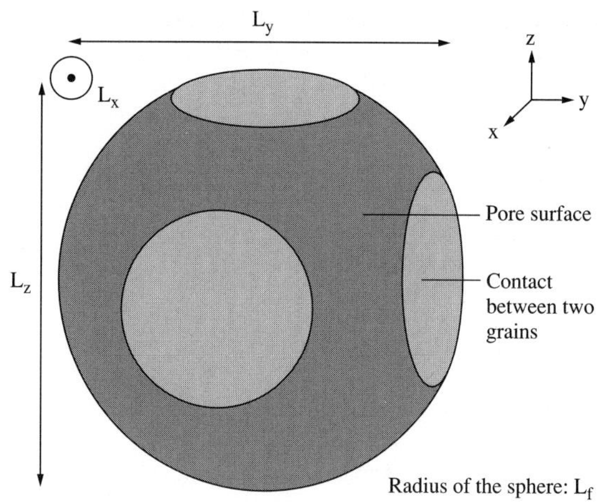


**Fig. 2.** Detailed description of the truncated spheres of Fig. 1a. The dark grey region is the grain surface adjacent to the pore whereas the lighter grey regions correspond to areas in contact with other grains. Each contact represents the circular composite contact of Fig. 1b. Four length variables, the radius of the sphere $L_f$, and the truncations $L_x$, $L_y$, and $L_z$, allow modifications of the grain shape, the rock texture, and the porosity to be modelled during deformation by pressure solution.

On a contact surface, the normal stress approaches lithostatic stress, whereas on the pore surface, the stress is equivalent to hydrostatic pressure. During deformation, the surface area of the contact increases (Fig. 3), thus the normal stress acting on it and the rate of deformation by pressure solution decreases. This mechanism couples stress acting on the rock and rock texture during deformation.

To permit pressure solution, fluid must be able to escape from the pore during compaction to allow a difference of normal stress between the contact and the pore. The geometric model considers deformation at a grain scale. If the system is open at a pore scale and if fluids can circulate, then as compaction reduces total porosity, the fluid circulates less because of the concomitant loss of permeability. Thus fluid pressure may increase during compaction. This process also couples stress and texture because in this case, local fluid pressure can reach lithostatic pressure, thus halting pressure solution.

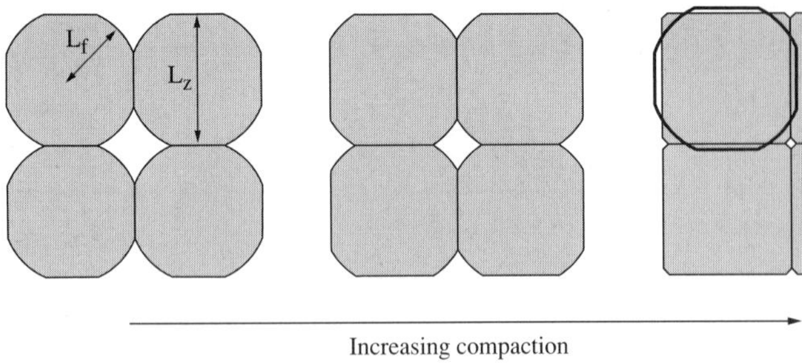


**Fig. 3.** A cross-section view of a cubic-packed network of truncated spheres. The grain shape evolves due to pressure solution as shown. The grain radius $L_f$ increases, to take account of overgrowths in the pore, whereas the grain flattens ($L_z$ decreases) resulting in a decrease in porosity and pore surface area.

**Geometry of the surface of dissolution**

North Sea sandstones contain quartz, micas, and potassic feldspars. Observations of thin sections of these rocks indicate the occurrence of pressure solution: detrital quartz grains dissolved at the contact with other minerals. Observations by SEM indicate that the contact surface between two grains of quartz can locally contain fluid inclusions with a hexagonal shape (Fig. 4a). These inclusions may be localized on dislocations in the mineral. Therefore the surface of contact can be rough and can be locally subdivided into contact points. Grain contacts should sustain a normal stress and should trap a water film, as well as containing inclusions where the fluid is free and where the stress is equivalent to the fluid pressure. Contacts between the grains can thus be described by geometry at two spatial scales: at a nanometre scale, a water film is trapped inside the actual contacts; at a microscale, the contact contains fluid inclusions (Spiers *et al.*, 1990).

A thick section from a sample of a metamorphic quartz vein from the Main Central Thrust in the Himalayas (Fig. 4b) also shows that an array of inclusions at the contacts between two grains can exist. These two natural examples verify that, in both diagenetic sandstones and metamorphic rocks, the contact surface between grains contains a region of actual contacts and fluid inclusions. In the case of the Himalayan sample, some of the fluid inclusions are connected and form small channels.

The distance of diffusion inside the contact may be the mean distance between two fluid inclusions, and not the radius of the full contact. As a consequence, a model of pressure solution should take into account the presence of fluid inclusions and regions of true contact at the interface between the grains. Grain contacts, as modelled in Fig. 2, should contain an array of fluid inclusions giving the composite contact geometry described in Fig. 1.b (Spiers *et al.*, 1990).

**Consequences for the mechanism of pressure solution**

All the preceding observations indicate that there are different sites with different free energies on the grain surface and silica is exchanged between them: the pore surface where silica can precipitate, the region of true contact, and fluid inclusions. Thus, the overall mechanism of pressure solution can be described in three steps (Fig. 5). First, dissolution occurs in the regions of true contact because of a high normal stress; secondly aqueous silica is transported by diffusion along the adsorbed water film. This second step takes into account diffusion from the actual contacts to the pore fluid and diffusion from the actual contacts to the inclusions. Inside the inclusions, precipitation or dissolution can occur, decreasing or increasing their volume and their reactive surface area. Then diffusion of the solute from the inclusions to the pore occurs. This second step of diffusion is made significantly more effective if inclusions are connected and form channels. At the third step, silica precipitation can take place on the pore surface (Fig. 5).

Pressure solution experiments using a metallic needle that indents a crystal of quartz or halite (Gratier, 1993a, 1993b) or on compaction of aggregates (Rutter, 1976, 1983; Gratier & Guiguet, 1986; Cox & Paterson, 1991; Schutjens, 1991) indicate that the mechanism of pressure solution can be limited by diffusion of silica along the trapped water film inside the contact (step 2 in Figs 1a & 5). In such a case, the factor controlling pressure solution and deformation is the stress acting on the water film, which also controls the driving force of reaction.

However, experiments on the deformation of fluid inclusions show that the limiting process is dissolution on the walls of the inclusions. In this case, diffusion is fast because solutes are transported in a free fluid (Gratier & Jenatton, 1984). Following observations of sandstone thin-sections, it was concluded that at low temperature

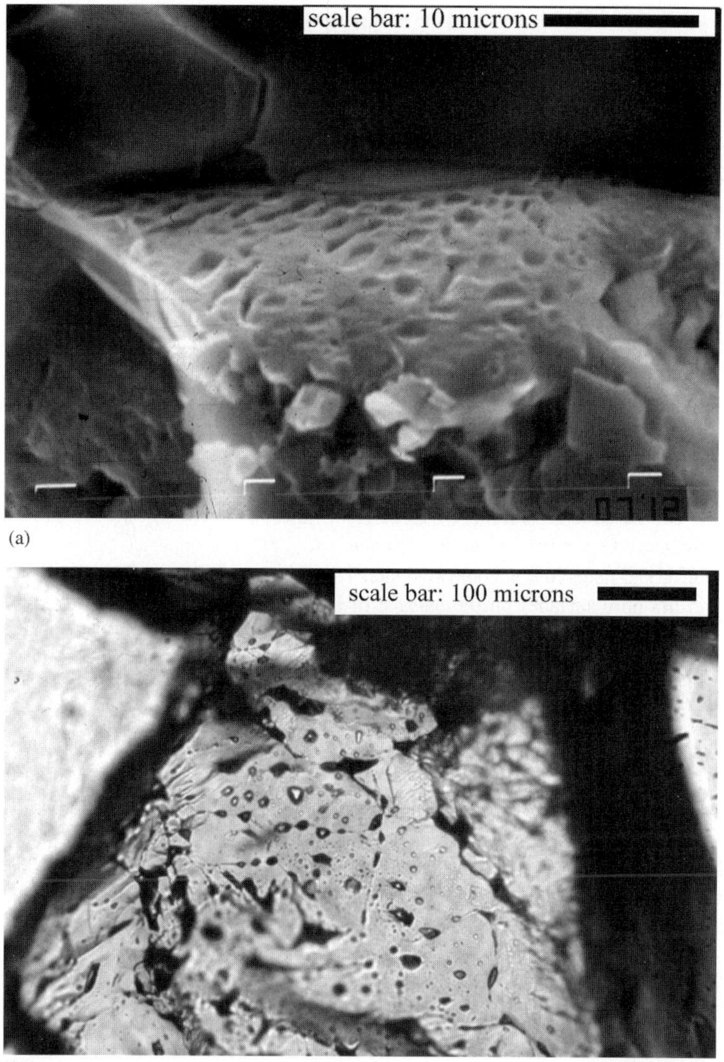


**Fig. 4.** (a) SEM picture of the contact surface between two grains of quartz in a North Sea sandstone. The surface contains flat actual contacts and fluid inclusions. (b) Thin section in a metamorphic Himalayan quartz. The surface of contact between the grains contains fluid inclusions that can be connected and form small channels. In both examples, fluid inclusions should be localized on dislocations of the crystal.

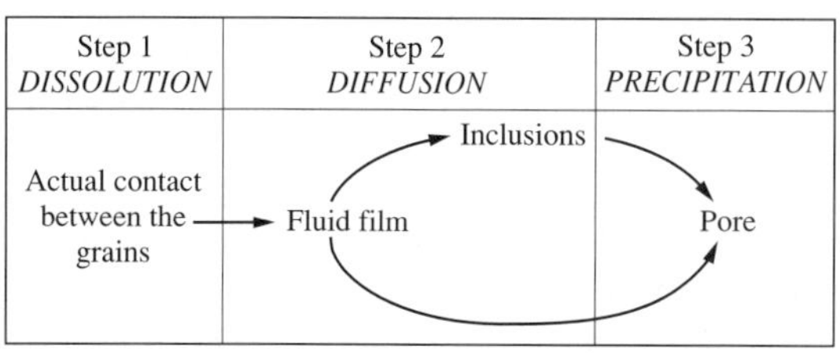


**Fig. 5.** The mechanism of pressure solution can be divided into three steps: (i) dissolution at grain interface; (ii) diffusion along the trapped water film from the actual contacts to the pore or to the inclusions inside the contact; (iii) precipitation on the pore surface. The slower step controls the overall rate of deformation.

(0–2500 m), the mechanism of sandstone deformation is also temperature controlled through the kinetics of quartz precipitation (step 3 in Figs 1a & 5) and that the effect of stress is not important (Bjørkum, 1996; Oelkers *et al.*, 1996). In this case, the limiting step for deformation is the rate of dissolution or precipitation of silica (steps 1 or 3 in Figs 1a & 5) for which the kinetics are slower than those for the rate of diffusion along the interface (Renard *et al.*, 1997).

All these experiments and natural observations indicate that pressure solution is complex with different processes interacting (Table 2). Depending on the conditions

**Table 2.** The different variables involved in pressure solution and the processes where they appear. The numbers in parenthesis correspond to the three steps of the mechanism of pressure solution described in Figs 1a & 5

| Stress | Temperature | Grain geometry |
|---|---|---|
| Driving force for pressure solution (1 to 3) | Kinetics of reaction (1, 3) | Pore surface area (3) |
| Diffusion inside the water film (2) | Diffusion (2) | Porosity and permeability (3) |
| Water film thickness (2) | Silica solubility (1, 3) | Surface area of grain–grain contact (1, 2) |
| | | Path length for diffusion (2) |

(stress, temperature, fluid and mineral compositions, grain size, and rock texture) several domains can be defined in which these variables have a different influence on the rate of deformation.

## MODEL FOR PRESSURE SOLUTION AT A GRAIN SCALE

Our model takes into account two sites (the contact area and the pore surface) on the surface of a grain and the exchange of matter between them to describe pressure solution at a grain scale (Lehner, 1995; Shimizu, 1995; Renard *et al.*, 1997).

### Geometrical model and compaction

For simplicity of the presentation, we will focus on the vertically directed grain contacts (normal to the $z$ axis) (Fig. 2). Results are identical for the horizontally directed contacts (normal to $x$ and $y$ contacts) because we consider an isotropic loading of the system, which is a reasonable assumption for a sedimentary basin setting. To complete the textural descriptions, different quantities are introduced:

$A_z$ = surface area of the contact between two grains ($m^2$)
$A_p$ = surface area of the grain facing the pore ($m^2$)
$P_z$ = lithostatic pressure (bar)
$p_p$ = pore fluid pressure (bar)
$p_a$ = normal stress across a grain contact (bar)
$\Delta_a$ = thickness of the water film inside a grain contact (m)
$D_a$ = diffusion coefficient for migration along a grain contact ($m^2 \cdot s^{-1}$)
$c_p$ = concentration of aqueous silica in the pore fluid ($mole \cdot m^{-3}$)
$c_a$ = concentration of aqueous silica in a grain contact ($mole \cdot m^{-3}$).

Also, there are two distinct rates, $G_a$ and $G_p$ ($m.s^{-1}$), for the evolution of grain texture during deformation. The rate $G_a$ describes grain indentation by dissolution at the contacts whereas $G_p$ takes account of the evolution of the grain radius, $L_f$, and represents the thickness of overgrowth on the pore surface. Given these rates, the textural evolution of the grains can be characterized by the two equations

$$\frac{dL_z}{dt} = G_a(c^a, p_a, P_z) \tag{7}$$

$$\frac{dL_f}{dt} = G_p(c^p, p_p) \tag{8}$$

The variables $G_a$ and $G_p$ give the rate of change of the length variables characterizing the two sites along the grain surface (Lehner, 1995). Rates are positive/negative for precipitation/dissolution, respectively. In these two equations, the two rates $G_a$ and $G_p$ and the two concentrations $c_a$ and $c_p$ are unknown. To solve Equations 7 and 8, four relationships are necessary. The first equation takes into account the conservation of mass in a closed system: all the matter that dissolves inside the contact precipitates in the pore.

$$6A_zG_a + A_pG_p = 0 \tag{9}$$

In this equation, the factor 6 represents the six contact areas on each grain (Fig. 2). A second relationship characterizes the diffusion of aqueous silica from the contact to the pore. In this case $c_a$ is related to $c_p$ by a steady state assumption (Shimizu, 1995); letting $l_a$ be the radius of a grain contact, Fick's Law gives:

$$\frac{2\pi l_a \Delta_a (c_a - c_p)}{l_a} = -\frac{A_z}{\overline{V}_s} G_a \tag{10}$$

where $\overline{V}_s$ is the molar volume of quartz ($22.688 \times 10^{-6}$ $m^3 \cdot mole^{-1}$) and $\Delta_a$ the thickness of the fluid film. $D_a$ is the coefficient of diffusion of aqueous silica along the grain contact. If the fluid inclusions have connections between them and to the pore, this coefficient should be higher than in a flat grain contact.

The rate $G_p$ represents the precipitation on the pore surface and can be described by a kinetic law:

$$G_p = k_p \left(1 - \frac{c_p}{K_{eq}}\right) \tag{11}$$

where $k_p$ is the rate of quartz precipitation, as measured in laboratory conditions (Brady & Walther, 1990; Dove, 1994; Rimstidt 1997) or estimated by Walderhaug (1994) from fluid inclusions microthermometry, and $K_{eq}$ is the equilibrium constant for quartz in the pore. The ratio in Equation 11 represents the degree of saturation.

The kinetic constant for quartz dissolution–precipitation can vary by several orders of magnitude depending if the rates are estimated under laboratory conditions or in nature. We have chosen to use the kinetic data measured by Brady & Walther (1990) because they give intermediate values.

The last relationship is based on an assumption about the limiting step of deformation. If diffusion inside the contact is the limiting step, the concentration in the contact $c_a$ is equal to the equilibrium constant at the contact. If the limiting step is precipitation in the pore, the concentration in the pore $c_p$ is equal to the concentration at the contact. In the numerical model, the diffusion step and the precipitation step are estimated and the slowest is chosen to be the limiting step for deformation. Depending on depth, the limiting step can vary (see Fig. 7): kinetics of precipitation at shallow depth, kinetics of diffusion under 2.5 to 3 km for geological conditions similar to the Norwegian margin.

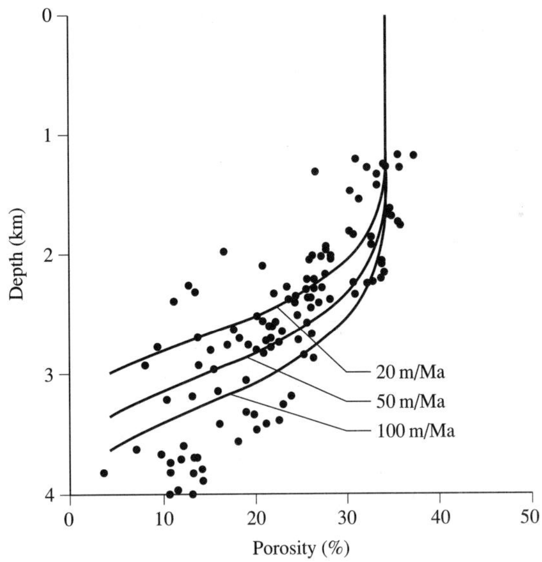


**Fig. 6.** Porosity–depth variation due to pressure solution. The porosity data are given in Ramm (1992). In the numerical model, initial grain size is 600 μm; the temperature gradient is chosen to be 40°C·km$^{-1}$, and a hydrostatic pressure gradient is assumed. The calculations were done for three sedimentation rates, 25, 50, and 100 m per million years. These conditions are similar to those for Jurassic sandstones of the Norwegian margin (Ramm, 1992). The theoretical curves fit the data at depths greater than 2 km. Between 0 and 2 km, porosity variations are essentially due to mechanical deformation. Below 2 km, pressure solution could be the dominant mechanism of porosity decrease.

## Normal stress inside the grain contacts

For the monomineralic system modelled as a periodic array of truncated spheres, the following relationship between the macroscopic vertical pressure $P_z$ (equivalent to a lithostatic pressure), the normal stress on a grain contact, $p_a$, and the pore pressure, $p_p$, allows a feedback between pressure and rock texture (Dewers & Ortoleva, 1990).

$$-L_xL_yP_z = A_zp_a + (L_xL_y - A_z)p_p \tag{12}$$

## Pressure solution and porosity variation with depth

Ramm (1992) has pointed out that mechanical compaction, due to grain rearrangements for example, is not sufficient to explain the porosity–depth trend in sandstones of the Norwegian continental shelf. We have applied our model of compaction due to pressure solution (Fig. 3; Equations 9–11) in the case of a subsiding sandstone, with similar conditions as those that have been found in the Norwegian shelf, namely, temperature gradient of 40°C·km$^{-1}$; lithostatic and hydrostatic pressure gradients and sedimentation rates varying from 20 to 100 m per million years (Fig. 6). For simplicity we consider that there are no overpressured compartments in the sediment, even if this assumption is far from reality (Ortoleva, 1994).

The following scenario seems to emerge. At shallow depth (lower than 2500 m in a sandstone), porosity variations are essentially due to grain sliding and rearrangement and brittle deformation. At greater depth, pressure solution becomes more efficient because of an increase of temperature and stress. Most of the compaction can be described by this mechanism of deformation.

## The limiting step for deformation: diffusion or precipitation

We have calculated the slowest step for deformation in conditions identical to Fig. 6. Between 0 and 2.5 to 3 km, the slowest step is quartz precipitation (Fig. 7). At these depths, the water film inside the grain contacts is thick and diffusion is fast, whereas kinetics of precipitation are

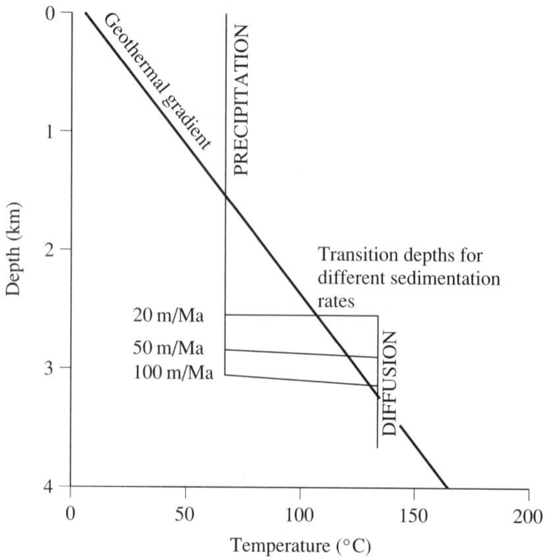


**Fig. 7.** The limiting step for pressure solution. The physical and textural conditions are the same as in Fig. 6 (the bold line represents a temperature gradient of 40°C·km$^{-1}$). Between 0 and 2.5 km the slowest step is precipitation in the pore. Below this limit, diffusion at the contact becomes the slowest step and controls the deformation.

slow because of a low temperature. Under the transition depth, kinetics of quartz precipitation are activated by an increase of temperature and the limiting step becomes silica diffusion along the grain contacts.

## DISCUSSION

### Effects of stress and temperature on deformation

An increase in effective stress on the grains has two consequences. First it decreases the water film thickness; second it increases the driving force for diffusion (Rutter, 1976; Renard *et al.*, 1997) (Table 2). The first factor decreases the pressure solution rate, whereas the second factor increases it.

Temperature plays a key role in quartz solubility and the rate of dissolution, both of which increase with temperature. Kinetics of the surface reactions are the limiting processes when there is a small effective stress because the water film is thick enough to allow the solute to diffuse easily (Fig. 5 & Table 2). In contrast, at high effective stress diffusion becomes the slowest process that limits the rate of the whole mechanism of deformation. Note that for burial depths between 0 and 3 km, the precipitation rate is so slow that regardless of the stress, the limiting factor would be the kinetics of quartz dissolution/precipitation (Oelkers *et al.*, 1996).

### Effect of clays on the rate of pressure solution

Clay minerals can promote mineral dissolution through 'water film diffusion' because a thick water film can be preserved at their surface. Clays can have structural electric charges spread on their surface which stabilize the water film (Fig. 1c), allowing the solutes to diffuse easily to the pore (Renard & Ortoleva, 1997).

In contrast, some authors have shown that clay coatings can retard pressure solution (Tada *et al.*, 1987; McBride, 1989; Ramm, 1992). In this case, quartz precipitation on the pore surface is hindered because of the loss of clean grain surfaces (step 3 in Fig. 5). The pore fluid at the site of precipitation becomes supersaturated in aqueous silica, so that the gradient of concentration between the contact and the pore should disappear thus diminishing the driving force for pressure solution.

### Case of an open system

In the model presented here, it is assumed that aqueous silica that goes outside the grain contact precipitates on the local grain surface opposite an open pore; the system being closed at a grain scale. If one considers a closed system at a decimetric to metric scale, transport phenomena should not be neglected. Silica can diffuse from a region where pressure solution is inhibited due to coating effects to a region where grains are cleaner and overgrowths can occur. Such a process could operate in sandstones where beds with abundant precipitated silica alternate with stylolite-rich regions where quartz dissolves (Oelkers *et al.*, 1996). In such systems, the driving force for pressure solution is the same in the clean layers and in the stylolites, whereas in the latter, precipitation is slower, due to clay coatings.

In the case of an open system at a grain scale where an aqueous fluid can circulate in the pores, fluid saturation is an important parameter. If the fluid is undersaturated with respect to silica, dissolution can occur directly from the grain surface adjacent to an open pore and not only at grain contacts. In contrast, if the fluid is supersaturated with respect to silica, the gradient of concentration between the contact and the pore decreases and the rate of pressure solution is diminished. And if the supersaturation is greater than a nucleation threshold, precipitation can occur directly in the pore.

## CONCLUSIONS

**1** The difference of silica solubility at a grain contact and in the pore fluid creates a concentration gradient, allowing solutes to diffuse along the grain–grain interface and precipitate in the pore. This overall process consists of several steps, the slowest of which controls the rate of deformation and quartz cementation. In laboratory pressure solution experiments, the effective stress is high and the water film trapped between the grains is very thin. Therefore, the rate limiting step is diffusion, especially at high temperature where the kinetics of dissolution and reaction are high. In contrast, geological observations suggest that when the effective stress is lower, diffusion is relatively faster than dissolution or precipitation, even if the driving force for solute transport is lower. Therefore, the limiting step can be the rate of precipitation on the pore free face. In clay-rich sandstones, especially at low temperatures, the kinetics of precipitation of quartz can limit the rate of deformation (Oelkers *et al.*, 1996), whereas at greater depths, diffusion controls the rate of deformation (Rutter, 1976).

**2** In some conditions, the pressure solution rate can be diminished. When the pore pressure reaches the lithostatic pressure, in an overpressured compartment for example (Ortoleva, 1994), the gradient of solubility between the grain contacts and the pore disappears. Fluid composition can also modify the rates of quartz dissolution, precipitation and transport. For example, an elevated concentration in $Na^+$ of the pore fluid, as well as a basic pH, activate quartz dissolution and increase quartz solubility (Dove, 1994).

**3** Finally, a crucial variable is rock fabric (grain shape and size, contact geometry, porosity and permeability). This variable has the same properties as a tensor: it can vary in the three directions of space. Therefore, a macroscopic model of pressure solution must take into account the evolution of rock texture to estimate the rate of compaction.

## ACKNOWLEDGEMENTS

This study was supported by the Institut Français du Pétrole and the CNRS (GdR Géomécanique des Roches Profondes). We would like to thank Anne-Marie Boullier and Christian France-Lanord who gave the sample of quartz from Himalayas. The reviewers, E. McBride and J. Parnell made valuable comments and insights. The last improvements of the manuscript, by R. Worden, have helped us to make the final corrections.

## REFERENCES

Anderson, G.M. & Crerar, D.A. (1993) *Thermodynamics in Geochemistry. The Equilibrium Model.* Oxford University Press, 588pp.

Applin, K.R. (1987) The diffusion of dissolved silica in dilute aqueous solution. *Geochimica and Cosmochimica Acta* **51**, 2147–2151.

Bjørkum, P.A. (1996) How important is pressure in causing dissolution of quartz in sandstones? *Journal of Sedimentary Research* **66** (1), 147–154.

Brady, P.V. & Walther, J.V. (1990) Kinetics of quartz dissolution at low temperatures. *Chemical Geology* **82**, 253–264.

Cox, S.F. & Paterson, M.S. (1991) Experimental dissolution–precipitation creep in quartz aggregates at high temperatures. *Geophysical Research Letters* **78** (8), 1401–1404.

Dewers, T. & Ortoleva, P. (1990a) A coupled reaction/transport/mechanical model for intergranular pressure solution stylolites, and differential compaction and cementation in clean sandstones. *Geochimica et Cosmochimica Acta* **54**, 1609–1625.

Dove, P.M. (1994) The dissolution kinetics of quartz in sodium chloride solutions at 25°C to 300°C. *American Journal of Science* **294**, 665–712.

Freer, R. (1981) Diffusion in silicate minerals and glasses: a data digest and guide to the literature. *Contributions to Mineralogy and Petrology* **76**, 440–454.

Gibbs, J.W. (1961) *The collected works of J. Willard Gibbs. Volume 1: Thermodynamics.* Yale University Press.

Gratier, J.P. & Jenatton, L. (1984) Deformation by solution-deposition and reequilibration of fluid inclusions in crystals depending on temperature, internal pressure, and stress. *Journal of Structural Geology* **6**, 189–200.

Gratier, J.P. & Guiguet, R. (1986) Experimental pressure solution-deposition on quartz grains: the crucial effect of the nature of the fluid. *Journal of Structural Geology* **8**, 845–856.

Gratier, J.P. (1993a) Experimental pressure solution of halite by an indenter technique. *Geophysical Research Letters* **20**, 1647–1650.

Gratier, J.P. (1993b) Le fluage des roches par dissolution-cristallisation sous contrainte dans la croûte supérieure. *Bulletin de la Société Géologique de France* **164**, 267–287.

Heald, M.T. (1955) Stylolites in sandstones. *Journal of Geology* **63**, 101–114.

Heidug, W.K. (1995) Intergranular solid–fluid phase transformations under stress: the effect of surface forces. *Journal of Geophysical Research* **100** (B4), 5931–5940.

Horn, R.G., Clarke, D.R. & Clarkson, M.T. (1988) Direct measurements of surface forces between sapphire crystals in aqueous solutions. *Journal of Materials Research* **3** (3), 413–416.

Horn, R.G., Smith, D.T. & Haller, W. (1989) Surface forces and viscosity of water measured between silica sheets. *Chemical Physics Letters* **162**, 404–408.

Lehner, F.K. (1995) A model for intergranular pressure solution in open systems. *Tectonophysics* **245**, 153–170.

Low, P.F. (1992) Interparticle forces in clay suspensions: Flocculation, viscous flow and swells. In: *Clay–Water Interface and its Rheological Implications, Vol. 4.* (eds Guven, N. & Pollastro, R.M.) pp. 158–186. Clay Minerals Society Workshop Lectures.

McBride, E.F. (1989) Quartz cement in sandstones: a review. *Earth Science Reviews* **26**, 69–112.

Mullis, A.M. (1993) Determination of the rate-limiting mechanism for quartz pressure dissolution. *Geochimica et Cosmochimica Acta* **57**, 1499–1503.

Oelkers, E.H., Bjørkum, P.A. & Murphy, W.M. (1996) A petrographic and computational investigation of quartz cementation and porosity reduction in North Sea sandstones. *American Journal of Science* **296**, 420–452.

Ortoleva, P.J. (1994) *Geochemical Self-organization*. Oxford University Press, 412pp.

Paterson, M.S. (1973) Nonhydrostatic thermodynamics and its geologic applications. *Reviews of Geophysics and Space Physics* **11**, 355–389.

Peschel, G. & Aldfinger, K.H. (1971) Thermodynamic investigations of thin liquid layers between solid surfaces. *Zeitschrift für Naturforschung* **26**, 707–715.

Poirier, J.P. (1985) *Creep of Crystals*. Cambridge University Press, 260pp.

Ramm, M. (1992) Porosity–depth trends in reservoir sandstones: theoretical models related to Jurassic sandstones offshore Norway. *Marine and Petroleum Geology* **9**, 553–567.

Renard, F. & Ortoleva, P. (1997) Water film at grain–grain contacts: Debye–Hückel, osmotic model of stress, salinity and mineralogy dependence. *Geochimica et Cosmochimica Acta* **61**, 1963–1970.

Renard, F., Ortoleva, P. & Gratier, J.P. (1997) Pressure solution in sandstones: influence of clays and dependence on temperature and stress. *Tectonophysics* **280**, 257–266.

Rimstidt, J.D. (1997) Quartz solubility at low temperature. *Geochimica et Cosmochimica Acta* **61**, 2553–2558.

Rutter, E.H. (1976) The kinetics of rock deformation by pressure solution. *Philosophical Transactions of the Royal Society of London* **283**, 203–219.

Rutter, E.H. (1983) Pressure solution in nature, theory and experiment. *Journal of the Geological Society of London* **140**, 725–740.

Schutjens, P.M.T. (1991) Experimental compaction of quartz sand at low effective stress. *Journal of the Geological Society of London* **148**, 527–539.

Shimizu, J. (1995) Kinetics of pressure solution creep in quartz: theoretical considerations. *Tectonophysics* **245**, 121–134.

Spiers, C.J., Schutjens, P.M., Brzesowsky, R.H. *et al.*, (1990) Experimental determination of constitutive parameters governing creep of rocksalt by pressure solution. In: *Deformation Mechanisms, Rheology and Tectonics* (eds Knipe, R.J. & Rutter, E.H.) pp. 215–227. Special Publication of the Geological Society of London 54.

Sposito, G. (1992) The diffuse-ion swarm near smectite particles suspended in 1:1 electrolyte solutions: Modified Gouy–Chapman theory and quasicrystal formation. In: *Clay–Water Interface and its Rheological Implications, Vol. 4.* (eds Guven, N. & Pollastro, R.M.) pp. 128–152. Clay Minerals Society Workshop Lectures.

Tada, R., Maliva, R. & Siever, R. (1987) A new mechanism of pressure solution in porous quartzose sandstone. *Geochimica et Cosmochimica Acta* **51**, 2295–2301.

Walderhaug, O. (1994) Precipitation rates for quartz cement in sandstones determined by fluid-inclusion microthermometry and temperature-history modeling. *Journal of Sedimentary Research* **A64** (2), 324–333.

Weyl, P.K. (1959) Pressure solution and the force of crystallization—a phenomenological theory. *Journal of Geophysical Research* **69**, 2001–2025.

Worden, R.H., Oxtoby, N.H. & Smalley, P.C. (1998) Can oil replacement prevent quartz cementation in sandstones? *Petroleum Geoscience* **4**, 129–138.

*Spec. Publs int. Ass. Sediment.* (2000) **29**, 79–88

# A test of hypotheses regarding quartz cementation in sandstones: a quantitative image analysis approach

C. M. PRINCE[1] *and* R. EHRLICH[2]

[1]*Petro Image, LLC, P.O. Box 12152 Columbia, SC 29211, USA; and*
[2]*Energy and Geoscience Institute, University of Utah, 423 Wakara Way, Salt Lake City, UT 84108, USA*

## ABSTRACT

Quartz cementation in sandstones is driven by the redistribution of silica. Three different processes have been proposed as having a major effect on this redistribution: (i) thermodynamic surface energy; (ii) diffusive transport; and (iii) advective transport. We have used quantitative image analysis procedures to test these hypotheses and have found all three to be lacking. Using a suite of quartz-rich sands and sandstones, as well as a suite of artificial sands made from progressively sintered glass spheres, we have found that in every case, the cementation in quartz-rich sands is a simple process that is indistinguishable from the mechanical compaction and plastic deformation found in the artificial sands.

## INTRODUCTION

Many diagenetic changes involve changes in porosity and so the nuances of porosity evolution can serve as an indicator of diagenetic state. Porosity consists of a three dimensional complex of pores connected by pore throats. The effectiveness of pigmented epoxy (a wetting phase) to invade microporosity and of mercury (a strongly non-wetting phase) to invade porosity at high injection pressures indicate that other than fluid inclusions, essentially all of the porosity is interconnected at some level. Diagenesis in quartz sands is usually accompanied by the loss of intergranular porosity. New porosity types may emerge during diagenesis (moulds, intragranular porosity, microporosity in clay-rich aggregates) but it is unlikely that this involves a net increase in porosity rather than a redistribution of existing porosity (Ehrenburg, 1990). The products of diagenesis may vary considerably over short-length scales (e.g. patchy cement), but in our observations diagenesis in many quartzose sandstones is marked by a pervasive and uniform loss of porosity coupled with quartz cementation. It is the diagenesis of the latter type of sandstone that is the subject of this paper.

Diagenesis most commonly concerns the transfer of material at a variety of length scales. Development of quartz overgrowths, for example, involves the transfer of silica on to grain surfaces accompanied by the displacement of the fluid occupying those locations. The details of this transfer are a major concern of diagenetic theory. Quartz overgrowth development can radically reduce porosity percent producing sandstones with five or less porosity (Heald & Anderegg, 1960). This must be accompanied by dewatering of a volume of water equivalent to about 30% or more of the original volume of sand as pore space is replaced by quartz. The volume of silica needed creates a difficulty, in that quartz is relatively insoluble. Meteoric water is normally supersaturated with respect to quartz (Blatt, 1979), but at depths where meteoric flushing is significant the temperature is generally too low to bring about the precipitation of quartz (Bjørlykke & Egeberg, 1993). At greater pressures and temperatures where silica is precipitated as quartz, cementation is limited by the availability of silica in solution. In order to replace 30% of the rock volume with quartz, we must infer a process in which the silica is either dissolved from within the formation, or silica is transported from an external source. An alternative assumes that the rock behaves plastically and that displacement of porosity is coupled to continuing compaction and dewatering, with the silica, supplied by pressure solution, filling adjacent pore space.

Silica migration may be due to advection or diffusion. Transport by advection is modulated by Darcy's law and can transfer fluids over very long length scales, scales large enough that the source of the silica need not be local. One problem associated with advective transport is the maintenance of a fluid that is initially oversaturated with respect to $SiO_2$ over long length scales. In a quartz-rich sand it is a question of maintaining a state of oversaturation in a rock matrix that is actively undergoing cementation and can readily precipitate silica.

Another problem with advection concerns the very low solubility of quartz over the range of pH values commonly encountered in sandstones. In conditions of neutral or acid pH, transport by advection thus requires the movement of enormous pore volumes of water. Estimates of the pore-water flux needed by the advective model are in the order of $10^6$–$10^8$ $cm^3/cm^2$, high enough to be unlikely in all but exceptional circumstances (Dutton & Diggs, 1990; Bjørlykke & Egeberg, 1993; Bjørlykke, 1997).

Diffusive transport is very slow and so must be considered to operate over very short length scales (< 1 m), scales short enough that the source of the silica must be local. The pressure solution mechanism has been invoked as a way to produce overgrowths concomitant with dissolution. Pressure solution can be considered a special case of diffusion controlled by local stresses. As overburden increases, so does the stress at the contact points between sand grains. In the case of a quartz-rich sand, this stress brings about localized dissolution at the contact points followed by precipitation in adjacent pore space. The net result is grain interpenetration and an increase in the area of contact, a process that is in many respects the same as the compaction of plastic grains (Bryant *et al.*, 1993).

Another diffusive mechanism concerns the tendency to reduce surface energy by depletion of small grains (Williams & Parks, 1985). To a first approximation, surface energy and surface area are proportional so long as the mineralogy is simple. This is a weak effect and should be important under conditions where all other thermodynamic potentials were essentially nil, the conditions one might expect for a quartz arenite. This type of reaction can bring about the complete destruction of silt and clay-size quartz fragments to serve as a source for silica. The reverse is also true in that if we start with a solution oversaturated with respect to silica, the result can bring about the selective cementation of fine-grained regions having the greater surface area per unit volume compared with nearby coarser-grained regions (e.g. Heald & Renton, 1966).

Can the various possible modes of quartz diagenesis produce diagnostic signatures in sandstone? We argue in this paper that hypotheses concerning the mode of quartz overgrowth emplacement can be determined using quantitative image analysis procedures. The key to this is the verification by Prince *et al.* (1995) of the conjecture of Graton & Fraser (1935) that all sandstones consist of juxtaposed close-packed and loose-packed domains. As shown in Fig. 1, an image of the porosity in a Cretaceous sandstone from the North Sea, most of these loose-packed domains are only a few grains wide, but they have larger pores and pore throats. They form a network of fluid flow pathways responsible for at least 80% of the fluid flow in unconsolidated sands and virtually all of the flow in lithified sandstones (Ehrlich *et al.*, 1991; McCreesh *et al.*, 1991; Riggert, 1994; Prince *et al.*, 1995; Carr, 1996). Close-packed domains consist of clusters of several grains that are in a close-packed configuration. These grain clusters have smaller pores and pore throats than the associated loose-packed domains. They account for virtually none of the advective flux in lithified sandstones and are accessible primarily by diffusion.

Previous efforts have shown that pore types can be derived objectively and precisely using image analysis procedures on binary images of porosity versus the solid matrix (Ehrlich *et al.*, 1991). In addition, the relationship between pore type and throat size has been demonstrated by relating pore type proportions to saturation values from mercury porosimetry (McCreesh *et al.*, 1991). Using 2-D Fourier analysis to isolate loose-packed porosity (Prince & Ehrlich, 1990; Prince *et al.*, 1995), further analysis of the same sandstones has shown that in every case there is a specific pore type associated with loose packed circuits and that the variations in permeability are associated with the abundance of this pore type. Such analysis indicates that by the time porosity has been reduced to approximately 25%, which is a modest degree of overgrowth development, the pore throats in the close packed domains are too small to support significant flow, and that most of the advected flow is constrained to the loose packed circuits (Carr, 1996; Murray *et al.*, 1994; Murray, 1996).

The objective of this study is to examine the systematic decline in porosity associated with the precipitation of quartz overgrowths and to test a number of hypotheses concerning overgrowth development. Does surface energy exert a strong influence upon quartz cementation? Is advective transport important for the introduction of silica or is this accomplished by diffusion? The first goal is to assess the surface area of porosity and examine the effect of surface energy with progressive diagenesis. The second goal is to examine progressive cementation within close-packed and loose-packed domains as a means to assess the effect of each mode of transport.

## SAMPLING

We have assembled two suites of samples: a suite of artificial sands composed of well-sorted soda-lime glass

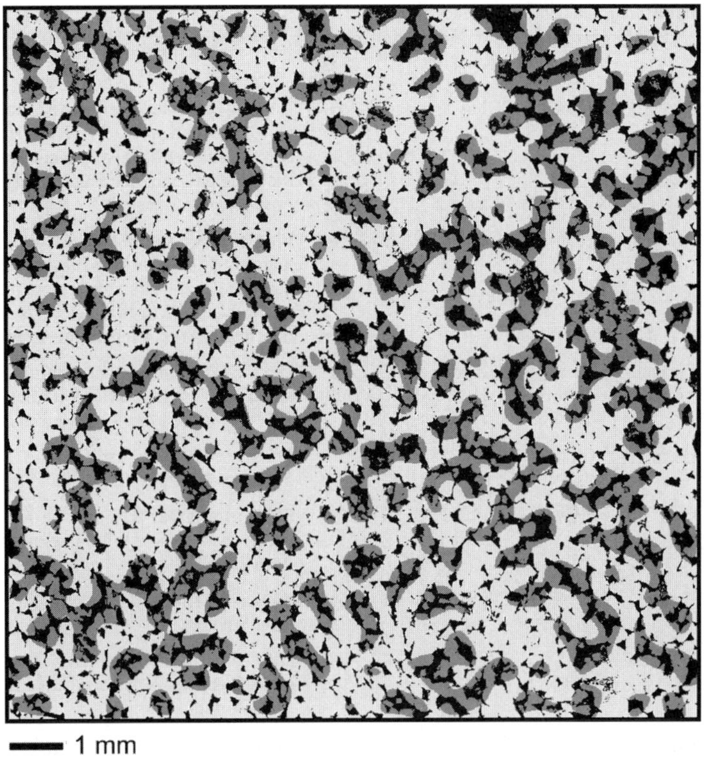


**Fig. 1.** A filtered binary image of a Cretaceous sandstone from the North Sea. Each pixel of the image is either 'pore' (black = porosity) or 'not pore' (white = sand grains and cement). The grey overlay is a filtered version of the binary image highlighting loose-packed ('expanded') porosity.

beads, and a suite of quartz-rich samples of sand and sandstone from a variety of basins and ages (Table 1). The beadpacks were progressively sintered at 600°C at ambient pressure for periods ranging from a few minutes to several hours, producing a set of artificial sands with porosity ranging from 35% down to 5%. The natural sandstones are taken from a suite of samples that range in age from Mississippian to Recent (Table 1). While there is a significant variation in both composition and textural parameters (size, sorting) within each sample set, the selected samples are all well sorted and in each case essentially all of the cementation is associated with the precipitation of quartz. Other than mechanical diagenesis (e.g. compaction), the progressive decline in porosity (43% down to 5%) is primarily a result of quartz cementation. At the outset, it was believed that such a heterogeneous assemblage of sands could produce scatter due to the variation in age, basin depths and temperatures, and geochemical environments of the samples. However, as we shall demonstrate below, the resulting data are surprisingly well-behaved.

## IMAGE ANALYSIS

Thin sections of both suites of samples were digitized, creating binary images of porosity. Using 2-D Fourier analysis, the images were examined both to determine grain size and to define the scale at which loose-packed porosity exists (Prince *et al.*, 1995). The images were filtered using the inverse Fourier transform, splitting porosity into two classes: 'expanded' porosity associated with loose-packed domains, and 'well packed' porosity associated with close-packed domains. The two phases were then examined separately, using image analysis to measure the size, perimeter, and area of each 'porel' (POROsity ELement, an individual pore as expressed in thin section).

### Image acquisition

The images presented in this paper were digitized using a Symbiotic Concepts MIS-386 video digitizing system attached to a petrographic microscope. The system uses

**Table 6.1.** Sand and sandstone samples

| Name | Age | Location | Porosity | Composition |
|---|---|---|---|---|
| Berea | Miss. | Ohio | 15–19% | Loosely cemented, well-sorted sublitharenite containing lithic fragments, feldspars (< 2%), and micas (< 1%). Rock fragments, feldspars, and micas all exhibit dissolution. While there is a minor amount of patchy carbonate cement, silica is the principal cementing agent and quartz overgrowths can be seen rimming original quartz clasts. |
| Perry | Penn. | Oklahoma | 5–23% | Fine- to medium-grained quartz arenite, ranging in porosity from 23% down to less than 5%. Cementation is primarily quartz overgrowth with variable amounts of patchy intergranular calcite. The diagenetic fabric varies from homogeneous quartz overgrowths to a heterogeneous fabric where quartz overgrowth interfingers with remnant intergranular calcite. |
| Cardium | Cret. | Alberta | 5–12% | Deltaic marine sandstone, sublitharenite, moderately to well-sorted, with minor amounts of chert, argillaceous metamorphic rock fragments, micas and clays. Feldspars (< 1%) are in advanced stages of dissolution. |
| Helm | Cret. | North Sea | 19–25% | Loosely cemented quartz arenite from the Netherlands. Primary cement is quartz overgrowth with trace amounts of pre-existing intergranular calcite. |
| Gulf | Pleist. | Gulf of Mexico | 21–26% | Pleistocene grain-flow sands from the Gulf of Mexico. Sublitharenite containing varying amounts of lithic fragments and minor amounts of feldspar. Porosities range from 21 to 26%. They have undergone compaction, but are relatively uncemented. |
| Beach | Modern | Virginia | 33–36% | Modern quartz sands from the Virginia coast. Box cores of unconsolidated beach samples dried and impregnaed with blue-dyed epoxy. |
| Artificial | — | — | 42–43% | Artificially sedimented using samples of quartz sand from the South Carolina coastal plain. Samples were sieved and the 250–300 μm size range was used to create a 'well-sorted' sample. |
| Beads | — | — | 5–33% | Sintered glass spheres ranging from 180 to 210 μm in diameter. Samples were sintered at 600°C under ambient pressure for periods of a few minutes to several hours. |

a digital filter to precisely define porosity (Crabtree *et al.*, 1984), producing 'binary' images. Each digital image consists of a 640 × 480 array of pixels, but seamless mosaics of up to 120 images can be constructed producing an image that combines relatively high resolution (small pixel size) with large coverage area (several square centimetres). The resulting images measure up to 7000 × 7000 pixels and often represent an entire thin section.

## 2-D Fourier analysis

A 2-D Fourier transform is a well-known signal analysis tool that can be used to decompose an image into a series of discrete spatial frequencies (harmonics). The output of the Fourier transform is conventionally reported in terms of power spectra, representing the squares of the harmonic amplitudes (Prince & Ehrlich, 1990). The radial power spectra presented in this paper are standardized to represent power as a relative contribution to the aggregate image (Prince, 1991). Using a radial power spectrum it is possible to determine the median grain size of the sand and define the scale at which expanded porosity exists (Prince *et al.*, 1995).

Grain size defines the minimum centre-to-centre distance at which grains can pack together and thus defines the fundamental spatial density of porosity (pore-to-pore separation). Pore features that exist at scales less than grain size are associated with pore/grain shape, whereas those that exist at scales greater than grain size are associated with the spatial interrelationship of grains–grain packing and sedimentary fabric. Figure 2 shows two power spectra, one typical of an artificial beadpack (Fig. 2a) and another typical of a natural sandstone (Fig. 2b). A comparison of the two illustrates the difference between the two suites of sands. The uniform shape of the glass beads creates a specific 'shape signature'—arcuate and cuspate porels—that is manifest as an increase in power at scales equal to two-thirds and one-third that of grain size. This is in marked contrast to natural sands where pore shapes are essentially random and the power declines at wavelengths less than grain size without any evidence of a shape signature.

The grain-size peak actually represents the contribution of grains (or associated intergranular porosity) arranged in a close-packed configuration. Loose-packed zones enhance the power at spatial wavelengths greater than the

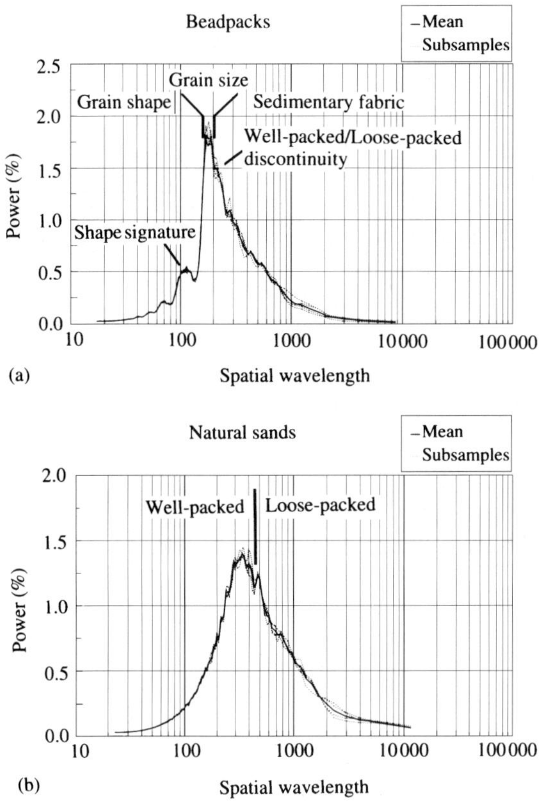


**Fig. 2.** Radial power spectra of (a) a typical beadpack and (b) a typical sand. Both exhibit a well-defined grain-size spike and a well-defined loose-packed discontinuity at spatial frequencies 20% greater than the modal grain size. The primary difference between the two is in the shape signature of the beadpack resulting from the uniform shape of the glass spheres.

modal grain size, creating a discontinuity at wavelengths approximately 20% greater than the modal grain size. This discontinuity is used as a cut-off for imaging expanded porosity.

One of the most desirable features of a 2-D Fourier transform is its reversibility; an image can be rebuilt from its Fourier components. By removing all spatial frequencies equal to or less than a specified length scale and inverting the remainder to generate a filtered image, it is possible to visualize any structure indicated in the spectrum. The grey overlay in Fig. 1 is a filtered version of the underlying image generated by removing all Fourier components with wavelengths less than the loose-packed threshold while retaining the components associated with the size, shape, and spatial arrangement of loose- and well-packed domains. Using this type of filter it is possible to segregate the image into 'expanded' and 'well-packed' porosity and examine each class of porosity separately (Fig. 3).

### Petrographic image analysis

The expanded and well-packed images were processed using the petrographic image analysis (PIA) system to extract information relating to the size and shape of individual porels. This information was then transformed into a series of variables detailing the pore size and perimeter.

One of the simplest variables to define is the amount of porosity in the image. The ratio of porel area to image area is defined as total optical porosity (TOP). Due to the resolution limit of the optics, TOP is usually less than the porosity measured through other means (i.e. helium porosity), but TOP can be considered to be an accurate estimate of macroporosity.

Total porel perimeter divided by total image area defines 'specific surface' which is proportional to the surface area of porosity per unit volume of rock (Kendall & Moran, 1963). A related variable, herein 'specific perimeter' is defined as the ratio of porosity perimeter to porosity area. Specific surface, as an estimator of surface area per unit volume is sensitive to thermodynamic effects that tend to conserve or decrease surface area. Specific perimeter is related to the size/shape complexity of the porosity, independent of the total rock volume in which it is found. That is, in general, a rock with low porosity may possess relatively low values of specific surface but high values of specific perimeter. Thus specific perimeter is sensitive to the pattern of spatial organization of the porosity independent of the amount of porosity.

## RESULTS

### Domainal porosity

The first and most straightforward analysis was in the examination of the proportions of each class of porosity. The amount of porosity associated with each type of domain provides insight into the progression of mechanical compaction and cementation. As total porosity declines from 43% down to approximately 30%, both expanded and well-packed porosity decline, but the amount of well-packed porosity declines at a faster rate (Fig. 4). In the range of 25% to 5% TOP, the amount of expanded porosity declines at a rate of 0.7% per unit decrease in total porosity. This is approximately twice the rate for well-packed porosity. The question of which class of porosity survives at low porosity values is a function of the initial contrast in pore sizes between loose-packed

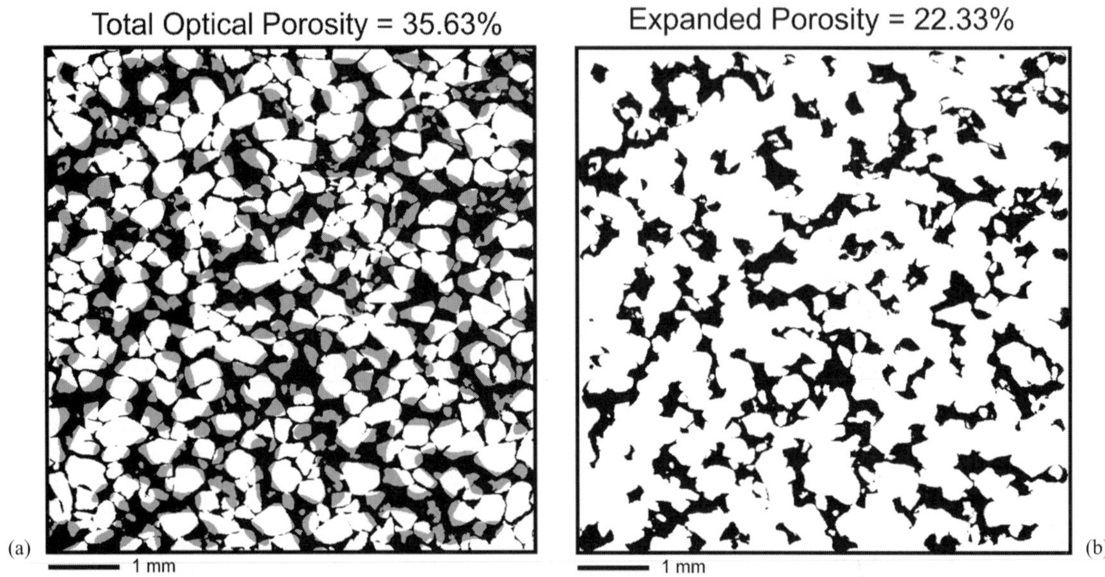


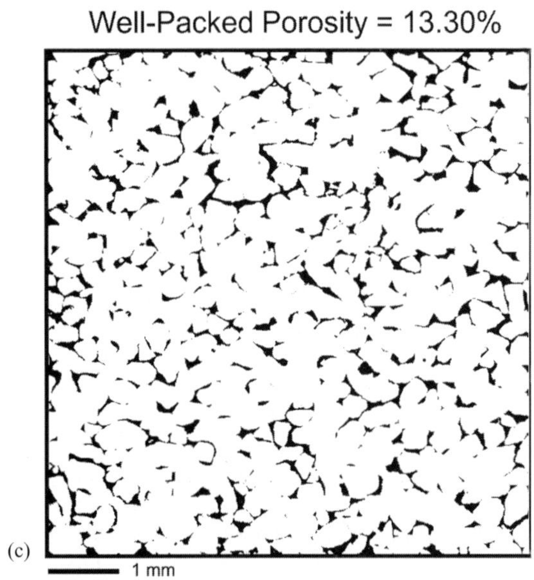


**Fig. 3.** (a) Modern beach. Binary images are filtered using the inverse transform to remove all image components at less than the well-packed/loose-packed discontinuity. The filtered image is used to subdivide porosity into (b) 'expanded' porosity associated with loose-packed domains, and (c) 'well-packed' porosity associated with close-packed domains.

and close-packed domains. Figure 4 shows that well-packed porosity 'loses the race' and accounts for almost none of the porosity at low porosity values.

At high porosity values (> 30%) in unconsolidated, uncompacted sands, well-packed porosity declines at a faster rate while expanded porosity remains relatively stationary. This suggests that there is an upper limit to expanded porosity that is governed by mechanical instability. Above that limit the grain matrix becomes unstable and compacts to fill the void. As porosity decreases from 43% down to 36% TOP, the major loss in porosity is within the well-packed domains. This loss may be a manifestation of the difference between artificially packed sands and the natural sedimentation process. However, the close agreement with the beadpack samples suggests that it may represent a mechanical realignment and dewatering within close-packed domains. The change in domainal porosity as a function of TOP represents a

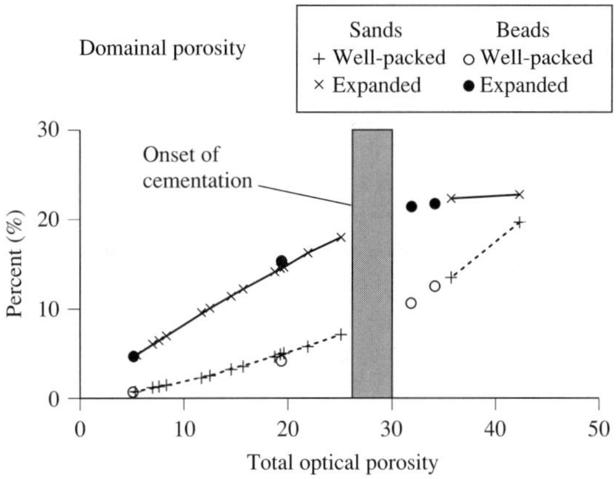


**Fig. 4.** The amount of each class of porosity as a function of total optical porosity (TOP). The graph is separated into two zones: a high porosity zone (30–43%) where mechanical compaction is the dominant lithification process, and lower porosity zone where cementation predominates.

simple trend, which suggests a simple underlying mechanism for porosity reduction.

### Specific surface

One of the hypotheses we set out to test was the assertion that surface energy exerts a strong influence upon quartz cementation. If it is true that surface energy exerts a strong influence on quartz cementation, its effect should be evident in the behaviour of specific surface during porosity decline. Close-packed domains have a higher surface area per unit volume than loose-packed domains. All other factors being equal, there should be a preference for cementation within close-packed domains until the surface area per unit volume approximates that of loose-packed domains. At that point the two should equilibrate with further cementation (Fig. 5a).

The results presented in Fig. 5(b) do not follow this theoretical trend. The trend in specific surface is strongly linear throughout the measured porosity range (43–5%). Not only does it not equilibrate, it remains at approximately the same rate throughout the range of TOP. The inverse correlation between well-packed and expanded porosity was to be expected in a binary system, but the linearity of the trend was not. This suggests that the compactional processes modifying specific perimeter at 40% TOP are the same as the processes modifying specific perimeter at 5% TOP. The idea that compactional processes are associated with this trend in specific perimeter is further supported by the fact that, at high porosity values (> 30%) we do not see much difference in the amount of expanded

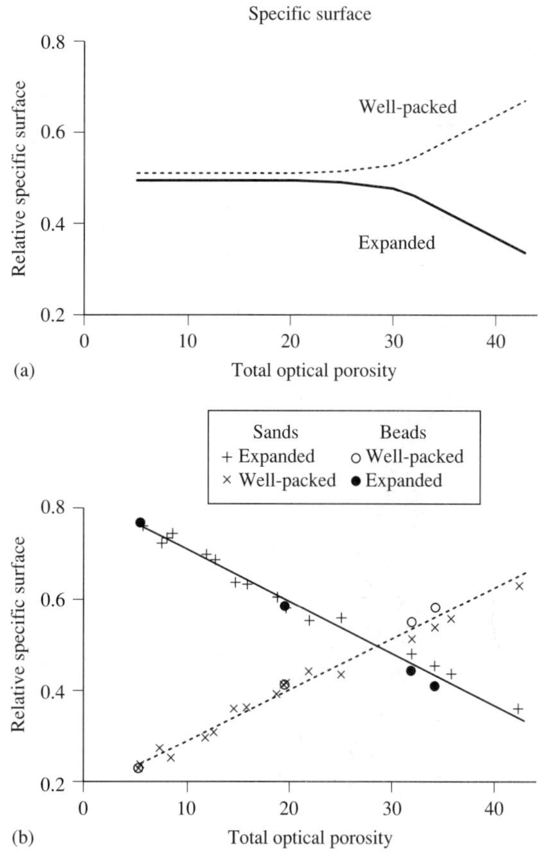


**Fig. 5.** The relative proportion of the total specific surface associated with each class of porosity. At the onset of diagenesis the porosity in close-packed domains has a smaller pore size and a greater surface area. (a) If surface energy exerts a strong influence upon cementation, we would expect to see preferential cementation within loose-packed domains until the surface area per unit volume approximates that of well-packed domains. At that point the two should remain in an equilibrium state. (b) The results of this investigation show a strong linear relationship. The linearity of the relationship argues against surface area as the controlling influence upon quartz cementation.

porosity (Fig. 4), yet we see a linear response in specific perimeter. This suggests that the compactional processes modifying the pore network in well-packed domains are part of the trend in specific perimeter.

Secondly, Fig. 5 does not show any indication of levelling to a constant surface. If surface energy were exerting a thermodynamic control on cementation, we would expect to see specific perimeter approach an equilibrium state, but there is no indication of any such trend. In fact, the reverse is true. The linearity of the trend suggests that if surface area were exerting a control, then the rate of cementation within expanded porosity should increase as

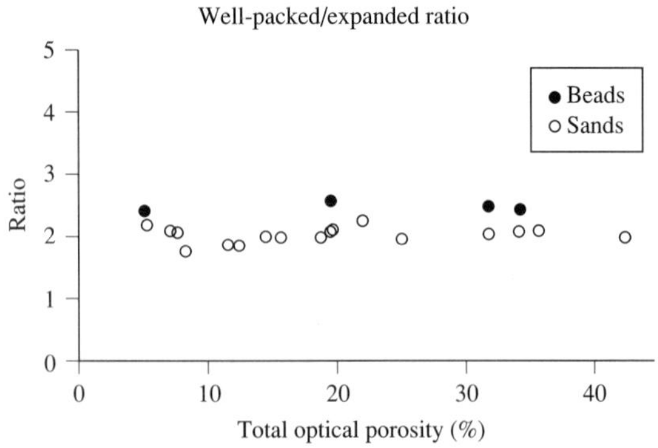


**Fig. 6.** Specific perimeter is a measure of porel shape. If we want to examine size/shape relationships between the two classes of porosity, we can examine the ratio of their specific perimeters or the 'well-packed/expanded ratio'.

specific surface increases and TOP falls from 25% down to 5%. The manner of the increase (linear, log-linear, etc.) is immaterial, but we should see an increase. As is shown in Fig. 4, the rate of cementation within expanded porosity remains relatively invariant in the 5–25% porosity range.

### Well-packed/expanded ratio

Specific perimeter is sensitive to the shape complexity of porosity, providing a measure of the magnitude of surface area per unit porosity. By extracting the specific perimeter from both classes of porosity we can examine the relationship between each class of porosity by calculating the ratio of their specific perimeters—herein the 'well-packed/expanded ratio'. The well-packed/expanded ratio is strongly modulated by the mode of quartz precipitation. In the case of advection, flow is concentrated in loose-packed circuits and so transported silica should be preferentially taken up on quartz grains either within or immediately adjacent to loose-packed domains. In contrast, if cementation is modulated by diffusion, then all quartz grains regardless of domain should acquire equal amounts of cement. Any cementation within either domain will modify the size and shape of porels, altering the specific perimeter within that domain, which would, in turn, alter the well-packed/expanded ratio.

As shown in Fig. 6, the well-packed/expanded ratio for both beadpacks and natural sands remains relatively invariant throughout the range of observed TOP. The ratio is slightly greater for the beadpacks (~2.5) than the sands (~2). Given the previous discussion of the shape differences between the beadpacks and natural sands, this slight difference is not surprising. What is surprising is the invariance of the ratio for both the beadpacks and the sands. As the following discussion will show, the stability of this ratio argues against both a diffusive model and an advective model.

If we assume that diffusion is the controlling process, then virtually all parts of the pore network are accessible. An argument can be made that accessibility is modulated by factors such as porosity and tortuosity within close-packed domains. However, the fluid-flow pathways are at most only a few grain diameters in length, which means that these effects should be minor. With virtually all parts of the pore network accessible, we would expect to see equal cementation in both classes of porosity.

We tested this assumption using a series of simple shapes; a square, circle, and triangle to represent porosity. We created two classes of 'porosity'; large (loose packed) pores and smaller (well-packed) pores, and added equal layers of cement to both classes of porosity, calculating the well-packed/expanded ratio with each iteration. As is shown in Fig. 7(a), the addition of equal layers, the diffusive model, brings about an exponential increase in the well-packed/expanded ratio as cementation increases. Equal cementation produces a result that is not present in either natural sands or beadpacks.

If we assume that advection is the controlling process, then we would have to assume that silica is preferentially available to expanded porosity. In other words, for every layer of cement added to well-packed porosity, a thicker layer should be added to expanded porosity. We repeated the simple shapes experiment using a ratio of 1/2, 1/3, and 1/4 (for every layer in well-packed porosity we added 2, 3, and 4 times the thickness to expanded porosity). The result, shown in Fig. 7(b), shows that the only way to keep the ratio constant is to add twice the thickness of cement to expanded porosity.

The results presented in Fig. 7(b) suggest that the advective model is valid and that the flow of mineralizing fluids tends to be concentrated within expanded porosity. However, if advection exerts a significant control on cementation, then cementation should be the integral of transport (fluid-flow and silica concentration) over time. The only way that the well-packed/expanded ratio could remain invariant throughout the sample set would be if all of the included formations had identical transport histories. The chance that all the units represented in the sample set had similar histories is essentially nil. More importantly, the well-packed expanded ratio remains at approximately 2 even in those samples with porosities above 30%, samples where we know that cementation is not occurring. Again, this implies that the cementation process occurring in sands with less than 25% TOP is indistinguishable from the compaction process occurring in high porosity sands.

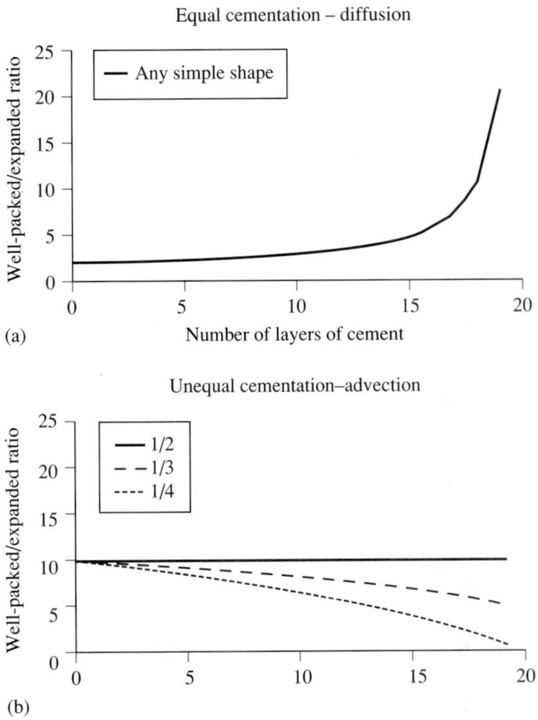


**Fig. 7.** Results of modelling using simple geometric shapes. Cementation was modelled by using two classes of pores: large (expanded) pores and smaller (well-packed) pores. Several experiments were run assuming equal cementation (a) within both classes of porosity (the diffusive model) and unequal cementation (b) (the advective model). The well-packed/expanded ratio was calculated for each iteration. The advective experiment was carried out three times using a ratio of 1/2, 1/3, and 1/4 (for every layer of cement within well-packed porosity, a layer 2, 3, and 4 times thicker was added to loose-packed porosity.

## CONCLUSIONS

In this investigation we have used quantitative image analysis techniques to test a series of hypotheses concerning the progression of quartz cementation in sandstones. We have examined a suite of samples with porosity ranging from 5 to 43%. The highest porosity values are represented by unconsolidated sands that have not undergone compaction. The sandstones in the range of 5–25% TOP are lithified and virtually all of the cementation is in the form of quartz overgrowths. These sandstones come from a wide range of basins and ages. Presumably they have been subjected to a variety of temperature, pressure, and pore-water chemistry regimes. In spite of this range in environmental factors, plots of each variable described in this paper exhibit smooth trends over the range of porosity, implying that these environmental factors might not play as significant a role in quartz cementation as previously theorized.

Throughout this investigation it has been noted that, in every case, the observed trends lie along the same trends as those derived from progressively sintered beadpacks. The trends in domainal porosity during cementation, the trend in specific surface, and the trend in specific perimeter do not suggest any significant differences between natural sandstones and sintered beadpack samples that are not subject to any geochemical interaction.

The only model that can accommodate both suites of samples is that of pressure solution. Geometrically, progressive sintering is a close model for the pressure/solution process in sandstones. Sintering brings about a plastic deformation at the contact points of the beads in the beadpack. As sintering progresses grains begin to interpenetrate, the contacts between grains become larger, more mechanically stable, and occlude adjacent pore space. A similar process occurs at the contact points of adjacent grains during pressure solution. As cementation proceeds, grains begin to interpenetrate, the contacts between grains become larger, more mechanically stable, and occlude adjacent pore space. The behaviour of our image variables as a function of porosity strongly suggests that the other hypothetical models for progressive quartz cementation are, in general, unlikely.

With further research along these lines, we expect to find exceptions to the rule, where one or the other cementation mechanisms may prove to be important. However, on the basis of the information available, we have to conclude that compaction involving the pressure solution mechanism is the best candidate for explaining the progressive loss of porosity by quartz cement in quartz-rich sandstones.

## ACKNOWLEDGEMENTS

The authors wish to thank the following people and corporations for their contributions to this research:

Tom Plona of Schlumberger-Doll Research provided the beadpack samples. Ross Clark and UNOCAL, provided samples of the Helm. Thomas Fate and British Petroleum provided thin sections from the Gulf of Mexico. Donald J.P. Swift provided the samples of unconsolidated beach sand. Conoco Inc., and the participants in the Conoco Borehole Test Project provided samples of the Perry Formation. Canadian Hunter Ltd. provided samples of the Cardium Formation and The Society of Core Analysts provided thin sections of the Berea

Sandstone. Additional thanks go to Richard Worden. His comments have significantly improved this manuscript.

## REFERENCES

BJØRLYKKE, K.O. (1989) *Sedimentology and Petroleum Geology*. Springer Verlag, New York, 363pp.

BJØRLYKKE, K.O. (1997) Mineral–water interaction, fluid flow, and Frio Sandstone diagenesis: evidence from the rocks: discussion. *Association of American Petroleum Geologists Bulletin* **81** (9), 1534–1535.

BJØRLYKKE, K.O. & EGEBERG, P.K. (1993) Quartz cementation in sedimentary basins. *Association of American Petroleum Geologists Bulletin* **77** (9), 1538–1548.

BLATT, H. (1979) Diagenetic processes in sandstones. In: *Aspects of Diagenesis*. (eds SCHOLLE, P.A. & SCHLUGER, T.H.) pp. 141–158. Society of Economic Paleontologists and Mineralogists Special Publication 26.

BRYANT, S., CADE, C. & MELLOR, D. (1993) Permeability prediction from geologic models. *Association of American Petroleum Geologists Bulletin* **77** (8), 1338–1350.

CARR, M.B. (1996) *Quantification of microstructural elements in natural porous media with correlation to single phase and multiphase flow parameters*. PhD Dissertation, University of South Carolina, 349pp.

CRABTREE, S.J., EHRLICH, R. & PRINCE, C.M. (1984) Evaluation of strategies for segmentation of blue-dyed pores in thin sections of reservoir rocks. *Comparative Vision Graphics and Image Processing* **28**, 1–18.

DUTTON, S.P. & DIGGS, N.T. (1990) History of quartz cementation in the Lower Cretaceous Travis Peak Formation, east Texas. *Journal of Sedimentary Petrology* **60**, 191–202.

EHRENBERG, S.N. (1990) Relationship between diagenesis and reservoir quality in sandstones of the Gam Formation, Haltenbanken, mid-Norwegians continental shelf. *American Association of Petroleum Geologists* **74**, 1538–1558.

EHRLICH, R., CRABTREE, S.J., HORKOWITZ, K.O. & HORKOWITZ, J.P. (1991) Petrography and reservoir physics I: Objective classification of reservoir porosity. *Association of American Petroleum Geologists Bulletin* **75**, 1547–1562.

GRATON, L.C. & FRASER, H.J. (1935) Systematic packing of spheres with particular relation to porosity and permeability. *Journal of Geology* **43** (8), 785–909.

HEALD, M.T. & ANDEREGG, R.C. (1960) Differential cementation in the Tuscarora Sandstone. *Journal of Sedimentary Petrology* **30** (4), 568–577.

HEALD, M.T. & RENTON, J.J. (1966) Experimental study of sandstone cementation. *Journal of Sedimentary Petrology* **36** (4), 977–991.

KENDALL, M.G. & MORAN, P.A.P. (1963) *Geometrical Probability*. Hafner, New York, 121pp.

MCCREESH, C.A., EHRLICH, R. & CRABTREE, S.J. (1991) Petrography and reservoir physics II: relating thin section porosity to capillary pressure, the association between pore types and throat size. *Association of American Petroleum Geologists Bulletin* **75** (10), 1563–1578.

MURRAY, C.J. (1996) The use of *nuclear magnetic resonance $T_2$ relaxation measurements in the characterization of the porous microstructure and reservoir properties of sandstone*. PhD Dissertation, University of South Carolina, 171pp.

MURRAY, C.J., EHRLICH, R., MASON, E. & CLARK, R. (1994) Evaluation of the diagenetic and structural influences on hydrocarbon entrapment in the Cardium Formation, Deep Basin, western Alberta. *Bulletin of Canadian Petroleum Geology* **42** (4), 529–544.

PRINCE, C.M. (1991) DECODE and DFOUR: 2-D Fourier processing of petrographic images. *Computers and Geosciences* **17** (4), 505–525.

PRINCE, C.M. & EHRLICH, R. (1990) Analysis of spatial order in sandstones I. Basic principals. *Mathematical Geology* **22** (3), 333–359.

PRINCE, C.M., EHRLICH, R. & ANGUY, Y. (1995) Analysis of spatial order in sandstones II. Grain clusters, packing flaws, and the small-scale structure of sandstones. *Journal of Sedimentary Research A* **65** (1), 13–28.

RIGGERT, V.L. (1994) *Petrophysical relationships of pores and pore throats to spatial fabric elements in sandstones and their implications for fluid and electrical flow*. PhD Dissertation, University of South Carolina, 192pp.

WILLIAMS, L.A. & PARKS, G. (1985) Silica diagenesis. I: solubility controls. *Journal of Sedimentary Petrology* **55** (3), 301–311.

*Spec. Publs int. Ass. Sediment.* (2000) **29**, 89–101

# Quantification of detrital, authigenic and porosity components of the Fontainebleau Sandstone: a comparison of conventional optical and combined scanning electron microscope-based methods of modal analyses

M. R. COOPER[1], J. EVANS[2], S. S. FLINT[3], A. J. C. HOGG[4] *and* R. H. HUNTER[3]

[1]*Geological Survey of Northern Ireland, 20 College Gardens, Belfast, Northern Ireland, BT9 6BS, UK;*
[2]*BP Exploration Operating Co. Ltd, Sherwood House, Holton Heath Trading Park, Poole, Dorset, BH16 6LS, UK;*
[3]**Department of Earth Sciences, Jane Herdman Laboratories, University of Liverpool, L69 3BX, UK; and*
[4]*BP Kuwait Ltd, PO Box 29335, Safat 13039, Kuwait*

## ABSTRACT

Textural and mineralogical maturity make the Fontainebleau Sandstone an ideal quartz arenite in which to investigate porosity reduction which has occurred by compaction and quartz cementation. However, measurement of quartz overgrowth volumes, using optical point counting methods, is problematic due to difficulties in distinguishing between detrital grains and overgrowths.

Optical point counting and scanning electron microscope (SEM)-based modal analysis methods have been undertaken on samples to independently measure the percentages of detrital quartz, quartz overgrowths, porosity and minus-cement porosity. Total core-plug porosity was measured using helium porosimetry. The results of each technique are compared in order to highlight the advantages, possible errors and/or limitations imposed by each method.

Comparison of total porosity measured using point counting with helium porosity shows that point counting results are at fault. Total porosities measured using the SEM and helium porosimetry methods are almost identical. It is proposed that the SEM-based method could be used to correct optical data sets, producing quantitative data at relatively low cost.

## INTRODUCTION

Quartz arenites are sandstones that contain > 95% detrital quartz (terminology of Williams *et al.*, 1954 and McBride, 1963). They are generally well-sorted, and individual grains typically show high sphericity and are well rounded (Pettijohn *et al.*, 1987). These compositional and textural characteristics make quartz arenites the most mature of the sandstones; they are therefore the simplest natural systems in which to investigate the effects of compaction and cementation.

Together, compaction and cementation by quartz account for most porosity reduction in quartz arenites (e.g. Rittenhouse, 1971a, 1971b, 1973; Girard & Deynoux, 1991; Dutton & Diggs, 1992). Porosity and pore-system interconnectivity are two very important controls on sandstone reservoir quality as they determine the rock's ability to store and to transmit fluids. It is therefore important to quantify how and to what degree compaction and quartz cementation affect porosity. Mechanical compaction and pressure solution are the dominant processes which result in porosity reduction and an overall bulk rock volume reduction in quartz arenites. Quartz cement, precipitated as syntaxial overgrowths (i.e. in crystallographic and optical continuity), is one of the most common cements found within quartz arenites (McBride, 1989). However, using conventional optical microscopy, the quantification of overgrowth volumes is commonly difficult, due to the absence of identifiable boundaries between the detrital and authigenic quartz (e.g. Thiry *et al.*,

*Please see the Acknowledgements, p. 100.

1988; Hogg *et al.*, 1992; Evans *et al.*, 1994). Textural information is also masked by the cement; collection of data regarding detrital grain size, sorting, grain-to-grain contacts and minus-cement porosity is therefore problematic.

The Fontainebleau Sandstone (early Oligocene, Paris Basin, France) was noted by previous workers (Bourbie & Zinszner, 1985; Thiry *et al.*, 1988) for its almost constant composition (pure quartz), grain size (fine sand) and sorting (well-sorted). It has also been noted for its large variations of porosity and permeability (Jacquin, 1964; Bourbie & Zinszner, 1985). The high degree of textural and mineralogical maturity makes it an ideal quartz arenite in which to investigate the occlusion of porosity by compaction and quartz cementation. Quartz cement has been recognized as the main cause of porosity occlusion within the sandstone (Thiry *et al.*, 1988). However, no study previous to this has been able to quantify how much quartz cement is present for a range of residual intergranular porosities. The recognition of quartz overgrowths within the Fontainebleau Sandstone, using standard optical point counting methods, is problematic due to difficulties in distinguishing between detrital grains and their overgrowths. The calculation of minus-cement porosity is useful for determining the degree of compaction and depth of burial. Minus-cement porosity is equivalent to intergranular volume as described by Houseknecht (1987). Its measurement is dependent upon the quantification of quartz cement and has therefore not previously been quantified either.

Optical modal analysis has advantages including wide availability of equipment (a reflection of low purchase and running costs) and rapid sample analysis. At the same time it has disadvantages in that the results are only semi-quantitative and, to a degree, operator-dependent. The above highlights the need for a technique other than the optical method in order to gain quantitative data. The SEM-based method of Evans *et al.* (1994) was chosen and will be described in detail.

The main aims of this work were to quantify the effects of compaction and quartz cementation on porosity within the Fontainebleau Sandstone. In order to achieve these aims, a range of analytical techniques were applied:

**1** measurement of core-plug helium porosity;

**2** measurement of grain size and sorting, determination of grain shape and sphericity;

**3** measurement of detrital, authigenic and porosity components using conventional optical point counting and SEM-based, combined backscattered scanning electron microscopy (BSEM), cathodoluminescence (CL) and image analysis;

**4** statistical and cross-plot analysis of data sets;

**5** comparing and contrasting conventional optical microscopy with SEM-based combined BSEM, CL and image analysis.

## GEOLOGICAL SETTING, STRATIGRAPHY AND SEDIMENTOLOGY

The Fontainebleau Sandstone forms part of the Paris Basin, which has a relatively simple 'layer-cake' stratigraphy. The basin has been a site of almost continuous subsidence and sedimentation from the Permian to the Palaeogene (Tertiary) (Mégnien, 1980a, 1980b; Cavelier & Lorenz, 1987). However, as a consequence of Palaeogene compression and associated uplift of the basin margins, the sea gradually retreated and sedimentation ceased. No substantial deposition has occurred within the basin since the Miocene. The maximum thickness of Palaeogene sediments deposited within the basin is ~150 m; the burial depth of the Fontainebleau Sandstone is therefore no greater than 150 m.

The Fontainebleau Sandstone was deposited as a 50–60 m thick laterally extensive sheet during the early Oligocene (Stampian, Thiry *et al.*, 1988). It rests upon Eocene–Oligocene marls and is overlain by Oligocene–Miocene limestones (Fig. 1). The sandstone is interpreted as a marine shoreface deposit, capped by subaerial aeolian dune-bedded sands (Alimen, 1936), suggesting progradation of a coastline.

Present day exposure of the Fontainebleau Sandstone covers the central part of the Paris Basin in the Ile de France region around Paris. The sandstone forms the scarps of the dissected Beauce Plateau to the west and south of Paris; north and east of Paris the sandstone forms residual buttes.

### Distribution of quartz cement

Quartz cement occurs within flat-lying lenses of variable thickness, but typically of the order of 2–8 m. A regional study of drill-hole data (Thiry *et al.*, 1988) shows that the lenses pinch out rapidly under Oligocene–Miocene limestone cover, extending generally less than 500 m. Cement volumes within individual lenses were seen to be variable, and appeared to increase from the outer margins to the centres.

### Sample collection and preparation

Comparative sample sets were collected from two working quarries approximately 1 km apart and located either side of Bonnevault, approximately 5 km west of Nemours

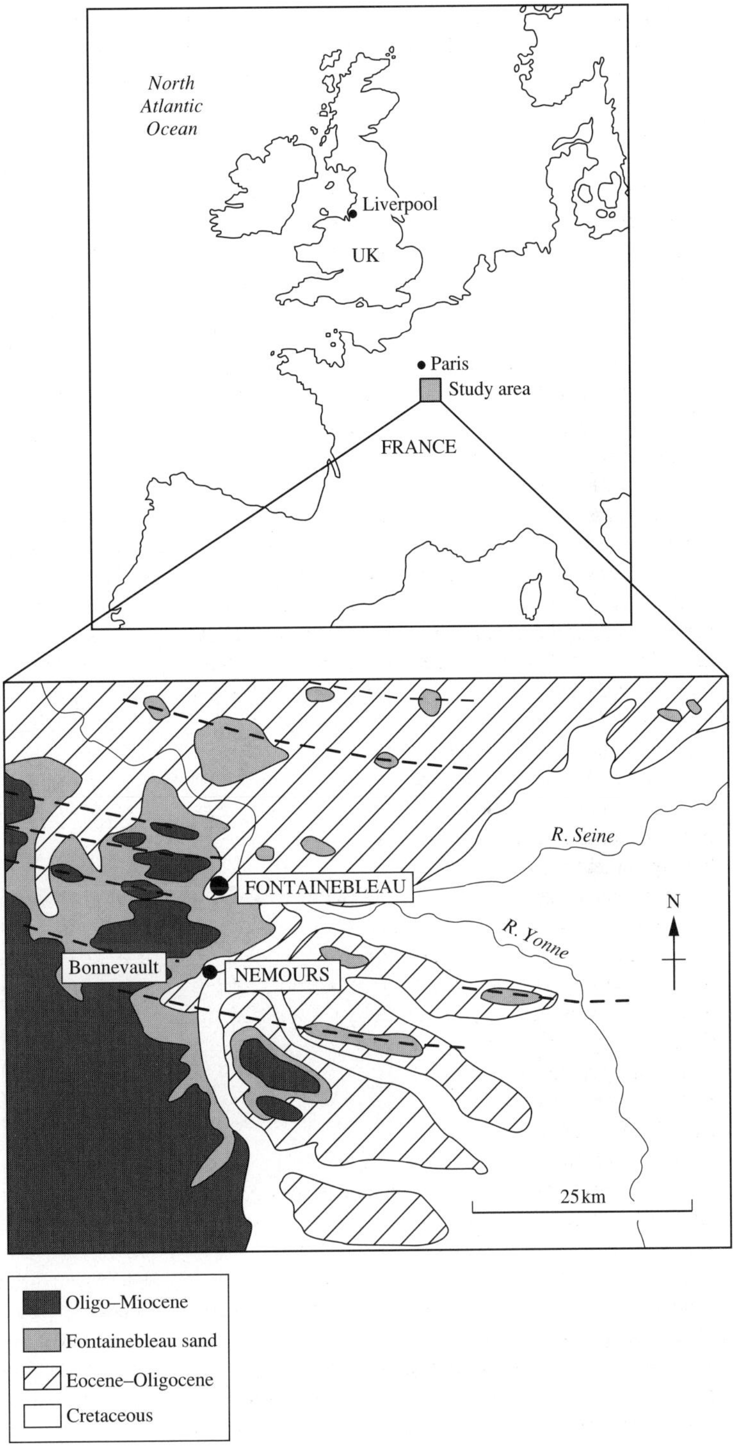


**Fig. 1.** Geological location map of the area around Fontainebleau and Nemours within the Paris Basin, France. (Modified after Thiry *et al.*, 1988.)

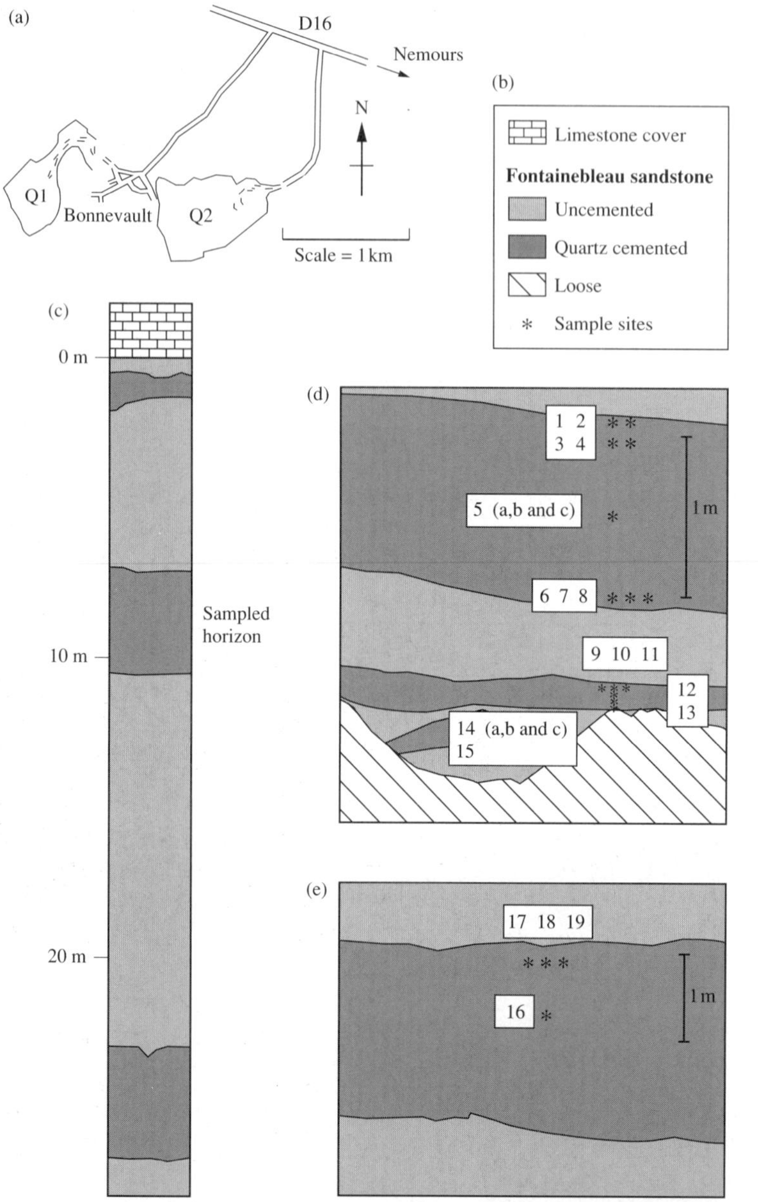


**Fig. 2.** (a) Map showing the location of the sampled quarries (Q1 and Q2) in relation to Bonnevault. (b) Key. (c) Schematic vertical section showing the approximate positions at which cemented horizons occur (including sampled horizon) in relation to the top of the sandstone. (d) Sample sites and numbers within Q1. (e) Sample sites and numbers within Q2.

(shown as Q1 and Q2 in Fig. 2a). Samples of suitable size for laboratory core-plugging were collected from the top, middle and base of two lenses in Q1, and from a lens occurring at a similar stratigraphic level in Q2 (Fig. 2c–e).

Core-plugs were cut parallel to lens boundaries or parallel to lamination if visible. Vertical orientation was marked on the cores for thin sectioning purposes. Core-plugging proved extremely difficult, friable samples tended to crumble, whereas extremely consolidated samples proved difficult to cut. However, 19 core-plugs were successfully cut. Core-plugs 1–15 are from Q1 (the Q1 sample set) and core-plugs 16–19 are from Q2 (the Q2 sample set). Figures 2(d) and 2(e) show the position of sample sites within cemented lenses. The samples when grouped are referred to as the total sample set. Trace fossil and petrographic palynological evidence (see sponge spicules shown Fig. 3c) confirm that all of the samples collected were of shallow marine origin.

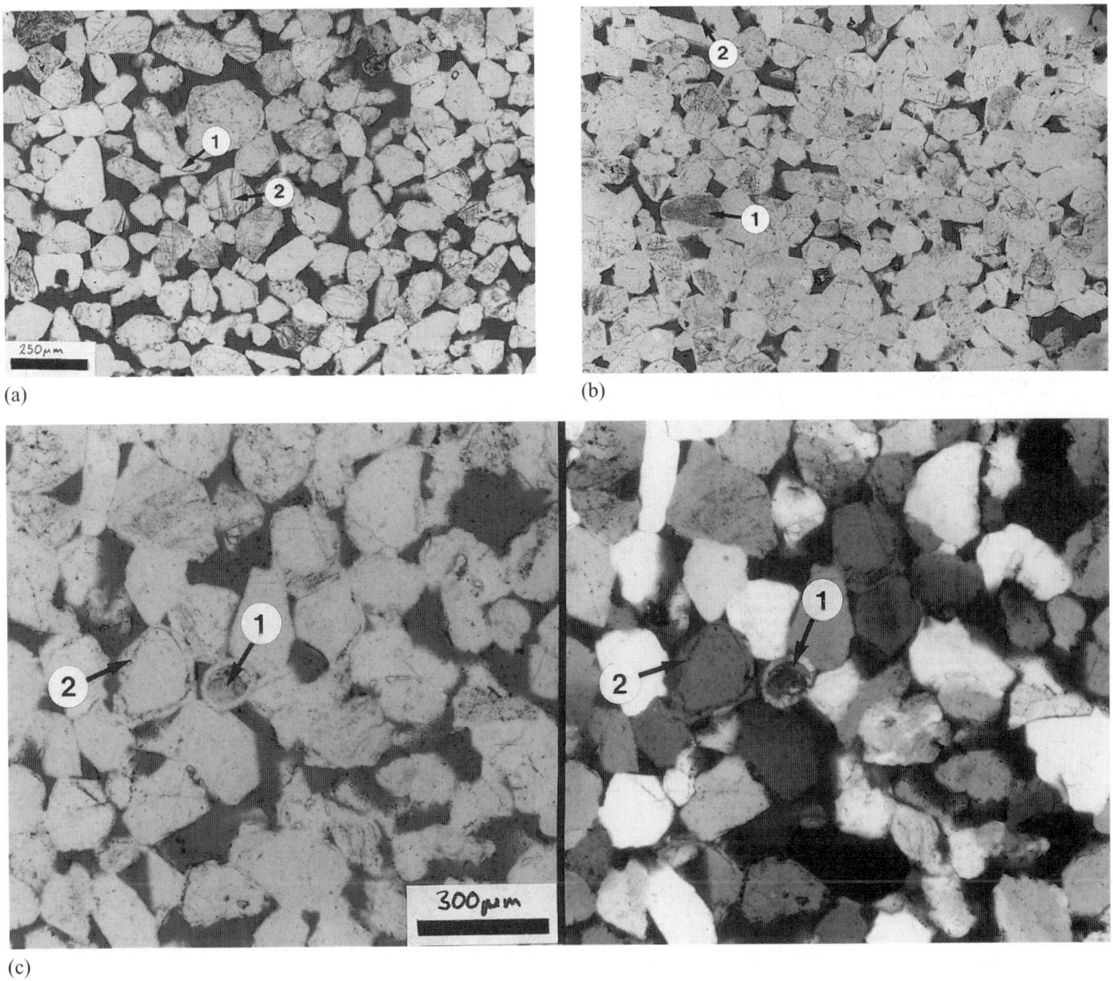


**Fig. 3.** (*Also reproduced in colour, see Plate 1, facing p. 98.*) Thin section photomicrographs of Fontainebleau Sandstone. (a) Plane polarized light thin section photomicrograph showing a high porosity sample (impregnated with blue epoxy resin) containing ~12% quartz as syntaxial overgrowths. Some of the criteria used to distinguish the detrital grains from the overgrowths include well-defined 'dust lines' (1) and cracks with inclusions (2) which terminate against clear overgrowths (Sample 4). (b) Plane polarized light thin section photomicrograph showing a low porosity sample containing ~27% authigenic quartz; note the reduction in pore interconnectivity. Scattered inclusions (1) within detrital grains giving a cloudy appearance and overgrowths projecting into open pore space (2) were also used as criteria for the recognition of overgrowths during point counting (Sample 5). (a) and (b) scale bar = 250 μm. (c) Plane and crossed polarized light thin section photomicrographs (left and right, respectively) showing sponge spicules (1), indicative of a marine depositional environment. Also indicated is a quartz grain showing an obvious dust line (2). (Sample 15: scale bar = 300 μm.)

## DETRITAL TEXTURE

### Grain size and sorting

Twenty-five measurements of detrital grain size and sorting were made from polished thin sections using conventional optical petrography. For each sample, the long axes of 100 grains were measured using a mechanical stepping stage; a stage interval of 1 mm and track spacing of 2 mm were used to ensure that a minimum of four complete traverses were made for each section. Where possible quartz overgrowths were excluded from the measurements using the criteria described for recognition of overgrowths (see Fig. 3a–c; Plate 1, facing p. 98).

Results indicate only small variations in grain size and sorting of samples used for this investigation. Figure 4 shows two bar charts of sample versus grain size and sample versus sorting. The Q1, Q2 and total sample set

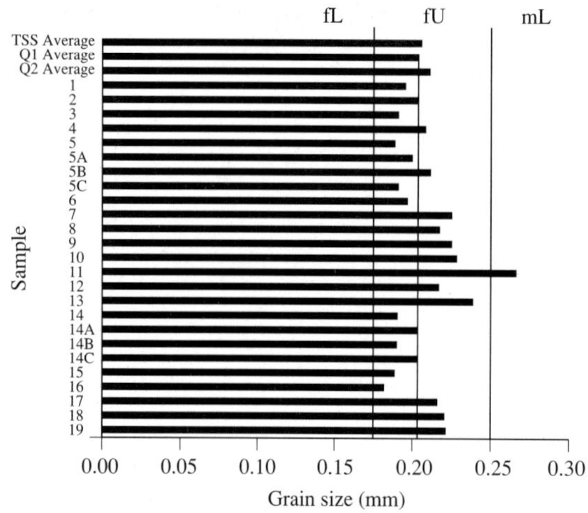


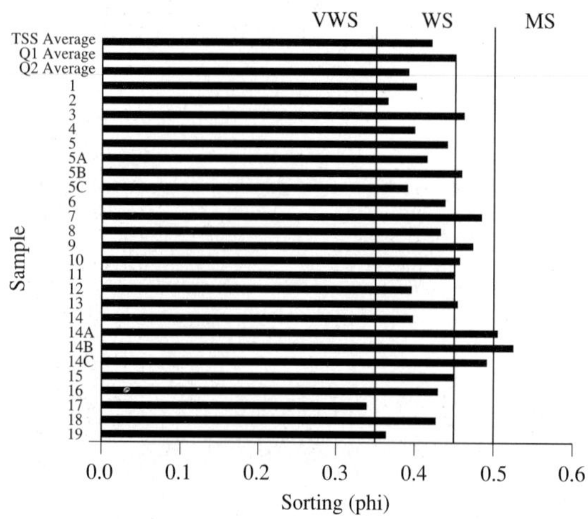


**Fig. 4.** Grain size and sorting bar graphs for the total sample set (TSS) and for the comparative sample sets (Q1 and Q2). Individual samples (1–19) are also shown. Grain size divisions: fL = fine lower sand; fU = fine upper sand; mL = medium lower sand. Sorting divisions: VWS = very well-sorted; WS = well-sorted; MS = medium sorting.

average values are displayed at the top of each chart. Most samples have a grain size within fine upper sand (fU: 0.177–0.25 mm) and are well sorted (WS: 0.35–0.5 phi). The mean grain size for the total data set is 0.208 mm, this compares favourably with 0.2 mm reported by Bryant *et al.* (1993) who used CL to distinguish quartz overgrowths from detrital grains. Other values include 0.2 mm by Cayeux (1929) and 0.25 mm reported by Bourbie & Zinszner (1985). The ranges and standard deviations for the total sample set are very small relative to the average values. Results indicate that the samples used during this investigation are consistently texturally mature.

### Sphericity and roundness

Categories of sphericity and roundness were visually assessed using the comparative chart of Powers (1953). When viewed in thin-section, quartz overgrowths essentially mask the shape and roundness of the detrital grains (Fig. 3; Plate 1, facing p. 98). The CL images shown in Fig. 5aii, bii and cii show best these characteristics unobscured by cement. Grains appear rounded to subrounded and have low to high sphericity. True grain sphericity is difficult to determine from two-dimensional sections and can vary if grains in the plane of the section lie parallel or perpendicular to a fabric. Observation of uncemented sand using a binocular microscope confirms that there is a mixture of low and high sphericity grains within the Fontainebleau Sandstone.

## MODAL ANALYSIS—METHODS

Conventional optical point counting and SEM-based, combined BSEM, CL and image analysis were undertaken on samples of Fontainebleau Sandstone in order to independently measure the percentages and percentage bulk volumes (%bv) of detrital quartz, quartz overgrowths,

**Fig. 5.** (*Opposite.*) Backscattered scanning electron microscopy and cathodoluminescence images. (ai) BSEM image, quartz appears light grey, detrital and authigenic quartz cannot be separated from this type of image. The grains have numerous euhedral boundaries and projections (1), indicating the presence of quartz cement occurring as syntaxial overgrowths. Porosity that has been impregnated with epoxy resin appears black, the dominant porosity type is intergranular (2), although some oversized intergranular porosity (3) was identified. (aii) CL image, detrital quartz (1) shows a range of grey-level both within and between individual grains. Quartz cement (2) is dominantly black and porosity (intergranular porosity 3, oversized intergranular porosity 4) is white. Detrital grain boundaries (b) are readily identifiable using this type of image. (Sample 4: total porosity = 20.5%bv; authigenic quartz = 12.2%bv; minus-cement porosity = 32.7%bv). (bi) Intergranular porosity within this sample shows relatively good interconnectivity (1), although pore throats (2) are much smaller than those of the previous example (a). (bii) The CL image reveals the true 2-D packing of detrital grains and the nature of detrital grain-to-grain contacts; floating grains are observed (1), touching grains show dominantly point (tangential) contacts (2). (Sample 13: total porosity = 12.2%bv; authigenic quartz = 19.3%bv; minus-cement porosity = 31.5%bv). (ci) Intergranular porosity within this sample shows relatively poor interconnectivity (1), pore throats (2) are much smaller than those of the previous example. (cii) Again floating grains are observed (1), touching grains show dominantly point (tangential) contacts (2). (Sample 5: total porosity = 8.4%bv; authigenic quartz = 27.2%bv; minus-cement porosity = 35.6%bv). Accelerating voltage 15 kV; scale = 200 μm.

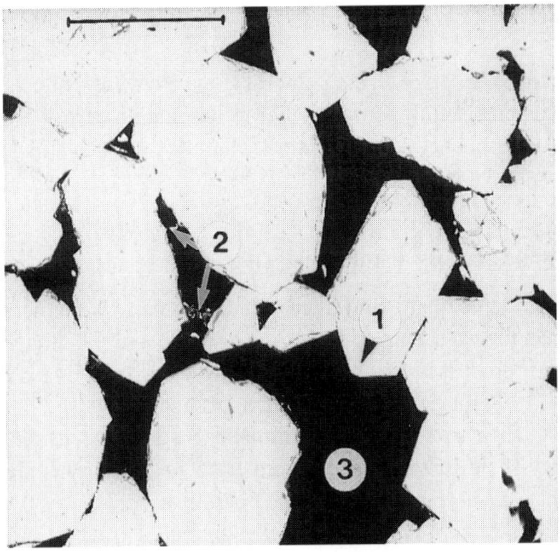

(ai)

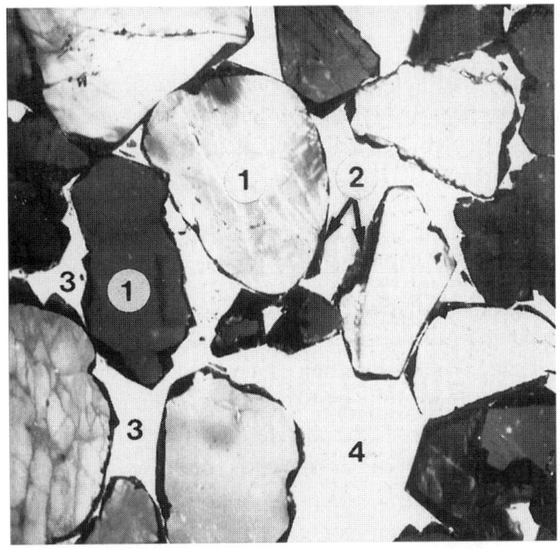

(aii)

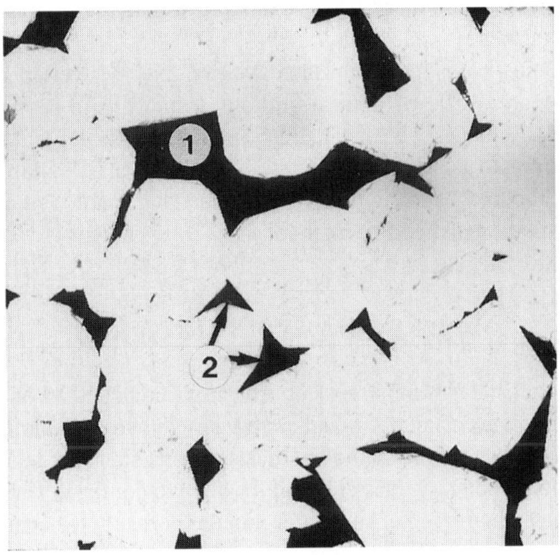

(bi)

(bii)

(ci)

(cii)

porosity and minus-cement porosity. High quality porosimetry was undertaken on core-plugs using a Boyle's Law shared volume helium porosimeter prior to thin section preparation.

### Optical point counting

Samples were impregnated with blue epoxy resin prior to preparation of polished thin sections. Data points were collected at a 0.66-mm stage interval along tracks spaced at 2 mm giving at least six traverses of the polished thin sections. Analyses were made consistently at a magnification of ×100 using plane and crossed polarized light. A test involving a 500 point count of a representative section showed that for greater than 300 points the errors are small relative to the modes measured. Therefore 300 points were considered to be sufficient to achieve consistent results during optical point counting.

### SEM-based combined BSEM, CL and image analysis

This method, described in detail by Evans *et al.* (1994), involved the collection and analysis of pairs of BSE and CL images from polished thin sections. The number of fields of view required to give accurate results depends upon magnification, grain size and inhomogeneity of the sample. A strategy for determining the required number of fields of view is given by Evans *et al.* (1994). A subset of seven sections was selected for analysis. The high degree of textural homogeneity meant that porosity range alone determined which samples were chosen. The analytical equipment used comprised a JEOL JSM-840 A SEM with a LINK BSE detector and an Oxford Instruments CL detector. The SEM was connected to a KONTRON IBAS-20 image analysis system. Thin sections were carbon coated prior to analysis using a Polaron (BIO-RAD) E6100 carbon coater.

The CL detector had a retractable parabolic mirror through which BSE and CL images were collected using a $LaB_6$ electron gun with an accelerating voltage of 15 kV. A probe current of 1.9 nA was used at a working distance of 39 mm. The beam diameter under these conditions was approximately 25 nm. The geometry of the CL collection system meant that a minimum magnification of ×100 was available. This magnification was used for all the samples during this investigation, allowing approximately 10 grains to be imaged for each field of view.

Analogue output from the CL and BSE detectors was converted to digital format by the image analysis system. Both CL and BSE images were acquired at 512 × 512 pixel (picture elements) resolution with 256 grey levels per pixel. Ten pairs of images were collected at equal spacing perpendicular to lamination across each of the polished thin sections. Examples of such images are shown in Fig. 5.

### Image processing and analysis

Authigenic quartz and intergranular porosity were measured for each pair of images. The BSE and CL images were processed using the following steps (see Evans *et al.*, 1994 for details of a similar approach):

**1** Quartz was separated from the BSE image using a grey-scale histogram and converted to a binary image (this was used to measure total porosity).

**2** The binary image was then subtracted from the CL image to produce a CL image containing only quartz.

**3** Authigenic quartz was separated from the CL image using a grey-scale histogram.

In most cases, however, there was a degree of overlap in the grey-scale of detrital and authigenic quartz shown by CL images. This meant that separation was rarely fully complete using stage 3, and was aided by manual editing of the images using a digitizing pad. This is an obvious limitation to the SEM method.

## MODAL ANALYSIS—RESULTS

A statistical summary of optical point count, SEM and helium porosimetry modal analyses is shown in Table 1. Regardless of the method of measurement, standard deviation (st.dev.) values are high for the two dominant variables (quartz cement and intergranular porosity), whereas the st.dev. are low for the constants (detrital grains volume and minus-cement porosity). All values in subsequent sections are expressed as a percentage of bulk volume (%bv).

### Detrital mineralogy

Both the optical and SEM methods show that quartz constitutes > 99.5% of the detrital composition for all samples used during this study. The optical method allowed division of detrital (and authigenic) quartz into three groups on the basis of grain structure. This differentiation could not be made using the SEM method. Grains composed of a single crystal, which showed uniform extinction under crossed polarized light, are termed 'unstrained monocrystalline'. Single crystal grains, showing sweeping or undulose extinction, are termed 'strained monocrystalline'. Grains composed of two or more crystals are termed polycrystalline. The average values of

**Table 1.** Statistical summary of point count (A); SEM-based BSEM, CL and image analysis (B); and helium porosimetry (C) results

**A**

| Detrital mineralogy | Average bulk volume | Standard deviation | Minimum | Maximum |
|---|---|---|---|---|
| Unstrained monocrystalline quartz | 35.75 | 2.48 | 29.98 | 41.9 |
| Strained monocrystalline quartz | 18.85 | 1.73 | 15.45 | 22.14 |
| Polycrystalline quartz | 3.52 | 0.77 | 2.16 | 4.52 |
| Heavy minerals | 0.25 | 0.66 | 0 | 2.35 |
| **Authigenic mineralogy** | **Average** | **Standard deviation** | **Minimum** | **Maximum** |
| Unstrained monocrystalline quartz | 15.8 | 5.45 | 5.99 | 27.7 |
| Strained monocrystalline quartz | 4.87 | 1.4 | 2 | 7.8 |
| Polycrystalline quartz | 0.38 | 0.36 | 0 | 1.2 |
| **Porosity** | **Average** | **Standard deviation** | **Minimum** | **Maximum** |
| Intergranular | 19.75 | 4.86 | 9.64 | 27.1 |
| Oversized intergranular | 0.82 | 0.67 | 0 | 2.53 |
| Total porosity | 20.57 | 5.19 | 9.64 | 28.06 |
| Minus-cement porosity | 41.62 | 3.41 | 33.17 | 48.92 |

**B**

| | Average | Standard deviation | Minimum | Maximum |
|---|---|---|---|---|
| Total detrital | 66.21 | 1.76 | 63.9 | 68.5 |
| Total authigenic quartz | 18.2 | 5.79 | 9.9 | 27.24 |
| Total porosity | 15.59 | 5.39 | 8.42 | 23.3 |
| Minus-cement porosity | 33.79 | 1.76 | 31.5 | 36.1 |

**C**

| | Average | Standard deviation | Minimum | Maximum |
|---|---|---|---|---|
| Helium porosity | 18.35 | 5.69 | 8.6 | 30.56 |

detrital quartz of unstrained and strained monocrystalline and polycrystalline quartz grains from point counting are 35.75%bv, 18.85%bv and 3.52%bv, respectively. Detrital grains other than quartz (rutile, tourmaline and zircon) accounted for 0.25%bv. The average total detrital grain volume from optical point counting was calculated as 58.37%bv. This compares with the higher average value of 66.21%bv determined from SEM-based analyses.

### Authigenic mineralogy

When viewed using plane polarized light, both detrital and authigenic quartz appear typically colourless and structureless (Fig. 3). Overgrowths are seen to be in optical and structural continuity with detrital grains when viewed with crossed polarized light; both therefore share the same interference colours and extinction characteristics. The boundaries between detrital grains and overgrowths were not consistent within samples of Fontainebleau Sandstone; differentiation of detrital and authigenic quartz was therefore difficult using standard optical petrography.

The following criteria were used for the recognition of quartz overgrowths during optical point counting (see Fig. 3 for examples):

**1** Euhedral grain boundaries and projections are observed where overgrowths have grown into open pore spaces.

**2** Inclusion clouds or trails (cracks) within detrital grains which terminate against clear overgrowths.

**3** Lines of impurities or dust lines on the detrital grain surface.

Criteria 2 and 3 define actual detrital grain/overgrowth boundaries and are of most use during optical point counting. Criterion 1 indicates only the presence of overgrowths and is of limited value during point counting without the presence of criteria 2 and/or 3. Examples of these criteria are indicated in Fig. 3. Note, however, that many of the grains in these photomicrographs lack the criteria outlined above.

Three main forms of authigenic quartz cement were identified within the samples investigated. These match the detrital groupings, the overgrowths having grown in structural and optical continuity with their detrital hosts. The average values of authigenic quartz of unstrained and strained monocrystalline and polycrystalline quartz, from point counting, are 15.8%bv, 4.87%bv and 0.38%bv, respectively (Cooper, 1995). The average total authigenic quartz, from optical point counting, was calculated as 21.05%bv. This compares to the lower average value of 18.2%bv determined from SEM-based analyses.

## POROSITY TYPES AND EVOLUTION

Total porosity was measured from core-plugs (helium porosity) and from thin sections using conventional optical point counting and SEM-based BSEM, CL and image analysis. The helium values of total porosity are accurate to ~0.1%bv. Table 1 includes a statistical summary of porosity values obtained using the three methods. The average total porosity as determined by optical point counting was 20.57%bv, which is significantly higher than that of the SEM-based method, which gave a value of 15.59%bv.

Numerous studies have obtained average values for the depositional intergranular porosity of beach sands, e.g. Pryor (1973) (49%bv); Beard & Weyl (1973) (43%bv); Palmer & Barton 1986 (47%bv); Atkins & McBride, 1992 (47%bv). The spread in values probably relates to true depositional variations, e.g. variable sorting, bioturbation. Problems associated with sample collection, handling and measurement must also be considered. Atkins & McBride (1992) by conducting control experiments have overcome sampling and handling errors, and have avoided natural variations, where possible, by sampling undisturbed recent deposits. Assuming that the depositional porosity of the Fontainebleau Sandstone was close to 47%bv. The average minus-cement porosity of ~34%bv, as determined via the SEM method, equates to a ~13%bv reduction of intergranular porosity as a result of mechanical compaction. The small standard deviation of minus-cement porosity values indicates that compaction occurred homogeneously throughout the samples studied.

Measurement of post-cementation intergranular porosity was achieved using optical point counting; this allowed the differentiation of porosity types (not possible using SEM or helium porosimetry methods). Porosity was visually divided into intergranular porosity and oversized intergranular porosity (i.e. pores of a size close to that of detrital grains), examples of which are shown in Fig. 5. Intergranular porosity was found to be dominant (average 19.75%bv) and determines the pore network interconnectivity and therefore permeability. The degree of interconnectivity is related to the amount of quartz cement and is therefore extremely variable between the samples. The dominant cement type (unstrained monocrystalline) has the greatest effect on porosity, followed by strained monocrystalline and then polycrystalline quartz. Oversized intergranular porosities were low (average 0.82%bv) and individual pores are isolated and unlikely to enhance permeability, as the connections between pores are controlled by the geometry of the intergranular porosity.

## COMPARISON OF CONVENTIONAL OPTICAL POINT-COUNTING, BSEM, CL AND IMAGE ANALYSIS AND HELIUM POROSIMETRY METHODS

Efficiency (in terms of speed and cost) and accuracy are the main objectives of modal analysis. During this investigation, modes have been obtained using conventional optical point counting, SEM-based BSEM, CL and image analysis and helium porosimetry methods. Comparison of total porosity measured using point counting, with that of helium porosimetry shows that point counting results are significantly different (Fig. 6a). The known accuracy of the helium porosity values indicates that point counting is at fault. However, total porosities measured using the SEM and helium porosimetry methods are almost identical (Fig. 6b). The agreement of values, shown by Fig. 6(b), indicates that all modes calculated using the SEM method, in the range ~8–24%bv (and probably outside this range), will be accurate.

Comparison of total porosity measured using the SEM with those measured optically (Fig. 7a) shows the difference between the results of the two methods, i.e. the errors associated with optical point counting. A similar comparison for quartz cement is shown graphically in Fig. 7(b). These cross-plots show that both total porosity and quartz cement have been overestimated during optical point counting. They also show that point count errors increase with increasing total porosity and with quartz cement, and that errors are systematic and cumulative.

The errors associated with optical point counting of total porosity and quartz cement, when combined during the calculation of minus-cement porosity, are large (25% greater than the real values obtained using SEM-based BSEM, CL and image analysis). Errors originate during

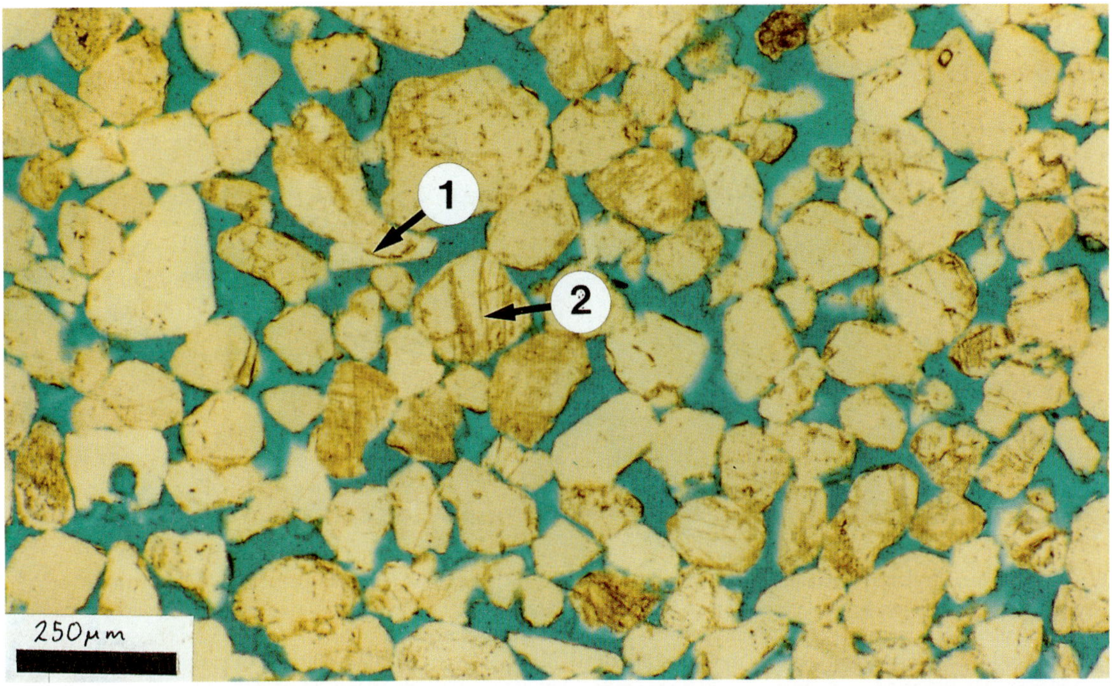


(a)

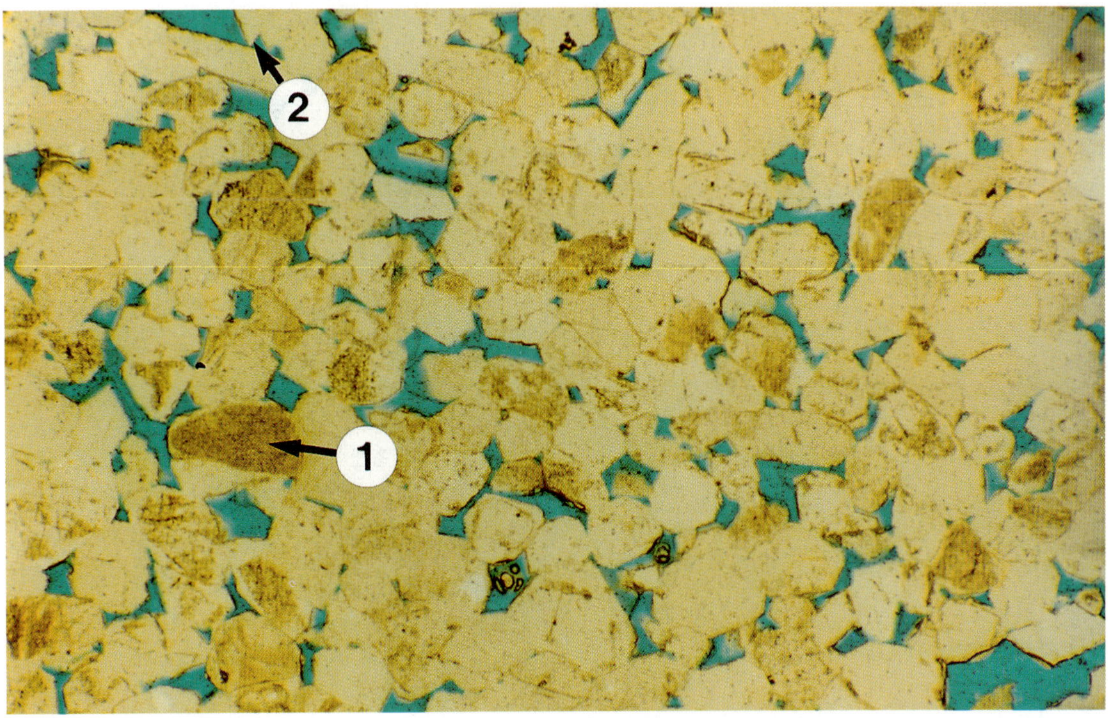


(b)

**Plate 1.** Thin section photomicrographs of Fontainebleau Sandstone. (a) Plane polarized light thin section photomicrograph showing a high porosity sample (impregnated with blue epoxy resin) sample containing ~12% quartz as syntaxial overgrowths. Some of the criteria used to distinguish the detrital grains from the overgrowths include well-defined 'dust lines' (1) and cracks with inclusions (2) which terminate against clear overgrowths (Sample 4). (b) Plane polarized light thin section photomicrograph showing a low porosity

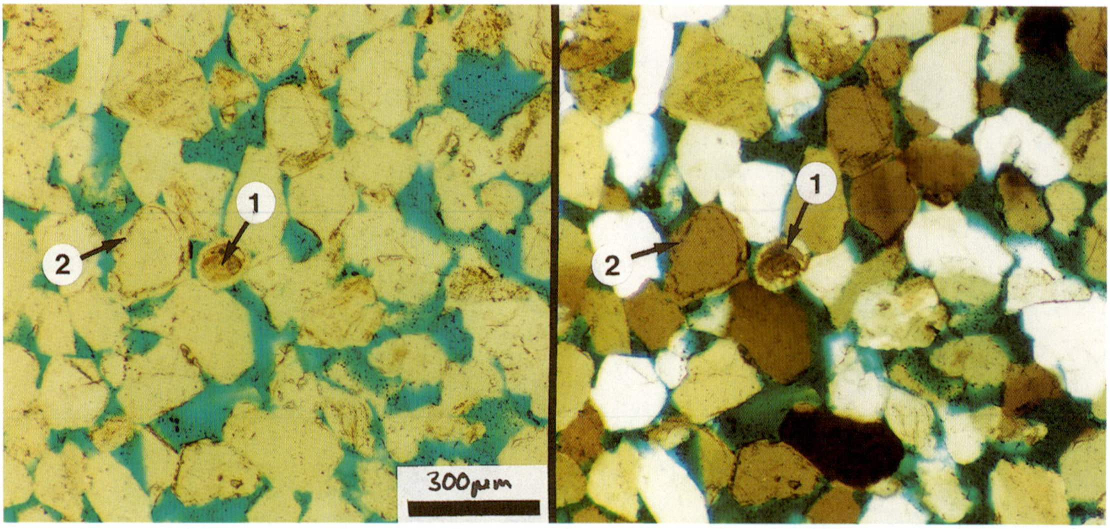


(c)

**Plate 1.** (*Continued*) sample containing ~27% authigenic quartz; note the reduction in pore interconnectivity. Scattered inclusions (1) within detrital grains giving a cloudy appearance and overgrowths projecting into open pore space (2) were also used as criteria for the recognition of overgrowths during point counting (Sample 5). (a) and (b) scale bar = 250 μm. (c) Plane and crossed polarized light thin section photomicrographs (left and right, respectively) showing sponge spicules (1), indicative of a marine depositional environment. Also indicated is a quartz grain showing an obvious dust line (2). (Sample 15 scale bar = 300 μm.)

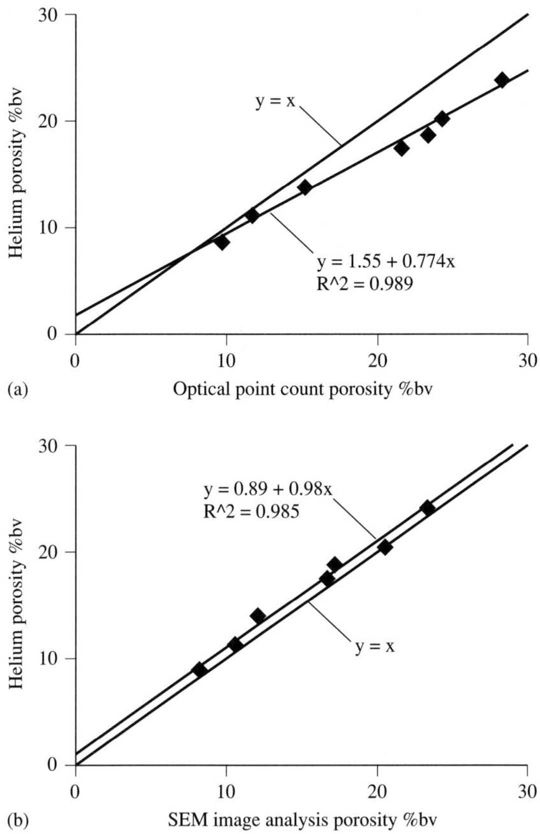


**Fig. 6.** (a) Relationship between the results of total porosity as measured using core-plug helium porosimetry and total porosity measured using optical point counting. Y = X represents 100% agreement, however, the graph indicates that the results of the optical method are overestimated and that the error is increasing with increasing porosity. (b) The results of the scanning electron microscope method are in close agreement with those of helium porosimetry. This indicates that the method produces results representative of core-plug and thin section.

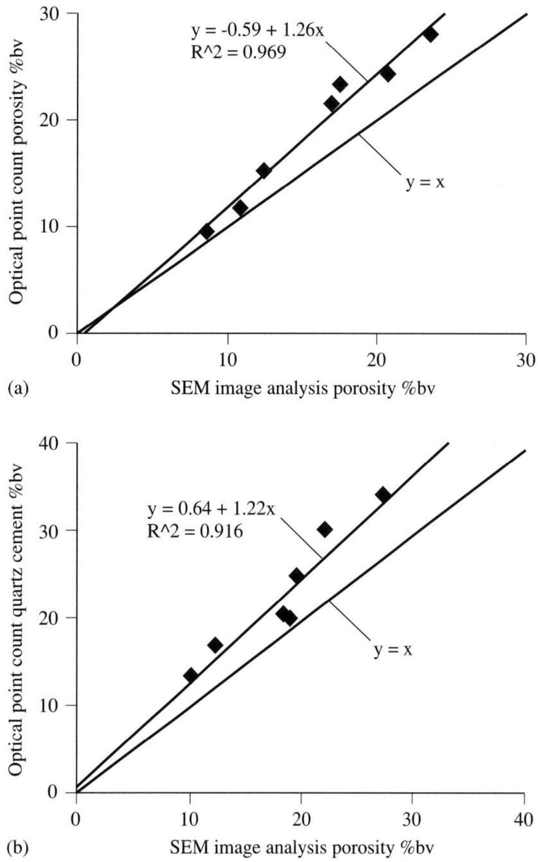


**Fig. 7.** (a) Cross-plot of total porosity measured using optical point counting versus total porosity measured using image analysis. It is evident that values obtained from point counting are higher than those of image analysis. (b) Cross-plot for quartz cement; it can be seen that errors associated with point counting increase with increasing quartz cement content.

manual operation and are also a fundamental statistical limitation of the optical technique. However, most of the discrepancy probably relates to difficulties in distinguishing boundaries between detrital grains and quartz overgrowths. Total porosity may have been overestimated as a result of the operator being influenced by blue resin when a point was counted on an inclined quartz/porosity boundary; even a 30 μm thin section can create numerous problematic situations during a standard point count.

The strong linear correlations shown in Figs 6, 7 and 8 indicate that corrections could be applied to the Fontainebleau optical data, which would bring the semi-quantitative results into line with the quantitative SEM data. Such corrections could be applied to other similar and perhaps more complex data sets, given carefully chosen samples (i.e. samples which describe the range of lithologies and textures being studied); corrections will also differ for each operator.

Figure 8 shows the relationship between quartz cement and total porosity for the Fontainebleau Sandstone. Figure 8(a) shows the results from optical point counting and Fig. 8(b) those from SEM-based analyses. The relationship in both cases is linear, showing that quartz cement is the dominant control on intergranular porosity. Correlation lines for both methods extrapolate into minus-cement porosity values calculated for each technique, again confirming that compaction acted homogeneously. It can be seen that the slopes of the correlations are very

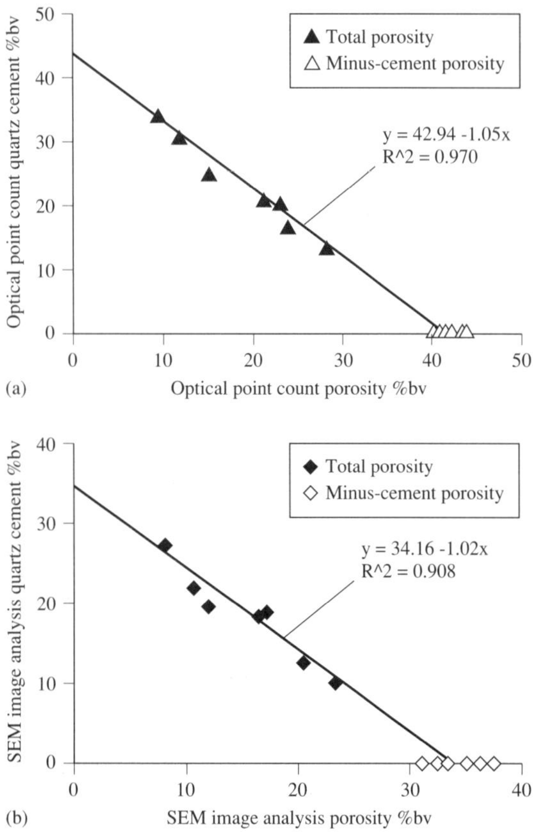


**Fig. 8.** (a) Cross-plot of the results of point count and (b) image analysis showing a strong linear relationship between total porosity and quartz cement. The correlation lines for both data sets extrapolate into the minus-cement porosity values calculated for each technique.

similar for the two techniques, however, the results are very different and would lead to quite different interpretations.

## CONCLUSIONS

**1** Analyses indicate that the Fontainebleau Sandstone is extremely mature in terms of mineralogy and texture.

**2** Initial porosity reduction is related to mechanical compaction. The minus-cement porosity values show little variation, indicating that compaction acted homogeneously throughout the samples examined. It is likely that cementation occurred prior to completion of mechanical compaction; this is reflected by high minus-cement porosity values and the nature of grain contacts.

**3** The precipitation of quartz cement has resulted in the loss of intergranular porosity.

**4** Quartz cement volume is the sole control on intergranular porosity in the samples examined.

**5** Measurements of detrital quartz, quartz cement, minus-cement porosity and post-cementation intergranular porosity using optical point counting are subject to large errors when applied to sandstones which lack easily visible boundaries between detrital and authigenic minerals.

**6** This study revealed that both total porosity and quartz cement were overestimated during optical point counting. Point count errors increase with increasing total porosity and quartz cement and appear systematic and cumulative. Errors are combined when minus-cement porosity is calculated.

**7** SEM-based BSEM, CL and image analysis provided quantitative results if a large enough area of sample surface is analysed. Use of the SEM-based method requires a determinable contrast in CL response between detrital and authigenic quartz.

**8** Using quantitative SEM data points which describe the range of lithologies and textures, semiquantitative optical data could be corrected; this would provide quantitative data more economically. The corrections will be data set and operator specific.

**9** Optical and SEM methods of modal analysis are both vital to the understanding of diagenetic evolution of sandstones and similar geological materials. They should be used in conjunction to optimize progress in quantitative diagenetic analysis.

## ACKNOWLEDGEMENTS

This work represents part of a University of Liverpool based PhD project undertaken by Mark R. Cooper. It was generously sponsored by B.P. Exploration, who also made available the necessary facilities located at the B.P. Sunbury Research Centre. Thanks to Jon Gluyas, Tim Primmer and Ed Warren for their interest and input at various stages of this research; Mark Hopkins, Phil Mitchell, Kevin Jauqes, for providing training and technical assistance. Thanks to the reviewers including Dr Tony Bazley (Geological Survey of Northern Ireland), R. Ehrlich, F. Walgenwitz and the Editor of this volume, Richard Worden, for constructive advice. Jane Cooper, thank you for significantly contributing towards this publication.

Bob Hunter, co-author, has sadly passed away since submission of this paper.

## REFERENCES


ALIMEN, H. (1936) Étude sur le stampien du Bassin de Paris. *Société Géologique de France Mémoire* **31**, 1–304.

ATKINS, J.E. & MCBRIDE, E.F. (1992) Porosity and packing of Holocene river, dune and beach sands. *American Association of Petroleum Geologists Bulletin* **76** (3), 339–355.

BEARD, D.C. & WEYL, P.K. (1973) Influence of texture on porosity and permeability of unconsolidated sand. *American Association of Petroleum Geologists Bulletin* **57**, 349–369.

BOURBIE, T. & ZINSZNER, B. (1985) Hydraulic and acoustic properties as a function of porosity in Fontainebleau Sandstone. *Journal of Geophysical Research* **90** (B)13, 11524–11532.

BRYANT, S., CADE, C. & MELLOR, D. (1993) Permeability prediction from geologic models. *American Association of Petroleum Geologists Bulletin* **77** (8), 1338–1350.

CAVELIER, C. & LORENZ, J. (1987) Aspect et Évolution géologiques du Bassin Parisien. *Bulletin d'Information Dès Géologiques Du Bassin de Paris. Mémoire Hors-série* **6**, 1–271.

CAYEUX, L. (1929) Les roches sédimentaires de France, roches siliceuses In:. *Mémoire Carte Géologique Détaillée de la France*, 774pp.

COOPER, M.R. (1995) *Diagenetic evolution and implications for the reservoir properties of selected shallow marine and aeolian sandstones*. PhD Thesis, The University of Liverpool.

DUTTON, S.P. & DIGGS, T.N. (1992) Evolution of porosity and permeability in the Lower Cretaceous Travis Peak Formation, East Texas. *American Association of Petroleum Geologists Bulletin* **76** (2), 252–269.

EVANS, J., HOGG, A.J.C., HOPKINS, M.S. & HOWARTH, R.J. (1994) Quantification of quartz cements using combined SEM, CL and image analysis. *Journal of Sedimentary Research A* **64** (2), 334–338.

GIRARD, J.-P. & DEYNOUX, M. (1991) Oxygen isotope study of diagenetic quartz overgrowths from the upper Proterozoic quartzites of Western Mali, Taoudeni Basin: implications for conditions of quartz cementation. *Journal of Sedimentary Petrology* **61** (3), 406–418.

HOGG, A.J.C., SELLIER, E. & JOURDAN, A.J. (1992) Cathodoluminescence of quartz cements in Brent Group sandstones, Alwyn South, UK North Sea. In: *Geology of the Brent Group*. (eds MORTON, A.C., HAZELDINE, R.S., GILES, M.R. & BROWN, S.) pp. 421–440. Special Publications of the Geological Society of London 61.

HOUSEKNECHT, D.W. (1987) Assessing the relative importance of compactional processes and cementation to the reduction of porosity in sandstones. *American Association of Petroleum Geologists Bulletin* **71** (6), 633–642.

JACQUIN, C. (1964) Corrélations entre la perméabilité et les caractéristiques géométriques du grés de Fontainebleau. *Review of the Institute of French Petrology* **19**, 921–937.

MCBRIDE, E.F. (1963) A classification of sandstones. *Journal of Sedimentary Petrology* **33**, 664–669.

MCBRIDE, E.F. (1989) Quartz cement in sandstone: A review. In: *Earth Science Reviews* **26**, 69–112. Elsevier Science, Amsterdam.

MÉGNIEN, C. (1980a) Tectogenèse du Bassin de Paris: Étapes de le évolution du bassin. *Société Géologique de France Bulletin* **12**, 669–680.

MÉGNIEN, C. (1980b) *Synthèse Géologique du Bassin de Paris, 2. Atlas*. Bureau de Recherches Géologiques et Minières Mémoire 102.

PETTIJOHN, F.J., POTTER, P.E. & SIEVER, R. (1987) *Sand and Sandstone*. Springer Verlag, New York, 553pp.

POWERS, M.C. (1953) A new roundness scale for sedimentary particles. *Journal of Sedimentary Petrology* **23**, 117–119.

PRYOR, W.A. (1973) Permeability–porosity patterns and variations in some Holocene sand bodies. *American Association of Petroleum Geologists Bulletin* **57** (1), 162–189.

RITTENHOUSE, G. (1971a) Pore space reduction by solution and cementation. *American Association of Petroleum Geologists Bulletin* **55** (1), 80–91.

RITTENHOUSE, G. (1971b) Mechanical compaction of sands containing different percentages of ductile grains: a theoretical approach. *American Association of Petroleum Geologists Bulletin* **55** (1), 92–96.

THIRY, M., AYRAULT, M.B. & GRISONI, J.-C. (1988) Groundwater silicification and leaching in sands: Example of the Fontainebleau Sand (Oligocene) in the Paris Basin. *Geological Society of America Bulletin* **100**, 1283–1290.

WILLIAMS, H., TURNER, F.J. & GILBERT, C.M. (1954) *Petrography*. Freeman, San Francisco, 406pp.

DUTTON & DIGGS (1992).

*Spec. Publs int. Ass. Sediment.* (2000) **29**, 103–117

# Effects of reservoir wettability on quartz cementation in oil fields

S. A. BARCLAY *and* R. H. WORDEN*

*School of Geosciences, Queen's University of Belfast, Belfast, BT7 1NN, Northern Ireland*

## ABSTRACT

The wettability of an oil field sandstone has a great impact upon the influence of oil emplacement on quartz cementation. Quartz cementation is most inhibited in reservoirs filled with low maturity, biodegraded and/or water-washed oil from NSO-rich source rocks (types IIS and III) containing quartz, pre-oil filling carbonate cement, swelling clays, kaolinite booklets and iron minerals. Conversely quartz cementation will be least inhibited in oil fields filled with high maturity, non-biodegraded/water washed oil from NSO-poor source rocks (types I and II), composed of quartz, illite, unaltered feldspar, finely crystalline kaolinite and devoid of early carbonate cements. Wettability and thus quartz cementation may also be affected by oil column height, temperature and water chemistry. Predicting the exact wettability of a reservoir is not yet feasible. However, with the general trends and patterns identified, it is possible to assess, and potentially predict, the relative degree of inhibition of quartz cementation in oil field sandstones.

## INTRODUCTION

Quartz cement is responsible for significant porosity and permeability reduction in many deeply buried, quartz-rich reservoir sandstones. The widely held assumption that quartz cementation ceases once oil has been emplaced into a sandstone has been questioned recently. A review by Worden *et al.* (1998) demonstrated that total cessation will not necessarily occur. The degree of inhibition of quartz cementation was shown to depend significantly upon the wettability of the sandstone–water–oil system. This paper builds upon these findings and attempts to define the controls on wettability and thus on quartz cementation. In this paper we will first introduce wettability, and the nature of grain surfaces, we will then synthesize the general role of wettability on quartz cementation. The specific effects of oil chemistry, formation water chemistry, reservoir mineralogy, pressure and temperature on wettability will then be assessed in turn. We finally present our findings as a series of general rules governing the controls on wettability and thus on quartz cementation in oil fields.

*Contact for reprints r.worden@qub.ac.uk.

### Introduction to wettability

Wettability is the tendency of a fluid to spread along a solid surface in the presence of another (immiscible) fluid (Crocker & Marchin, 1988). Common siliciclastic reservoir minerals (both framework grains and cements) have differing wettabilities which partly depend on their composition (Anderson, 1986; Robin *et al.*, 1995). Homogeneous wettability is when a rock surface has a uniform molecular affinity for a particular liquid, whereas heterogeneous wettability is when surfaces in a rock exhibit different affinities for the reservoir liquids (Kovscek *et al.*, 1993). Schematic diagrams for fluid distribution in oil- and water-wet sandstones are shown in Fig. 1.

Wettability can be measured in a number of ways (Table 1). The physical form of the sample used for wettability determination varies between methods. Some employ polished surfaces of isolated mineral crystals for wettability determination (e.g. Treiber *et al.*, 1972; Brown & Neustadter, 1980; Chilingar & Yen, 1983; González & Moreira, 1991). Others use broken or polished rock or core samples (e.g. Wolcott *et al.*, 1991; de Pedroza *et al.*, 1994; Grattoni *et al.*, 1995; Sayyouh & Al-Blehed, 1995). One of the most commonly used methods

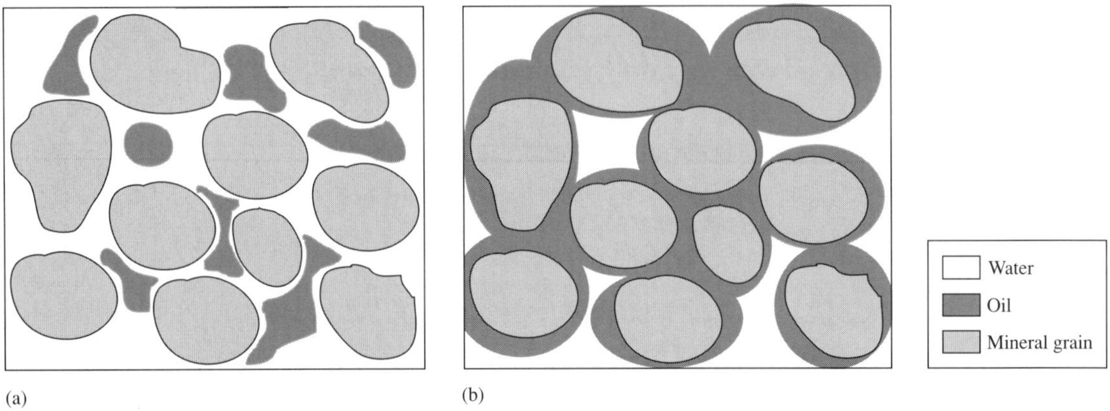


**Fig. 1.** Schematic diagram showing (a) a water-wet reservoir; and (b) an oil-wet reservoir.

**Table 1.** Tabulation of the variety of methods used to determine wettability

| Method | Author |
|---|---|
| Thermodynamic definition | Johansen & Dunning (1953) |
| Contact angle measurement | Treiber *et al.* (1972) |
| Interfacial tension and displacement pressure measurements | Slobod & Blum (1952) |
| Shape of relative permeability curves | Raza *et al.* (1968) |
| Shape of recovery curves | Raza *et al.* (1968) |
| Permeability and saturation measurements | Frehse (1973) |
| Spontaneous imbibition experiments | Gatenby & Marsden (1957) |
| Nuclear magnetic relaxation measurements | Brown & Fatt (1956) |
| Dye adsorption | Holbrook & Bernard (1958) |
| Flotation experiments | Anderson (1985) |
| Electrical resistivity well logs | Swanson (1980) |
| Capillary pressure curves | Donaldson *et al.* (1969) |
| Imbibition and displacement experiments | Amott (1959) |
| Nuclear magnetic resonance $T_1$ distributions | Howard (1994) |

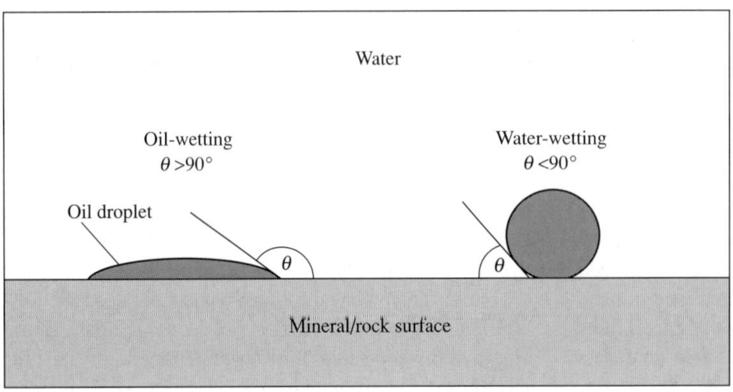


**Fig. 2.** Schematic diagram to demonstrate the relationship between contact angle and wettability. (After Cuiec, 1987.)

is an assessment of the contact angle between the fluid (oil or water) and the mineral surface (Fig. 2).

Laboratory wettability results obtained using single, isolated minerals cannot be assumed to represent the actual reservoir wettability, because these will give a wettability value for that mineral under unrealistic conditions ignoring heterogeneous wettability caused by reservoir rocks that are typically polymineralic (Anderson, 1986; Robin *et al.*, 1995). Probably the best way to obtain realistic laboratory wettability measurements is to use whole core samples. It is possible to define the overall wettability state of a core sample using techniques such as measurement of the extent of spontaneous imbibition of the non-wetting phase into the core sample (Grattoni *et al.*, 1995; Jadhunandan & Morrow, 1995) or the USA Bureau of Mines capillary pressure curve method (Wang & Guidry, 1994).

## GEOCHEMICAL CONTROLS ON QUARTZ CEMENTATION

The occurrence and extent of quartz cementation in a sandstone reservoir is controlled by relative rates of the supply of silica to the system, the transport rate from point of supply to point of quartz cement growth and the rate of precipitation of quartz (Worden *et al.*, 1998).

### Silica source

End-member sources of silica for cement are either (i) internal or (ii) external to the reservoir. Internal sources include locally dissolved detrital quartz grains, feldspars, sponge spicules as well as stylolite seams. External sources include silica from compacting and reacting mudrocks and diagenetic reactions occurring in other sandstone bodies. The wettability of the reservoir potentially has a strong effect on internally sourced silica. In oil-wet reservoirs any silica source is probably isolated from the rest of reservoir by the coat of oil; thus a geochemical feasible supply of silica may exist that is unavailable for cementation elsewhere in the reservoir. If the source of silica is external to the reservoir then the fundamental supply of silica will be unaffected by the wettability state of the reservoir itself (Worden *et al.*, 1998).

### Silica transport

Transport of silica from source to sink may involve two possible modes of transport: either (i) diffusion along a concentration gradient, or (ii) advection along a pressure gradient. Diffusion rates in oil legs of hydrocarbon reservoirs in comparison to the diffusion rate in water legs are reflected by the water saturation ($S_w$) and the saturation exponent ($n$). By direct analogy to electrical conductivity, Archie's Law has been used (Keller, 1953; Worden *et al.*, 1998) to show that:

$$\frac{\text{diffusion rate in water leg}}{\text{diffusion rate in oil leg}} = S_{w^n}$$

In water-wet oil fields, the saturation exponent is relatively low ($1 > n < 2$) reflecting the continuous film of residual water clinging to grain surfaces. As a system becomes increasingly oil-wet, the saturation exponent increases as the electrical, and by analogy diffusive, circuit becomes progressively more convoluted (Raza *et al.*, 1968; Søndenå *et al.*, 1991). Thus wettability strongly affects the value of 'n' and thus the rate of aqueous diffusion in oil fields. If there is an internal source of silica available in a reservoir, and assuming the mode of transport of silica within a reservoir is by diffusion (Worden *et al.*, 1998); then the rate of diffusion of silica in an oil-wet reservoir will be vastly reduced compared with a water-wet reservoir.

The rate of water flow in oil fields is strongly controlled by relative permeability (Archer & Wall, 1994). Imbibition of a non-wetting phase into oil fields (e.g. water into an oil-wet sandstone) is significantly more inhibited (i.e. lower relative permeability) than imbibition of a wetting phase (e.g. water into a water-wet sandstone. Thus wettability strongly effects the rate of flow of potentially silica-laden water into oil fields.

### Silica precipitation

The feasibility of quartz precipitation is strongly affected by the wettability state of the reservoir (Worden *et al.*, 1998). The fundamental rate of precipitation is strongly reduced in oil-wet reservoirs but presumably unaffected in water-wet reservoirs. To precipitate quartz cement, the silica-containing fluid must have access to quartz grain surfaces. Assuming that silica is transported as a hydrophilic aqueous complex, then if the surfaces of the mineral grains are coated with oil (as they would be in an oil-wet reservoir), then precipitation of quartz cement would not occur (Worden *et al.*, 1998). Note that the coat of oil must be continuous to totally halt quartz precipitation (Fig. 3).

A fundamental conclusion from this discussion is that oil-wet reservoirs are likely to have much slower rates of silica supply and transport and quartz precipitation than water-wet reservoirs (Table 2).

## PHYSICAL AND CHEMICAL NATURE OF (QUARTZ) SURFACES

The nature of the interaction between a surface and a fluid determines whether the surface is water-wet or oil-wet. From a purely thermodynamic perspective, surfaces try to attain their lowest possible surface energy in a particular fluid phase (Stumm, 1992). For example, at typical oilfield pH levels, negatively charged hydroxyl groups project from the surface of quartz. In formation water, these surface hydroxyl groups form hydrogen bonds with the positively charged dipoles of water molecules

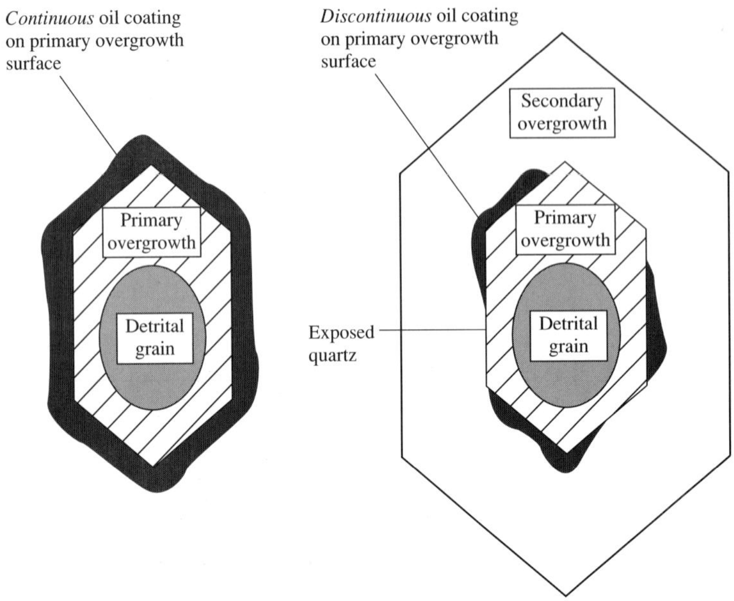


**Fig. 3.** The effect of a continuous and discontinuous oil film on the growth of quartz cement.

**Table 2.** Simplified list of the influence of wettability upon the component processes involved in quartz cementation

| Process | | Wettability effect |
|---|---|---|
| Source | Internal | Supply of silica retarded if source is oil-wet |
| | External | Supply of silica retarded if source is oil-wet |
| Transport | Diffusion | Transport retarded if rock is oil-wet |
| | Advection | Transport retarded if rock is oil-wet |
| Precipitation | | Retarded if site of precipitation is oil-wet |

(Schlangen *et al.*, 1995). In a relatively polar-free petroleum fluid, the formation of hydrogen bonds between quartz and petroleum is unlikely to occur and the hydroxyl groups on the surface of the quartz are forced to project into a fluid for which they have no affinity. The lowest (surface) energy state of quartz in most mixed oil–water systems is thus when the quartz is in contact with the formation water. Thus quartz is considered to be preferentially water-wet (Schlangen *et al.*, 1995; Fig. 4a–c) although we will investigate other complicating factors subsequently.

The addition of surface-active agents (surfactants) to a water–oil–quartz system can change the wetting preference of the quartz surface by reducing the surface energy between the quartz surface and the non-wetting fluid. This can theoretically cause the quartz to be wetted by the (polar-rich) oil (Brown & Neustadter, 1980; Schlangen *et al.*, 1995). The type of compounds that can act as surfactants for the water–oil–quartz system are polar compounds which typically contain one or more of the elements N, S or O (Fig. 4d and Fig. 5). Most surfactants are negatively charged and are thus not attracted to negatively charged mineral surfaces.

The type and strength of interaction between the mineral surface and the surfactant depends upon the mineral surface charge, the physical and chemical form of the surfactant molecule and the presence or absence of metal cations and water films on the mineral surface (Brown & Neustadter, 1980; Crocker & Marchin, 1988; González & Moreira, 1991; Poirier & Cases, 1991; Levinson *et al.*, 1994; Rixey & Fuerstenau, 1994; Schlangen *et al.*, 1995). For example, the relatively planar shape of asphaltene molecules (Tissot & Welte, 1984; González & Moreira, 1991) means that, when present in high concentrations, they can adsorb irreversibly on to mineral surfaces, by forming large numbers of lateral hydrogen bonds with surface hydroxyl groups (Dubey & Waxman, 1989; González & Moreira, 1991). Carboxylic acids in oils typically contain short to medium length alkane chains ($C_1$ to $C_{20}$, from Tissot & Welte, 1984). These adsorb perpendicular to the mineral surface, forming single hydrogen bonds with the surface hydroxyl groups (Schlangen *et al.*,

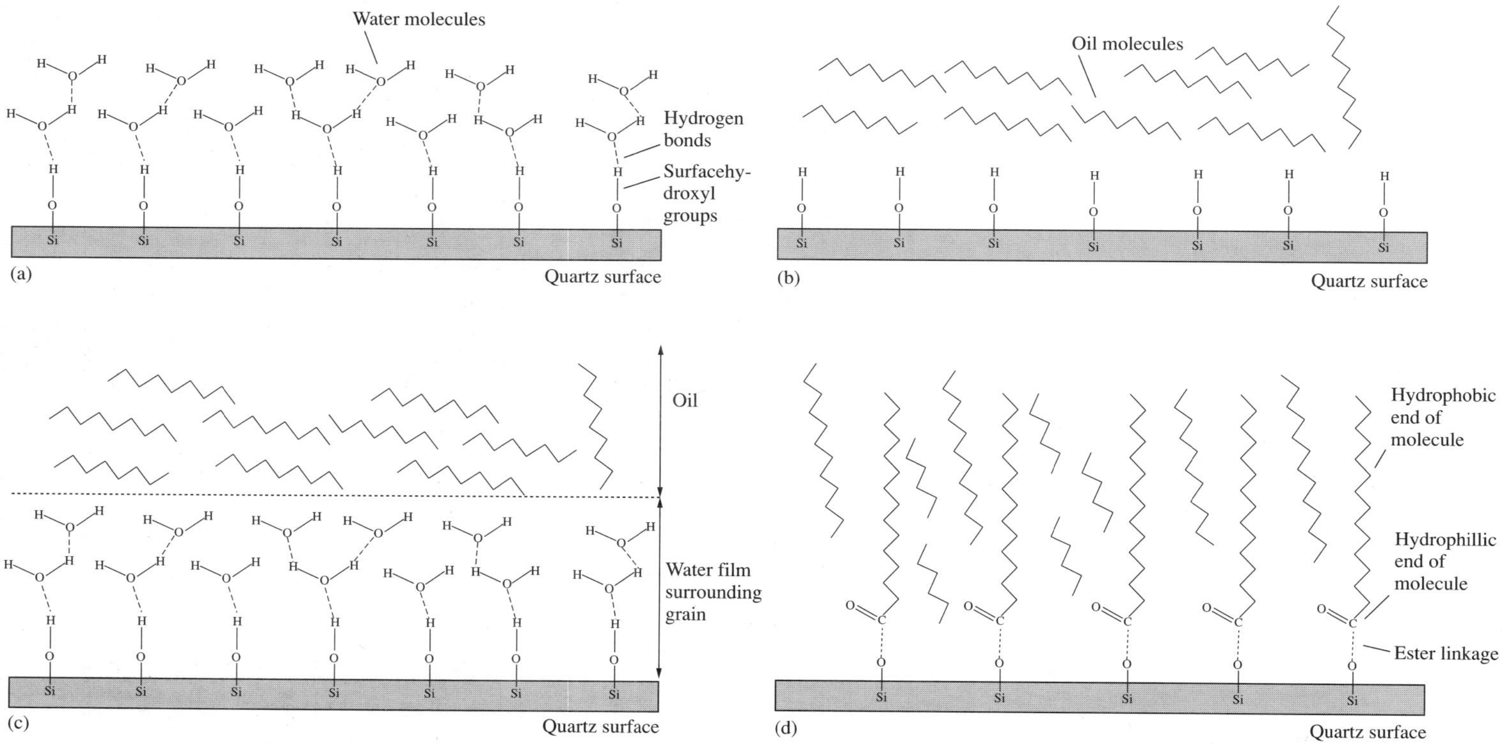


**Fig. 4.** Schematic representation of different types of surface–oil–water molecular arrangements. (a) Quartz surface in water (at normal pH) with a negative surface charge attracting positively charged oxygen atoms in water. (b) Quartz surface in oil (devoid of polar species) illustrating the lack of interaction between the grain surface and the liquid. (c) Quartz surface in mixed oil–water with water preferentially coating the grain surface due to the interaction between quartz and water. (d) Quartz surface shown interacting with surfactant (carboxylic acid) leading to the possibility of an oil-wet quartz surface.

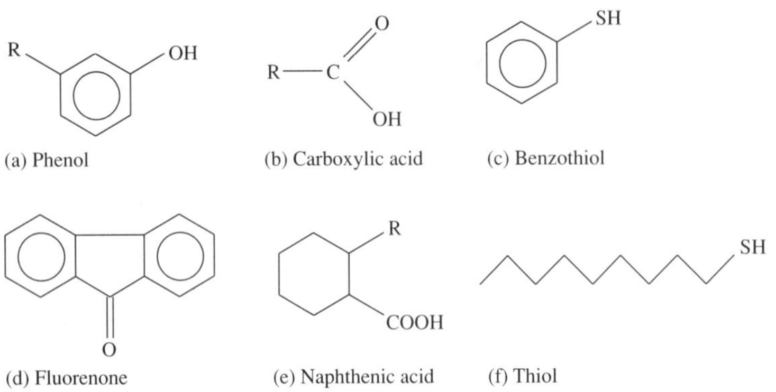


**Fig. 5.** Molecular structure of some common polar compounds.

1995). Therefore the strength of the interaction between a mineral surface and an asphaltene is typically stronger (due to the greater number of hydrogen bonds), than that between the surface and a carboxylic acid. However, longer chain carboxylic acids also show some evidence of lateral interaction with the mineral surface (Poirier & Cases, 1991), although carboxylic acids with chain lengths > $C_{20}$ are rare in oils.

The presence of adsorbed metal cations on mineral surfaces facilitates the formation of complexes between the mineral and the surfactant (Brown & Neustadter, 1980; Poirier & Cases, 1991; Stumm, 1992; Wang & Guidry, 1994) and thus can promote oil-wetting.

## CONTROLS ON WETTABILITY

### Petroleum-composition effects

Petroleum geochemistry has a major impact upon wettability. The source rock mineralogy, organic source material and temperature of petroleum generation all have major impacts upon the presence and abundance of surfactants in the oil. Post-petroleum emplacement influences may also affect the abundance and availability of surfactants.

#### *Petroleum source considerations and polar compounds*

Petroleum contains a number of different generic groups of compounds including: alkanes, aromatics and polars (Killops & Killops, 1993; Miles, 1994). Minerals have little or no affinity for the covalently bonded alkane group of compounds. Polar compounds have the greatest influence of all the generic groups on the wettability state of a reservoir (Fig. 5). The nitrogen–sulphur–oxygen (NSO) containing polar compounds can alter reservoir wettability from water-wet to oil-wet by a combination of (i) complex formation with mineral surfaces (Bennett & Siegel, 1987; Bennett *et al.*, 1988) and (ii) electrostatic attraction (Wang & Guidry, 1994), thus giving a mineral surface a hydrophobic layer (Fig. 4d).

The quantity of polar compounds in an unaltered (i.e. non-biodegraded and non-water-washed) oil is primarily controlled by the source rock that initially generates the oil. The mineralogy of the source rock may be critical. Marine carbonate-rich (and therefore typically iron-poor) source rocks (known hereafter as IIS source rocks) tend to generate oils with greater quantities of sulphur compounds than marine clastic source rocks (known as II; Tissot & Welte, 1984; Miles, 1994). The availability of sulphur to form organic compounds is limited by the presence of any iron oxides/oxyhydroxides allowing the formation of iron sulphide minerals.

The type of organic matter (i.e. kerogen) in the source rock has a major impact upon the polar-content of an oil. Oxygen-containing polar compounds, such as phenols, methoxy and alkyl phenols, are typical of the early maturation products produced from terrestrially derived source rocks (known hereafter as type III). Type IIS source rocks release the greatest amount of sulphur-containing organic compounds, such as alkylthiophenes and alkylbenzothiophenes. The overall conclusion being therefore that type III kerogens should generate oils rich in oxygen-containing polar compounds and type IIS kerogens should generate oils rich in sulphur-containing polar compounds. Lacustrine source rocks (type I) have the least potential for generating wettability-altering polar compounds although type II source rocks also tend to have relatively limited potential to generate oils rich in polar compounds.

The thermal maturity of the source kerogen generating the oil has a major effect upon the polar content of the oil. Initially kerogen has a disordered structure which evolves to a more ordered structure by the loss of organic functional groups with increasing thermal maturity (Killops &

Killops, 1993). The types of functional groups that are lost from the kerogen first are the most weakly bonded ones (e.g. carboxyl and carbonyl groups, both of which contain oxygen; Rouxhet & Robin, 1978). Therefore at low kerogen thermal maturities, any liquid oil generated will be enriched in oxygen-containing polar compounds, relative to more mature, later generated oils.

*Asphaltenes*

Asphaltenes are high molecular weight compounds composed of poly-aromatic nuclei linked by: aliphatic chains, NSO functional groups and metal cations. In a reservoir, asphaltenes can either be in solution or as an exsolved coat on mineral grains (Miles, 1994; Ehrenberg *et al.*, 1995). Exsolved asphaltenes can enhance the oil-wettability of a reservoir by being adsorbed on to mineral surfaces (González & Middea, 1987; Crocker & Marchin, 1988; Dubey & Waxman, 1989; González & Moreira, 1991; Wilhelms & Larter, 1994; Ehrenberg *et al.*, 1995). The solubility of asphaltenes in oil is controlled by a number of processes. (i) Asphaltenes tend to be exsolved from oil when the gas/oil ratio (GOR) increases (Milner *et al.*, 1977; Wilhelms & Larter, 1995). The GOR is itself controlled by source rock maturity and kerogen type (see earlier). Mature and type III source rocks tend to generate higher GOR oils than immature source rocks (Killops & Killops, 1993). (ii) Biodegradation and water-washing of an oil preferentially removes the lighter liquid components of an oil. This potentially oversaturates the remaining oil in asphaltenes (Palmer, 1993; Miles, 1994). (iii) Aerobic bacterial action can generate asphaltenes by oxidizing hydrocarbons and leading to the incorporation of sulphur into the petroleum liquid (North, 1985). (iv) *In situ* maturation of oil leading to gas-generation by oil cracking in a reservoir during burial and heating will also increase the GOR and thus cause asphaltenes to precipitate (Wilhelms & Larter, 1994). (v) Mixing of two compositionally distinct oils may result in asphaltene precipitation if, for example, one oil is enriched in light *n*-alkanes, or has a higher GOR relative to the other oil (Larter *et al.*, 1990; Wilhelms & Larter, 1994).

Asphaltenes in oil are in the form of micelles in solution (i.e. a colloidal solution; Tissot & Welte, 1984; Miles, 1994). Addition of a high ionic strength formation water may cause asphaltene micelles to coagulate and precipitate out of the oil (Wilhelms & Larter, 1994). Thus the salinity of formation water may have an indirect effect on the wettability state of the reservoir by controlling the availability of surface active compounds in the oil. The mechanism behind this cation-induced asphaltene precipitation is the reduction of electrostatic repulsion between the micelles by the cations in aqueous solution (Hunter, 1987; Atkins, 1990; Stumm, 1992). The precipitated asphaltenes may then alter wettability from water-wet to oil-wet.

**Table 3.** Tabulation of the effects of biodegradation and water-washing upon oil composition and thus wettability

| Process | Effects |
|---|---|
| Biodegradation | Generation of polar compounds<br>Generation of asphaltenes<br>Removal of *n*-alkanes<br>API gravity reduced |
| Water-washing | Removal of < $C_{15}$ polar compounds<br>Precipitation of asphaltenes<br>Introduction of aerobic bacteria |

*Post-petroleum emplacement processes*

Petroleum composition in a reservoir can be altered by two closely related mechanisms that tend to lead to the alteration of reservoir wettability towards being oil-wet.

Water-washing involves long-term advection of water under an oil accumulation that removes water-soluble compounds in the petroleum (North, 1985; Miles, 1994). Water in stagnant aquifers reaches saturation with the water-soluble compounds and thus has little effect on the overall properties of the petroleum. Water-washing tends to remove the lighter alkane liquids and other moderately water-soluble petroleum compounds causing precipitation of asphaltenes from the liquid (Tissot & Welte, 1984). Asphaltene exsolution from oil tends to lead to enhanced oil-wettability.

Water-washing by meteoric water can also introduce aerobic bacteria into a reservoir, thus providing the trigger for biodegradation and generation of wettability-altering polar compounds (Palmer, 1993). The initial composition of the oil and the supply of inorganic nutrients affect the occurrence and products of biodegradation (North, 1985; Miles, 1994). Water is an essential component for biodegradation of oil (Palmer, 1993). Unless the water saturation of the reservoir is relatively high, biodegradation and the production of wettability-altering compounds will be limited to the oil-water transition zone (Connan, 1984). The effects of post-emplacement processes are summarized in Table 3.

## Water–rock effects

Reservoir mineralogy has a profound effect upon wettability due to mineral-specific surface charge, the iron content of the surface layer of minerals, the adsorption

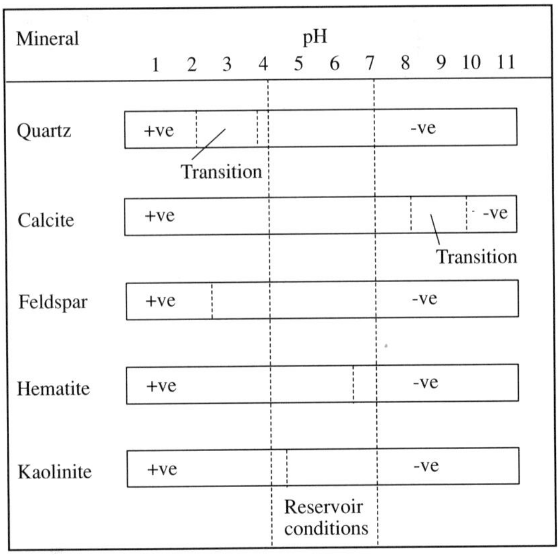


**Fig. 6.** Illustration of the influence of pH on the surface charge of common minerals.

of organic molecules into certain minerals and crystal shape/surface charge anisotropy. Consequently the overall mineralogy of a sandstone will affect the rate of quartz cementation, through its control on wettability.

### *Mineral-specific surface charge pH*

The pH of the formation water can indirectly control the wettability of a reservoir because the surface charge of a mineral is generally sensitive to pH. For example, the surface charge of quartz is positive at very low pH, and negative at high pH. The change from positive surface charge to a negative surface charge occurs over the pH range of 2–3.7 (Fig. 6; Stumm & Morgan, 1970). Thus, under typical oil field conditions (pH range 4–7; Hanor, 1994), the surface of quartz has a negative surface charge and is preferentially water-wet through hydrogen-bonding with water molecules (Fig. 7). Polar compounds typically have a negative charge (dipole) and are thus attracted to minerals with a positive surface charge. At atypically low pH levels (i.e. < 3.7), a quartz surface can adsorb negatively charged polar compounds (Fig. 7), which act as surfactants thus making the surface of the quartz oil-wet.

In comparison, carbonate minerals, such as calcite and dolomite, change from a positive surface charge to a negative surface charge over the pH range of 8–9.5 (Fig. 6; Somasundaran & Agar, 1967; Tabatabal *et al.*, 1993). Carbonate reservoirs are commonly oil-wet (Treiber *et al.*, 1972; Chilingar & Yen, 1983; Anderson, 1986; Cuiec, 1987), because under typical reservoir conditions (pH range 4–7, from Hanor, 1994), carbonate minerals have a positive surface charge. Carbonate reservoirs will be able to adsorb relatively more negatively charged wettability-altering compounds than siliciclastic reservoirs and are more likely to be oil-wet. Thus siliciclastic reservoirs that contain carbonate cements will have heterogeneous wettability, increased resistivity in oil legs and reduced diffusion rates from silica source to quartz precipitation point. The effects of pH on surface charge are summarized in Fig. 6.

### *Clay mineral effects*

Clay minerals have varying wettabilities which depend on their chemical composition, morphology and physical structure. Kaolinite ($Al_2Si_2O_5(OH)_4$) in the form of small, separate crystals is preferentially water-wet. However, in the form of 10 µm wide booklets it can be oil-wet (Robin *et al.*, 1995).

Hydroxyl groups project from the surface of kaolinite at right-angles to the Si–Al–O sheets in kaolinite (Fig. 8). When kaolinite is in the form of small crystals, the surface of kaolinite is dominated by these hydroxyl-coated (001) surfaces which attract charged water molecules, also the capillary sizes associated with small crystals of kaolinite (typically 0.1 µmm) means that

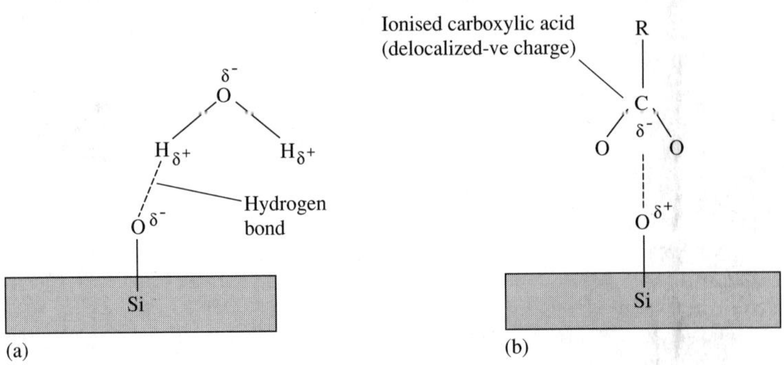


**Fig. 7.** Schematic illustration of the influence of pH on surface charge and wettability. (a) At pH > 3, quartz has a negative charge and is attracted to the positive dipole in water. (b) At pH < 3, quartz has a positive charge and is attracted to negative dipoles common in polar compounds.

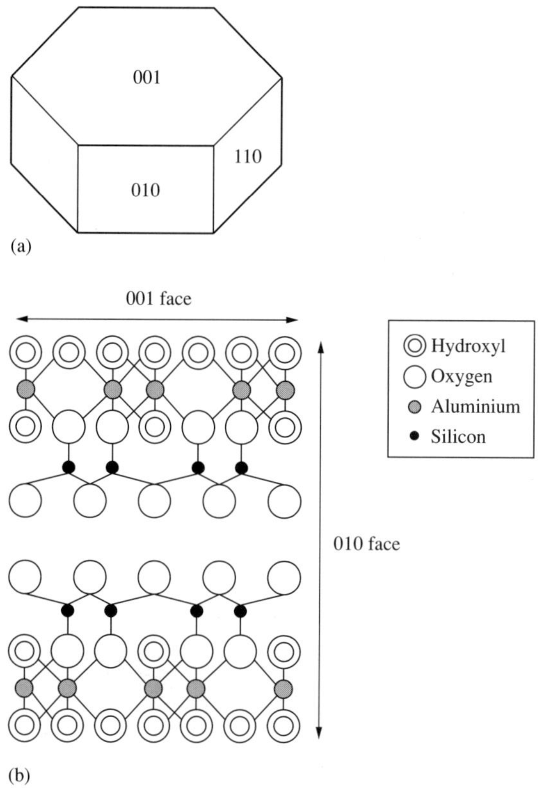


**Fig. 8.** Effect of kaolinite morphology on its wettability state. (a) Booklet of kaolinite showing the crystal faces exposed. (b) Cross section through kaolinite (viewed facing the 110 crystal face) showing the greater density of hydrophilic hydroxyl groups on the 001 face as compared with the 010 face.

the molecules in oil are not physically able to approach the surface of the kaolinite (Rezaee & Lemon, 1996). The combined effect is thus to render small crystals of kaolinite water-wet. When kaolinite is in the form of larger booklets, the surface of the kaolinite is dominated by the edges of the Si–Al–O sheets (110 and 010 surfaces). These have fewer hydroxyl groups and are thus more able to attract organic compounds than the hydroxyl-coated (001) surfaces. The oil-wetting character of kaolinite booklets was observed by Robin *et al.* (1995).

Expandable clays (the smectite group of minerals) are able to absorb larger quantities of polar compounds than non-expandable clays (analogous to gylcolation—used to identify expandable clay minerals). This makes expandable clays more likely to be oil-wet than non-expandable clays (Robin *et al.*, 1995). Smectite typically undergoes diagenetic transformation during burial (as defined by the illite/smectite ratio) so that diagenetic reactions in the reservoir may lead to evolution in wettability (Powers, 1967).

### *The effects of iron and other transition elements*

Iron oxide (haematite, $Fe_2O_3$) may be strongly oil-wet because it can adsorb large quantities of asphaltenes (González & Moreira, 1991). Wang & Guidry (1994) suggested two possible mechanisms for the strong oil-wetting nature of haematite. Firstly, transition metal cations on the mineral surface may act as adsorption sites for carboxylic acid anions which alter the wettability of the mineral surface. Secondly, catalytic oxidation of crude oil components by highly charged metal cations may form polar compounds which then alter wettability by adsorption on to metal cations. Surdam *et al.* (1993) suggested that hydrocarbons may be oxidized by reaction with the $Fe^{3+}$ and oxygen present in haematite-rich sandstone to form carboxylic acids. Sandstones tend to display increasing oil-wetting as their iron oxide content increases.

In oils and formation waters there are a variety of transition metals in the form of either aqueous complexes, organo-metallic complexes or metallo-porphyrins (Collins, 1975; Saunders & Swann, 1990; Barth, 1991; Filby, 1994; Giordano & Kharaka, 1994; Hanor, 1994; Shock & Koretsky, 1995). Trace quantities of aqueous transition metals can alter the wettability state of a reservoir (Treiber *et al.*, 1972; Brown & Neustadter, 1980; Chilingar & Yen, 1983; Wang & Guidry, 1994). Transition metal ions, from either aqueous solution or crude oil compounds may become adsorbed on to negatively charged mineral surfaces (Anderson, 1986; Stumm, 1992) thus acting as a bridge between the mineral surface and negatively charged polar compounds.

The redox state of the reservoir fluids can control indirectly the wettability state of the reservoir. Wettability can be altered from water- to oil-wet when iron is present in the mineralogy of the reservoir, but only when iron is in the $Fe^{3+}$ state (Wang & Guidry, 1994). The oxidation state of other transition metals is also likely to be an important influence on their effect on wettability although iron is the most volumetrically significant transition element.

Iron-rich silicates and carbonates (such as smectite, chlorite, siderite or ankerite) are also often oil-wet (Treiber *et al.*, 1972; Chilingar & Yen, 1983; Anderson, 1986; Cuiec, 1987), perhaps due to the presence of iron. The mechanism may be either catalytic oxidation or adsorption of polar compounds, analogous to the oil-wetting behaviour of haematite.

### *Feldspars*

Feldspars can display mixed wettability, probably controlled by two factors. The presence of partially coordinated, positively charged aluminium atoms on feldspar

**Table 4.** Tabulation of the probable wettabilities of different reservoir minerals

| Oil-wet minerals | Water-wet minerals |
|---|---|
| Calcite | Quartz |
| Dolomite | Illite |
| Ferroan dolomite | Kaolinite single crystals |
| Kaolinite booklets | Unweathered feldspar |
| Fe-rich smectite | |
| Haematite | |
| Weathered feldspar | |
| Fe-rich chlorite | |

**Table 5.** The effects of varying pressure and temperature upon wettability

| Increasing temperature | Increasing pressure |
|---|---|
| Destruction of carboxylic acids—water-wet | Increase in gas solubility, decrease in asphaltene solubility—oil-wet |
| Cracking of oil, increase in GOR—precipitation of asphaltenes—oil-wet | Increase in capillary pressure, water films around grains are thinner/non-existent—oil-wet |
| Reduction in gas solubility, increase in asphaltene solubility—water-wet | |
| pH decreases, quartz has a +ve surface charge, can attract surfactants—oil-wet | |

surfaces may allow the adsorption of wettability-altering polar compounds (Ehrenberg *et al.*, 1995). The generation of thin clay rims around feldspar grains during diagenetic reactions may allow adsorption of wettability-altering compounds by hydrogen bonding to hydroxylated aluminium atoms on the clay surface (Stumm, 1992).

Typical wettability states of different reservoir rock-forming minerals (at typical pH values) reveals that many minerals are oil-wet (Table 4). Most sandstone reservoirs are dominated volumetrically by quartz, feldspar and illite. However, surface coats of Fe-oxides, minor smectite or chlorite clay or a few per cent carbonate minerals can lead to oil-wet grain surfaces.

### *Salinity and major cation chemistry*

Formation water chemistry has been shown experimentally to influence wettability. Decreasing Na/Ca ratios tend to lead to increasing oil-wetting (Jadhunandan & Morrow, 1995) although the mechanism for this effect was not reported. Brown & Neustadter (1980) suggested that polyvalent cations (e.g. $Ca^{2+}$) may act as a 'bridge' between the mineral surface and a carboxylic acid anion, thus altering the wettability of the mineral surface from water-wet to oil-wet.

### *Heterogeneous mineralogy/wettability and quartz cementation*

It seems clear that the reservoir mineralogy is a major factor in determining its wettability. At a given pH, different minerals have different surface charges and therefore different preferences for water versus charged (polar) organic molecules. Thus typically polymineralic reservoir rocks have heterogeneous wettability on the grain size scale (Kovscek *et al.*, 1993). Quartz has a negative surface charge and is thus water-wet at typical reservoir pH values although diffusion may be hindered in quartzose sandstones containing oil-wet minerals due to the overall increase in diffusion path-length.

## Pressure and temperature effects

Reservoir pressure and temperature have a range of indirect influences upon wettability (Table 5).

### *Carboxylic acid stability*

Increasing the temperature above 120°C (i.e. during burial) causes the destruction of carboxylic acids (Carothers & Kharaka, 1978; Surdam *et al.*, 1984; Surdam & MacGowan, 1987). This may lead to increased water-wetting since carboxylic acids probably enhance the oil-wettability of a reservoir.

### *Asphaltene solubility and GOR*

Asphaltene solubility is inversely proportional to the GOR of an oil (Milner *et al.*, 1977; Wilhelms & Larter, 1995). The gas solubility and thus GOR of an oil is dependent partially upon the pressure and temperature of the reservoir (England *et al.*, 1987; Killops & Killops, 1993). Exsolved asphaltenes on mineral surfaces alter wettability from water-wet to oil-wet. Therefore increasing pressure and/or decreasing temperature is likely to lead to enhanced oil-wetting. Increasing temperature also causes oil-cracking which generates methane (Tissot & Welte, 1984) resulting in elevated GORs and thus increased oil-wetting.

### *Capillary pressure and petroleum column height*

The effect of high capillary pressure (e.g. at the top of the oil column) on the stability of the water films surrounding mineral grains in an oil-filled water-wet reservoir can alter wettability (Kovscek *et al.*, 1993). High capillary pressures can cause a reduction in the thickness of the water films surrounding grains from ~100 Å to ~5 Å (Renard & Ortoleva, 1997). This may permit asphaltenes and polar compounds to adsorb on to the surface of the mineral thus altering wettability from water-wet to oil-wet.

### *Formation water chemistry*

The pH of the aqueous reservoir fluids is partially determined by the temperature of the reservoir (Hanor, 1994). pH decreases with increasing temperature causing more minerals to develop a negative surface charge (Somasundaran & Agar, 1967; Anderson, 1986; Tabatabal *et al.*, 1993). This may lead to enhanced oil-wetting at increasing temperature if charged polar organic compounds are present. Formation waters tend to have increasing Ca/Na ratios at increasing depths of burial so that elevated temperature (caused by increasing burial depth) may also lead to more oil-wet behaviour.

## DISCUSSION

An oil-wet reservoir sandstone is less likely to be cemented by post-oil emplacement quartz than a water-wet reservoir. Oil-wetting behaviour inhibits diffusion of silica away from the source and advective transport into the reservoir is controlled by the degree of reservoir oil-wetting. Dissolution of silica and precipitation of quartz cement are inhibited if mineral surfaces are coated with oil. Thus, extent of retardation of quartz cementation depends strongly upon the wettability.

The degree of oil-wetting should be higher, and thus the rate of quartz cementation should be lower in reservoirs containing heavily biodegraded, high API gravity or high GOR oil generated from a low maturity, low iron source rock containing type IIS or III kerogen than in reservoirs containing an unbiodegraded, low API gravity or low GOR oil generated from a high maturity, high iron source rock containing type I or II kerogen.

We can also state that low pH, low salinity, reducing and transition metal-poor formation water is more likely to produce a water-wet reservoir and thus be quartz cemented than a high pH, high salinity, oxidizing and transition metal-rich water.

The degree of water-wetting and rate of quartz cementation will be lower in sandstone reservoirs containing carbonate cement, smectite, chlorite, kaolinite booklets or iron oxides because these encourage oil-wetting.

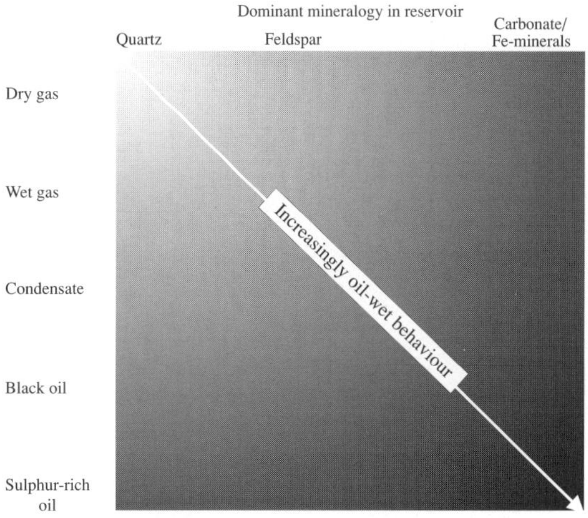


**Fig. 9.** Schematic diagram to reinforce the primary importance of reservoir mineralogy and oil type on wettability.

Reservoirs dominated by quartz, illite, finely crystalline kaolinite and unaltered feldspar are more likely to be water-wet and so the rate of quartz cementation will tend to be relatively high.

Wettability and quartz cementation are also affected by pressure and temperature, by source rock maturity, the phase behaviour of the oil (i.e. GOR and asphaltene exsolution) and reservoir pH. It is likely that quartz cementation will be more inhibited in deeply buried sandstone than shallow sandstones, although each case must be judged in turn in terms of source maturity, reservoir mineralogy and thus pH, and phase behaviour of the petroleum.

It is not possible to predict the exact rate of retardation of quartz cementation by oil emplacement. However, it is possible to establish some general rules to assess the relative rate of quartz cementation. In the first place we can assume that reservoir mineralogy and oil geochemistry are the most important and immutable factors that influence reservoir wettability (Fig. 9). Mineralogy will largely control water chemistry through water–rock interaction. Rock mineralogy is a primary and largely fixed part of the reservoir system and is therefore the first control that must be considered (Figs 9 and 10).

Sandstones can be classified according to their detrital mineralogy. Quartz arenites and unaltered arkosic sandstones will have a tendency to be water-wet and so have a limited capacity to retard quartz cementation in oil legs relative to water legs (Fig. 9). Lithic sandstones and altered arkoses contain potentially oil-wet minerals such as smectite, chlorite, iron oxides, etc., and thus may lead

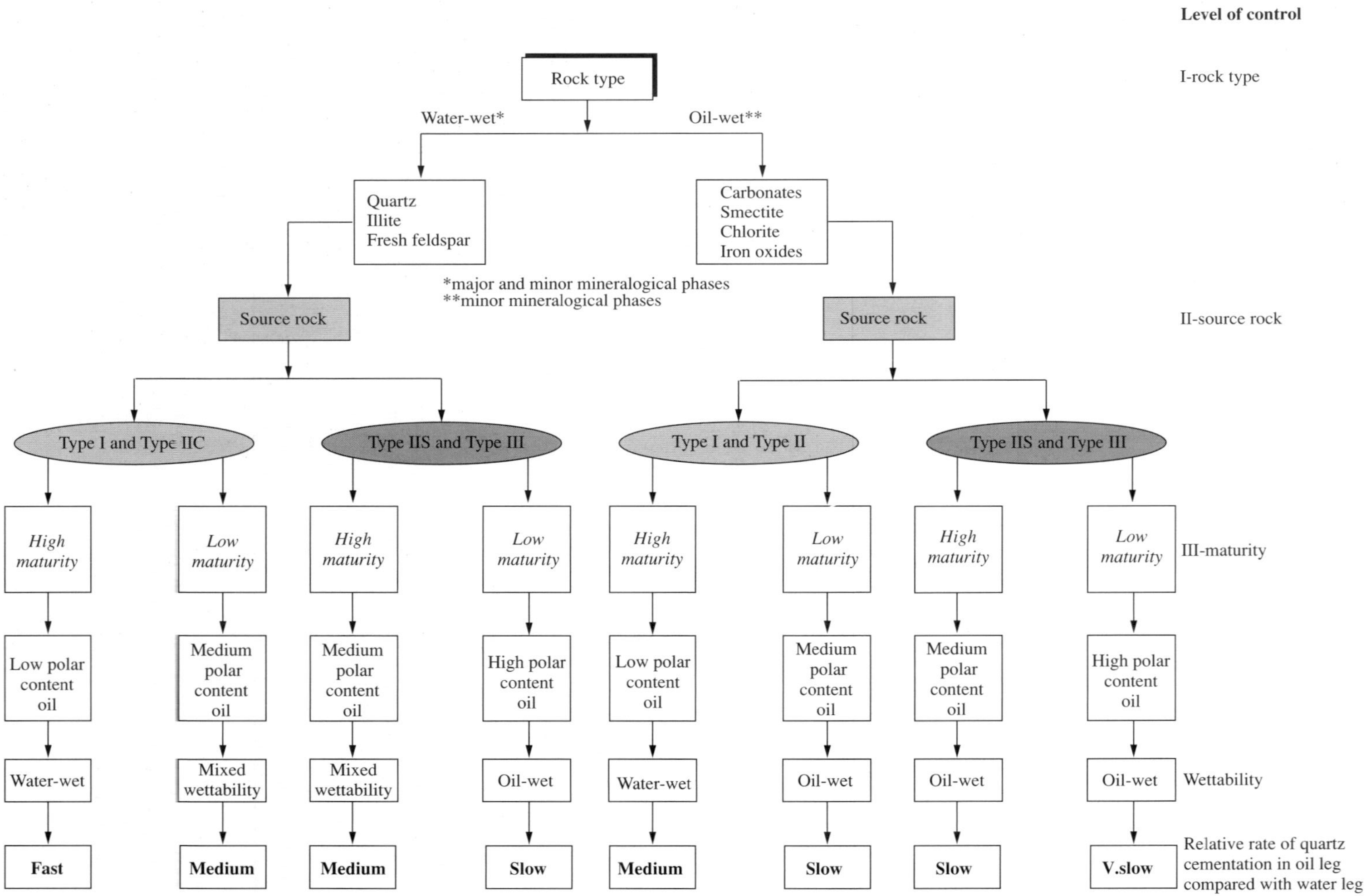


**Fig. 10.** Diagram to enable the estimation of the relative rate of quartz cementation in oil legs under different reservoir mineralogy, source types and source maturities. Note that post-emplacement processes may alter the primary scheme.

to radically reduced rates of quartz cementation in oil legs compared with water legs. Sandstones may also be classified according to their pre-emplacement diagenetic minerals. Sandstones with quartz, illite, feldspar and finely crystalline kaolinite cements will tend to be water-wet and the rate of quartz cementation in oil legs will only be a little slower than the rate in the attendant water legs. Sandstones with carbonate, smectite, chlorite, kaolinite booklet or iron oxide cements (e.g. red beds) will tend to be oil-wet and have a tendency to retard quartz cementation in the extreme in oil legs relative to water legs. This is due to the consequence of the reduced diffusion rates of silica in the oil leg compared with the water leg.

This initial mineralogical subdivision can be subdivided further by consideration of the source rock type and maturity. Type I and type IIS source rocks will produce oils with lower polar contents than type IIC or type III source rocks. The latter will lead to greater oil-wetting and relatively severe retardation of quartz cementation in oil legs. Reservoirs filled with low maturity oils will have larger quantities of polar compounds and will thus have relatively slower quartz cementation than those filled with high maturity oils (Fig. 10).

On top of this primary scheme there are other secondary considerations. Unusual sources of transition metals in formation waters (e.g. Hanor, 1994) elevate the likelihood of the reservoir being oil-wet and thus reduce quartz cementation rates in oil legs. Processes such as biodegradation or water washing may also enhance the oil-wetting state of the reservoir and lead to reduced quartz cementation rates in oil legs. Mixing of petroleum from different sources may lead to changes in asphaltene solubility and thus alter the wettability of the reservoir. Formation waters that are unusually acid or have high Ca/Na ratio can also lead to oil-wet reservoirs and thus reduced quartz cementation rates in oil fields. These secondary effects must be considered in order to derive a full understanding of the probable degree of retardation of quartz cementation in the reservoir.

## CONCLUSIONS

**1** The extent of the reduction of the rate of quartz cementation in oil legs relative to water legs depends upon the wettability of the reservoir.

**2** Wettability, and thus quartz cementation rates, reflect the polar content of the oil. Oils sourced from type I and II source rocks will tend to have low polar compound volumes, leading to water-wet reservoirs and relatively fast quartz cementation in oil fields. Oils sourced from type III and IIS source rocks will tend to have higher polar compound volumes, leading to oil-wet reservoirs and relatively slow quartz cementation in oil fields.

**3** Low maturity oils will tend to have higher polar volumes than high maturity oils and thus lead to more oil-wetting and slower rates of quartz cementation.

**4** Wettability and quartz cementation rates are also affected by reservoir mineralogy. Quartz, unaltered feldspars, illite and finely crystalline kaolinite are water-wet. Carbonates, smectite, chlorite, Fe-oxides and kaolinite booklets tend to be oil-wet. The degree of quartz cement inhibition will depend on the proportions of these minerals in the reservoir.

**5** Water-washing, biodegradation and mixing of petroleum from different sources will tend to cause an increase in the quantity of polar compounds and asphaltenes in the oil. These lead to an increased likelihood of oil-wetting behaviour and so will lead to inhibited quartz cementation rates.

**6** Ultra-low pH, high pressure, high formation water Ca/Na ratios and Eh values will all maximize the oil-wettability of the reservoir and thus minimize quartz cementation rates.

**7** The prediction of the relative rate of quartz cementation in an oilfield can be assessed by reference to all the above factors but cannot be exactly quantified.

## REFERENCES


Amott, E. (1959) Observations relating the wettability of porous rocks. *Petroleum Transactions of the AIME* **216**, 156.

Anderson, W.G. (1985) Wettability literature survey: Part 2. Wettability measurement. *SPE unsolicited paper 13933*.

Anderson, W.G. (1986) Wettability literature survey: Part 1. Rock/oil/brine interactions and the effects of core handling on wettability. *Journal of Petroleum Technology* **38**, 1125–1149.

Archer, J.S. & Wall, C.G. (1994). *Petroleum Engineering: Principles and Practice*. Graham & Trotman, London.

Atkins, P.W. (1990). *Physical Chemistry*, 4th edn. Oxford University Press, Oxford.

Barth, T. (1991) Organic acids and inorganic ions from petroleum reservoirs, Norwegian continental shelf: a multivariate statistical analysis and comparison with American reservoir formation waters. *Applied Geochemistry* **6**, 1–15.

Bennett, P.C., Melcer, M.E., Siegel, D.I. & Hassett, J.P. (1988) The dissolution of quartz in dilute aqueous solutions of organic acids at 25°C. *Geochimica et Cosmochimica Acta* **52**, 1521–1530.

Bennett, P.C. & Siegel, D.L. (1987) Increased solubility of quartz in water due to complexing by organic compounds. *Nature* **326**, 684–686.

Brown, R.J. & Fatt, I. (1956) Measurement of fractional wettability on oil field rocks by the nuclear magnetic relaxation method. *Petroleum Transactions of the AIME* **207**, 262.

Brown, C.E. & Neustadter, E.L. (1980) The wettability of oil/water/silica systems with reference to oil recovery. *Journal of Canadian Petroleum Technology* **19**, 100–110.

CAROTHERS, W.W. & KHARAKA, Y.K. (1978) Aliphatic acid anions in oilfield waters – implications for the origin of natural gas. *Bulletin of the American Association of Petroleum Geologists* **62**, 2431–2441.

CHILINGAR, G.V. & YEN, T.F. (1983) Some notes on wettability and relative permeabilities of carbonate reservoir rocks, II. *Energy Sources* **7**, 67–75.

COLLINS, A.G. (1975). *Geochemistry of Oilfield Waters*. Elsevier, Amsterdam, 496pp.

CONNAN, J. (1984) Biodegradation of crude oils in reservoirs. In: *Advances in Petroleum Geochemistry, Vol. 1.* (eds BROOKS, J. & WELTE, D.H.) pp. 300–330. Academic Press, London.

CROCKER, M.E. & MARCHIN, L.M. (1988) Wettability and adsorption characteristics of crude-oil asphaltenes and polar fractions. *Journal of Petroleum Technology* **40**, 470–474.

CUIEC, L. (1987) Wettability and oil reservoirs. In: *North Sea Oil and Gas Reservoirs.* (ed. KLEPPE, J.) pp. 193–207. Graham & Trotman, London.

DONALDSON, E.C., THOMAS, R.D. & LORENZ, P.B. (1969) Wettability determination and its effects on recovery efficiency. *Society of Petroleum Engineers Journal* **9**, 13.

DUBEY, S.T. & WAXMAN, M.H. (1989) Asphaltene adsorption and desorption from mineral surfaces. *SPE International Symposium on Oilfield Chemistry*. Houston, Texas, pp. 51–62.

EHRENBERG, S.N., SKJEVRAK, I. & GILJE, A.E. (1995) Asphaltene-rich residues in sandstone reservoirs of Haltenbank province, mid-Norwegian continental shelf. *Marine and Petroleum Geology* **12**, 53–69.

ENGLAND, W.A., MACKENZIE, A.S., MANN, D.M. & QUIGLEY, T.M. (1987) The movement and entrapment of petroleum fluids in the subsurface. *Journal of the Geological Society of London* **144**, 327–347.

FILBY, R.H. (1994) Origin and nature of trace element species in crude oils, bitumens and kerogens: implications for correlation and other geochemical studies. In: *Geofluids: Origin, Migration and Evolution of Fluids in Sedimentary Basins.* (ed. PARNELL, J.) pp. 203–219. Special Publications of the Geological Society of London 78.

FREHSE, W. (1973) Method for the evaluation of the wettability of reservoir rock. *Zeitschrift für Angewandte Geologie*. 19.

GATENBY, W.A. & MARSDEN, S.S. (1957) Some wettability characteristics of synthetic porous media. *Producers Monthly* **22**, 5.

GIORDANO, T.H. & KHARAKA, Y.K. (1994) Organic ligand distribution and speciation in sedimentary basin brines, diagenetic fluids and related ore solutions. In: *Geofluids: Origin, Migration and Evolution of Fluids in Sedimentary Basins.* (ed. PARNELL, J.) pp. 175–202. Special Publications of the Geological Society of London 78.

GONZÁLEZ, G. & MIDDEA, A. (1987) Asphaltenes adsorption by quartz and feldspar. *Journal of Dispersion Science Technology* **8**, 525–548.

GONZÁLEZ, G. & MOREIRA, B.C. (1991) The wettability of mineral surfaces containing adsorbed asphaltenes. *Colloids and Surfaces* **58**, 293–302.

GRATTONI, C.A., CHIOTIS, E.D. & DAWE, R.A. (1995) Determination of relative wettability of porous sandstones by imbibition studies. *Journal of Chemical and Technical Biotechnology* **64**, 17–24.

HANOR, J.S. (1994) Origin of saline fluids in sedimentary basins. In: *Geofluids: Origin, Migration and Evolution of Fluids in Sedimentary Basins.* (ed. PARNELL, J.) pp. 151–174. Special Publications of the Geological Society of London 78.

HOLBROOK, O.C. & BERNARD, G.G. (1958) Determination of wettability by dye adsorption. *Petroleum Transactions of the AIME* **213**, 261.

HOWARD, J.J. (1994) Wettability and fluid saturations determined from NMR $T_1$ distributions. *Magnetic Resonance Imaging* **12**, 197–200.

HUNTER, R.J. (1987). *Foundations of Colloid Science, Vol. 1.* Oxford Science Publications, Oxford.

JADHUNANDAN, P.P. & MORROW, N.R. (1995) Effect of wettability on waterflood recovery for crude oil/brine/rock systems. *Society of Petroleum Engineers: Research Engineering* **SPE22597**, 40–46.

JOHANSEN, R.T. & DUNNING, H.N. (1953) Relative wetting tendencies of crude oils by the capillarimetric method. *Producers Monthly* **17**, 20.

KELLER, G.V. (1953) Effect of wettability on the electrical resistivity of sand. *Oil and Gas Journal* **51**, 62–65.

KIANIPAY, S.A. & DONALDSON, E.C. (1986) Mechanism of oil displacement by microorganisms. Paper SPE 15601. *Society of Petroleum Engineers 61st International Technical Conference and Exhibition, New Orleans, Louisiana.*

KILLOPS, S.D. & KILLOPS, V.J. (1993). *An Introduction to Organic Geochemistry*. Longman, Harlow.

KOVSCEK, A.R., WONG, H. & RADKE, C.J. (1993) A pore-level scenario for the development of mixed wettability in oil reservoirs. *Journal of the American Institute of Chemical Engineers* **39**, 1072–1085.

LARTER, S.R., BJØRLYKKE, K.O., KARLSEN, D.A. *et al.* (1990) Determination of petroleum accumulation histories: examples from the Ula field, Central Graben, Norwegian North Sea. In: *North Sea Oil and Gas Reservoirs, Vol. 2.* (ed. BULLER, A.) pp. 319–330, Graham & Trotman, London.

LEVINSON, P., VALIGNAT, M.P., FRAYSSE, N. & CAZABAT, A.M. (1994) Adsorption isotherms of alkanes on silica: the role of preadsorbed layers. *Colloids and Surfaces A: Physicochemical and Engineering Aspects* **85**, 127–134.

MILES, J.A. (1994). *Illustrated Glossary of Petroleum Geochemistry*. Oxford University Press, Oxford.

MILNER, C.W.D., ROGERS, M.A. & EVANS, C.R. (1977) Petroleum transformations in reservoirs. *Journal of Geochemistry and Exploration* **7**, 101–153.

NORTH, F.K. (1985). *Petroleum Geology*. Allen & Unwin, Boston.

PALMER, S.E. (1993) Effect of biodegradation and water washing on crude oil composition. In: *Organic Geochemistry.* (eds ENGEL, M.H. & MACKO, S.A.) pp. 511–533, Plenum Press, New York.

DE PEDROZA, T.M., CALDERÓN, G.R., RICO, A.A. & REY, F.V. (1994) Influence of asphaltenes on some properties of Venezuelan reservoir rocks. *Vision Tecnologica* **2**, 59–65.

POIRIER, J.E. & CASES, J.M. (1991) Anionic surfactant adsorption onto silicate minerals: The role of the cations. *Colloids and Surfaces* **55**, 333–344.

POWERS, M.C. (1967) Fluid-release mechanisms in compacting marine mudrocks and their importance in oil exploration. *American Association of Petroleum Geology Bulletin* **51**, 1240–1254.

RAZA, S.H., TREIBER, L.E. & ARCHER, D.L. (1968) Wettability of reservoir rocks and its evaluation. *Producers Monthly* **32**, 2.

RENARD, F. & ORTOLEVA, P. (1997) Water films at grain–grain contacts: Debeye–Hückel, osmotic model of stress, salinity and mineralogy dependence. *Geochimica et Cosmochimica Acta* **61**, 1963–1970.

REZAEE, M.R. & LEMON, N. (1996) Controls on pore geometry in the Tirrawarra Sandstone Reservoir, Cooper Basin, Australia. *Petroleum Exploration Society of Australia Journal* **24**, 125–146.

RIXEY, W.G. & FUERSTENAU, D.W. (1994) The Young equation and the effect of surfactants on the wettability of minerals. *Colloids and Surfaces. A: Physicochemical and Engineering Aspects* **88**, 75–89.

ROBIN, M., ROSENBURG, E. & FASSI-FIHRI, O. (1995) Wettability studies at the pore level: a new approach by use of Cyro-SEM. *Society of Petroleum Engineers Formation and Evaluation* **10**, 11–19.

ROUXHET, P.G. & ROBIN, P.L. (1978) Infrared study of the evolution of kerogens of different origins during catagenesis and pyrolysis. *Fuel* **57**, 533–540.

SAUNDERS, J.A. & SWANN, C.T. (1990) Trace-metal content of Mississippi oil field brines. *Journal of Geochemistry and Exploration* **37**, 171–183.

SAYYOUH, M.H. & AL-BLEHED, M.S. (1995) Effect of microorganisms on rock wettability. *Journal of Adhesion Science and Technology* **9**, 425–431.

SCHLANGEN, L.J.M., KOOPAL, L.K., COHEN STUART, M.A. & LYKLEMA, J. (1995) Thin hydrocarbon and water films on bare and methylated silica: vapour adsorption, wettability, adhesion and surface forces. *Langmuir* **11**, 1701–1710.

SHOCK, E.L. & KORETSKY, C.M. (1995) Metal–organic complexes in geochemical processes: estimation of standard partial molal thermodynamic properties of aqueous complexes between metal cations and monovalent organic acid ligands at high pressures and temperatures. *Geochimica et Cosmochimica Acta* **59**, 1497–1532.

SLOBOD, R.L. & BLUM, R.A. (1952) Method for determining the wettability of reservoir rocks. *Petroleum Transactions of the AIME* **195**, 1.

SOMASUNDARAN, P. & AGAR, G.E. (1967) The zero point of charge of calcite. *Journal of Colloid Interface Science* **24**, 433–440.

SØNDENÅ, E., BRATTELI, F., NORMANN, H.P. & KOLLTVEIT, K. (1991) The effect of reservoir conditions and wettability on electrical resistivity. Paper SPE 22991. *Society of Petroleum Engineers Asia-Pacific Conference, Perth, Western Australia*, pp. 409–422.

STUMM, W. (1992). *Chemistry of the Solid–Water Interface*. Wiley, New York.

STUMM, W. & MORGAN, J.J. (1970). *Aquatic Chemistry*. Wiley, New York.

SURDAM, R.C., BOESE, S.W. & CROSSEY, L.J. (1984) The chemistry of secondary porosity. In: *Clastic Diagenesis*. (eds MCDONALD, D.A. & SURDAM, R.C.) AAPG Memoir 37, pp. 127–151. American Association of Petroleum Geologists.

SURDAM, R.C., JIAO, Z.S. & MACGOWAN, D.B. (1993) Redox reactions involving hydrocarbons and mineral oxidants: a mechanism for significant porosity enhancement in sandstones. *Bulletin of the American Association of Petroleum Geology* **77**, 1509–1518.

SURDAM, R.C. & MACGOWAN, D.B. (1987) Oilfield waters and sandstone diagenesis. *Applied Geochemistry* **2**, 613–619.

SWANSON, B.F. (1980) Rationalizing the influence of crude wetting on reservoir fluid flow with electrical resistivity behaviour. *Journal of Petroleum Technology* **32**, 1459.

TABATABAL, A., GONZALEZ, M.V., HARWELL, J.H. & SCAMEHORN, J.F. (1993) Reducing surfactant adsorption in carbonate reservoirs. *Society of Petroleum Engineers Reservoir Engineering* **8**, 117–122.

TISSOT, B.P. & WELTE, D.H. (1984). *Petroleum Formation and Occurrence*, 2nd edn. Springer-Verlag, Berlin.

TREIBER, L.E., ARCHER, D.L. & OWENS, W.W. (1972) Laboratory evaluation of the wettability of 50 oil producing reservoirs. *Society of Petroleum Engineers Journal* **12**, 531–540.

WANG, F.H.L. & GUIDRY, L.J. (1994) Effect of oxidation–reduction condition on wettability alteration. *Society of Petroleum Engineers Formation and Evaluation* **9**, 140–148.

WILHELMS, A. & LARTER, S.R. (1994) Origin of tar mats in petroleum. Part II: formation mechanisms for tar mats. *Marine Petroleum Geology* **11**, 442–456.

WILHELMS, A. & LARTER, S.R. (1995) Overview of the geochemistry of some tar mats from the North Sea and USA: implications for tar mat origin. In: *The Geochemistry of Reservoirs*. (eds CUBITT, J.M. & ENGLAND, W.A.) pp. 87–101. Special Publications of the Geological Society of London 86.

WOLCOTT, J.M., GROVES, F.R., JR, TRUJILLO, D.E. & LEE, H.G. (1991) Investigation of crude-oil/mineral interactions: factors influencing wettability alteration. Paper SPE 21042. *Society of Petroleum Engineers International Symposium on Oilfield Chemistry, Anaheim, California*, pp. 411–420.

WORDEN, R.H., OXTOBY, N.H. & SMALLEY, P.C. (in press) Can oil emplacement prevent quartz cementation in sandstones?. *Petroleum Geoscience* **4**, 129–138.

*Spec. Publs int. Ass. Sediment.* (2000) **29**, 119–127

# Experimental and field constraints on the role of silica–organic complexation and silica–microbial interactions during sediment diagenesis

J. B. FEIN

*University of Notre Dame, Civil Engineering and Geological Sciences, Notre Dame, IN 46556, USA*

## ABSTRACT

This paper reviews the results of experiments and field studies that examine the role of silica–organic interactions that may influence Si transport during sediment diagenesis. Three types of influences are reviewed: (i) aqueous silica–organic complexation; (ii) mineral surface silica–organic complexation; and (iii) microbial influences. The available evidence suggests that aqueous silica–organic complexation does not exert a significant effect on Si mobilities in the deep subsurface. Mineral surface silica–organic complexation can lead to enhanced quartz dissolution rates, but not at the temperatures, or for the organic acid anions, currently known to exist in sedimentary basin formation waters. The role of microorganisms on Si transport during diagenesis is just beginning to be investigated, although it is possible that microbial influences such as enhanced quartz dissolution due to microbial exudates, or bacterial-induced silicate mineral precipitation, can significantly affect Si mobilities. Although microorganisms have been found at extreme depths within sedimentary basins, more research must be conducted to determine the importance of microbial processes during deep sedimentary diagenesis.

## INTRODUCTION

The effect of aqueous organic molecules on the rate and extent of quartz dissolution within sedimentary basin fluid–rock systems has been the subject of considerable research. This interest stems from the importance of the Si budget in numerous geochemical processes that occur during diagenesis, such as porosity redistribution and the creation and destruction of secondary porosity within quartz- and/or aluminosilicate-rich sandstones. Without a quantitative understanding of the chemical controls on Si mobilities in diagenetic systems, our ability to accurately model the development of these systems is severely limited. The purpose of this article is to review the extent of understanding of silica–organic interactions, and to draw conclusions concerning the likelihood for such interactions to significantly influence Si mobilities during sediment diagenesis. The convention employed in this review is to refer to 'silica' complexation, but 'Si' transport. If aqueous and/or surface complexation occurs, it seems most likely that Si would be in the form of silica (that is, with associated oxygen atoms). When discussing transport, however, the chemical neighbourhood of the atom is irrelevant, and the generic 'Si' is employed.

Carboxylic acids are the dominant type of dissolved organic molecule found in sedimentary basin formation waters (Carothers & Kharaka, 1978; MacGowan & Surdam, 1988). The primary control on the importance of the carboxylic acids and their anions is the thermal stability of the molecule. All carboxylic acids are metastable under diagenetic conditions, and in general larger carboxylic acids exhibit faster thermal decarboxylation rates. For example, acetate, oxalate, and EDTA (ethylenediamine tetraacetic acid) all display first-order decarboxylation rate kinetics, with corresponding half-lives at 100°C on the order of millions of years, thousands of years, and hours, respectively (e.g. Motekaitis *et al.*, 1982; Kharaka *et al.*, 1983; Palmer & Drummond, 1986; Fein *et al.*, 1994). Although aqueous metal–organic anion complexation can increase the thermal stability of the organic anion significantly (e.g. Motekaitis *et al.*, 1982; Fein *et al.*, 1994; Fein *et al.*, 1995), only relatively short-chained carboxylic acid anions have been observed in formation waters to date. The maximum concentrations reported for dissolved carboxylic acids, compiled by MacGowan & Surdam (1988), are shown in Table 1. Due

**Table 1.** Maximum observed concentrations for carboxylic acid anions in oilfield waters

| | ppm | log molality |
|---|---|---|
| Monofunctional anions | | |
| Formate | 63 | −2.85 |
| Acetate | 10 000 | −0.77 |
| Propionate | 4400 | −1.22 |
| Butanate | 44 | −3.30 |
| Pentanate | 32 | −3.50 |
| Difunctional anions | | |
| Oxalate | 494 | −2.25 |
| Malonate | 2540 | −1.60 |

Data in this table were compiled by MacGowan & Surdam (1988).

to the scarcity of appropriate measurements of the aqueous organic geochemistry of formation waters, it is unclear whether the values shown in Table 1 are representative or anomalous (Kharaka *et al.*, 1993). However, the anions shown in Table 1 are clearly the most likely ones to potentially affect Si mobilities.

If either aqueous or mineral surface silica–carboxylate complexation occurs, then it would be possible that such complexation could augment Si mobility. Aqueous silica–carboxylate complexation would increase the solubility of quartz and other silicate minerals, and hence could directly increase the porosity and permeability of subsurface rocks. Mineral surface silica–carboxylate complexation, or the adsorption of carboxylate anions on to silica sites on silicate minerals, could significantly increase the *rate* of mineral dissolution by polarizing the crystallographic bonds holding a surface Si atom into a mineral structure (e.g. Furrer & Stumm, 1986; Stumm & Morgan, 1996). For systems in which the dissolution rate of quartz or silicate minerals is the rate-limiting step for Si mobility (that is, for systems in which the fluid flow is rapid enough for the aqueous phase to remain undersaturated with respect to quartz), an increase in the dissolution rate would directly cause an increase in Si mobility. Therefore, this review will examine the experimental and field constraints on the thermodynamic stabilities of both aqueous and surface silica–organic complexes in order to determine the role played by organic molecules in the distribution and transport of Si in diagenetic systems. Although silica precipitation also exerts a significant influence on Si mobility (e.g. Rimstidt & Barnes, 1980), there is virtually no work conducted to investigate the role of organic molecules on silica precipitation, and precipitation mechanisms and kinetics will not be reviewed here.

## AQUEOUS SILICA–ORGANIC COMPLEXATION

Aqueous silica–organic complexation has been investigated both in the field and with laboratory experiments using spectroscopy, solubility measurements, and potentiometry. The field studies (Bennett & Siegel, 1987; Bennett *et al.*, 1991) provide compelling evidence that enhanced Si mobilities are possible in organic-rich fluid–rock systems. However, although it is postulated that silica–organic aqueous complexation causes the enhanced Si mobilities that are observed, the evidence is circumstantial. Bennett & Siegel (1987) document a strong correlation between dissolved Si and dissolved organic carbon (DOC) concentrations in groundwater down gradient from a crude petroleum pipeline rupture site. The Si concentrations were significantly higher than the inorganic solubility of quartz, and aquifer quartz grains from areas affected by the enhanced DOC concentrations exhibited significant etching, an extremely unusual occurrence in near-neutral pH waters. Bennett & Siegel (1987) propose that the best explanation for these observations is that the biodegradation products of the crude petroleum (a variety of aliphatic and aromatic acids and hydroxy- and keto-acids not described in detail) enhance the extent of quartz dissolution through aqueous silica–organic complexation. Although Bennett & Siegel (1987) also observe a correlation between the zone of highest DOC concentrations and the spatial distribution of chemical etching of aluminosilicate minerals, dissolved Al concentrations were relatively low and constant throughout the study area. The only explanation for this is the precipitation of Al-bearing minerals subsequent to aluminosilicate mineral dissolution, a fact suggesting a lack of Al-organic aqueous complexation. Because of the marked decrease in solubility of Al-bearing minerals as a function of pH from acidic to neutral conditions, and because of the extent of hydrolysis that occurs involving dissolved Al, Al–organic aqueous complexes exert a minimal influence on Al-bearing mineral solubilities in near-neutral pH fluids (unless Al–OH–organic complexation occurs; e.g. Fein & Hestrin, 1994). Conversely, the solubility of quartz is independent of pH below the $pK_a$ of silicic acid (9.8; Iler, 1979), and aqueous Si does not hydrolyse to the extent of aqueous Al. Therefore, because silica–hydroxyl aqueous complexation would not compete with silica–organic complexation, silica–organic complexes, if they exist, would not exhibit the same pH dependency as Al–organic complexes, and could remain important in near-neutral pH fluids.

In a similar study, Bennett *et al.* (1991) examined an uncontaminated peat bog system for evidence of naturally occurring silica–organic complexation. As was the case for the contaminated site described by Bennett & Siegel (1987), the bog displayed Si groundwater concentrations in excess of quartz saturation, high DOC concentrations, relatively low Si concentrations in the bulk solid phase, and marked etching of quartz in near-neutral pH waters. Although both the studies by Bennett & Siegel (1987) and by Bennett *et al.* (1991) examine near-surface phenomena, both studies offer evidence that silica–organic complexation accounts for the enhanced Si mobilities documented at both sites. This evidence, too, is circumstantial, and the exact mechanisms, or silica–organic species, are not identified definitively. However, if such complexation affects near-surface Si budgets, then it is possible that aqueous silica–organic complexation is significant during deeper sediment diagenesis as well.

Laboratory experiments offer a means for isolating specific silica–organic interactions, and results from these experiments constrain which interactions are important in realistic systems, and the conditions under which they become important. Evidence for silica–organic aqueous complexation comes in the form of UV-difference spectroscopy, NMR data, and quartz and amorphous silica solubility measurements (Sjöberg *et al.*, 1985; Bennett *et al.*, 1988; Bennett, 1991; Öhman *et al.*, 1991). The spectroscopy results of Bennett *et al.* (1988) indicate a rapidly forming, pH- and concentration-dependent redshifted UV-difference peak for solutions containing silica and either citrate, oxalate, or pyruvate. The peak is apparent for solutions at pH 7 only, and not at the lower pH solutions studied. The experiments were conducted at relatively high Si and organic concentrations (both above 0.3 mmolal), and they do not quantify the thermodynamic stability of the proposed complexes. The solubility measurements by Bennett (1991) are partially consistent with the spectroscopic study. Bennett (1991) measured the solubility of quartz and of amorphous silica (separately) in citrate- and oxalate-bearing electrolytes from 25 to 70°C. At all temperatures at a pH of 7.0, the presence of 0.020 molal citrate increased quartz solubility by approximately 0.2 log molality units, and at 25°C, the same concentration of citrate increased the solubility of amorphous silica by 0.04 log molality units (a smaller effect presumably due to the higher Si/citrate ratios present in the amorphous silica systems). Both of the solubility experiments suggest the presence of a silica–citrate aqueous complex. However, no significant solubility enhancements were observed for analogous acetate or oxalate systems.

The lack of silica–oxalate aqueous complexation found by Bennett (1991) is consistent with spectroscopic and $^{29}$Si NMR experiments conducted by Marley *et al.* (1989) and by Tait *et al.* (1993). Marley *et al.* (1989) claim their spectroscopic results provide evidence for an aqueous silica–oxalate complex, but Tait *et al.* (1993) correct an interpretation error in the Marley *et al.* (1989) study, and both studies are consistent in showing that silica–oxalate complexation is negligible between pH values of 2 and 13, with large excess oxalate concentrations, and to temperatures of 125°C. Tait *et al.* (1993) draw the same conclusions for silica–citrate complexation as well.

The only studies to provide thermodynamic properties for silica–organic aqueous complexes are those of Sjöberg *et al.* (1985) and Öhman *et al.* (1991). Sjöberg *et al.* (1985), using potentiometry and $^{29}$Si-NMR data, demonstrate that Si (not silica) and tropolonate (an organic acid abundant in higher plant life) form a stable aqueous complex, and that the $SiL_3^+$ species (where L represents the tropolonate ligand) can dominate aqueous Si budgets at tropolonate concentrations above 0.1 M at pH 4 (although the predominance field is pH-dependent). The data not only constrain the stoichiometry of the complex, but they define its thermodynamic stability by quantifying its formation constant.

Öhman *et al.* (1991) use potentiometry and solubility measurements to examine aqueous Si–citrate, –oxalate, –sulphate, and –pyrocatechol complexation. Although they find evidence for each type of these complexes, the thermodynamic stability of the Si–pyrocatechol complex is the only one high enough to enable accurate determination of its stoichiometry and formation constant. The $SiL_3^{2-}$ pyrocatechol complex is the dominant Si species for relatively high pH silica systems containing more than 0.01 M total pyrocatechol. Although the presence of citrate and oxalate cause a measurable increase in the solubility of amorphous silica, the effect is small, and only significant at concentrations of citrate and oxalate much higher than those found in sedimentary basin formation waters thus far. For example, for both citrate and oxalate, ligand concentrations as high as 0.1 M are required to yield only a 10% increase in the solubility of amorphous silica at neutral pH.

These results are consistent with those of Fein & Hestrin (1994), who measured quartz solubility and amorphous silica solubilities in oxalate-bearing solutions as a function of pH (from 1.9 to 7.4) and oxalate concentration (from 0 to 0.1 molal) at 80°C (Fig. 1). In these experiments, the solubility of both silica phases was unaffected by the presence of oxalate, to oxalate concentrations of at least 0.1 molal, indicating no significant silica–oxalate aqueous complexation. This was also the conclusion arrived at by Franklin *et al.* (1994), who measured quartz solubility at 100°C, 347 bars, in a 0.07-molal acetate –0.005-molal oxalate solution (with a pH of approximately 4.4), in a 0.07-molal acetate-only solution

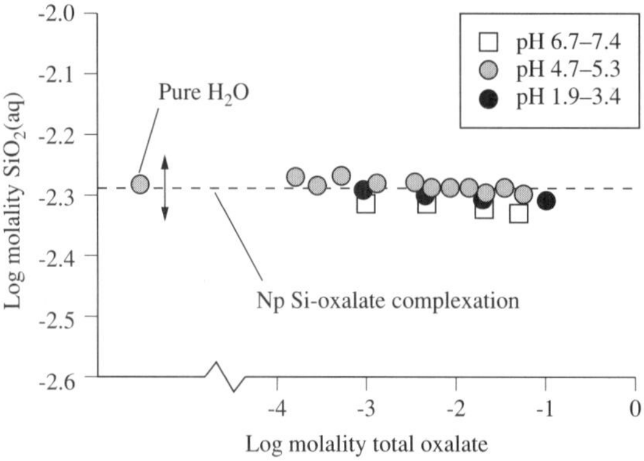


**Fig. 1.** Amorphous silica solubility measurements from Fein & Hestrin (1994). The double-headed arrow represents experimental uncertainties. The dashed line represents theoretical amorphous silica solubilities in the experimental fluids, calculated assuming that aqueous silica–oxalate complexation does not occur.

(with a pH of approximately 4.8), and in distilled water. Franklin *et al.* (1994) observed no significant increase in quartz solubility in either of the organic-bearing solutions relative to that in the distilled water. Poulson *et al* (1997) conducted titration experiments involving aqueous $Si(OH)_4$ and potassium oxalate, and compared the results with blank titrations. They also conclude that aqueous silica–oxalate complexation is negligible.

Although field studies suggest that enhanced quartz dissolution and Si mobility (relative to the dissolution and mobility controlled by purely inorganic processes) can occur in near-surface organic-rich environments, the experimental measurements of silica–organic aqueous complexation suggest that aqueous complexes are not likely to cause similar effects in deep sedimentary basin systems. The only ligands that form aqueous silica–organic complexes with stabilities high enough to significantly affect mineral solubilities and Si mobilities are catechol-like molecules. The ability of these molecules to complex with Si likely arises because the adjacent hydroxyl groups on the benzene ring are held in planar relation in a position such that the oxygen–oxygen distance is exactly that required between oxygen atoms in octahedral coordination with a Si atom (Iler, 1979). Although these ligands can be found in natural and/or contaminated waters, and although structures similar to them may be found on humic and fulvic acids, their concentrations in any natural fluid–rock system, if they are present at all, is orders of magnitude lower than would be required for them to significantly affect Si mobilities. A similar argument can be applied to citrate, which appears to form a weak aqueous complex with Si. It has not been identified in any sedimentary basin formation water, and therefore should not be important in the Si budget. Mono- and di-carboxylic acids such as acetate and oxalate, respectively, are found in sedimentary basin systems, but either they have no significant effect on Si mobilities (as is the case with acetate), or they are present at much lower concentrations than would be required even to observe a minor effect on the extent of quartz dissolution. Therefore, it is reasonable to conclude that aqueous complexation between Si and dissolved organic acid anions is not a significant process during deep sedimentary diagenesis.

## MINERAL SURFACE SILICA–ORGANIC COMPLEXATION

Poulson *et al.* (1997) observed negligible adsorption of oxalate on to quartz (at pH 7, from 5 to 21 μM oxalate). However, oxalate is the only organic acid anion whose adsorption behaviour on to quartz has been examined. For oxalate, and for other aqueous organic molecules, information concerning mineral surface silica–organic complexes comes indirectly, by measuring their influence on mineral dissolution rates. Although there are numerous studies of the effects of organic molecules on silicate mineral dissolution rates, this review will be limited to those few that only involve quartz.

Bennett *et al.* (1988) measured 25°C quartz dissolution rates in pH-buffered solutions containing either no other dissolved constituent (pure $H_2O$), NaCl, citrate, oxalate, salicylate, or acetate. Of these electrolytes, the NaCl solutions exhibited similar dissolution rates to those observed for the pure $H_2O$ solutions. The presence of acetate slowed quartz dissolution rates slightly. Salicylate, oxalate, and citrate each increased the dissolution rate of quartz in neutral pH solutions, with the magnitude of the effect strongly correlated to the concentration of organic acid anion in the system. For citrate, even 1 mmolal of the anion significantly increased the dissolution rate; for oxalate, the 1 mmolal solution did not cause a significant effect, but enhanced dissolution rates were observed for the 2 mmolal and higher concentrations. This result is not inconsistent with the findings of Poulson *et al.* (1997) who observed no adsorption of oxalate on to quartz under similar pH conditions, but at much lower oxalate concentrations (up to 21 μM). It is possible that a silica–oxalate mineral surface complex is only stable under the higher oxalate concentrations examined by Bennett *et al.* (1988).

In a follow-up study to Bennett *et al.* (1988), Bennett (1991) measured quartz dissolution kinetics in a variety

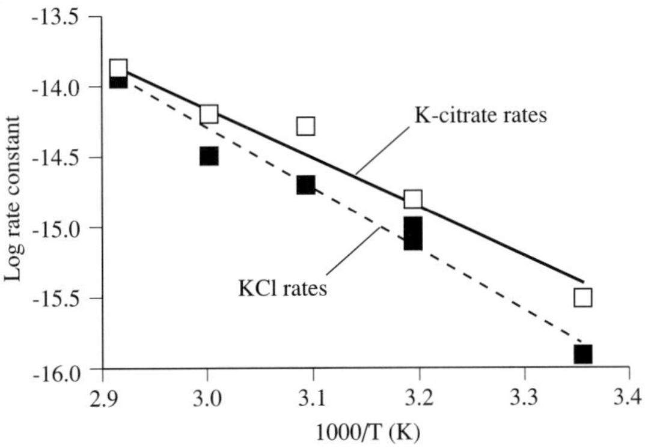


**Fig. 2.** Data from Bennett (1991), showing the temperature dependence of the quartz dissolution rate constant for 0.02 m K-citrate solutions, and for 0.05 m KCl solutions, both buffered to pH 7.0, over the temperature range of 25–70°C. Note the decreasing influence of citrate with increasing temperature.

of inorganic salt solutions, as well as organic acid anion solutions, as a function of temperature, pH, and electrolyte concentration. Consistent with the results of Bennett *et al.* (1988), citrate and oxalate were observed to enhance the dissolution rate of quartz, with the magnitude of the effects increasing from pH 3–7, and decreasing with increasing temperature from 25 to 70°C (Fig. 2). The temperature effect cannot be attributed to thermal degradation of citrate, because Bennett (1991) found no significant decrease in citrate concentration during the experimental time interval. The temperature effect is especially pertinent to an examination of the role of organic acid anions during sediment diagenesis. For example, at 25°C, a 20 mmolal citrate solution enhances the quartz dissolution rate by approximately 0.6 log rate units (moles $cm^{-2}$ $s^{-1}$), but only approximately 0.1 log rate units at 70°C. A similar decrease in the magnitude of the effect of oxalate was observed over the same temperature range. These data suggest that the effects of citrate and oxalate on the dissolution rate of quartz become negligible at temperatures much above 70°C.

The effects of simple carboxylic acid anions on quartz dissolution rates have also been examined by Manning *et al.* (1991), by Franklin *et al.* (1994), and by Knauss & Copenhaver (1995). Manning *et al.* (1991) measured quartz dissolution rates at 150°C at pH 6 in potassium acetate solutions. The results of those experiments indicate that the quartz dissolution rate decreases slightly with increasing acetate concentration, a result consistent with the findings of Bennett *et al.* (1988). The cause of the decrease is not known, and is inconsistent with the findings of Franklin *et al.* (1994), who observed a factor of four *increase* in the dissolution rate of quartz in acetate-bearing solutions relative to that measured in distilled water at 100°C, 345 bars. The experimental solutions of Franklin *et al.* (1994) were more acidic than those of Bennett *et al.* (1988) (4.8 versus. 7.0), but it is unclear whether the acidity can account for the inconsistency between datasets. Franklin *et al.* (1994) also observed that their acetate–oxalate experiments displayed slower quartz dissolution than the acetate-only experiments, a result also in apparent contrast to those reported by Bennett *et al.* (1988) and Bennett (1991). Poulson *et al.* (1997) re-evaluated the data of Bennett *et al.* (1988) and Bennett (1991) to illustrate that the observed quartz dissolution rate enhancement that Bennett *et al.* (1988) and Bennett (1991) claim is caused by the presence of oxalate may, in fact, be attributed to the concentration of $Na^+$ in solution. Until these issues are resolved, it remains uncertain whether acetate or oxalate exert any significant effects on quartz dissolution rates.

Conversely, the results of Knauss & Copenhaver (1995) unambiguously indicate that the di-carboxylic acid anion malonate can significantly affect quartz dissolution rates. These experiments were conducted using pH-buffered solutions at 70°C as a function of pH and malonate concentration. In the mildly acidic–near neutral pH range, 0.1 molal malonate solutions enhance quartz dissolution kinetics by approximately 0.5 log rate units. Because of the high concentration of malonate required to exert this effect, and because of the temperature effects observed by Bennett (1991), Knauss & Copenhaver (1995) conclude that while malonate may influence quartz dissolution rates in near-surface environments, it is unlikely to play a significant role in petroleum reservoir systems.

Experimental studies of the effects of dissolved organic acid anions on the dissolution rate of quartz indicate that it is possible that these molecules significantly affect Si mobilities during diagenesis. The experimental data are ambiguous concerning the effects of acetate on quartz dissolution rates, and the effects attributed to di- and tri-functional molecules may be caused by aqueous Na instead. The role that the di- and tri-functional molecules play, if any, at elevated temperatures, however, is limited by their thermal stabilities, and it is unlikely that tri-functional molecules exist, or exist for long enough, to significantly alter the Si budget in formation waters. Oxalate has been found in formation waters in concentrations that significantly enhance quartz dissolution rates in the laboratory between 25 and 70°C. However, as noted by Bennett (1991), the effect of organic acid anions on quartz dissolution rates decreases with increasing temperature, and has only a small effect at 70°C. Therefore, if oxalate influences the Si budget through enhanced quartz dissolution rates, its role is likely limited to low temperature (less than 70°C) regions of sedimentary basins.

## MICROBIAL INFLUENCES

Although our knowledge of microbial influences on fluid–rock interactions is rudimentary at best, because of the presence of viable microbial populations at depth within sedimentary basins, it is possible that microbial processes affect quartz dissolution and Si mobility during diagenesis. One means for microbes to affect mineral dissolution is through the secretion of organic molecules that are capable of chelating rock-forming cations in solution (e.g. Duff *et al.*, 1963; Webley *et al.*, 1963; McMahon & Chapelle, 1991; Ullman *et al.*, 1996). There are numerous studies that have quantified the effects of microbial exudates on the dissolution of minerals (e.g. see Ehrlich, 1981; Barker *et al.*, 1997; for reviews), however, there are few involving quartz directly. Quartz grains have been observed to be strongly etched in the presence of sponges (Bavestrello *et al.*, 1995), most likely due to the secretion of ascorbic acid by the organisms. Hiebert & Bennett (1992), in a related study to that of Bennett & Siegel (1987), performed *in situ* groundwater experiments to examine the dissolution of quartz and feldspar. The mineral grains were placed in a hydrocarbon-contaminated, organic-rich groundwater system for 14 months. After removal, a bacterial consortium was found, the mineral surfaces exhibited significant etching in the regions of bacterial attachment, and the minerals displayed enhanced dissolution rates relative to their theoretical rates in inorganic solutions. Hiebert & Bennett (1992) postulate that these results, and those of Bennett & Siegel (1987), can best be explained by the secretion of organic acids by the subsurface bacteria as a result of microbial metabolism of the contaminant hydrocarbons, leading to enhanced quartz dissolution rates.

Welch & Vandevivere (1994), using laboratory observations of feldspar dissolution rates in the presence of a range of bacterial exudates, find that the exudates can either enhance (presumably through ligand-promoted dissolution) or inhibit (through surface site blockage) mineral dissolution rates, depending on the type of organic exudate and the fluid conditions. Vandevivere *et al.* (1994) conducted similar studies, involving quartz, Ca-rich plagioclase (bytownite), and several other silicate minerals. They placed the minerals in contact with bacterial isolates and added glucose as a nutrient under near-neutral pH conditions. The experimental results (Fig. 3) show a strong correlation between the observed mineral

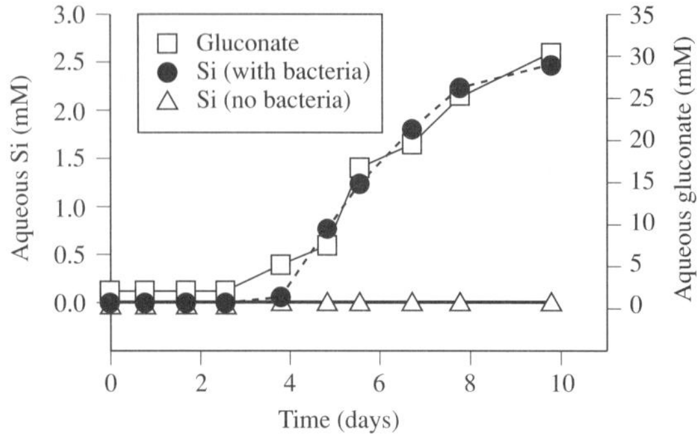


**Fig. 3.** Experimental data from Vandevivere *et al.* (1994), showing bytownite dissolution (in terms of Si release) and gluconate production from bacterial metabolic oxidation of glucose. The amount of Si release for bacteria-free control experiments is also depicted. The strong correlation between gluconate production and Si release, clearly represented in this graph, was observed for quartz as well.

dissolution rates and the concentration of gluconate, which was produced metabolically by the bacteria through the oxidation of glucose. All of these studies suggest that metabolic bacterial exudates have the capability of significantly influencing mineral dissolution rates. Whether these microbial processes are important in deep subsurface microbial communities remains an open question. It is often argued that 'high' concentrations of secreted organic acids can form in the immediate area of the bacteria, causing enhanced mineral etching (e.g. Hiebert & Bennett, 1992; Vandevivere *et al.*, 1994). Although such localized enhanced dissolution is possible, significant Si *transport* away from the site of mineral etching requires not only secretion of organic molecules capable of chelating Si and enhancing dissolution, but also the secretion of enough of these molecules to significantly raise their concentrations in the *bulk solution*. If the concentrations are enhanced only in the neighbourhood of the organisms, then Si mobility will also only be enhanced in that region.

Other possible means for microbes to affect Si mobilities are through direct silica–bacteria interactions such as aqueous Si adsorption on to bacterial surfaces, and bacterially induced silicate precipitation. Very little is known concerning these processes, and it is conceivable that Si mobilities are either enhanced or inhibited, depending on the fluid compositions and mineral substrate present. It is well established that the organic functional groups exposed on bacterial cell walls exhibit a high affinity for aqueous metal cations (Beveridge & Murray, 1976, 1980; Beveridge & Koval, 1981; Crist *et al.*, 1981; Harvey & Leckie, 1985; Goncalves *et al.*, 1987; Daughney *et al.*,

1998; Fein *et al.*, 1997). It is possible that through cationic bridging (the attachment of an aqueous cation to a negatively charged bacterial cell wall functional group), or through direct adsorption to positively charged cell wall amino groups, negatively charged $HSiO_3^-$ could become attached to bacterial cell walls (Urrutia & Beveridge, 1993, 1994). Because Si is present in solution dominantly as neutrally charged $SiO_{2(aq)}$ except at relatively high pH values, direct Si adsorption is unlikely to be a common occurrence in geological systems. However, ionization of $SiO_{2(aq)}$ to an anionic form can occur in the presence of strong-base surfaces (Iler, 1979), resulting in Si adsorption on to positively charged bacterial surface sites. In addition, Si has a strong chemical affinity for Fe (e.g. Iler, 1979), and it is possible that coadsorption of Fe and Si from solution occurs. Silica–bacterial adsorption is directly analogous to aqueous silica–organic complexation, and if it occurs, could tie the fate of Si to that of the bacteria. That is, if the bacteria are mobile in the subsurface, then Si mobilities could be enhanced. Conversely, if the bacteria themselves are adsorbed to mineral surfaces and are immobile, then the interaction with Si could dramatically limit the transport of Si.

Bacterial surfaces also have the ability to enhance silicate precipitation (Konhauser *et al.*, 1993; Urrutia & Beveridge, 1993, 1994). The precipitated silicate minerals appear to be retained surrounding the bacteria, so similar arguments apply concerning the ultimate fate of the precipitated Si. The process of bacterially-induced mineral precipitation can greatly influence element mobilities in surface water systems (Konhauser *et al.*, 1993); it is possible that the same processes also occur to some extent in deep sedimentary basin fluid–rock systems.

## CONCLUSIONS

A survey of experimental and field studies of the effects of organic molecules on the extent and rate of quartz dissolution suggests that solubility enhancement through direct aqueous silica (or Si) chelation is not likely to control Si mobilities in deep sedimentary basin environments. There is some evidence that silica–organic aqueous complexes do exist, although some datasets are inconsistent with this conclusion. Molecules such as catechol form relatively strong aqueous silica–organic complexes, but the concentration of catechol, or related molecules, in deep formation waters is too low to significantly affect Si transport. Citrate also appears to form a weak silica–organic complex, yet its thermal stability and its abundance in the subsurface are both likely to be too low to make citrate an important mobilizing agent. Most evidence indicates that oxalate and acetate, while much more abundant in the subsurface, do not significantly affect the transport of Si.

The effect of surface silica–organic complexation is to increase the dissolution rate of quartz and other silicate minerals. If fluid flow in a fluid–rock system is high enough for the fluid to remain undersaturated with respect to quartz, a change in the quartz dissolution rate would directly influence the mobility of Si in the subsurface. Experimental evidence is ambiguous concerning the effect of acetate on the dissolution rate of quartz, but it appears that di- and tri-functional carboxylic acid anions can significantly enhance the rate of dissolution, at least at relatively low temperatures. As temperature increases, the influence of the organic acid anions on the dissolution rate of quartz appears to decrease, and therefore it is unlikely that the anions significantly affect Si mobilities in deep sedimentary basin systems.

The effect of microbial organisms on Si mobilities is just beginning to be investigated. In near-surface environments, there is ample evidence that bacteria can strongly influence Si budgets through metabolite-promoted (or metabolite-inhibited) dissolution, and through bacterially-induced silicate precipitation. Future research should focus more closely not only on identifying the range of possible habitats for microorganisms in the subsurface, but also on identifying and quantifying the mechanisms by which the microorganisms affect geochemical processes during diagenesis.

## ACKNOWLEDGEMENTS

Acknowledgement is made to the Donors of the Petroleum Research Fund, administered by the American Chemical Society, for partial support of this research. Thorough reviews by P. Dove, D.A.C. Manning, and R. Worden were extremely helpful in improving the manuscript.

## REFERENCES

Barker, W.W., Welch, S.A. & Banfield, J.F. (1997) Biogeochemical weathering of silicate minerals. In: *Geomicrobiology: Interactions Between Microbes and Minerals.* (eds Banfield, J.F. & Nealson, K.H.) *Reviews in Mineralogy* **35**, 391–428. Mineralogical Society of America, Washington, DC.

Bavestrello, G., Arillo, A., Benatti, U. *et al.* (1995) Quartz dissolution by the sponge *Chondrosia reniformis* (Porifera, Demospongiae). *Nature* **378**, 374–376.

Bennett, P.C. (1991) Quartz dissolution in organic-rich aqueous systems. *Geochimica et Cosmochimica Acta* **55**, 1781–1797.

BENNETT, P. & SIEGEL, D.I. (1987) Increased solubility of quartz in water due to complexing by organic compounds. *Nature* **326**, 684–686.

BENNETT, P.C., MELCER, M.E., SIEGEL, D.I. & HASSETT, J.P. (1988) The dissolution of quartz in dilute aqueous solutions of organic acids at 25°C. *Geochimica et Cosmochimica Acta* **52**, 1521–1530.

BENNETT, P.C., SIEGEL, D.I., HILL, B.M. & GLASER, P.H. (1991) Fate of silicate minerals in a peat bog. *Geology* **19**, 328–331.

BEVERIDGE, T.J. & KOVAL, S.F. (1981) Binding of metals to cell envelopes of *Escherichia coli* K-12. *Applied and Environmental Microbiology* **42**, 325–335.

BEVERIDGE, T.J. & MURRAY, R.G.E. (1976) Uptake and retention of metals by cell walls of *Bacillus subtilis*. *Journal of Bacteriology* **141**, 876–887.

BEVERIDGE, T.J. & MURRAY, R.G.E. (1980) Sites of metal deposition in the cell wall of *Bacillus subtilis*. *Journal of Bacteriology* **141**, 876–887.

CAROTHERS, W.W. & KHARAKA, Y.K. (1978) Aliphatic acid anions in oilfield waters—implications for the origin of natural gas. *American Association of Petroleum Geologists Bulletin* **62**, 2431–2441.

CRIST, R.H., OBERHOLSER, K., SHANK, N. & NGUYEN, M. (1981) Nature of bonding between metallic ions and algal cell walls. *Environmental Science and Technology* **15**, 1212–1217.

DAUGHNEY, C.J., FEIN, J.B. & YEE, N. (1998) A comparison of the thermodynamics of metal adsorption onto two common bacteria. *Chemical Geology* **144**, 161–176.

DUFF, R.B., WEBLEY, D.M. & SCOTT, R.D. (1963) Solubilization of minerals and related materials by 2-ketogluconic acid-producing bacteria. *Soil Science* **95**, 105–114.

ERLICH, H.L. (1981) The geomicrobiology of silica and silicates. In: *Geomicrobiology*. (ed. EHRLICH, H.L.) Marcel Dekker, New York.

FEIN, J.B. & HESTRIN, J.E. (1994) Experimental studies of oxalate complexation at 80°C: Gibbsite, amorphous silica, and quartz solubilities in oxalate-bearing fluids. *Geochimica et Cosmochimica Acta* **58**, 4817–4829.

FEIN, J.B., YANE, L. & HANDA, T. (1994) The effect of aqueous complexation on the decarboxylation rate of oxalate. *Geochimica et Cosmochimica Acta* **58**, 3975–3981.

FEIN, J.B., MARSHALL, D., GORE, N. *et al.* (1995) The effect of aqueous complexation and gibbsite surface sites on the decarboxylation rate of malonate. *Geochimica et Cosmochimica Acta* **59**, 5071–5080.

FEIN, J.B., DAUGHNEY, C.J., YEE, N. & DAVIS, T. (1997) A chemical equilibrium approach to modeling metal adsorption onto bacterial surfaces. *Geochimica et Cosmochimica Acta* **61**, 3319–3328.

FRANKLIN, S.P., HAJASH, A. JR, DEWERS, T.A. & TIEH, T.T. (1994) The role of carboxylic acids in albite and quartz dissolution: an experimental study under diagenetic conditions. *Geochimica et Cosmochimica Acta* **58**, 4259–4279.

FURRER, G. & STUMM, W. (1986) The coordination chemistry of weathering: I. Dissolution kinetics of [Reinsert]-$Al_2O_3$ and BeO. *Geochimica et Cosmochimica Acta* **50**, 1847–1860.

GONCALVES, M.L.S., SIGG, L., REUTLINGER, M. & STUMM, W. (1987) Metal ion binding by biological surfaces: Voltammetric assessment in the presence of bacteria. *Science of the Total Environment* **60**, 105–119.

HARVEY, R.W. & LECKIE, J.O. (1985) Sorption of lead onto two gram-negative marine bacteria in seawater. *Marine Chemistry* **15**, 333–344.

HIEBERT, F.K. & BENNETT, P.C. (1992) Microbial control of silicate weathering in organic-rich groundwater. *Science* **258**, 278–281.

ILER, R.K. (1979). *The Chemistry of Silica*. Wiley, New York.

KHARAKA, Y.K., CAROTHERS, W.W. & ROSENBAUER, R.J. (1983) Thermal decarboxylation of acetic acid: Implications for origin of natural gas. *Geochimica et Cosmochimica Acta* **47**, 397–402.

KHARAKA, Y.K., AMBATS, G. & THORDSEN, J.J. (1993) Distribution and significance of dicarboxylic acid anions in oil field waters. *Chemical Geology* **107**, 499–510.

KNAUSS, K.G. & COPENHAVER, S.A. (1995) The effect of malonate on the dissolution kinetics of albite, quartz, and microcline as a function of pH at 70°C. *Applied Geochemistry* **10**, 17–33.

KONHAUSER, K.O., FYFE, W.S., FERRIS, F.G. & BEVERIDGE, T.J. (1993) Metal sorption and mineral precipitation by bacteria in two Amazonian river systems: Rio Solimoes and Rio Negro, Brazil. *Geology* **21**, 1103–1106.

MACGOWAN, D.B. & SURDAM, R.C. (1988) Difunctional carboxylic acid anions in oilfield waters. *Organic Geochemistry* **12**, 245–259.

MANNING, D.A.C., RAE, E.I.C. & SMALL, J.S. (1991) An exploratory study of acetate decomposition and dissolution of quartz and Pb-rich potassium feldspar at 150°C, 50 Mpa (500 bars). *Mineralogical Magazine* **55**, 183–195.

MARLEY, N.A., BENNETT, P., JANECKY, D.R. & GAFFNEY, J.S. (1989) Spectroscopic evidence for organic diacid complexation with dissolved silica in aqueous systems. I. Oxalic acid. *Organic Geochemistry* **14**, 525–528.

MCMAHON, P.B. & CHAPELLE, F.J. (1991) Microbial production of organic acids in aquitard sediments and its role in aquifer geochemistry. *Nature* **349**, 233–235.

MOTEKAITIS, R.J., COX, X.B., III, TAYLOR, P. *et al.* (1982) Thermal degradation of EDTA in aqueous solution. *Canadian Journal of Chemistry* **60**, 1207–1213.

ÖHMAN, L.-O., NORDIN, A., SEDEH, I.F. & SJÖBERG, S. (1991) Equilibrium and structural studies of silicon (IV) and aluminium (III) in aqueous solution. 28. Formation of soluble silicic acid–ligand complexes as studied by potentiometric and solubility measurements. *Acta Chemica Scandinavia* **45**, 335–341.

PALMER, D.A. & DRUMMOND, S.E. (1986) Thermal decarboxylation of acetate, Part I. The kinetics and mechanism of reaction in aqueous solution. *Geochimica et Cosmochimica Acta* **50**, 813–823.

POULSON, S.R., DREVER, J.I. & STILLINGS, L.L. (1997) Aqueous Si–oxalate complexing, oxalate adsorption onto quartz, and the effect of oxalate upon quartz dissolution rates. *Chemical Geology* **140**, 1–7.

RIMSTIDT, J.D. & BARNES, H.L. (1980) The kinetics of silica–water reactions. *Geochimica et Cosmochimica Acta* **44**, 1683–1699.

SJÖBERG, S., INGRI, N., NENNER, A.-M. & ÖHMAN, L.-O. (1985) Equilibrium and structural studies of silicon (IV) and aluminium (III) in aqueous solution. 12. A potentiometric and $^{29}$Si-NMR study of silicon tropolonates. *Journal of Inorganic Biochemistry* **24**, 267–277.

STUMM, W. & MORGAN, J.J. (1996). *Aquatic Chemistry*. Wiley, New York.

TAIT, C.D., JANECKY, D.R. & MUSGRAVE, J.A. (1993) *Direct speciation of metal and metalloid ions by optical spectroscopies: An update*. Office of Performance Assessment Program Review and DOE Workshop: Sedimentary Systems, Aqueous and Organic Geochemistry. Lawrence Berkeley Laboratory, Berkeley, CA. July 12–16, 1993.

ULLMAN, W.J., KIRCHMAN, D.L., WELCH, S.A. & VANDEVIVERE, P.V. (1996) Laboratory evidence for microbially mediated silicate mineral dissolution in nature. *Chemical Geology* **132**, 11–17.

URRUTIA, M.M. & BEVERIDGE, T.J. (1993) Mechanism of silicate binding to the bacterial surface (*Bacillus subtilis*). *Journal of Bacteriology* **175**, 1936–1945.

URRUTIA, M.M. & BEVERIDGE, T.J. (1994) Formation of fine-grained metal and silicate precipitates on a bacterial surface (*Bacillus subtilis*). *Chemical Geology* **116**, 261–280.

VANDEVIVERE, P., WELCH, S.A., ULLMAN, W.J. & KIRCHMAN, D.L. (1994) Enhanced dissolution of silicate minerals by bacteria at near-neutral pH. *Microbial Ecology* **27**, 241–251.

WEBLEY, D.M., HENDERSON, M.E.F. & TAYLOR, E.F. (1963) The microbiology of rocks and weathered stones. *Journal of Soil Science* **14**, 102–112.

WELCH, S.A. & VANDEVIVERE, P. (1994) Effect of microbial and other naturally occurring polymers on mineral dissolution. *Geomicrobiology Journal* **12**, 227–238.

*Spec. Publs int. Ass. Sediment.* (2000) **29**, 129–146

# Microstructures of deformed and non-deformed sandstones from the North Sea: implications for the origins of quartz cement in sandstones

Q. J. FISHER[1], R. J. KNIPE[1] *and* R. H. WORDEN[2]

[1]*Rock Deformation Research Group, Department of Earth Sciences, University of Leeds, Leeds LS2 9JT, UK; and* [2]*School of Geosciences, The Queen's University, Belfast, BT7 1NN, UK*

## ABSTRACT

Microstructural analysis of approximately 400 faults and fractures, from 28 gas and oil fields within the North Sea, has shown that mesocrystalline quartz cement is not present in significant quantities within any deformation features that acted as conduits for fluid flow. Pyrite, barite, anhydrite, dolomite, calcite and siderite were, however, identified as cements within parts of deformation features that experienced deformation-induced dilation. These cements probably precipitated from fluids focused along the deformation features. This supports isochemical models of quartz cementation, which suggest that it is unlikely that quartz cement in reservoirs originated from fluids flowing along faults.

Cataclastic faults with quartz cement are relatively common within clean sandstones that have been buried to > 90°C. However, these faults experienced pervasive grain fracturing resulting in a rapid decrease in porosity and permeability and are therefore unlikely to have been conduits for fluid flow. Instead, quartz cementation probably occurred because newly fractured, clean grain surfaces generated during cataclastic deformation provided favourable sites for the precipitation of quartz cement. The quartz cement was generated by local processes (occurring within, and up to a distance of ~1 m from, the fault) such as quartz grain-to-grain dissolution or quartz dissolution along stylolites.

In many reservoir sandstones, the distribution of quartz cement is influenced by the quantity and textural distribution of detrital and authigenic phyllosilicates. The presence of minor clays at quartz grain-to-grain contacts enhances silica solubility and therefore increases its supply for localized quartz cementation. However, grain-coating and pore-filling clays suppress the growth of authigenic quartz by reducing the surface area available for its precipitation. Small variations in the quantity and textural distribution of clays therefore dictate whether a particular part of a sandstone experiences quartz grain-to-grain dissolution and/or quartz cementation. As a consequence of this, the maximum quantity of quartz cement is found within sandstones containing small amounts of detrital clays concentrated at grain-to-grain contacts or stylolites where enhanced quartz grain contact dissolution is possible and precipitation sites are not occluded by too many clays. Fault-rocks experience enhanced grain-contact quartz dissolution relative to their host sediments, when clays are inserted at grain contacts during deformation.

## INTRODUCTION

Quartz cement is one of the major factors which reduce the porosity and permeability of reservoir sandstones during deep burial (e.g. McBride, 1989; Bjørlykke & Egeberg, 1993; Bjørlykke, 1994). The origin of this cement has been studied for over a century (e.g. Sorby, 1880) yet there still exists considerable debate as to which diagenetic processes are responsible for its distribution and abundance. Opinion is divided as to whether the quartz cement is derived locally (e.g. Bjørlykke & Egeberg, 1993; Dutton, 1993; Bjørlykke, 1994) or is imported into sandstones (e.g. Gluyas, 1985; McBride, 1989; Dutton & Diggs, 1992; Gluyas & Coleman, 1992).

Local mechanisms for silica cement generation include: quartz grain-to-grain dissolution (e.g. Waldschmidt, 1941; Heald, 1955; Fuchtbauer, 1974; Ramm, 1992; Bjørlykke, 1994), quartz dissolution along stylolites (Dewers & Ortoleva, 1990; Oelkers *et al.*, 1996), feldspar dissolution (Sorby, 1880; Bjørlykke, 1983; Bjørlykke & Egeberg, 1993) and the illitization of kaolin (Bjørlykke, 1980). Integration of petrographic observations with

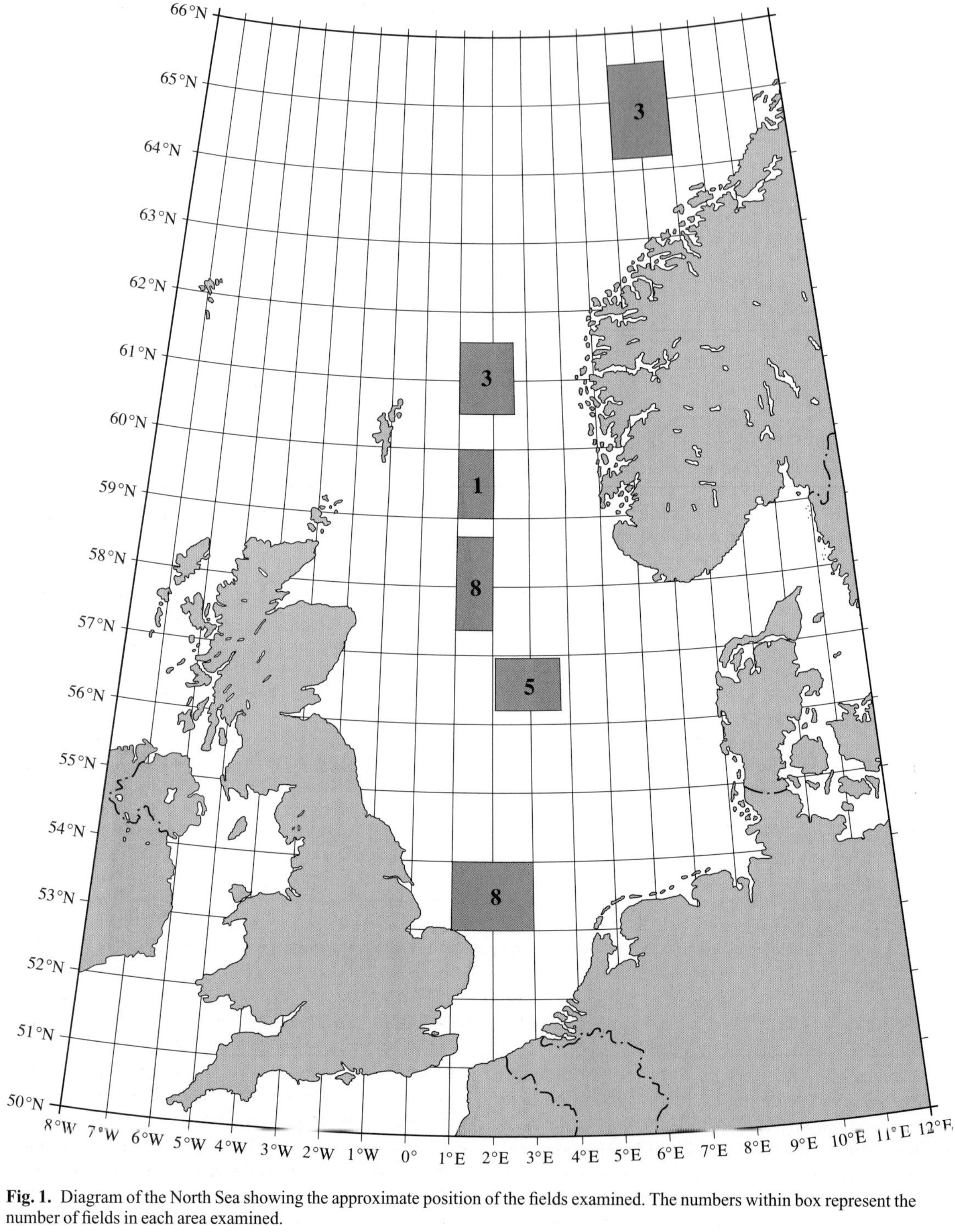


**Fig. 1.** Diagram of the North Sea showing the approximate position of the fields examined. The numbers within box represent the number of fields in each area examined.

geochemical modelling has shown that dissolution of quartz along stylolites can account for the distribution of quartz cement in some Jurassic sandstones of the North Sea and Norwegian continental shelf (Oelkers *et al.*, 1996).

Advocates of an external source of silica have suggested that fluids were derived from compacting mudrocks (Scotchman *et al.*, 1989; Burley, 1993) or from underlying sandstones (Wood & Hewett, 1984). Density-driven convection (Wood & Hewett, 1984) and fault-related fluid flow (Haszeldine *et al.*, 1984; Jourdan *et al.*, 1987; Scotchman *et al.*, 1989; Burley, 1993) have both been suggested as mechanisms for the movement of silica-bearing fluids. Geochemical models suggest that it is very unlikely that significant quantities of dissolved silica could be transported on a regional scale, even with the help of extensive fracture networks (Pedersen & Bjørlykke, 1994; Giles, 1997). However, it has been argued that our understanding of silica complexing and mobility in the subsurface is incomplete and that mechanisms for the redistribution of silica by advection may yet be found (Gluyas & Coleman, 1992). Geochemical models can be validated if their predictions match observed quartz cement distributions. Systematic microstructural studies have not, however, yet been presented to elucidate whether or not fault and open-fracture networks have indeed acted as conduits for the flow of silica-bearing fluids in sandstones.

The data presented here were gathered during studies of fault sealing within hydrocarbon reservoirs of the North Sea. The data set is unique in that the microstructures of a large number of both undeformed and deformed sediments have been analysed. The fields studied are diverse both in terms of the reservoir intervals present and their burial and deformation histories. The data gathered are particularly valuable for demonstrating the role of deformation features (a general term used here for faults and mode-I fractures) as conduits for the movement of silica-bearing fluids in hydrocarbon reservoirs. In particular, we discuss whether the microstructural data are consistent with fault-related fluid flow as a key mechanism for quartz cementation.

## SAMPLES EXAMINED

The data presented in this paper were gathered from 28 oil and gas fields in the North Sea. The approximate field locations are given in Fig. 1 and the stratigraphic units which were examined are shown in Table 1. The database includes a large number of samples from major North Sea oil reservoirs (the Brent Group and equivalents) as well as the major gas reservoirs (the Rotliegende). The samples examined have experienced large differences in their maximum depth of burial (1.6 km to 4.9 km).

**Table 1.** The total number of reservoirs of each age examined. All of the Permian reservoirs are Rotliegendes from the southern North Sea, whereas the Middle Jurassic are Brent and equivalents from the northern North Sea

| Age | Reservoirs |
|---|---|
| Cretaceous | 1 |
| Upper Jurassic | 4 |
| Middle Jurassic | 9 |
| Lower Jurassic | 4 |
| Triassic | 2 |
| Permian | 8 |

Unlike many previous studies, the sandstone samples discussed in this paper were collected primarily to study fault sealing and are therefore unlikely to be biased towards rocks with better reservoir quality. All samples of deformed and undeformed sandstone were collected during routine core description. Specimens were collected to represent the full range of faults and mode-I fractures (hereafter referred to as fractures) and lithofacies within the cored sections of wells. The size of deformation features sampled ranged from the occasional seismic scale fault, through to small-scale faults (< 1 cm off-set) within the damage zone of seismic faults to isolated deformation features situated many kilometres from seismic scale faults. Small-scale faults within the damage zone of larger faults were by far the most common type of feature analysed.

## METHODS

Polished blocks and stubs of broken surfaces were prepared for each specimen to be examined using secondary electron (SE), cathode luminescence (CL) and back-scattered electron (BSE) microscopy. Samples were examined with a CAMSCAN CS44 high performance scanning electron microscope (SEM) equipped with a secondary electron detector, a high-resolution solid-state four-quadrant back-scattered electron detector, a cathode luminescence detector (CL) and an EDAX energy dispersive X-ray spectrometry (EDS) system. The EDS detector had an ultra-thin window to allow the detection of low energy X-rays emitted from light elements (B, C, N, O and F).

The paragenetic history of both undeformed and deformed sediments was determined to establish the timing of deformation in relation to the major diagenetic events. The microstructures of the faults and fractures were used to assess/identify the deformation mechanisms.

Representative pairs of BSE and CL images were taken from both deformed and undeformed sandstones. These were transferred as 8-bit images (256 grey-levels) to an image analysis package (NIS-Image 1.52). As the brightness (grey-level) of a phase in a BSE image is dependant on its mean atomic number, all grains of the same composition have the same brightness level and grains with different compositions will generally have a different brightness. A binary image can therefore be made of each mineral within the specimen by thresholding of a BSE image. These binary images can then be processed to determine parameters such as the total area covered by that particular constituent and total number of grains/pores.

Several minerals were present in some samples with similar mean atomic mass (e.g. quartz and albite, K-feldspar and dolomite, K-feldspar and muscovite). As these could not be differentiated using BSE images a number of techniques were employed to determine their abundance. X-ray dot maps of Na and Mg distributions were gathered, which with careful image analysis allowed the determination of albite and dolomite abundances, respectively. K-feldspar and muscovite could not be distinguished using X-ray dot maps therefore they were manually differentiated based on their texture. In samples with a complex clay minerals assemblage, point-counting, based on a sample population of 300, was used. This was undertaken on the SEM using an automatic stage that allows the sample to be moved over a square grid at fixed intervals. The total clay mineral abundance quoted in this paper includes soft lithoclasts and detrital mica grains, but does not include microporosity between the clay particles.

The amount of authigenic quartz present within each undeformed sandstone was obtained in a similar manner to that described in Evans *et al.* (1994). The BSE and CL images were obtained from the same area of the sandstone. A binary image of the quartz was produced by thresholding the BSE image at the appropriate grey-level. The binary image was then combined with the CL image to produce a quartz-only CL image. The quartz overgrowths were then selected by either choosing the correct grey level which represents the quartz overgrowth or manually choosing authigenic quartz.

The pore-throat size distribution and threshold pressure were determined using mercury injection porosimetry capable of measuring values of up to 64 000 psi. Permeability of most deformed and undeformed sandstones was determined using a purpose-built, water-flow permeameter capable of measuring down to < 1 μD.

## MICROSTRUCTURE OF NORTH SEA FAULTS

### Fault sealing processes

Faults and fractures that were examined formed at a range of depths. They include early, often slump-related features, extensional faults commonly formed during the main rifting stage of the basin, as well as faults and fractures formed or reactivated during inversion. Such deformation features commonly act as barriers to fluid flow as a result of a number of porosity-reduction processes including: (i) porosity-collapse by disaggregation and deformation-induced mixing; (ii) cataclasis; (iii) diffusive mass transfer (DMT); (iv) clay-smearing; and (v) cementation (Knipe, 1992; Knipe *et al.*, 1997; Fisher & Knipe, 1998).

Porosity-collapse by disaggregation and deformation-induced mixing is a process which affects sandstones containing a heterogeneous distribution of coarse and fine grains (Fisher & Knipe, 1998). During deformation, grains of different size become intimately mixed and packed more efficiently. This reduces porosity and permeability. Sandstones particularly susceptible to porosity-reduction by this process are those that contain abundant (14–40%), heterogeneously distributed, fine-grained phyllosilicates.

Cataclasis is the process whereby grains become fractured during deformation (e.g. Borg *et al.*, 1960; Engelder, 1974; Knipe, 1989). In porous sandstones, a reduction in porosity and permeability occurs because the grain-size and grain-sorting of the fault rock is reduced. This causes the collapse of macroporosity (e.g. Engelder, 1974; Blenkinsop, 1991) and allows the grain fragments to be more tightly packed than the undeformed host sandstone. Cataclasis is the main process responsible for the reduction in porosity and permeability of faults within relatively clean sandstones (hereafter arbitrarily defined as rocks with < 14% phyllosilicates at the time of deformation).

Clay-smearing occurs when phyllosilicates become aligned on a fault surface during deformation. Lindsay *et al.* (1993) have identified a number of clay-smearing mechanisms including the shear deformation of shale horizons, faulting of shale horizons past other lithologies and injection of clay-rich gouges into fault zones. Cases have also been identified whereby the deformation of very clay-rich sandstones has resulted in the formation of clay-smears (Fisher & Knipe, 1998). Coherent clay-smears are formed by the deformation of sediments containing > 40% clays at the time of deformation.

Diffusive mass transfer (DMT) occurs when material is dissolved at grain-to-grain contacts and is transported by diffusion to grain surfaces adjacent to free pore-space where it precipitates (Rutter, 1983; Spiers & Schutjens, 1990). A large number of the studied fault rocks have experienced enhanced DMT relative to their host sandstone (Fisher & Knipe, 1998).

The degree of diagenetic cementation is often relatively enhanced within faults and fractures. This may occur either when mineralizing fluids are focused along dilatant deformation features or when deformation actually renders a fault rock more susceptible to cementation. Cemented faults and fractures are defined as those that contain significant concentrations of mineral cements. Approximately 10% of fault rocks examined contain more cement than their host sediment.

## Cements in faults and fractures

Cements concentrated within faults and fractures include anhydrite, ankerite, barite, calcite, dolomite, microcrystalline quartz, mesocrystalline quartz, pyrite, siderite and sphalerite (Fig. 2a). Microcrystalline quartz is defined here as that with a grain-size of < 1 μm, whereas mesocrystalline quartz usually has a grain-size of > 10 μm. The cemented deformation features may be divided into three broad groups; fluid escape structures, veins and cataclastic fault rocks (Fig. 2b).

Anhydrite, barite, calcite, dolomite, pyrite, siderite and sphalerite were all found as vein-filling cements. Calcite cemented veins are particularly common within calcite cemented sandstones. The other cements occur within faults and fractures formed in sandstones that were lithified by deep burial processes or along reactivated cataclastic faults. Anhydrite, dolomite/ankerite and barite are abundant within faults and fractures in the Rotliegende of the southern North Sea that were formed or reactivated during inversion.

Microcrystalline quartz and pyrite cement are concentrated within deformation features formed during very early burial due to soft-sediment fluid escape. These deformation features have a very high minus-cement porosity (> 40%). Microcrystalline quartz-cemented deformation features occur within sandstones containing large concentrations of biogenic silica (such as sponge spicules) suggesting that the cement is a product of its dissolution.

Mesocrystalline quartz is concentrated within faults that experienced cataclasis. These fault rocks often have very low porosities (<< 3%), and when viewed using BSE have the appearance of quartz-cemented veins (Fig. 3a). The CL imagery revealed that these faults are composed of fractured detrital grains cemented by mesocrystalline quartz (Fig. 3b). However, detrital fragments are not supported by the cement. Instead, the quartz appears to have preferentially precipitated between fragments in mutual contact and on fractured grain surfaces produced during deformation. Quartz cemented cataclastic faults occur within very clean sandstones (< 5% clay at the time of deformation) in reservoirs buried to depths > 3 km. All of the quartz-cemented cataclastic faults examined have lower minus-cement porosities than their host sandstone suggesting that they experienced a deformation-induced porosity-collapse.

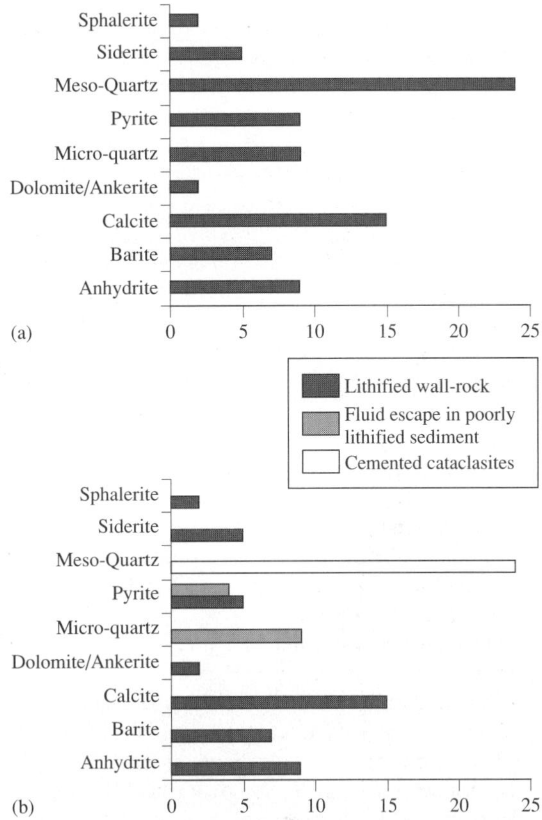


**Fig. 2.** Histograms showing (a) the number of deformation features containing each cement type; (b) the type of deformation feature containing each cement. Total number of samples: 76.

## Enhanced diffusive mass transfer within faults

A large number of the fault rocks contain highly sutured grain contacts formed due to post-faulting DMT (e.g. Fig. 4). The extent of post-deformation DMT has an enormous effect on the sealing capacity of faults. For example, the permeability of clay-rich (5–25% clay) fault rocks with similar grain-size distributions varies by as much as two orders of magnitude depending on the extent of DMT (Fisher & Knipe, 1998). Fault rocks buried to less than 3 km have not experienced DMT, suggesting

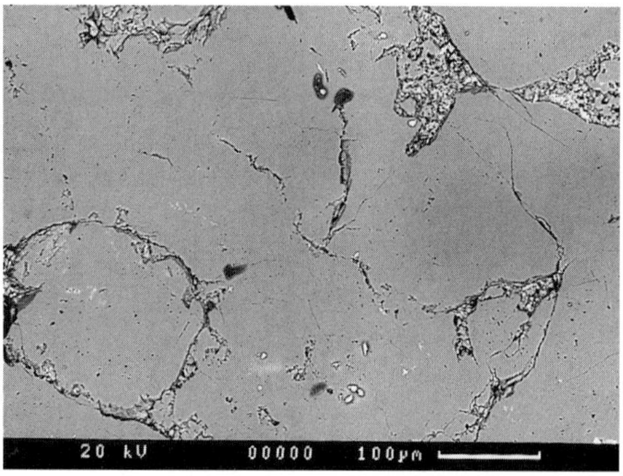


(a)

(b)

**Fig. 3.** Micrographs showing (a) BSE image of a quartz cemented cataclastic fault; and (b) CL image showing the same area. The quartz cement (arrow) is fracture filling and appears dark on the CL image. Note also the highly fractured and angular nature of the quartz fragments produced during cataclastic deformation.

(a)

(b)

**Fig. 4.** BSE micrographs showing (a) the undeformed sandstones close to (b) fault rock that has experienced enhanced quartz grain-to-grain dissolution. Note that quartz grain contacts are very rarely sutured in the host sandstone but are highly sutured within the fault rock.

that there is a critical depth or temperature before DMT becomes significant. Fault rocks formed at any stage during burial can experience DMT, in some cases, long after the initial deformation episode.

Fault rocks that have experienced enhanced DMT commonly formed in sandstones containing small to moderate concentrations of clays (5–25%) at the time of deformation. However, fault rocks formed adjacent to sandstones containing high concentrations of clays (> 25%) often experience less DMT than those formed from sandstones containing low to moderate concentrations of clays. Evidence of enhanced DMT was not observed within faults formed within sandstones containing very low concentrations of clay (< 5%) at the time of deformation. Instead, as discussed above, these often experienced enhanced quartz cementation. Non-horizontal stylolites were observed in some of the more deeply buried reservoirs, and may represent the end-product of DMT within phyllosilicate-rich faults. These observations suggest that the specific clay content is of critical importance to the extent a fault rock experiences post-faulting DMT.

## THE DISTRIBUTION AND MORPHOLOGY OF QUARTZ WITHIN UNDEFORMED SANDSTONES OF THE NORTH SEA

### Quartz cement content and burial depth

Sandstones buried to less than 2.7 km typically have negligible quartz cement (Fig. 5). However, sandstones

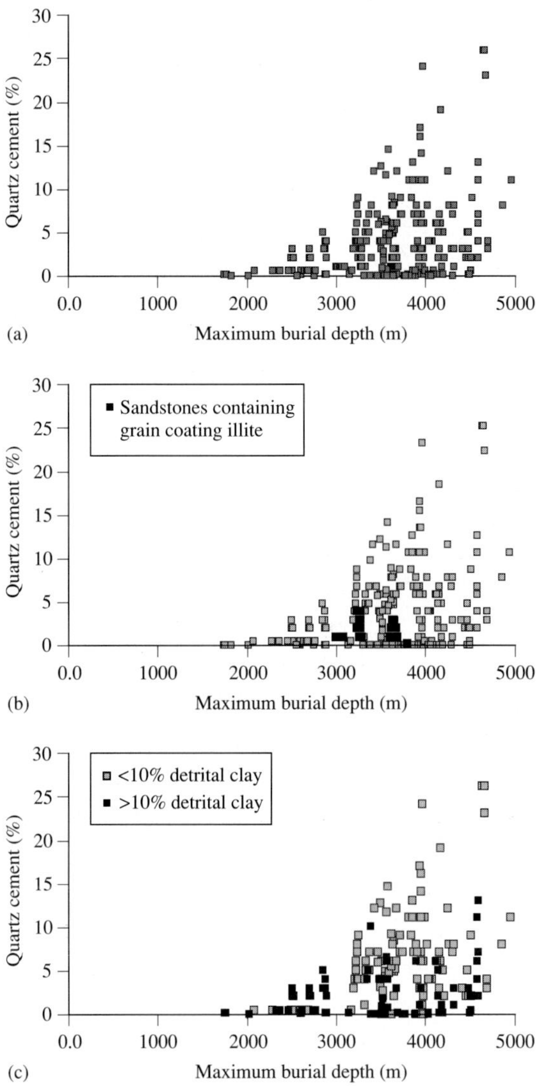


**Fig. 5.** Plots showing (a) maximum burial depth against the amount of quartz cement within the sandstones from the North Sea examined during this study; samples have been highlighted that (b) contain abundant grain coating illite; and (c) contain > 10% detrital clay. Total number of points included: 285.

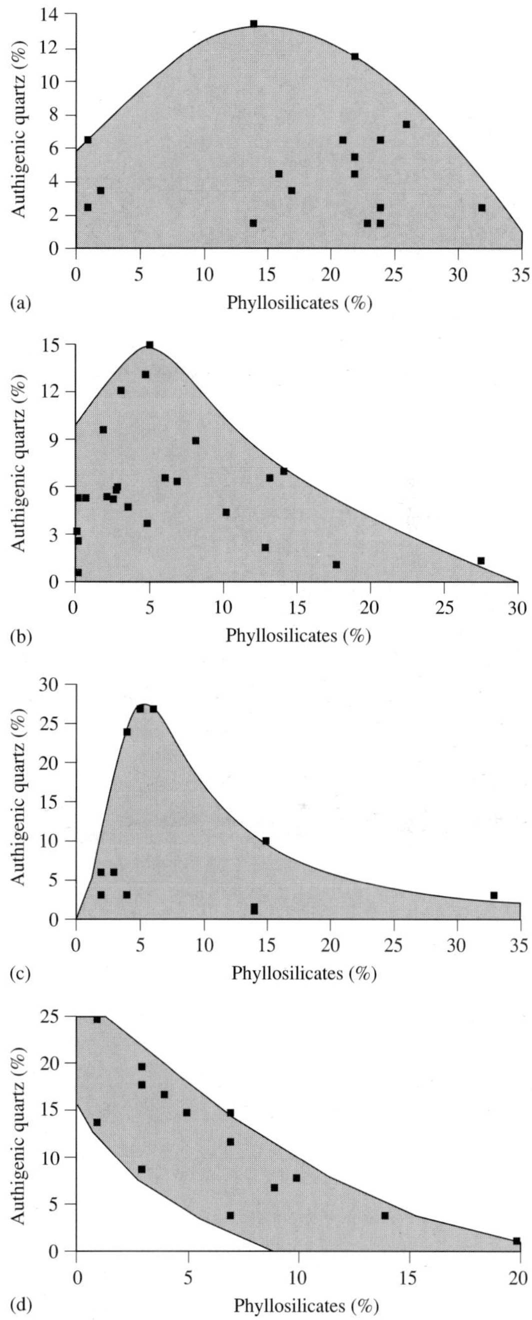


**Fig. 6.** Plots of the amount of authigenic quartz against the quantity of detrital or early authigenic clays for Middle Jurassic reservoir sandstones from four different fields (reservoirs a–d receptively). These have been buried to average burial depths of (a) 3700 m; (b) 3600 m; (c) 4500 m; (d) 4100 m.

buried deeper than 2.7 km have a huge range in authigenic quartz volumes (Fig. 6); although the maximum amount of authigenic quartz increases dramatically with depth. Similar observations have also been made by other workers (e.g. Ehrenberg, 1990; Bjørlykke *et al.*, 1992; Giles *et al.*, 1992; Ramm & Ryseth, 1996). An understanding of the factors which control the variation in the amount of quartz cement at any given depth is consequently important

**Fig. 7.** (a) BSE, and (b) combined CL–BSE micrographs showing a clay-rich region in the sample containing most authigenic quartz in reservoir C. (c) BSE, and (d) combined CL–BSE micrographs showing a relatively clay-free area within the sandstone containing most authigenic quartz in reservoir C. Note that grain-to-grain dissolution is common within the clay-rich area, and quartz overgrowths are common in the clay-free sandstone.

both in order to predict the distribution of authigenic quartz and to identify the sources of the silica cement.

## Quartz cement volume versus clay content

To further investigate the cause for the variation in quartz cement content in Fig. 5(a), samples have been highlighted which contain authigenic pore-lining clays (Fig. 5b) and high concentrations of detrital phyllosilicates (Fig. 5c). The results show that the sandstones with the lowest concentrations of quartz at each depth tend to contain either pore-lining clay or large quantities of detrital phyllosilicates. This observation is also consistent with previous studies of quartz cement distribution in hydrocarbon reservoirs (e.g. Tada & Siever, 1989; Ramm & Ryseth, 1996).

To further investigate the relationship between quartz cement and clay minerals, plots of the volume of detrital clay against the amount of authigenic quartz have been constructed for some of the Jurassic sandstones from individual hydrocarbon fields (e.g. Fig. 6a–d). These plots show considerable variation. In two of the fields (Fig. 6b,c) the maximum amount of authigenic quartz was found in samples containing small quantities of (~5%) detrital clays. For example, in one field, clay free samples contain a maximum of ~5% authigenic quartz and samples containing > 15% detrital clays contain less than ~2% authigenic quartz (Fig. 6b). In contrast, samples containing 6% detrital clay contain up to 14% authigenic quartz (Fig. 7). Another field contained the maximum amount of authigenic quartz in sandstones containing high concentrations of clay (~10–25%; Fig. 6a). Yet another reservoir contained the maximum amount of quartz in samples containing virtually no clay (Fig. 6d). The observation that the maximum amount of authigenic quartz in some reservoirs is within sandstones containing low to moderate concentrations of detrital clays is consistent with the work of Dewers & Ortoleva (1991) and suggests that minor quantities of clay can actually exacerbate quartz cementation.

A combination of core observation and detailed microstructural analysis shows that differences in the distribution of clays exist between these four reservoirs. In the sandstones containing high concentrations of authigenic quartz in Fig. 6(b–c), detrital clays are homogeneously distributed. Sandstones containing high concentrations of authigenic quartz in Fig. 6(a) experienced very early soft sediment deformation which resulted in extensive redistribution of the clays due to the fluid escape. The resulting sediment has a very heterogeneous distribution of clays and the authigenic quartz is concentrated in the clay-free regions (Fig. 7). The samples containing high concentrations of authigenic quartz in Fig. 6(d) were invariably situated adjacent to stylolite horizons (see also Oelkers *et al.*, 1996).

### The relationship between the permeability of undeformed sandstones and their authigenic quartz content

The maximum amount of authigenic quartz is concentrated within samples which have permeabilities mid-way between the maximum and minimum for those examined (Fig. 8a–d). The relatively low permeability of the most quartz-cemented sandstones is the consequence of quartz cementation. However, quartz cement is not concentrated within the sandstone horizons that are, or ever were, the most permeable in the reservoir (Fig. 9a–d). This suggests that fluid flow is not a primary control upon quartz cementation.

### Quartz morphology

Quartz cement in Middle Jurassic reservoirs of the northern North Sea usually occurs as overgrowths on detrital quartz grains (Fig. 10a,b). The only exception to this are samples containing large quantities of detrital clays, or more rarely grain-coating authigenic clays and/or microcrystalline quartz. In these cases, quartz tends to occur as outgrowths (Fig. 10c).

The majority of Rotliegende sandstones contained either authigenic illite or chlorite/berthierine grain-coating clays. In these samples, quartz outgrowths and pore-filling cements (Fig. 10d) are far more common than overgrowths.

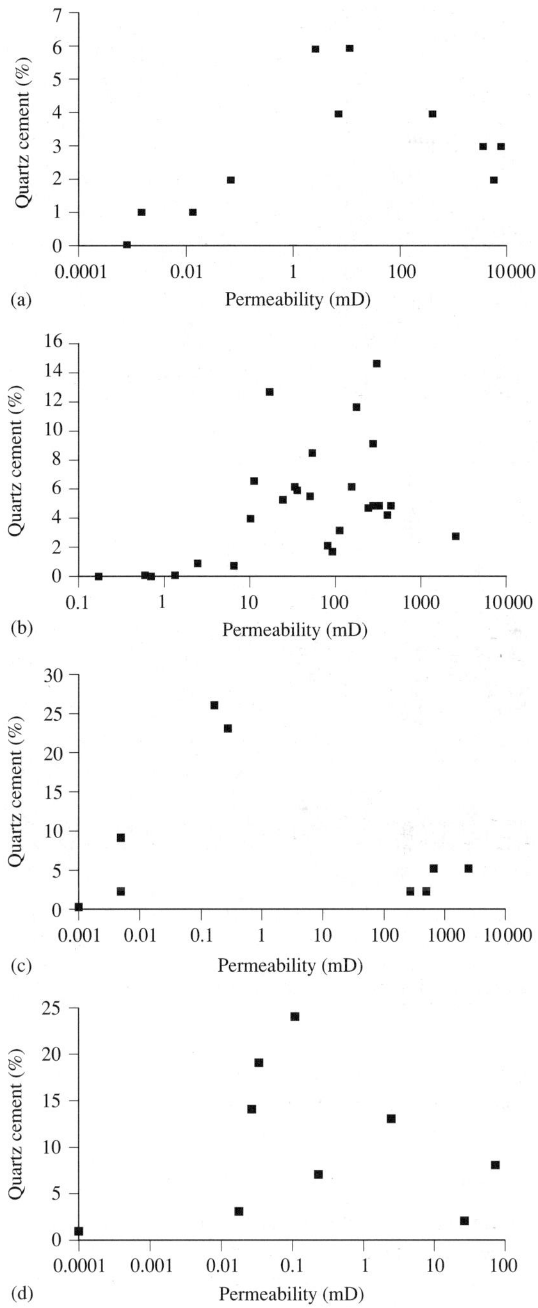


**Fig. 8.** Plots of permeability against the amount of authigenic quartz for the four reservoirs shown in Fig. 6. These have been buried to average burial depths of (a) 3700 m; (b) 3600 m; (c) 4500 m; (d) 4100 m. Note that quartz cement is not concentrated within the most permeable horizons.

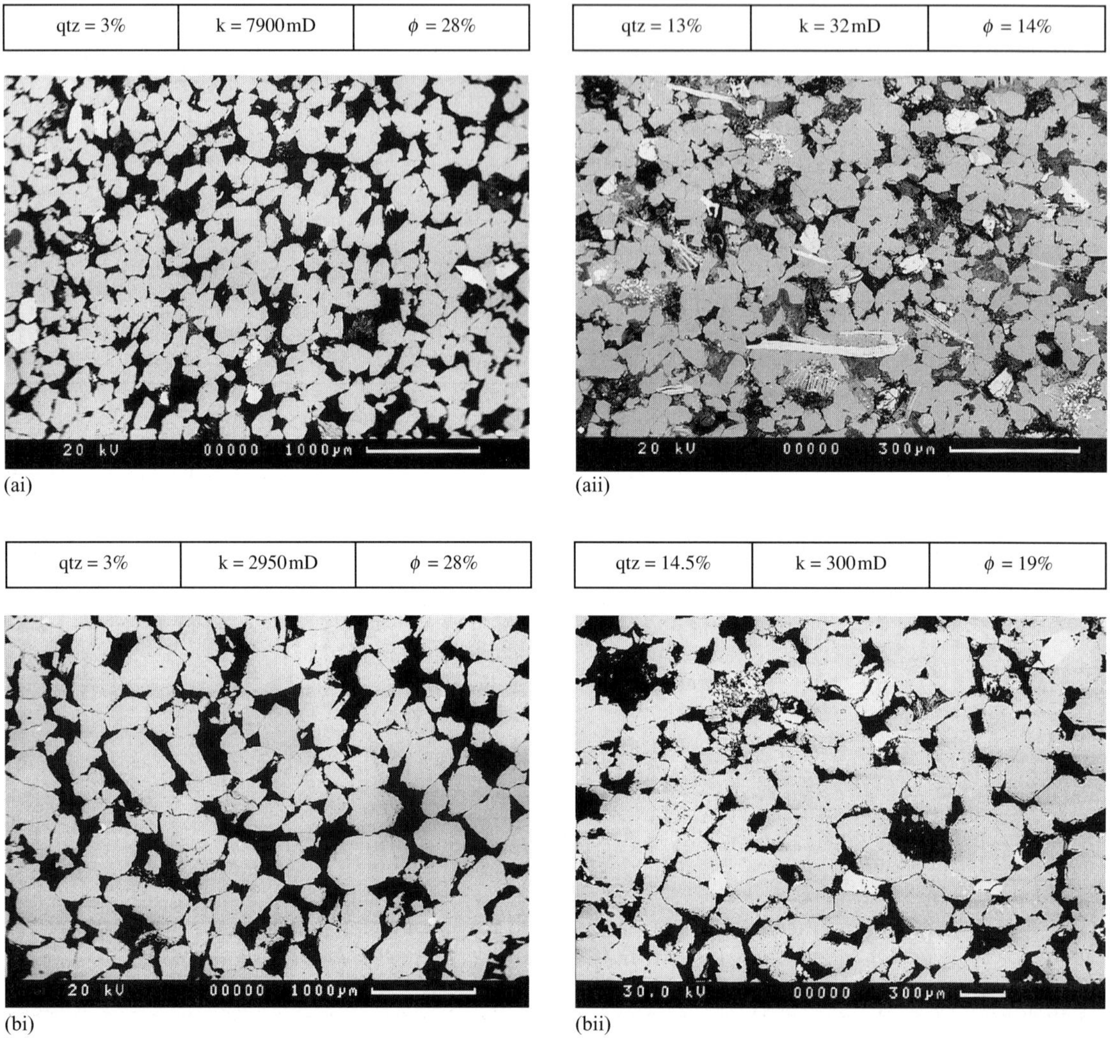


Some Rotliegende sandstones lacked grain-coating cements, instead the dominant authigenic clay is pore-filling kaolin. In these samples the authigenic quartz tended to occur as overgrowths similar to those observed within the northern North Sea sandstones.

## DISCUSSION

### Origin of cements within faults and fractures

Deformation features need to be more permeable than their host rock (i.e. dilated) for them to act as conduits for fluid flow. In general, for cementation to occur from fluids flowing along a fault, the wall rock at contact bridges needs to be sufficiently strong, or the fluid pressure sufficiently high, for dilation to be maintained (Knipe, 1992; Bjørlykke, 1994; Clennell *et al.*, 1998). Fundamental differences exist between both the extent of dilation and the dilation mechanism for the different types of cemented deformation features.

#### *Vein-filling cements*

The faults and fractures in which vein-filling cements were identified appear to have formerly had far higher permeability than their host sandstone and thus may have acted as conduits for fluid flow. Usually the wall rocks

| qtz = 5% | k = 2450mD | $\phi$ = 23% |
|---|---|---|

(ci)

| qtz = 26% | k = 0.19mD | $\phi$ = 7% |
|---|---|---|

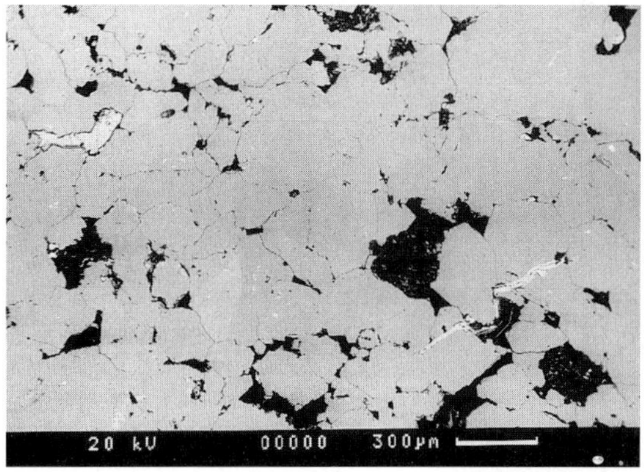


(cii)

| qtz = 8% | k = 98mD | $\phi$ = 20% |
|---|---|---|

(di)

| qtz = 24% | k = 0.12mD | $\phi$ = 8% |
|---|---|---|

(dii)

**Fig. 9.** (*Opposite and above.*) Micrographs contrasting the most permeable (micrographs on the left) and most quartz cemented (micrographs on the right) sandstones for the four reservoirs from which the data presented in Fig. 6 and Fig. 8 were derived. These reservoirs have been buried to average depths of (a) 3700 m; (b) 3600 m; (c) 4500 m; (d) 4100 m. The permeability (k), porosity (ø) and the quartz cement content (qtz) of the sandstones have been placed above the micrographs. Note that in each of the reservoirs, quartz cement is not concentrated within the most permeable and porous horizons. Instead, quartz cement is concentrated within sandstones with a smaller grain-size and/or slightly more detrital clays.

adjacent to cemented veins were sufficiently strong (either due to early cementation or later lithification) to permit fractures to remain open following deformation. Occasionally, high fluid pressures may also have allowed dilation to be maintained.

Many of the vein cements identified do not have obvious local sources suggesting that the solutes were transported into the reservoir. For example, there is no obvious local source of solutes for the precipitation of anhydrite, ankerite, barite and sphalerite cements in the Rotliegende of the southern North Sea. External sources of these cements can, however, be identified, such as the overlying Zechstein evaporites or the underlying Carboniferous shales (Sullivan *et al.*, 1990; Gaupp *et al.*, 1993; Leveille *et al.*, 1997). Precipitation of most of the cement types found within veins may be explained by invoking a decrease in pressure/temperature or the mixing of two fluids (e.g. Knipe, 1992).

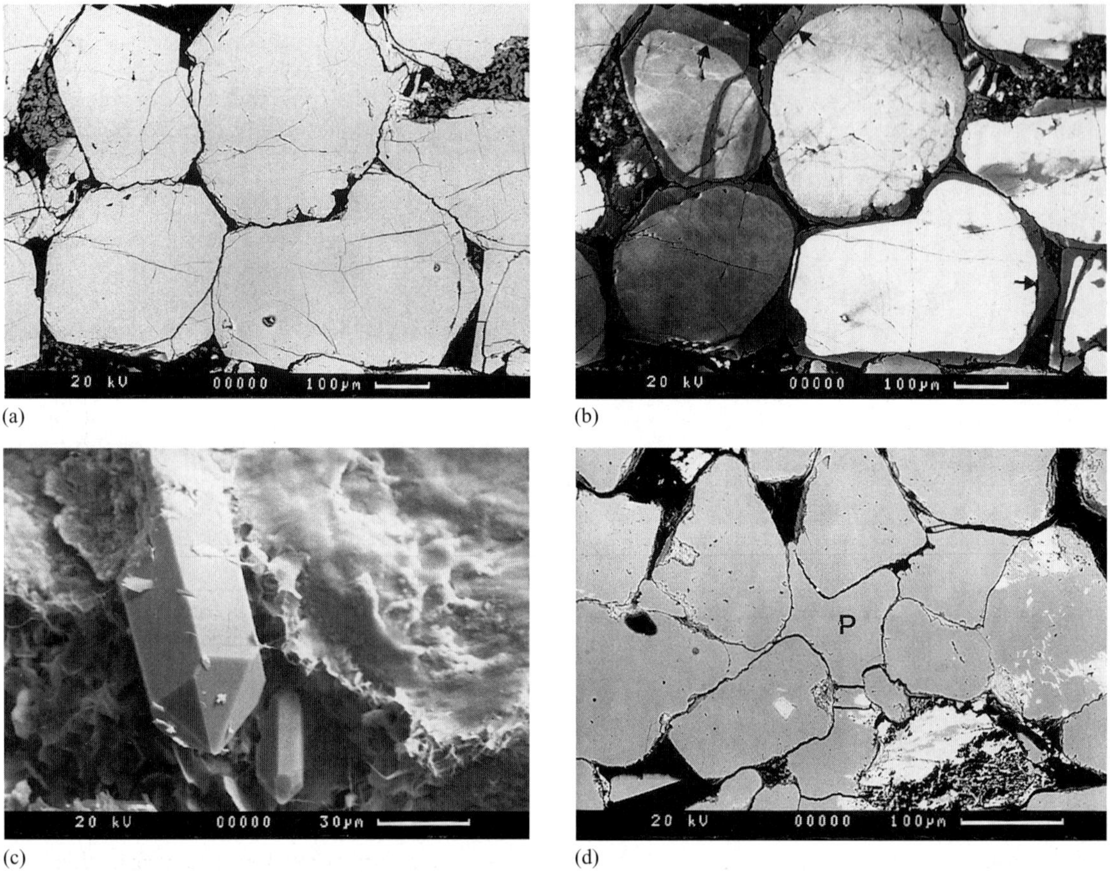


**Fig. 10.** Micrographs showing (a) BSE image of quartz grains in a sandstone that does not contain grain-coating clays; (b) CL image of the same area shown in (a), comparison of the two micrographs highlights the quartz overgrowths (arrows); (c) SE image of a quartz outgrowth; (d) BSE image of pore-filling quartz cement (P).

The concept that deformation of rocks may result in fluid conduits is well-established. Gold mineralized veins (Sibson, 1987; Sibson *et al.*, 1988), fractures within chalk reservoirs (Brown, 1987; Price, 1987) and faults within crystalline rock (Black, 1987) are examples where deformation of lithified sediments or crystalline material has resulted in the production of high permeability conduits.

### *Cemented water-escape structures*

Quartz cemented water-escape structures often have very high minus-cement porosities suggesting that they experienced significant dilation. In these cases, the host sandstone was poorly lithified and dilation was not maintained by the presence of a mechanically strong wall rock. Instead, elevated hydrostatic pressures probably caused dilation. As with the veins, cementation in fluid escape structures may result from decreases in temperature and pressure or due to the mixing of fluids. This type of cemented deformation feature formed during early burial and tells us nothing about quartz cementation in oil fields.

### *Quartz cemented cataclastic faults*

Quartz cemented cataclastic faults are fundamentally different from the cemented veins and water escape structures: their minus-cement porosity is much lower than their host sandstone. Cataclastic deformation resulted in a decrease in grain-size and grain-sorting of the fault rock causing a collapse of macroporosity and a decrease in permeability relative to the host sandstone.

These observations are supported by published accounts of high pressure deformation experiments on sands. For example, it has been shown that the permeability of sands decreases dramatically as the strain and effective stress are increased during triaxial deformation experiments (Zoback & Byerlee, 1976). As strain is increased minor pore-volume dilatancy occurs. However, even close to failure, permeability continues to decrease. It is possible that pore dilatancy occurred due to strain localization, which may have increased the permeability parallel to the fault. Indeed, Logan (1992) showed that at high effective stresses (equivalent to overburdens of > 2000 m) and/or high strains, localization resulted in the formation of discrete fractures which provided channels for fault-parallel fluid flow. The quartz cemented cataclastic faults examined do not appear to have experienced localization along discrete shear surfaces and their permeability is likely to have evolved in a similar manner to the sands during the triaxial experiments discussed above.

Small amounts of dilation may occur during the initial stages of cataclastic deformation at low effective stresses, due to grain-rotation and grain-rolling (Antonellini & Aydin, 1994; Menéndez *et al.*, 1996). If the effective stress is sufficient for cataclastic deformation to occur this dilation is short-lived because stress becomes heterogeneously distributed between grains resulting in their brittle deformation (Antonellini & Aydin, 1994). Indeed, Lambe & Whitman (1969) have shown that, at low pressures, dilatancy may occur due to large-scale grain movements, but as the pressure increases this effect is dramatically reduced.

Microstructural analysis and high pressure deformation experiments therefore strongly suggest that the quartz cemented cataclastic faults had far lower permeabilities than their host sandstone both during and following deformation. It is thus argued that they are unlikely to have acted as long-term conduits for fluid flow and that the quartz cement in both the cataclasites and the adjacent undeformed sandstones did not precipitate from silica-rich fluids flowing along the faults.

### *Implications for the origin of quartz cement within host sandstones*

The described deformation features that unequivocally acted as conduits for mineralizing fluids do not contain large volumes of quartz cement. The evidence also suggests that the quartz cemented cataclastic faults were not significant conduits for fluid flow (except during periods of reactivation following quartz cementation) and therefore the quartz cement associated with cataclasites did not precipitate from exotic fault-transported fluids.

It has been argued that fractures could act as conduits for silica-bearing fluids without necessarily becoming quartz cemented (Rimstidt & Barnes, 1980; Giles, 1997). However, many of the cemented fractures examined during this study were formed by the reactivation of cataclastic faults. These are typically composed of very small grains with very high reactive surface areas. We therefore believe that if these reactivated cataclastic faults had been conduits for silica-rich fluids then they should have become preferentially quartz cemented.

These microstructural observations agree with the geochemical modelling of Pedersen & Bjørlykke (1994) which indicates that, under diagenetic conditions, the fluid flux required to completely cement a fracture with quartz would be unrealistically high. The agreement between microstructural and theoretical calculations suggests that there are not unknown mechanisms to transport high concentrations of dissolved silica along faults and fractures within sedimentary basins.

Furthermore, geochemical modelling suggests that if faults had acted as conduits for silica-rich fluids the adjacent undeformed sandstone would be highly quartz cemented in comparison to the reservoir as a whole (Giles, 1997). This was not found to be the case in any of the samples analysed during this study.

The microstructural observations suggest that quartz cement within the reservoir sandstones could not have been supplied by the movement of fluids along faults and fractures and that other processes must be invoked to explain its presence.

## The effect of clay minerals on diffusive mass transfer and quartz cementation

The effect of clay minerals on quartz cementation and DMT is well documented in the literature (e.g. Heald, 1955, 1959; Thompson, 1959; De Boer *et al.*, 1977; Dewers & Ortoleva, 1991; Oelkers *et al.*, 1996). Diffusive mass transfer is a three stage process, involving dissolution, diffusive transport of solutes and precipitation (e.g. Rutter, 1983; Gratier & Guiguet, 1986). It is now recognized that the rate of DMT in sandstones may be controlled by the dissolution or precipitation kinetics of quartz (Dewers & Hajash, 1995) or the rate of diffusion (Worden *et al.*, 1998). Furthermore, the position of clay minerals within a sandstone is a critical control on which of the main steps in the DMT processes is rate determining (Dewers & Ortoleva, 1991).

Small concentrations of clays at grain-contacts may increase the rate of grain-to-grain quartz dissolution (e.g. Heald, 1955, 1959; Thompson, 1959; de Boer *et al.*, 1977). This may occur because traces of clay enhance the rate of silica diffusion from the sites of grain dissolution (Weyl, 1959; Dewers & Ortoleva, 1991) due to the presence of a thicker layer of absorbed water at clay-to-quartz contacts than at quartz–quartz contacts (Tada & Siever, 1989). Alternatively, clay may increase the pH of pore fluids adjacent to quartz grains and locally increase quartz solubility (Thompson, 1959; Oelkers *et al.*, 1996).

It has also been recognized that grain-coating clay minerals may suppress quartz cementation in sandstones (Cecil & Heald, 1971; Tada & Siever, 1989; Dutton & Diggs, 1992; Ehrenberg, 1993). Indeed, Mullis (1991) showed that the precipitation rate constant for quartz may be over three orders of magnitude lower in natural sandstones than was measured in laboratory experiments using clean sand. A number of mechanisms have been proposed to account for the suppression of quartz cementation by grain-coating clays. Dewers & Ortoleva (1991) suggested that grain-coating clays may retard solute diffusion to the grain surface and thereby suppress the quartz precipitation kinetics. On the other hand, Oelkers *et al.* (1996) have argued that the local pH is increased around grain-coating clays which increases silica solubility thereby suppressing quartz precipitation. The problem with both of these mechanisms is that they rely on the layer of grain-coating clays to be entirely continuous and this is unlikely to be the case. An alternative possibility is that the grain-coating clays simply reduce the reactive quartz surface area thereby reducing the rate of quartz precipitation.

### *Enhanced quartz dissolution and precipitation within faults*

It has been shown that the rate of quartz grain-to-grain dissolution may be enhanced by the presence of minor clays at grain contacts but quartz precipitation may be arrested by grain-coating clays. The proportion of clays that are present at grain-to-grain contacts compared with those which coat quartz grains can vary significantly. Clays in many sandstones occur either as clay-rich lithic grains or as pore-filling and grain-coating cements and therefore have a tendency to retard quartz precipitation without increasing the rate of quartz dissolution at grain contacts. However, during deformation, these clays can become intimately mixed with the framework grains resulting in an overall increase in the proportion of grain-contact clays which will increase the susceptibility to quartz grain contact dissolution. This is probably the main reason why faults in sandstones that contain small or moderate quantities of clay often experience enhanced DMT.

In terms of cataclastic faults, small quantities of grain-coating pollutants (clays or adsorbed ions) within the host sandstone may restrict quartz overgrowth development (e.g. Cecil & Heald, 1971; Heald, 1956; Iller, 1979). However, cataclastic deformation of relatively clay-free sandstones generates unpolluted fracture surfaces which provide ideal nucleation sites for silica precipitation. The distribution of quartz cemented cataclastic faults within reservoirs also provides important clues as to the source of the cement. If the quartz cement within cataclasites did precipitate from fault-related fluids it would be expected that quartz cementation of faults as well as reservoir sandstones would be at first approximation independent of structural level. The contrary was found to be true in every field examined during the present study—quartz cemented cataclastic faults were only identified in reservoirs that have at one time been buried to more than 3000 m. This depth is also similar to that which defines the onset of quartz cementation within reservoir sandstones of the North Sea (Ehrenberg, 1990; Bjørlykke *et al.*, 1992; Bjørlykke, 1994; Walderhaug, 1994; Ramm & Ryseth, 1996).

Quartz cementation and dissolution in very clay-rich sandstones is probably arrested due to the suppressed quartz precipitation kinetics. Deformation of such sandstones may increase the proportion of quartz–clay grain contacts, but does not affect the overall rate governing process—in this case precipitation—and therefore the resulting faults do not experience enhanced grain-to-grain dissolution.

### *Clay distribution and the origin of quartz cement within sandstones*

The maximum amount of authigenic quartz within some reservoirs was found within essentially clay-free intervals close to stylolites. However, in most reservoirs, the maximum amount of authigenic quartz was found in sandstones containing small quantities of detrital clays. The quartz cement variations described here are consistent with the lack of consensus in the literature as to whether quartz cement is concentrated in coarse grained (e.g. McBride, 1989) or fine grained (e.g. Walderhaug, 1996) sandstones.

The new data presented here support the idea that authigenic quartz is supplied by local quartz dissolution at quartz–clay contacts and that clay minerals at grain-contacts enhance quartz precipitation whereas grain-coating clays suppress quartz precipitation. In the case of the clean sandstones adjacent to stylolites, quartz is dissolved at the stylolite boundaries and precipitates in the interstylolite region where detrital clays do not suppress its precipitation. In reservoirs containing dispersed clays, there appears to be an optimum concentration of detrital clays for quartz overgrowth development. The extent of quartz cementation may be limited by quartz dissolution kinetics in sandstones containing lower concentrations of clays than this optimum and by quartz precipitation kinetics in more clay-rich lithologies. In particular, the rate of silica supply for quartz cementation in the very clean sandstones may be limited by the low abundance of clay–quartz contacts along which quartz dissolution occurs. However, whilst there are likely to be abundant clay–quartz contacts in the clay-rich sandstones (which could provide excess dissolved silica) the surfaces of quartz grains are likely to be clay-coated which will reduce the rate of quartz precipitation.

The relationship between quartz cement volume and clay content described above is far from simple. This is because it is the textural *distribution* of clay, not their absolute concentration, which controls the rate of quartz precipitation and dissolution. This is emphasized by the observation that whilst very small quantities (< 3%) of grain-coating authigenic illite are needed to suppress quartz cementation in the Rotliegende of the southern North Sea, low concentrations of detrital clays are required in fault rocks for them to experience enhanced quartz dissolution. The distribution of clays in the reservoirs from which the data shown in Fig. 6(a–d) was obtained varies dramatically. In the reservoir depicted in Fig. 6(d), most of the clays are concentrated along stylolites and are therefore ideally situated to increase the rate of quartz dissolution without decreasing the rate of quartz precipitation. In the reservoirs shown in Fig. 6(a–c), clays are present both as clay-rich lithic grains and as detrital particles situated between quartz grain-contacts, the former suppressing quartz precipitation, the latter enhancing dissolution.

### The effect of temperature/burial depth on quartz dissolution and precipitation

Quartz cementation and dissolution are not significant in reservoirs that have not achieved burial of ⩾ 2700 m. This observation does not, however, provide evidence as to whether the onset of these processes was governed by pressure or temperature. It is possible to elucidate whether temperature or pressure is the most important control onset of quartz precipitation and dissolution by comparing the results with those obtained from basins with different geothermal gradients. The North Sea has a geothermal gradient of ~35°C (Egeberg & Aagaard, 1989), therefore the depth of 2700 m which marks the onset of quartz dissolution and precipitation is equivalent to a temperature of ~90°C. A review of published data from sedimentary basins with varying geothermal gradients shows, however, that the onset of quartz cementation invariably occurs at ~90°C (Giles *et al.*, this volume, pp. 21–38). These results are compatible with the experimental results of Rimstidt & Barnes (1980) which show that the rate of quartz cementation is exponentially dependent upon temperature but virtually independent of pressure in the range of 0–500 bars. The same conclusion was emphasized by the modelling work of Oelkers *et al.* (1996) which suggests that the rate of quartz precipitation in sandstones is extremely low below 87°C but increases by two orders of magnitude between 87°C and 157°C. Based on the results of Rimstidt & Barnes (1980) and Oelkers *et al.* (1996) it is likely that temperature is more important than pressure in controlling quartz cementation in both cataclasites and undeformed sandstones.

## CONCLUSIONS

**1** Authigenic mesocrystalline quartz is not concentrated within faults and fractures that acted as conduits for fluid flow within the hydrocarbon reservoirs of the North Sea. This suggests that quartz cement within undeformed reservoir sandstones is not supplied by fluid movement along fracture networks.

**2** Enhanced quartz cementation of cataclastic fault rocks in the reservoir sandstones of the North Sea occurs because cataclastic deformation generates quartz fracture surfaces. These provide kinetically favourable sites for quartz cementation. The process is most effective if deformation occurs at temperatures close to, or greater than, 90°C.

**3** Fault rocks in clay-rich sandstones are often sites for enhanced quartz grain contact dissolution due to the deformation-induced insertion of clay between framework grains. Non-horizontal stylolites frequently form in deeply buried reservoirs; these are probably end-products of enhanced grain contact quartz dissolution along such deformation bands.

**4** Quartz cement distribution is not simply controlled by thermodynamic equilibrium, it can in different situations be controlled by precipitation and dissolution kinetics. In particular, grain-coating clays decrease the rate of quartz precipitation, whilst phyllosilicate–quartz grain contacts increase the rate of quartz dissolution.

**5** Quartz cement is not concentrated in the most permeable horizons suggesting that solutes for quartz precipitation were not transported by advective processes. Instead, quartz cement is concentrated within sandstones containing stylolites or small quantities (~5%) of detrital phyllosilicates. The distribution of quartz cement is therefore consistent with the model in which silica was supplied by processes such as quartz dissolution at phyllosilicate–quartz grain contacts and was transported short distances (< 1 m) by diffusion.

## ACKNOWLEDGEMENTS

We would first of all like to thank our industrial sponsors (Agip, Amoco, Arco, British Gas, British Petroleum, Conoco, Chevron, Elf, JNOC, Mobil, Phillips, Texaco and Statoil) who provided financial support and core material for this study. We would also like to thank Sadoon Morad, Jim Hendry and Mogens Ramm for providing thoughtful reviews which greatly improved this manuscript. Comments by Einar Sverdrup on another manuscript by QJF and RJK also improved this paper. Ben Clennell, Barbara Kidd and Alex Harrison are thanked for conducting the physical property analyses.

## REFERENCES

Antonellini, M. & Aydin, A. (1994) Effect of faulting on fluid flow in porous sandstones: petrophysical properties. *Bulletin of the American Association of Petroleum Geologists* **78**, 335–377.

Bjørlykke, K. (1980) Clastic diagenesis and basin evolution. *Revista Del Instituto de Investigaciones Geológicas, Diputacion Provincial, Universidad de Barcelona* **34**, 21–44.

Bjørlykke, K. (1983) Diagenetic reactions in sandstones. In: *Proceedings of the NATO Advanced Study Institute on Sediment Diagenesis.* (eds Parker, A. & Sellwood, B.) pp. 169–213. Reidel, Holland.

Bjørlykke, K. (1994) Fluid-flow processes and diagenesis in sedimentary basins. In: *Geofluids: Origin, Migration and Evolution of Fluids in Sedimentary Basins.* (ed. Parnell, J.) pp. 127–140. Special Publications of the Geological Society of London 78.

Bjørlykke, K. & Egeberg, P.K. (1993) Quartz cementation in sedimentary basins. *Bulletin of the American Association of Petroleum Geologists* **77**, 1538–1548.

Bjørlykke, K., Nedkvitne, T., Ramm, M. & Saigal, G. (1992) Diagenetic processes in the Brent Group (Middle Jurassic) reservoirs of the North Sea—an overview. In: *Geology of the Brent Group.* (eds Morton, A.C., Haszeldine, R.S., Giles, M.R. & Brown, S.) pp. 263–287. Special Publications of the Geological Society of London 61.

Black, J.H. (1987) Flow and flow mechanisms in crystalline rock. In: *Fluid Flow in Sedimentary Basins and Aquifers.* (eds Goff, J.C. & Williams, B.P.J.) pp. 185–200. Special Publications of the Geological Society of London 34.

Blenkinsop, T.G. (1991) Cataclasis and processes of particle size reduction. *Pure and Applied Geophysics* **136**, 60–86.

Borg, I., Friedman, M., Handin, J. & Higgs, D.V. (1960) Experimental deformation of St. Peter sand: a study of cataclastic flow. In: *Rock Deformation.* pp. 133–192. Geological Society of America Memoirs 79.

Brown, D.A. (1987) The flow of water and displacement of hydrocarbons in fractured reservoirs. In: *Fluid Flow in Sedimentary Basins and Aquifers.* (eds Goff, J.C. & Williams, B.P.J.) pp. 201–218. Special Publications of the Geological Society of London 34.

Burley, S.D. (1993) Models of burial diagenesis for deep exploration plays in Jurassic fault traps of the Central and Northern North Sea. In: *Petroleum Geology of Northwest Europe.* (ed. Parker, J.R.) pp. 1353–1375. Geological Society of London.

Cecil, C.B. & Heald, M.T. (1971) Experimental investigation of the effects of grain-coatings on quartz over-growth. *Journal of Sedimentary Petrology* **41**, 582–584.

Clennell, M.B., Knipe, R.J. & Fisher, Q.J. (1998) Fault zones as barriers to, or conduits for, fluid flow in argillaceous formations: a microstructural and petrophysical perspective. In: *Fluid Flow through Faults and Fractures in Argillaceous Formations.* Organisation for Economic Co-operation and Development Publications, Paris, pp. 125–140.

De Boer, R.D., Nagtegaal, P.J.C. & Duyvus, E.M. (1977) Pressure solution experiments on quartz sand. *Geochimica et Cosmochimica Acta* **41**, 249–256.

Dewers, T. & Hajash, A. (1995) Rate laws for water-assisted compaction and stress-induced water–rock interaction in sandstones. *Journal of Geophysical Research* **100**, 13093–13112.

Dewers, T. & Ortoleva, P.J. (1990) Interaction of reaction, mass transport, and rock deformation during diagenesis: mathematical modelling of intergranular pressure solution, stylolites, and differential compaction/cementation. In: *Prediction of Reservoir Quality Through Chemical Modelling.* (eds Meshri, I.D. & Ortoleva, P.J.) pp. 147–160. Memoirs of the American Association of Petroleum Geologists 49.

Dewers, T. & Ortoleva, P.J. (1991) Influences of clay minerals on sandstone cementation and pressure solution. *Geology* **19**, 1045–1048.

Dutton, S.P. (1993) Influence of provenance and burial history on diagenesis of Lower Cretaceous Frontier Formation sandstones, Green River Basin, Wyoming. *Journal of Sedimentary Petrology* **63**, 665–677.

Dutton, S.P. & Diggs, T.N. (1992) Evolution of porosity and permeability in the Lower Cretaceous Travis Peak Formation, East Texas. *Bulletin of the American Association of Petroleum Geologists* **76**, 252–269.

Egeberg, P.K. & Aagaard, P.A. (1989) Origin and evolution of formation waters from oil fields on the Norwegian shelf. *Applied Geochemistry* **4**, 131–142.

Ehrenberg, S.N. (1990) Relationship between diagenesis and reservoir quality in sandstones of the Garn Formation, Haltenbanken, mid-Norwegian continental shelf. *Bulletin of the American Association of Petroleum Geologists* **74**, 1538–1558.

Ehrenberg, S.N. (1993) Preservation of anomalously high-porosity in deeply buried sandstones by grain-coating chlorite—examples from the Norwegian continental-shelf. *Bulletin of the American Association of Petroleum Geologists* **77**, 1260–1286.

Engelder, J.T. (1974) Cataclasis and the generation of fault gouge. *Bulletin of the Geological Society of America* **85**, 1515–1522.

Evans, J., Hogg, A.J., Hopkins, M.S. & Howarth, J.R. (1994) Quantification of quartz cements using combined SEM, CL, and image-analysis. *Journal of Sedimentary Research* **64**, 334–338.

Fisher, Q.J. & Knipe, R.J. (1998) Microstructural controls on the petrophysical properties of deformation features. In: *Faulting and Fault Sealing in Hydrocarbon Reservoirs.* (eds Jones, G., Fisher, Q.J. & Knipe, R.J.). pp. 117–134. Special Publications of the Geological Society of London **147**, 117–134.

Fuchtbauer, H. (1974) *Sediments and Sedimentary Rocks 1.* Wiley, New York.

Gaupp, R., Matter, A., Platt, J., Ramseyer, K. & Walzebuck, J. (1993) Diagenesis and fluid evolution of deeply buried Permian (Rotliegende) gas reservoirs, Northwest Germany. *Bulletin of the American Association of Petroleum Geologists* **77**, 1111–1128.

Giles, M. (1997) *Diagenesis and its Impact on Rock Properties. A Quantitative Perspective.* Kluwer, Amsterdam, 520pp.

Giles, M.R., Indrelid, S.L., Beynon, G.V. & Amthor, J. (2000) (This volume). The origin of large-scale quartz cementation: evidence from large data sets and coupled heat–fluid mass transport modelling. In: *Quartz Cementation in Sandstones.* (eds Morad, S. & Worden, R.H.) pp. 21–38. Special Publications of the International Association of Sedimentology 29. Blackwell Science, Oxford.

Giles, M.R., Stevenson, S., Martin, S.V. *et al.* (1992) The reservoir properties and diagenesis of the Brent Group: a regional perspective. In: *Geology of the Brent Group.* (eds Morton, A.C., Haszeldine, R.S., Giles, M.R. & Brown, S.) pp. 289–327. Special Publications of the Geological Society of London 61.

Gluyas, J.G. (1985) Reduction and prediction of sandstone reservoir potential, Jurassic, North Sea. *Philosophical Transactions of the Royal Society of London, Series A* **135**, 187–202.

Gluyas, J.G. & Coleman, M. (1992) Material flux and porosity changes during sediment diagenesis. *Nature*, 356, 52–54.

Gratier, J.P. & Guiguet, R. (1986) Experimental pressure solution—precipitation on quartz grains: The crucial effect of the nature of the fluid. *Journal of Structural Geology* **8**, 845–856.

Haszeldine, R.S., Samson, I.M. & Cornford, C. (1984) Quartz diagenesis and convective fluid movement; Beatrice Oilfield, UK North Sea. *Clay Mineralogy* **19**, 391–402.

Heald, M.T. (1955) Stylolites in sandstones. *Journal of Geology* **63**, 101–114.

Heald, M.T. (1956) Cementation of Simpson and St. Peter sandstones in parts of Oklahoma, Arkansas, and Missouri. *Journal of Geology* **64**, 16–30.

Heald, M.T. (1959) Significance of stylolites in permeable sandstones. *Journal of Sedimentary Petrology* **29**, 251–253.

Iller, R.K. (1979) *The Chemistry of Silica.* John Wiley, Chichester.

Jourdan, A., Thomas, M., Brevart, O., Robson, P., Sommer, F. & Sullivan, M. (1987) Diagenesis as the control of the Brent sandstone reservoir properties in the Greater Alwyn area (East Shetland Basin). In: *Petroleum Geology of North West Europe.* (eds Brooks, J. & Glennie, K.) pp. 951–961. Graham & Trotman, London.

Knipe, R.J. (1989) Deformation mechanisms—recognition from natural tectonites. *Journal of Structural Geology* **1**, 127–146.

Knipe, R.J. (1992) Faulting processes and fault seal. In: *Structural and Tectonic Modelling and its Application to Petroleum Geology.* (eds Larsen, R.M., Brekke, H., Larsen, B.T. & Talleraas, E.) pp. 325–342. Special Publications of the NPF 1.

Knipe, R.J., Fisher, Q.J., Jones, G. *et al.* (1997) Fault seal prediction methodologies, applications and successes. In: *Hydrocarbon Seals—Importance for Exploration and Production* (ed.) pp. 15–38. Special Publications of the NPF.

Lambe, T.W. & Whitman, R.V. (1969) *Soil Mechanics.* John Wiley, New York.

Leveille, G.P., Knipe, R.J., More, C. *et al.* (1997) Compartmentalisation of Rotliegendes gas reservoirs by sealing faults, Jupiter Fields Area, Southern North Sea. In: *Hydrocarbon Potential of the Southern North Sea.* (eds Ziegler, K., Turner, P. & Daines, S.R.) pp. 97–104. Special Publications of the Geological Society of London 123.

Lindsay, N.G., Murphy, F.C., Walsh, J.J. & Watterson, J. (1993) Outcrop studies of shale smears on fault surfaces. *Special Publications of the International Association of Sedimentology* **15**, 113–123.

Logan, J.M. (1992) The influence of fluid flow on the mechanical behaviour of faults. In: *Rock Mechanics.* (eds Tillerson, J.R. & Wawersik, W.R.) pp. 141–149. Balkema, Rotterdam.

McBride, E.F. (1989) Quartz cement in sandstones—a review. *Earth Science Review* **26**, 69–112.

Menéndez, B., Zhu, W. & Wong, T.-F. (1996) Micromechanics of brittle faulting and cataclastic flow in Berea Sandstone. *Journal of Structural Geology* **18**, 1–16.

Mullis, A.M. (1991) The role of silica precipitation kinetics in determining the rate of quartz pressure solution. *Journal of Geophysical Research* **96**, 10007–10013.

Oelkers, E.H., Bjørkum, P.A. & Murphy, W.M. (1996) A petrographic and computational investigation of quartz cementation and porosity reduction in North Sea sandstones. *American Journal of Science* **296**, 420–452.

Pedersen, T. & Bjørlykke, K. (1994) Fluid flow in sedimentary basins: a model of water flow in a vertical fracture. *Basin Research* **6**, 1–16.

Price, M. (1987) Fluid flow in the Chalk of England. In: *Fluid Flow in Sedimentary Basins and Aquifers.* (eds Goff, J.C. & Williams, B.P.J.) pp. 141–156. Special Publications of the Geological Society of London 34.

Ramm, M. (1992) Porosity–depth trends in reservoir sandstones: offshore Norway. *Marine and Petroleum Geologists* **9**, 553–567.

Ramm, M. & Ryseth, A.E. (1996) Reservoir quality and burial diagenesis in the Statfjord Formation. *Petrological Geoscience* **2**, 313–324.

Rimstidt, J.D. & Barnes, H.L. (1980) The kinetics of silica–water reactions. *Geochimica et Cosmochimica Acta* **44**, 1683–1699.

Rutter, E.H. (1983) Pressure solution in nature, theory and experiment. *Journal of the Geological Society of London* **140**, 725–740.

Scotchman, I., Johnnes, L.H. & Miller, R.S. (1989) Clay diagenesis and oil migration in Brent Group sandstones of NW Hutton field, UK North Sea. *Clay Mineralogy* **24**, 339–374.

Sibson, R.H. (1987) Earthquake rupturing as a mineralising element in hydrothermal systems. *Geology* **15**, 701–704.

Sibson, R.H., Robert, F. & Poulsen, K.H. (1988) High-angle reverse faults, fluid pressure cycling and mesothermal gold–quartz deposits. *Geology* **16**, 551–555.

Sorby, H.C. (1880) On the structure and origin of non-calcareous stratified rocks. *Quarterly Journal of the Geological Society of London* **37**, 49–92.

Spiers, C.J. & Schutjens, P.M.T.M. (1990) Densification of crystalline aggregates by fluid phase diffusional creep. In: *Deformation Processes in Minerals, Ceramics and Rocks.* (eds Barber, D.J. & Meredith, P.G.) pp. 334–353. Special Publications of the Mineralogical Society 1.

Sullivan, M.D., Haszeldine, R.S. & Fallick, A.E. (1990) Linear coupling of carbon and strontium isotopes in Rotliegend Sandstone, North Sea: Evidence for cross-formational fluid flow. *Geology* **18**, 1215–1218.

Tada, R. & Siever, R. (1989) Pressure solution during diagenesis: a review. *Annual Review of Earth and Planetary Science* **17**, 89–118.

Thompson, A. (1959) Pressure solution and porosity. In: *Silica in Sediments.* (ed. Ireland, H.A.) pp. 92–110. Special Publications of the Society of Economic Paleontology and Mineralogy 7.

Walderhaug, O. (1994) Temperatures of quartz cementation in Jurassic sandstones from the Norwegian continental shelf–evidence from fluid inclusions. *Journal of Sedimentary Research* **64**, 311–323.

Walderhaug, O. (1996) Kinetic modelling of quartz cementation and porosity loss in deeply buried sandstone reservoirs. *Bulletin of the American Association of Petroleum Geologists* **80**, 731–745.

Waldschmidt, W.A. (1941) Cementing materials in sandstones and their influence on the migration of oil. *Bulletin of the American Association of Petroleum Geologists* **25**, 1839–1879.

Weyl, P.K. (1959) Pressure solution and the force of crystallisation—a phenomenological theory. *Journal of Geophysical Research* **64**, 2001–2025.

Wood, J.R. & Hewett, T.A. (1984) Reservoir diagenesis and convective fluid flow. In: *Clastic Diagenesis.* (eds McDonald, D.A. & Surdam, R.C.) pp. 99–110. *Memoirs of the American Association of Petroleum Geologists* 37.

Worden, R.H., Oxtoby, N.H. & Smalley, P.C. (in press) Can oil emplacement stop quartz cementation in sandstones? *Petrology and Geoscience* **42**, 129–138.

Zoback, M.D. & Byerlee, J.D. (1976) Effect of high-pressure deformation on permeability of Ottawa Sand. *Bulletin of the American Association of Petroleum Geologists* **60**, 1531–1542.

*Spec. Publs int. Ass. Sediment.* (2000) **29**, 147–161

# Petrophysical and petrographical analysis of quartz cement volumes across oil–water contacts in the Magnus Field, northern North Sea

S. A. BARCLAY *and* R. H. WORDEN

*School of Geosciences, The Queen's University of Belfast, Belfast, BT7 1NN, Northern Ireland*

## ABSTRACT

Quartz cement is a significant porosity-reducing mineral cement in many sandstones and thus affects flow-rate (through its effect on permeability) and calculation of reserves. The presence of oil in a reservoir is commonly assumed to retard quartz cement precipitation and thus early oil emplacement is often thought to preserve porosity and permeability. A combined petrographic and wireline log approach was utilized to investigate whether quartz cement volumes and the total quantity of quartz varies across the oil–water contact in a sandstone reservoir. Thin-section point-count data and bulk density, neutron porosity and sonic transit time wireline log data were obtained across the oil–water contact from three wells in the Magnus Field, an Upper Jurassic turbidite sandstone reservoir in the northern North Sea. Oil filled inclusions in quartz overgrowths in this reservoir show that quartz cementation occurred either during or after oil emplacement. Point-count data were used to determine quartz cement and total quartz volumes across the oil–water contact, whilst wireline data were transformed to estimate the total quantity of quartz across the oil–water contact. Results show that the volume of quartz cement and the total volume of quartz show little or no variation across the oil–water contact. These data imply that the presence of oil in the reservoir had no appreciable effect on the component processes involved in quartz cementation in this field: a paradox that needs to be further investigated in other reservoirs.

## INTRODUCTION

The distribution of quartz cement can be a major control on sandstone reservoir quality (Coskun *et al.*, 1993). Quartz cement can reduce porosity by occluding pores, thus reducing oil volume. It can also have an effect on permeability by the general reduction in porosity and specifically by reducing the diameter of pore throats.

One of the major factors that has widely been interpreted to either retard, or halt quartz cementation in a reservoir is the emplacement of oil prior to cementation (e.g. Glasmann *et al.*, 1989; Robinson & Gluyas, 1992). It is commonly thought that replacing pore water by oil would halt inorganic geochemical processes, including those involved in quartz cementation. This interpretation, however, does not take account of either the preferred wetting state of the reservoir or the source of the silica in the quartz cement. Although a reservoir may have reached maximum oil saturation, it can be water-wet and the pore network can be filled with approximately 20% water. There exists therefore the possibility of continued silica transport, quartz dissolution and silica precipitation after oil emplacement (Walderhaug, 1990).

The possibility of quartz cementation continuing in the presence of oil is strongly influenced by the water saturation of the reservoir and its wettability (Worden *et al.*, 1998). Oil inclusions are not uncommon in quartz cements (Kvenvolden & Roedder, 1971) although these are often considered to form early in oil-filling history before maximum oil saturation has been achieved (Larter & Aplin, 1995). The oil trapped in the inclusions is often less mature than the oil in the reservoir (Larter & Aplin, 1995). In areas where the oil source rock was progressively being buried and heated (e.g. the North Sea), oil migrating to the reservoir could become more mature with time. This suggests that oil inclusions typically are formed during the early stages of reservoir filling and do not permit the implication of continued quartz cementation at maximum oil saturation.

Gluyas *et al.* (1993) illustrated an inverse relationship between quartz cement volume and oil inclusion abundance, with the greatest abundance of oil inclusions at the crest of the field. Assuming reservoirs fill from the crest to the flank (England *et al.*, 1987), this inverse relationship allows us to infer that early migrated oil was trapped to form oil inclusions at the crest and that as the oil saturation of the reservoir decreased toward the flanks, the growth of quartz cement was less inhibited towards the flanks of the field (Gluyas *et al.*, 1993).

Quartz is usually preferentially water-wet (Schlangen *et al.*, 1995; Barclay & Worden, 1997). One of the key problems with the formation of oil inclusions is the mechanism of trapping a non-wetting fluid. Macleod *et al.* (1993) showed that oil in inclusions is more enriched in polar compounds compared with the oil in the reservoir. Brown & Neustadter (1980) and Schlangen *et al.* (1995) have both demonstrated that polar compounds can act as surfactants for the water–oil–quartz system, altering the wetting preference of quartz from hydrophilic to oleophilic. This might permit quartz to trap droplets of oil, thus forming oil inclusions, although how an oil-wet system allows quartz cementation remains to be understood.

The evidence presented above implies that oil inclusions form in conditions that are potentially unrepresentative of the reservoir at maximum oil saturation with a mature oil, and therefore should not be used as proof that quartz cementation continues throughout reservoir filling. However, oil inclusions may still be used to show that oil emplacement and quartz cementation occurred synchronously in a reservoir, as oil inclusions cannot form in the absence of oil in the reservoir.

In this paper, we describe the distribution of quartz cement across the oil–water contact (OWC) of a submarine fan sandstone hydrocarbon reservoir: the Magnus Field in the northern North Sea, United Kingdom Continental Shelf (UKCS). We have used a combination of point-count data and wireline analysis methods to examine the distribution of quartz cement and the bulk distribution of quartz in the reservoir. The results obtained allow us to draw conclusions on (i) the possibility of quartz cementation continuing after oil emplacement in Magnus; (ii) the likely preferred wetting-state of the reservoir during oil emplacement; (iii) the potential source of the silica in the quartz cement; and (iv) the use of wireline logs to obtain information about total quartz and quartz cement volumes in reservoirs.

## THE MAGNUS FIELD

### Regional setting

The Magnus Field lies 160 km north-east of the Shetland Islands and is within UKCS exploration blocks 211/12a and 211/7a (Fig. 1). The field occurs at the southern margin of the North Shetland Basin, typified by easterly dipping fault blocks and Upper Jurassic reservoir sandstones (Figs 1 & 2; De'Ath & Schuyleman, 1981). The main reservoir, the Magnus Sandstone Member, occurs stratigraphically between the Lower and Upper Kimmeridge Clay Formations (Fig. 3).

### Structural and stratigraphic evolution

Deposition of the Magnus Sandstone and Upper Kimmeridge Clay Formation (Fig. 3) was followed by a depositional hiatus during the early Cretaceous (late Cimmerian) when the reservoir was faulted, uplifted, exposed and eroded. Deposition of the overlying Cretaceous sediments was further interrupted by a second break in deposition in the mid-Cretaceous (Cenomanian–Turonian) with erosion locally removing Lower Cretaceous and Jurassic sediments from the northern area of the field. Rapid subsidence and deposition occurred during the Tertiary and Quaternary, with tilting of the reservoir towards the north-east during the early Tertiary (De'Ath & Schuyleman, 1981).

### Reservoir characteristics and oil source

Maximum closure of the Magnus Sandstone Member is 350 m and the field covers an area of 34 $km^2$, the maximum vertical thickness of gross oil sand is 140 m, although it thins both eastwards and westwards. The mean porosity of the Magnus Sandstone ranges from 25% in the western half of the field to 19% in the eastern half of the field (De'Ath & Schuyleman, 1981).

Oil was sourced from the Kimmeridge Clay Formation, the source is thought to be downflank to the north and east of the reservoir (Emery *et al.*, 1993). Timing of the major phase of oil generation from the source was during the Late Cretaceous, *c.*75 Ma (Shepherd *et al.*, 1990). The oil is contained within a combined stratigraphic and structural trap. The seal is a combination of the overlying unconformable Cretaceous marls and the conformable Upper Kimmeridge Clay Formation mudstones (De'Ath & Schuyleman, 1981; Emery *et al.*, 1993).

## GEOLOGICAL CHARACTERISTICS OF THE MAGNUS SANDSTONE MEMBER

### Sedimentology

The Magnus sandstones are predominantly submarine fan, subarkosic to arkosic (average feldspar content = 15%), fine-to coarse-grained and generally poorly sorted sediments (De'Ath & Schuyleman 1981). Typical

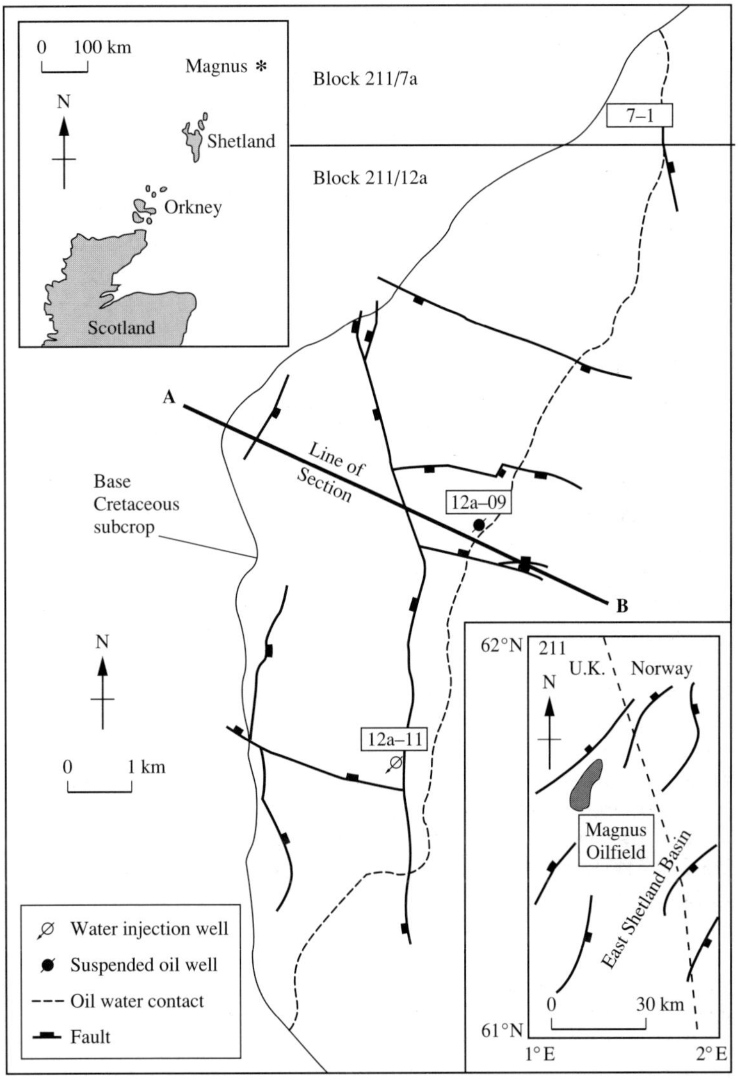


**Fig. 1.** Maps showing location of Magnus field in northern North Sea and well locations.

composition of the Magnus Sandstone is as follows: detrital quartz ~50 vol.%; detrital feldspar 10–20 vol.%; mica and detrital clay ~1 vol.%; authigenic minerals (see next section) 9–19 vol.%; and porosity ~20 vol.%. Sedimentological analysis of the Magnus Sandstone Member revealed that five distinct depositional lithofacies are present (Table 1). In terms of the volume of oil in the reservoir, lithofacies IV (see Table 1 for description) is the most important (Emery *et al.*, 1993). All of the data presented in this paper are from lithofacies IV, thus negating facies-dependent controls on cementation.

## Diagenetic history

Early diagenesis of the Magnus Sandstone Member occurred during and shortly after deposition, with the precipitation of non-ferroan calcite and pyrite (Emery *et al.*, 1993). The next phase of diagenesis occurred when the Magnus Sandstone Member was subaerially exposed during the Early Cretaceous, allowing entry of meteoric water causing K-feldspar dissolution, and precipitation of kaolinite (Emery *et al.*, 1990). The Magnus Sandstone Member was subsequently reburied and compacted,

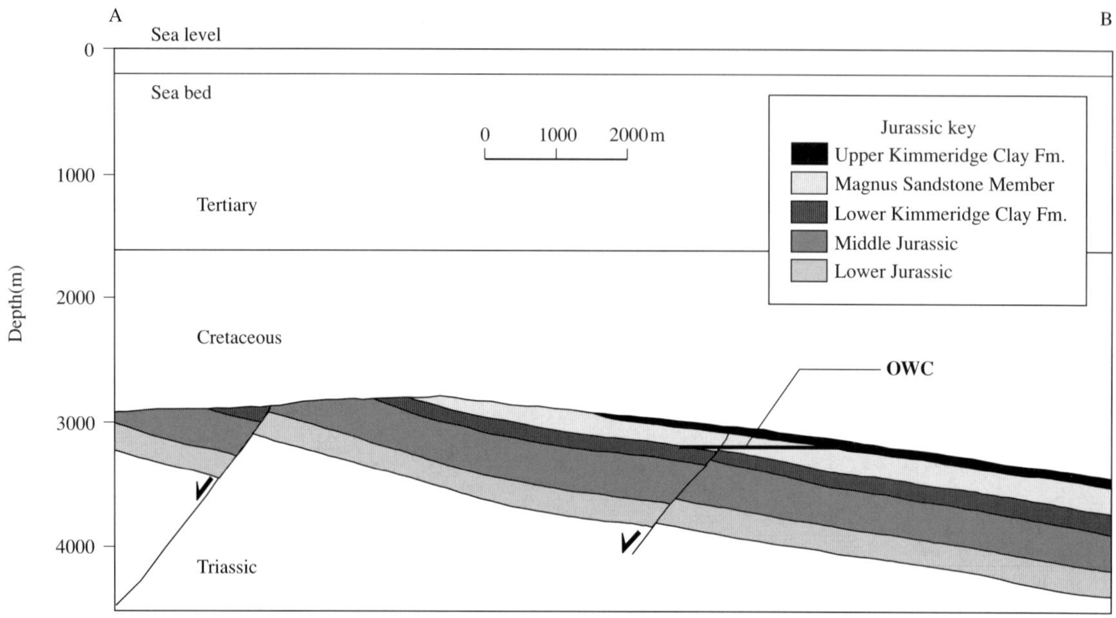


**Fig. 2.** Geological cross-section of the Magnus Field (line of section AB shown on Fig. 1).

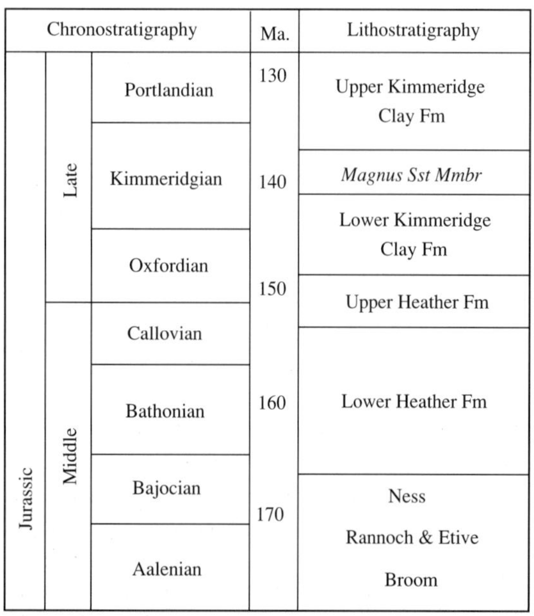


**Fig. 3.** Stratigraphic column showing the Magnus Sandstone Member.

although negligible cementation occurred until the first oil reached the sandstone. The movement of oil into the Magnus sandstones during the Late Cretaceous (*c.*80 Ma) coincided with the onset of deep (> 2.5 km) burial diagenesis (Emery *et al.*, 1993). The cements formed during deep burial diagenesis are (in paragenetic order)—K-feldspar, quartz, kaolinite, illite, siderite and ankerite (Emery *et al.*, 1993). K-feldspar and siderite cements are of very minor significance (> 0.5 vol.%) and as such no variations across the oil–water contact have been noted. Shepherd *et al.* (1990) reported that the volume of illite in the water leg of the reservoir was much greater than the volume in the oil leg, this implies that the presence of oil in the reservoir halted illite cementation. The volume

**Table 1.** Reservoir lithofacies in the Magnus Field. Lithofacies IV is the most important in terms of the volume of trapped petroleum

| Lithofacies | Description | Depositional environment |
|---|---|---|
| 0 | Laminated mudstones of hemipelagic turbiditic origin | Basin plain |
| I | Various sediment remobilized during mass flow processes | Outer-fan and basin plain |
| II & III | Interlaminated thin mudstone and very fine/fine grained sandstones | Mid-fan (interchannel, lobe) or outer fan |
| IV | Thickly bedded fine- to coarse-grained sandstones | Channelled or unchannelled fan lobes in mid-fan |

of ankerite increases around intra-reservoir mudstone bands and the enclosing Kimmeridge Clay Formation (Macauley *et al.*, 1992). This indicates that ankerite distribution is controlled by possible sources of $CO_2$ and metal cations (i.e. from mudstones) rather than the presence or absence of oil.

## SAMPLES AND METHODS

Wireline and point-count data were obtained from across the oil–water contact (OWC) for three wells in the Magnus Field: 211/12a-11 (depth range = 3140 m to 3220 mRKB); 211/12a-09 (depth range = 3180 m to 3380 mRKB); and 211/7–1 (depth range = 3160 m to 3220 mRKB). The locations of these wells are shown on Fig. 1.

### Wireline data

Sonic transit time, neutron porosity and bulk density wireline log data for each well were used to derive three mineral components ('quartz', 'clay' and 'dolomite') and porosity using the methodology introduced by Savre (1963), and described by Doveton (1994) and Hearst & Nelson (1985) for each depth interval. There are clearly more than these three mineral components present in the reservoir (e.g. various feldspar types), but these three were chosen because previous work on the Magnus Sandstone has shown that these three components are volumetrically dominant (De'Ath & Schuyleman, 1981), and the three wireline logs used could differentiate between them successfully. Dolomite was substituted for ankerite when calculating the wireline-log-derived mineralogy, as more reliable wireline-tool responses were available for dolomite than ankerite.

The rationale behind the mineralogy derivation method is that different minerals have different characteristic responses to the sonic, neutron and density tools. The signals from the sonic transit time, neutron porosity and bulk density logs can be integrated and resolved for three mineral types and total porosity using three algorithms relating each separate log signal at any given depth to solid grain volume (occupied by the three minerals) and the assumption that the sum of the three mineral fractions plus porosity equals unity (Equation 1). This approach also assumes a linear relationship between mineral proportions and their contribution to the response on any of the logs used. Therefore, with four equations and four unknowns (the proportions of the three minerals and porosity), the following algorithms (Equations 1–4) can be solved simultaneously at each depth.

**Table 2.** Definition of terms used in Equations 1 to 4

| Term | Definition |
|---|---|
| $\Delta t$ | Sonic transit time recorded by log ($\mu$sec ft$^{-1}$) |
| $\Delta t_{minx}$ | Sonic transit time of mineral X ($\mu$sec ft$^{-1}$) |
| $\Delta t_{\Phi}$ | Sonic transit time of fluid in pore space ($\mu$sec ft$^{-1}$) |
| $\rho$ | Density recorded by log (g cm$^{-3}$) |
| $\rho_{minx}$ | Density of mineral X (g cm$^{-3}$) |
| $\rho_{\Phi}$ | Density of fluid in pore space (g cm$^{-3}$) |
| $\Phi n$ | Neutron porosity recorded by log (porosity units) |
| $\Phi n_{minx}$ | Neutron porosity of mineral X (porosity units) |
| $\Phi n_{\Phi}$ | Neutron porosity of fluid in pore space (porosity units) |
| minX | Proportion of mineral X (as a fraction of total rock volume) |
| $\Phi$ | Porosity (as a fraction of total rock volume) |

$$\text{unity} = \text{min1} + \text{min2} + \text{min3} + \phi \quad (1)$$

$$\rho = \text{min1} \cdot \rho_{\text{min1}} + \text{min2} \cdot \rho_{\text{min2}} + \text{min3} \cdot \rho_{\text{min3}} + \rho \cdot \phi \quad (2)$$

$$\phi n = \text{min1} \cdot \phi n_{\text{min1}} + \text{min2} \cdot \phi n_{\text{min2}} + \text{min3} \cdot \phi n_{\text{min3}} + \phi n \cdot \phi \quad (3)$$

$$\Delta t = \text{min1} \cdot \Delta t_{\text{min1}} + \text{min2} \cdot \Delta t_{\text{min2}} + \text{min3} \cdot \Delta t_{\text{min3}} + \Delta t \cdot \phi \quad (4)$$

The terms used in Equations 1–4 are defined in Table 2. The sonic transit time, neutron porosity and bulk density wireline responses for the three minerals were taken from Rider (1986).

The value of 'quartz' derived from the wireline data used includes all types of detrital and diagenetic quartz as well as feldspars. This is because the three wireline logs used are insensitive to the difference between detrital and authigenic quartz, also quartz and feldspar have a very similar response on these logs. Therefore the wireline 'quartz' data actually represents a 'pseudo-quartz' value which may be defined:

'pseudo-quartz' = mono- and polycrystalline detrital quartz + quartz cement + feldspar (5)

The gamma log has often been used in the past to quantify the clay content of reservoir sandstones (Hearst & Nelson, 1985). The composite gamma log records the total potassium, thorium and uranium content of the rock; the spectral gamma log differentiates between the gamma radiation from the three elements. Apart from illite, most clay minerals do not contain any potassium, the composite gamma log and the potassium spectral gamma log indiscriminately record the total abundance of potassium feldspar, illite and mica. Potassium feldspar is not uncommon in the Magnus Sandstone (classified as subarkosic; De'Ath & Schuyleman, 1981) which also contains ~2 vol.% of mica and potassium-bearing clays such as illite (typically 5–8 vol.%). Using the composite gamma log to attempt to estimate the clay content of the Magnus Sandstone will produce artificially high estimates of the clay content. The problem is compounded because many detrital feldspars contain variable and unpredictable quantities of potassium.

The thorium gamma signal (derived from the spectral gamma log), has been used in the past to identify and quantify the amount of kaolinite in sandstones (Serra *et al.*, 1980; Quirein *et al.*, 1982). However, Hurst & Mildowski (1996) have recently shown that the thorium gamma signal reflects the abundance of thorium-bearing heavy minerals (e.g. monazite) within the sandstone, and bears no genetic relationship to the amount of kaolinite.

It was therefore not possible to use the gamma wireline logs to quantify potassium feldspar or clay in the reservoir, because of the ambiguities inherent in the allocation of the radioactive potassium or thorium signal to feldspar and clay minerals.

### Petrography

Thin-sections were prepared using blue-dyed epoxy impregnation and stained for feldspars and carbonates using standard techniques (Hayes & Klugman, 1959; Dickson, 1965). Petrographic data were originally collected by British Petroleum on the basis of 200 solid grain counts (with porosity counted on a separate channel) per section. The point count data were also converted into values of point-count pseudo-quartz using Equation 5 to facilitate comparison between wireline and petrographic data.

## RESULTS

The point-count quartz cement volume, point-count pseudo-quartz volume and wireline-derived pseudo-quartz volume are plotted as percentages as a function of measured depth (mRKB) for wells 211/12a-11, 211/12a-09 and 211/7–1 (Figs 4, 5 and 6, respectively). The other lithological data derived from the wireline logs (i.e. 'dolomite', 'clay' and 'porosity') are omitted to simplify the data presentation. Mean and standard deviation of the wireline pseudo-quartz percentage, point-count pseudo-quartz percentage and quartz cement percentage are presented in Table 3.

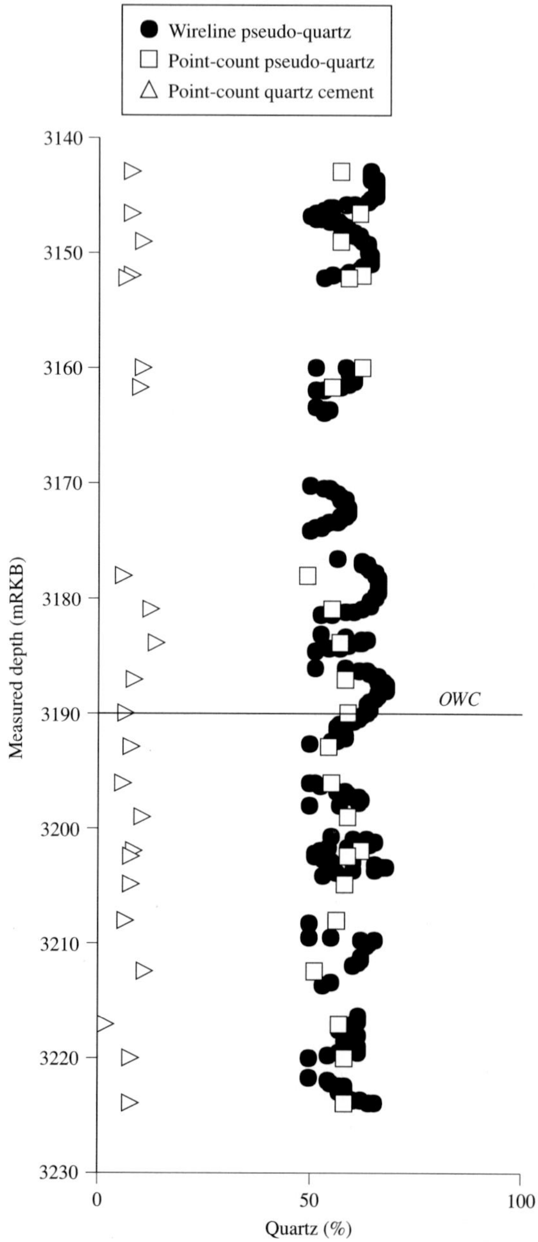


**Fig. 4.** Variation of point-count quartz cement percentage (open triangles), point-count pseudo-quartz percentage (open squares) and wireline pseudo-quartz percentage (black circles) across the oil–water contact (OWC) for 211/12a-11.

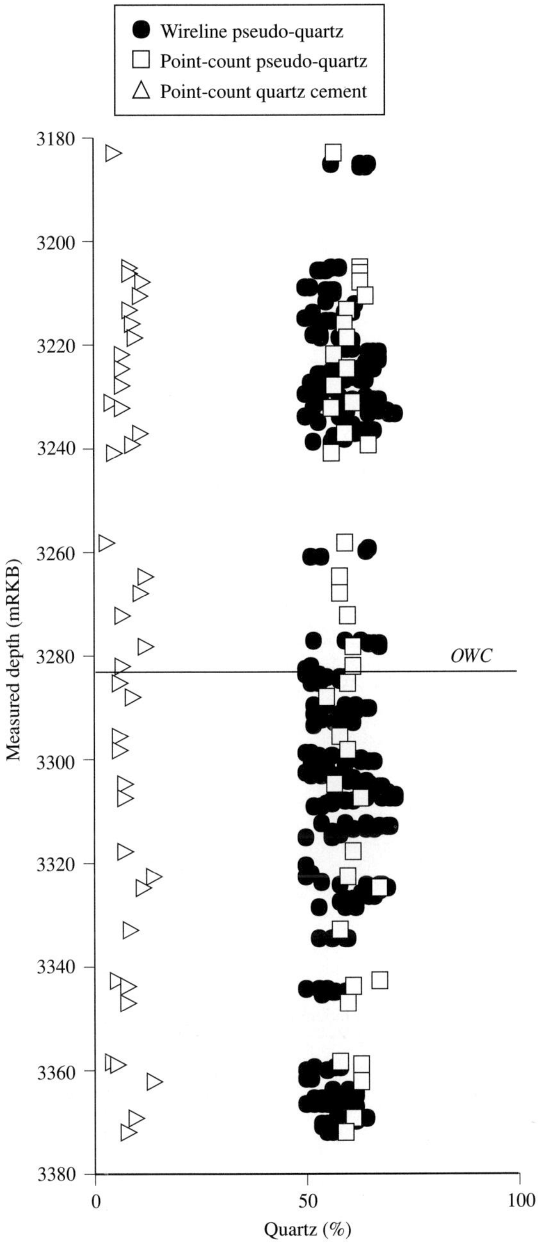


**Fig. 5.** Variation of point-count quartz cement percentage (open triangles), point-count pseudo-quartz percentage (open squares) and wireline pseudo-quartz percentage (black circles) across the oil–water contact (OWC) for 211/12a-09.

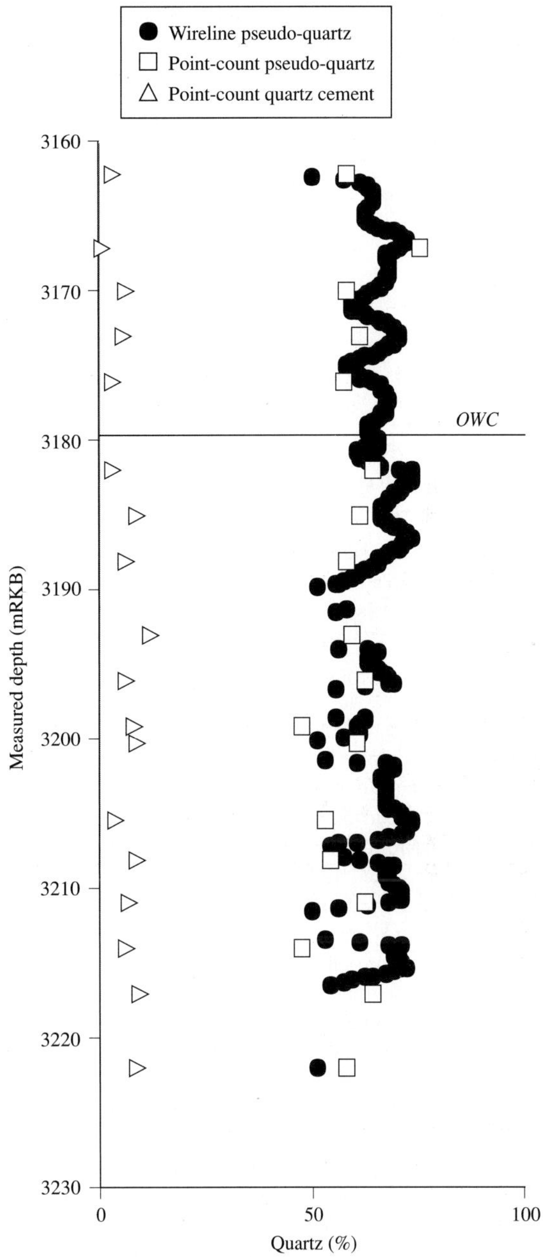


**Fig. 6.** Variation of point-count quartz cement percentage (open triangles), point-count pseudo-quartz percentage (open squares) and wireline pseudo-quartz percentage (black circles) across the oil–water contact (OWC) for 211/7–1.

**Table 3.** Calculated means (± standard deviations) for the point-count quartz cement (%), point-count pseudo-quartz (%) and wireline pseudo-quartz (%) in each of the studied wells

| Well | Position | Point-count quartz cement (%) | Point-count pseudo-quartz (%) | Wireline pseudo-quartz (%) |
|---|---|---|---|---|
| 211/12a-09 | Oil leg | 8 (± 3) | 60 (± 3) | 60 (± 5) |
| | Water leg | 8 (± 3) | 61 (± 3) | 59 (± 6) |
| 211/12a-11 | Oil leg | 9 (± 2) | 57 (± 4) | 60 (± 5) |
| | Water leg | 8 (± 3) | 58 (± 3) | 59 (± 4) |
| 211/7-1 | Oil leg | 4 (± 2) | 62 (± 6) | 65 (± 4) |
| | Water leg | 7 (± 2) | 58 (± 6) | 65 (± 5) |

## DISCUSSION

### The timing of oil emplacement on quartz cementation—previous work on Magnus

Oil emplacement and quartz cementation are thought to have occurred synchronously in Magnus (Emery *et al.*, 1993). The burial and thermal history of the Magnus Field source rock, the Kimmeridge Clay Formations, indicates that oil started migrating into the Magnus reservoir at approximately 80 Ma. Primary aqueous fluid inclusions reveal a relatively restricted range of homogenization temperatures (approximately 90–120°C; Emery *et al.*, 1993). This range of temperatures was equated to quartz cementation occurring between 80 and 65 Ma for quartz cementation (from thermal history models of the Magnus Field; Emery *et al.*, 1993). The coincidence of oil emplacement and quartz cementation was used by Emery *et al.* (1993) to infer that the two events were synchronous. The presence of primary oil inclusions in quartz cement within Magnus was used by Emery *et al.* (1993) as supporting evidence for the coincidence of oil emplacement and quartz cementation.

### Distribution of quartz in the Magnus Field

The mean amounts of wireline and point-count pseudo-quartz do not vary by a significant amount across the oil–water contact (OWC) in any of three wells (Figs 7–9; Table 3). This suggests that the total amount of quartz is uniformly distributed about the OWC in Magnus. The mean amount of point-count quartz cement also does not vary significantly across the OWC (Table 3). There is a slight difference in well 211/7–1, where the mean amount of point-count quartz cement increases from 4% in the oil

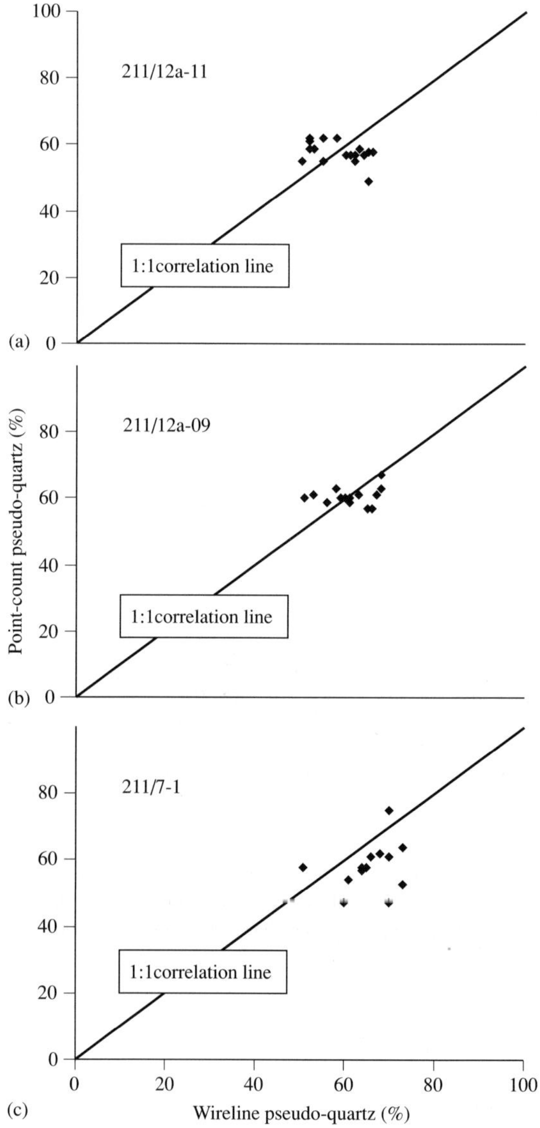


**Fig. 7.** (*Right.*) (a–c) Cross-plots of point-count pseudo-quartz and wireline pseudo-quartz for wells 211/12a-11, 211/12a-09 and 211/7–1, respectively.

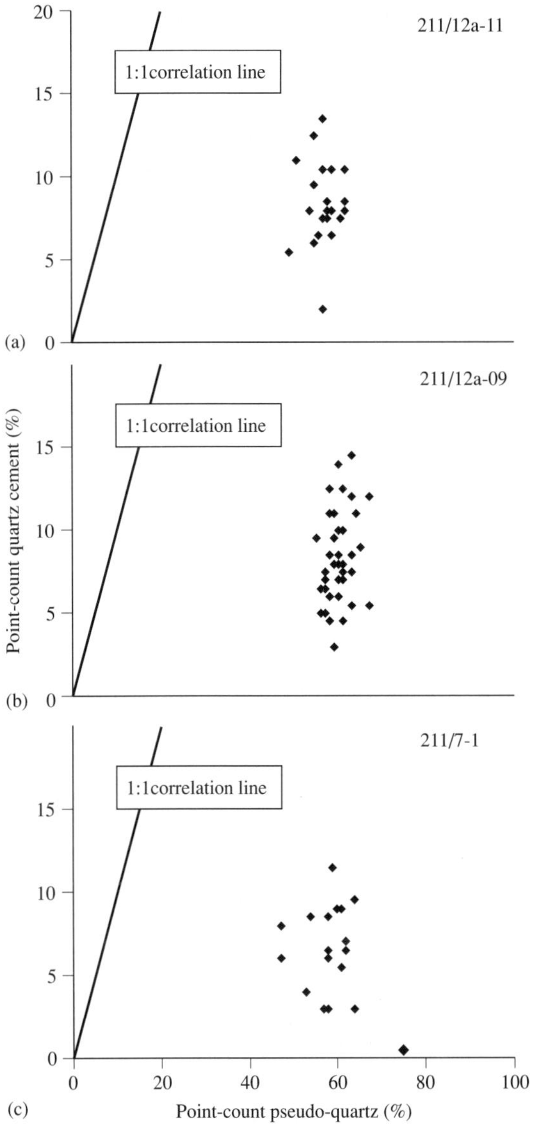


**Fig. 8.** (a–c) Cross-plots of point-count quartz cement and point-count pseudo-quartz for wells 211/12a-11, 211/12a-09 and 211/7–1, respectively.

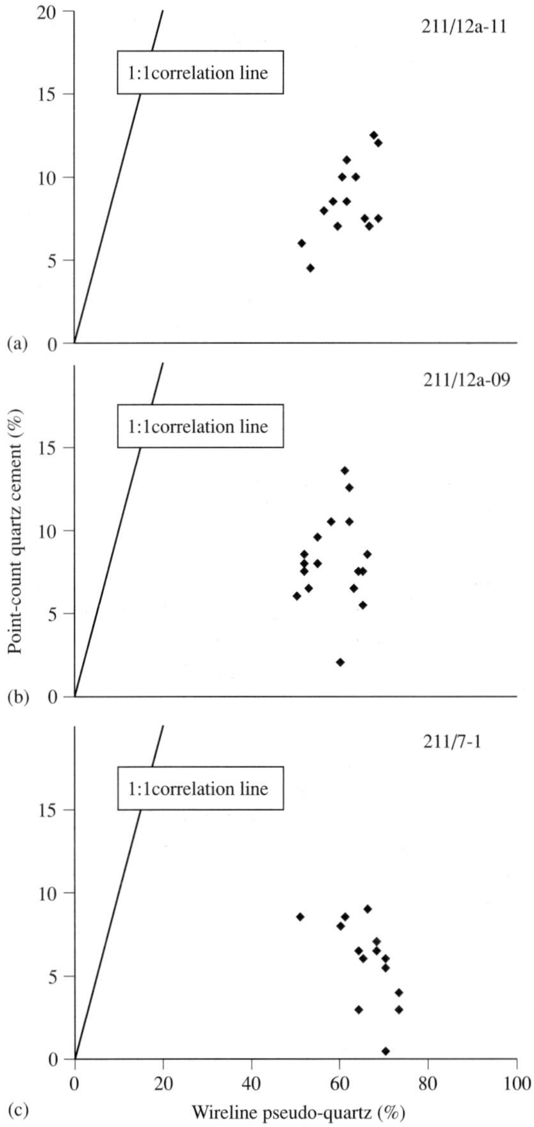


**Fig. 9.** (a–c) Cross-plots of point-count quartz cement and wireline pseudo-quartz for wells 211/12a-11, 211/12a-09 and 211/7–1, respectively.

leg to 7% in the water leg (both values on a scale of 0–1), this may suggest an increase in the mean amount of quartz cement in the water leg relative to the oil leg.

## Comparison of point-count pseudo-quartz and wireline pseudo-quartz data

The point-count pseudo-quartz and wireline pseudo-quartz data for wells 211/12a-11, 211/12a-09 and 211/7–1 are plotted in Fig. 7(a–c). The data for all three wells falls on, or near to, the 1 : 1 correlation line on each plot, suggesting that there is a reasonably good correlation between petrographically and petrophysically determined total quartz volumes. These results indicate that the wireline pseudo-quartz value can be used to estimate the total quartz volume in a reservoir.

### Comparison of point-count and wireline pseudo-quartz with point-count quartz cement data

The point-count and wireline pseudo-quartz versus point-count quartz cement data for wells 211/12a-11, 211/12a-09 and 211/7–1 are plotted in Fig. 8(a–c) and Fig. 9(a–c). The data for these three wells do not plot near to the 1 : 1 line and do not correlate. The pseudo-quartz values are composed of more than one quartz type (quartz cement, mono- and polycrystalline detrital quartz) and includes feldspar (see Equation 5), whilst the point-count quartz cement value represents quartz cement only, with no contribution from detrital quartz or feldspar. However, these data (Fig. 8a–c and Fig. 9a–c) are useful because they show that the total quartz volume does not increase with increasing quartz cement contents. This suggests that the occurrence of quartz cement in Magnus was not associated with an increase in the overall amount of quartz in the reservoir.

### Effects of oil emplacement on quartz cementation in the Magnus Field

There are three major controls on quartz cementation in oil fields:

**1** the source of the cement—external or internal to the reservoir;

**2** the specific effects of oil emplacement on quartz cementation—whether oil emplacement halts quartz cementation (e.g. Gluyas *et al.*, 1993), or allows continued quartz cementation (e.g. Bjørlykke & Egeberg, 1993);

**3** the relative timing of quartz cementation and oil emplacement control the distribution of quartz cement across an OWC (Emery *et al.*, 1993).

If the timing of quartz cementation could be fixed relative to the timing of oil emplacement, then it should be possible to quantify the likely effects of oil emplacement on quartz cementation in the three wells studied. Firstly, assuming that the rate of quartz cementation is *adversely* affected by emplacement of oil within a reservoir, then theoretically there are three possible distributions of quartz cement across the OWC:

**1** early oil emplacement would halt quartz cementation in the oil leg and have no effect on the water leg, resulting in an abrupt change in quartz cement volume (Fig. 10a);

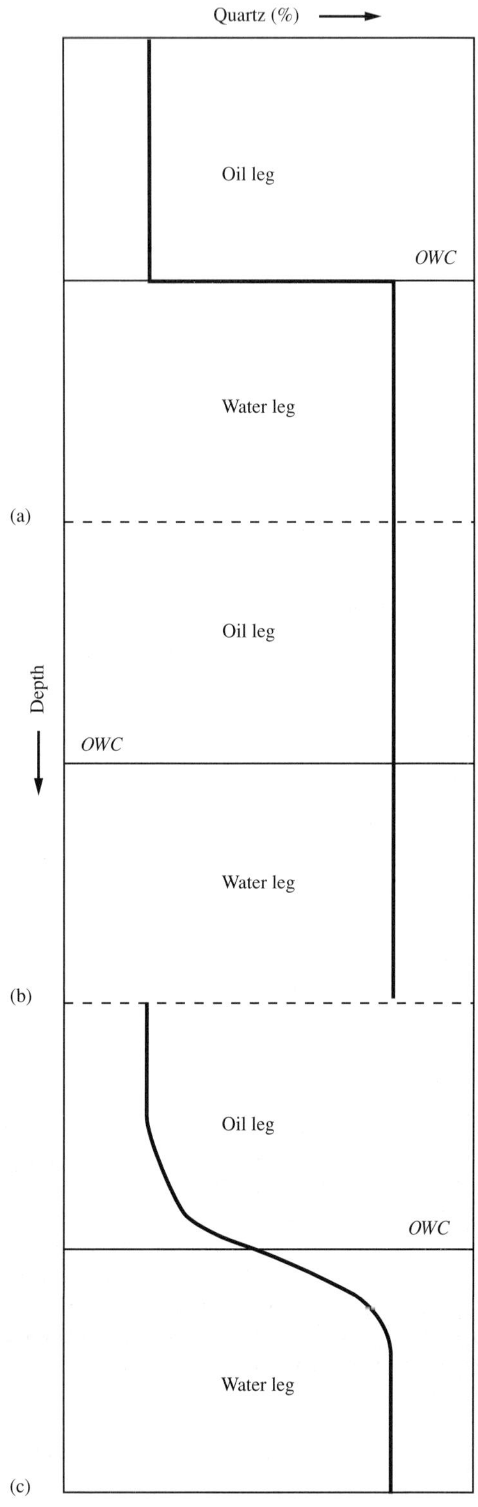


**Fig. 10.** (*Right.*) (a–c) Effect of petroleum emplacement on cement distribution across the oil–water contact: (a) before quartz cementation; (b) after quartz cementation; (c) synchronous with quartz cementation. Assuming quartz cementation is halted by oil emplacement.

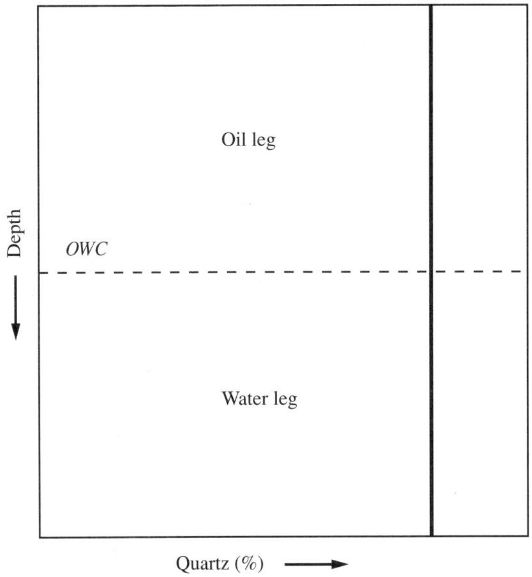


**Fig. 11.** Distribution of quartz cement across the oil–water contact is unaffected by the relative timing of oil emplacement and quartz cementation (assuming that quartz cementation is unaffected by oil emplacement).

**2** oil emplacement after quartz cementation would have no affect on quartz cement volumes in either the oil or the water legs (Fig. 10b);

**3** oil emplacement during quartz cementation would lead to progressively less quartz cement passing up into the oil leg (Fig. 10c).

Secondly, assuming that the rate of quartz cementation is not affected by the emplacement of oil in the reservoir, then whenever quartz cementation occurred, there should be equal amounts of quartz cement in the oil and water legs. In this last case, cement distribution should be independent of the relative timings of oil emplacement and quartz cementation (Fig. 11).

Emery *et al.* (1993) asserted that quartz cementation and oil emplacement occurred synchronously in Magnus at 80 Ma through the use of fluid inclusion petrography and burial history modelling. Thus in Magnus we should witness either (i) progressively less quartz cement passing up into the oil leg if quartz cementation is halted by oil emplacement (i.e. as shown by Fig. 10c); or (ii) uniform quartz cement volumes in the oil and water legs if quartz cementation is not affected by oil emplacement (i.e. as shown by Fig. 11).

The results from two wells: 211/12a-11 and 211/12a-09 (Figs 4 & 5; Table 3) seem to show that the volumes of quartz cement, point-count pseudo-quartz and wireline pseudo-quartz do not change significantly across the OWC. The results from the remaining well: 211/7-1 (Fig. 6; Table 3) show that the volume of quartz cement does show an increase from a mean of 4% in the oil leg to 7% in the water leg. This increase could be explained by the adverse effect of oil emplacement on quartz cementation or by the difference in the number of samples above and below (Fig. 5) the OWC which makes any potential interpretation less reliable.

Thus, despite previous assertions about inhibition of quartz cementation during and following oil emplacement within the Magnus Field (Emery *et al.*, 1993), the majority of the data seem to show that the distribution of quartz cement in Magnus was unaffected by the presence of oil in the reservoir.

## Possible sources of quartz for cementation in the Magnus Field

In some previous studies it has been concluded that silica is imported into reservoirs from an external source to supply the quartz cementation process (e.g. Gluyas & Coleman, 1992). An alternative source of quartz cement is from within the reservoir itself. Possible internal sources postulated include pressure solution between quartz grains in the reservoir (e.g. Saigal *et al.*, 1992), dissolution of quartz grains at stylolites (e.g. Oelkers *et al.*, 1992; Walderhaug, 1994) and dissolution of silicate sponge spicules (e.g. Vagle *et al.*, 1994).

If the source of quartz is external to the reservoir, then transport of silica into the reservoir must occur by advective processes (Worden *et al.*, 1998). Advection of fluids in the subsurface is driven by fluid potential (in this case water potential, England *et al.*, 1987). To set-up a water potential difference between two points requires a pressure gradient. The velocity of water flow (v) is given by Darcy's law (Equation 6).

$$v = (k_{eff}/\mu) \cdot (\Delta P/L) \quad (6)$$

Where $k_{eff}$ is the effective permeability of the rock, $\mu$ is the viscosity of the fluid and $\Delta P/L$ is the pressure gradient. Effective permeability can be defined as:

$$k_{eff} = k_r \cdot k \quad (7)$$

Where $k_r$ is the relative permeability (Fig. 12) and k is the intrinsic permeability of the rock. Relative permeability reflects the permeability of a rock to two or more immiscible fluid phases (Archer & Wall, 1994). Therefore assuming a uniform pressure gradient, constant fluid viscosity and constant intrinsic permeability in the oil-filled sandstone and the aquifer, then the flow velocity of water

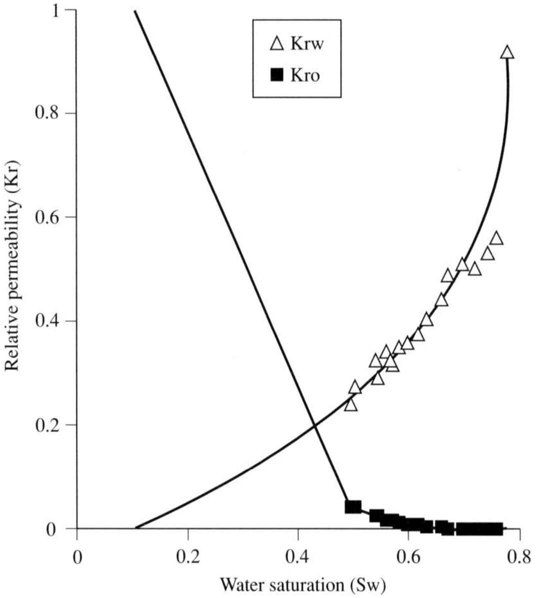


**Fig. 12.** Results of waterflood tests on preserved Magnus core plugs, demonstrating that the relative permeability of the reservoir to water at low water saturations is very low.

into the reservoir is effectively controlled by the $k_r$ of the rock to water. Assuming quartz is transported as an aqueous complex, the relative velocity of transport of quartz into the reservoir versus the underlying aquifer is controlled by the $k_r$ of the reservoir to water.

Relative permeability of reservoir sandstones to aqueous and non-aqueous fluids is usually assessed by the use of waterflood tests on core samples, the results being presented as a function of fractional water saturation (Archer & Wall, 1994). Waterflood tests have been carried out on preserved core from Magnus (Gamble & Brooking, 1989), and a representative example of the results obtained is shown in Fig. 12. At low values of water saturation ($S_w$) the relative permeability of the Magnus Sandstone Member to water ($k_{rw}$) is very low. This implies that when the reservoir contains oil, flow of water and influx of silica into the Magnus reservoir will be negligible. If we accept the assertion of Emery *et al.* (1993) that oil emplacement and quartz cementation were synchronous, then this seems to rule out the possibility of the quartz cement in Magnus being externally sourced, as the large volumes of fluid required to precipitate the amount of quartz cement observed could not have gained access to the reservoir. This conclusion is corroborated by the lack of correlation between the quartz cement and total quartz data (Figs 8 & 9), which showed that silica appears not to have been imported into the sandstone.

With an internal source of quartz in the reservoir, transport of silica in the reservoir probably occurs dominantly by diffusion (Worden *et al.*, 1998). The diffusion rate of silica in solution is governed by Fick's law:

$$J = D \cdot (dc/dx) \cdot \Phi/\theta^2 \quad (8)$$

Where J is the diffusional flux of quartz (i.e. the rate of diffusion of silica), D is the diffusion coefficient of silica, dc/dx is the concentration gradient, $\Phi$ is the porosity and $\theta^2$ is the tortuosity. Water saturation ($S_w$) and wettability exert controls on the rate of the component processes involved in internally sourced quartz cementation by influencing the amount of the porosity available for diffusion (i.e. that part that is filled with water) and the tortuosity of the remaining water (Worden *et al.*, 1998). Quartz cementation will be least inhibited when water saturation is highest and less inhibited in water-wet than oil-wet reservoirs.

Thus patterns of quartz cementation will likely be influenced by $S_w$ and wettability of the reservoir at the time of quartz cementation. If the value of $S_w$ is relatively high in a water-wet reservoir, i.e. reservoir-wide $S_w$ of 98% during the first stages of oil migration into a reservoir (England *et al.*, 1987), then the transport, dissolution and precipitation rates of silica will be relatively unaffected and the volume of quartz cement above and below the OWC should be effectively identical.

After oil filling, the $S_w$ value is usually < 20–25% (Hearst & Nelson, 1985). In water-wet reservoirs, this residual water exists in the form of grain-coating films. Transport of silica in the oil-leg will be adversely affected by the reduced water volume and the increased tortuosity of the water film so that the amount of quartz cement precipitated above the OWC should be recognizably reduced relative to the aquifer (all other things being equal).

In oil-wet reservoirs, the rate of silica diffusion and the access of the aqueous medium to the sites of would-be dissolution and precipitation should be radically reduced from the early stages of oil filling and should be effectively zero at maximum oil saturations. Oil-wet reservoirs should present differences in quartz cement content about the OWC even at the earliest stages of oil emplacement for synchronous quartz cementation and oil filling (Fig. 13).

If we concur that oil emplacement and quartz cementation were synchronous in Magnus, then we can also conclude that the quartz cementation must have occurred at the earliest stages of oil filling and that Magnus was water-wet at the time of cementation. Any other scenario

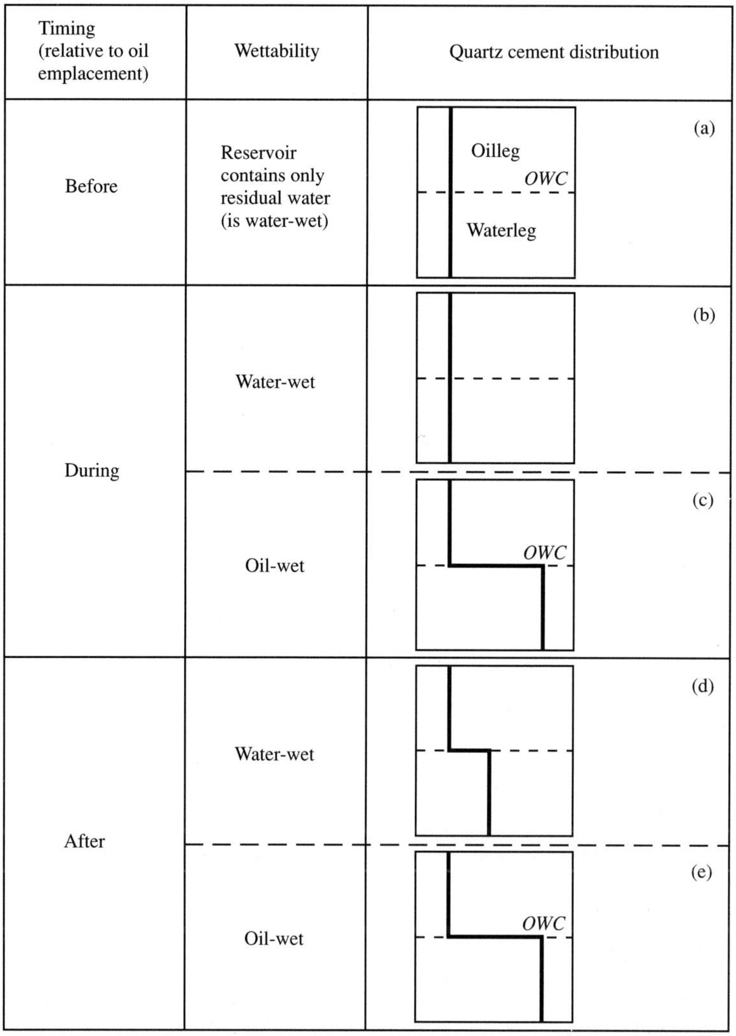


**Fig. 13.** (a–e) Possible distributions of quartz cement across the oil–water contact assuming that the source for quartz cement is internal to the reservoir, showing the dependence on reservoir wettability during quartz cementation.

of wettability or $S_w$ (i.e. exact timing of oil emplacement) would produce differences in quartz cement content not observed in the Magnus Field.

## CONCLUSIONS

**1** The wireline (petrophysical) pseudo-quartz values correlate reasonably well with the point-count (petrographic) pseudo-quartz values. Both techniques can thus be used to examine the distribution of bulk quartz across the oil–water contact in the Magnus reservoir.

**2** Neither the wireline pseudo-quartz nor the point-count pseudo-quartz values correlate with the point-count quartz cement values and are not good indicators of quartz cement values in the Magnus reservoir.

**3** The lack of correlation of total quartz content and quartz cement implies that silica was not imported into the reservoir.

**4** The point-count quartz cement, point-count pseudo-quartz and the wireline pseudo-quartz appear not to change significantly across the oil–water contact in the Magnus Sandstone Member.

**5** The fact that point-count quartz cement, point-count pseudo-quartz and wireline pseudo-quartz do not change significantly across the oil–water contact shows that quartz cementation in the Magnus reservoir was largely unaffected by the emplacement of oil.

**6** The reported presence of oil-filled fluid inclusions in quartz cement and the reported simultaneous oil generation and quartz cementation suggest that quartz cementation occurred in the presence of oil. However, the lack of correlation between total quartz and quartz cement and the uniform quartz cement volumes in the oil and water legs seem to indicate that the reservoir was water-wet at the time of cementation, the water saturation must still have been very high during quartz cementation and the silica forming the cement was locally sourced.

## ACKNOWLEDGEMENTS

The authors would like to thank British Petroleum Ltd. (and Jon Rowse in particular) for supplying the petrographic data and the wireline log data, and also the reviewers, Dr Harry Shaw and Morten Bergan for their comments.

## REFERENCES

ARCHER, J.S. & WALL, C.G. (1994). *Petroleum Engineering: Principles and Practice*. Graham & Trotman, London.

BARCLAY, S.A. & WORDEN, R.H. (1997) Reservoir wettability and its effect upon cementation in oil fields. In: *Geofluids II '97: Contributions to the Second International Conference on Fluid Evolution, Migration and Interaction in Sedimentary Basins and Orogenic Belts. Belfast, Northern Ireland, 10–14 March.* (eds HENDRY, J., CAREY, P., PARNELL, J., RUFFELL, A. & WORDEN, R.H.) pp. 264–267.

BJØRLYKKE, K. & EGEBERG, P.K. (1993) Quartz cementation in sedimentary basins. *Bulletin, American Association of Petroleum Geologists* **77**, 1538–1548.

BROWN, C.E. & NEUSTADTER, E.L. (1980) The wettability of oil/water/silica systems with reference to oil recovery. *Journal of Canadian Petroleum Technology* **19**, 100–110.

COSKUN, S.B., WARDLAW, N.C. & HAVERSLEW, B. (1993) Effects of composition, texture and diagenesis on porosity, permeability and oil recovery in a sandstone. *Journal of Petroleum Science and Engineering* **8**, 279–292.

DE'ATH, N.G. & SCHUYLEMAN, S.F. (1981) The geology of the Magnus oilfield. In: *Petroleum Geology of the Continental Shelf of North-West Europe.* (eds ILLING, L.V. & HOBSON, G.D.) pp. 342–351. Heyden, London.

DICKSON, J.A.D. (1965) A modified staining technique for carbonates in thin section. *Nature* **205**, 587.

DOVETON, J.H. (1994) *Geologic Log Analysis using Computer Methods, AAPG Computer Applications in Geology* 2. American Association of Petroleum Geologists, Tulsa.

EMERY, D., MYERS, K.J. & YOUNG, R. (1990) Ancient subaerial exposure and freshwater leaching in sandstones. *Geology* **18**, 1178–1181.

EMERY, D., SMALLEY, P.C. & OXTOBY, N.H. (1993) Synchronous oil migration and cementation in sandstone reservoirs demonstrated by quantitative description of diagenesis. *Philosophical Transactions of the Royal Society of London A* **344**, 115–125.

ENGLAND, W.A., MACKENZIE, A.S., MANN, D.M. & QUIGLEY, T.M. (1987) The movement and entrapment of petroleum fluids in the subsurface. *Journal of the Geological Society of London* **144**, 327–347.

GAMBLE, I.J.A. & BROOKING, M.R.A. (1989) Magnus: wettability and reservoir condition waterflood study on wells 211/12a-M 12 (A:5) and 211/12a-M 13 (C:7). *BP Internal Report*. BP Research, Exploration and Production Division, London.

GLASMANN, J.R., CLARK, R.A., LARTER, S.R., BRIEDIS, N.A. & LUNDEGARD, P.D. (1989) Diagenesis and Hydrocarbon Accumulation, Brent Sandstone (Jurassic), Bergen High Area, North Sea. *American Association of Petroleum Geologists Bulletin* **73**, 1341–1360.

GLUYAS, J.G. & COLEMAN, M.L. (1992) Material flux and porosity changes during sandstone diagenesis. *Nature* **356**, 52–54.

GLUYAS, J.G., ROBINSON, A.G., EMERY, D., GRANT, S.M. & OXTOBY, N.H. (1993) The link between petroleum emplacement and sandstone cementation. In: *Petroleum Geology of Northwest Europe: Proceedings of the 4th Conference.* (ed. PARKER, J.R.) pp. 1395–1402. Geological Society of London.

HAYES, J.R. & KLUGMAN, M.A. (1959) Feldspar staining methods. *Journal of Sedimentary Petrology* **29**, 227–232.

HEARST, J.R. & NELSON, P.H. (1985). *Well Logging for Physical Properties*. McGraw-Hill, New York.

HURST, A. & MILDOWSKI, A. (1996) Thorium distribution in some North Sea sandstones: implications for petrophysical evaluation. *Petroleum Geoscience* **2**, 59–68.

KVENVOLDEN, K.A. & ROEDDER, E. (1971) Fluid inclusions in quartz crystals from South-West Africa. *Geochimica et Cosmochimica Acta* **35**, 1209–1229.

LARTER, S.R. & APLIN, A.C. (1995) Reservoir geochemistry: methods, applications and opportunities. In: *The Geochemistry of Reservoirs.* (eds CUBITT, J.M. & ENGLAND, W.A.) pp. 5–32. Special Publications of the Geological Society of London 86.

MACAULEY, C.I., HASZELDINE, R.S. & FALLICK, A.E. (1992) Diagenetic pore waters stratified for at least 35 million years: Magnus Oil Field, North Sea. *Bulletin, American Association of Petroleum Geologists* **76**, 1625–1634.

MACLEOD, G., PETCH, G.S., LARTER, S.R. & APLIN, A.C. (1993) Investigations on the composition of hydrocarbon fluid inclusions. *205th American Chemical Society Meeting, Abstracts*. American Chemical Society.

OELKERS, E.H., BJØRKUM, P.A. & MURPHY, W.M. (1992) The mechanism of porosity reduction, stylolite development and quartz cementation in North Sea sandstones. In: *Proceedings of the 7th International Symposium on Water–Rock Interaction, Park City, Utah, 13–18 July.* (eds KHARAKA, Y.K. & MAEST, A.S.) pp. 1183–1186.

QUIREIN, J.A., GARDNER, J.S. & WATSON, J.T. (1982) Combined natural gamma ray spectral litho-density measurements applied to complex lithologies. *57th Annual Fall Meeting, New Orleans*. Society of Petrological Engineering Paper 11143, pp. 1–14.

RIDER, M.H. (1986). *The Geological Interpretation of Well Logs*. Blackie, Glasgow.

ROBINSON, A. & GLUYAS, J. (1992) Duration of quartz cementation in sandstones, North Sea and Haltenbank Basins. *Marine and Petroleum Geology* **9**, 324–327.

SAIGAL, G.C., BJØRLYKKE, K. & LARTER, S. (1992) The effects of oil emplacement on diagenetic processes—examples from the Fulmar reservoir sandstones, Central North Sea. *American Association of Petroleum Geologists Bulletin* **76**, 1024–1033.

SAVRE, W.C. (1963) Determination of a more accurate porosity and mineral composition in complex lithologies with the use of sonic, neutron and density surveys. *Journal of Petroleum Technology* **15**, 945–959.

SCHLANGEN, L.J.M., KOOPAL, L.K., COHEN STUART, M.A. & LYKLEMA, J. (1995) Thin hydrocarbon and water films on bare and methylated silica: vapour adsorption, wettability, adhesion and surface forces. *Langmuir* **11**, 1701–1710.

SERRA, O., BALDWIN, J. & QUIREIN, J. (1980) Theory, interpretation and practical application of natural gamma ray spectoscopy. *SPWLA Transactions of the 21st Annual Logging Symposium*, Paper Q.

SHEPHERD, M., KEARNEY, C.J. & MILNE, J.H. (1990) Magnus Field. In: *Treatise of Petroleum Geology, Atlas of Oil and Gas Fields: Structural Traps II.* (eds BEAUMONT, E.A. & FOSTER, N.H.) pp. 95–125. American Association of Petroleum Geologists.

VAGLE, G.B., HURST, A. & DYPVIK, H. (1994) Origin of quartz cements in some sandstones from the Jurassic of the Inner Moray Firth (UK). *Sedimentology* **41**, 363–377.

WALDERHAUG, O. (1990) A fluid inclusion study of quartz-cemented sandstones from offshore Mid-Norway—Possible evidence for continued quartz cementation during oil emplacement. *Journal of Sedimentary Petrology* **60**, 203–210.

WALDERHAUG, O. (1994) Temperatures of quartz cementation in Jurassic sandstones from the Norwegian continental shelf—evidence from fluid inclusions. *Journal of Sedimentary Research* **64**, 311–323.

WORDEN, R.H., OXTOBY, N.H. & SMALLEY, P.C. (1998) Can oil emplacement prevent quartz cementation in sandstones? *Petroleum Geoscience* **4**, 129–138.

*Spec. Publs int. Ass. Sediment.* (2000) **29**, 163–182

# Quartz cementation in Cretaceous and Jurassic reservoir sandstones from the Salam oil field, Western Desert, Egypt: constraints on temperature and timing of formation from fluid inclusions

R. MARFIL[1], C. ROSSI[1], R. P. LOZANO[1], A. PERMANYER[2] *and* K. RAMSEYER[3]

[1]*Departmento de Petrología y Geoquímica, Universidad Complutense, 28040 Madrid, Spain;*
[2]*Departmento de Geoquímica, Petrología y Prospección Geológica, Universidad de Barcelona, 08028 Barcelona, Spain; and*
[3]*Geologisches Institut, Universität Bern, CH 3012 Bern, Switzerland*

## ABSTRACT

Middle Jurassic and Lower Cretaceous reservoir sandstones in the Salam oil field (in Egypt's Western Desert), are mostly quartz-arenites with abundant quartz overgrowths. Most overgrowths precipitated relatively early in the diagenetic history, only pre-dated by mechanical- and minor chemical compaction. Integration of the diagenetic sequence with the geological and maturation history of the basin, indicates that quartz cementation occurred prior to oil filling the reservoirs during the Eocene, probably during the Late Cretaceous, related to high subsidence and heating rates.

Homogenization temperatures ($T_h$) of aqueous fluid inclusions that occur along the boundary between the detrital quartz grains and their overgrowths indicate that the quartz cements started to grow at temperatures from about 116°C to 135°C. These palaeotemperatures are similar to, or slightly higher than, present-day formation temperatures, and show a positive correlation with present depth of burial. The palaeotemperatures obtained from fluid inclusions are higher than predicted from thermal history models. There are several possible explanations for these relatively high palaeotemperatures: (i) the Late Cretaceous burial depths were higher than initially presumed; (ii) the Late Cretaceous palaeogeothermal gradients were considerably higher than present gradients; (iii) migrating hot fluids were responsible for the growth of quartz cements, at temperatures above those resulting from the contemporary conductive geothermal gradient; or (iv) fluid inclusions were trapped at lower temperatures and re-equilibrated during subsequent burial. Although thermal re-equilibration of the grain-boundary fluid inclusions is hypothetically possible, and would resolve the inconsistencies between $T_h$ data and the thermal and migration histories, there is no clear evidence supporting re-equilibration, given the narrow ranges of $T_h$ in each sample and the lack of a correlation between $T_h$ and parameters such as inclusion salinity or size.

## INTRODUCTION

Homogenization temperatures of fluid inclusions have been used to estimate minimum temperatures of precipitation of quartz overgrowths and palaeo-fluid composition in sandstones (e.g. Roedder, 1984; Burley *et al.*, 1989; Walderhaug, 1990, 1994a). Most reported fluid inclusions in quartz cement occur along the boundary between detrital grains and overgrowths, and thus their homogenization temperatures could be indicative of the minimum temperatures at the onset of quartz cementation (Walderhaug, 1994a; Wilson & Stanton, 1994). In sandstones of different localities, ages and burial histories, the temperatures of onset of quartz cementation, deduced from grain-boundary fluid inclusions, seem to be highly variable, usually within the 75–150°C range (Walderhaug, 1994b). Combined with models of the thermal history of the basin, the homogenization temperatures of inclusions at grain–cement boundaries have been used to estimate the date of initial quartz cementation (Burley *et al.*, 1989; Walderhaug, 1990, 1994b).

In certain oil-field sandstones, the reported homogenization temperatures of grain-boundary, aqueous inclusions in quartz overgrowths, increase with burial depth and show a positive correlation with maximum burial temperatures (Walderhaug, 1990; Grant & Oxtoby, 1992;

Gluyas *et al.*, 1993; Haszeldine & Osborne, 1993; Osborne & Haszeldine, 1993). Two possible explanations have been proposed for this: (i) the sandstones experienced rapid burial, and cementation by quartz only started at peak temperature (Robinson *et al.*, 1992; Walderhaug, 1994b); or (ii) the inclusions were trapped at lower temperatures and re-equilibrated as a result of overheating during subsequent burial (e.g. Osborne & Haszeldine, 1993, 1995a,b).

Although thermal re-equilibration is common in some diagenetic minerals (e.g. carbonates), fluid inclusions in quartz are usually believed to be resistant to non-elastic deformation, making their re-equilibration unlikely under diagenetic conditions (Robinson *et al.*, 1992; Goldstein & Reynolds, 1994, p. 61; Worden *et al.*, 1995). In spite of this, Osborne & Haszeldine (1993) have proposed that re-equilibration of fluid inclusions in authigenic quartz at diagenetic temperatures could be a common process in certain subsiding basins, supporting this hypothesis with data from reservoir sandstones from the North Sea (Haszeldine & Osborne, 1993; Osborne & Hazeldine 1993, 1995a,b). However, the hypothesis and supporting evidence presented by Osborne & Haszeldine (1993) have been questioned and the presumed re-equilibrated inclusions were explained alternatively as unaltered inclusions formed at high temperatures (Goldstein & Reynolds, 1994, p. 61; Worden *et al.*, 1995; Walderhaug, 1995).

This paper describes the characteristics and homogenization temperatures for fluid inclusions located at the boundaries of quartz grains and their overgrowths, in deeply buried (2400–3500 m depth) Cretaceous and Jurassic reservoir sandstones from an oil field in the Western Desert of Egypt. We will discuss: (i) the timing of quartz cement regarding the diagenetic sequence, burial history and hydrocarbon generation and migration; (ii) the implications of the palaeotemperatures obtained from fluid inclusions regarding the age of quartz cement and thermal history of the basin; and (iii) the likelihood of thermal re-equilibration of the fluid inclusions.

## GEOLOGICAL SETTING

The samples studied were taken from cores of reservoir sandstones of the Middle Jurassic Khatatba Formation and the Lower Cretaceous Alam El Bueib (AEB) Formation in the Salam Oil Field. Salam Field lies in the northern part of the Egypt's Western Desert, approximately 80 km to the south of Mersa Matruh (Fig. 1). The northern Western Desert represents a passive margin of the African plate, in which numerous rift-style basins developed from the early Mesozoic in association with the opening of the Tethys ocean (El Shazly, 1977). These basins were mainly filled with Jurassic, Cretaceous and Palaeogene sediments. The pattern of basins and structural highs (Fig. 1) is masked under gently dipping, Miocene blanket deposits, which cover most of the surface of the northern Western Desert (El Ayouty, 1990; Kerdany & Cherif, 1990).

Two large Mesozoic grabens, the Matruh and Shusham Basins, intersect in the Salam Field area (Fig. 1). The Matruh basin has a north–south trend and was particularly active during the Early Cretaceous and earlier (Hantar, 1990). The Shusham Basin is a north-east trending half graben system bounded to the north and west by the Faghur-Mamura high (Fig. 1). Salam Field lies in the eastern border of the Shusham Basin, in a relative structural high bounded by deep-seated normal faults. The Salam ridge ends to the south in a major east–west orientated palaeoridge (Ras Qattara ridge), which represents a large fault-block tilted to the north (Fig. 1) (El Ayouty, 1990).

In the northern Western Desert, the Palaeozoic basement is formed by rocks of Cambrian to Carboniferous age. Along the edges of the Shusham Basin, however, the Palaeozoic is strongly eroded and only Cambrian rocks are preserved. The Mesozoic succession in the Salam area (Fig. 2) comprises four major tectono-sedimentary cycles separated by unconformities: Lower–Middle Jurassic, Lower Cretaceous, Upper Cretaceous and Eocene to Miocene. Each cycle begins with fluvio-deltaic siliciclastics and terminates with marine carbonates (May, 1991).

The Early to Middle Jurassic syn-rift deposits of the Ras Qattara and Khatatba Formations are composed of fluvial to coastal plain sandstones and shales, grading upwards into marine shales and finally into the marine limestones of the Masajid Formation (Keeley *et al.*, 1990). This phase of Jurassic extensional faulting ended with the late Kimmeridgian uplift. The deltaic coals and carbonaceous shales of the Khatatba Formation represent the main oil source rock of the area (Bagge & Keeley, 1994). Additionally, the interbedded fluvial channel sandstones form high-quality reservoirs. In the Salam Field, the Khatatba Formation is found at depths from 3300 to 3700 m. Rift reactivation during the Lower Cretaceous is recorded by the fluvial to shallow marine, quartz-arenitic sandstones, shales and carbonates of the AEB Formation, which are capped by the transgressive, lower Aptian carbonates of the Alamein Formation. The Early Cretaceous rifting ends with the Aptian uplift and erosion (Nashaat *et al.*, 1994). In the Salam Field, the AEB Formation is found at depths from 2400 to 3300 m and has four reservoir intervals (Fig. 2) mainly formed by massive, fluvial-channel sandstones. The Upper Cretaceous was a period of regional subsidence, represented by the sedimentation

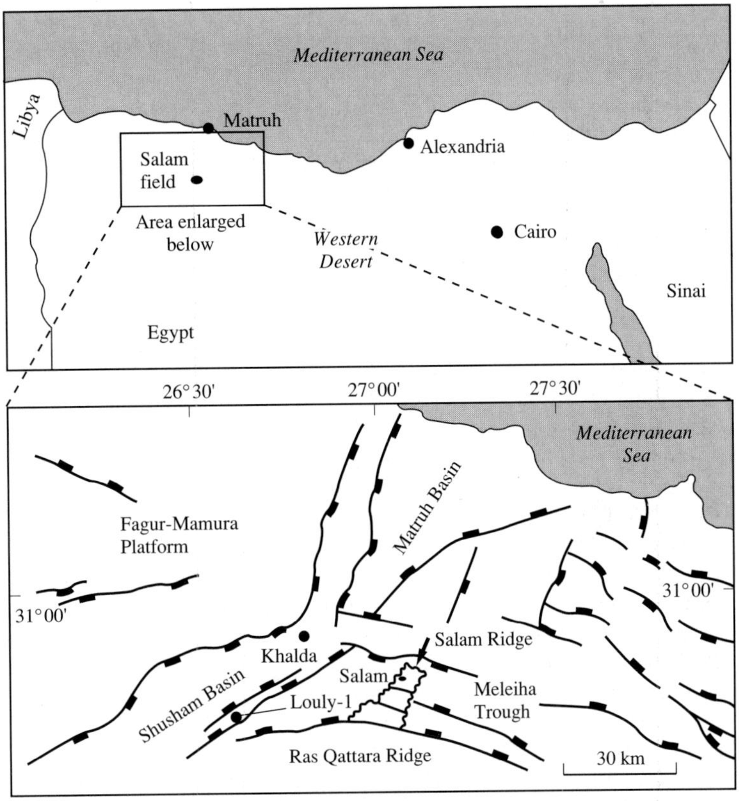


**Fig. 1.** Location map of the Salam oil field in the Western Desert, Egypt. The enlarged area shows the pattern of Mesozoic basins and main faults.

of the fluvial to shallow marine sandstones of the Kharita and Bahariya Formations, which are oil-bearing in the Salam Field, and the open marine shales and carbonates of the Abu-Roash and Khoman Formations (Kerdany & Cherif, 1990).

From the latest Cretaceous to the middle Eocene, intense Alpine tectonic activity occurred related to the closing of the Tethys as a result of the movement of the African plate toward Eurasia. It resulted in folding and faulting, and the elevation of major portions of the north-western Desert (Hantar, 1990; Kerdany & Cherif, 1990). Alpine shear and compression caused reversal of the north-east trending faults and most former rift basins disappeared or even became highs as a result of basin inversion (Kerdany & Cherif, 1990; Nashaat *et al.*, 1994). The majority of the Western Desert oil field traps, including the Salam Field, are faulted anticlines or fault blocks that were formed during the Alpine orogeny as a result of tectonic inversion of Mesozoic extensional structures (El Ayouty, 1990).

A significant hiatus, covering the Palaeocene and early Eocene, exists in the Salam area, which was probably subjected to erosion in this period (Said, 1990). The shallow-water facies and reduced thickness of the Apollonia Formation in the Salam area suggest that this area was a relative structural high during the Eocene (Hantar, 1990; Said, 1990). From the middle Eocene to the Middle Miocene, the area acted as an intracratonic, marginal marine basin (Nashaat *et al.*, 1994), with deposition of the Apollonia Limestone, the Dabaa and Moghra Shales and Sandstones and, finally, the Middle Miocene Marmarica Limestone, which is the surface outcrop today. The Middle Miocene uplift and emergence caused the sedimentation to cease in the entire Western Desert, and the beginning of an erosional period which is still active.

## ANALYTICAL METHODS

This study is based on the petrographic analysis of 70 samples selected from the cored sections of four oil wells. After elimination of oil from the porosity by organic solvents, epoxy-impregnated, polished thin sections underwent detailed petrographic analysis. Mineralogy

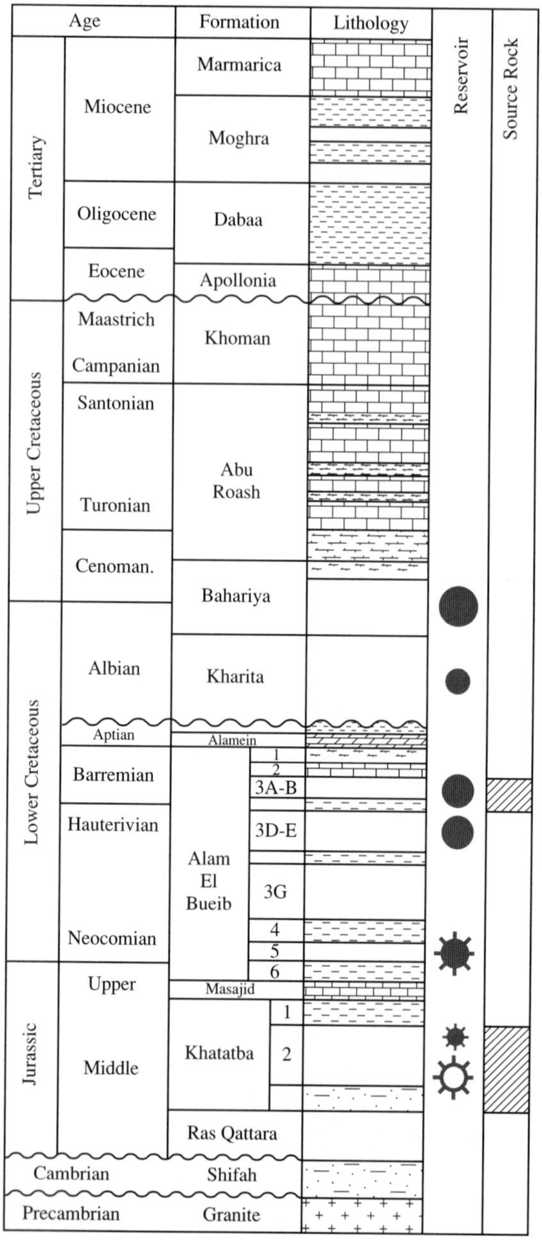


**Fig. 2.** Generalized stratigraphic diagram of the Salam Field area. Vertical scale diagrammatic only.

was confirmed by X-ray diffraction. The samples were also studied under scanning electron microscopy (SEM), in secondary electron mode and backscattered electron imaging (BSE) using a JEOL JSM 6400, equipped with a Link System energy dispersive X-ray microanalyser (EDX). Cathodoluminiscence (CL) petrography was carried out using a hot cathode luminiscope (Ramseyer *et al.*, 1989). Vitrinite reflectance ($R_0$) of dispersed vitrinite particles, concentrated from siltstones and coaliferous shales, were performed using a Leitz–Leica reflected-light microscope equipped with a MPV C2 photometry tube (photometer and photomultiplier).

The microthermometric studies were performed on a petrographic microscope equipped with a Chaixmeca heating–cooling stage. The accuracy of a homogenization temperature ($T_h$) measurement is estimated to be around ±1°C. Microthermometric measurements were performed in rock slices, 60–100 μm thick, polished on both sides and detached from their glass mount. Samples were examined in a standard polarizing microscope equipped with a video-camera, and fluid inclusions selected for microthermometry were recorded with a video-printer. The recorded samples were then observed under a fluorescence microscope in order to identify possible petroleum inclusions. $T_h$ measurements were made on 150 inclusions. Due to the small size of the inclusions (usually less than 4 μm), it was only possible to record reproducible ice-melting for 14 larger (> about 10 μm) inclusions. In most inclusions, their small diameter prevented the detection of clathrates.

## COMPOSITIONAL AND DIAGENETIC FEATURES

The sandstones of both Khatatba and AEB Formations are fine to medium grained, texturally and compositionally mature quartz-arenites. Minor components are K-feldspar (< 5%, commonly kaolinitized), micas and clay aggregates (< 5%) (Table 1). The interbedded shales, coaly shales and mudstones show similar bulk mineralogical compositions but variable clay composition. In the Khatatba Formation, the predominant detrital clay is kaolinite, with occasional presence of mixed-layer smectite–illite (S–I) and rare illite. In the AEB Formation, clay minerals in the interbedded mudstones and shales are kaolinite, illite, S–I and chlorite (Marfil *et al.*, 1997). Siderite and framboidal pyrite are abundant in the mudstones and shales of both formations.

Diagenetic fabrics and authigenic mineral phases and their mutual relationships are similar in the Jurassic and Cretaceous reservoir sandstones (Fig. 3), with minor differences related to variations in the depositional environment (Rossi *et al.*, 1996). Most primary porosity was lost during burial diagenesis, initially reduced by compaction and later occluded by quartz cement and ferroan dolomite, ankerite and siderite pore-filling/grain-replacing cements. In some samples from marine-

**Table 1.** Data on petrographic point counts for some representative samples used in this study

| Sample | BO-67 | BO-38 | BO-68 | BO-76 | BO-78 | BO-40 | BO-80 | BO-82 | BO-41 | BO-84 | BO-90 | BO-96 | BO-97 | BO-98 | BO-99 |
|---|---|---|---|---|---|---|---|---|---|---|---|---|---|---|---|
| Depth (m) | 2525 | 2527 | 2533 | 2637 | 2645 | 2655 | 2668 | 3076 | 3088 | 3092 | 3476 | 3525 | 3529 | 3531 | 3535 |
| Formation | AEB | AEB | AEB | AEB | AEB | AEB | AEB | AEB | AEB | AEB | KHAT | KHAT | KHAT | KHAT | KHAT |
| Quartz | 59.6 | 66.1 | 64.2 | 50 | 48.6 | 70.3 | 77.8 | 55.6 | 68.3 | 63.6 | 57.3 | 64 | 59 | 54 | 63.6 |
| Feldspar | 2 | 4.9 | 3.8 | 3 | 0.3 | — | — | 2.3 | — | 0.3 | 1 | 1.5 | — | 0.3 | 1.3 |
| Ductile rock fragment | 3.3 | 2.4 | — | — | — | — | — | 0.3 | — | 0.3 | 0.6 | — | — | 0.3 | — |
| Dolomite replacive | — | — | — | 5.6 | — | 1.4 | — | — | 0.3 | — | — | — | — | — | — |
| Pyrite replacive | — | 3.4 | 1 | — | — | — | 0.5 | — | — | — | — | — | — | — | — |
| IGV | 20.1 | 22.8 | 29.5 | 28.1 | 27.3 | 8.4 | 21.9 | 36.6 | 31.4 | 24.6 | 20.8 | 17.5 | 19.8 | 38.9 | 35.1 |
| COPL | 24.1 | 22.7 | 14.9 | 16.5 | 4.2 | 34.5 | 16.8 | 5.4 | 12.5 | 20.4 | 24.2 | 27.3 | 25.2 | 1.8 | 7.6 |
| Quartz cement | 9.3 | 4.4 | 7.9 | 8.6 | 14.6 | 2.1 | 4.1 | 11.3 | 10.4 | 9.3 | 11.1 | 12 | 14.6 | 24.3 | 10.6 |
| Dolomite cement | 0.6 | — | — | 7.3 | 0.6 | 5.8 | 0.3 | 1.1 | 8.5 | 0.6 | — | — | — | — | — |
| Kaol–dickite cement | 1 | 0.1 | 0.5 | — | 6.3 | 0.1 | — | 13.3 | 0.2 | 9 | 2.3 | 1 | 2.3 | 1.6 | 23.3* |
| Illite cement | 1 | — | — | — | — | — | 0.5 | — | 0.2 | — | — | — | — | 1.1 | — |
| Pyrite cement | 3.6 | 2.3 | 1.5 | 2.3 | 10.6 | 0.2 | 0.7 | 3.1 | 0.4 | 0.6 | — | 1 | 1.3 | 1.6 | — |
| Siderite cement | 0.6 | — | — | 6.6 | 0.6 | — | — | 1.8 | — | 2 | — | 1.5 | — | — | — |
| K-feldspar cement | — | 0.1 | — | — | — | — | — | — | — | — | — | — | — | — | — |
| Other cements | — | 0.1 | 1.3 | — | — | — | — | — | — | — | 0.3 | — | — | — | — |
| Primary porosity | 4 | 15.8 | 0.9 | 3.3 | 4.6 | — | — | 6 | — | 3.1 | 7.1 | 2 | 1.6 | 10.3 | 1.2 |
| Secondary porosity | 15 | 0.8 | 18.8 | 13 | 13.3 | 20 | 16.3 | 5.3 | 11.8 | 11.2 | 20.2 | 17 | 21.2 | 6.3 | — |

IGV, intergranular volume; COPL, compactional (mechanical and chemical) porosity loss (%); * Infiltrated detrital kaolinite. Data of BO-38; BO-68; BO-40 and BO-41 samples, after Rossi *et al.* (1996).

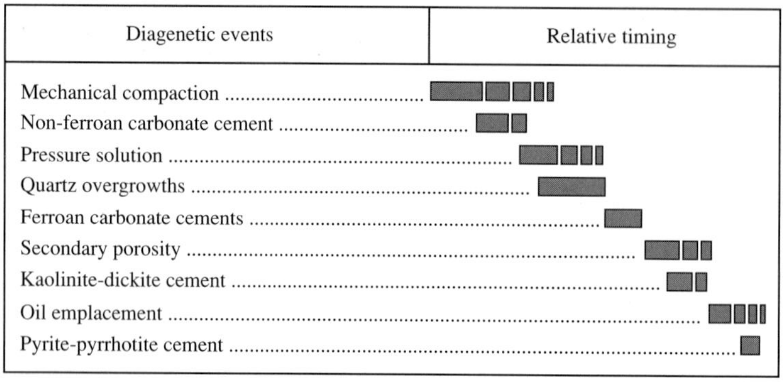


**Fig. 3.** Summary chart of the diagenetic sequence of the Khatatba and AEB sandstones in Salam Field.

influenced facies, an early generation of non-ferroan, zoned poikilotopic dolomite occurs. Where present, this dolomite occluded most primary porosity and inhibited the formation of quartz overgrowths.

Quartz overgrowths are typically non-luminescent, non-zoned and presumably formed in a single phase. In thin section, the overgrowths are usually distinguished by the presence of dust rims crowded with primary aqueous fluid inclusions. In the more deeply buried sandstones of the Khatatba Formation, there is a thin, second phase of dull green luminescent quartz. In most samples, CL shows that moderate pressure-solution pre-dated the formation of the quartz overgrowths. The silica required for quartz cementation was probably derived locally from intergranular pressure dissolution and stylolitization. Although pressure dissolution at grain contacts is moderate, stylolites are common in the studied sandstones, initiated at detrital kaolinite- and organic-rich laminae (Fig. 4a). Within the stylolites, the relatively high titanium content (up to 8% of $TiO_2$, as revealed by EDX), suggest that a relatively large amount of detrital quartz could have been dissolved along the stylolites (cf. Stone & Siever, 1996). The interbedded organic-rich mudstones could have provided additional silica by dissolution of silt-size quartz by organic acids released during organic matter diagenesis (Bennet *et al.*, 1988; Crossey & Larsen, 1992; Marfil *et al.*, 1997). The BSE images of the Khatatba source rocks show that quartz grains are indeed strongly corroded in the presence of organic matter (Fig. 4b).

Ferroan dolomite, ankerite and siderite cements occur in patches as non-luminescent poikilotopic crystals that post-date and corrode the quartz overgrowths (Fig. 5a). In most samples, abundant intergranular secondary porosity was formed after extensive dissolution of the carbonate cements, giving porosities up to 20%. In the sandstones with secondary porosity, partially dissolved relics of the replacive carbonate cement are commonly preserved, and the quartz grains and overgrowths adjacent to intergranular pores usually exhibit strongly corroded margins and v-shaped re-entrants (Rossi *et al.*, 1996). Under CL, the v-shaped re-entrants are commonly seen affecting both the overgrowths and the detrital grains, excluding an origin related to the growth surfaces of the quartz cement. The sandstones with secondary porosity typically show heterogeneous packing, coexisting in the same thin-section areas of denser packing still retaining abundant quartz cement, and areas of oversized pores after cement dissolution.

Late cements, kaolinite–dickite, and pyrite–pyrrhotite cements are interpreted to partially fill secondary porosity. Both types of cement seem to be ubiquitous in all the studied sandstones. The kaolinite–dickite cement typically has a fresh appearance and post-dates the quartz cement and quartz corrosion (Fig. 5b & c). Rarely, in the deeper samples of the Khatatba Formation, kaolinite–dickite crystals may show incipient partial intergrowth with the second phase of quartz overgrowth. The pyrite–pyrrhotite cements, which also show a fresh appearance as seen under the SEM (Fig. 5d), are interpreted to be the final authigenic phase, possibly related to the emplacement of oil in the reservoir.

## BURIAL AND THERMAL HISTORY OF THE STUDIED SANDSTONES

### Burial history

Published burial history curves of areas adjacent to Salam Field (Fig. 6) indicate similar trends: widespread rapid subsidence during the Cretaceous rifting stage, interrupted in the Palaeocene by the Alpine Orogeny, and a phase of regional subsidence during the Tertiary. In constructing these curves, the unconformity between the Late Cretaceous and the early Tertiary was represented either by a hiatus or by 500 m of erosion. In both cases, the max-

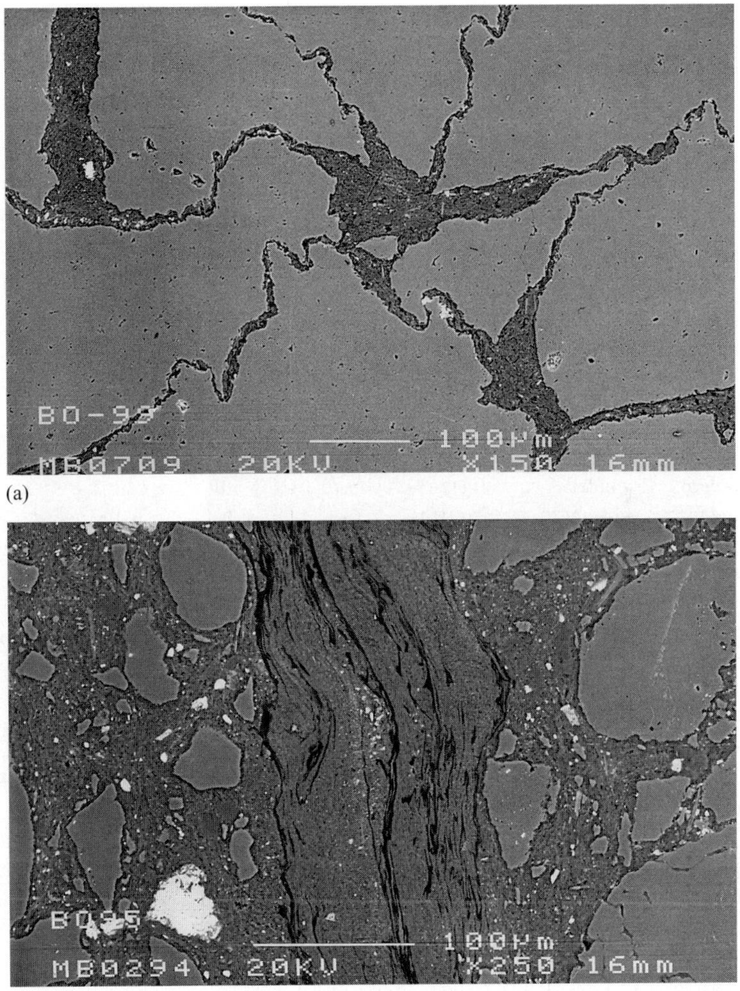


**Fig. 4.** (a) BSE image of stylolitized quartz-arenite. The stylolites are mainly composed of detrital kaolinite, with a high content of $TiO_2$. (b) BSE image of a Khatatba mudstone source rock showing bands of algal filaments and strongly corroded quartz grains floating in a matrix composed of detrital kaolinite, organic matter and some pyrite.

imum burial depth for the AEB and Khatatba Formations was considered to have occurred in the Middle Miocene. Erosion post-dating Middle Miocene times has been assumed to represent at least 60 m of uplift in the Melehia area (Taher *et al.*, 1988), to 120–500 m of uplift along the Shusham Basin axis (Lotfy, 1994; Nashaat *et al.*, 1994).

Figure 7 shows a calculated burial curve for a representative Salam Field well. Taking into account the previously expressed uncertainties regarding the magnitude of the Palaeocene and post-Miocene erosion events, this curve must be considered as a first approximation. Highly conservative estimates of the erosive gaps (50 m in the Palaeocene, and 100 m since the Middle Miocene) have been used for this curve. This picture would change drastically if we assumed that the Salam area acted as a structural high during the early Palaeogene and was therefore subjected to intense erosion. In this case, maximum or near-maximum depths of burial could have been reached by the end of the Cretaceous.

## Maturation history

At the present time in the Shusham Basin, the top of the oil window lies between 2700 and 3200 m of depth. In most areas, the Khatatba source rock is still in the oil generation phase, but in the basin centres it has passed into the gas window (Lotfy, 1994; Nashaat *et al.*, 1994). In the Salam Field, the top of the oil window ($R_o = 0.6\%$) is encountered at around 2450 m (Table 2), and the Khatatba source rocks are actually at the onset of peak oil generation and migration ($R_o = 0.8$–$1.2\%$). Similar maturity–depth relationships have been reported from

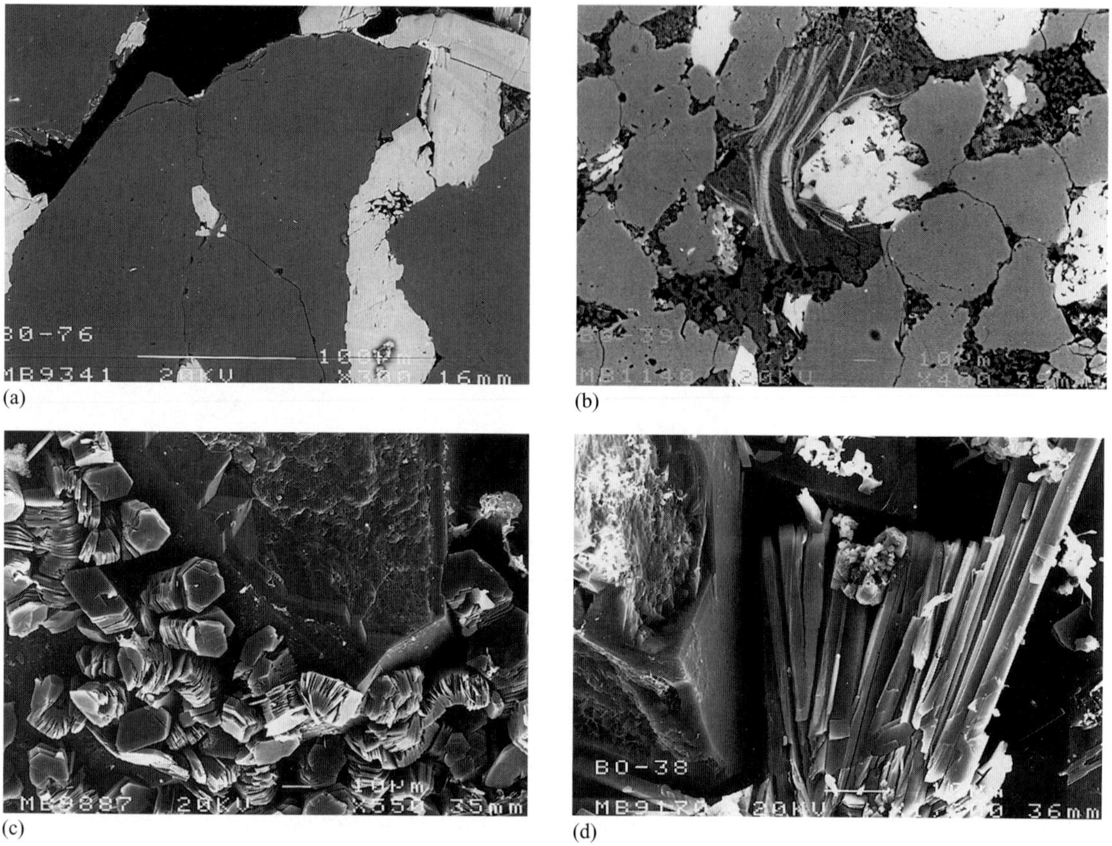

**Fig. 5.** (a) BSE image of replacive ferroan carbonate cement post-dating quartz overgrowths. Secondary porosity is also shown in the upper left of the photomicrograph. Pore-filling kaolinite–dickite that post-dates quartz cement and the dissolution of replacive ankerite is shown in the upper right of the image. (b) BSE image showing extensive pore-filling of kaolinite–dickite post-dating the corrosion of authigenic and detrital quartz. (c) SEM image of quartz cement post-dated by kaolinite–dickite. (d) Prismatic crystals of pyrrhotite post-dating quartz cement, and kaolinite–dickite pore-fillings.

the nearby Melehia area, were the top of the oil window is encountered between 2300 and 2400 m (Taher *et al.*, 1988).

In the Shusham and Matruh Basin centres, the Khatatba Formation entered the oil window during the Cenomanian–Turonian, and the onset of peak oil generation and migration was during the Campanian–Maastrichtian, continuing through the Palaeogene (Lotfy, 1994; Nashaat *et al.*, 1994). In structural highs, such as the Salam ridge, the Khatatba Formation reached the oil window during the Eocene (Lotfy, 1994). Oil migration from the Shusham Basin into the Salam structure presumably has occurred since the Palaeocene (Nashaat *et al.*, 1994). In addition, vertical migration from the local Kahatba source in the Salam Field has taken place since the Eocene. As the Salam trap formed as a result of the Palaeocene–Eocene tectonism (El Ayouty, 1990), the deeper reservoirs in Salam (Khatatba and AEB) were probably filled after the Eocene.

## Thermal history

The present-day geothermal gradient increases on a regional scale from about 29.1°C km$^{-1}$ in the Shusham Basin axis to more than 34.6°C km$^{-1}$ over the Qattara ridge, the Fagur platform and other structural highs (Taher *et al.*, 1988; Lotfy, 1994; Nashaat *et al.*, 1994). In Salam Field, the present-day geothermal gradient, calculated from borehole temperature measurements, is 34°C km$^{-1}$. This is a relatively high value, compared with the typical geothermal gradients of the northern Western Desert, which are normally about 20°C km$^{-1}$. In order to

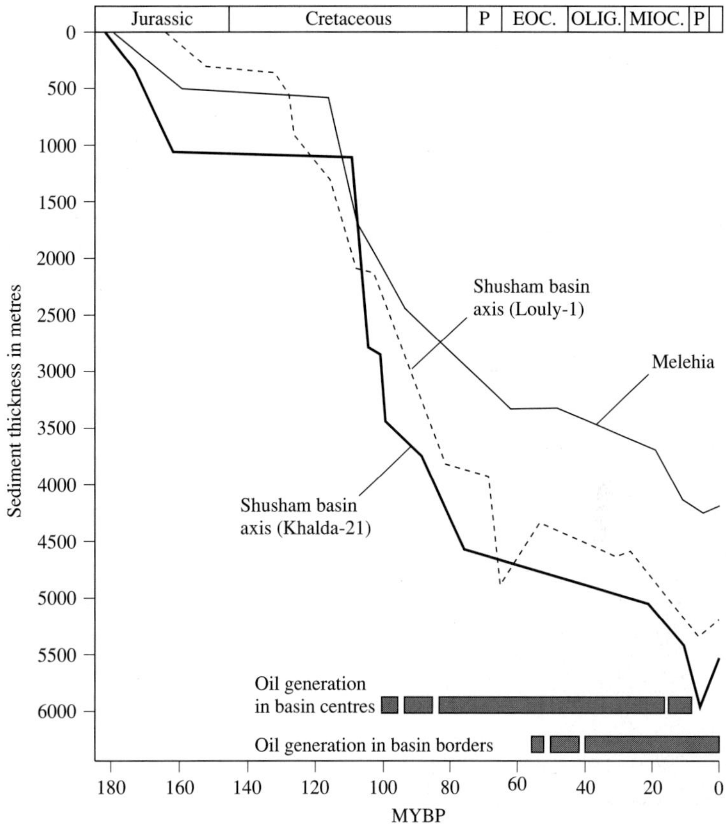


**Fig. 6.** Representative burial histories for the Khatatba Formation in areas adjacent to the Salam Field. Data from Nashaat *et al.* (1994) (Louly-1); Lotfy (1994) (Khalda-21); and Taher *et al.* (1988) (Melehia). Khalda-21 and Louly-1 wells are located along the axis of the Shusham Basin, 20–40 km to the SW and NW of Salam. Melehia field lies 15 km to the NE of Salam. See Fig. 1 for location of the wells.

bring measured and calculated maturities into agreement, Taher *et al.* (1988) proposed that the relatively high temperature gradients in the area originated about 5 million years ago, and that lower palaeogradients prevailed earlier: 25.5°C km$^{-1}$ between 5 and 22 million years with 20°C km$^{-1}$ before 22 million years ago.

According to the thermal history model developed by Lotfy (1994), in the Shusham Basin centres, the temperature increased rapidly from the Aptian to the Maastrichtian, when it reached its maximum value (142°C at the top of Khatatba). According to Lofty's model (1994), in the Salam ridge the temperatures for the top of the Khatatba Formation during the Maastrichtian varied from 120°C in the southern part of the ridge to 90°C in Salam. During the Palaeogene, the model shows a pronounced cooling, and the Oligocene temperatures in Shusham Basin centres were presumably up to 23°C lower than those during the Maastrichtian. In the Salam ridge, this temperature difference varied from a pronounced drop of 30°C in the southern part of the ridge to a slight increase of 2°C in Salam Field.

## $R_0$ palaeothermometry

Vitrinite reflectance ($R_0$) may be used as an indication of the maximum temperature reached by the rock, as $R_0$ may be converted to temperatures using empirical calibrations (e.g. Barker & Pawlewicz 1986; Barker & Pawlewicz, 1994). In Salam (Table 2) a trend of increasing $R_0$-deduced temperature with depth can be observed, with some gaps in the samples which show higher standard deviations. Taking into account the recent burial history, present-day temperatures must be lower than maximum temperatures. $R_0$-deduced temperatures are, however, lower than present-day formation temperatures. Therefore, $R_0$-deduced temperatures are considered of limited utility as indicators of peak temperatures in this case. $R_0$ values obtained from non-fluorescent vitrinite are higher than those obtained from fluorescent vitrinite in the same sample, and thus they may be a better indication of the maximum temperature, although their deduced temperatures are still lower than present-day formation temperatures. The lower reflectance of the fluorescent vitrinite may be

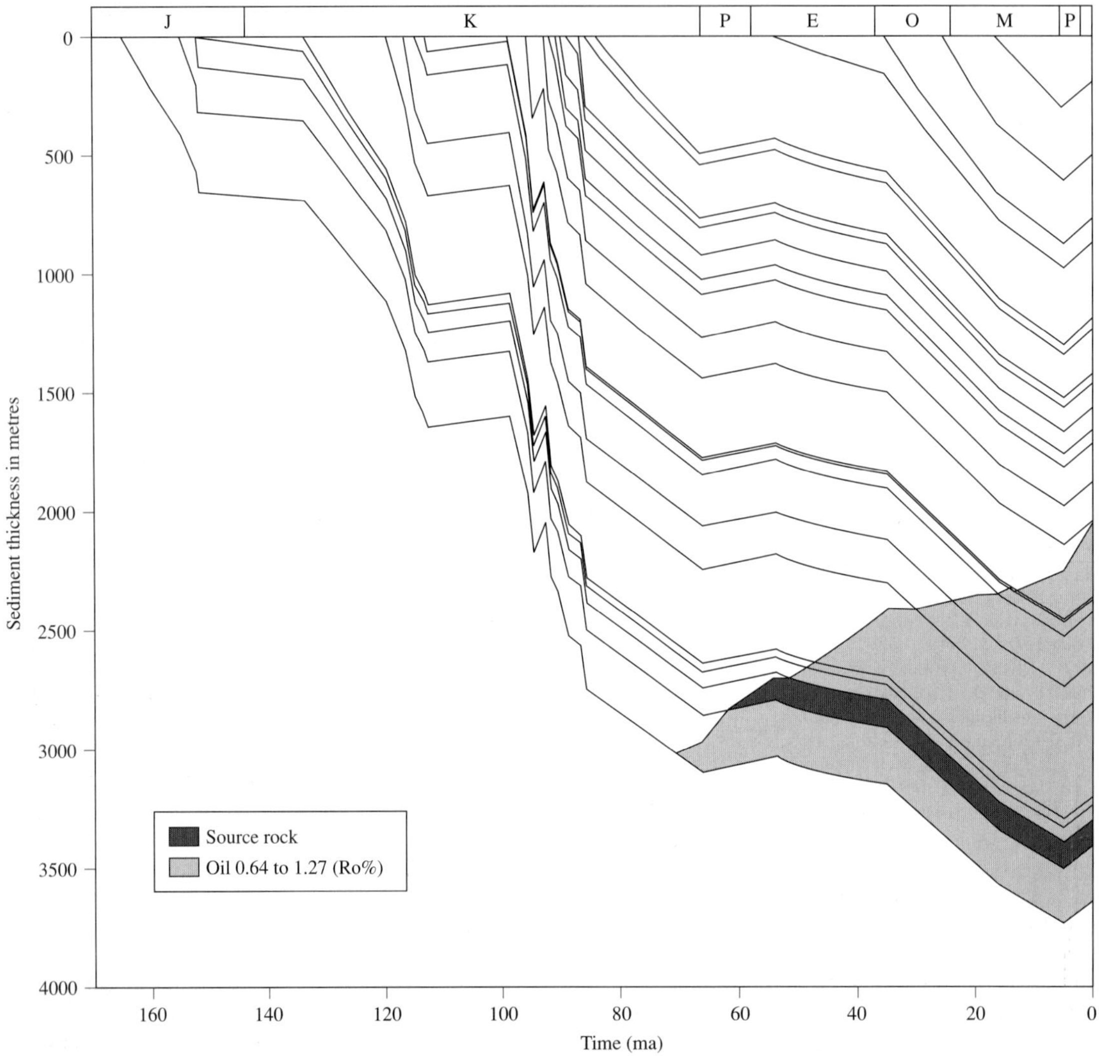


**Fig. 7.** Burial history curve for the studied well. For the Khatatba Formation source rocks, we have assumed an average of 3.5%wt organic carbon and 40–50% type II and type III kerogen, respectively. The vitrinite values used correspond to Table 2. The $T_{max}$ of the rock-eval pyrolysis is taken into consideration (Permanyer *et al.*, 1995).

related either to an initial hydrogen-rich composition (Newman, 1997), or to increases in the hydrogen index caused by bacterial degradation of algal kerogen (Thompson *et al.*, 1994; Marfil *et al.*, 1997).

## FLUID INCLUSION ANALYSES

Three samples from the same well were selected for fluid-inclusion microthermometry, representing depths of 2654, 3092 (AEB) and 3525 m (Khatatba). All the inclusions studied in these samples are aqueous. No petroleum inclusions have been observed in quartz overgrowths in these samples or in other samples from the AEB and Khatatba reservoirs.

Sixty-one homogenization temperatures were measured in inclusions located at quartz grain–overgrowth boundaries (Fig. 8a,b,d,f). These inclusions have been termed Q1 A. They are abundant along the 'dust-rims', delineating the rounded surfaces of the detrital grains. The Q1 A inclusions are commonly small (less than 3 μm), but can locally reach up to 15 μm in diameter. They show

**Table 2.** Vitrinite reflectance ($R_o$%) results and maximum burial temperatures (Tpeak) calculated from $R_o$ using equivalences proposed by Barker and Pawlewicz (1986, 1994)

| Sample depth | Formation | *n* | % $R_o$ (mean) | Fluorescent vitrinite | St. dev. | Tpeak (°C) |
|---|---|---|---|---|---|---|
| 2429 | AEB | 48 | 0.51 | — | 0.07 | 82 |
| 2436 | AEB | 51 | 0.57 | — | 0.06 | 91 |
| 2446 | AEB | 20 | 0.75 | 0.55 | 0.08 | 113 |
| 2608 | AEB | 38 | 0.61 | 0.50 | 0.08 | 96 |
| 2609 | AEB | 42 | 0.63 | 0.58 | 0.05 | 99 |
| 2642 | AEB | 26 | — | 0.58 | 0.04 | 92 |
| 2991 | AEB | 20 | — | 0.55 | 0.04 | 88 |
| 3076 | AEB | 39 | 0.81 | 0.70 | 0.11 | 119 |
| 3381 | Khatatba | 20 | — | 0.49 | 0.07 | 78 |
| 3383 | Khatatba | 50 | 0.71 | — | 0.10 | 108 |
| 3383-2 | Khatatba | 50 | — | 0.66 | 0.06 | 102 |
| 3435 | Khatatba | 50 | 0.77 | — | 0.11 | 115 |
| 3535-2 | Khatatba | 50 | 0.74 | — | 0.07 | 112 |
| 3442 | Khatatba | 14 | 0.67 | 0.62 | 0.07 | 104 |
| 3480 | Khatatba | 50 | 0.77 | 0.66 | 0.03 | 115 |
| 3480-2 | Khatatba | 50 | 0.80 | 0.61 | 0.03 | 119 |
| 3530 | Khatatba | 45 | 0.95 | 0.55 | 0.16 | 132 |

variable shapes, from equant (Fig. 8f) to highly irregular (Fig. 8c). They show uniform ratios of liquid-to-vapour of about 10 : 1. Locally, some monophase inclusions have been observed in the dust rims in sample 2654, but they are very small and were interpreted as metastable. The overall range of $T_h$ in Q1 A inclusions is 110–145°C, but with a marked increase in the ranges, means and modes of $T_h$ with increasing depth (Fig. 9). The limited data on final ice melting correspond to a range of salinity variation (according to the equation of Hall *et al.*, 1988), between 0.3 and 6.6 wt% NaCl equivalent (Table 3).

Fifty-eight homogenization temperatures correspond to inclusions whose petrographic location is unclear. These have been termed Q1 B. They apparently occur along the boundaries of grains and overgrowths (Fig. 8e), but the absence of clearly defined dust rims diminish the confidence of this assignation. It is plausible for this group to include inclusions of different origins: inclusions of grain–overgrowth boundaries, inclusions in secondary planes located near grain edges, or even inclusions isolated within overgrowths. Q1 B inclusions seem to be larger than Q1 A (Table 3), and their $T_h$ ranges are similar (sample 3525) or slightly higher than Q1 A (samples 2654 and 3092) (Fig. 9). The salinities of Q1 B inclusions (2.3–5.8 wt% NaCl) are similar to those displayed by Q1 A inclusions. The liquid-to-vapour ratios of Q1 B are commonly similar to those in Q1 A, but rare low-density inclusions may occur (one inclusion in sample 3092) in which the presence of $CO_2$ is indicated by the melting point of solid $CO_2$ at –56.5°C (Shepherd *et al.*, 1985).

Additionally, we have studied secondary inclusions located in planes within the detrital grains and showing ranges in homogenization temperatures similar to those of Q1 A and Q1 B inclusions. These secondary inclusions have been termed Q2 (Fig. 8c). Commonly, Q2 inclusions show consistent liquid-to-vapour ratios, similar to Q1 A and Q1 B, but some planes contain abundant all-liquid inclusions (in sample 2654) or inclusions with varied densities, from low-density to all-liquid (in sample 3092). The $T_h$ values for Q2 are therefore variable (Fig. 9), with modes that may be similar (sample 3525) or higher than Q1 (sample 2654). Some secondary planes have inclusions that homogenize well above 170°C, commonly at 230–270°C. Given their elevated $T_h$ ranges (240–270°C), we interpret these inclusions to have been inherited from the source materials of the detrital quartz grains and therefore they will not be discussed here.

### Basic interpretations of the fluid-inclusion data

In each sample, Q1 A fluid inclusions represent a consistent assemblage in which 90% of $T_h$ values are within a range of 10–15°C. As Q1 A inclusions are located at detrital grain–overgrowth boundaries, they must have been trapped during the first stages of quartz overgrowth development. It is assumed that the pressure correction for $T_h$ is negligible due to the probable presence of dissolved methane. This is a reasonable assumption taking into account the abundance of organic matter in the shales interbedded with the reservoir sandstones. If we also assume that the inclusions have not leaked or stretched since entrapment, then their $T_h$ values would be indicative of the temperatures at the onset of quartz cementation. If this is true, then authigenic quartz began to grow at around 135°C in sample 3525, 123°C in sample 3092 and 116°C in sample 2654.

The interpretation of Q1 B inclusions is ambiguous because their petrographic location is unclear. In sample 3525, the $T_h$ modes of Q1 B inclusions overlap with those of Q1 A inclusions, suggesting common origins and therefore most Q1 B inclusions in this sample could represent true grain-boundary inclusions. In sample 3092, the broad range in $T_h$ of the Q1 B inclusions suggests considerable mixture of different fluid inclusion assemblages, perhaps some of the grain-boundary type and some of the Q2-type inclusions. In sample 2654, the $T_h$ distribution of Q1 B inclusions is more consistent, unimodal and within a narrow range, but around five degrees hotter than Q1 A inclusions. In this sample, Q1 B inclusions could be interpreted either as part of the Q1 A

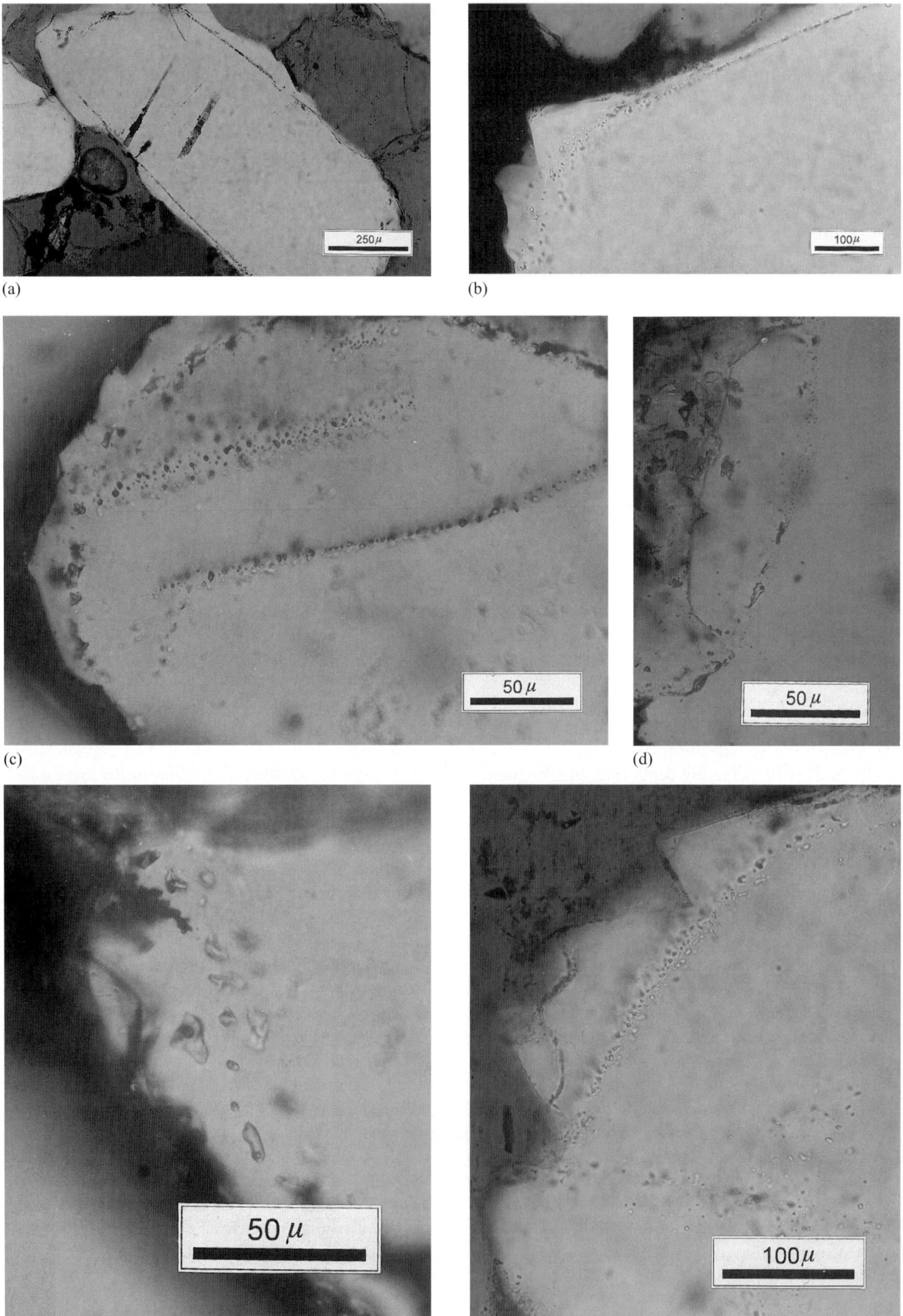
(a)
250μ
(b)
100μ
(c)
50 μ
(d)
50 μ
(e)
50 μ
(f)
100μ

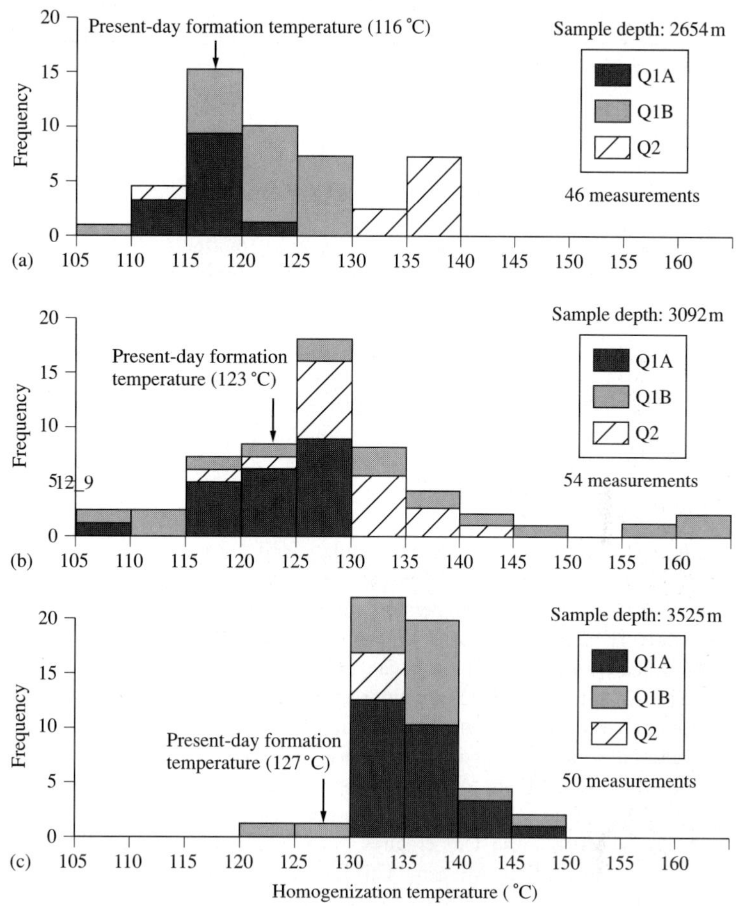


**Fig. 9.** Homogenization temperature measurements: Q1 A represents inclusions at the boundary between the quartz grains and overgrowths; Q1 B represents inclusions whose petrographic location is unclear (see text for explanation); Q2 represents secondary inclusions aligned along planes within the detrital grains.

assemblage which for some reason yields higher $T_h$, or as a later-formed assemblage, located within the overgrowths, and trapped at slightly higher temperatures than Q1 A inclusions.

The Q2 inclusions are located along microfracture planes within the detrital grains. As no Q2 inclusion planes have been observed to cross-cut overgrowths, we cannot say whether they are inherited from the source material of the detrital grain, or the inclusion-bearing microfractures formed *in situ* during burial. The Q2 inclusions, however, in samples 3092 and 3525 show similar $T_h$ ranges as the neighbouring Q1 A (grain-boundary) inclusions, suggesting entrapment (or re-equilibration) under similar conditions (cf. Morad *et al.*, 1991). In sample 2654, the range of Q2 inclusions is 15–20°C hotter than that of Q1 A, suggesting entrapment at higher temperatures. In this sample, however, the presence in the Q2 assemblage of inclusions with variable densities may indicate that their $T_h$ values could be altered by necking-down processes (Goldstein & Reynolds, 1994).

**Fig. 8.** (*Opposite.*) Thin-section photomicrographs showing some of the studied fluid inclusions. (a) The boundary between the detrital grain and its overgrowth is marked by a discontinuous line of Q1 A fluid inclusions. Note that the quartz cement is strongly corroded, and this corrosion can affect the detrital quartz. Note also the presence of partially dissolved ankerite cement (lower left). The dissolution of this ankerite has created abundant secondary porosity. Sample depth: 2654 m. (b, f) Details of the grain-boundary (Q1 A) fluid inclusions. Note the corrosion re-entrants affecting the quartz overgrowths. (c) Large Q1 A fluid inclusions, located at the grain–cement boundary, and smaller secondary Q2 inclusions aligned along planes located within the detrital grain. Sample depth: 3525 m. (d) Quartz overgrowth, marked by a line of Q1 A fluid inclusions, and post-dated by ankerite cement (left) which is partially replacing the authigenic quartz. The Q1 fluid inclusions mark the boundary between the quartz grain and the quartz cement. The quartz cement is post-dated by replacive ankerite cement (to the left). (e) Detail of Q1 B fluid inclusions. See text for explanation. Sample depth: 3525 m.

**Table 3.** Summary of fluid inclusion data

| Sample depth | Present-day temperature | Inclusion location | *n* | Range $T_h$ | Mean $T_h$ | Modal interval | Mean size | Salinity (wt% NaCl) |
|---|---|---|---|---|---|---|---|---|
| 2654 m | 116°C | Q1A (grain boundary) | 13 | 110–122°C | 115.8°C | 115–120°C | 3.2 µm | 0.3 |
| | | Q1B (unclear) | 23 | 108–128°C | 121.8°C | 120–125°C | 5.5 µm | — |
| | | Q2A (secondary planes) | 10 | 113–140°C | 132.6°C | 135–140°C | 4.8 µm | — |
| 3092 m | 123 °C | Q1A (grain boundary) | 21 | 116–127°C* | 122.7°C | 125–130°C | 3.8 µm | — |
| | | Q1B (unclear) | 16 | 106–162°C | 132.8°C | 130–135°C | 5.7 µm | 2.3, 2.3, 2.3 |
| | | Q2A (secondary planes) | 17 | 118–140°C | 129.5°C | 125–130°C | 3.7 µm | 2.6 |
| 3525 m | 127 °C | Q1A (grain boundary) | 27 | 130–145.5°C | 135.2°C | 130–135°C | 5.3 µm | 2, 3.3, 4.6, 4.6, 6.6 |
| | | Q1B (unclear) | 19 | 121–145°C | 135.7°C | 135–140°C | 7.3 µm | 2.6, 3.3, 4.5, 5.8 |
| | | Q2A (secondary planes) | 4 | 133–134°C | 133.8°C | 130–135°C | 3.5 µm | — |

* After supressing two extreme values of 108 and 129.

## DISCUSSION

Assuming that the $T_h$ values of grain-boundary inclusions are representative of the onset of quartz cementation and if we convert these palaeotemperatures into time using the available thermal and burial history models (Lotfy, 1994), then quartz cementation began in the Miocene, when the Khatatba and AEB Formations approached near maximum burial depths. In the studied samples, the mean $T_h$s values of the grain-boundary inclusions are similar to the present-day reservoir formation temperatures (Fig. 10). The sandstones represented by these samples could have also reached these temperatures (116, 123, and 135.7°C) during the Early Miocene. This is possible if we assume, for example, a palaeogeothermal gradient of 28°C km$^{-1}$ and erosion post-dating Middle Miocene times of 200 m of sediments.

There are, however, some possible inconsistencies between this Miocene age assignation for the onset of quartz cementation and other independent constraints, derived from the diagenetic sequence and the timing of oil migration into the reservoirs. In addition, the burial–thermal history models that support a Miocene age for the quartz cement must be treated with caution, as there are considerable uncertainties in the input data and theoretical assumptions of these models, especially regarding the Late Cretaceous palaeogeothermal gradients and the early Tertiary erosive event.

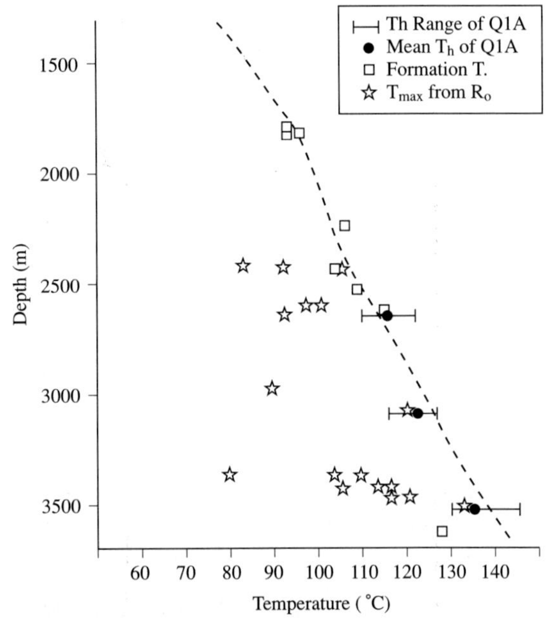


**Fig. 10.** Present depth versus temperature plot, showing the means and ranges of $T_h$ values of Q1 A inclusions. $T_{max}$ obtained from $R_o$ are included also. The dashed line represents the present-day temperature curve deduced from corrected borehole measurements. The deepest temperature point (128°C at 3623 m) was uncorrected for fluid circulation and the real value is estimated to be 15°C higher.

### Quartz cementation versus oil migration

Oil emplacement probably post-dated quartz cementation in the Khatatba and AEB reservoirs. Quartz cementation is unlikely to be taking place today in Salam sandstones, as the late-stage cements of kaolinite–dickite and pyrite–pyrrhotite, which clearly post-date the quartz overgrowths, have a fresh appearance and in most cases are not intergrown with authigenic quartz. Also, oil actually fills secondary porosity that resulted from the leaching of carbonate cements (that post-dated quartz overgrowths).

Oil emplacement can halt quartz cementation in sandstone oil reservoirs, especially in oil-wet reservoirs with low water saturations (Worden *et al*., 1998). Production

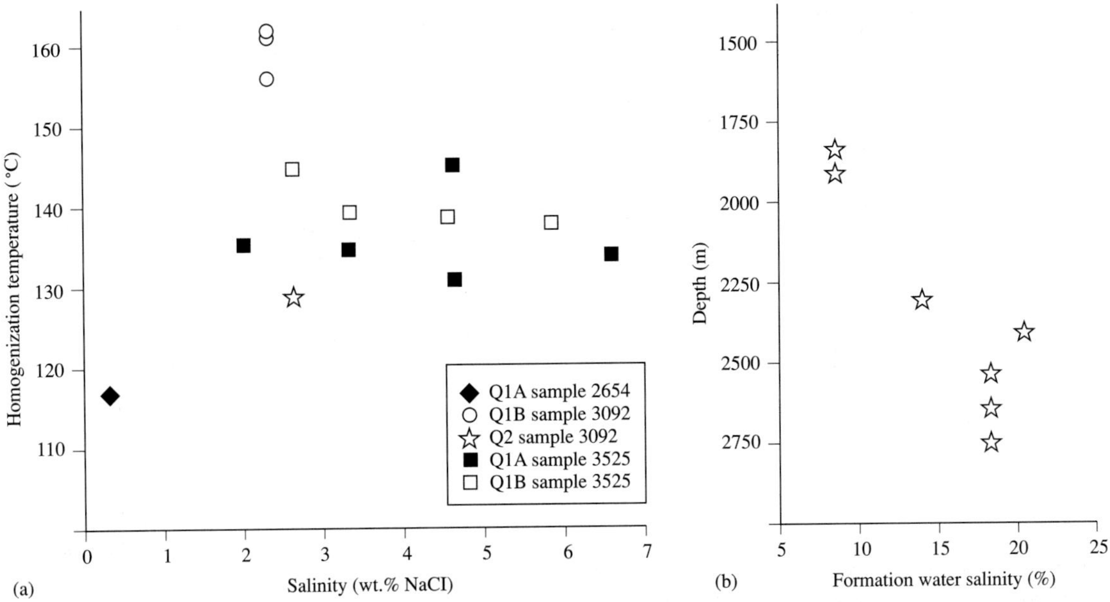


**Fig. 11.** (a) Plot of salinity versus $T_h$ data. (b) Plot of present-day salinity of the formation water versus depth.

data show that water saturations in the AEB and Khatatba reservoirs decrease with depth and are relatively low (14% in the AEB 3E reservoir, at 2650 m). These low water saturations make it improbable that the lower reservoirs are water-wet. Moreover, high porosity–permeability sandstones, such as those of the AEB and Khatatba reservoirs, commonly lead to oil-wet behaviours (Worden *et al.*, 1998). Taking into account these data and also the relationship observed between the authigenic phases, it is unlikely that quartz cementation has continued up to the present in Salam sandstones.

No petroleum inclusions were found in the quartz overgrowths of Khatatba and AEB sandstones. Yellow fluorescent petroleum inclusions are, however, abundant in grain–overgrowth boundaries in sandstones of the shallower Bahariya reservoir (1800–2000 m depth). Therefore, it appears that either quartz cementation largely pre-dated petroleum filling of the Khatatba and AEB reservoirs, or for some unknown reason, petroleum was trapped in the quartz overgrowths of the shallower reservoirs but *not* in the deeper Khatatba and AEB reservoirs. This later explanation seems unlikely, taking into account that the deeper reservoirs, being closer to the source rocks, filled earlier. This is demonstrated by the trend of increasing water saturations up-section (reaching up to values up to 46% in the upper Bahariya reservoir), and by the fact that the most shallow reservoirs such as Bahariya are not filled to overflowing.

The limited salinity data of the Q1 A and Q1 B fluid inclusions indicate that the salinities of the aqueous fluid involved in quartz precipitation were low to moderately saline, ranging 0.3–6.6 wt% NaCl and usually around 2–4 wt% NaCl (Table 3). This contrasts with the high salinities displayed by the present-day formation water in the AEB reservoirs, which are above 18%. No salinity data are available from Khatatba reservoir, but the trend of increasing salinity with depth in the Salam Field (Fig. 11b) suggests that even higher salinities could be present in the Khatatba reservoir. These data indicate that the water from which quartz precipitated was markedly different from oil-related aqueous brines, also supporting the likelihood that quartz precipitation pre-dated oil filling.

## Timing of quartz cementation

As discussed previously, oil traps formed during the early Tertiary and oil started to fill the reservoirs of the Salam structure following the Eocene. The vertical distribution of gas–oil ratios in the Salam Field indicates that later generated light oils and gas have displaced heavier oils downward to spill points, where they were able to migrate further up-section, to the upper AEB and Bahariya reservoirs (Nashaat *et al.*, 1994). This suggests that oil migration is not a recent phenomenon. Instead, a relatively long history of filling must have occurred, especially in the Khatatba and deeper AEB reservoirs, as they are close to the oil source rocks which supplied progressively more mature (lighter) oils during progressive burial and heating.

If oil filled the Khatatba and AEB reservoirs in the Eocene, and oil filling post-dated quartz cementation, then quartz cement must be of pre-Eocene age. In studies on the timing of quartz cementation in extensional basins world-wide, quartz appears to precipitate during and/or immediately after periods of rapid sedimentation and high heating and fluid-flow rates, presumably related to hydrocarbon generation and migration (Robinson & Gluyas, 1992; Bjørlykke & Egeberg, 1993; Gluyas *et al.*, 1993; Walderhaug, 1994a). In the Salam Field area, the Late Cretaceous was a period of rapid subsidence, high heating rates and hydrocarbon generation. It is therefore conceivable that, in the Salam Field, the quartz cements precipitated during the rapid burial at the end of the Cretaceous.

Although the palaeotemperatures obtained from the Q1 A fluid inclusions match with those expected to have prevailed in the Miocene, we believe that it is unlikely that the quartz cement started to grow in the Miocene, when the traps of the Salam structure were already formed and mostly filled with oil. This would mean that the deeper, highly porous and permeable reservoirs were water-wet during the Miocene, despite their location close to the mature source rock, and for some reason no oil inclusions were trapped in the quartz cement. A Miocene age for the quartz would imply also that a drastic increase in the salinity of the formation water has occurred in recent times. If quartz precipitated during the Miocene, then the diagenetic events that post-date authigenic quartz (ferroan carbonate cementation, carbonate leaching, kaolinite–dickite and pyrite–pyrrhotite precipitation), all would have occurred very recently, and the sandstones would have suffered almost no diagenetic alterations in a period of 100–170 million years characterized by several burial and uplift cycles.

### Interpretation of the diagenetic sequence

If we assume that oil filled the reservoirs in the Eocene, and quartz overgrowths probably grew in the Late Cretaceous, then we can try to place temporally the other recognized diagenetic events.

Because oil fills secondary porosity, the generation of secondary porosity must be Eocene or earlier. As the secondary porosity formed by dissolution of carbonate that post-dated quartz overgrowths, then the secondary porosity generation must have taken place during the latest Cretaceous or later. An important tectonic event occurred in the area between the latest Cretaceous and the middle Eocene, leading to basin inversion, uplift and erosion. This situation may have lead to large-scale movement of waters that might have been capable of dissolving carbonates. These waters could have been cooling, hot compactional waters (Burley *et al.*, 1989) enriched in organic acids derived from organic matter maturation (Keeley *et al.*, 1990). However, it is also feasible that they were meteoric waters, whose regional flow was triggered by the generalized uplift and by the basinward migration from the discharge areas, resulting in higher hydraulic gradients and therefore deeper meteoric water circulation. Basinal or meteoric in origin, it is feasible that these low-pH waters were also responsible for the widespread kaolinite–dickite cements, which precipitated in secondary porosity.

The formation water from which the kaolinite–dickite precipitated were progressively displaced by oil from the middle Eocene onwards. As kaolinite–dickite shows a marked tendency to be oil-wet (Worden *et al.*, 1998), it is very likely that the ingress of oil into the reservoir stopped its precipitation. Finally, the reduction of sulphate in the highly saline oil brines by petroleum potentially resulted in the precipitation of the pyrite–pyrrhotite cement, which is the latest diagenetic event recorded in the studied sandstones.

### Temperatures of quartz cementation

If we assume that quartz overgrowths precipitated during the Late Cretaceous, then the palaeotemperatures obtained from fluid inclusions are higher than expected for the Salam Field using the thermal history model of Lotfy (1994). If we assume that the inclusions are not re-equilibrated (see discussion below), then there are several possible explanations for the higher palaeotemperatures:

**1** The Late Cretaceous burial depths were higher than presumed, with a significant portion of the Maastrichtian section being removed by erosion during the early Tertiary.

**2** The Late Cretaceous palaeogeothermal gradient was considerably higher than the present gradient. This may not be surprising, as geothermal gradients in modern extensional basins can be as high as 45°C km$^{-1}$. Because of the rapid Late Cretaceous subsidence, quartz cementation could span a relatively short period of time but cover a broad depth interval. This would explain the correlation observed between $T_h$ and burial depth.

**3** Migrating hot fluids were responsible for the precipitation of the quartz cements, at temperatures higher than those resulting from the contemporary conductive geothermal gradient. This could explain the high palaeotemperatures given by the fluid inclusions and the correlation observed between $T_h$ and burial depth, as the quartz cement could have been in this case synchronous over a broad range of burial depths. The hot fluids could have originated in the adjacent Shusham graben, where the Khatatba mudrocks were at burial depths exceeding 3.5 km during Late Cretaceous (Fig. 6) times and undergoing depth-related reactions involving dehydration, maturation of organic matter and compactional expulsion of aqueous fluids. Furthermore, the Salam structure is bounded by deep-seated faults, which could have facilitated the vertical migration of fluids along the sand-prone sedimentary succession.

### Re-equilibration of the fluid inclusions

Assuming the quartz overgrowths to be Late Cretaceous, another possible explanation for the relatively high palaeotemperatures yielded by the fluid inclusions is that these inclusions were trapped at lower temperatures and re-equilibrated during further burial (cf. Osborne & Haszeldine, 1993). This could resolve the inconsistencies between the data and thermal history models, as no significant early Tertiary erosion and/or Late Cretaceous overheating would be required. According to Lotfy (1994), in the Salam Field the highest Khatatba temperatures in the Maastrichtian were around 90°C, i.e. 38°C lower than the temperature obtained from the regression of mean $T_h$ values of Q1 A fluid inclusions. As stated previously, the temperatures marked by Q1 A fluid inclusions are similar to present-day downhole values (Fig. 10). If re-equilibration has occurred, the most obvious interpretation of the $T_h$ data is that they would reflect Miocene or younger temperatures, when the reservoirs were presumably at maximum or near-maximum burial depths and temperatures.

Re-equilibration could also explain the increase in the $T_h$ ranges and mean values with burial depth and their correlation with present-day temperatures (Osborne & Haszeldine, 1993), as well as the absence of oil inclusions in the quartz overgrowths and the differences in salinity between the present-day oil brines and the fluid inclusions.

Although fluid inclusions in quartz are generally considered not prone to re-equilibration at diagenetic temperatures (Robinson *et al.*, 1992; Goldstein & Reynolds, 1994, p. 61), the case of grain-boundary inclusions, such as the Q1 A assemblage, must be treated with caution. When aqueous fluid inclusions are overheated during burial above their entrapment temperatures, they develop internal overpressures that can cause re-equilibration of fluid inclusions by either stretching (irreversibly expand through plastic deformation) or leakage and refilling. As quartz has a slight tendency to plastic deformation (Prezbindowski & Tapp, 1991), the most likely mechanism of fluid inclusion re-equilibration in quartz is stretching. In this case, the stretched inclusions would retain their original salinities. In the case of grain-overgrowth boundaries, their relative physical weakness may allow the boundary to deform or even crack (Goldstein & Reynolds, 1994, p. 15) as a result of the combined action of the numerous overpressured inclusions that lay along these boundaries. In the samples studied here, the mechanical strength of the quartz overgrowths may be locally reduced by the abundant v-shaped corrosion re-entrants that commonly affect not only the authigenic but also the detrital quartz.

However, while re-equilibration is hypothetically possible and would resolve the inconsistencies between $T_h$ data and the thermal and migration histories, the narrow and unimodal $T_h$ distributions of Q1 A inclusions (Fig. 9) do not suggest resetting, as 90% of $T_h$ values are within a range of 10–15°C (cf. Goldstein & Reynolds, 1994). When looking at the combined distribution of Q1 A and Q1 B, the $T_h$ ranges increase, but this may mean nothing, given the uncertainties regarding the petrographic location of Q1 B inclusions. In a re-equilibrated assemblage, one would expect some correlation between $T_h$ and salinity, and between $T_h$ and size, as larger and less saline inclusions are more susceptible to be reset (Bodnar *et al.*, 1989; Osborne & Haszeldine, 1993, 1995a). However, there is no correlation between the available salinity data and $T_h$ (Fig. 11a). In terms of the $T_h$ versus size pattern, there is a weak tendency for the larger inclusions to yield slightly higher $T_h$ values, specially in sample 3525 (Fig. 12). This tendency is clearer when considering together Q1 A and Q1 B types (Fig. 12), but, as stated, the significance of Q1 B inclusions is doubtful. We must therefore conclude that, although remaining plausible, there is no unequivocal evidence supporting re-equilibration, at least with the available data.

## CONCLUSIONS

Quartz overgrowths in Jurassic and Lower Cretaceous reservoir sandstones from Salam Field precipitated relatively early in the diagenetic history, pre-dated by mechanical and chemical compaction, and post-dated by ferroan carbonate cements, secondary porosity generation, and kaolinite–dickite and pyrite–pyrrhotite cement precipitation. Quartz cementation occurred prior to oil filling the reservoirs during the Eocene, probably during the rapid burial period that occurred in the area in the Late Cretaceous. Secondary porosity generation was probably related to the ingress of acid waters during the Palaeocene uplift. These waters, which were also responsible for the

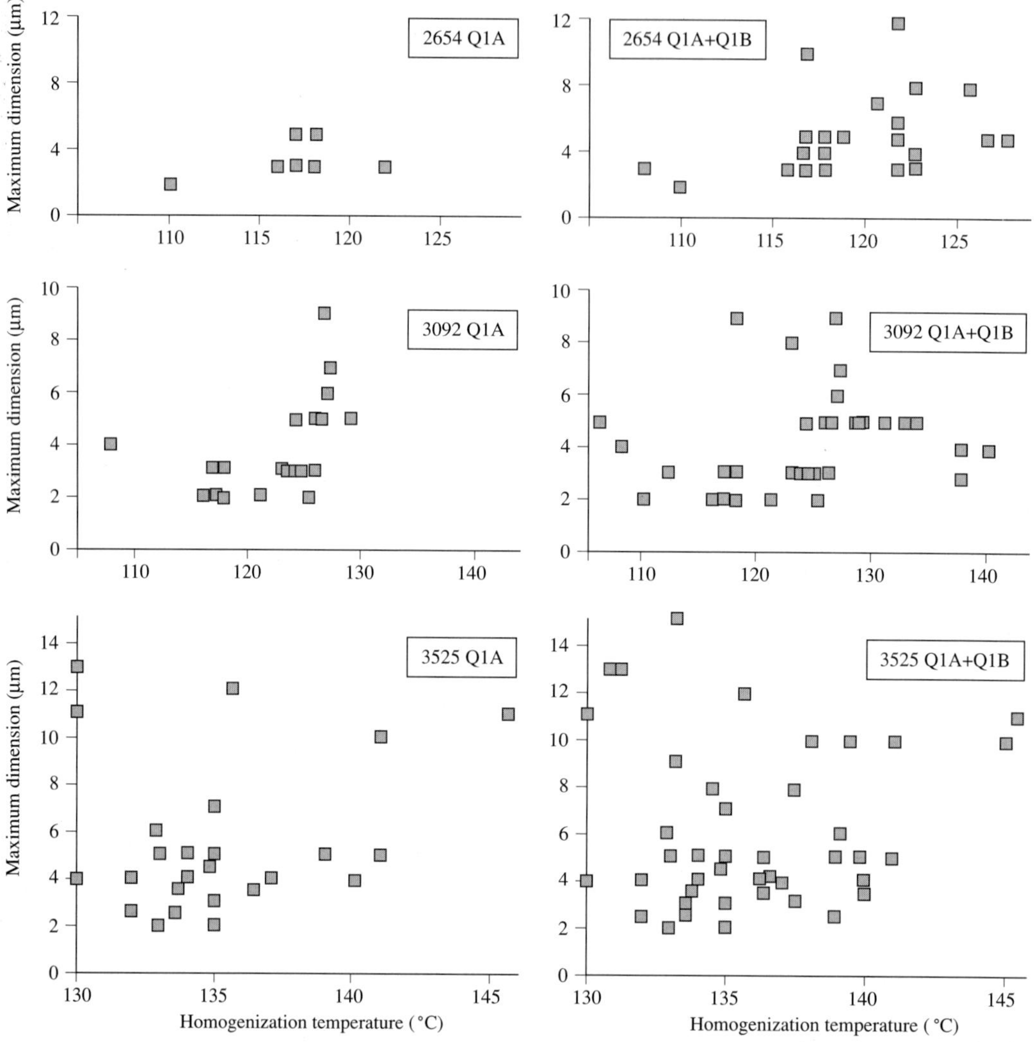


**Fig. 12.** Plots of $T_h$ versus estimated inclusion size.

widespread kaolinite–dickite cements, were progressively displaced by oil after the Eocene. The oil-filling of the reservoir presumably halted the kaolinite–dickite precipitation and resulted in the precipitation of the pyrite–pyhrrotite cement.

The homogenization temperatures of fluid inclusions located at grain-overgrowth boundaries indicate that quartz cements started to grow at temperatures ranging from about 116°C to 135°C. These palaeotemperatures are similar to present-day formation temperatures, and show a positive correlation with present depth of burial. The palaeotemperatures obtained from fluid inclusions are higher than predicted for the Late Cretaceous in the Salam area by the available thermal history models. There are several possible explanations for these relatively high palaeotemperatures: (i) the Late Cretaceous burial depths were higher than initially presumed; (ii) the Late Cretaceous palaeogeothermal gradients were considerably higher than present gradients; (iii) migrating hot fluids were responsible for the precipitation of the quartz cements at temperatures above those resulting from the contemporary conductive geothermal gradient; or (iv) fluid inclusions were formed at lower temperatures and re-equilibrated during subsequent burial and heating.

Although thermal re-equilibration of the grain-boundary fluid inclusions is hypothetically possible and would resolve the inconsistencies between $T_h$ data and the thermal and migration histories, there is no unequivocal evidence supporting re-equilibration, given the unimodal $T_h$ distributions (90% of $T_h$ values are within a range of 10–15°C in each sample) and the lack of a clear correlation between $T_h$ and parameters such as inclusion salinity or inclusion size.

## ACKNOWLEDGEMENTS

This work was funded by DGICYT PB-96–1236-CO2-O2 Project. The authors wish to thank Khalda Petroleum Co. and Repsol Exploracion Egipto for facilitating sampling and providing geological and well data, especially to J. Garcia Mallo, P. Cámara, F. Martínez-Fresneda, R. Ferrando and J. Suárez. Invaluable help during sampling in Salam Base was provided by A. Morsi and D. Perez. J. Mullis and C. Casquet made helpful suggestions regarding the interpretation of fluid inclusion data. The constructive and critical reviews by S. Morad, R.H. Worden and J. Parnell are gratefully acknowledged.

## REFERENCES


BAGGE, M.A. & KEELEY, M.L. (1994) The oil potential of Mid-Jurassic coals in Northern Egypt. In: *Coal and Coal-bearing Strata as Oil-Prone Source Rocks?* (eds SCOTT, A.C. & FLEET, A.J.) pp. 183–200. Special Publications of the Geological Society of London 77.

BARKER, CH.E. & PAWLEWICZ, M.J. (1986) The correlation of vitrinite reflectance with maximum temperature in humic organic matter. In: *Paleogeothermics.* (eds BUNTERBARTH, G. & STEGENA, L.) pp. 79–93. Springer-Verlag, Berlin.

BARKER, CH.E. & PAWLEWICZ, M.J. (1994) Calculation of vitrinite reflectance from thermal histories and peak temperatures. In: *Reevaluation of Vitrinite Reflectance.* (eds MUKHOPADHYAY, P.K. & DOW, W.G.) pp. 216–229. American Chemical Society Symposium Series 570.

BENNETT, P., MELCER, M.E., SIEGEL, D.I. & HASSETT, J.P. (1988) The dissolution of quartz in dilute aqueous solutions of organic acids at 25°C. *Geochimica et Cosmochimica Acta* **52**, 1521–1530.

BJØRLYKKE, K. & EGEBERG, P.K. (1993) Quartz cementation in sedimentary basins. *American Association of Petroleum Geologists Bulletin* **77**, 1538–1548.

BODNAR, R.J., BINNS, P.R. & HALL, D.L. (1989) Synthetic fluid inclusions. VI. Quantitative evolution of the decrepitation behaviour of fluid inclusions in quartz at one atmosphere confining pressure. *Journal of Metamorphic Geology* **7**, 229–242.

BURLEY, S.D., MULLIS, J. & MATTER, A. (1989) Timing diagenesis in the Tartan reservoir UK, North Sea; Constraints from combined cathodoluminiscence microscopy and fluid inclusion studies. *Marine and Petroleum Geology* **6**, 98–120.

CROSSEY, L.J. & LARSEN, D. (1992) Authigenic mineralogy of sandstones intercalated with organic-rich mudstones: Integrating diagenesis and burial history of the Mesaverde Group, Piceance Basin, NW Colorado. *Origin, Diagenesis and Petrophysics of Clay Minerals in Sandstones.* (eds HOWSEKNECHT, D.W. & PITTMAN, E.D.) pp. 125–144. Special Publications of the Society of Economic Paleontologists and Mineralogists 47.

EL AYOUTY, M.K. (1990) Petroleum geology. In: *The Geology of Egypt.* (ed. SAID, R.) pp. 567–599. A.A. Balkema, Rotterdam.

EL SHAZLY, E.M. (1977) The geology of the Egyptian region. In: *The Ocean Basins and Margins, Vol. 4.* (eds NAIRN, A.E.M., KANES, W.H. & STEHLI, F.G.) pp. 379–444. Plenum, New York.

GLUYAS, J.C., GRANT, S.M. & ROBINSON, A.G. (1993) Geochemical evidence for a temporal control on sandstone cementation. In: *Diagenesis and Basin Development.* (eds HORBURY, A.D. & ROBINSON, A.G.) pp. 23–33. American Association of Petroleum Geologists Studies in Geology 36.

GOLDSTEIN, R.H. & REYNOLDS, T.J. (1994) *Systematics of Fluid Inclusions in Diagenetic Minerals.* Society of Economic Paleontologists and Mineralogists Short Course 31.

GRANT, S.M. & OXTOBY, N.H. (1992) The timing of quartz cementation in Mesozoic sandstones from Haltenbanken, offshore Mid-Norway: fluid inclusion evidence. *Journal of Geological Society of London* **149**, 479–482.

HALL, D.L., STERNER, S.M. & BODNAR, R.J. (1988) Freezing point depression of aqueous sodium chloride solutions. *Economic Geology* **83**, 197–202.

HANTAR, G. (1990) North Western Desert. In: *The Geology of Egypt.* (ed. SAID, R.) pp. 293–321. A.A. Balkema, Rotterdam.

HASZELDINE, R.S. & OSBORNE, M. (1993) Fluid inclusion temperatures in diagenetic quartz reset by burial: Implications for oil field cementation. In: *Diagenesis and Basin Development.* (eds HORBURY, A.D. & ROBINSON, A.G.) pp. 35–46. American Association of Petroleum Geologists Studies in Geology 36.

KEELEY, M.L., DUNGWORTH, G., FLOYD, C.S. *et al.* (1990) The Jurassic System in northern Egypt. Regional stratigraphy and implication for hydrocarbon prospectivity. *Journal of Petroleum Geology* **13**, 397–420.

KERDANY, M.T. & CHERIF, O.H. (1990) Mesozoic. In: *The Geology of Egypt.* (ed. SAID, R.) pp. 407–438. A.A. Balkema, Rotterdam.

LOTFY, A. (1994) Jurassic source rock maturity and thermal history modelling of the Khalda West area. North Western Desert, Egypt. *12th Exploration and Production Seminar, Egyptian General Petroleum Corporation, Cairo.* pp. 217–233.

MARFIL, R., DORRONSORO, C., ROSSI, C. & PERMANYER, A. (1997) Controles composicionales, diagenéticos, y de microfábrica sobre las características de la roca madre jurásica del petróleo del campo de Salam (Western Desert, Egipto). *Geogaceta* **21**, 159–162.

MAY, R.M. (1991) The Eastern Mediterranean Mesozoic basin: Evolution and oil habitat. *American Association of Petroleum Geologists Bulletin* **75**, 1215–1223.

MORAD, S., BHATTACHARYYA, A., AL-AASM, I.S. & RAMSEYER, K. (1991) Diagenesis of quartz in the Upper Proterozoic Kaimur sandstones, Son Valley, central India. *Sedimentary Geology* **73**, 209–225.

NASHAAT, M., GAMAL EL DIN, A. & GAD, M. (1994) Geochemical material balance and entrapment mechanics in Shusham/Matruh basins, Western Desert, Egypt. *12th*

*Exploration and Production Seminar, Egyptian General Petroleum Corporation, Cairo*. pp. 198–216.

NEWMAN, J. (1997) New approaches to detection and correction of suppressed vitrinite reflectance. *APPEA Journal* **37** (2), 524–535.

OSBORNE, M. & HASZELDINE, R.S. (1993) Evidence for resetting of fluid inclusion temperatures from quartz cements in oil fields. *Marine and Petroleum Geology* **10**, 271–278.

OSBORNE, M. & HASZELDINE, R.S. (1995a) Reply to a discussion of 'Evidence for resetting of fluid inclusion temperatures from quartz cements in oilfields' by Osborne and Haszeldine (1993). *Marine and Petroleum Geology* **12**, 561–565.

OSBORNE, M. & HASZELDINE, R.S. (1995b) Reply to a discussion of 'Evidence for resetting of fluid inclusion temperatures from quartz cements in oilfields' by Osborne and Haszeldine (1993). *Marine and Petroleum Geology* **12**, 570–575.

PERMANYER, A., MARFIL, R., ROSSI, C., DORRONSORO, C. & LAHCINI, A. (1995) Oil features in reservoirs in Western Desert, Egypt. In: *Organic Geochemistry: Developments and Applications to Climate, Energy, Environment and Human History*. (eds GRIMALT, J.O. & DORRONSORO, C.) pp. 344–347. AIGOA, The Basque Country, Spain.

PREZBINDOWSKI, D.R. & TAPP, J.B. (1991) Dynamics of fluid inclusion alteration in sedimentary rocks: a review and discussion. *Organic Geochemistry* **17**, 131–142.

RAMSEYER, K., FISHER, J., MATTER, A., EBERHARDT, P. & GEISS, J. (1989) A cathodoluminiscence microscope for low intensity luminescence. *Journal of Sedimentary Petrology* **59**, 619–622.

ROBINSON, A. & GLUYAS, J. (1992) Duration of quartz cementation in sandstones, North Sea and Haltenbanken basins. *Marine and Petroleum Geology* **9**, 324–327.

ROBINSON, A., GRANT, S. & OXTOBY, N. (1992) Evidence against natural deformation of fluid inclusions in diagenetic quartz. *Marine and Petroleum Geology* **9**, 568–572.

ROEDDER, E. (1984) Fluid inclusions. In: *Reviews in Mineralogy 12*. (ed. Ribbe, P.H.) 644pp. Mineralogical Society of America.

ROSSI, C., MARFIL, R. & PERMANYER, A. (1996) Control diagenético sobre la porosidad y permeabilidad en areniscas-almacén cretácicas (Área de Salam, Western Desert, Egipto). *Geogaceta* **20**, 172–175.

SAID, R. (1990) Cenozoic. In: *The Geology of Egypt*. (ed. SAID, R.) pp. 451–486. A.A. Balkema, Rotterdam.

SHEPHERD, T.J., RANKIN, A.H. & ALDERTON, D.H.M. (1985). In: *A Practical Guide to Fluid Inclusions Studies*. Blackie, London, 239pp.

STONE, W.N. & SIEVER, R. (1996) Quantifying compaction, pressure solution and quartz cementation in moderately— and deeply—buried quartzose sandstones from the Greater Green River Basin, Wyoming. In: *Siliciclastic Diagenesis and Fluid Flow: Concept and Applications*. (eds CROSSEY, L.J., LOUCKS, R. & TOTTEN, M.W.) pp. 129–150. Special Publication of the Society of Economic Paleontologists and Mineralogists 55.

TAHER, M., SAID, M. & EL-AZHARY, T. (1988) Organic geochemical study in Meleiha area, Western Desert, Egypt. *9th Exploration and Production Seminar, Egyptian General Petroleum Corporation, Cairo*. pp. 1–28.

THOMPSON, S., COOPER, B.S. & BARNARD, P.C. (1994) Some examples and possible explanations for oil generation from coals and coaly sequences. In: *Coal and Coal-bearing Strata as Oil-Prone Source Rocks?* (eds SCOTT, A.C. & FLEET, A.J.) pp. 119–137. Special Publications of the Geological Society of London 77.

WALDERHAUG, O. (1990) A fluid inclusion study of quartz-cemented sandstones from offshore Mid-Norway. Possible evidence for continued quartz cementation during oil emplacement. *Journal of Sedimentary Research* **60**, 203–210.

WALDERHAUG, O. (1994a) Temperatures of quartz cementation in Jurassic sandstones from the Norwegian continental shelf. Evidence from fluid inclusions. *Journal of Sedimentary Research A* **64**, 313–323.

WALDERHAUG, O. (1994b) Precipitation rates for quartz cement in sandstones determined by fluid-inclusion microthermometry and temperature-history modelling. *Journal of Sedimentary Research A* **64**, 324–333.

WALDERHAUG, O. (1995) Discussion of 'Evidence for resetting of fluid inclusion temperatures from quartz cements in oilfields' by Osborne and Haszeldine (1993). *Marine and Petroleum Geology* **12**, 559–561.

WILSON, M.D. & STANTON, P.T. (1994) Diagenetic mechanisms of porosity and permeability reduction and enhancement. In: *Reservoir Quality Assessment and Prediction in Clastic Rocks*. (ed. WILSON, D.) pp. 59–119. Society of Economic Paleontologists and Mineralogists Short Course 30.

WORDEN, R.H., OXTOBY, N.H. & SMALLEY, P.C. (1998) Can oil emplacement prevent quartz cementation in sandstones? *Petroleum Geoscience* **4**, 129–137.

WORDEN, R.H., WARREN, E.A., SMALLEY, P.C., PRIMMER, T.J. & OXTOBY, N.H. (1995) Discussion of 'Evidence for resetting of fluid inclusion temperatures from quartz cements in oilfields' by Osborne and Haszeldine (1993). *Marine and Petroleum Geology* **12**, 566–570.

*Spec. Publs int. Ass. Sediment.* (2000) **29**, 183–197

# Regional loss of $SiO_2$ and $CaCO_3$, and gain of $K_2O$ during burial diagenesis of Gulf Coast mudrocks, USA

L. S. LAND *and* K. L. MILLIKEN

*Department of Geological Sciences, University of Texas at Austin, Austin TX 78713, USA*

## ABSTRACT

Ten of 11 wells from the onshore Texas portion of the Tertiary Gulf of Mexico sedimentary basin for which whole-rock analyses are available exhibit loss of $SiO_2$ relative to $Al_2O_3$, and in three of the wells $SiO_2/Al_2O_3$ ratios decrease with increasing depth. At least some mudrocks lose $SiO_2$ during burial diagenesis, and mudrocks seem to be a sufficient source for the quartz cement in Gulf Coast Tertiary sandstones. Most wells are also characterized by loss of Ca (as $CaCO_3$) and some display gain in $K_2O$. Burial diagenesis of mudrocks is an open chemical process in which all mudrocks lose some components, and some mudrocks gain components as well.

## INTRODUCTION

The source of $SiO_2$ for quartz cement in sandstones is uncertain, and in his review, McBride (1989) lists 23 proposed silica sources. Silica released by the transformation of smectitic to illitic I/S (illite/smectite) is an attractive source of $SiO_2$ in mud-rich settings because of the immense volume of smectite which is deposited in most sedimentary basins, and because there is consensus that smectitic clay is richer in $SiO_2$ (relative to $Al_2O_3$) than illitic clay. At issue is the fate of the $SiO_2$ mobilized during the clay mineral transformation. The classic works of Hower *et al.* (1976) and Yeh & Savin (1977) suggested that quartz precipitation in shales served as a sink for the $SiO_2$, even though Hower (1983) subsequently suggested that some $SiO_2$ was exported from the shales.

Land *et al.* (1997) analysed a single well from the south Texas Gulf Coast for several presumably immobile elements ($Al_2O_3$, $TiO_2$, HREE, Zr, Hf, and Th). All the presumably immobile elements remained constant in their relative proportions with respect to each other with increasing depth in the well, whereas $SiO_2$ decreased relative to all of them with increasing depth. This observation, coupled with the demonstration of oxygen isotopic inhomogeneity of modern quartz silt, similar to silt from the analysed well, led Land *et al.* (1997) to conclude that approximately 6 g of $SiO_2$ was lost per 100 g of mudrock in that single well.

In the course of our studies of Gulf Coast diagenesis, we have accomplished whole-rock analyses on mudrock cuttings from nine additional wells using the procedures outlined in Land *et al.* (1997). Table 1 presents relevant data on the 11 wells for which analyses are available. The well locations are shown in Fig. 1. Previously unpublished data for eight wells are presented in the Appendix (see p. 195).

## CHEMICAL EVIDENCE FOR $SiO_2$ LOSS

Gains and losses of components are typically assessed by comparison against a presumably immobile component. Of the elements in a conventional whole-rock analysis, $Al_2O_3$ and $TiO_2$ have been accepted by many authors as being the least mobile (e.g. Krauskopf, 1979). Indeed, in the samples they analysed, Land *et al.* (1997) found that $Al_2O_3$ and $TiO_2$ behaved conservatively during burial with respect to each other, as well as with respect to minor elements such as Zr, Hf, HREE, and Th, which were analysed to a similar level of precision using isotope dilution. High precision chemical analyses of a large number of samples for minor immobile elements such as Zr, Hf, Th, and HREE is a daunting task, and accurate concentrations of these elements are not available except for the single well analysed by Land *et al.* (1997). Choosing $Al_2O_3$ rather than $TiO_2$ as a 'normalizing' element from a conventional whole-rock analysis is preferred because of the small relative abundance of $TiO_2$.

**Table 1.** Gulf Coast wells for which whole-rock chemical analyses are available

| Well designation | | Well name | County | Formation | Depth | No. of samples |
|---|---|---|---|---|---|---|
| 1 | R-30819 | Shell Oil Co. No. 1 Carroll | Dewitt | Wilcox | 3082–19 495 | 26 |
| 2 | R-36527 | Shell Oil Co. No. 1 L. H. Penwell | Duval | Wilcox | 2435–15 438 | 22 |
| 3 | CWRU No. 6 | Mobil No. 1 Hall's Bayou Ranch | Galveston | Frio | 6070–18 045 | 13 |
| 4 | R-45572 | Amerada No. 1 Tally | Victoria | Frio | 14 477–23 881 | 8 |
| 5 | R-36512 | Walter Van Norman No. 1 Powderhorn Ranch | Calhoun | Frio | 8010–16 005 | 15 |
| 6 | R-44207 | Atlantic No. 1 Mustang Island | Nueces | Frio | 12 640–16 990 | 8 |
| 7 | R-47731 | Mobil No. 406 State Tract | Kenedy | Frio | 7145–18 085 | 23 |
| 8 | R-36334 | Shell No. 1 Continental and Fee | Cameron | Frio | 3005–14 984 | 20 |
| 9 | R-17494 | Amerada No. 1 T&N | Hidalgo | Vicksburg | 2511–11 020 | 11 |
| 10 | R-36720 | Tenneco No. 36 McAllen | Hidalgo | Vicksburg | 9485–14 835 | 12 |
| 11 | R-44220 | Shell Oil Co. No. 1 Lehman | Starr | Wilcox | 2154–14 198 | 20 |

Data for well No. 1 are from Awwiller (1993); data from well No. 3 are from Hower *et al.* (1977); and data from well No. 7 are from Land *et al.* (1997). Data for the other eight wells are reported in the Appendix.

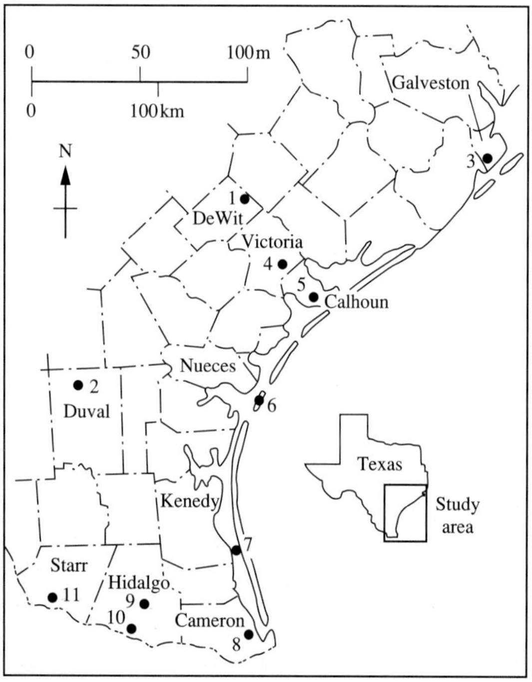


**Fig. 1.** Location map for the 11 wells listed in Table 1.

We used two methods to assess elemental mobility relative to $Al_2O_3$ in this regional data set. In the first, weight/weight scatter plots were constructed with $Al_2O_3$ plotted on the abscissa. If two components initially are distributed uniformly, and both behave conservatively during subsequent alteration, then a regression line through their concentrations should pass (statistically) through the origin. If both elements behave conservatively, their ratio remains constant (see additional discussion in Land *et al.*, 1997). Figure 2(a) shows a plot of weight percentage $TiO_2$ (the component being tested) against weight percentage $Al_2O_3$ (the presumed immobile component) for all 11 wells. A least squares regression line through the data passes close to, but not through the origin, and the range of extrapolated $TiO_2$ contents at 0% $Al_2O_3$ which encompasses 95% confidence also does not include the origin. There are several possible reasons for the apparent non-conservative behaviour of $TiO_2$ relative to $Al_2O_3$ in this regional data set. A systematic analytical error, such as incomplete sample digestion, cannot be ruled out, and the two components may actually behave conservatively. Alternatively, there may have been small losses of $Al_2O_3$ from the mudrocks, to form clay mineral cements in sandstones, for example. The degree of non-conservative behaviour of $Al_2O_3$ and $TiO_2$ using this technique is not large, however, when compared with the behaviour of the other components of interest.

If a component is selectively lost relative to an immobile component, then the intercept of a regression line through the data on the ordinate will be positive. The initial composition can be considered to be a mixture of the final, modified composition, and some composition which was lost and which contained none of the immobile component. The reasons we reject other possible explanations for the trend (e.g. differential sorting during deposition, or temporal changes in provenance) have been discussed by Land *et al.* (1997).

Figure 2(b) shows a plot of weight percentage $SiO_2$ versus weight percentage $Al_2O_3$ for all 11 wells, and demonstrates that $SiO_2$ has been lost relative to $Al_2O_3$. This kind of procedure was accomplished for each well

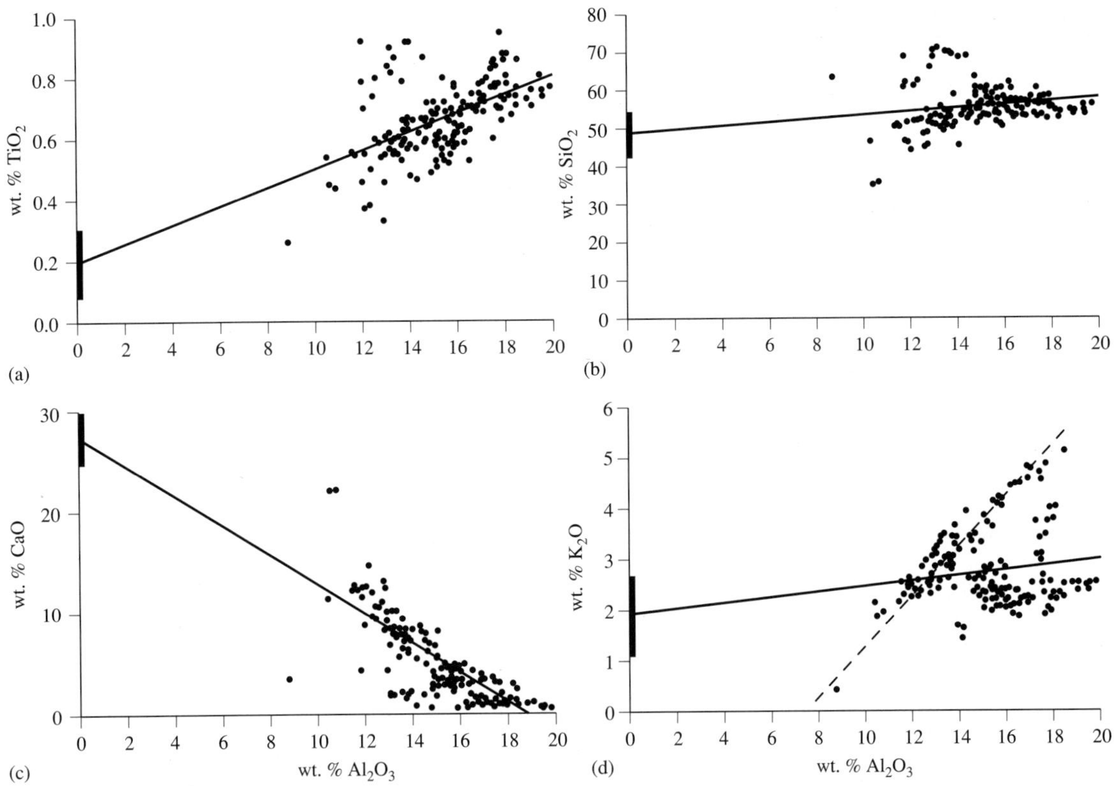


**Fig. 2.** Weight–weight scatter plot of components versus $Al_2O_3$. Solid bar on the ordinate shows the interval of 95% confidence for the extrapolation to 0% $Al_2O_3$. In the case of $TiO_2$ (a), nearly conservative behaviour is indicated, whereas $SiO_2$ (b) and CaO (c) have clearly been lost relative to $Al_2O_3$. In the case of $K_2O$ (d), there appear to be two populations of data, although the entire data-set suggests net $K_2O$ loss. One population displays approximately constant $K_2O$ content irrespective of $Al_2O_3$ content, suggesting that some mudrocks have lost $K_2O$. A second population can be extrapolated to negative $K_2O$ at 0% $Al_2O_3$ (dashed line), suggesting $K_2O$ gain by those mudrocks.

individually, and the results are summarized in Table 2 (and in Table 3 using $TiO_2$ as the immobile component). All wells except well number 3 indicate $SiO_2$ loss relative to $Al_2O_3$, using this procedure.

The second method of assessing elemental mobility is to plot component/$Al_2O_3$ ratio versus depth. $TiO_2$ (Fig. 3a) shows the expected relation of no significant trend with increasing depth, as the slopes of the lines which encompass the 95% confidence envelope are of different sign. Using this method, the $SiO_2/Al_2O_3$ ratio (Fig. 3b) also does not change significantly with depth for the entire data set. Three of the 11 wells do show a clear decrease in $SiO_2/Al_2O_3$ with increasing depth, however (Table 2), in which the slopes of both lines which define the 95% confidence envelope, as well as the slope of the regression line, are negative. No well displays increased $SiO_2/Al_2O_3$ ratios with increasing depth.

We conclude that considerable evidence exists in all wells, and unequivocal evidence in three, for $SiO_2$ loss relative to $Al_2O_3$ during burial diagenesis. In addition to regional $SiO_2$ loss during burial, most of these wells (Tables 2 & 3) also exhibit loss of CaO (as $CaCO_3$) (Figs 2c & 3c) and some wells exhibit gain of $K_2O$ (Figs 2d & 3d).

## PETROGRAPHIC EVIDENCE FOR $SiO_2$ GAIN/LOSS

During burial of mudrocks, some authors have proposed that $SiO_2$ precipitates locally (e.g. Yeh & Savin, 1977), whereas other authors (e.g. Füchtbauer, 1978; Boles & Franks, 1979) have proposed that $SiO_2$ dissolves. These conclusions were reached largely without the benefit of

**Table 2.** Summary of gain/loss of four components relative to $Al_2O_3$

| | Component versus $Al_2O_3$ | | | | Component/$Al_2O_3$ ratio versus depth | | | |
|---|---|---|---|---|---|---|---|---|
| Well | $TiO_2$ | $SiO_2$ | CaO | $K_2O$ | $TiO_2$ | $SiO_2$ | CaO | $K_2O$ |
| 1 | | L | G | | I | | | I |
| 2 | | L | L | L | | D | D | D |
| 3 | | | L | G | | | D | I |
| 4 | | L | L | G | | | | |
| 5 | L | L | | | I | | D | |
| 6 | | L | | | | | D | I |
| 7 | | L | L | G | D | D | D | I |
| 8 | L | L | | | I | | D | |
| 9 | L | L | L | G | D | D | D | I |
| 10 | L | L | L | | | | | |
| 11 | | L | L | L | I | | | |
| All wells | L | L | L | L | | | D | I |

For the four components plotted versus $Al_2O_3$, loss (L) relative to $Al_2O_3$ is indicated when the intersection of the 95% confidence envelope with the ordinate at 0% $Al_2O_3$ is positive. Gain (G) is indicated when the intersection is negative. Conservation is indicated (no symbol) when the intersection encompasses the origin (see Fig. 2). When component ratios are plotted versus depth, increase (I) relative to $Al_2O_3$ is indicated when the slope of both 95% confidence envelopes is positive. Decrease (D) relative to $Al_2O_3$ is indicated when the slope of both 95% confidence envelopes is negative. When the slopes of the 95% confidence envelope are of opposite sign, no significant change in component/$Al_2O_3$ ratio occurs with increasing depth (no symbol, see Fig. 3). Statistics were calculated using Statview II on a Macintosh computer.

**Table 3.** Summary of gain/loss of four components relative to $TiO_2$

| | Component versus $Al_2O_3$ | | | | Component/$Al_2O_3$ ratio versus depth | | | |
|---|---|---|---|---|---|---|---|---|
| Well | $TiO_2$ | $SiO_2$ | CaO | $K_2O$ | $TiO_2$ | $SiO_2$ | CaO | $K_2O$ |
| 1 | L | L | | G | D | D | | I |
| 2 | | L | L | L | | D | D | |
| 3 | L | L | L | | | | D | I |
| 4 | | L | L | | | | | |
| 5 | L | L | L | L | D | D | D | D |
| 6 | L | L | L | | | | | |
| 7 | | L | L | | I | D | D | I |
| 8 | L | L | L | L | D | D | D | D |
| 9 | G | | L | G | I | | D | I |
| 10 | | L | L | | | | | |
| 11 | L | L | L | L | D | | D | |
| All wells | L | L | L | L | | | D | |

direct petrographic evidence. Detrital quartz in mudrocks is sufficiently coarse (typically in the coarse silt range of 40–60 μm) so as to be amenable to petrographic observation. But the characteristically pure chemical composition of quartz does not permit identification of detrital versus authigenic quartz using the same methods that have been used successfully with other non-clay minerals in mudrocks (back-scattered electron imaging and electron microprobe analysis, for example). We used two approaches for petrographic examination of detrital quartz in mudrocks. First, we ultrasonically disaggregated the rocks and physically removed the silt fraction by gravity settling in distilled water. The grain surfaces were then examined with a scanning electron microscope (SEM)

**Fig. 3.** (*Opposite.*) The ratios of $TiO_2/Al_2O_3$ (a) and $SiO_2/Al_2O_3$ (b) do not change significantly with increasing depth because the slopes of the lines representing the 95% confidence envelope (dotted lines) are of different sign. In the case of the $CaO/Al_2O_3$ ratio (c) and $K_2O/Al_2O_3$ ratio (d), the slopes of the lines representing the 95% confidence envelopes have the same sense as the least squares regression line, indicating regional loss of CaO (as $CaCO_3$) and gain of $K_2O$.

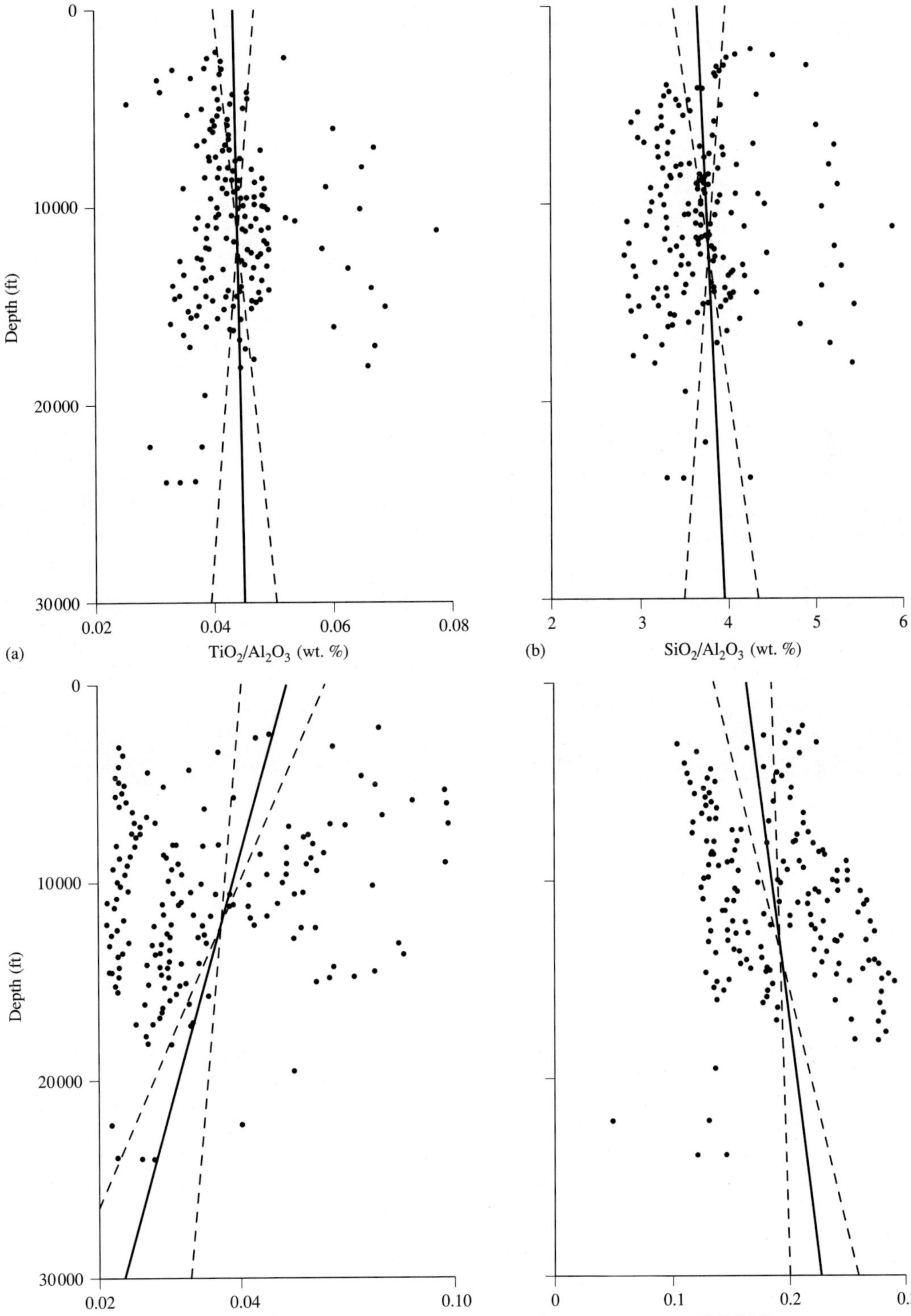
0
10000
20000
30000
Depth (ft)
0.02
0.04
0.06
0.08
(a)
$TiO_2/Al_2O_3$ (wt. %)
2
3
4
5
6
(b)
$SiO_2/Al_2O_3$ (wt. %)
0
10000
20000
30000
Depth (ft)
0.02
0.04
0.10
(c)
$CaO/Al_2O_3$
0
0.1
0.2
0.3
(d)
$K_2O/Al_2O_3$

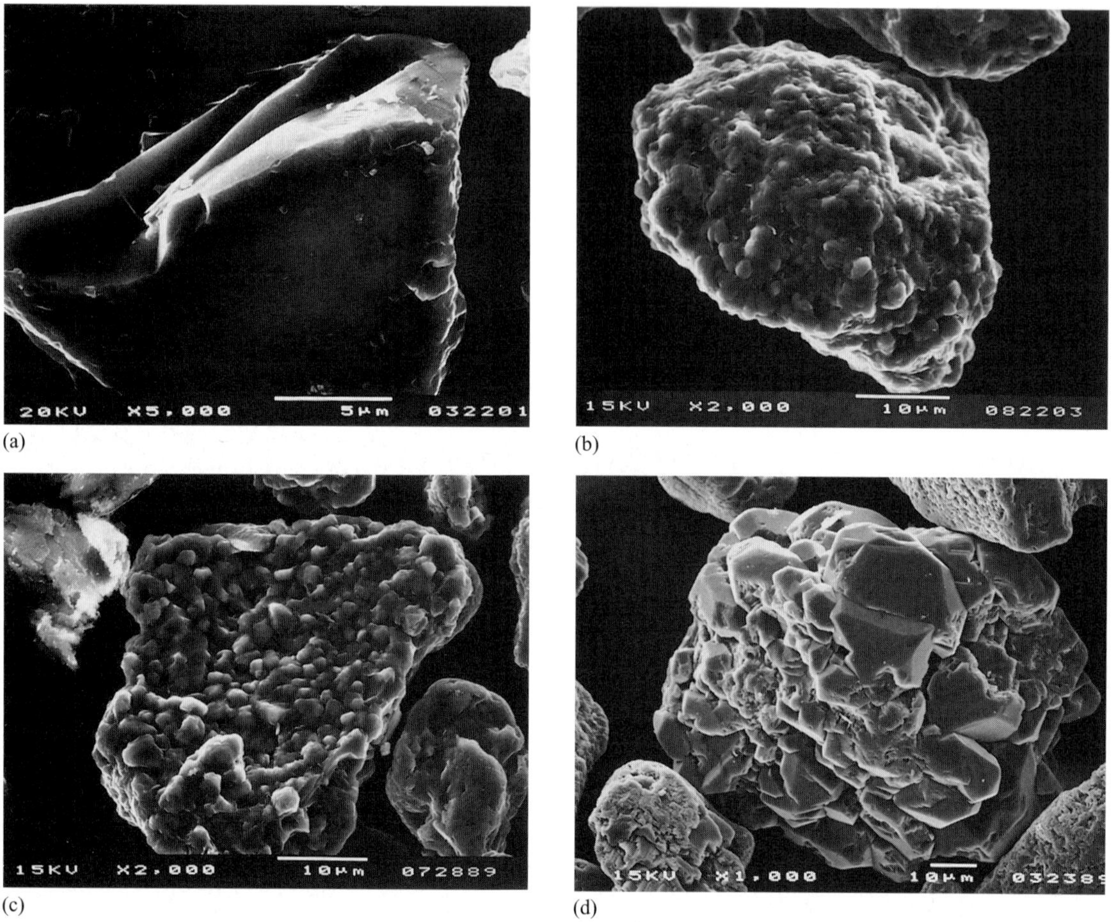


**Fig. 4.** Surface textures of quartz silt grains from Frio Formation (Oligocene) mudrocks. (a) Smooth chonchoidal surface showing no evidence of either dissolution or precipitation (well 8, 14 984 ft). (b) An example of vaguely defined 1–2 µm lumps which could be due to *in situ* precipitation of $SiO_2$, or which could be inherited (well 8, 12 285 ft). (c) Highly distinct 1–2 µm lumps with subhedral shapes (well 8, 12 850 ft). (d) Distinctly euhedral crystals on the surface of a quartz silt particle, which could be interpreted as an *in situ* feature (well 8, 3586 ft).

operating in conventional secondary electron mode. Second, we examined carbon-coated thin sections for cathodoluminescence with the SEM. Both methods were applied to a variety of mudrocks from a broad range of depths in both the Oligocene Frio and in Eocene formations (Fig. 1 and Table 1). As a partial model for the preburial character of the Frio Formation, silt from the modern Rio Grande river also was examined.

Contrasting observations were obtained from the two Cenozoic Gulf Coast units. Quartz silt grains from Frio samples display a wide range of surface textures. Some grains have smooth surfaces characterized by chonchoidal fracture (Fig. 4a). In addition to these simple and probably unmodified surfaces, other Frio silt grains show a broad diversity of features. Surface 'decorations' range from small (1–2 µm), ill-defined equant lumps (Fig. 4b), to more distinct, but still small and subhedral lumps (Fig. 4c), to larger (5–10 µm) distinct euhedra (Fig. 4d). There is no depth trend in the occurrence of these various surface fabrics, and all can be found both above and below the depth characteristic of the transformation of smectitic I/S to illitic I/S (see also Milliken, 1994a). Surfaces with varying degrees of decoration were observed on silt grains from all samples irrespective of depth. Significantly, both simple and highly decorated surfaces were also observed on quartz silt grains from the modern Rio Grande river. It

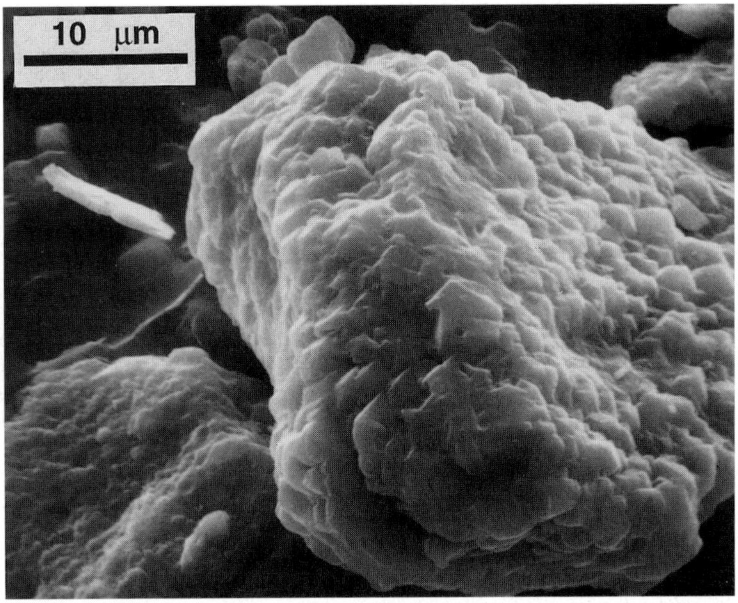


**Fig. 5.** Euhedral coatings on grain surfaces of typical quartz silt grains from deeply buried samples from the Eocene (well 11, 10 398 ft).

is not clear whether the surface features on Frio quartz silt are inherited from the source rocks (primarily Oligocene volcanics of west Texas and northern Mexico) or produced by processes during weathering and transport, analogous to the irregular features observed on fine quartz particles from the Mississippi River (Leschak & Ferrell, 1988) and the Rio Grande. It does seem clear, however, that none of these features can be correlated in any meaningful way with the strongly depth-dependent diagenesis of Frio mudrocks, either in terms of progressive quartz precipitation or quartz dissolution. In the Frio Formation, quartz silt grains appear to be unreactive, based both on examination of grain surfaces and on cathodoluminescence textures.

Cathodoluminescence imaging of quartz silt from Frio mudrocks as well as silt separated from the modern Rio Grande confirms the conclusions reached on the basis of grain surface textures. A wide range of cathodoluminescence textures are observed in both suites of samples, and no plausible connection between these textures and diagenesis can be found (Milliken, 1994a).

In contrast, examination of silt extracted from Eocene mudrocks reveals many cases in which authigenic quartz apparently has precipitated as overgrowths on detrital silt particles. In Eocene samples, surfaces of quartz silt are commonly modified by euhedra (2–5 μm) that uniformly cover the surfaces of many grains in a given sample (Fig. 5). Quartz silt grains from relatively shallow Eocene samples do not show these euhedra, and thus these features are plausibly interpreted as having resulted from diagenesis. Cathodoluminescence imaging again supports findings from secondary electron imaging of grain surfaces, suggesting clear evidence for quartz overgrowth development in Eocene mudrocks (Fig. 6).

## DISCUSSION: $SiO_2$ LOSS

Petrographic observations of samples from Oligocene and Eocene units reveal that quartz precipitation is not uniformly observed in Gulf Coast mudrocks. Quartz silt from the Frio Formation (Oligocene) manifests a wide diversity of surface textures and cathodoluminescence fabrics that are most plausibly interpreted as features that pre-date deposition of the mudrocks and were subsequently unmodified during diagenesis. Unequivocal evidence of mobilization of $SiO_2$ by pressure dissolution of quartz silt in Frio mudrocks was never observed. In contrast, Eocene mudrocks buried to depths greater than approximately 6000 ft do contain convincing evidence for quartz overgrowth modification of detrital quartz silt, visible both on grain surfaces and in cathodoluminescence images. Like the Frio, however, there is no evidence for mobilization of $SiO_2$ from detrital quartz by pressure dissolution in Wilcox mudrocks.

Reasons for the differences between Oligocene and Eocene mudrocks with regard to authigenic quartz are uncertain, but differences in petrographic character of

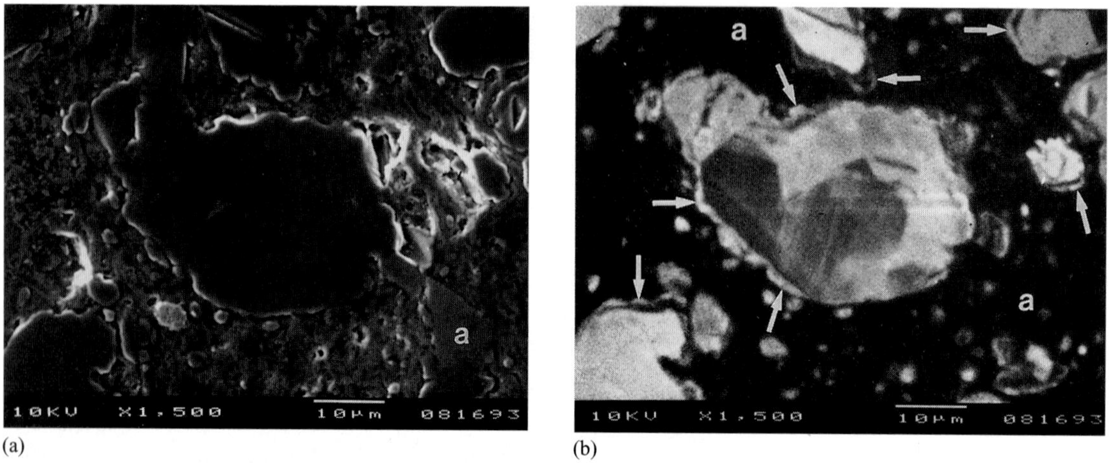


**Fig. 6.** Paired cathodoluminescence and BSE images of an Eocene mudrock. Displaying clear evidence (arrows) of a detrital overgrowths which may contribute toward explaining the $^{18}O$-rich nature of quartz silt (Blatt, 1987), as well as probable *in situ* quartz (well 11, 17 000 ft).

quartz do not correlate with differences in bulk elemental trends observed in these formations (Awwiller, 1993; Land *et al.*, 1997), as both formations appear to have experienced $SiO_2$ and $CaCO_3$ loss in addition to other elemental transfers. Percentages of quartz cement in Frio (2.6%) and Wilcox (4.6%—Land & Macpherson, 1992) sandstones are similar, and sandstones from the Eocene formations are somewhat more mature than those of the Frio Formation. In addition, the Eocene formations, as well as being older, contain a larger component of cratonic debris. None of the differences between the two units suggest simple explanations for the different behaviour of quartz in the mudrocks, however.

Even though some $SiO_2$ seems to be sequestered as authigenic quartz in some Eocene mudrocks, the whole-rock chemical data demonstrate that silica released from clay reactions in both Eocene and Oligocene units has been lost from the mudrocks. Tables 2 and 3 indicate that there is strong evidence for loss of $SiO_2$ relative to both $Al_2O_3$ and $TiO_2$, based on weight–weight scatter plots. Only wells 3 and 9 do not display significant $SiO_2$ loss relative to both $Al_2O_3$ and $TiO_2$. Changes in $SiO_2/Al_2O_3$ or $SiO_2/TiO_2$ ratio with depth are less convincing. Wells 2 and 7 display decreased $SiO_2$/immobile component ratios with increasing depth, whereas three other wells display a decrease relative to one immobile component but not with respect to the other. No other well displays a significant change with increasing depth, and no well displays increasing $SiO_2$ with increasing depth. When all the data are combined, $SiO_2$ is lost relative to both $Al_2O_3$ and $TiO_2$ (Fig. 2), but neither the $SiO_2/Al_2O_3$ nor the $SiO_2/TiO_2$ ratio changes significantly with increasing depth (Fig. 3).

We believe these data support the contention that $SiO_2$ is commonly lost from mudrocks as burial progresses. The conclusion based on the chemical data is in accord with the petrographic observations, at least for the Frio Formation, that authigenic $SiO_2$ does not become an obvious major component of the rocks as burial proceeds. It is difficult to assess the magnitude of $SiO_2$ loss from the available data. If 100 g of sediment initially contains approximately 14.0 weight percentage $Al_2O_3$, and losses produce a sediment containing approximately 18 weight percentage $Al_2O_3$ (e.g. Fig. 2), then the loss of mass of about 28 g ($14 \times 1.28 = 18$) is approximately half accounted for by $SiO_2$ ($55.7 \times 1.28–57.6$). Such a large value seems excessive, and Land *et al.* (1997) found a loss of only 6 g of $SiO_2$/100 g of sediment in the well they studied. No matter what the actual amount of $SiO_2$ loss may be, it is certainly sufficient to account for the approximately 3% of $SiO_2$ quartz overgrowth cement in Tertiary Gulf Coast sandstones (Land & Macpherson, 1992), recognizing the large mud/sand ratio which characterizes the basin. The fact that the sand–mud system may be out of balance, and losses by mudrocks may exceed gains by sandstones, is annoying, but much more data will be required to determine if a satisfactory balance exists or if a net loss from the system is actually taking place.

## DISCUSSION: CaO LOSS

CaO loss during burial is even more convincing than loss of $SiO_2$. Only well 1 does not display significant loss of CaO relative to either $Al_2O_3$ or $TiO_2$, and that well contains such a small amount of CaO that it is possible that loss of $CaCO_3$ has proceeded to completion. Three other wells display loss relative to one immobile component but not the other, whereas all other wells, in addition to the entire set of data, display clear loss of CaO. In the case of changes in CaO/immobile component ratio with increasing depth, six wells, and the entire data set, display decrease in both the $CaO/Al_2O_3$ and $CaO/TiO_2$ ratio, two wells display a decrease with respect to only one immobile component, and three wells exhibit no significant change in CaO/immobile component ratio with increasing depth. No well displays an apparent increase in CaO with increasing depth, as would be expected if the data were random. X-ray diffraction data reported by Freed (1980a, 1980b) provide additional evidence for $CaCO_3$ loss from Gulf Coast Tertiary mudrocks. Of the wells Freed analysed, four display a clear decrease in the percentage of detrital (mostly skeletal) calcite with increasing depth, whereas one exhibits no significant change with increasing depth.

As in the case of $SiO_2$, the magnitude of $CaCO_3$ loss as burial proceeds is difficult to quantify. If 28 g are lost per 100 g of mud (see above), then the loss of CaO derived from Fig. 2(c) is about 8 g of CaO (or about 20 g of $CaCO_3$) per 100 g of mud, nearly twice the value found by Land *et al.* (1997). As in the case of $SiO_2$, the loss of CaO (as $CaCO_3$) is clearly more than sufficient to supply the approximately 3% carbonate cement which characterizes Gulf Coast Tertiary sandstones (Land & Macpherson, 1992). There is an even stronger suggestion than is true in the case of $SiO_2$, that because of the large mud/sand ratio of the basin, a net loss from the sand–mud system may be taking place. The petrographic evidence relating to loss of $CaCO_3$ from Gulf Coast mudrocks is discussed by Milliken & Land (1993), and the source(s) of acid necessary for $CaCO_3$ removal are discussed by Land *et al.* (1997).

## DISCUSSION: $K_2O$ GAIN

The case for regional gain/loss of $K_2O$ is much less certain than is true of the loss of either $SiO_2$ or CaO. A case can be made for gain of $K_2O$ relative to either $Al_2O_3$ or $TiO_2$, or gain of $K_2O$ with increasing depth for wells 1, 3, 4, 6, 7, and especially 9. An equally strong case can be made for loss of $K_2O$ relative to either (or both) $Al_2O_3$ and $TiO_2$, or loss of $K_2O$ with increasing depth for wells 2, 5, 8, and 11. The entire data set (Fig. 2d) suggests that both gains and losses may be real, as the data appear to cluster along two trends, one indicating $K_2O$ loss and the other indicating $K_2O$ gain. Thus on a regional basis, there may have been little net gain or loss of $K_2O$ from the mudrocks during burial, merely an exchange between different localities. It is not at all clear from a mineralogic perspective why some mudrock sections seem to sequester $K_2O$, whereas others appear to lose $K_2O$, however. It is difficult to imagine that large local differences in clay mineralogy existed (smectite-rich versus smectite-poor areas, for example) due to the relatively uncomplicated provenance of the Gulf Coast basin. Likewise, it is hard to accept that smectite can transform to illitic I/S, losing $K_2O$ at the same time. The variable behaviour of $K_2O$ is easier to envision from a hydrologic perspective, however. If some mudrock sections were more open as the result of connected sands or faults, then two-way mass transfer could have taken place, whereas localities which were more isolated from the sands and/or faults which served as aquifers may have behaved as more nearly closed systems, or systems which only lost material as compaction proceeded. $K_2O$ is more difficult to relate to the degree of openness of the mudrock system because of the relatively small amount of $K_2O$ initially present in the sediment. On a regional basis, small amounts of $K_2O$ gain/loss appear to be taking place locally in an overall system which is characterized by loss of much larger amounts of $SiO_2$, $CaCO_3$, and $H_2O$.

## DISCUSSION: SOURCE OF QUARTZ CEMENT IN ASSOCIATED SANDSTONES

The $SiO_2$ for quartz cementation of Gulf Coast sandstones is clearly not derived from within the sandstones themselves, as advocated by Bjørlykke & Egeberg (1993). Pressure dissolution is not a significant factor in either Frio or Wilcox sandstones. Rare concavo-convex boundaries at contacts between quartz grains in deep Frio sandstones do occur, but are revealed by cathodoluminescence to arise primarily from brittle deformation followed by re-cementation of the grain fragments (Fig. 7), similar to brittle features observed in a wide range of other cases of apparent pressure dissolution (Milliken, 1994b; Dickinson & Milliken, 1995). Quartz precipitation precedes feldspar dissolution in both Frio and Wilcox sandstones, because secondary pores formed as a result of feldspar dissolution rarely contain quartz cement (Fisher & Land, 1986; Land & Fisher, 1987). Thus neither *in situ* pressure dissolution of quartz nor feldspar dissolution provides a sufficient local source for $SiO_2$ in Gulf Coast

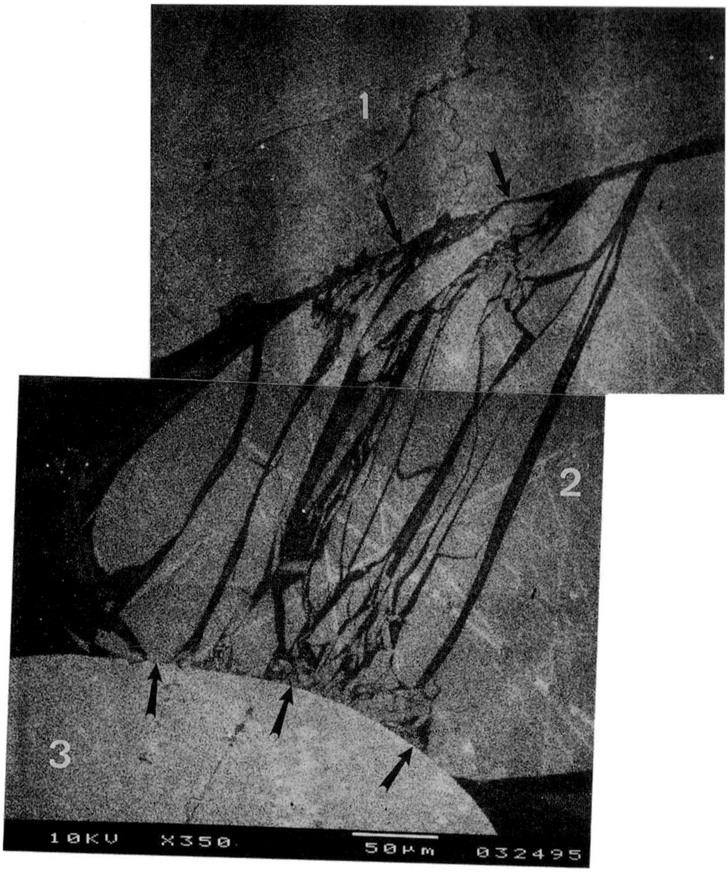


**Fig. 7.** Cathodoluminescence images of concavo-convex boundaries (arrows) between three quartz grains (bright luminescence) in a Frio Formation sandstone from 15 620.5 feet. The field of view includes only quartz, based on the BSE image. The curvilinear contacts between grains seen in the BSE image (or in thin section) might be interpreted to arise from pressure dissolution. But cathodoluminescence imaging demonstrates that these contacts arise from crushing of the central grain, the pieces of which have been cemented by authigenic quartz (dark luminescence). Volume loss due to grain interpenetration as interpreted from this image is much less than would be interpreted from a conventional light microscope image.

sandstones, and a source of $SiO_2$ external to the sandstones must exist. Additionally, a transport mechanism must exist to move the dissolved components both out of (some products of feldspar dissolution) and into ($CaCO_3$ and $SiO_2$) the sandstones.

## CONCLUSIONS

During burial, the reactions which characterize mudrocks do not take place in a closed system, or in a system solely characterized by compaction. Gulf Coast mudrocks lose $SiO_2$, $CaCO_3$, $H_2O$ and hydrocarbons (the latter two are not discussed herein). In some cases $K_2O$ is lost as well, whereas in other instances $K_2O$ is added as burial diagenesis proceeds. It is important to recognize that the mass potentially lost from mudstones is large, up to about 20%. Such large chemical losses are not conventionally considered in basin models involving compaction. Loss of such large amounts of material must have important and ongoing structural consequences (Cartwright & Lonergan, 1996), especially if, as appears to be the case, the losses are not uniform in time and space.

The conclusion that mudrock diagenesis is an open system process raises two important questions. First, 'Are the gains/losses evidenced in the mudrocks balanced by losses/gains in nearby sandstones?' The answer to that question appears to be 'no', as the loss of $CaCO_3$ and $SiO_2$ from mudrocks appears to be larger than can be balanced by chemical cements in Gulf Coast Tertiary sandstones. In at least one well (number 7, Land *et al.*, 1997) a source of $K_2O$ external to the sand–mud system is required. Arguably, available data may simply be insufficient to support such a conclusion on a basinal scale.

The answer to the second question, 'How does the mass transfer take place?', is also unclear. Insufficient chemical gradients exist in formation waters to enable diffusive transport over the large distances that characterize the sand–mud system. Convective flow within the overpressured mud-rich rock section is an attractive, albeit unproven, transport mechanism to move both solutes and hydrocarbons. Fractures generated *in situ* (Fig. 7) suggest that intergranular fluid flow is not the sole pathway in this complex hydrodynamic system. Additionally, a slow, through-going, vertically orientated flow system resulting from metamorphic dewatering of 'basement' units cannot be ruled out.

'More study of mudrocks is warranted' is a trite, but appropriate conclusion of our studies. The 'scale of system closure' is not yet clear in thick sedimentary prisms such as occur in the Gulf of Mexico sedimentary basin.

## ACKNOWLEDGEMENTS

This research was begun with support from the National Science Foundation (EAR-8904489) and continued under DE-FG05–92ER14249 from the Department of Energy and TX-131 and TX-534 from the Texas Advanced Research Program. ICP analyses were made by Dr Cassia Wolfson, and several undergraduate students, including Wendy Donaldson, Colby Drexler and Cori Lambert who carefully and cheerfully (almost always) picked cuttings.

## REFERENCES

AWWILLER, D.N. (1993) Illite/smectite formation and potassium mass transfer during burial diagenesis of mudrocks: a study from the Texas Gulf Coast Paleocene–Eocene. *Journal of Sedimentary Petrology* **63**, 501–512.

BJØRLYKKE, K. & EGEBERG, P.K. (1993) Quartz cementation in sedimentary basins. *American Association of Petroleum Geologists Bulletin* **77**, 1538–1548.

BLATT, H.A. (1987) Oxygen isotopes and the origin of quartz. *Journal of Sedimentary Petrology* **57**, 373–377.

DICKINSON, W.W. & MILLIKEN, K.L. (1995) The diagenetic role of brittle deformation in compaction and pressure solution, Etjo Sandstone, Namibia. *Journal of Geology* **103**, 339–347.

FISHER, R.S. & LAND, L.S. (1986) Diagenetic history of Eocene Wilcox sandstones, South-Central Texas. *Geochimica et Cosmochimica Acta* **50**, 551–561.

FREED, R.L. (1980a) Clay minerals as indicators of depositional environment in the South Hallettsville Field, Lavaca County, Texas. *Transactions of the Gulf Coast Association Geological Society* **30**, 347–350.

FREED, R.L. (1980b) Shale mineralogy and burial diagenesis in four geopressured wells, Hidalgo and Brazoria counties, Texas. In: *Factors Controlling Reservoir Quality in Tertiary Sandstones and their Significance to Geopressured Geothermal Production.* (eds LOUCKS, R.G., RICHMAN, D.L. & MILLIKEN, K.L.) 188pp. Annual Contract Report for U.S. Department of Energy, Contract no. DE-AC08–79ET27111.

FÜCHTBAUER, H. (1978) Zur Herkunft Quarzzements; Abschätzung der Quartzauflösung in Silt- und Sandsteinen. *Geologische Rundschau* **67**, 991–1008.

HOWER, J. (1983) Clay mineral reactions in clastic diagenesis (Abstract). *American Association of Petroleum Geologists Bulletin* **67**, 486.

HOWER, J., ESLINGER, E.V., HOWER, M.E. & PERRY, E.A. (1976) Mechanism of burial metamorphism of argillaceous sediment: 1. Mineralogical and chemical evidence. *Bulletin of the Geological Society of America* **87**, 725–737.

KRAUSKOPF, K.B. (1979). *Introduction to Geochemistry.* McGraw-Hill, 617pp.

LAND, L.S. & FISHER, R.S. (1987) Wilcox Sandstone diagenesis, Texas Gulf Coast: A regional isotopic comparison with the Frio Formation. In: *Diagenesis of Sedimentary Sequences.* (ed. MARSHALL, J.D.) pp. 219–235. Special Publications of the Geological Society of London 36.

LAND, L.S. & MACPHERSON, G.L. (1992) Origin of saline formation waters, Cenozoic section, Gulf of Mexico sedimentary basin. *American Association of Petroleum Geologists Bulletin* **76**, 1344–1362.

LAND, L.S., MACK, L.E., MILLIKEN, K.L. & LYNCH, F.L. (1997) Burial diagenesis of argillaceous sediment, South Texas Gulf of Mexico sedimentary basin: a re-examination. *Bulletin of the Geological Society of America* **109**, 2–15.

LESCHAK, P. & FERRELL, R.E. JR (1988) Morphologies of suspended clay-sized quartz particles in the Mississippi River and their relation to sedimentary sources. *Geology* **16**, 334–336.

MCBRIDE, E.F. (1989) Quartz cement in sandstones: a review. *Earth-Science Reviews* **26**, 69–112.

MILLIKEN, K.L. (1994a) Cathodoluminescence textures and the origin of quartz silt in Oligocene mudrocks, south Texas. *Journal of Sedimentary Research A* **64**, 567–571.

MILLIKEN, K.L. (1994b) The widespread occurrence of healed microfractures in siliciclastic rocks: Evidence from scanned cathodoluminescence imaging. In: *Rock Mechanics: Models and Measurements, Challenges from Industry.* (eds NELSON, P.P. & LAUBACH, S.E.) pp. 825–832. First North American Rock Mechanics Symposium. A.A. Balkema, Rotterdam.

YEH, H.W. & SAVIN, S.M. (1977) Mechanism of burial metamorphism of argillaceous sediment: 3. Oxygen-isotopic evidence. *Bulletin of the Geological Society of America* **88**, 1321–1330.

# APPENDIX

## Well No. 2

| Depth (ft) | $SiO_2$ | $TiO_2$ | $Al_2O_3$ | $Fe_2O_3$ | MnO | MgO | CaO | $Na_2O$ | $K_2O$ | $P_2O_5$ | $H_2O$ | Total | Li | V | Zn | Sr | Ba |
|---|---|---|---|---|---|---|---|---|---|---|---|---|---|---|---|---|---|
| 2435 | 55.67 | 0.53 | 13.62 | 4.41 | 0.05 | 1.91 | 6.51 | 1.90 | 2.69 | 0.12 | | 87.40 | 46 | 76 | 71 | 218 | 285 |
| 3070 | 51.49 | 0.55 | 13.27 | 4.35 | 0.08 | 1.91 | 8.67 | 2.03 | 2.57 | 0.13 | 8.77 | 93.81 | 41 | 74 | 71 | 289 | 336 |
| 4240 | 55.66 | 0.68 | 14.97 | 4.80 | 0.04 | 2.18 | 3.80 | 1.92 | 2.63 | 0.13 | | 86.82 | 49 | 84 | 78 | 255 | 348 |
| 4795 | 56.65 | 0.71 | 16.59 | 5.42 | 0.02 | 1.97 | 0.94 | 1.42 | 2.12 | 0.09 | | 85.93 | 76 | 83 | 81 | 171 | 310 |
| 5385 | 53.45 | 0.73 | 17.92 | 6.39 | 0.02 | 1.93 | 1.17 | 1.13 | 2.24 | 0.10 | 10.36 | 95.43 | 87 | 107 | 106 | 174 | 332 |
| 5875 | 53.59 | 0.74 | 18.43 | 6.40 | 0.02 | 2.04 | 1.41 | 1.18 | 2.31 | 0.09 | 10.12 | 96.34 | 85 | 110 | 102 | 190 | 362 |
| 6385 | 56.40 | 0.71 | 16.66 | 5.56 | 0.02 | 1.86 | 1.61 | 1.08 | 2.26 | 0.09 | 8.36 | 94.60 | 81 | 93 | 100 | 165 | 346 |
| 6915 | 55.95 | 0.71 | 16.85 | 5.97 | 0.02 | 1.85 | 1.71 | 1.20 | 2.27 | 0.09 | 8.76 | 95.37 | 87 | 99 | 98 | 142 | 320 |
| 7485 | 56.84 | 0.69 | 17.08 | 5.16 | 0.03 | 1.76 | 1.61 | 1.21 | 2.54 | 0.10 | 8.16 | 95.17 | 73 | 98 | 91 | 129 | 337 |
| 8105 | 57.97 | 0.73 | 16.98 | 4.55 | 0.02 | 1.71 | 0.87 | 1.30 | 2.27 | 0.08 | 7.80 | 94.27 | 63 | 95 | 82 | 131 | 685 |
| 8690 | 57.65 | 0.74 | 17.15 | 5.14 | 0.02 | 1.71 | 1.09 | 1.22 | 2.24 | 0.08 | 7.92 | 94.96 | 68 | 97 | 83 | 154 | 1384 |
| 9200 | 56.41 | 0.78 | 18.03 | 5.20 | 0.02 | 1.73 | 0.79 | 1.26 | 2.33 | 0.09 | 8.76 | 95.38 | 69 | 103 | 85 | 130 | 804 |
| 9890 | 56.61 | 0.81 | 18.05 | 6.19 | 0.04 | 2.01 | 0.96 | 1.26 | 2.26 | 0.14 | 7.71 | 96.05 | 66 | 98 | 91 | 135 | 601 |
| 10 385 | 56.67 | 0.74 | 18.15 | 5.44 | 0.03 | 1.77 | 1.55 | 1.21 | 2.22 | 0.09 | 8.00 | 95.85 | 76 | 104 | 88 | 136 | 700 |
| 10 925 | 56.05 | 0.76 | 19.53 | 5.14 | 0.03 | 1.68 | 0.51 | 1.19 | 2.42 | 0.11 | 8.28 | 95.70 | 74 | 108 | 85 | 120 | 675 |
| 12 005 | 57.24 | 0.77 | 19.84 | 5.51 | 0.03 | 1.76 | 0.52 | 1.15 | 2.56 | 0.11 | 7.43 | 96.93 | 87 | 113 | 95 | 124 | 697 |
| 12 575 | 55.44 | 0.74 | 19.56 | 5.33 | 0.03 | 1.80 | 0.76 | 1.12 | 2.55 | 0.10 | 8.04 | 95.46 | 77 | 114 | 91 | 119 | 592 |
| 13 115 | 57.22 | 0.81 | 19.47 | 5.48 | 0.03 | 1.78 | 0.65 | 1.26 | 2.51 | 0.11 | 7.43 | 96.73 | 89 | 115 | 99 | 130 | 721 |
| 13 655 | 55.78 | 0.73 | 18.90 | 5.83 | 0.03 | 1.97 | 1.06 | 1.19 | 2.55 | 0.11 | 7.67 | 95.83 | 76 | 108 | 95 | 119 | 589 |
| 14 655 | 54.94 | 0.76 | 19.10 | 6.66 | 0.08 | 1.95 | 1.12 | 1.53 | 2.42 | 0.16 | 7.04 | 95.77 | 82 | 129 | 108 | 172 | 1081 |
| 15 135 | 55.40 | 0.77 | 18.49 | 7.59 | 0.15 | 2.06 | 0.91 | 1.56 | 2.52 | 0.19 | 7.06 | 96.71 | 79 | 140 | 131 | 154 | 611 |
| 15 438 | 55.38 | 0.71 | 19.11 | 7.58 | 0.16 | 2.11 | 1.11 | 1.48 | 2.54 | 0.20 | 4.41 | 94.78 | 77 | 139 | 125 | 117 | 573 |

## Well No. 4

| Depth (ft) | $SiO_2$ | $TiO_2$ | $Al_2O_3$ | $Fe_2O_3$ | MnO | MgO | CaO | $Na_2O$ | $K_2O$ | $P_2O_5$ | $H_2O$ | Total | Li | V | Zn | Sr | Ba |
|---|---|---|---|---|---|---|---|---|---|---|---|---|---|---|---|---|---|
| 14 477 | 61.31 | 0.60 | 17.48 | 4.21 | 0.04 | 1.58 | 0.58 | 1.34 | 3.12 | 0.10 | 6.52 | 96.88 | 73 | 97 | 112 | 153 | 514 |
| 14 466 | 64.37 | 0.65 | 14.83 | 4.60 | 0.05 | 1.46 | 0.60 | 1.38 | 2.44 | 0.11 | 5.96 | 96.45 | 74 | 78 | 194 | 179 | 417 |
| 14 451 | 61.31 | 0.62 | 15.96 | 6.32 | 0.07 | 1.77 | 0.58 | 1.40 | 2.63 | 0.12 | 6.20 | 96.97 | 59 | 87 | 140 | 159 | 598 |
| 22 125 | 61.41 | 0.62 | 16.37 | 5.30 | 0.05 | 1.52 | 0.60 | 1.69 | 2.11 | 0.09 | 6.28 | 96.04 | 30 | 85 | 42 | 104 | 513 |
| 22 129 | 63.95 | 0.26 | 8.82 | 8.72 | 0.18 | 2.78 | 3.53 | 1.13 | 0.42 | 0.10 | 3.92 | 93.81 | 34 | 41 | 65 | 72 | 126 |
| 23 858 | 59.28 | 0.66 | 17.86 | 5.73 | 0.07 | 1.76 | 0.93 | 1.73 | 2.58 | 0.19 | 5.84 | 96.65 | 6 | 93 | 51 | 130 | 990 |
| 23 865 | 59.61 | 0.48 | 14.00 | 8.16 | 0.25 | 2.60 | 2.19 | 1.38 | 1.68 | 0.11 | 5.16 | 95.62 | 6 | 67 | 54 | 101 | 410 |
| 23 881 | 57.72 | 0.53 | 16.51 | 5.99 | 0.14 | 2.07 | 1.98 | 1.62 | 2.39 | 0.14 | 7.64 | 96.73 | 19 | 101 | 50 | 115 | 608 |

**Well No. 5**

| Depth (ft) | $SiO_2$ | $TiO_2$ | $Al_2O_3$ | $Fe_2O_3$ | MnO | MgO | CaO | $Na_2O$ | $K_2O$ | $P_2O_5$ | $H_2O$ | Total | Li | V | Zn | Sr | Ba |
|---|---|---|---|---|---|---|---|---|---|---|---|---|---|---|---|---|---|
| 8010 | 56.54 | 0.65 | 15.86 | 4.54 | 0.04 | 1.98 | 3.48 | 1.42 | 2.43 | 0.10 | 9.38 | 96.42 | 46 | 69 | 71 | 198 | 676 |
| 8505 | 58.55 | 0.61 | 15.83 | 4.02 | 0.03 | 1.69 | 2.96 | 1.75 | 2.40 | 0.09 | 7.60 | 95.52 | 42 | 61 | 80 | 171 | 1631 |
| 9005 | 58.59 | 0.55 | 15.69 | 4.12 | 0.04 | 1.63 | 3.58 | 1.50 | 2.33 | 0.08 | 7.88 | 95.99 | 49 | 58 | 586 | 193 | 1623 |
| 9510 | 56.97 | 0.60 | 15.19 | 4.06 | 0.04 | 1.60 | 3.59 | 1.52 | 2.36 | 0.08 |  | 86.01 | 57 | 58 | 77 | 200 | 2144 |
| 10 532 | 57.50 | 0.58 | 15.52 | 3.96 | 0.04 | 1.55 | 3.32 | 1.44 | 2.39 | 0.09 | 8.76 | 95.14 | 60 | 57 | 74 | 211 | 2158 |
| 11 050 | 57.23 | 0.58 | 15.63 | 4.26 | 0.04 | 1.65 | 3.57 | 1.62 | 2.35 | 0.09 | 7.40 | 94.42 | 47 | 63 | 81 | 238 | 2402 |
| 11 555 | 58.57 | 0.60 | 15.45 | 4.65 | 0.04 | 1.58 | 2.88 | 1.50 | 2.20 | 0.08 | 7.24 | 94.79 | 55 | 65 | 80 | 221 | 2351 |
| 12 050 | 58.38 | 0.62 | 15.85 | 4.39 | 0.04 | 1.60 | 3.28 | 1.50 | 2.37 | 0.08 | 7.84 | 95.94 | 54 | 67 | 74 | 174 | 1237 |
| 12 547 | 58.54 | 0.59 | 15.86 | 4.63 | 0.04 | 1.59 | 3.12 | 1.39 | 2.30 | 0.09 | 7.52 | 95.67 | 57 | 68 | 82 | 177 | 1480 |
| 13 050 | 58.44 | 0.63 | 16.43 | 4.53 | 0.04 | 1.60 | 2.94 | 1.44 | 2.39 | 0.09 | 7.88 | 96.41 | 63 | 68 | 78 | 162 | 1127 |
| 13 547 | 58.47 | 0.64 | 16.16 | 4.69 | 0.03 | 1.64 | 2.87 | 1.45 | 2.44 | 0.08 | 7.72 | 96.19 | 63 | 71 | 82 | 160 | 1005 |
| 14 062 | 58.88 | 0.57 | 15.35 | 4.34 | 0.04 | 1.60 | 3.58 | 1.49 | 2.48 | 0.08 | 8.48 | 96.88 | 63 | 66 | 78 | 156 | 762 |
| 15 046 | 58.46 | 0.59 | 15.70 | 4.55 | 0.03 | 1.62 | 3.90 | 1.40 | 2.30 | 0.08 |  | 88.63 | 45 | 62 | 83 | 161 | 1417 |
| 15 540 | 58.41 | 0.58 | 15.95 | 4.79 | 0.03 | 1.69 | 3.50 | 1.35 | 2.26 | 0.07 | 8.48 | 97.12 | 40 | 66 | 89 | 149 | 936 |
| 16 005 | 56.85 | 0.62 | 16.00 | 4.99 | 0.03 | 1.71 | 4.09 | 1.49 | 2.18 | 0.08 | 7.60 | 95.65 | 47 | 69 | 94 | 208 | 3310 |

**Well No. 6**

| Depth (ft) | $SiO_2$ | $TiO_2$ | $Al_2O_3$ | $Fe_2O_3$ | MnO | MgO | CaO | $Na_2O$ | $K_2O$ | $P_2O_5$ | $H_2O$ | Total | Li | V | Zn | Sr | Ba |
|---|---|---|---|---|---|---|---|---|---|---|---|---|---|---|---|---|---|
| 12 670 | 59.51 | 0.53 | 15.44 | 5.22 | 0.05 | 1.68 | 3.14 | 1.12 | 2.52 | 0.07 | 7.36 | 96.64 | 52 | 67 | 108 | 115 | 335 |
| 13 375 | 61.40 | 0.53 | 15.11 | 5.03 | 0.05 | 1.70 | 3.07 | 1.26 | 2.62 | 0.07 | 6.80 | 97.64 | 44 | 61 | 108 | 100 | 305 |
| 13 960 | 59.48 | 0.52 | 15.59 | 4.92 | 0.04 | 1.70 | 3.17 | 1.32 | 2.72 | 0.07 | 7.16 | 96.69 | 50 | 61 | 100 | 99 | 337 |
| 14 545 | 61.24 | 0.51 | 15.14 | 5.17 | 0.04 | 1.76 | 2.73 | 1.42 | 2.73 | 0.07 | 6.16 | 96.97 | 52 | 64 | 112 | 100 | 430 |
| 15 230 | 60.28 | 0.55 | 15.34 | 5.25 | 0.05 | 1.78 | 2.88 | 1.53 | 2.82 | 0.07 | 6.40 | 96.95 | 51 | 63 | 113 | 106 | 409 |
| 15 835 | 61.68 | 0.49 | 14.87 | 5.11 | 0.04 | 1.69 | 3.00 | 1.57 | 2.66 | 0.07 | 6.16 | 97.34 | 55 | 58 | 117 | 103 | 461 |
| 16 435 | 61.52 | 0.54 | 15.40 | 5.05 | 0.04 | 1.72 | 2.79 | 1.58 | 2.89 | 0.08 | 6.20 | 97.81 | 60 | 61 | 155 | 112 | 810 |
| 17 035 | 61.26 | 0.57 | 15.81 | 5.21 | 0.04 | 1.76 | 2.45 | 1.61 | 2.96 | 0.08 | 6.68 | 98.43 | 55 | 66 | 113 | 103 | 626 |

**Well No. 8**

| Depth (ft) | $SiO_2$ | $TiO_2$ | $Al_2O_3$ | $Fe_2O_3$ | MnO | MgO | CaO | $Na_2O$ | $K_2O$ | $P_2O_5$ | $H_2O$ | Total | Li | V | Zn | Sr | Ba |
|---|---|---|---|---|---|---|---|---|---|---|---|---|---|---|---|---|---|
| 3005 | 47.45 | 0.46 | 11.96 | 4.10 | 0.04 | 2.84 | 12.59 | 1.49 | 2.64 | 0.14 | | 92.43 | 42 | 101 | 469 | 277 | 331 |
| 3585 | 46.88 | 0.37 | 12.10 | 3.97 | 0.04 | 2.77 | 12.65 | 1.45 | 2.49 | 0.13 | | 91.58 | 41 | 103 | 76 | 275 | 264 |
| 4192 | 44.79 | 0.38 | 12.23 | 4.12 | 0.04 | 2.40 | 14.78 | 1.40 | 2.41 | 0.13 | | 91.79 | 42 | 120 | 65 | 311 | 256 |
| 4768 | 45.69 | 0.33 | 12.84 | 4.33 | 0.04 | 2.43 | 13.19 | 1.37 | 2.46 | 0.13 | 6.76 | 98.18 | 40 | 125 | 79 | 308 | 316 |
| 5348 | 46.18 | 0.46 | 12.89 | 4.57 | 0.04 | 2.62 | 12.56 | 1.62 | 2.56 | 0.16 | 6.72 | 98.05 | 38 | 121 | 75 | 304 | 249 |
| 5876 | 49.11 | 0.54 | 12.75 | 4.96 | 0.05 | 2.51 | 11.17 | 1.81 | 2.53 | 0.14 | 6.86 | 97.99 | 36 | 118 | 87 | 312 | 353 |
| 6568 | 49.55 | 0.55 | 12.92 | 4.95 | 0.05 | 3.03 | 10.21 | 1.64 | 2.71 | 0.15 | 6.86 | 99.07 | 39 | 117 | 89 | 254 | 301 |
| 7085 | 50.20 | 0.58 | 13.60 | 5.28 | 0.05 | 2.94 | 9.41 | 1.57 | 2.86 | 0.15 | 7.34 | 99.72 | 43 | 126 | 96 | 257 | 306 |
| 7662 | 51.40 | 0.60 | 13.75 | 5.20 | 0.05 | 2.82 | 7.88 | 1.91 | 2.80 | 0.14 | 7.26 | 98.87 | 43 | 122 | 85 | 232 | 696 |
| 8180 | 53.39 | 0.59 | 13.71 | 4.82 | 0.04 | 2.72 | 7.19 | 1.76 | 2.99 | 0.14 | 7.38 | 99.09 | 48 | 121 | 80 | 216 | 519 |
| 8740 | 52.15 | 0.66 | 14.07 | 5.33 | 0.04 | 2.68 | 8.37 | 1.69 | 3.20 | 0.16 | 6.74 | 99.37 | 38 | 90 | 105 | 238 | 408 |
| 9376 | 51.20 | 0.66 | 13.73 | 5.35 | 0.04 | 2.51 | 8.40 | 1.75 | 2.88 | 0.16 | 6.98 | 98.20 | 44 | 87 | 84 | 259 | 841 |
| 9932 | 53.90 | 0.67 | 13.88 | 5.47 | 0.04 | 2.51 | 7.15 | 1.89 | 3.25 | 0.16 | 6.84 | 99.25 | 44 | 78 | 80 | 220 | 554 |
| 10 482 | 53.98 | 0.60 | 13.32 | 5.41 | 0.05 | 2.47 | 7.60 | 2.08 | 2.93 | 0.14 | 6.14 | 98.19 | 40 | 78 | | 301 | 2234 |
| 11 106 | 55.79 | 0.68 | 15.10 | 5.38 | 0.04 | 2.56 | 5.71 | 1.59 | 2.85 | 0.11 | 7.14 | 99.07 | 48 | 90 | 101 | 242 | 1012 |
| 11 703 | 54.51 | 0.64 | 14.73 | 5.39 | 0.05 | 2.56 | 6.96 | 1.54 | 2.59 | 0.11 | 6.96 | 99.74 | 49 | 89 | 102 | 270 | 1437 |
| 12 285 | 55.38 | 0.67 | 14.46 | 5.43 | 0.05 | 2.55 | 8.18 | 1.53 | 2.63 | 0.12 | 6.54 | 98.03 | 44 | 88 | 117 | 319 | 2634 |
| 12 850 | 56.06 | 0.66 | 14.60 | 4.97 | 0.04 | 2.64 | 7.95 | 1.80 | 3.53 | 0.15 | 6.00 | 98.00 | 44 | 95 | 88 | 274 | 799 |
| 14 488 | 54.24 | 0.56 | 13.32 | 5.31 | 0.08 | 2.48 | 10.26 | 1.82 | 3.46 | 0.14 | | 97.19 | 41 | 75 | 96 | 370 | 5101 |
| 14 984 | 54.13 | 0.62 | 14.34 | 5.96 | 0.05 | 2.60 | 8.74 | 1.28 | 3.96 | 0.15 | | 91.44 | 49 | 85 | 104 | 235 | 1428 |

**Well No. 9**

| Depth (ft) | $SiO_2$ | $TiO_2$ | $Al_2O_3$ | $Fe_2O_3$ | MnO | MgO | CaO | $Na_2O$ | $K_2O$ | $P_2O_5$ | Total | Li | V | Zn | Sr | Ba |
|---|---|---|---|---|---|---|---|---|---|---|---|---|---|---|---|---|
| 2511 | 47.24 | 0.54 | 10.45 | 4.12 | 0.57 | 2.50 | 11.49 | 1.00 | 2.14 | 0.16 | 68.72 | 34 | 86 | 110 | | 478 |
| 4535 | 52.27 | 0.55 | 12.05 | 4.17 | 0.51 | 2.11 | 8.85 | 1.10 | 2.26 | 0.16 | 75.19 | 38 | 82 | 116 | 360 | 220 |
| 5026 | 51.81 | 0.59 | 13.19 | 4.22 | 0.50 | 2.69 | 10.17 | 1.03 | 2.42 | 0.19 | 76.63 | 55 | 452 | 158 | 379 | 259 |
| 7023 | 55.29 | 0.55 | 12.88 | 4.31 | 0.67 | 1.92 | 8.35 | 1.63 | 2.32 | 0.19 | 79.77 | 32 | 94 | 597 | 255 | 355 |
| 8021 | 54.36 | 0.56 | 13.22 | 4.51 | 0.53 | 1.97 | 7.89 | 1.32 | 2.65 | 0.19 | 79.30 | 26 | 107 | 104 | 246 | 452 |
| 8515 | 51.02 | 0.55 | 13.48 | 4.69 | 0.67 | 2.33 | 8.52 | 1.29 | 3.04 | 0.18 | 77.25 | 40 | 209 | 125 | 349 | 371 |
| 9028 | 51.11 | 0.58 | 13.97 | 5.07 | 0.52 | 2.39 | 7.34 | 1.51 | 3.44 | 0.15 | 78.73 | 37 | 111 | 205 | 205 | 322 |
| 9520 | 53.46 | 0.57 | 13.05 | 4.63 | 0.49 | 2.57 | 6.83 | 1.34 | 3.22 | 0.14 | 79.46 | 31 | 120 | 105 | 214 | 217 |
| 10 020 | 54.42 | 0.62 | 15.23 | 5.10 | 0.43 | 2.41 | 4.42 | 1.30 | 3.75 | 0.14 | 83.40 | 30 | 122 | 100 | 176 | 454 |
| 10 520 | 53.61 | 0.61 | 15.06 | 5.04 | 0.45 | 2.33 | 5.61 | 1.39 | 3.88 | 0.14 | 82.51 | 33 | 122 | 109 | 205 | 817 |
| 11 020 | 53.27 | 0.59 | 14.58 | 5.11 | 0.57 | 2.40 | 7.28 | 1.18 | 3.35 | 0.14 | 81.19 | 32 | 116 | 141 | 225 | 275 |

**Well No. 10**

| Depth (ft) | $SiO_2$ | $TiO_2$ | $Al_2O_3$ | $Fe_2O_3$ | MnO | MgO | CaO | $Na_2O$ | $K_2O$ | $P_2O_5$ | Total | Li | V | Zn | Sr | Ba |
|---|---|---|---|---|---|---|---|---|---|---|---|---|---|---|---|---|
| 9485 | 51.10 | 0.55 | 11.74 | 2.85 | 0.04 | 1.21 | 12.33 | 1.25 | 2.30 | 0.14 | 71.17 | 29 | 70 | 53 | 278 | 448 |
| 9995 | 51.01 | 0.56 | 11.52 | 2.46 | 0.04 | 1.27 | 12.32 | 1.49 | 2.16 | 0.14 | 70.65 | 26 | 79 | 53 | 277 | 777 |
| 10 575 | 53.31 | 0.65 | 13.82 | 3.39 | 0.04 | 1.74 | 7.56 | 1.31 | 3.07 | 0.13 | 77.45 | 32 | 102 | 58 | 217 | 897 |
| 11 175 | 56.57 | 0.61 | 13.46 | 2.81 | 0.03 | 1.55 | 5.64 | 1.13 | 3.51 | 0.11 | 79.79 | 30 | 88 | 84 | 135 | 382 |
| 11 745 | 53.47 | 0.64 | 14.67 | 3.59 | 0.04 | 1.87 | 6.23 | 0.96 | 3.16 | 0.13 | 78.54 | 26 | 106 | 58 | 174 | 823 |
| 12 300 | 52.82 | 0.63 | 13.70 | 4.08 | 0.04 | 1.91 | 8.30 | 1.11 | 2.96 | 0.13 | 77.38 | 24 | 106 | 70 | 173 | 493 |
| 12 465 | 51.76 | 0.55 | 11.61 | 3.67 | 0.06 | 2.11 | 12.83 | 1.73 | 2.55 | 0.12 | 74.16 | 29 | 82 | 78 | 448 | 4491 |
| 13 065 | 52.85 | 0.59 | 12.65 | 3.76 | 0.04 | 1.93 | 10.61 | 1.44 | 2.98 | 0.13 | 76.37 | 36 | 89 | 78 | 265 | 1269 |
| 13 635 | 52.64 | 0.60 | 12.52 | 4.09 | 0.05 | 2.03 | 10.68 | 1.52 | 2.87 | 0.14 | 76.45 | 31 | 92 | 92 | 329 | 2119 |
| 14 235 | 51.94 | 0.62 | 13.07 | 3.47 | 0.04 | 1.97 | 8.59 | 1.25 | 3.14 | 0.14 | 75.63 | 33 | 108 | 103 | 321 | 2384 |
| 14 745 | 52.34 | 0.61 | 12.94 | 3.67 | 0.04 | 2.00 | 9.25 | 1.27 | 3.06 | 0.14 | 76.05 | 28 | 100 | 107 | 232 | 732 |
| 14 835 | 52.14 | 0.61 | 13.07 | 3.10 | 0.04 | 1.82 | 8.43 | 1.22 | 2.87 | 0.14 | 75.00 | 29 | 98 | 127 | 255 | 1243 |

**Well No. 11**

| Depth (ft) | $SiO_2$ | $TiO_2$ | $Al_2O_3$ | $Fe_2O_3$ | MnO | MgO | CaO | $Na_2O$ | $K_2O$ | $P_2O_5$ | $H_2O$ | Total | Li | V | Zn | Sr | Ba |
|---|---|---|---|---|---|---|---|---|---|---|---|---|---|---|---|---|---|
| 2154 | 52.72 | 0.50 | 12.35 | 5.08 | 0.07 | 2.60 | 9.66 | 1.70 | 2.58 | 0.14 |  | 87.38 | 65 | 73 | 78 | 305 | 532 |
| 2645 | 55.33 | 0.57 | 13.86 | 5.00 | 0.06 | 2.02 | 6.06 | 1.83 | 2.44 | 0.13 |  | 87.31 | 67 | 73 | 90 | 262 | 429 |
| 3322 | 57.26 | 0.60 | 14.63 | 4.66 | 0.04 | 1.94 | 4.95 | 1.80 | 2.36 | 0.12 | 8.04 | 96.40 | 69 | 68 | 87 | 270 | 404 |
| 4357 | 52.65 | 0.68 | 15.78 | 5.74 | 0.04 | 1.72 | 2.23 | 0.92 | 2.06 | 0.12 |  | 81.94 | 85 | 88 | 124 | 865 | 3688 |
| 5067 | 53.62 | 0.63 | 16.53 | 6.43 | 0.03 | 1.82 | 3.07 | 0.96 | 1.87 | 0.11 | 7.00 | 92.08 | 80 | 92 | 123 | 564 | 3127 |
| 5640 | 46.15 | 0.57 | 14.22 | 9.91 | 0.06 | 3.21 | 5.45 | 0.92 | 1.66 | 0.13 |  | 82.28 | 74 | 83 | 112 | 534 | 3067 |
| 6207 | 51.23 | 0.64 | 16.01 | 8.36 | 0.05 | 2.47 | 4.81 | 1.23 | 2.03 | 0.10 |  | 86.92 | 100 | 88 | 111 | 304 | 2593 |
| 6889 | 55.74 | 0.68 | 18.27 | 7.25 | 0.04 | 2.08 | 2.96 | 1.33 | 2.36 | 0.10 | 7.72 | 98.54 | 108 | 91 | 110 | 185 | 1073 |
| 7477 | 57.29 | 0.59 | 15.08 | 6.44 | 0.04 | 2.59 | 1.83 | 1.07 | 2.36 | 0.12 | 6.88 | 94.29 | 95 | 74 | 123 | 423 | 3548 |
| 8036 | 56.20 | 0.70 | 16.17 | 5.07 | 0.04 | 1.54 | 3.47 | 0.91 | 2.06 | 0.08 |  | 86.24 | 99 | 82 | 115 | 310 | 3396 |
| 8636 | 58.68 | 0.69 | 15.64 | 5.54 | 0.05 | 1.70 | 3.04 | 0.94 | 2.07 | 0.11 | 6.56 | 95.01 | 84 | 75 | 96 | 168 | 1139 |
| 9277 | 59.05 | 0.68 | 16.10 | 4.40 | 0.04 | 1.52 | 3.34 | 1.02 | 2.21 | 0.09 | 6.80 | 95.25 | 93 | 82 | 98 | 156 | 570 |
| 9858 | 56.99 | 0.72 | 15.44 | 5.98 | 0.04 | 1.79 | 3.07 | 0.91 | 1.98 | 0.14 | 6.44 | 93.50 | 94 | 76 | 119 | 305 | 3466 |
| 10 398 | 58.62 | 0.69 | 16.09 | 5.24 | 0.04 | 1.70 | 4.17 | 1.02 | 2.43 | 0.10 | 6.44 | 96.53 | 87 | 90 | 100 | 161 | 575 |
| 11 002 | 57.68 | 0.70 | 15.15 | 6.06 | 0.04 | 1.76 | 3.51 | 1.07 | 2.19 | 0.13 | 6.52 | 94.80 | 85 | 76 | 95 | 186 | 1043 |
| 11 580 | 58.61 | 0.66 | 15.63 | 4.79 | 0.04 | 1.68 | 4.22 | 1.10 | 2.21 | 0.11 | 6.40 | 95.44 | 88 | 77 | 97 | 176 | 785 |
| 12 143 | 55.92 | 0.67 | 14.68 | 5.25 | 0.04 | 1.90 | 6.37 | 1.25 | 2.36 | 0.13 | 5.96 | 94.54 | 64 | 80 | 104 | 226 | 448 |
| 12 697 | 59.03 | 0.66 | 14.91 | 5.43 | 0.04 | 1.78 | 4.19 | 1.09 | 2.27 | 0.10 | 6.36 | 95.85 | 76 | 80 | 95 | 163 | 456 |
| 13 599 | 59.90 | 0.69 | 14.90 | 7.09 | 0.04 | 1.88 | 2.46 | 1.05 | 2.30 | 0.19 | 6.48 | 96.99 | 69 | 85 | 85 | 136 | 553 |
| 14 198 | 59.95 | 0.66 | 15.59 | 5.57 | 0.03 | 1.62 | 2.70 | 1.28 | 2.44 | 0.15 | 6.36 | 96.35 | 72 | 92 | 89 | 133 | 511 |

*Spec. Publs int. Ass. Sediment.* (2000) **29**, 199–218

# Quartz cement: the Miller's Tale

J. GLUYAS[1]*, C. GARLAND[2]†, N. H. OXTOBY[3] *and* A. J. C. HOGG[4]†

[1]*BP Exploración de Venezuela SA, Edificio Seguros Sudamerica, Avenida Fransisco de Miranda, El Rosal, Caracas, Venezuela;*

[2]*BP Exploration Operating Company Ltd, Farburn Industrial Estate, Dyce, Aberdeen, AB2 0PB, UK;*

[3]*University of London, Department of Geology, Egham Hill, TW20 0EX, UK; and*

[4]*BP Exploration Operating Company Ltd, Blackhill Rd., Holton Heath, Poole, Dorset, BH16 6LS, UK*

## ABSTRACT

The Brae Formation sandstones of the Miller Field, UK Continental Shelf, are quartz cemented quartz arenites. The only parts of the sandstone not cemented by quartz lie within calcite concretions. These concretions, precipitated soon after deposition of the sand, preserve the original fabric and chemistry of the sand. Comparison of the calcite cemented rock with the quartz cemented rock reveals that the sandstone gained $SiO_2$ during subsequent diagenesis.

The Brae Formation sandstones are overlain and interbedded with mudstones of the Kimmeridge Clay Formation. However, the mudstones do not have a uniform chemical composition. Mudstones that are interbedded with the Brae Formation sandstones contain less $SiO_2$ than mudstones that occur above the Brae Formation. The contrast between mudstones which are relatively rich in silica and those which are relatively poor in silica cross-cuts chronostratigraphic surfaces. This indicates that those mudstones which are interbedded with the sandstones lost (exported) silica during diagenesis.

The volume of silica exported from the mudstones matches that imported by the sandstone. Moreover, the date of quartz cementation in the sandstone was much the same as the inferred timing of clay mineral transformations in the mudstones. Such clay transformations could have liberated silica for cementation.

The average size of the isochemical, diagenetic system with respect to quartz was small. Indeed the system is better described as short since the important parameters are a vector and a measure of net (sandstone) to gross (sandstone plus mudstone). The average distance travelled for a silica quantum was $1.15 \pm 2.02$ m, in a direction normal to the stratigraphy and in a period of 2–20 Myr. Thus the silica flux during cementation was in the order of $10^{-4}$–$10^{-2}$ moles $m^{-1}$ $yr^{-1}$.

## INTRODUCTION

Geoffery Chaucer (1400) told the Miller's Tale; a bawdy story of sexual intrigue. For our Miller's Tale, that of quartz cementation within the reservoir sandstones of the Miller Oilfield, we can promise much intrigue, but if you are in search of salacious gossip read no further. The aim of this paper is to investigate the mobility of $SiO_2$ during cementation of a sandstone. The Upper Jurassic Brae (sandstone) Formation of the South Viking Graben in the Miller Field, UK North Sea (Fig. 1) was chosen for this study because it:

**1** is quartzose with a quartz cement;
**2** is largely of uniform, medium grain size;
**3** is encased within the Upper Jurassic, Kimmeridge Clay Formation, which is both source and seal for the oil accumulation of Miller;
**4** has simple and well constrained burial and thermal histories;
**5** contains preserved remnants of the original sand as it was deposited.

The strategy for the study was to determine (i) when, at what temperature and at what depth the quartz

* Present address: Monument Oil and Gas plc, Kierran Cross, 11 Strand, London, WC2N 5H, UK.

† Present address: ADDAX Petroleum 12, Rue Michel Servet 1206 Geneva, Switzerland.

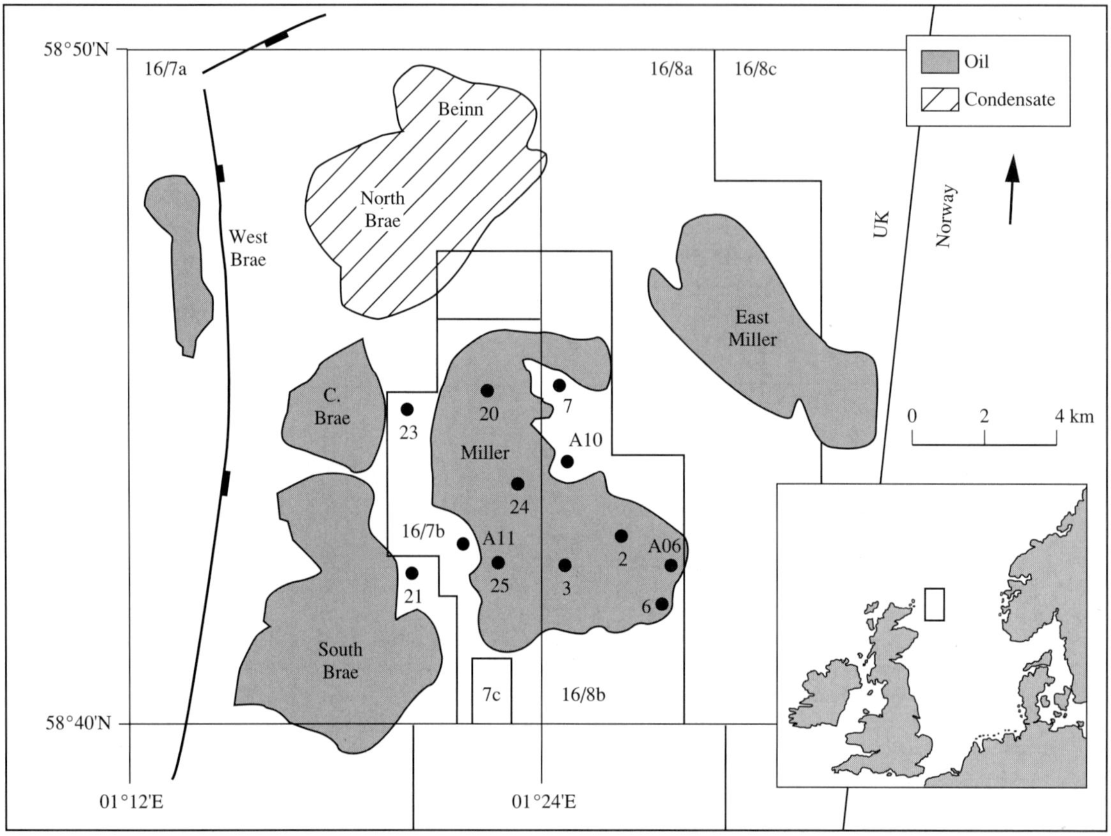


**Fig. 1.** Location map adapted from Garland (1993). West Brae has a Palaeocene reservoir. The remaining fields all have Upper Jurassic reservoir sections.

cementation occurred; (ii) what volume changes were associated with compaction and cementation; and (iii) what component oxides were imported into and exported from the sandstone.

A similar strategy was adopted for analysis of the Kimmeridge Clay Formation that overlies and is interbedded with the Brae Formation. Accurate quantification of mass transfer to and from the mudstone was difficult because the mudstones appear not to contain any preserved remnants of the original deposited mud.

Finally, the timing of diagenetic processes in the sandstone and mudstone formations and the volumes of material moved during diagenesis were compared. This was to investigate material balance and establish the size of the diagenetic system; a volume for which there was no net material gain or loss.

## GEOLOGICAL BACKGROUND

The Miller Field (Fig. 1) was discovered by BP Exploration well 16/7b-20 in December 1982. Early in 1983, well 16/8b-2 drilled by a Conoco/Saxon partnership penetrated the same accumulation in the adjacent licence. The field was the largest discovery of the 1980s for the North Sea, with an estimated original oil in place (STOOIP) of 519 million barrels (Garland, 1993).

The Miller Field occurs in the Brae Formation sandstones which were deposited during active rifting in the Late Jurassic (Kimmeridgian, Portlandian). The Upper Jurassic sediment pile is thick, more than 1000 m in Miller and thicker to the west. Deposition of the reservoir sandstones occurred in a sand-rich submarine fan close to the basin boundary, with sediment shed from the Fladen Ground Spur (Fig. 2). The oldest Upper Jurassic sediments in Miller comprise the toes of apron-fan sands and conglomerates that stacked up against the bounding faults

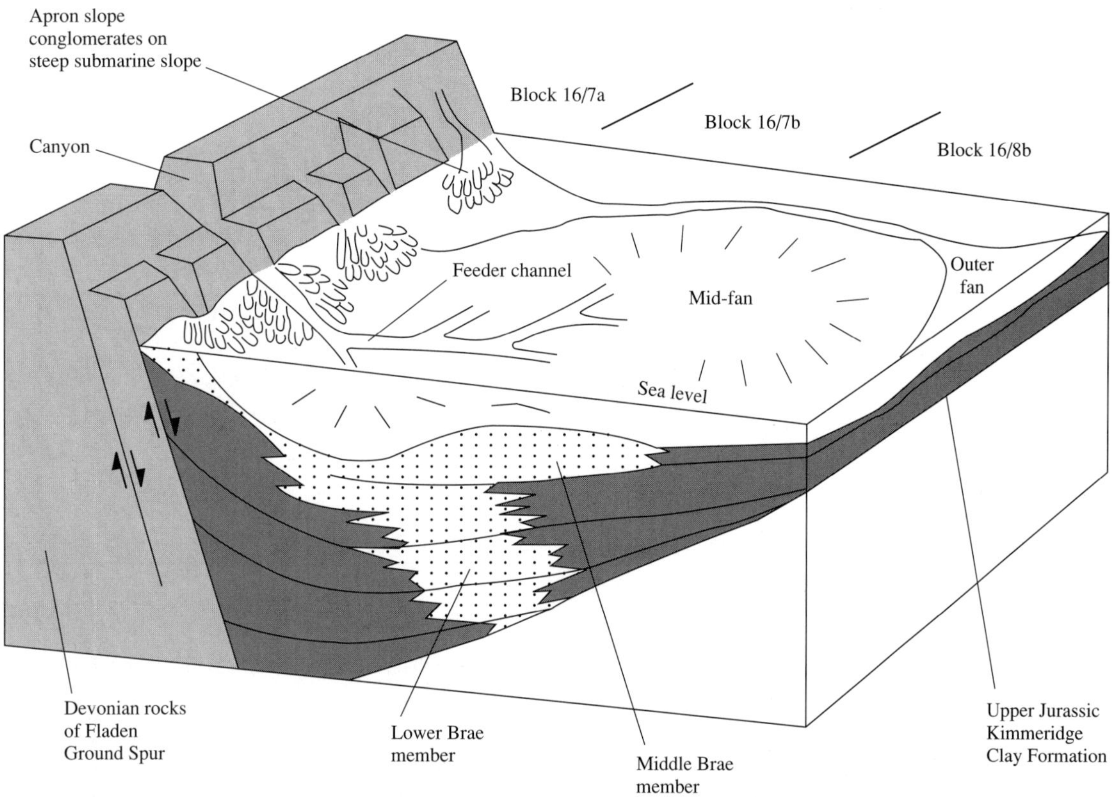


**Fig. 2.** Block diagram showing the depositional geometries of the reservoir intervals in the Brae and Miller fields.

within the area of block 16/7a. The main reservoir interval in Miller accumulated as a lobe-shaped fan closer to the basin centre, shed from a point-source located in the area of the South Brae Field. Thin, inter-bedded oil-bearing, sandstones and mudstones that overlie part of the Miller Field were deposited as part of a younger fan system, which developed to the north and which now forms the North Brae (condensate) Field. By Portlandian times the sediment sources had been denuded and the restricted basin filled with marine mudstones, what is now, the Kimmeridge Clay Formation. Subsequent sediment accumulation has been largely continuous until the present day. The burial rate peaked in the Early Eocene (Fig. 3).

The Kimmeridge Clay Formation overlies and is intimately interbedded with the Brae Formation. It acts as both seal and source to the Miller Field (Mackenzie *et al.*, 1987). The source is currently at peak production, expelling petroleum to the sandstone today. Generation of petroleum has probably been active since 10–15 Ma.

The Miller trap combines both structural and stratigraphic elements. The main component is compactional drape of the Kimmeridge Clay Formation over the reservoir sandstone lobes forming a low relief dome. Sandstone pinchout to the north-east forms the stratigraphic element to the trap (Rooksby, 1991).

The Brae and Kimmeridge Clay formations have been divided into six lithofacies (McClure & Brown, 1992; Fig. 4):

**1** Granular sandstone (lithofacies 2). This is the main lithofacies in the Miller reservoir. The sandstones are medium-grained and thick-bedded. Some contain dish structures but most are massive. These sandstones were deposited by high-density turbidity currents in what were probably unconfined or partly confined lobes and channels. Amalgamated beds are commonly several metres thick.

**2** Interlaminated mudstone and fine-grained sandstone (lithofacies 1). Commonly referred to as 'tiger stripes' (Stow *et al.*, 1982), these consist of alternations of thin mudstones, siltstones, and very fine- to fine-grained sandstones (Bouma sequences). Deposition was from low-density classical turbidity currents in relatively quiescent areas of the mid- to outer fan or in interchannel areas.

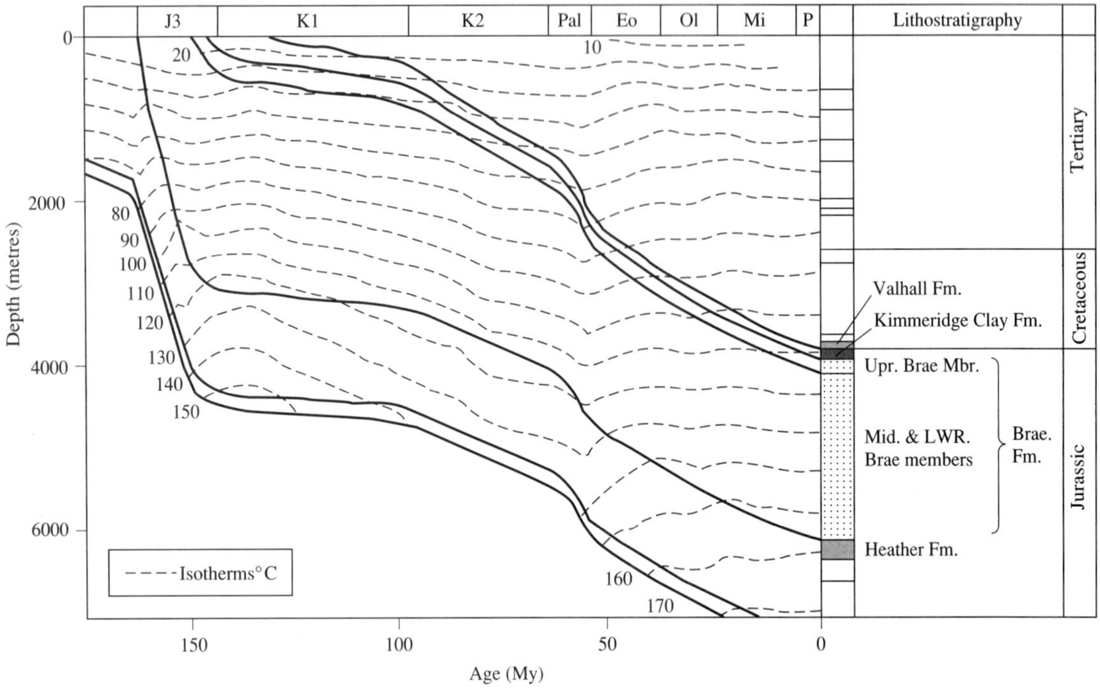


**Fig. 3.** Burial and thermal histories for the Miller Field, based on the thermal modelling of Mackenzie *et al.* (1987) for well 16/7b-20.

i. Lithofacies 2 ii. Lithofacies 4 iii. Lithofacies 1 iv. Lithofacies 6

**Fig. 4.** Core photographs of the main lithotypes in the Miller reservoir: (i) structureless medium-grained lithofacies 2 sandstones; (ii) argillaceous and carbonaceous medium-grained lithofacies 4 sandstones; (iii) thinly interbedded very fine sandstones, siltstones and mudstones (tiger stripe rock, lithofacies 1); (iv) laminated mudstones (lithofacies 6). The lower left corner of core slab (i) is part of a carbonate cemented concretion.

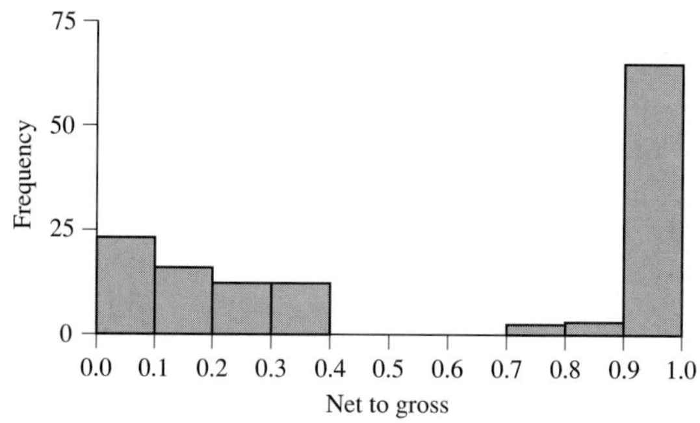


**Fig. 5.** Net to gross distribution in the cored interval of 16/7b-20. The graph was constructed on a metre by metre basis. Most intermediate values of net to gross result from partial sampling of high and low net to gross intervals in a single metre.

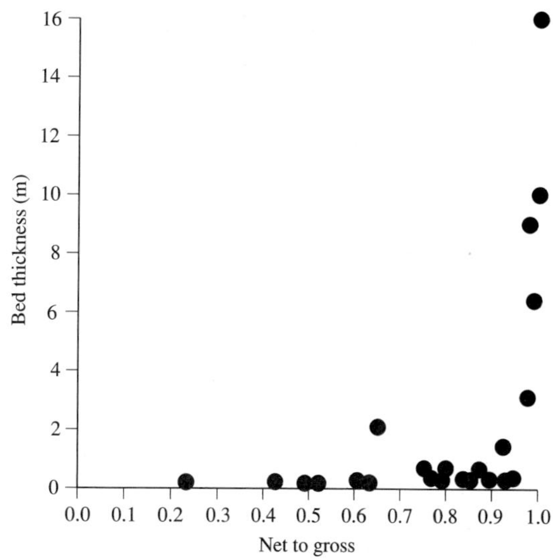


**Fig. 6.** Relationship between bed thickness and net to gross for the cored interval of 16/7b-20.

Individual beds of both sandstone and mudstone have a range in thickness from 0.001 to 0.1 m.

**3** Argillaceous sandstones (lithofacies 4). These are medium to fine-grained sandstones with abundant mudstone laminae and some carbonaceous plant debris. Bed thickness ranges from 0.05 to 1.00 m. The sands were deposited by high density turbidity currents. The presence of mudstone laminae in these sandstones indicates that deposition occurred in a marginal setting within the mid-fan and inner fan, away from the main axis of sand input.

**4** Mudstones (lithofacies 6). These are laminated mudstones with occasional siltstones and very fine-grained sandstone laminae (< 5 mm thick). These represent hemipelagic and dilute turbidite deposition on the outer fan and basin plain.

**5** Mud-supported conglomerates (lithofacies 5). These are rare in the Miller Field, only occurring in the proximal, western parts of the area. They comprise beds (0.1–1.0 m thick) of angular, mixed mudstone and sandstones clasts supported in a contorted mudstone matrix. Clast size is commonly 1–10 cm. Such sediment was deposited as debris flows.

**6** Sand-supported conglomerates (lithofacies 3). These too are rare in Miller in contrast to the adjacent Brae fields (Harmes *et al.*, 1981). Clast size in the Miller area is rarely greater than a few centimetres. Deposition of the few examples in Miller probably occurred from traction currents at the base of the high density turbidites.

The average net to gross for the oil bearing interval is about 0.8 (Abbots, 1991). These average net to gross values obscure a more fundamental subdivision of the sandstone distribution. Figure 5 was compiled from the cored interval of 16/7b-20. Net to gross was calculated on a metre by metre basis. The distribution is bimodal. The thinly interbedded, 'tiger stripe' low density turbidites commonly have a net to gross of less than 0.4. The intervals dominated by high density turbidite sandstones commonly have a net to gross greater than 0.8. Almost all of the samples with intermediate net to gross result from sampling across the boundary between a low net to gross interval and a high net to gross and so averaging two extreme values.

Bed thickness also shows a bimodal distribution. The 'tiger stripe' sandstones are commonly millimetres to centimetres thick while the main reservoir sandstones are rarely less than 1 m thick. Bed thickness and net to gross are loosely correlated (Fig. 6).

## DATABASE AND ANALYTICAL METHODS

Ninety-six samples from five wells were randomly selected for analysis from the extensively cored intervals for analysis. For all wells, the permeability data for the randomly selected set was compared with that for the whole well. Only when the permeability distribution of the selected samples was indistinguishable from that of the whole interval (99.9% confidence using student *t*-test) was the sample set used for further analyses. This approach was used to ensure that subsequent calculations of element mobility could be considered representative of the volume of rock within the 16/7b Block. Given the spatial distribution of wells (Fig. 1) and thickness of both the Brae and Kimmeridge Clay formations, we estimate the sampling to be representative of about a billion cubic metres ($10^9$ $m^3$) of rock.

**Table 1.** Mineralogy of the Miller reservoir sandstones

| | *n* | qtz | pqtz | af | pf | mic | irf | mrf | srf | bio | mx | org | cal | fecal | dol | ank | sid | qc | py | glau | ko | ill | heav | vpor |
|---|---|---|---|---|---|---|---|---|---|---|---|---|---|---|---|---|---|---|---|---|---|---|---|---|
| *Quartz cemented sandstone at 4000–4200 m* | | | | | | | | | | | | | | | | | | | | | | | | |
| Mean | 45 | 69.9 | 15.7 | 2.0 | 0.01 | 1.44 | 0 | 0.34 | 0.92 | 0.20 | 1.09 | 2.34 | 0.09 | 0.30 | 0 | 0 | 0.02 | 4.59 | 0.11 | 0.10 | 0 | 1.07 | 0.01 | 8.58 |
| St dev | 45 | 6.4 | 6.5 | 1.5 | 0.08 | 1.15 | 0 | 0.47 | 0.88 | 0.56 | 1.58 | 5.97 | 0.32 | 2.01 | 0 | 0 | 0.10 | 3.30 | 0.24 | 0.33 | 0 | 1.00 | 0.07 | 5.20 |
| *Quartz cemented sandstone at 4700 m* | | | | | | | | | | | | | | | | | | | | | | | | |
| Mean | 7 | 70.8 | 16.7 | 1.6 | 0 | 1.79 | 0 | 0.50 | 0.36 | 0.07 | 0.36 | 0.50 | 0 | 0 | 0 | 0 | 0 | 5.50 | 0.43 | 0 | 0 | 1.93 | 0 | 4.71 |
| St dev | 7 | 7.2 | 6.2 | 1.5 | 0 | 1.26 | 0 | 0.41 | 0.56 | 0.19 | 0.75 | 1.12 | 0 | 0 | 0 | 0 | 0 | 2.69 | 0.73 | 0 | 0 | 1.13 | 0 | 2.71 |
| *Carbonate concretions at 4000–4200 m* | | | | | | | | | | | | | | | | | | | | | | | | |
| Mean | 8 | 46.1 | 13.9 | 0.9 | 0.06 | 0.44 | 0 | 0.13 | 0.56 | 1.00 | 0 | 0.19 | 16.37 | 15.38 | 0 | 3.81 | 0 | 0 | 0.31 | 0 | 0 | 0.38 | 0 | 0.06 |
| St dev | 8 | 6.6 | 5.6 | 1.1 | 0.18 | 0.56 | 0 | 0.23 | 0.78 | 1.28 | 0 | 0.26 | 17.95 | 20.52 | 0 | 10.8 | 0 | 0 | 0.70 | 0 | 0 | 0.58 | 0 | 0.18 |

qtz: quartz; pqtz: polycrystalline quartz; af: alkali feldspar; pf: plagioclase feldspar; mic: mica (commonly muscovite); irf: igneous rock fragments (quartzo-feldspathic); mrf: metamorphic rock fragments; srf: sedimentary rock fragments (commonly siltstone); bio: bioclastic (shell) fragments; mx: matrix clay (undifferentiated); org: organic matter (commonly wood); cal: calcite cement; fecal: ferroan calcite cement; dol: dolomite cement; ank: ferroan dolomite and/or ankerite cement; sid: siderite cement; qc: quartz cement; py: pyrite cement; glau: glauconite grains; ko: kaolinite cement; ill: illite cement; heav: heavy/opaque minerals; vpor: visible porosity (percentage of rock plus void).

All of the sandstone samples and about half of the mudstone samples were examined petrographically using conventional and polished thin-sections in both polarized light and for cathodoluminescence (using a scanning electron microscope). The thin sections were point counted for mineralogy using 200 grain counts. Porosity was counted in addition to grain counts. Rock chips of the same samples were also viewed using secondary electron imaging on a scanning electron microscope.

For the whole-rock analysis, about a gram of rock was ground to a fine powder and homogenized. Approximately 0.25 g of the powder was then dissolved in a mixture of hot hydrofluoric and hydrochloric acid. Resultant solutions were diluted with distilled water before analysis. The solutions were analysed using inductively coupled plasma emission spectroscopy for silicon, aluminium, titanium, iron, manganese, calcium and magnesium (Gluyas & Coleman, 1992). Potassium and sodium were measured by flame photometry. Precision on whole-rock analysis is estimated to be within ±1% (Thompson & Walsh, 1983).

Fluid inclusion measurements were made on a different sample set from that used for chemical and petrographic analysis. This difference in sampling was an artefact of the commercial criteria that generated the projects from which this paper has been compiled. Fluid inclusions were identified using a combination of transmitted light, incident UV illumination, and microthermometry. Five hundred and twenty-one measurements were made on primary aqueous inclusions in quartz overgrowths on 32 samples from seven wells. Roedder (1984) and Burruss (1991) give details of the techniques and potential pitfalls. $T_h$ and $T_m$ were measured using a Linkam TH600 heating–cooling stage attached to a Leitz microscope with a ×40 LWD objective. Calibration of the stage against synthetic inclusions and bulk standards gave a measurement accuracy of $\pm < 1°C$ throughout the range of the Miller data.

## SANDSTONE MINERALOGY (Table 1)

Quartz and polycrystalline quartz are the major detrital grains. Together they typically form > 90% of the fabric grains in any one sample. Matrix clay content is low (< 3% in lithofacies 4) or absent (lithofacies 2). Most samples contain 1–2 vol.% alkali (potassium) feldspar but grains of plagioclase feldspar are rare. Minor quantities of mica, glauconite, sedimentary rock fragments (mudstone, siltstone) and metamorphic rock fragments (quartz and mica) have been recorded. Cellular wood fragments, shell debris (more common in carbonate cemented intervals) and a few siliceous sponge spicules are the only bioclastic matter.

The sandstones contain two major authigenic minerals: calcite and quartz. There are small quantities of illite and minor to trace amounts of ferroan dolomite, kaolinite and pyrite.

Ferroan and non-ferroan calcite occurs as concretions (0.05–2.0 m in diameter). The concretions contain about 35 vol.% calcite cement and 65 vol.% fabric grains (Fig. 7; Plate 1, facing p. 218). Individual calcite crystals are large (> 100 μm) and poikilotopic. Within the concretions the detrital quartz grains appear free of authigenic quartz overgrowths (Fig. 7; Plate 1, facing p. 218). Great care was taken to ensure that absence of quartz cement on petrographic evidence was real because the absence of quartz cement within the concretions is of quantitative importance in assessing the flux of silica during diagenesis. To this end, the calcite cement of the concretions was dissolved using a weak acid and the released grains examined using a scanning electron microscope. The grains showed no signs of syntaxial quartz cement and of equal importance they showed no evidence of partial dissolution. Feldspar trapped within the concretions is also well-preserved and free of the dissolution features seen in feldspars outside the concretions (Fig. 7; Plate 1, facing p. 218).

Both ferroan and non-ferroan calcite cemented concretions fluoresce bright yellow in ultra-violet light. The margins of many of the concretions are sharp but digitate and they commonly have a dull UV fluorescence. They may have undergone partial dissolution at their margins. Traces of pyrite are associated with the concretions. We conclude that precipitation of this calcite was passive, thus preserving the original fabric and chemistry of the sand.

Minor non-ferroan calcite also occurs as disseminated patches a few centimetres in diameter. Such intervals contain between 2 vol.% and 14 vol.% of poikilotopic calcite (Fig. 7; Plate 1, facing p. 218). Unlike the concretionary masses, framework quartz grains contained within these calcite cemented intervals have authigenic quartz overgrowths and the calcite lacks mineral fluorescence when illuminated with UV light.

Quartz cement occurs throughout the sandstones as syntaxial overgrowths on detrital grains (Fig. 7; Plate 1, facing p. 218), forming $5.8 \pm 1.6$ ($2\sigma$) vol.% (Gluyas, 1985). Traces of rhombic, ferroan dolomite (30 μm) occur in two samples from 16/7B-24 and 16/7B-25. Most samples that contain quartz cement also contain some fibrous illite cement (< 3 vol.%). Authigenic kaolinite was found in one sample (4103 m, 16/7B-24) during clay mineral separation. The extract contained 5% kaolinite from a sample with *c.* 0.5% clay (= 0.025% kaolinite in the whole rock). Kaolinite appears to be associated with

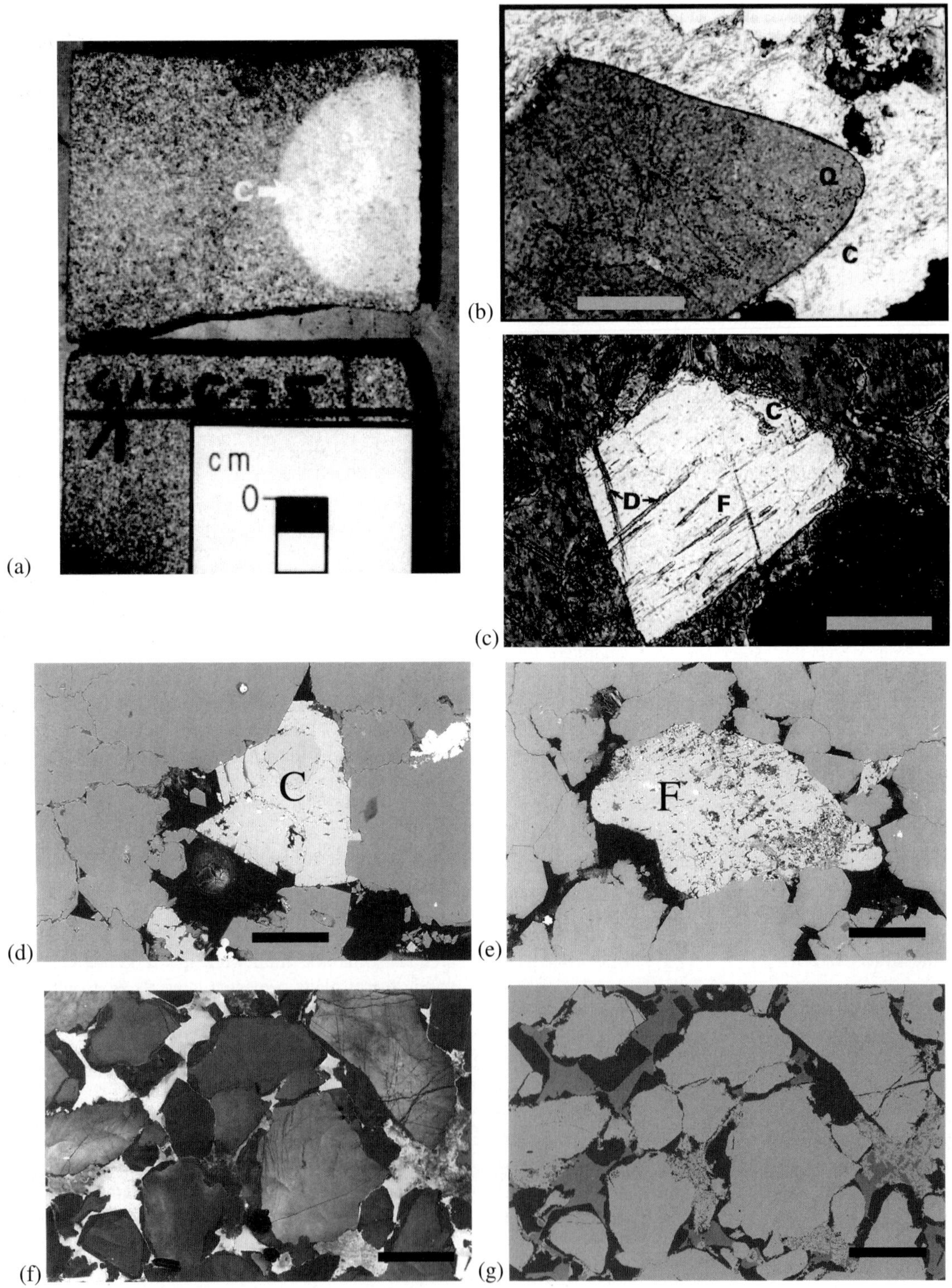
(a)
C
cm
0
(b)
Q
C
(c)
C
D
F
(d)
C
(e)
F
(f)
(g)

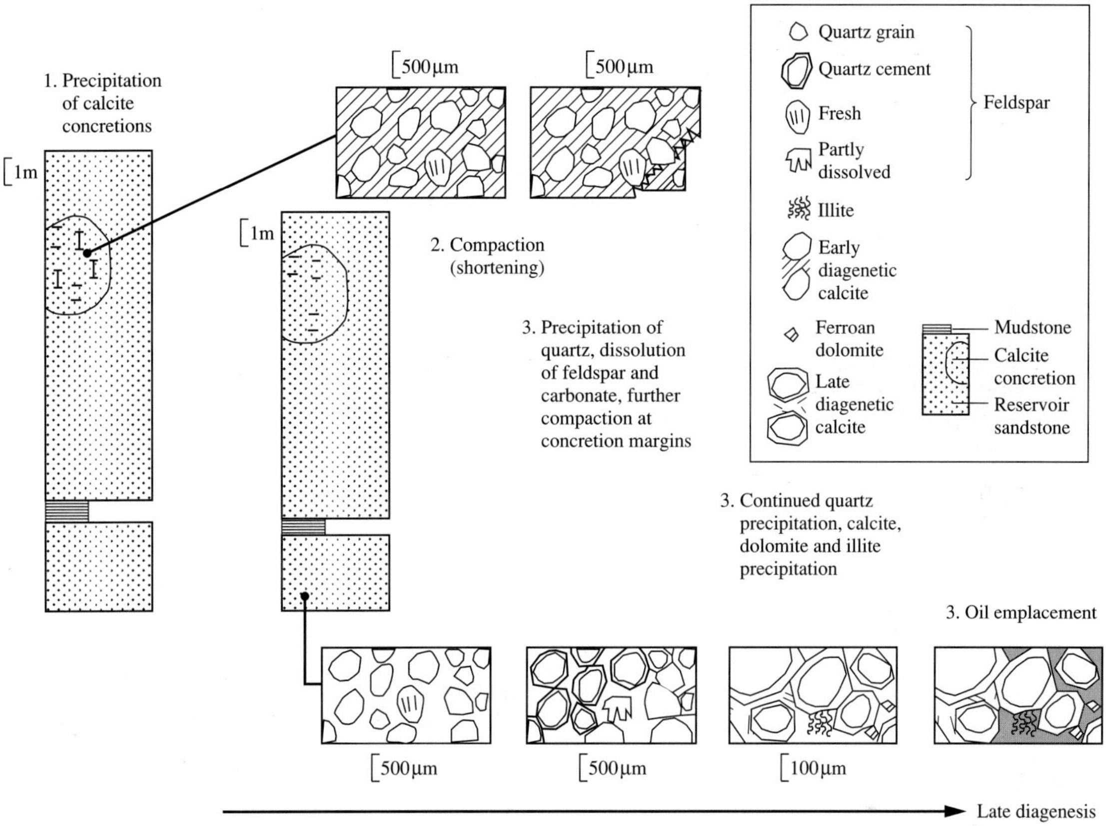


**Fig. 8.** Diagenetic history for the Brae Formation of the Miller Field.

feldspar dissolution. Considerable effort was made to identify quartz dissolution features both during core logging and petrographic analysis. However, no stylolites were observed during logging of over one kilometre of slabbed core and no grain to grain quartz dissolution features were observed during petrographic analysis.

## DIAGENETIC HISTORY

The diagenetic sequence was deduced from petrographic analysis (Fig. 8). Calcite precipitated as concretionary masses during earliest diagenesis, close to the sediment–water interface (Gluyas, 1985). Bacterial sulphate

**Fig. 7.** (*Opposite. Also reproduced in colour, see Plate 1, facing p. 218.*) (a) Core photograph of a calcite cemented concretion. This particular concretion is the smallest observed in the cored Brae Formation of the Miller Field. Most are 2–3 m in diameter. All the concretions are subspherical and contain trace quantities of pyrite typical of carbonate cementation occurring near to the sediment–water interface through the destruction of organic matter by sulphate-reducing bacteria (Gluyas, 1984). Small concretions such as this allowed sampling of both calcite and quartz cemented sandstones on the same depositional surface. This enabled us to show that the sites of concretion growth were in no way different from the remainder of the sediment. Sample from well 16/7b-23 at 4145.3 m below rotary table. (b) Photomicrograph, calcite cement within a carbonate concretion. The quartz grain is free of quartz cement and there appears to have been no dissolution of the quartz grain during calcite precipitation. Scale of photomicrograph 1.3 mm × 0.9 mm. Photomicrograph taken with cross-polars and $^1/_4\lambda$ plate. (c) Thin section photomicrograph of well preserved feldspar (F) within calcite concretion (C), the grain has only minor dissolution features (D). Photomicrograph taken with cross-polars and $^1/_4\lambda$ plate. Scale bar: 100 µm. (d) Late diagenetic calcite cement (C) infilling residual porosity after quartz cementation. Photomicrograph taken using back-scattered electron imaging on a scanning electron microscope. Scale bar: 100 µm. (d) Corroded feldspar (F) outwith a calcite cemented concretion. Photomicrograph taken using back-scattered electron imaging on a scanning electron microscope. Scale bar: 100 µm. (f) Quartz cemented Miller reservoir sandstone. The image combines a backscattered electron image and one obtained from cathodoluminescence. The dark grey quartz cement occurs as syntaxial rims on most grains and as fracture fills. Scale bar: 100 µm. (g) The same image as in (f) but with false colours used to highlight the grey levels contrast between pore space (blue; see Plate 1, facing p. 218), quart cement (red) and grains (green).

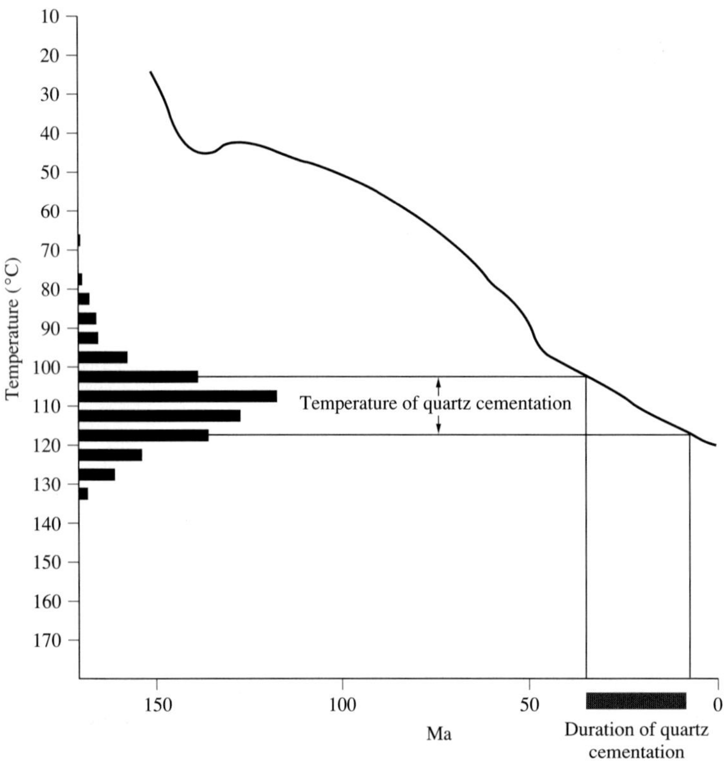


**Fig. 9.** Combined fluid inclusion histogram and thermal history of the Miller Field to yield precipitation dates for quartz.

reduction and associated oxidation of organic matter was responsible for both the bicarbonate and sulphide that now occur as calcite and pyrite (Gluyas, 1984; Gluyas, 1985). Diagenesis within the concretion was halted. Only at the concretion margins, which suffered some calcite dissolution, did a little quartz precipitate. Thus the concretions are believed to preserve both the original chemistry and fabric of the deposited sand. Moreover there is nothing to suggest that the sites of concretion growth were different from the remainder of the sand at the time of deposition (Fig. 7).

Only in the host sandstone unaffected by calcite precipitation, did diagenesis and compaction continue. Compaction reduced the near surface porosities of 40 down to 22%. This was followed by quartz precipitation and feldspar dissolution. There is insufficient evidence to determine which of these precipitation and dissolution processes occurred first or, alternatively, if they occurred at the same time. Subsequent to these events, minor amounts of calcite, ferroan dolomite and illite precipitated in the pore space before oil emplacement.

## TIMING OF QUARTZ CEMENTATION

It is not possible to date directly when quartz precipitated. Instead, we have used two indirect methods that use the burial and thermal history of the Miller reservoir to convert the measurements of temperature and depth to time. Aqueous fluid inclusion homogenization temperatures were used to estimate the temperature at which quartz precipitated. Of the 521 fluid inclusion homogenization temperatures measured, 70% indicated a precipitation temperature for quartz of between 102.5°C and 117.5°C (Fig. 9, Table 2). Using the thermal history diagram for the Miller Field (Fig. 9, Mackenzie *et al.*, 1987), these temperatures correspond to the age range 30–10 Ma (early Oligocene to Late Miocene) with the most likely being 25 Ma (late Oligocene/Early Miocene).

An alternative approach is to calculate the porosity of the sand before quartz cementation and then use this porosity to calculate the minimum depth the sand was at when quartz precipitated. Subsequently the burial history diagram (Fig. 10) can be used to convert depth of

**Table 2.** Fluid inclusion homogenization temperatures. The data are presented as average and standard deviation of temperature data taken from between 4 and 37 inclusions in any one sample

| Depth (ft) | Depth TVDss (m) | Aqueous $T_h$ (°C) Number | Average | Standard deviation |
|---|---|---|---|---|
| 16/8b-6 | | | | |
| −13404.0 | −4062.0 | 14 | 102.8 | 8.5 |
| −13410.2 | −4063.8 | 14 | 96.7 | 10.1 |
| −13439.9 | −4073.0 | 25 | 107.6 | 7.5 |
| −13474.0 | −4083.5 | 25 | 110.5 | 7.2 |
| 16/8b-A06 | | | | |
| −5874.0 | −4075.0 | 23 | 111.9 | 9.7 |
| −5881.3 | −4079.9 | 18 | 108.5 | 9.3 |
| −5909.3 | −4099.0 | 32 | 105.7 | 8.9 |
| −5923.8 | −4108.8 | 22 | 108.5 | 7.5 |
| 16/8b-7 | | | | |
| −14388.5 | −4102.0 | 28 | 116.7 | 9.0 |
| −14467.0 | −4127.5 | 25 | 114.1 | 9.5 |
| 16/8b-A10 | | | | |
| −14438.2 | −4080.1 | 26 | 110.5 | 5.9 |
| −14481.2 | −4093.4 | 27 | 114.5 | 8.6 |
| 16/8b-A11 | | | | |
| −17037.5 | −4088.8 | 38 | 109.7 | 6.5 |
| −17066.4 | −4096.3 | 28 | 114.9 | 6.1 |
| −17083.0 | −4100.6 | 29 | 109.2 | 6.5 |
| −17241.7 | −4141.8 | 37 | 113.9 | 11.1 |
| 16/8b-3 | | | | |
| −4011.5 | −3993.0 | 8 | 102.2 | 12.0 |
| −4016.4 | −3998.2 | 4 | 82.9 | 11.0 |
| −4027.3 | −4009.1 | 16 | 114.6 | 10.2 |
| −4031.0 | −4012.7 | 12 | 110.0 | 3.5 |
| −4039.5 | −4021.3 | 16 | 113.3 | 6.1 |
| 16/7b-25 | | | | |
| −4084.0 | −4062.1 | 9 | 104.4 | 7.0 |
| −4085.8 | −4063.9 | 11 | 101.9 | 6.3 |
| −4089.4 | −4067.5 | 9 | 103.1 | 5.2 |
| −4090.6 | −4068.7 | 6 | 105.9 | 5.0 |
| −4106.3 | −4084.4 | 16 | 109.3 | 8.0 |

cementation to time of cementation. The reservoir sands currently have a porosity of 16% (lithofacies 2 and 4, Gluyas, 1985; Rooksby, 1991) and quartz cement content of 5.8 ± 1.6% (Gluyas, 1985). Thus prior to cementation the porosity of the sand was about 22 ± 1.5%. Using the global equation of Gluyas & Cade (1997):

$$\Phi = 50\exp\left(\frac{-10^{-3}z}{2.4 + 5 \times 10^{-4}z}\right)$$

where porosity (Φ) is in percentage and depth (z) is in metres.

The depth at which cementation occurred is in the range 2.9 km to 3.8 km with a most likely cementation depth of 3.3 km. In time terms, these depths are equivalent to 42 Ma to 21 Ma with a most likely cementation date of 32 Ma. The similarity between the precipitation dates obtained from two different methods gives us confidence to assert that most cementation of the Miller reservoir sandstone occurred in late Oligocene and Early Miocene times.

## BRAE FORMATION GEOCHEMISTRY

### Sandstone whole-rock chemistry

A rudimentary measure of the quantity of quartz added to the sandstone during diagenesis was derived from petrographic point count data. However, such data alone are insufficient to be able to differentiate between possible internal sources for the silica (such as stylolitization, intergranular pressure dissolution or feldspar dissolution) and external sources such as the enclosing mudstone. However, data from whole-rock chemical analysis (Table 3) may be used to differentiate between these sources.

In order to calculate how much matter has been imported to or exported from the sandstones during diagenesis, the depositional and present chemical compositions of the sandstone must be known as must the porosity today and the porosity at the time of cementation. Present day porosity and chemistry were easily measured on the quartz cemented, host sandstones. The chemistry of the silicate fraction in the calcite cemented sandstones was taken to be the chemistry of the sandstone at the time of deposition. We believe this assumption to be justified on the basis of the observations made on the petrography of the silicate minerals both within and without the concretions and upon the selection procedure. We have already shown that there is no evidence of grain dissolution, alteration or cement precipitation (other than calcite) in the concretions and that the horizons on which the concretions grew were mineralogically the same as the adjacent horizons. The porosity at the time of cementation by quartz can be calculated in two ways: (i) the point counted cement volume can be added to the present porosity; or (ii) the degree of compaction and hence porosity loss may be calculated by comparing the abundance of an immobile phase inside and outside the concretions.

For the Brae Formation sandstones the two calculation methods give the same result. However, for many

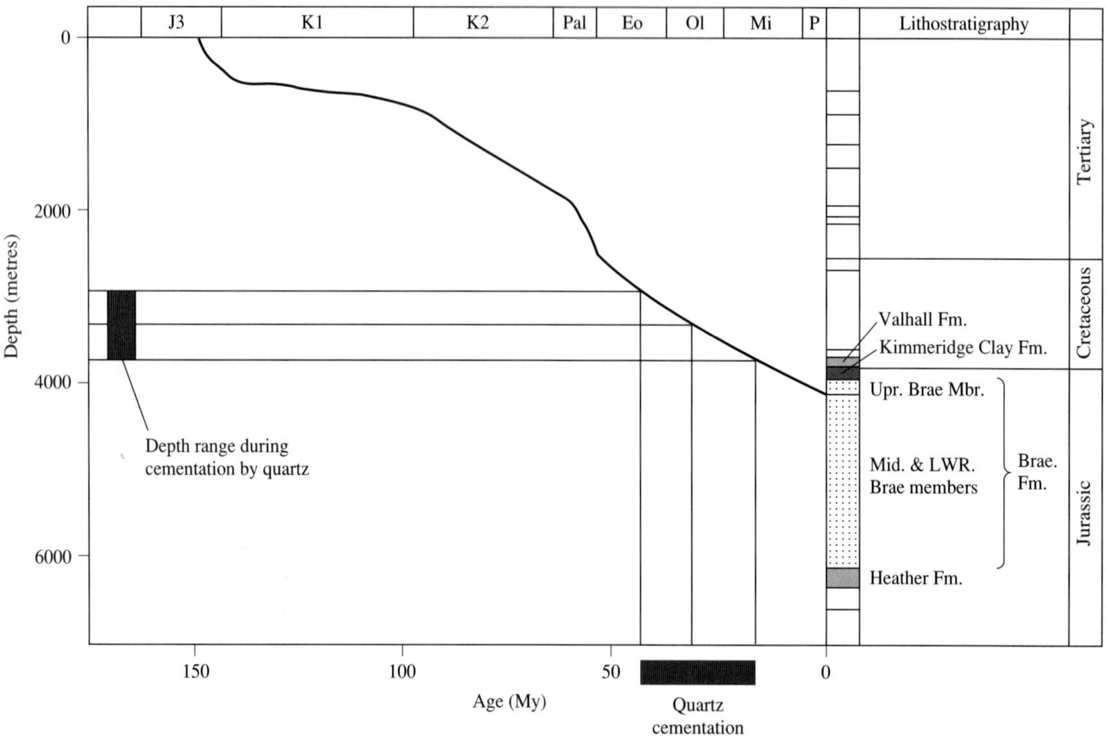


**Fig. 10.** Combined burial history and depth of cementation data to deliver precipitation dates for the quartz cement.

**Table 3.** Oxide chemistry for the Miller reservoir sandstones and associated mudstones

| | | Oxide (wt%) | | | | | | | | | | | | |
|---|---|---|---|---|---|---|---|---|---|---|---|---|---|---|
| Depth (m) | *n* | $SiO_2$ | $TiO_2$ | $Al_2O_3$ | FeO | MnO | CaO | MgO | $K_2O$ | $Na_2O$ | $CO_2$ | Lithofacies | Permeability (mD) | Porosity (%) |
| 4000–4200 | | | | | | | | | | | | | | |
| Average | 7 | 84.10 | 0.37 | 8.12 | 1.00 | 0.10 | 0.27 | 0.31 | 2.97 | 0.23 | 0.16 | 1 | 0.35 | 6.8 |
| St dev | 7 | 5.23 | 0.09 | 1.28 | 0.44 | 0.10 | 0.24 | 0.10 | 0.35 | 0.11 | 0.21 | 1 | 0.24 | 0.9 |
| Average | 38 | 94.47 | 0.10 | 1.88 | 0.16 | 0.04 | 0.33 | 0.05 | 0.75 | 0.07 | 0.21 | 2 | 236.98 | 14.3 |
| St dev | 38 | 6.79 | 0.04 | 0.64 | 0.14 | 0.05 | 0.68 | 0.03 | 0.30 | 0.04 | 0.52 | 2 | 428.08 | 2.6 |
| Average | 1 | 94.57 | 0.07 | 2.57 | 0.13 | 0.01 | 0.80 | 0.05 | 0.80 | 0.06 | 0.63 | 3 | | |
| Average | 7 | 71.55 | 0.28 | 6.14 | 0.74 | 0.03 | 0.92 | 0.25 | 2.18 | 0.16 | 0.22 | 4 | 37.57 | 11.0 |
| St dev | 7 | 29.88 | 0.26 | 4.53 | 0.78 | 0.02 | 1.69 | 0.22 | 1.39 | 0.13 | 0.17 | 4 | 45.27 | 4.3 |
| Average | 16 | 60.45 | 0.45 | 12.91 | 2.00 | 0.18 | 0.58 | 0.65 | 4.05 | 0.36 | 0.44 | 6 | | |
| St dev | 16 | 7.96 | 0.09 | 2.34 | 0.84 | 0.10 | 0.38 | 0.14 | 0.83 | 0.08 | 0.32 | 6 | | |
| 4000–4200 calcite cemented | | | | | | | | | | | | | | |
| Average | 14 | 56.10 | 0.07 | 1.29 | 0.36 | 0.10 | 18.23 | 0.15 | 0.54 | 0.04 | 14.31 | 2 | 0.07 | 2.1 |
| St dev | 14 | 8.72 | 0.06 | 0.79 | 0.50 | 0.06 | 2.68 | 0.08 | 0.19 | 0.02 | 2.10 | 2 | 0.03 | 0.9 |
| Average | 4 | 53.35 | 0.04 | 1.24 | 0.26 | 0.07 | 13.95 | 0.06 | 0.51 | 0.01 | 10.96 | 4 | | |
| St dev | 4 | 1.18 | 0.01 | 0.23 | 0.07 | 0.01 | 0.65 | 0.01 | 0.14 | 0.00 | 0.51 | 4 | | |
| 4700 | | | | | | | | | | | | | | |
| Average | 5 | 96.48 | 0.07 | 5.61 | 0.19 | 0.02 | 0.14 | 0.08 | 0.71 | 0.11 | 0.21 | 2 | 2.01 | 9.0 |
| St dev | 5 | 0.59 | 0.011 | 8.16 | 0.07 | 0.01 | 0.09 | 0.01 | 0.08 | 0.03 | 0.14 | 2 | 2.38 | 2.0 |
| Average | 2 | 55.43 | 0.52 | 17.63 | 2.09 | 0.14 | 0.11 | 0.49 | 6.44 | 0.40 | 0.00 | 6 | | |
| St dev | 2 | 0.60 | 0.02 | 0.63 | 0.21 | 0.06 | 0.05 | 0.04 | 0.08 | 0.01 | 0.00 | 6 | | |
| Average | 2 | 93.23 | 0.14 | 4.26 | 0.32 | 0.04 | 0.06 | 0.11 | 1.64 | 0.14 | 0.09 | 4 | 0.45 | 6.8 |
| St dev | 2 | 0.064 | 0.02 | 0.17 | 0.06 | 0.01 | 0.04 | 0.01 | 0.08 | 0.00 | 0.05 | 4 | 0.33 | 2.3 |

sandstones that contain appreciable quantities of soluble detrital grains, method (ii) gives more reliable and reproducible results (Gluyas & Coleman, 1992). The least mobile element in the Brae sandstones is titanium. It is significantly less mobile during chemical weathering and diagenesis (Staudigel & Hart, 1983; Morton, 1984) than the other elements analysed during this study. Traces of authigenic titanium oxides were seen during petrographic analysis (R. Garden, personal communication, 1987) but these were interpreted to have formed *in situ* from the decomposition of mica. In consequence, the titanium oxide abundance in each of the whole-rock analyses has been used as a datum against which the abundance of other elements has been compared.

In the following calculations, data from only the clean, massive lithofacies 2 sandstones are presented. Besides being the volumetrically most important sandstones, and the prime reservoir sandstones in the Brae Formation, they are the only ones that contain carbonate cemented concretions. All the following calculations are based on volume and include the present porosity of the sandstone. The sandstones that are not cemented by calcite contain about $1.23 \pm 0.90$ times more $TiO_2$ than do those which are cemented by calcite (where calcite volume = porosity at time of cementation). Therefore, because we define that titanium has remained immobile during diagenesis the apparent increase in $TiO_2$ concentration per unit volume of rock including porosity, can only be due to compaction (grain fracturing and rearrangement). That is, there are more grains and thus more $TiO_2$ in a unit volume of rock including porosity.

If the assumption of $TiO_2$ immobility is accepted (we will examine the consequences of not doing so later), then the difference between the percentage change in $TiO_2$ abundance and other oxide abundances from calcite concretion to surrounding sandstone is a measure of the quantity of oxide imported to or exported from the sandstone (Table 3).

The present potassium content of the sandstone is less than that at deposition. This is taken to indicate that potassium has been lost from the sandstone since deposition. The remaining three oxides: soda, silica and alumina are all enriched (aluminium marginally) in the sandstone today, relative to the sandstone at deposition. This is taken to indicate that matter has been imported to the sandstone during diagenesis.

The quantity of silica imported to the sandstone was 2–3 orders of magnitude greater than that for other element oxides. Hence most must have precipitated as quartz. This conclusion accords with petrographic evidence that quartz is the only volumetrically significant silicate cement in the Brae Formation sandstones.

### Mudstone whole-rock chemistry

The interbedded Kimmeridge Clay Formation is a possible source for the cements. Here we investigate the export capabilities of the local Kimmeridge Clay Formation. Unlike the Brae Formation sandstones, the interbedded and overlying Kimmeridge Clay mudstones do not contain early diagenetic calcite concretions. Thus it is not possible to compare directly the chemistry of the original silicate detritus with rock that has undergone further diagenesis. However, it is possible to examine the diagenetic evolution of the clay minerals and feldspar minerals using a plot of $K_2O/Al_2O_3$ against $Al_2O_3$ (Gluyas & Leonard, 1995).

In Fig. 11 potassium and aluminium data from lithofacies 1 (interbedded siltstone and mudstone) and lithofacies 6 (mudstone) samples are plotted. If each sample initially represented a simple mixture of silt and mud, and if subsequent diagenesis occurred in a closed system (on the scale of a 3-cm poroperm plug sample), then all the points should plot on a simple mixing line, $K_2O/Al_2O_3$ against $Al_2O_3$, of silt and mud. The best line that can be fitted to the data represents mixing a siltstone with about 90 wt% quartz and about 10 wt% potassium feldspar with an illite mudstone (line AA′). Although this mixing line accurately portrays the behaviour of lithofacies 6 samples selected from above the main reservoir sandstone (Fig. 11), it does not fit data obtained from mudstones that are interbedded with the main reservoir sandstones. All mudstones samples from within the main reservoir sandstones (group 1) appear to be enriched in both aluminium and potassium relative to the mudstones that occur above the main reservoir sandstones (group 2). This difference is unlikely to be depositional because the two groups of mudstones are in part time-equivalent (Fig. 12) and because the mudstones (group 1) which are interbedded with the highly quartzose sandstones contain less silica than do the mudstones (group 2) that are not associated with the sandstones. If depositional effects controlled the present differences in chemistry (mineralogy) between the two groups there should be a positive correlation between the abundance of silica in the mudstones and the abundance of sandstones in the same interval; that is, opposite to the above observation. The alternative possibility, that variations in the clay mineralogy at deposition can explain the current differences in chemistry between mudstones that overlie the Brae Formation and mudstones that are interbedded with the Brae Formation, is not tenable. For this case to be true would require that the clay minerals falling to the sea bed were able to predict whether they finally rest interbedded with or overlying Brae Formation sands. This possibility is clearly ridiculous.

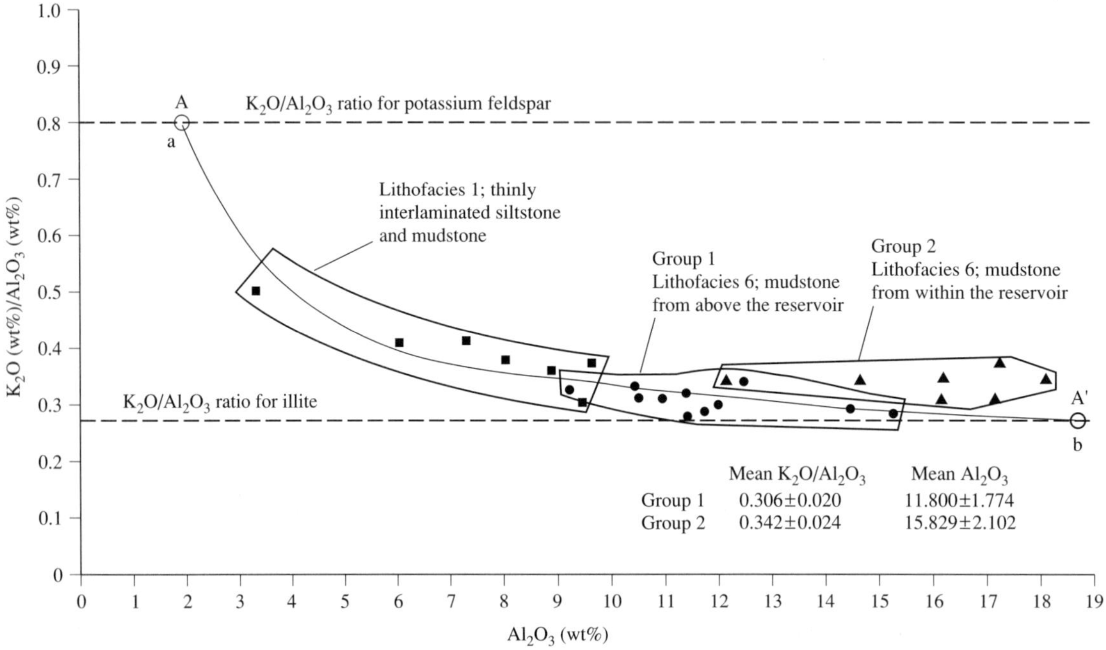


**Fig. 11.** Relationship between $K_2O/Al_2O_3$ ratio and $Al_2O_3$ content for the mudstone and tiger stripe lithofacies of the Miller reservoir. The mudstones have been divided into those which occur interbedded with the sandstones and those which occur above the sandstones (Fig. 13).

Hence the differences in chemistry between the two groups of mudrocks are regarded as diagenetic.

Observations on the sandstones indicated that aluminium was the least mobile of the components relative to defined immobility of $TiO_2$. Comparison of $TiO_2/Al_2O_3$ ratios from the two groups of mudstones shows them to be identical. We take this to indicate that alumina has also remained relatively immobile in both sets of mudstones during diagenesis. When compared with alumina (or $TiO_2$) the mudstones within the Brae Formation seem to have gained a little potassium and lost substantial silica.

The quantity of silica exported from the mudstones within the Brae Formation may be calculated from the whole-rock chemical data by normalizing to the alumina content for both sets of mudstones. The percentage export of silica from mudstone within the Brae Formation as a function of the original silica content (Table 4) is given by:

$$(1 - (75.38/(14.74/11.00 \times 79.95))) \times 100 = 29.64\%$$

and as a percentage of the total original rock = $79.95 \times 0.296 = 23.7\%$.

## TEMPERATURE OF MUDSTONE DIAGENESIS

We only have direct data on the chemistry of the Brae Formation and Kimmeridge Clay Formation mudstones. From this we deduced that the Brae Formation mudstones had exported silica. There seem to be two possible mechanisms that could deliver dissolved silica for export from the mudstones. The reactions most commonly implicated with release of silica are those including clay mineral reactions. Another possibility was suggested by Evans (1990). He observed up to 20% loss of detrital quartz from mudstones during diagenesis. No temperature data were presented by Evans (1990) and here we are unable to comment on whether such a situation might have existed in the Kimmeridge Clay Formation.

Silica is a product of clay mineral transformations (smectite to illite) in mudstones (Hower & Longstaffe, 1981; Curtis, 1985). We have no specific evidence to show that such a reaction occurred in the Kimmeridge Clay Formation around the Miller Field. There is, however, much evidence that such a transformation did take

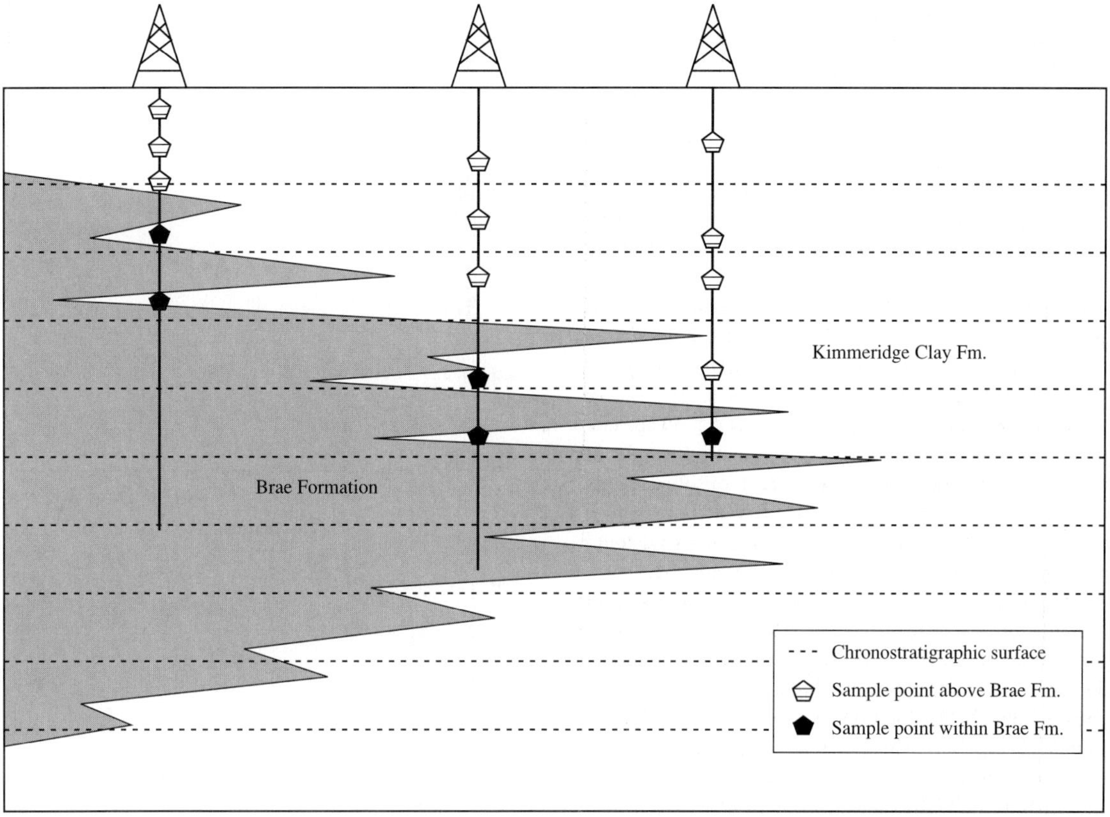


**Fig. 12.** Sampling strategy for the mudstones within and above the Miller Reservoir. The figure illustrates the temporal equivalent of some of the mudstones which are internal to the reservoir in the west but above the reservoir in the east.

**Table 4.** Average silica and alumina composition of mudstones within and above the Brae Formation

| | $SiO_2$ (wt%) | $Al_2O_3$ (wt%) |
|---|---|---|
| Mudstones within the Brae Fm. | 75.38 ± 2.10 | 14.74 ± 1.88 |
| Mudstones above the Brae Fm. | 79.95 ± 2.02 | 11.00 ± 1.77 |

place in the Kimmeridge Clay Formation throughout the more deeply buried parts (> about 2 km) of the North Sea and adjacent areas (Dypvik, 1983; Scotchman, 1987; Pearson & Small, 1988). Moreover, the presence of quartz cement within the peripheral parts of carbonate concretions (in the Kimmeridge Clay Formation of the North Sea) has been taken as evidence for the liberation of silica during smectite to illite transformation (Scotchman, 1993). We infer that the clay mineral reactions that dominate Kimmeridge Clay Formation diagenesis elsewhere in the North Sea were also important in the diagenesis of the Brae Formation and Kimmeridge Clay Formation mudstones in Block 16/7b. Dypvik (1983) reports the transformation of smectite to illite and mixed-layer illite–smectite to illite. Smectite converts to illite–smectite at temperatures below about 75°C, whereas conversion of mixed-layer clays with more than 70% illite layers to illite occurs in the range 70–100°C, that is over the same temperature range at which petroleum begins to be generated and expelled. The upper end of the temperature range is similar to the estimated precipitation temperature for quartz in the Brae Formation.

## MATERIAL BALANCE

We calculated from comparison of the silica contents both inside and outside calcite concretions that the quartz content of the sandstone increased by 5.57 vol.% during diagenesis. At a similar time the mudstones appear to have lost 23.7% of their silica mass. It is possible therefore that one volume of mudstone could supply 23.7/5.57 or 4.25 volumes of sandstone with cement. The net to gross of the Upper Member of the Brae Formation is 80%, that is a sandstone to mudstone ratio of 4 : 1. Thus there is agreement between the volume of silica that could have been supplied by the mudstones and that which has been received by the sandstones.

Further evidence that the mudstones played an important role in delivering silica for cementation comes from the relationships between bed thickness and porosity and between net to gross and porosity (Figs 13 & 14). For sandstone beds of less than 0.6 m there is a good correlation between bed thickness and porosity (inverse of cement volume). For thick beds this relationship does not exist. Net to gross and porosity are also positively correlated. Thus where mudstones are abundant the sandstones are more cemented than where mudstones are few.

To investigate the possible processes responsible for cementation we need to consider the volume of silica transported, the distance the silica was transported and the time available. From our fluid inclusion and compaction studies, the duration of quartz cementation was estimated at 20 Myr. This is perhaps an upper limit, much of the cementation could have taken place in a much shorter time (cf. Gluyas & Oxtoby, 1995). The petrographic and geochemical data suggest that an average 6% of the sandstone has been 'mobile' during cementation. The average distance for the silica to have travelled during diagenesis is $1.15 \pm 2.02$ m (Table 5). The greatest distance, based on the thickest sandstone, was about 8 m. For this particularly thick sandstone the porosity is about 14%, with an estimated 8% quartz cement based upon a post-compaction/pre-cementation porosity of 22%.

Using values of 2.65 for grain density, 14% for porosity and 60 g for a mole of $SiO_2$, the silica flux required to satisfy the observed cement volume is in the order of $10^{-4}-10^{-3}$ moles $m^{-1}$ $yr^{-1}$ for a 20-Myr cementation phase and an order of magnitude greater for a 2-Myr cementation period. Such rate demands are trivial compared with the quantities of silica that could be supplied by advection (Canals & Meunier, 1995) and similar to that which might be achieved through diffusion of silica from source to precipitation site (Bjølykke & Egeberg, 1993).

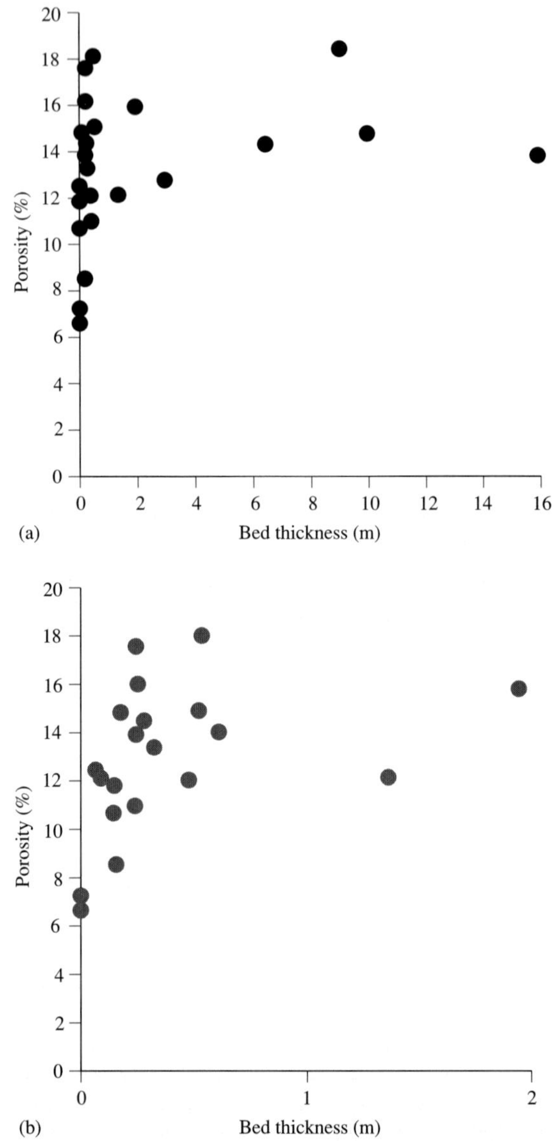


**Fig. 13.** Relationship between bed thickness and porosity for the cored section of 16/7b-20. (a) All beds; (b) thin beds (< 2 m).

## DISCUSSION

The preceding section on the mobility of silica and other components was based upon two important assumptions: (i) in part, the Brae Formation sandstones have retained their original chemistry; and (ii) titanium was essentially immobile during diagenesis. The argument for retention of original chemistry was based on deductive reasoning. It has already been presented in the sections on geological background and petrography. However, the statement that titanium was largely immobile was a conclusion reached by induction. The consequences of other interpretations are examined here.

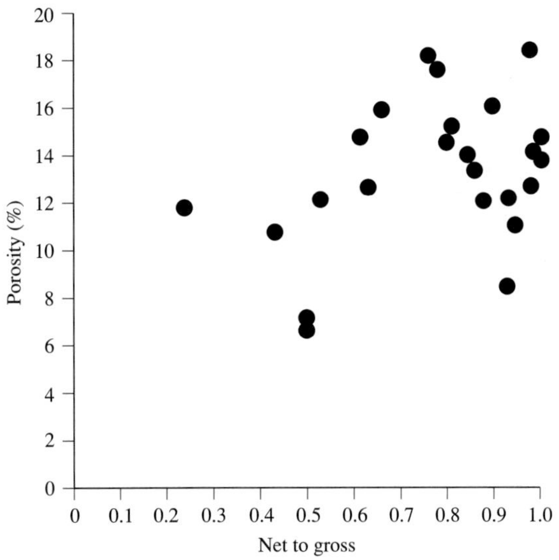


**Fig. 14.** Relationship between net to gross and porosity for the cored section of 16/7b-20.

We began with the observation that the $TiO_2$ content of the host sandstone was greater than the $TiO_2$ content of the concretion and where concretionary calcite was taken as a proxy for porosity of the sand soon after deposition. We defined this increase as a result of compaction. However, the absolute increase of $TiO_2$ from within to without the concretion was accompanied by a decrease in the $TiO_2$ to $SiO_2$ ratio. We argued that this was indicative of silica import during diagenesis of the host sandstone. Let us now examine other possibilities.

The two alternative possibilities are that silica was dissolved from the concretion site during concretion growth or that titanium was removed from the host sandstone during later diagenesis. The first possibility allows titanium to remain largely immobile. The second statement demands that the titanium is totally removed from the billion cubic metres of rock we sampled rigorously. We have already demonstrated from petrographic evidence that growth of the concretion was not accompanied by quartz dissolution. Complete removal of titanium from such a large volume of rock seems to us highly improbable. If these arguments alone are insufficient to demonstrate, beyond reasonable doubt, that silica was imported to the host sandstones during diagenesis, then let the relationship with the Kimmeridge Clay Formation be re-examined.

Much evidence points towards the Kimmeridge Clay Formation as a source for the quartz cement in the Brae Formation sandstone. Both sandstone and mudstone experienced their main phase of silicate mineral diagenesis at much the same time. Clay mineral reactions in the mudrock generated silica. Estimates of silica volume exported from the mudstones and that imported by the sandstones are similar. The relationships between porosity of the sandstone and both bed thickness and net to gross indicated that thinner sandstones nearer to mudstones were more likely to receive a substantial flux of silica than thick sandstones distant from mudstones. The data used to generate the conclusion that silica was exported from parts of the mudstone most intimately associated with the sandstone, could be reinterpreted as demonstrating aluminium and titanium import to the mudstone. In order to do this it would be necessary to ignore the net to gross, bed thickness, porosity relationships and the material balance between sandstone and mudstone. It would be necessary to look beyond the billion cubic metres to capture a balanced diagenetic system. Moreover it would be necessary to invoke massive transport distances for elements during diagenesis.

If the small-scale material balance between the Brae Formation and Kimmeridge Clay Formation is accepted then the data here indicate that the average 'quantum' of silica for quartz cement moved little more than 1 m in a period of up to 20 Myr. We initially asked, 'How big was the diagenetic system?' The data collected here lead us to the conclusion that the system, although it contains both sandstone and mudstone, is small. A simple calculation based upon the core diameter of 15 cm and the average and maximum silica transport distances of 1 m and 8 m yields volumes of $5 \times 10^{-3}$ m$^3$ – $5 \times 10^{-2}$ m$^3$. The volume figures although contrived, serve to illustrate the conclusion that silica has moved short distances vertically through the sediment column. There is no need to appeal

**Table 5.** Silica flux during diagenesis

| Quartz cement content (%) | Distance travelled (m) | Silica flux moles m$^{-1}$ yr$^{-1}$ (20 Myr cementation period) | Silica flux moles m$^{-1}$ yr$^{-1}$ (2 Myr cementation period) |
|---|---|---|---|
| 6 | 1.15 | $10^{-4}$ | $10^{-3}$ |
| 8 | 8 | $8.7 \times 10^{-4}$ | $8.7 \times 10^{-3}$ |

to long distance vertical transport of silica as a cementation mechanism. Nor is there a need to invoke imaginary stylolites or other features internal to the sandstone.

The relationship between net to gross and quartz cement abundance in the Brae Formation begs further investigation. If our interpretation of short distance vertical transport of silica is correct then reservoir sections with a net to gross which is either higher or lower than the 0.8–0.6 for Miller could show quite different cement abundances and distributions. Such high and low net to gross areas do exist. The Brae Fields West of Miller commonly have enormous thicknesses of sandstone and conglomerate where the net to gross is about one. Similarly the eastern sections of the Miller and East Miller Fields show sand pinchout where the net to gross approaches zero. Here one might predict that quartz cementation progressed to completion (zero porosity).

## CONCLUSIONS

The Brae Formation and the Kimmeridge Clay Formation of the Miller Field area in the North Sea acted as a coupled, closed system, during silicate diagenesis. Silica was exported from the mudstone to the sandstone. The Kimmeridge Clay Formation was capable of delivering an average 6% quartz cement to the sandstone in a period of less than 20 million years. On average the transport distance for the silica was little more than a metre. Diffusion alone is adequate to accomplish this. Indeed it might not be the limiting process.

## ACKNOWLEDGEMENTS

We thank BP and partners for giving permission for this work to be published. Mark Hopkins took the SEM photomicrographs and selected the best for publication. The whole-rock analyses were performed by Ken Liberty.

## REFERENCES

Abbots, I.L. (1991) *United Kingdom Oil and Gas Fields, 25 Years Commemorative Volume.* Geological Society of London Memoir 14, 573pp.

Bjølykke, K. & Egeberg, P.K. (1993) Quartz cementation in sedimentary basins. *American Association of Petroleum Geologists Bulletin* **77**, 1538–1548.

Burruss, R.C. (1991) Practical aspects of fluorescence microscopy of petroleum fluid inclusions. In: *Luminescence Microscopy, Quantitative and Qualitative Aspects.* (eds Barker, C.E. & Kopp, O.C.) Society of Economic Paleontologists and Mineralogists Short Course 25.

Canals, M. & Meunier, J.D. (1995) A model for porosity reduction in quartzite reservoirs by quartz cementation. *Geochimica et Cosmochimica Acta* **59**, 699–709.

Chaucer, G. (1400) *The Canterbury Tales.* World's Classics, Oxford University Press (1986), p. 482.

Curtis, C.D. (1985) Clay mineral precipitation and transformation during burial diagenesis. *Philosophical Transactions of the Royal Society of London A* **315**, 91–106.

Dypvik, H. (1983) Clay Mineral transformations in Tertiary and Mesozoic sediments from North Sea. *American Association of Petroleum Geologists Bulletin* **67**, 160–165.

Evans, I.J. (1990) Quartz dissolution during shale diagenesis, implications for quartz cementation in sandstones (abstract). In: *Geochemistry of the Earth's Surface and of Mineral Formation, 2nd International Symposium, July 2–8, 1990, Aix en Provence, France.*

Garland, C.R. (1993) Miller Field: reservoir stratigraphy and its impact on development. In: *Petroleum Geology of Northwest Europe: Proceedings of the 4th Conference.* pp. 401–414. (ed. Parker, J.R.) Geological Society of London.

Gluyas, J.G. (1984) Early carbonate diagenesis within Phanerozoic shales and sandstones of the NW European continental shelf. *Clay Minerals* **19**, 309–322.

Gluyas, J.G. (1985) Reduction and prediction of sandstone reservoir potential. Jurassic North Sea. *Philosophical Transactions of the Royal Society of London A* **315**, 187–202.

Gluyas, J.G. & Cade, C.A. (1997) Prediction of porosity in compacted sands. In: *Reservoir Quality Prediction in Sandstones and Carbonates.* (eds. Kupecz, J.A., Gluyas, J.G. & Bloch, S.) American Association of Petroleum Geologists Memoir 69 1997, 19–28.

Gluyas, J.G. & Coleman, M.L. (1992) Material flux and porosity changes during sediment diagenesis. *Nature* **356**, 52–53.

Gluyas, J.G. & Leonard, A.J. (1995) Diagenesis of the Rotliegend sandstone: The answer ain't blowin' in the wind. *Marine and Petroleum Geology* **12**, 491–497.

Gluyas, J.G. & Oxtoby, N.H. (1995) Diagenesis a short (2 million year) story—Miocene sandstones of Central Sumatra, Indonesia. *Journal of Sedimentary Research, A* **65**, 513–521.

Harmes, J.C., Tackenberg, P., Pickles, E. & Pollock, R.E. (1981) The Brae oilfield area. In: *Petroleum Geology of the Continental Shelf of North-West Europe.* (eds Illing, L.V. & Hobson, G.D.) pp. 352–357. Heyden, London.

Hower, J. (1981) Shale Diagenesis. In *Mineralogical Association of Canada, Short Course in Clays and the Resource Geologist.* (ed. Longstaffe, F.J.) pp. 60–80. Canadian Society of Petroleum Geologists, Calgary.

Mackenzie, A.S., Price, I., Leythaeuser, D., Müller, P., Radke, M. & Schaefer, R.G. (1987) The expulsion of petroleum from Kimmeridge clay source-rocks in the area of the Brae Oilfield, UK continental shelf. In: *Petroleum Geology of North West Europe.* (eds Brooks, J. & Glennie, K.) pp. 865–877. Graham & Trotman, London.

McClure, N.M. & Brown, A.A. (1992) Miller Field: A subtle Upper Jurassic submarine fan trap in the South Viking Graben, UK sector, North Sea. In: *Giant Oil and Gas Fields of the Decade 1978–88.* (ed. Halbouty, M.T.) American Association of Petroleum Geologists Memoir 54.

Morton, A.C. (1984) Stability of detrital heavy minerals in Tertiary sediments from the North Sea Basin. *Clay Minerals* **19**, 287–308.

PEARSON, M.J. & SMALL, J.S. (1988) Illite–smectite diagenesis and palaeotemperatures in Northern North Sea Quaternary to Mesozoic shale sequences. *Clay Minerals* **23**, 109–132.

ROEDDER, E. (1984) *Fluid Inclusions*. Reviews in Mineralogy 12, 644pp.

ROOKSBY, S.K. (1991) The Miller's Field, Blocks 16/7b, 16/8b, UK North Sea. In: *United Kingdom Oil and Gas Fields, 25 Years Commemorative Volume*. (ed. ABBOTS, I.L.) pp. 159–164. Geological Society of London Memoir 14.

SCOTCHMAN, I.C. (1987) Relationship Between Clay Diagenesis and Organic Matter Maturation in the Kimmeridge Clay Formation, Onshore UK. In: *Petroleum Geology of North West Europe.* (eds BROOKS, J. & GLENNIE, K.) pp. 251–262. Graham & Trotman, London.

SCOTCHMAN, I.C. (1993) Diagenetic pore fluid evolution in the Kimmeridge Clay Formation. From concretions to sandstone cements. In: *Geochemistry of Clay–Pore Fluid Interactions.* (eds MANNING, D.A.C., HALL, P.L. & HUGHES, C.R.) pp. 127–159. Chapman & Hall, London.

STAUDIGEL, H. & HART, S.B. (1983) Alteration of basaltic glass: Mechanisms and significance for the oceanic crust–seawater budget. *Geochimica et Cosmochimica Acta* **43**, 337–350.

STOW, D.A.V., BISHOP, C.D. & MILLS, S.J. (1982) Sedimentology of the Brae Oilfield, North Sea. Fan Models and Controls. *Journal of Petroleum Geology* **5**, 129–148.

THOMPSON, M. & WALSH, J.N. (1983) *A Handbook of Inductive Coupled Spectrometry*. Blackie, Glasgow, 250pp.

**Plate 1.** (*Opposite.*) (a) Core photograph of a calcite cemented concretion. This particular concretion is the smallest observed in the cored Brae Formation of the Miller Field. Most are 2–3 m in diameter. All the concretions are subspherical and contain trace quantities of pyrite typical of carbonate cementation occurring near to the sediment water interface through the destruction of organic matter by sulphate reducing bacteria (Gluyas, 1984). Small concretions such as this allowed sampling of both calcite and quartz cemented sandstones on the same depositional surface. This enabled us to show that the sites of concretion growth were in no way different from the remainder of the sediment. Sample from well 16/7b-23 at 4145.3 m below rotary table. (b) Photomicrograph, calcite cement within a carbonate concretion. The quartz grain is free of quartz cement and there appears to have been no dissolution of the quartz grain during calcite precipitation. Scale of photomicrograph 1.3 mm × 0.9 mm. Photomicrograph taken with cross-Polars and $^1/_4\lambda$ plate. (c) Thin section photomicrograph of well preserved feldspar (F) within calcite concretion (C), the grain has only minor dissolution features (D). Photomicrograph taken with cross-Polars and $^1/_4\lambda$ plate. Scale bar 100 μm (d) late diagenetic calcite cement (C) infilling residual porosity after quartz cementation. Photomicrograph taken using back-scattered electron imaging on a scanning electron microscope. Scale bar: 100 μm (d) corroded feldspar (F) outwith a calcite cemented concretion Photomicrograph taken using back-scattered electron imaging on a scanning electron microscope. Scale bar 100 μm (f) quartz cemented Miller reservoir sandstone. The image combines a backscattered electron image and one obtained from cathodoluminescence. The dark grey quartz cement occurs as syntaxial rims on most grains and as fracture fills. Scale bar: 100 μm (g) The same image as in (f) but with false colours used to highlight the grey levels contrast between pore space (blue), quart cement (red) and grains (green).

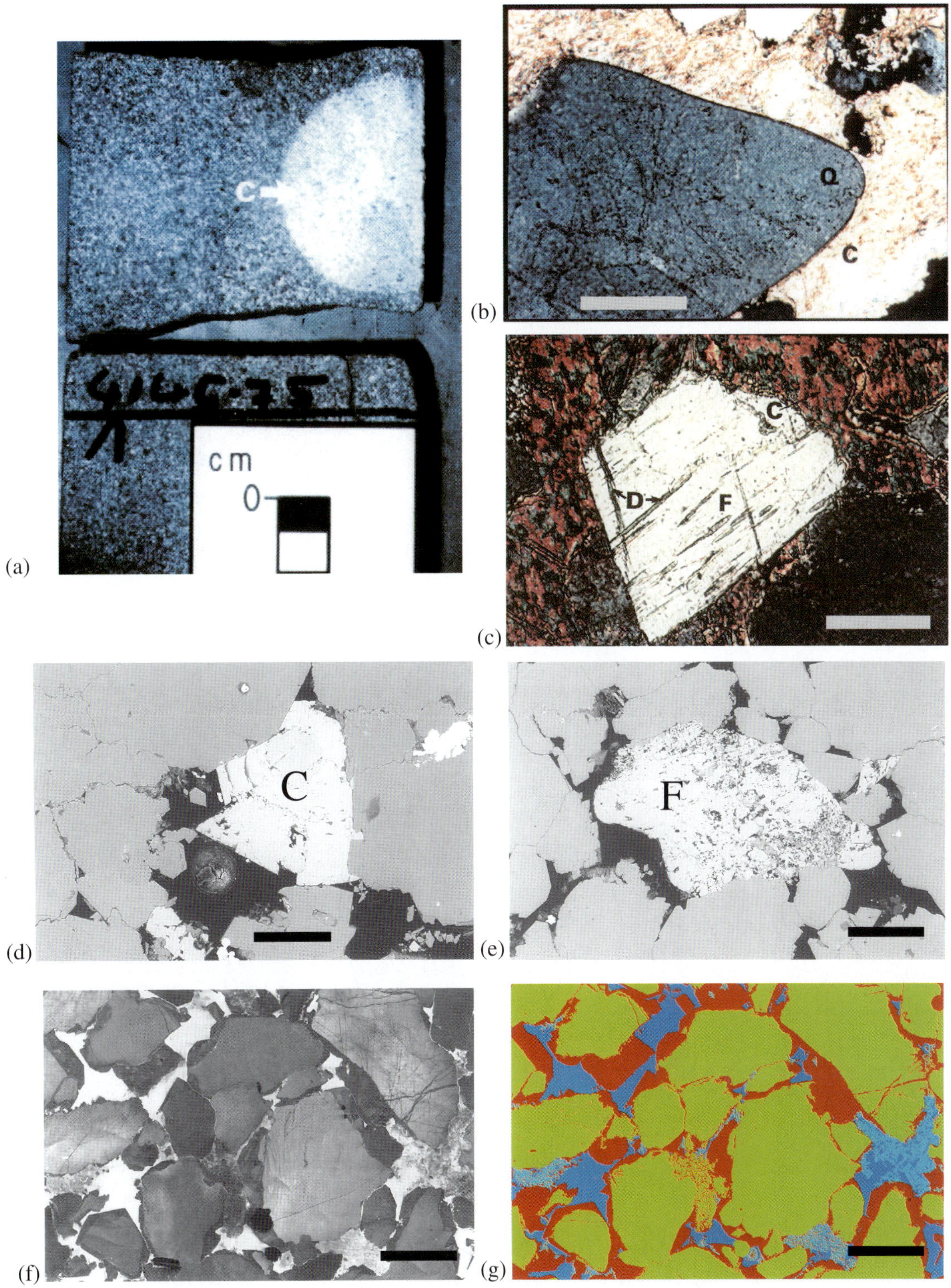
(a)
C
cm
0
(b)
Q
C
(c)
C
D
F
(d)
C
(e)
F
(f)
(g)

*Spec. Publs int. Ass. Sediment.* (2000) **29**, 219–229

# Quartz cement origins and budget in the Tumblagooda Sandstone, Western Australia

N. H. TREWIN[1] *and* A. E. FALLICK[2]

[1]*Department of Geology and Petroleum Geology, University of Aberdeen, Aberdeen, AB24 3UE, UK; and*
[2]*Scottish Universities Research and Reactor Centre, East Kilbride, Glasgow G75 0QF, UK*

## ABSTRACT

The Tumblagooda Sandstone (late Silurian) includes facies of mixed fluvial and aeolian sandsheet origin. It is over 1 km thick in the type outcrop area and comprises quartz-rich medium to coarse grained red-bed sandstones which are thought to rest unconformably on metamorphic basement. Cementation is dominantly by quartz with minor illite and Fe-rich grain coatings. Fluvial facies have extensive quartz overgrowths (8–24% rock volume) and low fabric compaction due to net introduction of quartz. Aeolian facies show extensive quartz dissolution at grain contacts and contain only a minor (up to 5%) quartz cement volume. Point-count data imply compactional volume losses due to grain-to-grain dissolution of up to 3% for fluvial and 8–19% for aeolian facies. Calculation of pre-cement porosity gives figures of 21–33% (av. 27.5%) for fluvial and 22–34% (av. 29.3%) for aeolian sandstones. Silica is generally conserved within the formation, some being transferred from aeolian to fluvial facies with migration distances in the range of centimetres to tens of metres. However, other sources of silica are required to account for the quantity of quartz cement present within the formation; such sources are considered to be the redistribution of products from abrasion of detritus during transport, and solution products from stylolite seams. Feldspar dissolution could only have supplied a minor quantity of silica for cementation, dissolution of minor feldspar present took place later than the development of quartz overgrowths.

Quartz cementation of fluvial sandstones was aided by relatively clean detrital grain surfaces, but grain coatings of clay and oxides on aeolian grains encouraged dissolution at grain contacts during burial and inhibited overgrowth formation. Macroporosity (primary intergranular) values in the two facies are broadly similar (fluvial 1–12%, av. 6.6%; aeolian 2–17%, av. 9.1%) but pore configuration and connectivity varies and would greatly affect gas or fluid production from the major facies.

Oxygen isotope data measured on separated overgrowths range from $\delta^{18}O$ –14.2‰ to 17.6‰ SMOW with no apparent relationship to locality, facies or stratigraphic position. For precipitation temperatures of 125°C and lower, calculated water $\delta^{18}O$ values are less than zero, implying a meteoric origin. It is concluded that silica transfer took place at depths of ≈ 3 km and a temperature of ≈ 100°C, and commenced when porosity was generally 25–30% in both major facies.

## INTRODUCTION

The Tumblagooda Sandstone is generally considered to be of late Silurian age, although ages from early Ordovician to Devonian have been suggested (Discussion in Trewin & McNamara, 1995). The sandstone is exposed in the gorges of the Murchison River and the sea cliffs at Kalbarri, Western Australia (Fig. 1). The sequence is about 1 km thick, there being uncertainty due to the presence of faults and the difficulty of compiling a succession from strata having a shallow dip, and using sections up to 50 km apart (Hocking, 1991). The sedimentology of the sequence has been described by Hocking (1991) and partly reinterpreted by Trewin (1993a;b).

Those features relevant to this contribution are briefly summarized below. The sandstone is a red-bed sequence which is presumed to rest unconformably on Precambrian basement. It consists almost entirely of sandstone; the only siltstone or mudstone present is in the form of very rare rip-up clasts in sandstone. The succession (Fig. 2) is dominated by two units of medium to coarse grained fluvial sandstone (Facies Assemblages FA1 and FA3 of

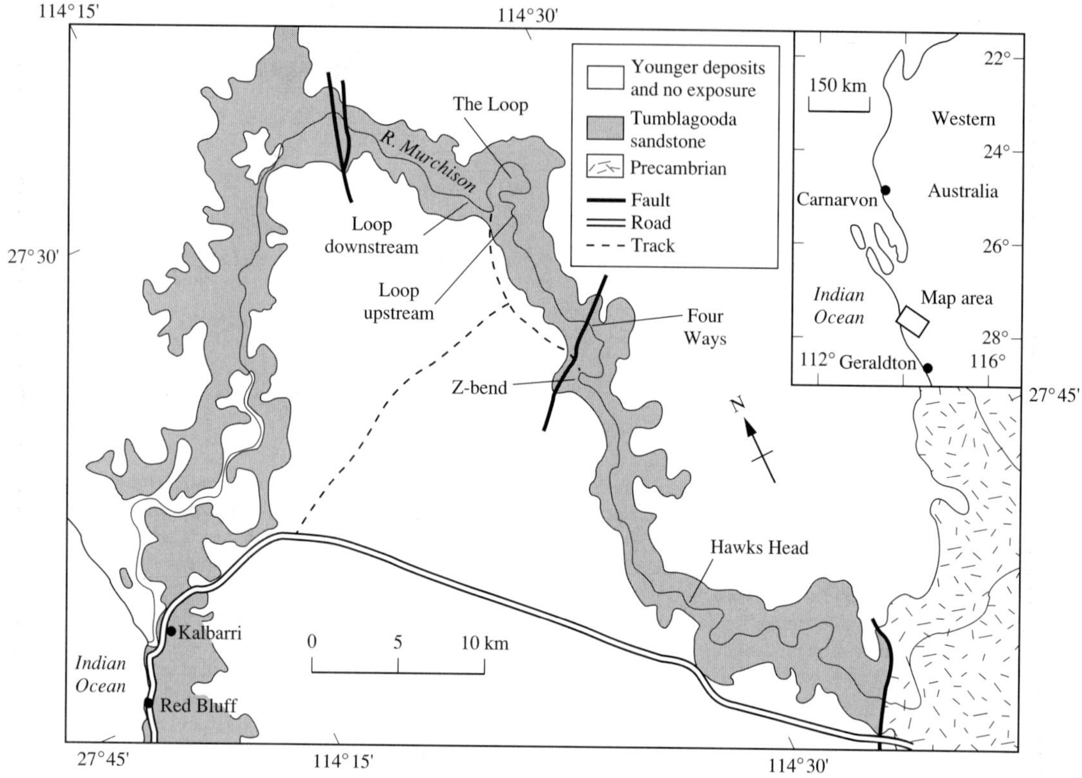


**Fig. 1.** Locality map, outcrop of Tumblagooda Sandstone in gorge of the Murchison River and at Kalbarri.

Hocking *et al.*, 1987) which are separated by a unit (Facies Assemblage FA2) which has been interpreted (Trewin, 1993a) as a mixed fluvial and aeolian sandsheet deposit. This unit had previously been interpreted as being tidally influenced (Hocking, 1991). The topmost unit is only exposed on the coast near Kalbarri (Facies Assemblage 4 of Hocking *et al.*, 1987) and was deposited with strong marine influence as shown by a spectacular *Skolithos*/*Diplocraterion* trace fossil assemblage (Trewin & McNamara, 1995).

The sandstone outcrops, particularly of the mixed fluvial and aeolian unit, show marked weathering contrasts which can be related to quartz cementation. It is readily apparent in hand specimen that fluvial trough cross-bedded sandstones have an extensive quartz-overgrowth cement, whereas grains in aeolian sandsheet and dune deposits generally lack overgrowths. Bioturbated and water-reworked sands with wave ripples show a variety of cementation features.

The object of this contribution is to demonstrate that quartz cementation has been facies selective, and that silica has been transferred in large volumes, and over distances of tens of metres within the formation. The problems of determining a quartz budget for the sandstone are discussed.

## SAMPLING AND METHODS

Samples for the petrographic study were taken from field sections in Facies Association 2 (FA2) of Hocking *et al.* (1987) which were logged for sedimentology in the Murchison Gorge from Hawks Head downstream to the area of The Loop (Fig. 1).

Specimens were chosen to represent the major lithofacies identified in FA2 (See Fig. 2) of Hocking *et al.* (1987) as interpreted by Trewin (1993a). These comprise:

**1** Trough cross-bedded and planar bedded sandstones of fluvial origin.

**2** Sub-parallel bedded sandsheets of dominantly aeolian origin with some reworking by waves in shallow water pools.

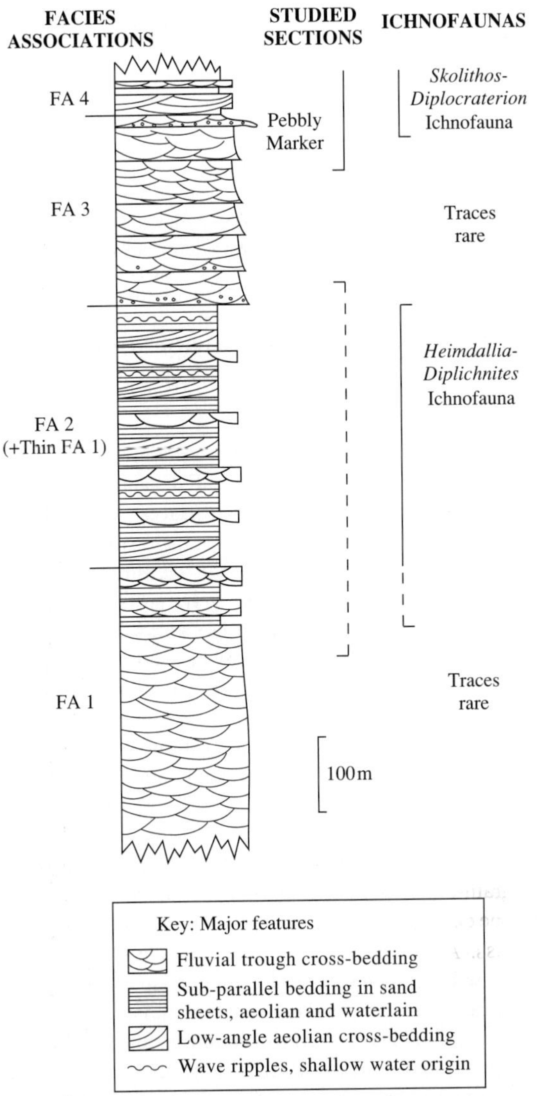


**Fig. 2.** Generalized stratigraphic section of the Tumblagooda Sandstone. Modified from Trewin & McNamara (1995).

**3** Bioturbated sandstones; beds formed by bioturbation by *Heimdallia* and occurring within aeolian sandsheet deposits, reworked by water.

**4** Low-angle cross-bedded sandstones of aeolian origin.

The rocks were examined petrographically in standard impregnated thin sections. SEM examination of polished surfaces utilized standard backscatter electron and cathodoluminescence (CL) images to confirm the nature of quartz overgrowth fabrics observed in thin section.

The grain size of fine to coarse-grained sandstone, lack of fines, and the clear preservation of original detrital grain shapes, as defined by dust-rims and discontinuities of inclusion trails, makes these rocks ideal for determination of quartz overgrowth percentage by traditional point-counting.

Four hundred points were counted for each of 50 slides, inclusive of grains, cement phases and both primary intergranular porosity and secondary porosity. In addition a record was made of the number of points that fell in probable areas of grain dissolution at grain-to-grain contacts. This is subjective, requiring a decision to be made as to whether any point lies in an area of 'grain overlap' as defined by Houseknecht (1984) and illustrated in Fig. 3. The percentage of overlap points in the grain counts provides a measure of silica lost from framework grains by chemical compaction, and an estimate can be made of porosity at the point in burial when grain-to-grain dissolution commenced. Similarly, estimates are obtained of the porosity at commencement of quartz overgrowth formation. This is the porosity of the rock at the time it acquired a rigid frame, and mechanical compaction ceased. In calculation of this porosity value, secondary porosity due to grain dissolution is counted as original grains following Shanmugam (1985). It should be stressed that this petrographic method has been adapted to investigate silica redistribution, and could only work in situations where overgrowths are easily seen and grain shape is reasonably predictable. The method is complementary to sandstone compaction studies (see Ehrenberg (1995) for discussion) which involve the assumption of a value for 'original porosity' based on sorting, and result in values for compactional porosity loss.

To estimate the $\delta^{18}O$ of authigenic silica we have employed the methods of physical separation of quartz overgrowth developed by Lee & Savin (1985) and adapted by Brint *et al.* (1991). This method attempts direct physical separation of quartz overgrowths from host detrital grains and employs repeated size fraction sieving, on the assumption that overgrowths are concentrated into the finer size fractions. The efficacy of the diagenetic versus detrital fraction separation is assessed qualitatively at each step by SEM–CL. A more quantitative assessment comes from the pattern of $\delta^{18}O$ with grain size: a $\delta^{18}O$ plateau at fine grain sizes is taken to indicate a relatively pure diagenetic component (on the argument that a constant proportional detrital fraction is relatively unlikely). Error estimates are deliberately wide to allow for less than 100% authigenic composition.

Quartz $\delta^{18}O$ was determined on $CO_2$ following oxygen release by reaction of 1 mg aliquots with $CF_3$ (Borthwick & Harmon, 1982) in a $CO_2$-laser heated system. The

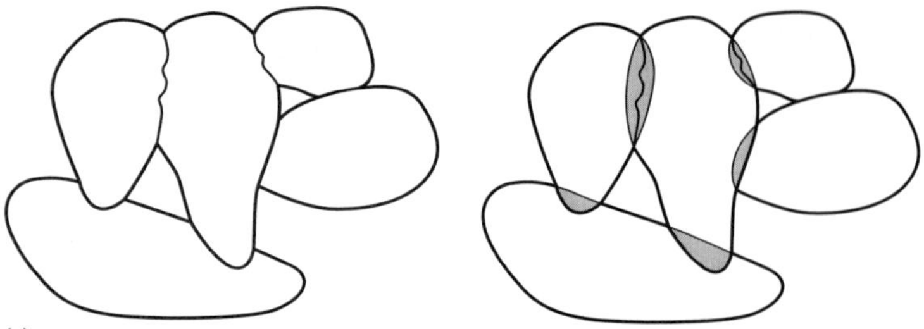

(a) (b)

(a) Compacted fabric with silica lost by dissolution at grain contacts

(b) Shaded areas represent grain overlap, and give a measure of silica lost by grain-to-grain dissolution when expressed as a % of total grain counts.

**Fig. 3.** Interpretation of silica loss by grain contact solution. Shaded areas in (b) are point counted to give an estimate of silica lost from grains during chemical compaction, and expressed as a percentage of grain overlap.

analytical precision for $\delta^{18}O$ measurement is better than $\pm 0.2‰$ ($1\sigma$), but estimates of overgrowth $\delta^{18}O$ are less precise and accurate to allow for effects described above.

## PETROGRAPHY

The detrital mineralogy of the sandstones is dominated by quartz (unstrained, strained and polycrystalline) which constitutes 53–84% (av. 72.3%) of the whole rock (Fig. 4) with minor K-feldspar 0–5.7% (av. 2.2%). The sum of all other detrital components does not exceed 1%. Quartz grain shapes range from subangular to well-rounded, frequently with a distinct population of well-rounded, spherical grains which show relics of abraded red grain coatings.

The remainder of the rock consists of cements, grain coatings, diagenetic phases (mainly quartz and illite) and porosity. The quartz grains display variable preservation of dark red haematite-bearing grain coatings which generally form less than 5% of the rock but exceptionally reach 16%. Illite (Fig. 4e) is also present (0–12.5%) growing from the grain coatings as a boxwork of flakes (10–50 μm) perpendicular to grain surfaces. Quartz overgrowths enclose and grow over illite on grain surfaces. It is difficult to accurately point count red grain coatings and illite, and commonly haematite intensely stains and invades illite boxworks. Red grain coatings contain Fe, K and Al when analysed by EDAX in the SEM. Overgrowths of quartz on the detrital grains are clearly seen in sandstones of all the categories listed (Fig. 4), and since the detrital grain boundaries can be seen, the point counting of quartz overgrowths is straightforward. Quartz overgrowths reach a maximum of 24.4% of the rock. Feldspar grains show variable dissolution, but do not have feldspar overgrowths. Feldspar dissolution postdates quartz overgrowth formation.

Porosity has been recorded as primary or secondary. Primary porosity in fluvial sandstones with quartz overgrowths is generally reduced to small disconnected pores bounded by smooth overgrowth surfaces. Thus, in the fluvial sandstones, primary porosity ranges from 1 to 12% with an average of 6.6%. In aeolian sandstones primary pores are surrounded by detrital grain surfaces covered with grain coatings from which boxworks of illite have grown. Primary porosity in aeolian sandstones ranges from 2 to 17%, averaging 9.1%. However, the presence of illite will give these sandstones very low permeability. Secondary porosity was identified by the presence of oversize pores from which grains, probably mainly feldspars, have been dissolved. Feldspar can be observed in all stages of disintegration.

Estimates from point-count data of volume loss due to grain-to-grain solution (Fig. 4) range up to 18.5% with one exception of a stylolitized texture with 31.5% volume loss. A cross plot (Fig. 5) of percentage quartz overgrowth against percentage of grain overlap produces two clear fields, one of waterlain, generally fluvial sandstones with abundant quartz overgrowths and another, of aeolian sandsheet and dune sandstones, with high compaction values due to grain-to-grain dissolution. The few sandsheet facies points that lie in the fluvial field relate to thin wave rippled sands deposited in interdune ponds. The bioturbated sands show a wide scatter but generally lie in the area where the aeolian and fluvial fields are in close proximity.

## ISOTOPIC DATA

Nine samples were selected for oxygen isotopic analysis of the quartz overgrowth cement; the methods used have been given above. The samples are from Hawks Head, The Loop and Red Bluff (Fig. 1) representing a wide geographical area, and including from Red Bluff

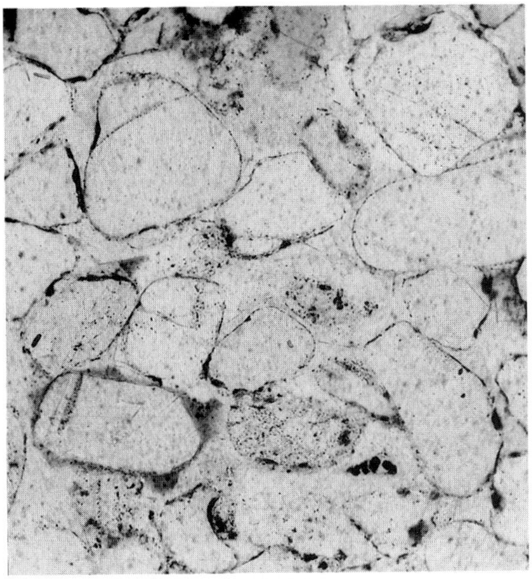

(a)

(b)

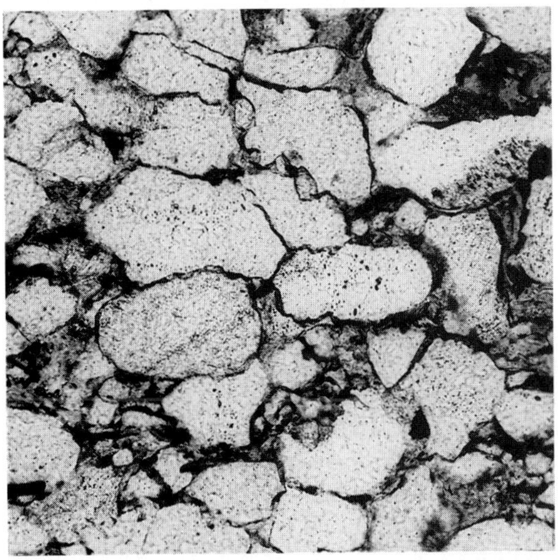

(c)

(d)

(e)

**Fig. 4.** Typical textures in the Tumblagooda Sandstone. (a, b) Fluvial sandstone facies with well-preserved original grain boundaries and extensive quartz overgrowth cement. (a) Plane light; (b) crossed polars. (c, d) Aeolian sandstone with solution contacts between coated grains and virtual absence of quartz overgrowths. (c) Plane light; (d) crossed polars. (e) Quartz overgrowths partly enclosing illite. Grain at top right has grain coating with illite and lacks quartz overgrowths.

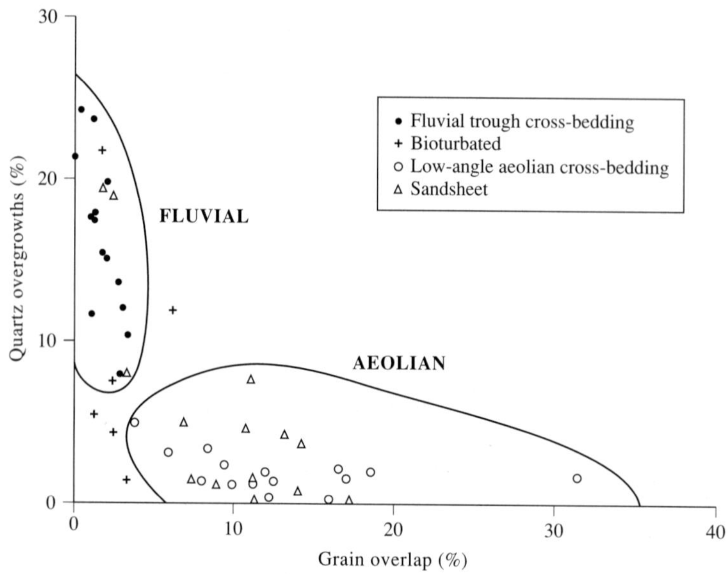


**Fig. 5.** Plot of percentage quartz overgrowths against percentage grain overlap to show distinctive fields for fluvial and aeolian sandstones. Sandsheet samples plotting within the fluvial field are generally water rippled sands from interdune areas which were periodically flooded. Bioturbated sands show no clear preference for either field. For individual data points uncertainty values (following Van Der Plas & Tobi, 1965) are generally about ±4% (95% confidence limit) for values of 20%. Hence the separation of the fields is significant.

**Table 1.** Measured $\delta^{18}O$ (‰ SMOW) values for sieve-separates of pure quartz produced by the Lee & Savin (1985) method for nine samples

(a) Fluvial to marginal marine sandstones

| Grain size | | < 53 μm | 53–85 μm | > 85 μm |
|---|---|---|---|---|
| Red Bluff | RB1 | 14.6; 14.6 | 14.5 | 10.3 |
| Red Bluff | RB5 | 14.3; 14.8 | 13.8 | 9.7 |
| Red Bluff | RB7 | 16.5; 16.9 | 15.5 | 10.5 |

(b) Fluvial sandstones in mixed fluvial/aeolian sequence

| Grain size | | 30 μm | 30–53 μm | 53–85 μm | > 85 μm |
|---|---|---|---|---|---|
| Hawks Head | HH8 | 15.8 | 16.3 | 13.9 | 11.9 |
| Hawks Head | HH13 | 16.1 | 16.6 | 13.9 | 10.5; 10.6 |
| Hawks Head | K34 | 16.8 | 15.1 | 13.0 | 10.6 |
| The Loop | LU2 | 14.2 | 10.0 | 11.3 | 9.9 |
| The Loop | LF3 | 17.6, 17.6 | 17.2 | 12.6 | 11.2, 11.5 |
| The Loop | KB | 15.3 | 15.1 | 13.8 | 10.3 |

The duplicate analyses all show good precision with a mean difference of 0.2‰. The range in the finest size fraction (dominated by detrital quartz) $\delta^{18}O$ and the weakly related coarse size fraction (dominated by detrital quartz) $\delta^{18}O$ are very interesting and worthy of further consideration from a sedimentological viewpoint, but this is outside the scope of the current paper.

**Table 2.** Calculation of $\delta^{18}O$ of water in equilibrium with silica of $\delta^{18}O$ 17.6‰ as a function of temperature

| Temperature (°C) | 25 | 50 | 100 | 125 | 150 | 200 |
|---|---|---|---|---|---|---|
| $\delta^{18}O$ water | −16.7 | −11.1 | −3.1 | −0.2 | 2.2 | 5.0 |

representatives of Facies Association 4 (FA4 on Fig. 2) of Hocking *et al.* (1987) which contains a trace fossil assemblage indicative of marine influence (Trewin & McNamara, 1995). At each of the three localities three samples were taken to test for local variation in isotopic values. It is readily apparent from Table 1 that reasonable $\delta^{18}O$ plateaux were attained for six of these samples (RB1, RB5, HH8, HH13, LF3 and KB). An error estimate of ±1‰ on the proposed overgrowth $\delta^{18}O$ appears appropriate. An accuracy-error reducing $\delta^{18}O$ by 1‰ corresponds to a 12.5% contribution of detrital quartz at 10‰ to the fine size fraction. The finest size fraction $\delta^{18}O$ values range from 14.2 to 17.6‰ with no apparent relationship to locality, facies or stratigraphic position. We note than an overgrowth estimate of 14.5 ± 1‰ would be reasonable for four samples (RB7, HH8, HH13 and K34). Since detrital quartz contamination of the fine size fractions resulted in a measured $\delta^{18}O$ lower than the true overgrowth value, there is suggestive evidence that LF3 has a slightly higher overgrowth $\delta^{18}O$ of around 17.6‰.

Using the low temperature extrapolation of the Matsuhisa *et al.* (1979) quartz–water oxygen isotope fractionation factor, we can calculate the $\delta^{18}O$ of water in equilibrium with silica of $\delta^{18}O$ 17.6‰ as a function of temperature as in Table 2.

Therefore, for temperatures of 125°C and below, calculated water $\delta^{18}O$ values are less than zero, strongly implying a meteoric origin. *A fortiori*, this argument also applies to the other samples with lower overgrowth $\delta^{18}O$ estimates; for the other extreme of 14.5‰, the temperature estimate for water with a $\delta^{18}O$ value of 0‰ is 160°C.

## DIAGENESIS DISCUSSION

Diagenesis in the Tumblagooda Sandstone is controlled by depositional facies. The most striking feature is the extensive development of quartz overgrowths and low compaction in the fluvial sandstones, contrasting with the very low percentage of overgrowths combined with high compaction in the aeolian sandstones.

It is likely that the very early diagenetic history was dominated by meteoric pore waters. It has been suggested (Trewin, 1993b) that the water table was high and that rivers flowed in broad shallow channels on an outwash area with a palaeoslope to the west-north-west. The lack of vegetation resulted in surface sand being reworked by the wind into aeolian sandsheets and dunes which were transported obliquely up the palaeoslope to the south-east.

In the dunes and sandsheets, perched above the water table, grain coatings developed on quartz grains. The coatings probably contained Fe and also K and Al derived from the dissolution of unstable detrital minerals and formation of clays under near-surface conditions. Red amorphous coatings in recent stabilized dunes in Western Australia contain Fe, K and Al. Aeolian dust washed in by rain was probably also a source of material for grain coatings (Pye & Tsoar, 1987).

The fluvial sandstones with subangular to rounded grains are interpreted to contain a mixture of reworked aeolian grains and material derived by fluvial processes from the sediment source area. Any coatings present on aeolian grains were abraded during fluvial transport and hence most deposited grains had relatively clean surfaces in comparison with grains in the aeolian deposits. Such processes can be observed today in the Cooper Creek area of central Australia, where red dune sands become white when reworked and abraded by water transport.

During early burial, with strong groundwater flow, any unstable detrital minerals would have been removed, leaving the clean quartzose sands with minor K-feldspar. In contrast, grain coatings on quartz grains were generally retained in aeolian deposits.

As burial proceeded the critical depth was reached where grain-to-grain 'pressure' dissolution began at grain contacts. The grain coatings in the aeolian sandstones provided a catalytic film which actively promoted dissolution. Illite grain coatings are known to promote pressure solution (Houseknecht, 1988), however, Bjørkum (1996) prefers to refer to this as a clay-induced dissolution process rather than 'pressure solution'. The general term pressure solution is useful to apply to the process whether it involves amorphous grain coatings, clays or organic matter. We envisage that the silica released by pressure solution was able to migrate in solution to the fluvial sandstones where it was deposited on the clean quartz grain surfaces as overgrowths, so providing a strong frame and preventing further compaction.

**Table 3.** Average values for feldspar and secondary porosity created by feldspar distribution in the four lithofacies. The total amount of feldspar present at the time of lithification was similar for all four lithofacies

| Lithofacies | | A Feldspar (%) | B Secondary porosity (%) | C Total (%) |
|---|---|---|---|---|
| Fluvial | ($n = 15$) | 1.40 | 4.05 | 5.45 |
| Aeolian | ($n = 15$) | 2.91 | 3.48 | 6.39 |
| Sandsheet | ($n = 14$) | 2.10 | 3.06 | 5.16 |
| Bioturbated | ($n = 6$) | 2.45 | 3.80 | 6.25 |

Illite developed from the grain coatings which contained the necessary K and Al for illite growth—a feature also seen in the sandstones of the Rotliegende in the Southern North Sea, where illite forms from apparently amorphous grain coatings which contain Al and K as well as Fe. The Permian sandstones in Arran (Van Panhuys-Sigler & Trewin, 1990) also have amorphous grain coating with Al and K but have not been deeply enough buried for this to be converted to illite. Quartz overgrowths in the Tumblagooda Sandstone (Fig. 4e) are locally observed to enclose diagenetic illite grown on grain surfaces, thus quartz overgrowth development continued after illite formation.

The range in K-feldspar content from 0 to 5.7% might represent slight provenance variation or reflect reworking of the sands prior to deposition, but some variation is due to dissolution of feldspar to give oversize pores. Where petrographic evidence is available it appears that feldspar dissolution post-dates quartz overgrowth and illite formation. There is no relationship between facies and original feldspar content (Table 3).

Whilst some feldspar loss may have been associated with the initial burial event, it is also apparent that some has been lost more recently. Partly dissolved feldspars are seen coated and partially replaced by Fe-oxides and similar oxides coat quartz overgrowth surfaces. Thus feldspar loss may be associated with the extensive Tertiary to recent lateritic weathering of the area.

The diagenetic history of these rocks is relatively simple, and mainly concerns mobility of silica.

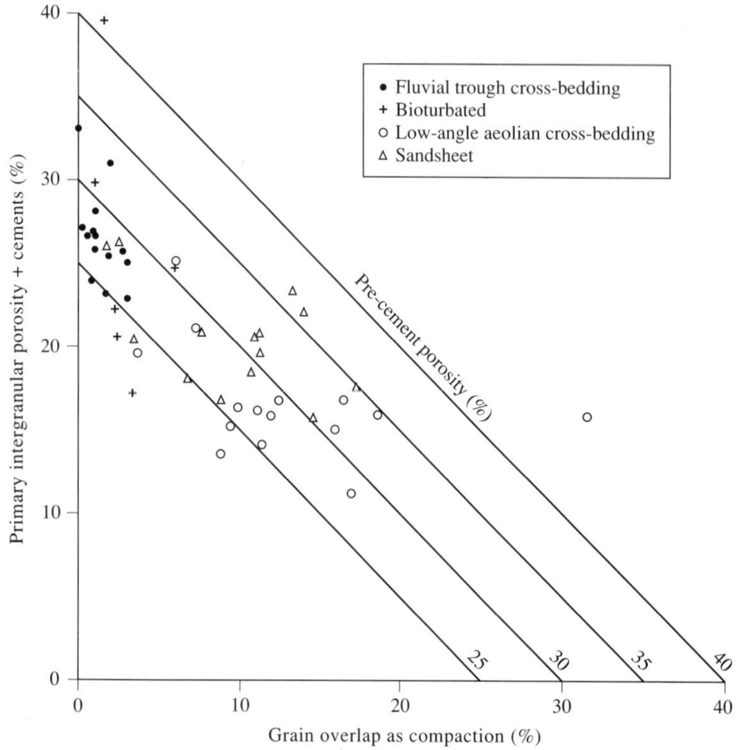


**Fig. 6.** Plot of primary porosity + cements against chemical compaction as expressed by grain overlap. Pre-cement porosities are similar for both fluvial and aeolian lithologies.

## POROSITY EVOLUTION

The question which arises from the basic petrographic observations and diagenetic interpretation relates to silica mobility. Is it possible that silica was conserved within the formation and that aeolian facies acted as a donor of silica for cementation of the fluvial facies? If this was the case, or even partly the case, silica dissolution in aeolian facies must occur at the same time as quartz overgrowth cementation of the fluvial facies. In order to test this hypothesis, a cross plot (Fig. 6) of primary porosity plus cements against chemical compaction was constructed. This can only, at best, be a rough guide, but the plot produces similar ranges and scatter of pre-cement porosity for both the aeolian and the fluvial facies. Thus it is reasonable to conclude that silica derived from the aeolian facies was redeposited in the fluvial facies as overgrowths, and that this process started when the average porosity of the fluvial sandstones was 27.5% (range 21–33%) and the aeolian 29.3% (range 22–34%). The similarity is remarkable, and probably reflects the well-sorted nature of the sands on deposition, giving them similar initial porosity values. This similarity in porosity values was retained until the onset of framework cementation following mechanical compaction. Some secondary porosity has been created by K-feldspar dissolution, but this was largely created after the illite and quartz overgrowths had formed, and in some cases appears to be due to leaching in the present uplift phase, during Miocene lateritization or when the sandstone was eroded in the Mesozoic before being covered by Cretaceous marine deposits. Hence, the secondary porosity observed is not considered to be connected with quartz cementation during the main burial phase.

## SOURCES OF SILICA FOR CEMENTATION

In a review of quartz cementation in sandstones, McBride (1989) listed 23 possible sources of silica for quartz cements previously proposed in the literature. In the present study many mechanisms relating to shales, organic matter and carbonate replacement can be eliminated on the basis of the field and stratigraphic evidence. The candidates which remain were all proposed over 30 years ago (references in McBride, 1989).

The possible sources are:

**1** Decomposition of feldspars.

**2** Silica from dissolution of aeolian quartz abrasion dust.

**3** Silica precipitated from descending meteoric water, $SiO_2$ derived from silicate minerals in the zone of weathering.
**4** Pressure solution at grain contacts.
**5** Pressure solution at stylolites.

It is not reasonable to derive all the silica required for cementation of the Tumblagooda Sandstone from grain-to-grain dissolution in the aeolian facies as described above. The aeolian rocks comprise, at most, only a third of the whole succession. The average value for quartz overgrowths in the fluvial facies is (17.7%) but the chemical compaction average 'donor potential' of the aeolian facies is only (12.1%). Thus another source of silica must be found. If the whole formation is considered as presently exposed some 30–50% of the quartz cement must be derived from other sources, but a balance could exist within FA2 (Fig. 2) if considered in isolation.

Oxygen isotopic values show that cementation was from meteoric fluids. This feature combined with the view that the sandstone rests on metamorphic basement, eliminates shale compaction fluids, or fluids migrating from a marine basin, and marine $SiO_2$-rich organisms as a source of $SiO_2$.

Feldspar and other silicate mineral dissolution in the subsurface zone of mechanical compaction prior to the rock acquiring a load-bearing frame could supply some of the silica shortfall. Whole-rock percentage averages for feldspar, and feldspar plus secondary porosity (which is considered due to feldspar dissolution) are very close for all four selected lithofacies (Table 3). This clearly shows that transfer of silica due to feldspar dissolution during chemical compaction is not a source of observed differences in quartz cements. Furthermore, feldspar dissolution post-dates both the formation of the load-bearing frame and deposition of illite and quartz overgrowths. Feldspar may have been leached at the time the sandstone was exposed prior to Mesozoic deposition, or in the present erosion cycle.

During both fluvial and aeolian transport attrition of grains produces abundant quartz chips of very fine grain size. Whalley *et al.* (1987) showed that experimental abrasion of quartz in an air stream initially produced coarse–medium silt as an abrasion product, but finer material was produced as rounding of grains increased. Some 60–80% of the abrasion product was 10–50 μm in size, and 300–1600 p.p.m. of water soluble silica was obtained from the abrasion detritus by water treatment.

Dust deposits in deserts tend to occur where the dust is washed into hollows and trapped at the vegetation margin (Pye & Tsoar, 1987), in the relatively wet Tumblagooda depositional environment, the abrasion dust would have concentrated in interdune ponds and on damp interdune surfaces, hence adding to the cementation potential of water reworked sands in such situations.

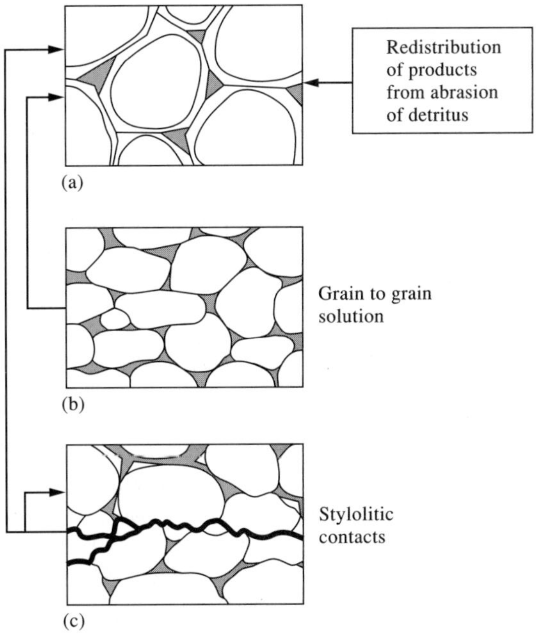


**Fig. 7.** Sources of cementing silica in fluvial sandstones: (a) derivation from aeolian sandstones; (b) derivation from stylolitic contacts; (c) derivation from redistribution of products of abrasion of detritus during transport.

Descending meteoric water might carry silica derived from dissolution of unstable silicates or from quartz abrasion dust. However, solubility of silica increases with increase in temperature, and silica deposition is unlikely by this method of introduction acting in isolation, but could provide pore waters close to silica saturation.

Another feature of the sandstone is the presence of stylolitic seams which are common at bedding planes but difficult to section and study. They are particularly common within the sandsheet facies. Since they are low amplitude stylolites following bedding and lamination it is impossible to assess their overall contribution of silica for cementation but it could be a significant quantity. One sample of aeolian sandstone with a stylolitized texture gave a compactional grain volume loss of 31.5% but it is not possible to quantify loss from low amplitude stylolites. Within quartz cemented fluvial sandstone beds, stylolites are not seen, but solution is probably associated with bed contacts and cannot be quantified. The possible sources of quartz cement for the sandstone are summarized in Fig. 7.

## TRANSFER OF SILICA WITHIN THE FORMATION

It is clear from textural considerations that silica has been lost from sections of dominantly aeolian facies tens of metres thick. Similarly, silica has been introduced to fluvial sandstones which are virtually continuous stratigraphically for up to 350 m. Local silica redistribution from grain contact area to adjacent porosity does not explain the observed textures. Giles & De Boer (1990) argued that products from dissolution during diagenesis are normally deposited in nearby pore space. This may be so when suitable substrates exist, but in the case of red-bed sandstones with coated grains, sites for quartz precipitation may be rare and silica can be carried to more distant sites, both by strong groundwater flow and vertical fluid movement due to compaction. Migration of silica over a minimum of tens of metres by fluid flow is required to satisfy the petrographic observations in the Tumblagooda Sandstone. The fact that thick (tens of metres) and thin (centimetre–decimetre) units or beds of sandstone appear to show (on field observation) no variation in the degree of cementation from their margins to centres, which might be associated with local diffusion gradients, supports the contention that large-scale fluid flow is implicated in the transfer of silica.

Silica transfer from aeolian to fluvial facies in the Tumblagooda Sandstone is difficult to constrain in terms of burial depth, temperature and timing. It is not possible to reconstruct a reliable burial curve for the Tumblagooda Sandstone due to lack of information in the area. The nearest significant borehole (Yaringa no. 1) is nearly 200 km to the north but contains *c.* 1500 m of Silurian–Devonian strata above the presumed Tumblagooda Sandstone and beneath the sub-Mesozoic unconformity (Hocking *et al.*, 1987). An offshore well (Edel. no. 1) 100 km to the north-west contains more than 2.5 km of Tumblagooda Sandstone. It is thus reasonable to suggest that the sandstone was buried to depths in excess of 2 km by the end of Devonian times. Application of the compaction curve for quartzose sand given by Robinson & Gluyas (1992) indicates that cementation possibly commenced at 2 km depth on the basis of pre-cement porosities. The observation that illite growth pre-dated cessation of quartz cementation implies a burial depth greater than 3 km and temperature of 100°C by the time quartz cementation was completed. These figures are based on an assumed 30°C $km^{-1}$ temperature gradient and illite formation at 100°C+, but considerable variation is possible from these frequently quoted figures as discussed by Chamley (1994). However, if the boxwork illite observed is a transformation of earlier illite–smectite, a shallower depth is possible. The oxygen isotope data (Table 1) vary from 14.2‰ to 17.6‰, with no apparent trend relating isotopic composition to the other parameters discussed. Assuming that the quartz precipitated in isotopic equilibrium with its parental fluid and that the quartz–water oxygen isotope fractionation equation of Matsuhisa *et al.* (1979) can be extrapolated to around 100°C, then the measured silica $\delta^{18}O$ values require a water $\delta^{18}O$ of less than 0‰ for all temperatures less than 125°C. This implies a strong meteoric water component, in agreement with the earlier suggestion of Trewin (1993b). Calculations by Robinson & Gluyas (1992) imply considerable flow rates (1–10 m $yr^{-1}$) maintained over 10 Myr to produce extensive quartz cementation, and they point out that such rates are generally characteristic of artesian flow through an aquifer connected to the surface. We know little of flow rates of meteoric water into basins at 2–3 km depth. However, evidence from quartz cementation as noted here, combined with observations of early meteoric diagenesis in deep water marine clastic sediments (Watson *et al.*, 1995) may indicate that meteoric water flow into basins with escape to the sea bed is more common than generally envisaged.

## CONCLUSIONS

This study demonstrates that large scale redistribution of silica can take place over distances in excess of 10 m within a sandstone formation. In this case aeolian sandstones have been silica donors and fluvial sandstones silica receptors.

Silica dissolution in aeolian sandstones was promoted by the presence of grain coatings which acted as catalytic films to give dissolution at grain contact points. The grain coatings also prevented the dissolved silica from being redeposited in the adjacent pore space of the aeolian sandstones since no suitable quartz substrate existed. However, clean abraded quartz grain surfaces in the fluvial sandstones acted as ready substrates for quartz cement.

There appears to be an excess, possibly 30–50%, of quartz cement above that supplied by simple transfer of quartz from chemical compaction at grain contacts. Other sources are from stylolites, which frequently modify bedding and lamination planes, and from the abrasion detritus produced during aeolian and fluvial transport.

The contribution of silica from K-feldspar dissolution could only supply a minor proportion of silica for cementation in the Tumblagooda Sandstone.

## ACKNOWLEDGEMENTS

NHT thanks the Carnegie Trust for the Universities of Scotland, and The Leverhulme Trust and the Royal Society for support for field work. Research facilities were kindly provided by the Western Australian Museum, and the Department of Conservation and Land Management allowed access to Kalbarri National Park. Overgrowth separation was by Brian Davidson. The drafting and secretarial services of Barry Fulton, Karen Chalmers and Emma Jones are gratefully acknowledged. The Isotope Geosciences Unit at SURRC is supported by the Consortium of Scottish Universities and by NERC. The authors also thank Mogens Ramm and Christoph Sp WP MultinationalA Romantl for interesting and constructive reviews of the submitted manuscript.

## REFERENCES

BJØRKUM, P.A. (1996) How important is pressure in causing dissolution of quartz in sandstones? *Journal of Sedimentary Research* **66**, 147–154.

BORTHWICK, J. & HARMON, R.S. (1982) A note regarding $ClF_3$ as an alternative to $BrF_5$ for oxygen isotope analysis. *Geochimica et Cosmochimica Acta* **46**, 1665–1668.

BRINT, J.F., HAMILTON, P.J., HASZELDINE, R.S., FALLICK, A.E. & BROWN, S. (1991) Oxygen isotopic analysis of diagenetic quartz overgrowths from the Brent sands: a comparison of two preparation methods. *Journal of Sedimentary Petrology* **61**, 527–533.

CHAMLEY, H. (1994) Clay mineral diagenesis. In: *Quantitative Diagenesis: Recent Developments and Applications to Reservoir Geology.* (eds PARKER, A. & SELLWOOD, B.W.) pp. 161–188. NATO ASI Series, Series C. Mathematical and Physical Sciences no. 453. Kluwer, Amsterdam

EHRENBERG, S.N. (1995) Measuring sandstone compaction from modal analyses of thin sections: how to do it and what the results mean. *Journal of Sedimentary Research A* **65**, 369–379.

GILES, M.R. & DE BOER, R.B. (1990). Origin and significance of redistributional secondary porosity. *Marine and Petroleum Geology* **7**, 378–397.

HOCKING, R.M. (1991) The Silurian Tumblagooda Sandstone, Western Australia. *Geological Survey of Western Australia Report* 27, 124pp.

HOCKING, R.M., MOORS, H.T. & VAN DE GRAAF, W.J.E. (1987) Geology of the Carnarvon Basin, Western Australia. *Geological Survey of Western Australia Bulletin* **133**.

HOUSEKNECHT, D.W. (1984) Influence of grain size and temperature on intergranular pressure solution, quartz cementation and porosity in a quartzose sandstone. *Journal of Sedimentary Petrology* **54**, 348–361.

HOUSEKNECHT, D.W. (1988) Intergranular pressure solution in four quartzose sandstones. *Journal of Sedimentary Petrology* **58**, 228–246.

LEE, M. & SAVIN, S.M. (1985) Isolation of diagenetic overgrowths on sand grains for oxygen isotopic analysis. *Geochimica et Cosmochimica Acta* **49**, 497–501.

MATSUHISA, Y., GOLDSMITH, J.R. & CLAYTON, R.N. (1979) Oxygen isotope fractionation in the system quartz–albite–anorthite–water. *Geochimica et Cosmochimica Acta* **43**, 1131–1140.

MCBRIDE, E.F. (1989) Quartz cement in sandstones: a review. *Earth Science Reviews* **26**, 69–112.

PYE, K. & TSOAR, H. (1987) The mechanics and geological implications of dust transport and deposition in deserts with particular reference to loess formation and dune sand diagenesis in the northern Negev, Israel. In: *Desert Sediments: Ancient and Modern.* (eds FROSTRICK, L. & REID, I.) pp. 139–156. Special Publications of the Geological Society of London 35.

ROBINSON, A.G. & GLUYAS, J.G. (1992) Duration of quartz cementation in sandstones, North Sea and Haltenbanken basin. *Marine and Petroleum Geology* **9**, 324–327.

SHANMUGAM, G. (1985) Significance of secondary porosity in interpreting sandstone composition. *American Association Petroleum Geologists Bulletin* **69**, 378–384.

TREWIN, N.H. (1993a) Mixed aeolian sandsheet and fluvial deposits in the Tumblagooda Sandstone, Western Australia. In: *Characterisation of Fluvial and Aeolian Reservoirs.* (eds NORTH, C.P. & PROSSER, D.J.) pp. 219–230. Special Publications of the Geological Society of London 73.

TREWIN, N.H. (1993b) Controls on fluvial deposition in mixed fluvial and aeolian facies within the Tumblagooda Sandstone (Late Silurian) of Western Australia. *Sedimentary Geology* **85**, 387–400.

TREWIN, N.H. & MCNAMARA, K.J. (1995) Arthropods invade the land: trace fossils and palaeoenvironments of the Tumblagooda Sandstone (?late Silurian) of Kalbarri, Western Australia. *Transactions of the Royal Society of Edinburgh: Earth Science* **85**, 177–210.

VAN DER PLAS, L. & TOBI, A.C. (1965) A chart for judging the reliability of point counting results. *American Journal of Science* **263**, 87–90.

VAN PANHUYS-SIGLER, M. & TREWIN, N.H. (1990) Authigenic sphene cement in Permian sandstones from Arran. *Scottish Journal of Geology* **26**, 139–144.

WATSON, R.S., TREWIN, N.H. & FALLICK, A.E. (1995) The formation of carbonate cements in the Forth and Balmoral fields; Northern North Sea; a case for biodegradation, carbonate cementation and oil leakage during early burial. In: *Characterisation of Deep Marine Clastic Systems.* (eds HARTLEY, A.J. & PROSSER, D.J.) pp. 175–198. Special Publications of the Geological Society of London 94.

WHALLEY, W.B., SMITH, B.J., MCALISTER, J.J. & EDWARDS, A.J. (1987) Aeolian abrasion of quartz particles and the production of silt-size fragments: preliminary results. In: *Desert Sediments: Ancient and Modern.* (eds FROSTICK, L. & REID, I.) pp. 129–138. Special Publications of the Geological Society of London 35.

*Spec. Publs int. Ass. Sediment.* (2000) **29**, 231–252

# Influence of uplift and magmatism on distribution of quartz and illite cementation: evidence from Siluro-Devonian sandstones of the Paraná Basin, Brazil

L. F. DE ROS[1], S. MORAD[2], C. BROMAN[3], P. DE CÉSERO[4] *and* D. GOMEZ-GRAS[5]

[1]*Universidade Federal do Rio Grande do Sul, Instituto de Geociências, Departamento de Mineralogia e Petrologia, Av. Bento Gonçalves, 9500, CEP 91501–970 Porto Alegre-RS, Brazil;*
[2]*Sedimentary Geology Research Group, Institute of Earth Sciences, Uppsala University, Norbyvägen 18 B, S–75236, Uppsala, Sweden;*
[3]*Institutionen för geologi och geokemi, Stockholms Universitet, Kungstensgatan 45, S-106 91 Stockholm, Sweden;*
[4]*Universidade do Estado do Rio de Janeiro, Departamento de Geologia/Geofísica, R. São Francisco Xavier, 524, Bloco 'A', Maracanã, 20 550–013, Rio de Janeiro, RJ, Brazil; and*
[5]*Department de Geologia, Universitat Autonoma de Barcelona, Edifici C (s), 08193-Bellaterra, Spain*

## ABSTRACT

Siluro-Devonian sandstones of the Furnas Formation in the intracratonic Paraná Basin, southern Brazil display heterogeneous distribution of quartz, kaolin and illite cementation and feldspar dissolution. Extensive dissolution and kaolinization of detrital feldspar, mica and clay pseudomatrix occurred under an eodiagenetic meteoric regime. The heterogeneous precipitation of quartz and illite is attributed to the circulation of hot fluids related to the uplift and magmatism along the regional Ponta Grossa Arch. This is supported by the high (up to 156°C) fluid inclusion homogenization temperatures in quartz overgrowths and by the intense kaolinite illitization in sandstones with maximum burial depths between 1200 and 2200 m, as well as by the high crystallinity and the K/Ar derived ages of illite that coincide with the timing of magmatism. The extensive destruction of detrital feldspar has generated diagenetic quartz arenites with poorly connected pore systems, and hence low reservoir potential. The heterogeneous patterns of diagenetic modifications displayed by the sandstones suggests that 'isochemical' diagenetic models of the system $SiO_2$–$Al_2O_3$–$K_2O$–$H_2O$ have limited validity for basins with complex burial histories and magmatic activity, and may apply only to some sandstones buried deeply in rapidly subsiding basins.

## INTRODUCTION

Quartz cementation is a major factor in porosity and permeability destruction, and hence reservoir-quality deterioration, during advanced burial diagenesis. Therefore, numerous authors have studied quartz cementation both in terms of sources and transportation mechanisms of silica as well as timing and temperature of precipitation (e.g. Haszeldine *et al.*, 1984; Leder & Park, 1986; Porter & James, 1986; Houseknecht, 1988; Dutton & Diggs, 1990; Girard & Deynoux, 1991; Morad *et al.*, 1991; Oelkers *et al.*, 1992; Gluyas *et al.*, 1993; Walderhaug, 1994a,b).

Most models of quartz diagenesis in sandstones are biased towards the large amounts of data available from oil-producing, rapidly subsiding basins, primarily the Gulf Coast and the North Sea. These models relate the diagenetic evolution of sandstones to a systematic evolution of formation-water chemistry and temperature during progressive burial, and generally consider quartz cementation to be important only below 2 or 3 km (e.g. Land *et al.*, 1987; McBride, 1989; Gluyas & Coleman, 1992; Bjørlykke & Egeberg, 1993; Walderhaug, 1994a,b).

Despite their homogeneous detrital composition and depositional facies, the Silurian–Devonian fluvial sandstones of the Furnas Formation of the intracratonic Paraná Basin (southern Brazil) are heterogeneously cemented by quartz and illite. This pattern contrasts with the diagenetic models developed for the Gulf Coast and North Sea basins. The aim of this study is to decipher the sources, timing and factors which control the distribution of quartz cement and associated diagenetic phases in the

**Fig. 1.** Location map of the Paraná Basin.

Furnas sandstones along the eastern portion of the Paraná Basin (Fig. 1).

In this paper, the terms *eo-*, *meso-* and *telodiagenesis* are applied for the diagenetic stages *sensu* Schmidt & McDonald (1979)), the term *kaolin* is applied to the polytypes dickite and kaolinite, and *illitic clays* refers to both pure illite and mixed-layer illite–smectite.

## SAMPLES AND ANALYTICAL METHODS

Epoxy-impregnated thin sections of 65 unweathered outcrop samples from the eastern border of the basin and 18 core samples from depths of 500–3900 m were examined with a standard petrographic microscope. The volumes of detrital and diagenetic components, and of different pore types, were determined by counting 300 points in each of 25 representative thin sections. The morphology and textural relationships among minerals were examined in 11 gold-coated samples with a JEOL JSM-T330 scanning electron microscope (SEM) at an accelerating voltage of 10 kV.

Thirteen thin-sections were carbon-coated and analysed with a Cameca Camebax SX50 electron microprobe

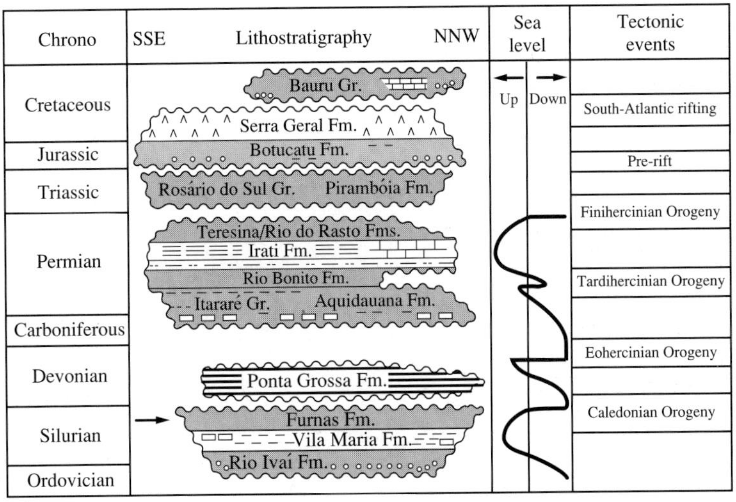


**Fig. 2.** Simplified stratigraphic column of the Paraná Basin; arrow indicates the stratigraphic position of the Furnas Formation (modified from Zalán *et al.*, 1990a).

(EMP) equipped with three spectrometers and a back-scattered electron detector (BSE). An acceleration voltage of 8 kV was used for the carbonates and 12 kV for the silicates, oxides and sulphides, at a beam current of 6 nA and 1–5 μm of diameter. The standards used were wollastonite (Ca and Si, 10 s), MgO (Mg, 10 s), $MnTiO_3$ (Mn, Ti, 10 s), haematite (Fe, 10 s), orthoclase (K, 5 s), albite (Na, 5 s), barite (S, Ba 10 s), strontianite (Sr, 5 s) and corundum (Al, 20 s).

Fifteen thin sections were also examined using a hot-cathode luminescence microscope (CL; Ramseyer *et al.*, 1989) at an acceleration voltage of 30 kV and a beam current density of 0.4 μA/mm$^2$.

Microthermometric analyses of the fluid inclusions were made on a Chaixmeca heating and freezing stage mounted on a modified microscope equipped with a 50/0.63 objective. The temperatures were measured using the procedure described in Shepherd *et al.* (1985). The stage was calibrated at regular intervals using pure standards. The accuracy of the temperatures below 0°C was ±0.5°C and ±1° for temperatures up to 200°C.

For the stable-isotope analysis of carbonate minerals, the bulk sample was powdered (< 200 mesh) and reacted with 100% phosphoric acid after 1 h at 25°C for calcite and after 6 days at 50°C for siderite. The evolved $CO_2$ gas was analysed for carbon and oxygen isotopes on a SIRA-12 mass spectrometer. The phosphoric acid fractionation factors used were 1.01025 for calcite at 25°C (Friedman & O'Neil, 1977) and 1.009082 for siderite (Rosenbaum & Sheppard, 1986). The carbonate isotope data are presented in the normal δ notation relative to PDB (Craig, 1957).

## GEOLOGICAL SETTING

The intracratonic Paraná Basin covers around 1 700 000 km$^2$ of southeastern South America, 1 100 000 km$^2$ of which are within Brazilian territory. During the Palaeozoic, prior to the Andean orogeny, the basin was connected with the Argentinean Paraná-Chaco Basin (Fig. 1). According to Zalán *et al.* (1990b), this wide intracratonic sag was formed due to the thermal subsidence of crust that accreted through multiple collisions during the Brasiliano–Panafrican orogeny that built the Gondwana supercontinent (Zalán *et al.*, 1990b). The basin was filled with more than 6000 m of sediments and volcanics, arranged in five major depositional sequences (Silurian, Devonian, Carboniferous–Permian, Triassic and Jurassic–Cretaceous; Fig. 2; Zalán *et al.*, 1990a,b), separated by basin-wide unconformities. The first four sequences are dominantly siliciclastic and the fifth comprises the most voluminous tholeiitic volcanic flows on Earth (Zalán *et al.*, 1990b). Associated dykes and sills of diabase are intruded in the pre-Cretaceous units.

The filling of the basin started during the Ordovician, and occurred in a large gulf facing the initially passive Pacific margin (Zalán *et al.*, 1987b). The Furnas Formation was dominantly deposited by a braided fluvial system, interpreted as the outwash of a periglacial fringe, during a regressive phase (Zalán *et al.*, 1987b). It is composed of white, cross-stratified, quartzose sandstones and conglomerates, arranged in amalgamated fining-up cycles. The formation is more than 260 m thick in the subsurface and more than 220 m in the eastern outcrop belt (Fig. 3). The Furnas sandstones overlie the Silurian

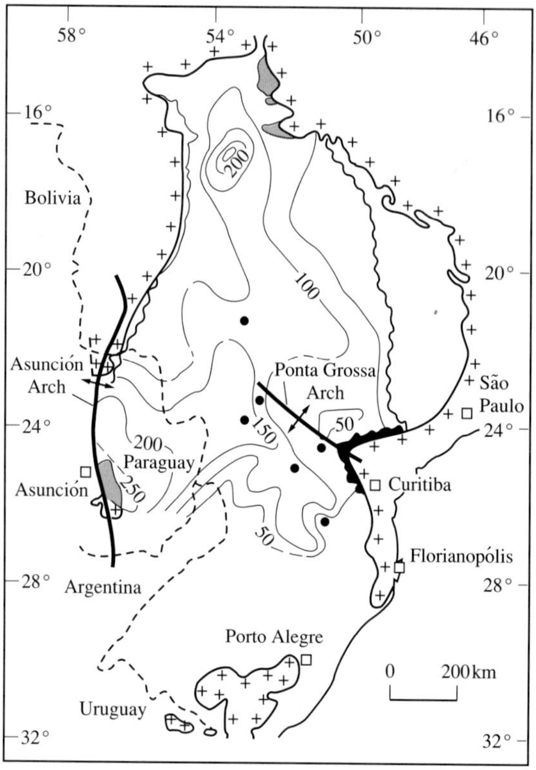


**Fig. 3.** Thickness of the Furnas Formation in the Paraná Basin (modified from Zalán *et al.*, 1987b). Eastern outcrop belt and sampled wells shown in black. Other outcrop areas in grey.

Vila Maria Formation, which is composed of periglacial sandstones, transgressive fossiliferous shales and diamictites. The contact of the Furnas with the overlying marine Devonian black shales of the Ponta Grossa Formation is considered gradational by Milani *et al.* (1994), Dino & Rodrigues (1995) and Dino *et al.* (1995), but is interpreted as abrupt, basin-wide and unconformable by Zalán *et al.* (1987b). The latter authors estimate a depositional hiatus of around 10 Myr between the two formations, and consider this unconformable contact to reflect a transition from a passive to a compressional Pacific margin (Zalán *et al.*, 1987a,b). The sandstones are extensively fractured in outcrop along NW–SE and NE–SW lines, and commonly are intruded by Jurassic diabase dykes and sills (Cruz, 1987). These structures are related to the Mesozoic uplift of the Ponta Grossa Arch (Sial *et al.*, 1987; Zalán *et al.*, 1990a).

The burial history of the Furnas Formation varies considerably in different parts of the basin, because of major differential vertical movements involved in subsidence of the depocentres and uplift of the regional arches, such as the Ponta Grossa Arch (Zalán *et al.*, 1990b). Three subsidence phases occurred at the basin centre, related to deposition of Silurian–Devonian and Permian–Carboniferous sequences and Jurassic–Cretaceous volcanics (Fig. 4). The Silurian and Permian sandstones in the basin centre contain hypersaline pore fluids (> 120 g/L tds), whereas dilute meteoric fluids dominate along the basin margins (Milani *et al.*, 1990). However, there are no detailed and up-to-date formation water analyses available.

The maximum burial depth of the Furnas Formation along the eastern outcrop belt is estimated from palinspastic reconstructions to be 1200–2200 m. The unit was raised to the surface as a result of the rise of the regional Ponta Grossa Arch (Zalán *et al.*, 1990b) that began during the Triassic, continued during the Jurassic to Early Cretaceous phase of maximum tectonism, and attained full development after the Serra Geral volcanism (Figs 3 & 5; Zalán *et al.*, 1990b). The uplift of the Ponta Grossa Arch was associated with the extrusion of huge amounts of tholeiitic lavas, and intrusion of dyke swarms with NW orientation (Fig. 3; Sial *et al.*, 1987; Zalán *et al.*, 1987a). The most likely uplift pathway of the unit along the eastern outcrop belt is shown in Fig. 6, which also depicts the burial evolution of an area with a marginal position, similar to that of the outcrop, but south of the uplifted Arch region. The exposure of the Silurian sequence probably occurred in Tertiary times and was related to the uplift of the coastal range, the Serra do Mar.

## DETRITAL TEXTURE AND COMPOSITION

The Furnas sandstones are mainly medium- to coarse-grained, with moderate to poor sorting, but locally have bimodal, very fine–medium or medium–[very]coarse grain size lamination. Some depositional cycles are topped by micaceous siltstones and mudstones, commonly eroded by the overlying cycles (cf. Cruz, 1987). Other cycles begin with a conglomeratic lag, or have conglomeratic intercalations. The original shape of the grains is difficult to define with the petrographic microscope, due to the precipitation of overgrowths and to intergranular pressure dissolution.

The sandstones are dominantly quartz arenites and quartzose subarkoses (Fig. 7). The quartz grains (23–62 bulk vol.%; average 50%; Table 1) are mainly monocrystalline (av. ≈40%). Polycrystalline grains are locally abundant (av. 10%; up to 27%). The feldspars (0–8%; av. 1%; Table 1) are mostly potassic, and extensively dissolved and replaced by kaolin, illite and albite.

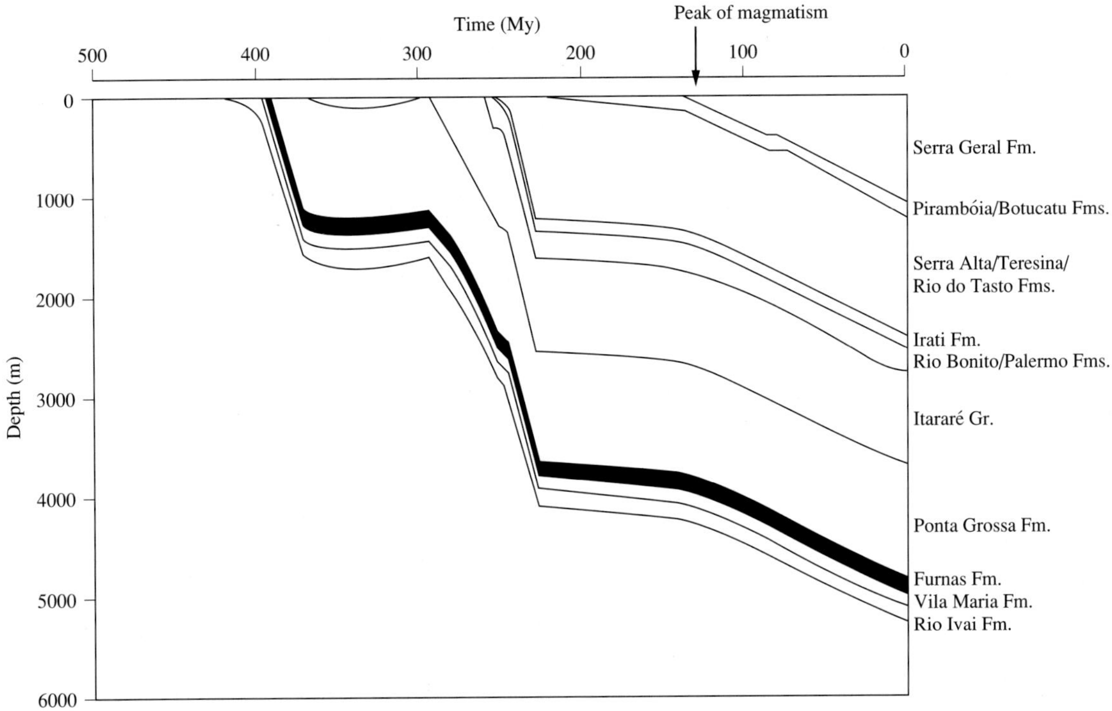


**Fig. 4.** Typical burial history of the Furnas Formation in the centre of the basin (modified from Milani *et al.*, 1990).

Microcline dominates over orthoclase/sanidine and perthite. The rock fragments (0–3%; av. 1%; Table 1) are essentially micaceous low-grade metamorphic, with schist and phyllite more common than sericitic meta-sandstone. Gneissic and granitic rock fragments, as well as hydrothermal quartz with abundant tourmaline, chlorite and fluid inclusions are significant in the conglomerates.

Micas are the dominant accessory grains (0–10%; av. 1%; Table 1), extensively kaolinized and concentrated in the fine sandstones and siltstones at the tops of the fluvial cycles. Tourmaline and rutile are the dominant heavy minerals. Remnants of the original, less stable heavy mineral assemblage, including epidote, sphene, apatite and Fe–Ti oxides are preserved in the micaceous fine sandstones and siltstones deposited in the cycle tops. Mud intraclasts eroded from cycle-top and interchannel fluvial plain sediments (original average ≈7%) were extensively compacted to pseudomatrix, kaolinized and illitized.

## DIAGENETIC CONSTITUENTS

### Quartz

Quartz cement occurs as overgrowths, outgrowths, and microfracture-healing of detrital quartz. Due to the absence of oxides, clay minerals, or fluid inclusions around the quartz grains, the overgrowths are difficult to recognize through conventional petrography. Therefore, cathodoluminescence (CL) imaging was used to quantify quartz cement (average 14.7%) and intergranular volume (IGV average 22.8%; Table 1) and to evaluate pressure dissolution. Packing measured from CL images varies from normal to tight (packing proximity index ≈50–70%; Kahn, 1956). CL examination and modal analysis revealed that quartz cementation is heterogeneously distributed, ranging from 1 to 26.7% (Table 1) even for texturally very similar samples, taken sometimes from the same layer and only a few metres apart. Overgrowths fill intergranular pores either in moderately compacted sandstones (Fig. 8a), or in sandstones that have suffered substantial compaction and pressure dissolution (Fig. 8b). The overgrowths display dull-brown luminescence and complex zonation (Fig. 8c). In a few samples, the overgrowths outline mouldic pores after dissolved feldspars. Thick overgrowths and outgrowths engulf remnants of dissolved feldspar, kaolin and illite (Fig. 8d,e), and fill large dissolution pores. CL images further reveal that authigenic quartz commonly heals fractured quartz grains (Fig. 8f; cf. Morad *et al.*, 1991; Milliken, 1994), and

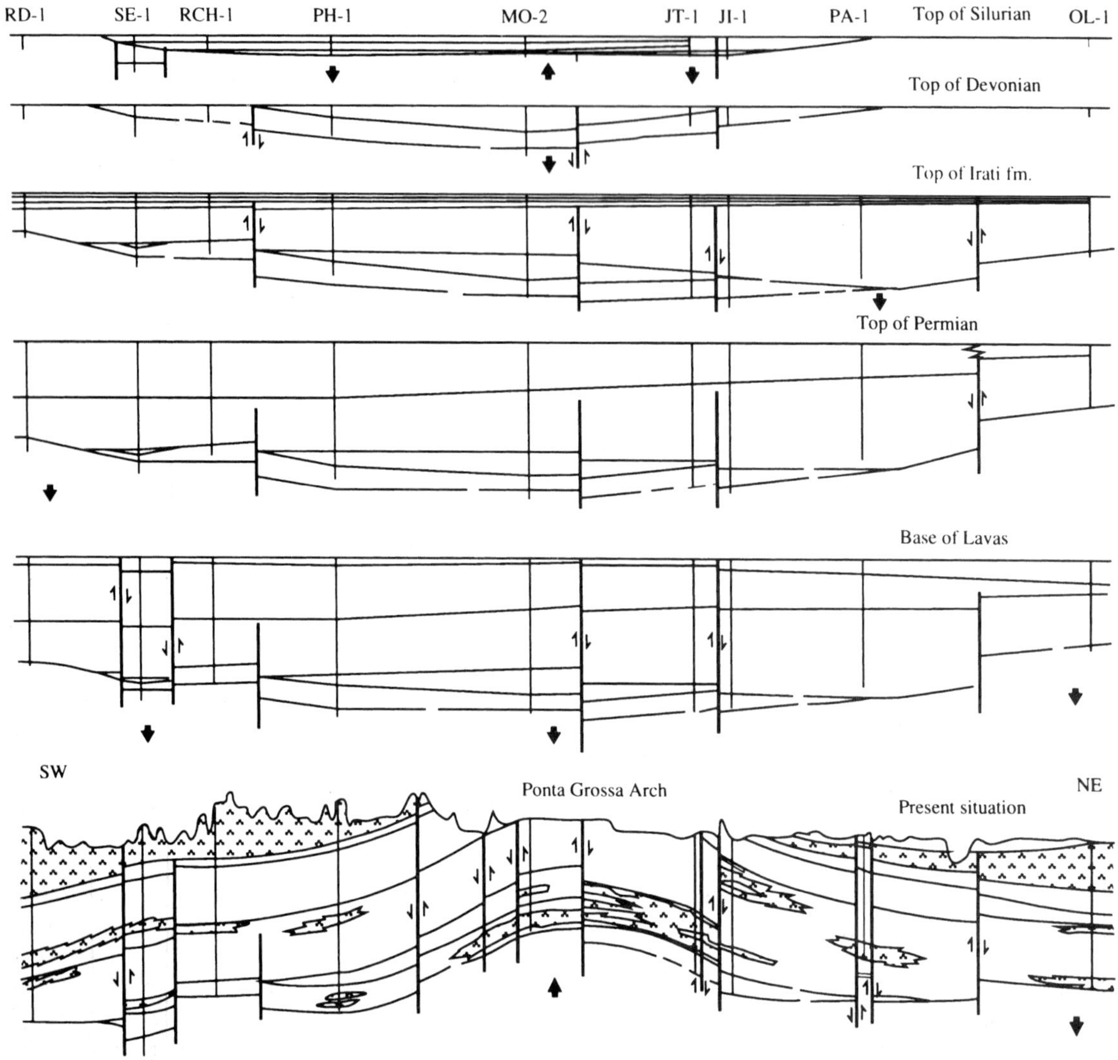


**Fig. 5.** Sections showing the structural evolution and maximum burial of the Silurian sequence during Late Permian times, followed by the uplift of the Ponta Grossa Arch in the Mesozoic with intense faulting, diabase intrusion and basalt extrusion (shown patterned; cf. Zalán *et al.*, 1990b).

locally replaces detrital quartz along the fractures, and along the contacts between overgrowths and host grains, as indicated by the corroded fracture margins and grains borders (Fig. 8c,f).

Prismatic quartz outgrowths extend into mouldic, intra- and intergranular pores (0–4%; av. 2%; Fig. 9a,b). Well-developed multiple overgrowths and outgrowths engulf illite and dickite which in turn cover earlier overgrowths (Figs 8d,e & 9b), suggesting the presence of at least two main quartz cement generations. The scarcity and diminutive size of fluid inclusions in the quartz overgrowths rendered the acquisition of homogenization temperatures difficult. The overgrowths in outcrop samples yielded a homogenization temperature range from 11° to 156°C and a salinity range between 16 and 28 eq. wt.% NaCl. In subsurface samples (1595.2–3812.1 m), Th ranges from 83 to 110°C and salinity between 16 and 19 eq. wt.% NaCl.

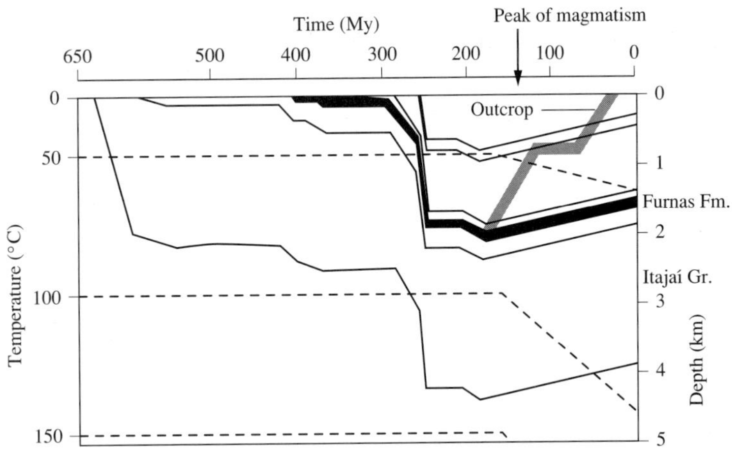


**Fig. 6.** Burial history diagram of the Furnas Formation south of the outcrop belt, suggesting possible hydrocarbon generation from Vendian shales (cf. Meister *et al.*, 1989), and illustrating the probable burial and uplift history of the outcropping sandstones (grey).

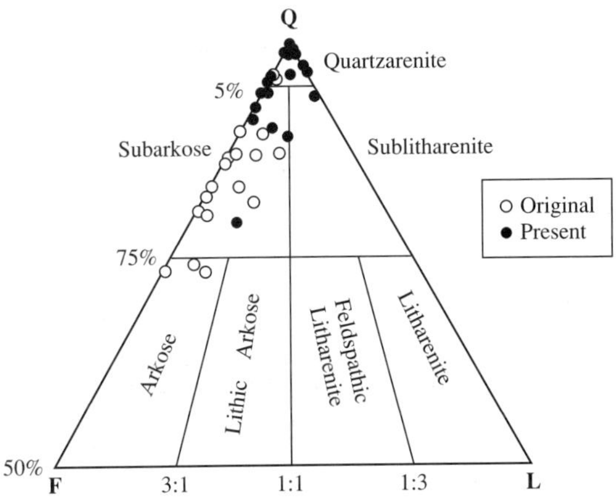


**Fig. 7.** Present and reconstructed detrital composition of 25 representative Furnas sandstones (plotted in the upper half of Folk, 1968 diagram).

### Authigenic clay minerals

Authigenic clay minerals include kaolin, illite and trace chlorite. Kaolin occurs both as kaolinite and dickite, as indicated by XRD and SEM analyses (cf. Ehrenberg *et al.*, 1993; Morad *et al.*, 1994). Kaolin is abundant (0–44 vol.%; av. 13%) in outcrop samples and in subsurface samples from moderate depths (< 2200 m; Ramos & Formoso, 1975), but is absent towards the deep centre of the basin.

Kaolinite occurs mostly as replacement of mica (0–19%; av. 3%; Table 1), feldspar (0–10%; av. 3%) and pseudomatrix (0–10%; av. 2%). Intergranular kaolinite is less abundant (0–9%; av. 2%). Kaolinized micas show pseudomorphic replacement of the lamellae and additional displacive kaolinite precipitation between the cleavage planes, resulting in characteristic fan-shaped flakes (Fig. 9c), and in large expanded 'accordions' (Fig. 9d). The loosely expanded aspect of most kaolinized micas and their close association with intergranular kaolinite cement indicate an *in situ* diagenetic origin.

Kaolinized pseudomatrix is riddled with dark inclusions of unreplaced detrital mixed-layer illite–smectite, haematite and Ti oxides, and shows a smaller crystal size (0.5–5 μm) than intergranular or feldspar-replacive kaolinite. Kaolin which replaced feldspar occurs mostly as tightly packed booklets or vermicules of platy crystals (5–40 μm). Intergranular kaolinite cement appears as loose, microporous booklets and vermicules (3–25 μm across; Fig. 8d). In most outcrop samples, secondary pores derived from feldspar dissolution contain less kaolin than adjacent intergranular pores which are filled either by kaolinite cement, kaolinized pseudomatrix, or illite pseudomorphically replacing both kaolin types (Fig. 8d).

Dickite occurs only in the subsurface samples of moderate present depth (≈1500 m to 2200 m), as blocky crystals (5–20 μm) replacing kaolinite, and covering or being engulfed by quartz overgrowths (Fig. 9e). Dickite occurs as vermicular aggregates containing etched remnants of kaolinite (Fig. 9E). Such textural features are indicative of kaolinite transformation into dickite via small scale dissolution–reprecipitation at temperatures of 80–130°C (cf. Ehrenberg *et al.*, 1993; McAulay *et al.*, 1993; Morad *et al.*, 1994).

Illite (0–42%; av. 7%; Table 1) occurs mostly as aggregates of lamellar crystals which replaced vermicular

**Table 1.** Modal and average composition and texture of 25 representative Furnas sandstones. Minima and maxima of the constituents are also shown

| Sample | FUR-7 | 8SD-27-1 | 8SD-25-2 | FUR-40 | P.G. 3 | S-22/74 | S-12/68 | P.G. 10 | 8SD-24 | 8SD-6/1A | 12SD-2/4 | Zn201 | M-04 | M-22 |
|---|---|---|---|---|---|---|---|---|---|---|---|---|---|---|
| **Detrital quartz** | **47.7** | **50.7** | **53.7** | **55.7** | **54** | **56** | **36.7** | **58.3** | **62** | **56.3** | **48.3** | **36.3** | **53** | **54.3** |
| Quartz monocrystalline | 39.3 | 29.3 | 35.3 | 38.7 | 38.3 | 43 | 33.7 | 46 | 33 | 47 | 31.3 | 34 | 42.3 | 44 |
| Quartz polycrystalline | 8 | 21.3 | 16 | 15.7 | 12.7 | 11 | 3 | 12.3 | 26.7 | 9.3 | 17 | 2.3 | 10.7 | 10 |
| Quartz in plutonic r.f. | 0.3 | — | 2.3 | 1.3 | 3 | 2 | — | — | 2.3 | — | — | — | — | 0.3 |
| **Detrital feldspar** | **0** | **0** | **0** | **3** | **1** | **0.3** | **0** | **3.3** | **0** | **0.7** | **0.3** | **0** | **0** | **0.3** |
| Orthoclase | — | — | — | 0.7 | 0.3 | — | — | 0.7 | — | 0.7 | 0.3 | — | — | 0.3 |
| Microcline | — | — | — | 1.7 | 0.7 | 0.3 | — | 2.3 | — | — | — | — | — | — |
| Perthite | — | — | — | 0.7 | — | — | — | 0.3 | — | — | — | — | — | — |
| K-feldspar in plutonic r.f. | — | — | — | — | — | — | — | — | — | — | — | — | — | — |
| **Plutonic rock fragments** | **0.3** | **0.3** | **2.3** | **1.3** | **3** | **2** | **0** | **0** | **2.3** | **0** | **0** | **0** | **0** | **0.3** |
| **Total fine lithics** | **0.3** | **1.3** | **0** | **0** | **1.7** | **0** | **1.3** | **0.3** | **0** | **0** | **3** | **0** | **0** | **0** |
| Micaceous metamorphic r.f. | 0.3 | 1 | — | — | 1.3 | — | 1.3 | 0.3 | — | — | 3 | — | — | — |
| Meta-sandstone rf. | — | — | — | — | — | — | — | — | — | — | — | — | — | — |
| Chert r.f. | — | 0.3 | — | — | 0.3 | — | — | — | — | — | — | — | — | — |
| Muscovite | 0.3 | 3.7 | — | 0.3 | 0.3 | 0.7 | 1 | 0.3 | — | 0.3 | 2 | 3 | 1.3 | 0.3 |
| Biotite | — | — | — | — | — | — | — | — | — | — | — | — | — | — |
| Mica in plutonic r.f. | — | 0.3 | — | — | — | — | — | — | — | — | — | — | — | — |
| Tourmaline | 0.3 | — | — | — | — | — | 1.3 | 0.3 | — | 0.3 | 0.7 | 1 | — | 0.3 |
| Rutile | 1 | — | — | — | 0.3 | — | — | — | — | — | — | 0.3 | — | — |
| Opaques | — | — | — | — | — | — | 0.3 | — | — | — | — | — | — | — |
| Other heavy minerals | 0.3 | — | — | — | — | — | 0.3 | — | — | — | — | — | — | 0.7 |
| Pseudomatrix | — | — | — | 1.7 | — | — | — | — | — | — | 4.3 | — | 5 | 0.3 |
| **Diagenetics total** | **32.3** | **33.7** | **38** | **24** | **30** | **35.3** | **59** | **24.3** | **34** | **36.3** | **39.7** | **44.7** | **36.3** | **37** |
| Clay coats | — | — | — | — | 0.3 | 1 | — | — | 0.3 | — | — | — | 0.7 | — |
| Iron oxides | — | 0.3 | — | 0.3 | 1 | 1 | 0.3 | 0.7 | — | — | — | 0.3 | — | — |
| Quartz overgrowth | 15.3 | 10 | 5 | 10.3 | 9.3 | 14.3 | 11.7 | 11.7 | 4.7 | 13.7 | 9 | 13.7 | 13.7 | 18.7 |
| Quartz discrete | 0.7 | 2.7 | 1.3 | 1.7 | 1.7 | 2 | 0.7 | 0.7 | 0.7 | — | 0.7 | 4 | 1 | 2 |
| Kaolin intergranular | 0.3 | 4.3 | — | 2 | 4.3 | 3.7 | 8.7 | 3.7 | — | 5.3 | 2 | 7.3 | — | 3.7 |
| Kaolin replacing detrital feldspar | — | — | — | 3 | 1.3 | 3.3 | — | 3 | — | 3.3 | 0.7 | — | — | 0.7 |
| Kaolin replacing unidentified grain | 1.3 | 4.7 | — | 1 | 0.7 | 5.3 | 11.7 | 1 | — | 8.3 | 7 | 4.3 | 0.7 | 5 |
| Kaolin replacing mica | 1.3 | 5.7 | — | 2 | 2 | 3.3 | 19 | 2.3 | — | 4 | 10.7 | 3.7 | 4.3 | 4.7 |
| Kaolin replacing pseudomatrix | 8 | 5 | 0.7 | 2.7 | 6 | — | 4.3 | 0.3 | — | 1 | 9.7 | 1.3 | 3.3 | 1.3 |
| Kaolin oversized patch | — | — | — | — | — | — | — | 1 | — | — | — | 1.3 | — | — |
| Kaolin in clay coats | — | — | — | — | — | 1 | — | — | — | — | — | — | — | — |
| Illite intergranular cement | — | — | 8.3 | — | 0.7 | — | — | — | 2.3 | — | — | 0.7 | 3 | — |
| Illite replacing feldspar | — | — | 10.3 | — | — | — | — | — | 9.7 | — | — | 0.3 | 1.3 | — |
| Illite replacing mica | — | — | — | — | — | — | — | — | — | — | — | 1.3 | — | — |

| | | | | | | | | | | | | | | |
|---|---|---|---|---|---|---|---|---|---|---|---|---|---|---|
| Illite replacing pseudomatrix | — | — | 11 | — | — | — | — | — | 11.3 | — | — | — | 4 | — |
| Illite replacing kaolin | 0.7 | 0.7 | 1 | 1 | 1.7 | — | — | — | 3 | — | — | 5 | 3.3 | — |
| Chlorite | — | — | — | — | — | — | — | — | — | — | — | — | — | — |
| Rutile/anatase intergranular | 1.7 | 0.3 | 0.3 | — | 0.3 | 0.3 | 1.7 | — | 0.7 | 0.7 | — | 1 | — | 0.3 |
| Rutile/anatase replacing grain | 1.3 | — | — | — | — | — | 0.7 | — | 1.3 | — | — | 0.3 | 0.3 | 0.7 |
| Opaque oxides | 1.7 | — | — | — | 0.7 | — | 0.3 | — | — | — | — | — | 0.7 | — |
| Albite replacing feldspar grain | — | — | — | — | — | — | — | — | — | — | — | — | — | — |
| Siderite replacing feldspar grain | — | — | — | — | — | — | — | — | — | — | — | — | — | — |
| Siderite intergranular | — | — | — | — | — | — | — | — | — | — | — | — | — | — |
| **Macroporosity total** | **17.7** | **10.3** | **8.3** | **15.3** | **12.7** | **7.7** | **0** | **13** | **4** | **6** | **1.7** | **14.7** | **4.3** | **6.7** |
| Intergranular | 5 | 2.3 | 2 | 3.7 | 2.3 | 4 | — | 4.3 | — | 2 | 0.7 | 3 | 1.7 | 5 |
| Intragranular in feldspar | — | — | — | 5.3 | 4 | 1.7 | — | 6 | 0.7 | 1.3 | — | — | — | 0.3 |
| Intragranular in mica | — | 1 | — | — | — | — | — | — | — | 2.3 | — | 0.7 | — | 0.3 |
| Intragranular in plutonic r.f. | — | — | — | 1.3 | 0.7 | — | — | — | — | — | — | — | — | 0 |
| Dissolution of pseudomatrix | — | — | — | — | — | — | — | — | — | — | — | — | — | 0 |
| Mouldic | 7 | 3 | 2 | 3 | 3 | 0.3 | — | 2 | — | — | — | 6 | — | 0 |
| Lamellar decompaction | 2.3 | — | 1.7 | 1 | 1 | 1.7 | — | 0.3 | 1.3 | 0.3 | — | 3.7 | — | 0 |
| Fracture | 0.3 | 0.3 | 1 | — | 0.7 | — | — | 0.3 | 0.7 | — | — | — | — | 0 |
| Shrinkage | 0.3 | 3 | 1.7 | 1 | 0.7 | — | — | — | 1.3 | — | 1 | 0.7 | 2.7 | 0.3 |
| Oversized | 2.7 | 0.7 | — | — | 0.3 | — | — | — | — | — | — | 0.7 | — | 0.7 |
| **Microporosity** | **Abundant** | **Abundant** | **Abundant** | **Common** | **Abundant** | **Abundant** | **Abundant** | **Abundant** | **Abundant** | **Abundant** | **Abundant** | **Abundant** | **Abundant** | **Abundant** |
| Granulometry | Coarse | Coarse | Coarse | Coarse | Coarse | Medium | Fine | Medium | Coarse | Very coarse | Very coarse | Coarse | Coarse | Coarse |
| Sorting | Poor | Poor | Poor | Moderate | Poor | Moderate | Poor | Moderate | Poor | Moderate | Poor | Good | Poor | Good |
| **Intergranular volume** | **27** | **20** | **18.7** | **19** | **21.7** | **29** | **23.3** | **21.3** | **10** | **22** | **12.3** | **33.7** | **20.7** | **29.7** |
| **Cement volume** | **19.7** | **17.7** | **15** | **14.3** | **18.3** | **22.3** | **23.3** | **16.7** | **8.7** | **19.7** | **11.7** | **27** | **19** | **24.7** |
| Authigenic quartz total | 16 | 12.7 | 6.3 | 12 | 11 | 16.3 | 12.3 | 12.3 | 5.3 | 13.7 | 9.7 | 17.7 | 14.7 | 20.7 |
| Grain volume original | 73 | 80 | 81.3 | 81 | 78.3 | 71 | 76.7 | 78.7 | 90 | 78 | 87.7 | 66.3 | 79.3 | 70.3 |
| Grain replacing/dissolution total | 22.7 | 23.7 | 26.7 | 20.3 | 20.3 | 14 | 35.7 | 15.7 | 27.3 | 20.3 | 29 | 25.7 | 20 | 14 |
| Feldspar dissolution total | 11.3 | 12.7 | 12.3 | 14.3 | 13.7 | 14.3 | 20.3 | 16.7 | 10.3 | 18.3 | 9.7 | 20 | 2 | 10.3 |
| Kaolin total | 11.7 | 20.3 | 1.7 | 11.7 | 16 | 16.7 | 43.7 | 11.3 | 3 | 22 | 30 | 23 | 11.7 | 15.3 |
| Grain-replacive kaolin | 10.7 | 15.3 | 0.7 | 8.7 | 10 | 12 | 35 | 6.7 | 0 | 16.7 | 28 | 9.3 | 8.3 | 11.7 |
| Grain-replacive clays | 11.3 | 16 | 23 | 9.7 | 11.7 | 12 | 35 | 7.7 | 24 | 16.7 | 28 | 17.3 | 17 | 11.7 |
| Illite total | 0.7 | 0.7 | 30.7 | 1 | 2.3 | 0 | 0 | 0 | 26.3 | 0 | 0 | 7.3 | 11.7 | 0 |
| Q | 98.6 | 97.4 | 95.7 | 92.6 | 90 | 95.9 | 96.5 | 94.1 | 96.2 | 98.8 | 93.6 | 100 | 100 | 98.8 |
| F | 0.7 | 0 | 4.4 | 7.4 | 7.1 | 4.1 | 0 | 5.4 | 3.8 | 1.2 | 0.7 | 0 | 0 | 1.2 |
| L | 0.7 | 2.6 | 0 | 0 | 2.9 | 0 | 3.5 | 0.5 | 0 | 0 | 5.8 | 0 | 0 | 0 |
| Qo | 96 | 89.4 | 80.2 | 86.7 | 86.9 | 83.1 | 73.8 | 87.1 | 83.3 | 82 | 81.5 | 85.8 | 96.4 | 89.5 |
| Fo | 3.4 | 8.2 | 19.8 | 13.3 | 10.2 | 16.9 | 23.5 | 12.4 | 16.7 | 18 | 13.5 | 14.2 | 3.6 | 10.5 |
| Lo | 0.7 | 2.4 | 0 | 0 | 2.8 | 0 | 2.7 | 0.5 | 0 | 0 | 5.1 | 0 | 0 | 0 |

*Continued p. 240*

**Table 1.** (*continued*)

| Sample | M-32 | M-41 | M-44 | M-45 | M-46 | M-52 | M-59 | CMI 1595.2 | MA 1822.8 | CSI 3811.6 | CSI 3812.1 | Average | Minimum | Maximum |
|---|---|---|---|---|---|---|---|---|---|---|---|---|---|---|
| **Detrital quartz** | **39.7** | **22.7** | **62.3** | **40.3** | **52** | **57** | **48.3** | **48.3** | **55.7** | **49.7** | **52.7** | **50.1** | **22.7** | **62.3** |
| Quartz monocrystalline | 33.3 | 21 | 41.7 | 36.7 | 42.7 | 49.7 | 42.3 | 43 | 53.7 | 42.3 | 46.7 | 39.5 | 21 | 53.7 |
| Quartz polycrystalline | 6.3 | 1.7 | 20 | 3.7 | 9.3 | 4 | 6 | 5.3 | 2 | 7.3 | 6 | 9.9 | 1.7 | 26.7 |
| Quartz in plutonic r.f. | — | — | 0.7 | — | — | 3.3 | — | — | — | — | — | 0.6 | 0 | 3.3 |
| **Detrital feldspar** | **8.3** | **0** | **5** | **2.7** | **1** | **0** | **2.3** | **0** | **0** | **0** | **0** | **1.1** | **0** | **8.3** |
| Orthoclase | 3.7 | — | 1.3 | 1.3 | 0.3 | — | 0.3 | — | — | — | — | 0.4 | 0 | 3.7 |
| Microcline | 4 | — | 3.3 | 1.3 | 0.7 | — | 2 | — | — | — | — | 0.7 | 0 | 4 |
| Perthite | 0.7 | — | — | — | — | — | — | — | — | — | — | 0.1 | 0 | 0.7 |
| K-feldspar in plutonic r.f. | — | — | 0.3 | — | — | — | — | — | — | — | — | 0 | 0 | 0.3 |
| **Plutonic rock fragments** | **0** | **0** | **1** | **0** | **0** | **3.3** | **0** | **0** | **0** | **0** | **0** | **0.7** | **0** | **3.3** |
| **Total fine lithics** | **2.3** | **0** | **0.3** | **2.3** | **1** | **0** | **0** | **0** | **0.3** | **0.7** | **1.7** | **0.7** | **0** | **3** |
| Micaceous metamorphic r.f. | 1.3 | — | — | 2.3 | 0.7 | — | — | — | 0.3 | 0.7 | 1.3 | 0.6 | 0 | 3 |
| Meta-sandstone r.f. | 1 | — | — | — | 0.3 | — | — | — | — | — | 0.3 | 0.1 | 0 | 1 |
| Chert r.f. | — | — | 0.3 | — | — | — | — | — | — | — | — | 0 | 0 | 0.3 |
| Muscovite | 1.7 | 7 | — | — | — | 1 | — | — | — | — | — | 0.9 | 0 | 7 |
| Biotite | — | 2.7 | — | — | — | — | — | — | — | — | — | 0.1 | 0 | 2.7 |
| Mica in plutonic r.f. | — | — | — | — | — | — | — | — | — |  | — | 0 | 0 | 0.3 |
| Tourmaline | — | 0.3 | — | 0.3 | — | — | 0.3 | 0.3 | — | — | — | 0.2 | 0 | 1.3 |
| Rutile | 0.3 | 0.3 | — | 0.7 | — | — | — | — | — | — | — | 0.1 | 0 | 1 |
| Opaques | — | 0.3 | — | — | — | — | — | — | — | 0.3 | — | 0 | 0 | 0.3 |
| Other heavy minerals | — | 0.3 | — | — | — | — | — | — | — | — | — | 0.1 | 0 | 0.7 |
| Pseudomatrix | 3.3 | — | 5 | — | 6 | 2.3 | 1.7 | — | — | 0.7 | 2.7 | 1.3 | 0 | 6 |
| **Diagenetics total** | **35** | **55.3** | **15.3** | **40.7** | **27.3** | **24.3** | **34.3** | **49** | **33.3** | **45.7** | **41.3** | **36.3** | **15.3** | **59** |
| Clay coats | — | — | — | — | — | — | — | 0.3 | 0.3 | — | — | 0.1 | 0 | 1 |
| Iron oxides | 0.3 | 1.7 | 2 | — | 1 | — | 0.3 | — | — | — | 0.3 | 0.4 | 0 | 2 |
| Quartz overgrowth | 7 | 0.7 | 4.7 | 14.7 | 13.3 | 11 | 23.7 | 23.3 | 22.7 | 19.7 | 21.7 | 12.9 | 0.7 | 23.7 |
| Quartz discrete | 1 | 0.3 | 2.3 | 2.3 | 1.3 | 2 | 3 | 2 | 1.3 | 4.7 | 4 | 1.8 | 0 | 4 |
| Kaolin intergranular | 1.7 | — | — | — | — | — | 1.3 | 3 | — | — | — | 2.1 | 0 | 8.7 |
| Kaolin replacing detrital feldspar | 1.3 | — | — | — | — | — | 0.7 | 7.7 | — | — | — | 1 | 0 | 7.7 |
| Kaolin replacing unidentified grain | 1 | 3.7 | — | 1 | — | — | — | — | 2.3 | — | — | 2.4 | 0 | 11.7 |
| Kaolin replacing mica | 1.3 | 5.3 | 1.3 | 2.7 | — | 3 | 3 | 3 | — | — | — | 3.3 | 0 | 19 |
| Kaolin replacing pseudomatrix | 2.3 | — | 4 | 1.7 | 2.3 | 3.7 | 2.3 | 7 | — | — | — | 2.7 | 0 | 9.7 |
| Kaolin oversized patch | 0.3 | — | — | — | — | — | — | 0.3 | — | — | — | 0.1 | 0 | 1.3 |
| Kaolin in clay coats | — | — | — | — | — | — | — | — | — | — | — | 0 | 0 | 1 |
| Illite intergranular cement | 1.7 | — | — | 2.3 | 2 | 1.3 | — | — | 2.7 | 2.7 | 2.3 | 1.2 | 0 | 8.3 |
| Illite replacing feldspar | 1.3 | 4.7 | — | — | — | — | — | — | 1 | 0.7 | — | 1.2 | 0 | 10.3 |
| Illite replacing mica | 1 | 6.3 | — | — | — | — | — | — | 0.7 | 1 | 1.3 | 0.5 | 0 | 6.3 |
| Illite replacing pseudomatrix | 6.7 | 30.7 | 1 | 9 | 1.7 | 3 | — | — | 1.3 | 3 | 1 | 3.4 | 0 | 30.7 |

| | | | | | | | | | | | | | | |
|---|---|---|---|---|---|---|---|---|---|---|---|---|---|---|
| Illite replacing kaolin | 0.7 | — | — | 4.3 | 3.7 | — | — | 1.3 | 0.7 | — | — | 1.1 | 0 | 5 |
| Chlorite | — | — | — | — | — | — | — | — | — | 4 | 2 | 0.2 | 0 | 4 |
| Rutile/anatase intergranular | 1 | 1 | — | 1 | — | — | — | 1 | 0.3 | 0.3 | — | 0.5 | 0 | 1.7 |
| Rutile/anatase replacing grain | 2.3 | 0.3 | — | 1 | — | — | — | — | — | 0.7 | 0.3 | 0.4 | 0 | 2.3 |
| Opaque oxides | 4 | 0.7 | — | 0.7 | 2 | 0.3 | — | — | — | 0.7 | — | 0.5 | 0 | 4 |
| Albite replacing feldspar grain | — | — | — | — | — | — | — | — | — | 7 | 6 | 0.5 | 0 | 7 |
| Siderite replacing feldspar grain | — | — | — | — | — | — | — | — | — | 0.3 | 1 | 0.1 | 0 | 1 |
| Siderite intergranular | — | — | — | — | — | — | — | — | — | 1 | 1.3 | 0.1 | 0 | 1.3 |
| **Macroporosity total** | **9.3** | **11** | **12** | **13** | **12.7** | **15.3** | **13** | **2.3** | **10.7** | **3** | **1.7** | **9.1** | **0** | **17.7** |
| Intergranular | 2.7 | — | 1.7 | 2.3 | 2.3 | 2 | 6.3 | 1.3 | 6.7 | 0.7 | 0.3 | 2.5 | 0 | 6.7 |
| Intragranular in feldspar | 4.3 | 0.7 | 5.7 | 9 | 4.3 | 1.7 | 4.3 | — | — | 0.3 | 0.7 | 2 | 0 | 9 |
| Intragranular in mica | — | — | — | — | — | — | — | 0.3 | — | — | — | 0.2 | 0 | 2.3 |
| Intragranular in plutonic r.f. | — | — | — | — | — | — | — | — | — | 0.3 | — | 0.1 | 0 | 1.3 |
| Dissolution of pseudomatrix | — | — | — | — | 2 | — | 0.3 | — | — | 1 | 0.7 | 0.2 | 0 | 2 |
| Mouldic | 0.7 | — | 1.3 | 0.7 | 1.7 | 7 | 1.7 | — | 3.7 | 0.7 | — | 1.8 | 0 | 7 |
| Lamellar decompaction | 0.3 | — | 1 | 0.3 | 1.3 | 2.7 | 0.3 | — | — | — | — | 0.8 | 0 | 3.7 |
| Fracture | — | — | 1.3 | — | — | 0.7 | — | — | — | — | — | 0.2 | 0 | 1.3 |
| Shrinkage | 1.3 | 10.3 | 1 | 0.7 | 1 | 1.3 | — | 0.7 | — | — | — | 1.2 | 0 | 10.3 |
| Oversized | — | — | — | — | — | — | — | — | 0.3 | — | — | 0.2 | 0 | 2.7 |
| **Microporosity** | **Abundant** | **Abundant** | **Common** | **Abundant** | **Common** | **Abundant** | **Common** | **Abundant** | **Common** | **Common** | **Rare** | **Abundant** | **Rare** | **Abundant** |
| Granulometry | Medium | Fine | Coarse cgl. | Medium | Coarse | Very coarse | Coarse | Medium | Medium | Medium | Medium | Coarse | Fine | V. coarse cgl. |
| Sorting | Good | Moderate | Very poor | Moderate | Good | Poor | Moderate | Poor | Moderate | Good | Good | Moderate | Poor | Good |
| **Intergranular volume** | **19.7** | **4.3** | **11.7** | **23.7** | **23.3** | **19.3** | **35** | **31** | **34** | **29.7** | **30** | **22.8** | **4.3** | **35** |
| **Cement volume** | **16.7** | **4.3** | **9** | **21** | **19.7** | **14.7** | **28.3** | **29.7** | **27.3** | **29** | **29.7** | **19.5** | **4.3** | **29.7** |
| Authigenic quartz total | 8 | 1 | 7 | 17 | 14.7 | 13 | 26.7 | 25.3 | 24 | 24.3 | 25.7 | 14.7 | 1 | 26.7 |
| Grain volume original | 80.3 | 95.7 | 88.3 | 76.3 | 76.7 | 80.7 | 65 | 69 | 66 | 70.3 | 70 | 77.2 | 65 | 95.7 |
| Grain replacing/dissolution total | 24.7 | 62 | 14.3 | 30 | 16.7 | 19.7 | 12.3 | 20.3 | 10 | 19 | 13 | 22.3 | 10 | 62 |
| Feldspar dissolution total | 10.7 | 9 | 7 | 10.7 | 8 | 8.7 | 8.3 | 11 | 7.3 | 10 | 8.3 | 11.4 | 2 | 20.3 |
| Kaolin total | 8.7 | 9 | 5.3 | 9.7 | 6 | 6.7 | 7.3 | 22.3 | 3 | 0 | 0 | 12.6 | 0 | 43.7 |
| Grain-replacive kaolin | 6 | 9 | 5.3 | 5.3 | 2.3 | 6.7 | 6 | 17.7 | 2.3 | 0 | 0 | 9.4 | 0 | 35 |
| Grain-replacive clays | 16 | 50.7 | 6.3 | 18.7 | 7.7 | 9.7 | 6 | 19.3 | 6 | 8.7 | 4.3 | 15.8 | 4.3 | 50.7 |
| Illite total | 11.3 | 41.7 | 1 | 15.7 | 7.3 | 4.3 | 0 | 1.3 | 6.3 | 7.3 | 4.7 | 7.3 | 0 | 41.7 |
| Q | 78.8 | 100 | 91.1 | 89 | 96.3 | 94.2 | 95.4 | 100 | 99.4 | 98.7 | 96.9 | 95.5 | | |
| F | 16.6 | 0 | 8.4 | 5.9 | 1.9 | 5.9 | 4.6 | 0 | 0 | 0 | 0 | 3.2 | | |
| L | 4.6 | 0 | 0.5 | 5.2 | 1.9 | 0 | 0 | 0 | 0.6 | 1.3 | 3.1 | 1.3 | | |
| Qo | 73 | 73.1 | 91.1 | 87.1 | 96.3 | 94.2 | 94.2 | 85.8 | 93.8 | 79.7 | 83.2 | 86.1 | | |
| Fo | 22.7 | 26.9 | 8.4 | 7.9 | 1.9 | 5.9 | 5.8 | 14.2 | 5.6 | 19.3 | 14.2 | 12.7 | | |
| Lo | 4.3 | 0 | 0.5 | 5 | 1.9 | 0 | 0 | 0 | 0.6 | 1.1 | 2.6 | 1.2 | | |

r.f.: rock fragment.

**Fig. 8.** (a) Cathodoluminescence (CL) photo of a sandstone with mostly tangential intergranular contacts due to early and pervasive quartz overgrowth cementation (arrows); bright grains are K-feldspars; outcrop; (b) CL photo of a sandstone with intense compaction through pressure dissolution and grain fracturing (arrows), followed by precipitation of dull-luminescing quartz overgrowths; outcrop; (c) CL photo of large, zoned quartz overgrowths partially filling dissolution pores and replacing detrital quartz (arrows); some fractured grains healed by diagenetic quartz; outcrop; (d) Backscattered electron (BSE) image of partially dissolved K-feldspar grains and intergranular vermicular kaolinite (darker grey, engulfed by quartz overgrowths; arrows); outcrop; (e) scanning electron microscope (SEM) image of multiple late quartz overgrowths partially engulfing lamellar illite 1822.3 m; (f) CL photo of intensely fractured quartz grains healed by authigenic quartz and cemented by quartz overgrowths; bright feldspar dissolution remnants; outcrop.

**Fig. 9.** (a) Optical photomicrograph of a sandstone with heterogeneous pore system of large secondary pores after feldspar dissolution (some with quartz outgrowths; arrows) poorly connected by lamellar throats reduced by quartz cementation and pressure dissolution; plane light; outcrop; (b) SEM image showing multiple overgrowths and outgrowths of late quartz partially engulfing blocky dickite, and being covered by fibrous illite; 1595.2 m; (c) BSE image of an expanded and kaolinized mica flake showing characteristic 'fanning' along the edges; vermicular kaolinite (darker grey) fills adjacent pores; outcrop; (d) BSE image of pervasively kaolinized and widely expanded mica flakes; outcrop; (e) SEM image of blocky dickite crystals partially engulfed by late quartz overgrowths; 1595.2 m; (f) SEM photo of illitized vermicular kaolinite; outcrop.

(a)

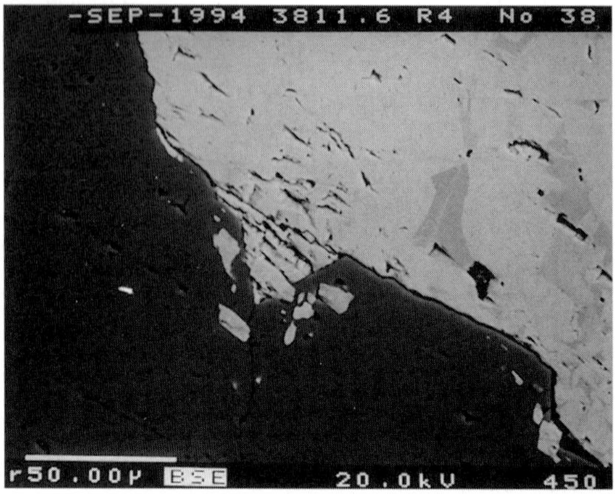


(b)

(c)

**Fig. 10.** (a) SEM image of lamellar pore-rimming illite overgrowing an illite–smectite coating (cc); outcrop; (b) BSE image of Mg-rich siderite (medium grey) filling fractures in Fe-rich siderite (bright grey), and late quartz overgrowths (dark grey) engulfing remnants of dissolution along the siderite borders; 3811.6 m; (c) optical photomicrograph of a sandstone with pervasive kaolinite replacement of feldspar and cementation; crossed polarizers; outcrop.

kaolin (Fig. 9f; 0–5%; av. 1%), micas (0–6%; av. 0.5%), feldspar grains (0–10%; av. 1%), and pseudomatrix (0–31%; av. 3%). Neoformed illites occur mostly as divergent aggregates of lamellar crystals, up to 40 μm wide (Fig. 10a; 0–8%; av. 1%). Rims and divergent aggregates of fibrous or lath-shaped crystals are less widespread, and associated with late quartz overgrowths and outgrowths. Mixed-layer illite–smectite (I–S) occurs as thin coatings with morphology typical of mechanically infiltrated clays, which are covered by neoformed illite (Fig. 10a; cf. Moraes & De Ros, 1990; 0–1%; av. 0.1%) and replacing pseudomatrix. Illite occurs at all depths, from outcrop to basin centre. Illite and subordinate chlorite are the only clays encountered below 2200 m of present-day depth (Ramos & Formoso, 1975). Illite shows high crystallinity, with Kübler (1968) index commonly below 2, and intense 5 Å peaks (Ramos & Formoso, 1975). The I–S clays are generally well-ordered and with a large proportion of illite. The lamellar illites show higher crystallinity than the fibrous type (Dos Santos & Bonhomme, 1991).

Authigenic chlorite occurs as rosette or cabbage-head aggregates associated with illite and quartz in subsurface samples deeper than 1440 m. Pyrophyllite occurs in a few subsurface samples (Ramos & Formoso, 1975).

## Accessory diagenetic minerals

Albite occurs only in the deep subsurface samples close to the basin centre (3800–3900 m), where it totally replaces detrital feldspars in sandstones massively cemented by quartz and illite. Grain-replacive albite is fine to coarsely crystalline (2–80 μm), and arranged in parallel aggregates. Small (< 1–20 μm) discrete crystals occur associated with fibrous illite. Albite is commonly rimmed by illite or engulfed by late siderite and quartz cements. Albitization of detrital potassium feldspars and plagioclase is pervasive in the deep subsurface samples.

Diagenetic carbonates are rare (0–2%; av. 0.1%), and occur only in subsurface samples. Some deep subsurface samples (3800–3900 m) contain coarse, blocky to poikilotopic, magnesian siderite ($MgCO_3$ = 8–28 mol%). Magnesian siderite is a common mesogenetic carbonate (Macaulay *et al.*, 1993; Morad *et al.*, 1994). The siderite precipitated in intergranular spaces reduced by intense pressure dissolution and engulfs quartz, albite and illite cements. BSE examination revealed a complex zonation pattern, with Mg-rich siderite filling intracrystalline pores after fracturing and partial dissolution of Mg-poor siderite ($MgCO_3$ ≈10%; Fig. 10b). $MnCO_3$ content varies from 1 to 6.8 mol% and $CaCO_3$ from 0 to 0.8 mol%. Siderite covers, and thus post-dates the continuous quartz overgrowths, but is engulfed and replaced by later euhedral overgrowths and outgrowths (Fig. 10b). The $\delta^{18}O_{PDB}$ and $\delta^{13}C_{PDB}$ values of siderite are –15.8‰ and –10.4‰, respectively. Assuming that siderite was formed from mesogenetic fluids highly modified by water–rock interaction with $\delta^{18}O$ of ≈ + 2 to + 5‰, calculated precipitation temperature is ≈150°C, which is reasonable for the burial history in the central area of the basin (Fig. 4). The $\delta^{13}C_{PDB}$ value of –10.4‰ indicates derivation of the carbon from the decarboxylation of organic matter.

Conversely, small amounts of calcite (< 1%) occur in shallow subsurface samples (≈500 m), close to the outcrop belt. Calcite is unzoned, poikilotopic and slightly ferroan (0.8–2.1% $FeCO_3$, 0–1.7% $MnCO_3$, 0–0.8% $MgCO_3$). The $\delta^{18}O_{PDB}$ and $\delta^{13}C_{PDB}$ values of calcite are –14.8‰ and –10.2‰, respectively. As calcite is restricted to areas that have been influenced by meteoric eodiagenesis, the $\delta^{18}O$ values suggest precipitation from high latitude meteoric waters at near-surface temperatures, with carbon derived from the oxidation of organic matter.

Authigenic titanium oxides are conspicuous, although scarce (av. ≈0.5%; up to 3%; Table 1). Prismatic rutile crystals (2–100 μm) are scattered in secondary pores, in kaolin replacing mica or pseudomatrix and are engulfed by quartz overgrowths, and typically replace detrital Ti–Fe minerals. Bipyramidal anatase crystals and cryptocrystalline leucoxene are less common. The textural evidence indicates that the Ti oxides are residual after the dissolution of detrital Fe–Ti minerals.

Tourmaline occurs as partial overgrowths on some detrital tourmaline grains. Pyrite occurs in small cubic crystals in pseudomatrix that has been replaced by kaolin and illite. Barite is present as small discrete crystals associated with altered feldspar and pseudomatrix.

## DISCUSSION

### Paragenetic sequence, compaction and porosity evolution

The paragenetic sequences which affected the sandstones in basin margin and basin centre settings are depicted in Fig. 11. The different diagenetic evolution pathways in basin margin and basin centre sandstones are also illustrated in Fig. 12. The main differences involve: (i) the much larger extent of meteoric flushing, and consequent production of feldspar, mica and smectite pseudomatrix along the basin margin, allowing the better preservation of feldspars in the basin centre which were later albitized; (ii) the obviously greater intensity of mechanical and chemical compaction in the basin centre; and (iii) the telogenetic leaching, oxidation and decompaction, limited to the sandstones cropping out on the basin margin. The paragenetic relationships have been interpreted from the integration of optical microscopy, SEM, CL and BSE observations. However, due to the complex diagenetic patterns and to the scarcity of precise geothermometric data, a precise timing cannot be achieved for all the diagenetic effects observed.

Mechanical compaction was, in general, relatively intense, as suggested by the widespread deformation of mud intraclasts into pseudomatrix, bending of mica plates and fracturing of quartz grains, particularly in the coarser sandstones. Most of the fractured grains are healed by diagenetic quartz (Fig. 8f). Some samples, however, display limited mechanical compaction due to early quartz cementation (Fig. 8a). Tight packing values (packing proximity index ≈60–70%; Kahn, 1956), sutured intergranular contacts and stylolitic surfaces occur mostly in fine-grained micaceous rocks. Intergranular volume varies from 4 to 35% (av. 23%). However, remaining primary intergranular porosity averages only 2% (0–6.7%), grain-dissolution porosity averages 3% (0–9%), and total macroporosity averages 9% (0–17.7%; Table 1). This suggests that cementation, although heterogeneous, was a major factor of porosity destruction and limited the extent of compaction.

A plot of cement vol.% versus intergranular volume percentage of the Furnas sandstones (Fig. 13; cf. Houseknecht, 1987) shows that compaction was nearly as important as cementation in porosity reduction, and that the samples with larger intergranular volume have a consistently larger volume of cement. This distribution pattern combined with the low average macroporosity suggest that a relatively shallow-burial cementation by quartz was the main constraint to compaction.

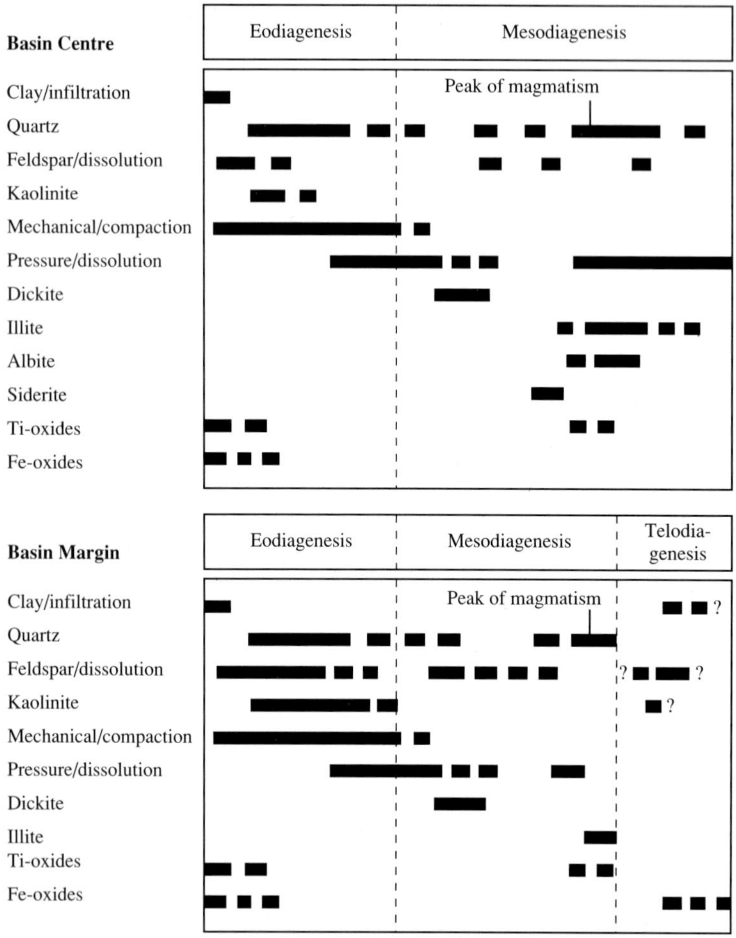


**Fig. 11.** Schematic paragenetic sequence of diagenetic processes in Furnas sandstones.

## Patterns of quartz and illite cementation

Several lines of evidence suggest that quartz cementation in the Furnas sandstones commenced at shallow burial and proceeded recurrently through mesodiagenesis (Fig. 11), including: (i) overgrowths which both engulf and are covered by diagenetic phases that range from early kaolinite to late dickite, illite and siderite (Figs 8d,e & 9b,e); (ii) widely variable volume of diagenetic quartz (1–26.7%; av. 14.7%) and intergranular volume (4–35%; av. 23%); (iii) CL images show that, in some samples, quartz overgrowths occur between the grains and before substantial compaction, while in others they only fill pores remaining after intense pressure dissolution (Fig. 8a,b,f); (iv) a wide range of homogenization temperatures of fluid inclusions (11° to 156°C), with, however, a dominance of high temperatures (> 100° to > 150°C) both in surface and subsurface samples.

This pattern of quartz cementation contrasts with most accepted models of quartz cementation, which are based on rapidly subsiding and deep basins, such as the U.S. Gulf Coast and the North Sea. In these basins, quartz cementation is substantial only at depths > 2–2.5 km, and fluid inclusion data suggest that most quartz precipitated at temperatures > 100°C. This has led to the generalization that quartz cementation would be important only under deep burial (> 2–3 km; e.g. Land *et al.*, 1987; McBride, 1989; Gluyas & Coleman, 1992; Bjørlykke & Egeberg, 1993; Walderhaug, 1994a;b). According to these models, silica would be internally derived by diffusion from pressure dissolution along intergranular contacts and stylolites (Bjørlykke & Egeberg, 1993; Walderhaug, 1994a;b). However, other authors consider that the homogenization temperatures from quartz fluid inclusions, and the associated illite isotopic data suggest that quartz cementation is episodic, related to the

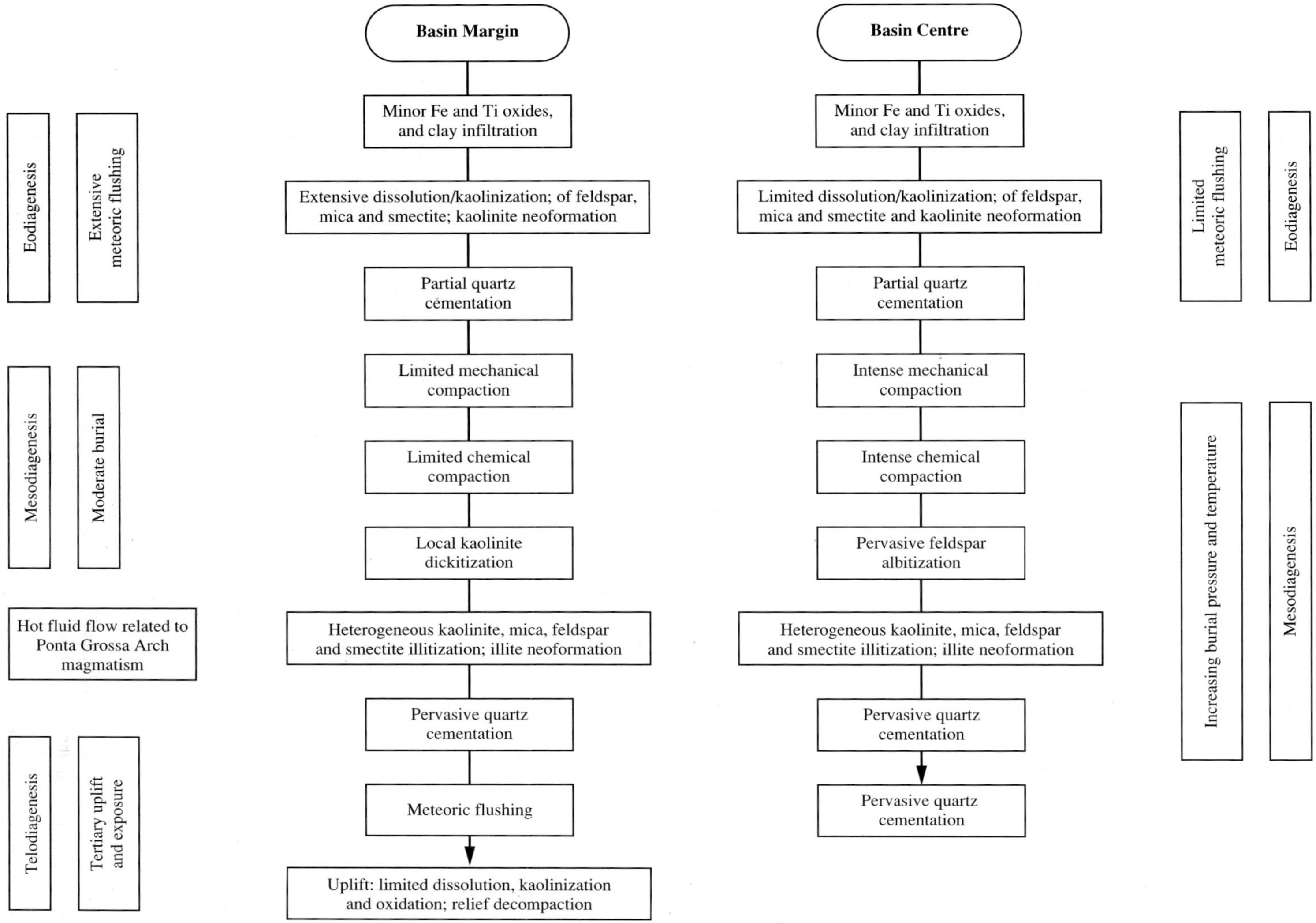


**Fig. 12.** Main diagenetic evolution pathways in Furnas sandstones in the margin and centre of the basin.

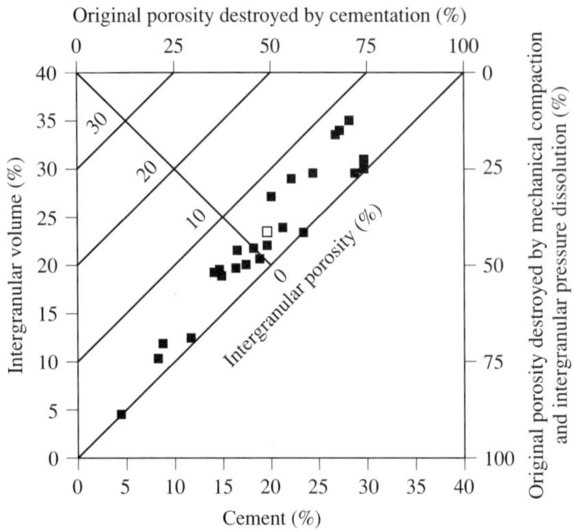


**Fig. 13.** Plot of cement percentage versus intergranular volume percentage (cf. Houseknecht, 1987) for 25 Furnas sandstones, showing the approximately equal importance of cementation and compaction processes in the destruction of intergranular porosity. Open square = average porosity.

circulation of hot fluids through faults during rifting and heating periods (Lee *et al.*, 1985; Glasmann *et al.*, 1989; Gluyas *et al.*, 1993). Detailed studies associating fluid inclusions and cathodoluminescence indicate a multi-episodic pattern of quartz cementation (Burley *et al.*, 1989; Rezaee & Tingate, 1997).

Plausible sources of silica for quartz cementation in the Furnas sandstones include meteoric kaolinization of feldspar (up to 20.3 of bulk vol.% feldspars were destroyed; av. 11.4%; Table 1). The destruction of 11.4 vol.% of K-feldspar would produce 5.2 vol.% of kaolinite and 4.8 vol.% of quartz. Alteration of 11.4 vol.% $An_{50}$ plagioclase yields 8.4 vol.% of kaolinite and 2.6 vol.% of quartz. It is therefore suggested that approximately one-third to one-sixth of the observed average 14.7 vol.% of quartz cement may have been derived from eodiagenetic dissolution and kaolinization of feldspar. An eodiagenetic meteoric setting for feldspar leaching is supported by the engulfment of kaolinite and feldspar dissolution remnants by quartz overgrowths, the wide expansion observed in contemporaneously kaolinized micas and the skeletal, subhedral morphology of the kaolinite (cf. McAulay *et al.*, 1993), although diagnostic isotopic data are not available.

According to Zalán *et al.* (1987a,b), the Furnas Formation remained buried at shallow depths for a period of 10 Myr, before being overlain by the transgressive marine deposits of the Devonian Ponta Grossa Formation, thus providing sufficient time for the meteoric alteration of the sandstones. The relatively rugged relief left after the Brasiliano–Panafrican orogeny along basin margins would provide a suitable hydraulic gradient for extensive meteoric flushing, and consequent dissolution and kaolinization of feldspars and micas. Relatively large volumes of water from seasonal melting could be available for infiltration under the periglacial palaeogeographic setting of the Paraná Basin during Silurian to Early Carboniferous time (Scotese *et al.*, 1979; Zalán *et al.*, 1987b). Good connectivity and permeability would be supplied by the amalgamated fluvial cycles and the average coarse grain size of the unit, respectively.

The extensive dissolution and kaolinization changed the composition of the Furnas sandstones from originally arkosic with an average of 12.5 vol.% of detrital feldspars, to diagenetic quartzarenites and quartzose subarkoses with an average of 1.1% of feldspars (0–8.3%; Fig. 7). As most feldspar destruction occurred relatively early and at shallow depths, the original amounts of feldspar are underestimated due to the collapse of secondary pores or deformation of replacive clay during compaction (Land *et al.*, 1987; Harris, 1989; Weedman *et al.*, 1992). Feldspar destruction was accompanied by precipitation of equivalent volumes of authigenic clay and quartz by-products, and porosity in the unit is thus dominantly redistributional (Figs 8d & 10c; cf. Giles & deBoer, 1990). However, some samples show no clay by-products after feldspar dissolution (Fig. 9a), indicating that alumina and silica were mobilized at least on a thin-section scale.

The salinity range of fluid inclusions within quartz overgrowths in subsurface Furnas sandstones (between 16 and 19 eq. wt.% NaCl) suggests that highly saline fluids were responsible for quartz cementation. Such fluids, which presently occur in the pores of Silurian and Permian sandstones of the basin centre (Milani *et al.*, 1990), were presumably evolved during long-range fluid–rock interaction or derived from the magmatism, as there are no evaporites in the basin. The present distribution of dilute fluids along the basin margins and hypersaline fluids in the centre agrees with models of fluid circulation in intracratonic basins which indicate persistence of an active meteoric regime along the margins, whereas stagnant conditions gradually develop towards the basin centre (cf. Bethke, 1985). Sandstones of cratonic settings, characterized by a large residence time at shallow depths, commonly show eodiagenetic quartz cementation which can limit compaction and preserve large intergranular volumes (e.g. Sibley & Blatt, 1976; Odom *et al.*, 1979; Thiry *et al.*, 1988; Longstaffe, 1993).

Another source for quartz cementation observed in Furnas sandstones is intergranular pressure dissolution. However, it is difficult to quantify the volumes of silica derived from this source. As stylolites are common only in mudstones and fine-grained sandstones, which are volumetrically subordinate in the unit, stylolitic pressure dissolution cannot account for large volumes of quartz cement. Similarly, the relatively small volume of the Silurian and Devonian shales limits the availability of compactional fluids and of silica from smectite illitization.

Therefore, the most likely explanation for the large volumes and the heterogeneous distribution of quartz cement in the Furnas sandstones is the transportation of silica by convecting thermobaric fluids related to the extensive magmatism along the Ponta Grossa Arch (Zalán *et al.*, 1990b). A similar connection between quartz cementation and magmatism was recognized elsewhere (Girard *et al.*, 1989; França & Potter, 1991; Summer & Verosub, 1994). Increased heat flow due to the intrusion of basaltic magma would induce active fluid convection through faults and extensional fractures related to the Arch uplift (Fig. 5). This is also suggested by the elevated Th of fluid inclusions trapped within quartz overgrowths in outcrop samples (up to 156°C). The possibility that these high homogenization temperatures may be due to heterogeneous gas trapping, instead of to hot fluids, cannot be dismissed, although it is very unlikely, considering the basin-margin setting and the very small size of the inclusions. Assuming maximum burial depths around 1220–2200 m in the eastern marginal area (Fig. 6) and geothermal gradients of $\approx$25–30°C km$^{-1}$ during Jurassic–Cretaceous magmatism and initial uplift of the Ponta Grossa Arch, maximum burial temperatures in the area would be around 90°C. The widespread microfracturing of quartz grains is due not only to burial overburden, but also to tectonic stresses related to arch uplift (cf. Morad *et al.*, 1991; Milliken, 1994). Convection of hot fluids related to the Ponta Grossa Arch magmatism and uplift are thus the probable source of late, well-developed quartz overgrowths and outgrowths, and of coeval illite as well.

The dominance of illite in Furnas samples of moderate to deep subsurface agrees with the models of illite generation by reaction of K-feldspar and kaolinite at elevated temperature under deep burial. However, this explanation is unsatisfactory for the heterogeneous distribution of illite in outcrop and shallow subsurface samples, considering that the maximum regional burial temperatures of the eastern marginal area were around 90°C. Furthermore, non-illitized kaolinite occurs together with detrital K-feldspar in several samples, indicating that the observed heterogeneous illitization was not constrained by lack of remaining K-feldspar (cf. Ehrenberg, 1991).

Evidence for the conditions of illitization in the Furnas Formation is provided by the Kübler index of illites which is around 1.5–2, even in shallow subsurface samples of areas not affected by substantial uplift. Such a high crystallinity suggests precipitation under fairly high temperatures (Dunoyer de Segonzac, 1970; Ramos & Formoso, 1975). Additionally, K/Ar ages of illites in the Furnas sandstones range between 138 and 102 Ma (Dos Santos & Bonhomme, 1991). The age of the older illites, which display higher crystallinity and lamellar habit, coincides with the peak of basaltic extrusion (cf. Dos Santos & Bonhomme, 1991). This age, the high crystallinity and the very heterogeneous distribution of illite in outcrop samples suggest that the precipitation of illite and associated late quartz occurred from hot fluids circulated through faults and extensional fractures and was related to the magmatism and uplift of the Ponta Grossa Arch (Fig. 5).

## Reservoir implications

The Furnas sandstones are considered as potential reservoirs for hydrocarbons generated from Silurian and Devonian shales in the deep central Paraná Basin (Milani *et al.*, 1990; Zalán *et al.*, 1990a), as well as from Vendian synorogenic black shales in the shallow eastern portion (Meister *et al.*, 1989). This interpretation is based on the proximity of these mature source rocks, on the regionally continuous seal provided by the Ponta Grossa shales, and on the widespread secondary porosity due to feldspar dissolution. However, this study indicates that feldspar dissolution was not accompanied by significant porosity enhancement, for two main reasons:

**1** Most secondary porosity that developed through feldspar dissolution during early meteoric diagenesis was subsequently destroyed by mechanical compaction. Nevertheless, in some samples, early quartz cementation allowed the preservation of these secondary pores (Fig. 9a). However, the large dissolution pores are poorly connected by thin lamellar throats which present a low hydrocarbon recovery efficiency (Wardlaw & Cassan, 1979), because a large residual hydrocarbon saturation remains trapped in the large pores after the throats are drained.

**2** Feldspar dissolution was not accompanied by significant porosity enhancement, because of local reprecipitation of kaolinite and quartz cements. Kaolinite authigenesis corresponded to the introduction of abundant microporosity, but no bulk increase in pore volume (Fig. 10c). Microporosity contained in kaolinite, dickite and illite cements may further diminish reservoir performance in terms of high residual water saturation and substantial permeability reduction.

Towards the basin centre, the Furnas sandstones were less affected by meteoric flushing than along the margins.

Larger amounts of feldspar had thus escaped dissolution and kaolinization. These feldspars were albitized and illitized during deep burial diagenesis. Sandstones of the deep central portions of the basin suffered a substantial decrease in porosity and permeability by intense pressure dissolution, and by precipitation of quartz overgrowths and illite.

The preservation of permeability and porosity is expected to be greater for sandstones at intermediate depths (< 2000 m) and away from the Ponta Grossa Arch region, where quartz and illite cementation was extensive. Gas could be trapped in low-permeability reservoirs sealed by Ponta Grossa shales, in compressional structures generated by the large volumes of intrusion of diabase sills (Zalán *et al.*, 1987a; 1990b).

The reservoir potential of Furnas sandstones and similar units that have been influenced by magmatism and complex burial and diagenetic histories, is low due to pervasive porosity destruction. However, factors such as the focusing of fluid flow along preferential permeability conduits and early saturation by hydrocarbons may generate and preserve high porosity values.

## CONCLUSIONS

The main implications of this study are:

**1** The diagenesis of sandstones in cratonic settings and/or under conditions influenced by regional magmatism generate particularly heterogeneous cementation and alteration patterns. The 'isochemical' diagenetic models of the system $SiO_2$–$Al_2O_3$–$K_2O$–$H_2O$ have a limited validity, and may apply only to sandstones deeply buried in rapidly subsiding basins.

**2** The development of diagenetic quartzarenites may involve successive stages of feldspar destruction, accompanied by the precipitation of clay minerals and quartz by-products rather than by porosity enhancement. Diagenetic quartzarenites are commonly developed by pervasive destruction of the feldspars during eodiagenetic meteoric flushing that is accompanied by abundant precipitation of kaolinite. Eodiagenetic secondary pores after dissolved feldspars were largely destroyed during progressive burial and compaction. Accordingly, the reservoir quality of such sandstones is fairly low, unless early quartz cementation provides effective framework support.

**3** The redistributional aspect of porosity evolution and the poorly connected pore-systems substantially limit the reservoir quality of Furnas sandstones. The best reservoir potential is expected for areas at intermediate burial depths and away from the Ponta Grossa Arch region.

**4** The heterogeneous distribution of authigenic quartz and illite in Furnas sandstones is attributed to the convection of hot fluids through fractures related to the uplift and magmatism along the Ponta Grossa Arch region. This is supported by high (up to 156°C) fluid inclusion homogenization temperatures in quartz overgrowths and by extensive kaolinite illitization in sandstones with maximum burial depths between 1200 and 2200 m, as well as by the high crystallinity and the K/Ar ages of the illite which coincide with the timing of magmatism.

## ACKNOWLEDGEMENTS

We thank L. Borghi, R.A. Gonçalves and F.E.G. Cruz for providing outcrop samples and information, and M.A.S. Moraes and PETROBRAS for access to subsurface samples, data, and permission to publish this work. We also thank K. Ramseyer for thoughtful discussions and for CL photographs. The financial support by the Brazilian National Council of Research (CNPq) and the Swedish National Research Foundation (NFR) is gratefully acknowledged. H. Harrysson is thanked for aiding with the microprobe analyses, C. Bäck and B. Giös for photographic work, and C. Wernström for drafting the figures. Comments by R. Worden and reviewers N. Oxtoby and R. Gaupp helped to improve the manuscript. The paper also benefited from fruitful discussions with Drs K. Ramseyer, A. Matter, K. Bjørlykke, P. Aagaard and S.N. Ehrenberg, among others, but the interpretations expressed here are solely our responsibility.

## REFERENCES

Bethke, C.M. (1985) A numerical model of compaction-driven groundwater flow and heat transfer and its application to the paleohydrology of cratonic sedimentary basins. *Journal of Geophysical Research* **90**, 6817–6828.

Bjørlykke, K. & Egeberg, P.K. (1993) Quartz cementation in sedimentary basins. *American Association of Petroleum Geologists Bulletin* **77**, 1538–1548.

Burley, S.D., Mullis, J. & Matter, A. (1989) Timing diagenesis in the Tartan Reservoir (UK North Sea): constraints from combined cathodoluminescence microscopy and fluid inclusion data. *Marine and Petroleum Geology* **6**, 98–120.

Craig, H. (1957) Isotopic standards for carbon and oxygen correction factors for mass spectrometric analysis of carbon dioxide. *Geochimica et Cosmochimica Acta* **12**, 133–149.

Cruz, F.E.G. (1987) Petrologia dos Arenitos da Formação Furnas na Região de Ponta Grossa—Paraná. PETROBRAS-CENPES. *Technical Report CENPES-DIGER-SEGEX 01/87*.

Dino, R., Bergamaschi, S., Pereira, E. *et al.* (1995) Biochronostratigraphic investigations of the Pragian and Emsian stages on the southeastern border of the Paraná Basin. *2° Simpósio Sobre Cronestratigrafia Da Bacia Do Paraná, Porto*

*Alegre, RS, Brazil.* UFRGS, Boletim de Resumos Expandidos pp. 19–25.

Dino, R. & Rodrigues, M.A.C. (1995) Palinomorfos Eodevonianos da Formação Furnas—Bacia do Paraná. *An Anois dos Academia Brasileira de Ciências* **67**, 107–116.

Dos Santos, R.P. & Bonhomme, M.G. (1991) K/Ar dating of clays associated with fluorite mineralizations along the Atlantic coast of South America—Relationships with South Atlantic Ocean opening. In: *Source, Transport and Deposition of Metals.* (eds Pagel, M. & Leroy, J.L.) pp. 381–384. Proceedings of the 25 Years Society of Geology Applied to Mineral Deposits Anniversary Meeting, Nancy, France, 30 August–3 September 1991, Rotterdam.

Dunoyer de Segonzac, G.D. (1970) The transformation of clay minerals during diagenesis and low-grade metamorphism: a review. *Sedimentology* **15**, 281–396.

Dutton, S.P. & Diggs, T.N. (1990) History of quartz cementation in the Lower Cretaceous Travis Peak Formation, East Texas. *Journal of Sedimentary Petrology* **60**, 191–202.

Ehrenberg, S.N. (1991) Kaolinized, potassium-leached zones in the contacts of the Garn Formation, Haltenbanken, mid-Norwegian continental shelf. *Marine and Petroleum Geology* **8**, 250–269.

Ehrenberg, S.N., Aagaard, P., Wilson, M.J., Fraser, A.R. & Duthie, D.M.L. (1993) Depth-dependent transformation of kaolinite to dickite in sandstones of the Norwegian Continental shelf. *Clay Mineralogy* **28**, 325–352.

Folk, R.L. (1968). *Petrology of Sedimentary Rocks.* Hemphills, Austin, TX.

França, A.B. & Potter, P.E. (1991) Stratigraphy and reservoir potential of glacial deposits of the Itararé Group (Carboniferous-Permian) Paraná Basin, Brazil. *American Association of Petroleum Geologists Bulletin* **75**, 62–85.

Friedman, I. & O'Neil, J.R. (1977) *Compilation of Stable Isotopic Fractionation Factors of Geochemical Interest.* United States Geological Survey. USGS Professional Paper 440-KK, 12pp.

Giles, M.R. & deBoer, R.B. (1990) Origin and significance of redistributional secondary porosity. *Marine and Petroleum Geology* **6**, 378–397.

Girard, J.-P. & Deynoux, M. (1991) Oxygen isotope study of diagenetic quartz overgrowths from the Upper Proterozoic quartzites of western Mali, Taoudeni Basin: implications for conditions of quartz cementation. *Journal of Sedimentary Petrology* **61**, 406–418.

Girard, J.-P., Deynoux, M. & Nahon, D. (1989) Diagenesis of the Upper Proterozoic siliciclastic sediments of the Taoudeni Basin (West Africa) and relation to diabase emplacement. *Journal of Sedimentary Petrology* **59**, 233–248.

Glasmann, J.R., Clark, R.A., Larter, S., Briedis, N.A. & Lundegard, P.D. (1989) Diagenesis and hydrocarbon accumulation, Brent Sandstones (Jurassic), Bergen High, North Sea. *American Association of Petroleum Geologists Bulletin* **73**, 1341–1360.

Gluyas, J. & Coleman, M. (1992) Material flux and porosity changes during sediment diagenesis. *Nature* **356**, 52–54.

Gluyas, J.G., Grant, S.M. & Robinson, A.G. (1993) Geochemical evidence for a temporal control on sandstone cementation. In: *Diagenesis and Basin Development.* (eds Horbury, A. & Robinson, A.) pp. 23–33. American Association of Petroleum Geologists Studies in Geology 36.

Harris, N.B. (1989) Diagenetic quartzarenite and destruction of secondary porosity: an example from the Middle Jurassic Brent sandstone of northwest Europe. *Geology* **17**, 361–364.

Haszeldine, R.S., Samson, I.M. & Cornford, C. (1984) Quartz diagenesis and convective fluid movement: Beatrice oilfield, UK North Sea. *Clay Mineralogy* **19**, 391–402.

Houseknecht, D.W. (1987) Assessing the relative importance of compaction processes and cementation to reduction of porosity in sandstones. *American Association of Petroleum Geologists Bulletin* **71**, 633–642.

Houseknecht, D.W. (1988) Intergranular pressure solution in four quartzose sandstones. *Journal of Sedimentary Petrology* **58**, 228–246.

Kahn, J.S. (1956) The analysis and distribution of the properties of packing in sand-size sediments: 1. On the measurement of packing in sandstones. *Journal of Geology* **64**, 385–395.

Kübler, B. (1968) Evaluation quantitative du metamorphisme par la cristalinite de l'illite; etate des progres realises ces derniers annees. *Bulletin du Centre de Reserche Pau SNPA* **2**, 385–397.

Land, L.S., Milliken, K.L. & McBride, E.F. (1987) Diagenetic evolution of Cenozoic sandstones, Gulf of Mexico sedimentary basin. *Sedimentary Geology* **50**, 195–225.

Leder, F. & Park, W.C. (1986) Porosity reduction in sandstone by quartz overgrowth. *American Association of Petroleum Geologists Bulletin* **70**, 1713–1728.

Lee, M., Aronson, J.L. & Savin, S.M. (1985) K/Ar dating of the time of gas emplacement in Rotliegendes sandstone, Netherlands. *American Association of Petroleum Geologists Bulletin* **69**, 1381–1385.

Longstaffe, F.J. (1993) Meteoric water and sandstone diagenesis in the western Canada sedimentary basin. In: *Diagenesis and Basin Development.* (eds Hornbury, A.D. & Robinson, A.G.) pp. 49–68. American Association of Petroleum Geologists Studies in Geology 36, 49–68.

Macaulay, C.I., Haszeldine, R.S. & Fallick, A.E. (1993) Distribution, chemistry, isotopic composition and origin of diagenetic carbonates: Magnus Sandstone, North Sea. *Journal of Sedimentary Petrology* **63**, 33–43.

McAulay, G.E., Burley, S.D. & Johnes, L.H. (1993) Silicate mineral authigenesis in the Hutton and NW Hutton fields: implications for sub-surface porosity development. In: *Petroleum Geology of Northwest Europe: Proceedings of the 4th Conference.* (ed. Parker, J.R.) pp. 1377–1394. The Geological Society London.

McBride, E.F. (1989) Quartz cement in sandstones; a review. *Earth Science Review* **26**, 69–112.

Meister, E.M., Astolfi, M.A.V. & Gonzaga, P.M. (1989) Gravimetria e radiometria integradas à prospecção de petróleo: um exemplo no flanco leste da Bacia do Paraná. *I Seminário Interno de Exporação—SINTEX, Rio de Janeiro, RJ*, PETROBRAS/DEPEX, pp. 259–267.

Milani, E.J., Kinoshita, E.M., Araújo, L.M. & Cunha, P.R.C. (1990) Bacia do Paraná: possibilidades petrolíferas da calha central. *Boletim de Geociências da PETROBRAS* **4**, 21–34.

Milani, E.J., França, A.B. & Scheider, R.L. (1994) 6—Bacia do Paraná. *Boletim de Geociências da PETROBRAS* **8**, 69–82.

Milliken, K.L. (1994) The widespread occurrence of healed microfractures in siliciclastic rocks: Evidence from scanned cathodoluminescence imaging. In: *Rock Mechanics.* (eds Nelson, X.X. & Laubach, X.X.) pp. 825–832. Balkema Rotterdam.

Morad, S., Bhattacharyya, A., Al-Aasm, I.S. & Ramseyer, K. (1991) Diagenesis of quartz in the Upper Proterozoic

Kaimur Sandstones, Son Valley, central India. *Sedimentary Geology* **73**, 209–225.

MORAD, S., BEN ISMAIL, H., DE ROS, L.F., AL-AASM, I.S. & SERRHINI, N.-E. (1994) Diagenesis and formation water chemistry of Triassic reservoir sandstones from southern Tunisia. *Sedimentology* **41**, 1253–1272.

MORAES, M.A.S. & DE ROS, L.F. (1990) Infiltrated clays in fluvial Jurassic sandstones of Recôncavo Basin, northeastern Brazil. *Journal of Sedimentary Petrology* **60**, 809–819.

ODOM, I.E., WILLAND, T.N. & LASSIN, R.J. (1979) Paragenesis of diagenetic minerals in the St. Peter sandstone (Ordovician), Wisconsin and Illinois. In: *Aspects of Diagenesis.* (eds SCHOLLE, P.A. & SCHLUGER, P.R.) pp. 425–443. Society of Economic Paleontologists and Mineralogists Special Publication 26.

OELKERS, E.H., BJØRKUM, P.A. & MURPHY, W.M. (1992) The mechanism of porosity reduction, stylolite development and quartz cementation in North Sea sandstones. In: *Water–Rock Interaction.* (eds KHARAKA, Y.K. & MAEST, A.S.) pp. 1183–1186. Proceedings of the 7th International Symposium on Water–Rock Interaction, Balkema Rotterdam, the Netherlands, 2—Moderate and high temperature environments.

PORTER, E.W. & JAMES, W.C. (1986) Influence of pressure, salinity, temperature and grain size on silica diagenesis in quartzose sandstones. *Chemical Geology* **57**, 359–369.

RAMOS, A.N. & FORMOSO, M.L.L. (1975) *Argilominerais das Rochas Sedimentares da Bacia do Paraná.* Ciência-Técnica-Petróleo, Seção Exploração e Produção de Petróleo 9, PETROBRAS—CENPES, Rio de Janeiro, RJ.

RAMSEYER, K., FISCHER, J., MATTER, A., EBERHARDT, P. & GEISS, J. (1989) A cathodoluminescence microscope for low intensity luminescence. *Journal of Sedimentary Petrology* **59**, 619–622.

REZAEE, M.R. & TINGATE, P.R. (1997) Origin of quartz cement in the Tirrawarra Sandstone, southern Cooper Basin, South Australia. *Journal of Sedimentary Research* **67**, 168–177.

ROSENBAUM, J.M. & SHEPPARD, S.M.F. (1986) An isotopic study of siderites, dolomites and ankerites at high temperatures. *Geochimica et Cosmochimica Acta* **50**, 1147–1150.

SCHMIDT, V. & MCDONALD, D.A. (1979) The role of secondary porosity in the course of sandstone diagenesis. In: *Aspects of Diagenesis.* (eds SCHOLLE, P.A. & SCHLUGER, P.R.) pp. 175–207. Society of Economic Paleontologists and Mineralogists Special Publication 29.

SCOTESE, C.R., BAMBACH, R.K., BARTON, C., VANDERVOO, R. & ZIEGLER, A.M. (1979) Paleozoic base maps. *Journal of Geology* **87**, 217–277.

SHEPHERD, T., RANKIN, A.H. & ALDERTON, D.H.M. (1985) *A Practical Guide to Fluid Inclusion Studies.* Blackie, Glasgow.

SIAL, A.N., OLIVEIRA, E.P. & CHOUDHURI, A. (1987) Mafic dyke swarms of Brazil. In: *Mafic Dike Swarms.* (eds HALLS, H.C. & FAHRIG, W.F.) pp. 467–481. *Geological Association of Canada Special Paper* 34.

SIBLEY, D.F. & BLATT, H. (1976) Intergranular pressure solution and cementation of the Tuscarora orthoquartzite. *Journal of Sedimentary Petrology* **46**, 881–896.

SUMMER, N.S. & VEROSUB, K.L. (1994) Diagenesis and organic maturation of sedimentary rocks under volcanic strata, Oregon. *American Association Petroleum Geologists Bulletin* **76**, 1190–1199.

THIRY, M., AYRAULT, M.B. & GRISONI, J.-C. (1988) Groundwater silicification and leaching in sands: Example of the Fontainebleau Sand (Oligocene) in the Paris Basin. *Geological Society of the American Bulletin* **100**, 1283–1290.

WALDERHAUG, O. (1994a) Temperatures of quartz cementation in Jurassic sandstones from the Norwegian Continental Shelf—evidence from fluid inclusions. *Journal of Sedimentary Research A* **64**, 311–323.

WALDERHAUG, O. (1994b) Precipitation rates for quartz cement in sandstones determined by fluid-inclusion microthermometry and temperature-history modelling. *Journal of Sedimentary Research A* **64**, 324–333.

WARDLAW, N.C. & CASSAN, J.P. (1979) Oil recovery efficiency and the rock-pore properties of some sandstone reservoirs. *Bulletin of Canadian Petroleum Geology* **27**, 117–138.

WEEDMAN, S.D., BRANTLEY, S.L. & ALBRECHT, W. (1992) Secondary compaction after secondary porosity: can it form a pressure seal? *Geology* **20**, 303–306.

ZALÁN, P.V., WOLFF, S., CONCEIÇÃO, J.C.J. *et al.* (1987a) Tectônica e sedimentação da Bacia do Paraná. *III Simpósio Sul-Brasileiro de Geologia, Curitiba, PR.* Sociedade Brasileira de Geologia, Atas **1**, pp. 441–477.

ZALÁN, P.V., WOLFF, S., CONCEIÇÃO, J.C.J. *et al.* (1987b) A divisão tripartite do Siluriano da Bacia do Paraná. *Revista Brasileira de Geociências* **17**, 242–252.

ZALÁN, P.V., WOLFF, S., CONCEIÇÃO, J.C.J. *et al.* (1990a) Bacia do Paraná. In: *Origen E Evolução de Bacias Sedimentares.* (eds RAJA GABAGLIA, G.P. & MILANI, E.J.) pp. 135–168. PETROBRAS Rio de Janeiro.

ZALÁN, P.V., WOLFF, S., ASTOLFI, M.A.M. *et al.* (1990b) The Paraná Basin, Brazil. In: *Interior Cratonic Basins* (eds LEIGHTON, M.W., KOLATA, D.R., OLTZ, D.F. & EIDEL, J.J.) pp. 683–708. American Association of Petroleum Geologists Memoir 51.

*Spec. Publs int. Ass. Sediment.* (2000) **29**, 253–270

# Polyphased quartz cementation and its sources: a case study from the Upper Palaeozoic Haushi Group sandstones, Sultanate of Oman

B. H. HARTMANN, K. JUHÁSZ-BODNÁR, K. RAMSEYER *and* A. MATTER

*Geologisches Institut, Universität Bern, Baltzerstrasse 1, CH-3012 Bern, Switzerland*

## ABSTRACT

The late Westphalian to early Artinskian Haushi Group in the Interior Oman Sedimentary Basin consists of the glaciogenic Al Khlata Formation and the Gharif Formation, which contains marginal marine, coastal plain and fluvial sediments. Due to a geographically varied subsidence history, the sediments presently range from outcrop in the south-east to almost 5000 m in the north-west of the study area, representing maximum burial. This favourable geological setting provides a detailed picture of the genesis of authigenic quartz over a wide depth range.

Authigenic quartz formed as zoned syntaxial overgrowths on detrital quartz. It varies from trace amounts in outcrop samples to almost 25 vol% in sandstones buried over 4000 m. Evidence from cathodoluminescence, fluid inclusion microthermometry, and stable isotopes supports a multistage origin of authigenic quartz.

A detailed study of the diagenetic sequence reveals that dissolution of aluminosilicates, pressure solution and stylolitization are the major silica-providing processes in the Haushi Group sandstones. Meteoric infiltration along the basin margins is responsible for intense dissolution of aluminosilicates during shallow burial, contributing to quartz precipitation. Transformation of smectite to illite in mixed-layer clay minerals, supplies silica with progressive burial. However, the composition–depth curve of the mixed-layer clay minerals shows an abrupt increase in illite content at 1500 m, supplying a pulse of silica at that depth. Pressure solution and stylolitization are responsible for most of the observed quartz cement. A marked increase in pressure solution at 2600 m corresponds with an increase in authigenic quartz at the same depth. Illitization of kaolinite supplies minor amounts of silica at temperatures above 100°C.

## INTRODUCTION

Precipitation of authigenic quartz has long been recognized as an important factor in reducing reservoir quality in siliciclastic sediments (Wilson & Stanton, 1994). A large number of possible sources for silica have been proposed, with pressure solution, illitization of clay minerals and dissolution and alteration of aluminosilicates thought to be the volumetrically most important ones (McBride, 1989). Since these diagenetic alterations are difficult to quantify, the relative importance of the different processes has seldom been documented adequately (Houseknecht, 1988; Bjørkum *et al.*, 1993; Bjørlykke & Egeberg, 1993; Wintsch & Kvale, 1994).

The wide present depth range, from surface to almost 5000 m, of the Upper Palaeozoic Haushi Group provides a unique setting for separately studying burial related diagenesis and the effects of the basin margins. Intense quartz cementation is the most important process reducing reservoir quality in the Haushi Group sandstones (Ramseyer, 1983; Hartmann, 1996; Juhász-Bodnár *et al.*, 1997).

The goals of this study are to identify processes that produce silica and link them to quartz cementation during burial diagenesis along a cross section from the margin towards the centre of the continuously subsiding basin.

## GEOLOGICAL SETTING

### Interior Oman Sedimentary Basin (IOSB)

The Sultanate of Oman occupies the south-eastern corner of the Arabian Peninsula. The Interior Oman Sedimentary Basin (IOSB) is a low-lying desert area bounded by the Huqf–Haushi axis, a NNE- to SSW-running anticlinorium in the east and by the NE—SW orientated Jabal Qara arch in the south. To the north, the basin is bordered by

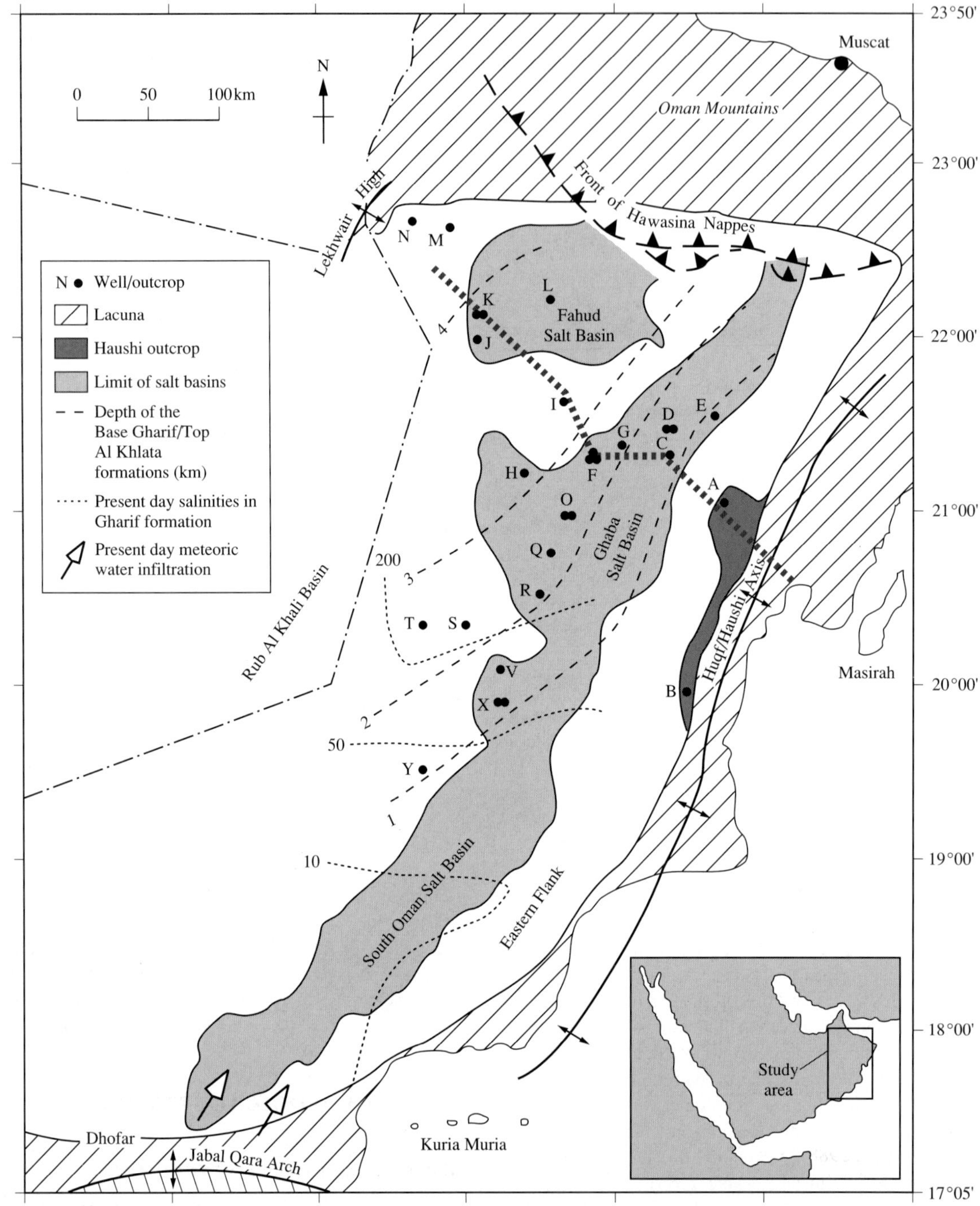


**Fig. 1** Locations of the studied wells and outcrops in the northern part of the Interior Oman Sedimentary Basin (modified after Al Lamki & Terken, 1996).

(a)

(b)

**Plate 1.** (a) Sandstone (sample A, Figs 13 & 14) with 4% of overlapping grain boundaries and 20% quartz cement acting as silica importer (Al Khlata Formation, well I, depth 3592.5 m; scale bar is 200 μm). (b) Sandstone (sample B, Figs 13 & 14) with 1.5% quartz cement and 32% pressure solution, acting as silica exporter (Al Khlata Formation, well I, depth 3593.9 m; scale bar is 200 μm).

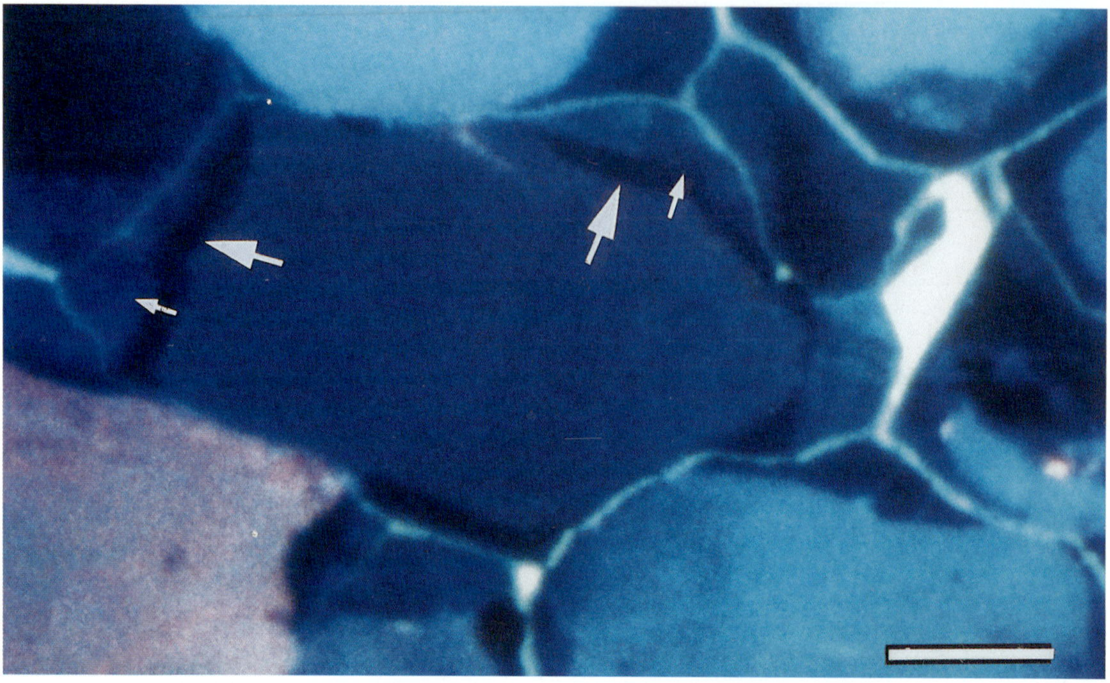

(c)

(d)

**Plate 1.** (*Continued*) (c) Early quartz I (large arrows) as non-luminescing overgrowth followed by later pore-filling, greenish-blue zoned quartz II (small arrows) (Gharif Formation, well F, depth 2738.8 m, CL-image; scale bar is 200 μm). (d) Early diagenetic framework-stabilizing calcite II with floating grains (arrows) (Gharif Formation, well X, depth 1550.2 m, CL-image; scale bar is 500 μm).

the Oman Mountains. To the west, the basin grades into the Rub al Khali Basin of Saudi Arabia (Fig. 1; Tschopp, 1967; Al-Marjeby & Nash, 1986; Hughes Clarke, 1988; Levell *et al.*, 1988; Visser, 1991; Loosveld *et al.*, 1996; Millson *et al.*, 1996).

Due to different tectonic histories (Loosveld *et al.*, 1996) and stratigraphy (Hughes Clarke, 1988), the IOSB can be subdivided into four areas: a southern area, the South Oman Salt Basin, a northern area, the Fahud and Ghaba Salt Basin, the Central Oman Platform and the foreland area of the Oman Mountains (Fig. 1). The southern and northern areas are also two distinct hydrocarbon habitats (Al-Marjeby & Nash, 1986). Most potential reservoir rocks in South Oman are presently at shallow depth, whereas in the northern part of the basin Late Palaeozoic reservoir units are at depths of almost 5000 m.

## Haushi Group sediments

The predominantly clastic Haushi Group is divided into two units, the Al Khlata Formation and the Gharif Formation (Fig. 2). The tillites, conglomerates, sandstones, siltstones and mudstones of the Al Khlata have been dated as late Westphalian to early Sakmarian (Braakman *et al.*, 1982; De la Grandville, 1982; Besems & Schuurman, 1987; Levell *et al.*, 1988). The uppermost unit of the Al Khlata Formation, the lacustrine Rahab shale (Hughes Clarke, 1988), contains rare marine foraminifera in the north, suggesting a marine connection to the basin. The Al Khlata is characterized by a highly variable sediment thickness, from 0 to > 1000 m.

The late Sakmarian to Artinskian Gharif Formation conformably overlies the Al Khlata Formation and is further subdivided into three informal members: the Lower, Middle and Upper Gharif (Focke & van Popta, 1989; Mercadier & Livera, 1993). In southern Oman the Lower Gharif contains locally massive diamictite horizons and laterally extensive cross-bedded sandstones. Towards the north, the siliciclastic sequence laterally grades into the marine Haushi Limestone facies. The Middle Gharif consists of a fining-upward sequence of sandstones, siltstones and mudstones. Its upper part is a uniform, massive mudstone, correlatable over most of the basin. The Upper Gharif is composed of sand sheets which expand stratigraphically and split into discrete or multistorey sand bodies, separated by red-brown mudstones. Except along the northern and south-eastern margins of the IOSB, the Gharif Formation is characterized by quite constant sediment thickness, varying from 200 to 350 m.

Palaeoenvironmental reconstructions (Braakman *et al.*, 1982; Besems & Schuurman, 1987) show that the Al Khlata sediments are of glacio-lacustrine, glacio-fluvial and locally glacio-marine origin, deposited during several advances and retreats of ice sheets during the Late Palaeozoic Gondwana glaciation. The Rahab Shale is interpreted as a basin-wide lacustrine deglaciation horizon (Hughes Clarke, 1988), marking the onset of general warming. The shallow marine, coastal plain and fluvial plain sediments of the Gharif were the products of glacio-eustatic sea level changes superimposed on a general eustatic sea level rise caused by Gondwana deglaciation (Guit *et al.*, 1994). The appearance of red-bed diagenesis, typical for semiarid to arid conditions (Walker, 1976), within the Middle Gharif is another clear indication of general warming during the Permian.

## Burial and thermal history

The burial and thermal history of the basin has been reconstructed using PetroMod® (Hermanrud, 1993; IES GmbH, 1994; Sachsenhofer, 1994; Büker *et al.*, 1995; Welte *et al.*, 1997) along a geological cross-section through the study area (Fig. 3). Input parameters for the model are described in Hartmann (1996).

The stratigraphic sequence of the IOSB reveals evidence of four erosional events of regional extent. However, Visser (1991) showed that none of those erosional events are connected to major inversions of the study area. Throughout the subsidence history, subsidence rates are much higher in the north-western part of the study area, compared with almost no subsidence close to the Huqf–Haushi axis. This led to the wide variation in the present-day depth of the Haushi Group sediments, which show a generally deepening trend from the outcrop area in the Huqf–Haushi region towards the north-west, where they reach burial depths of almost 5000 m.

General palaeo-heatflow values for Oman (Visser, 1991) have been modified to take into consideration the geological history of the northern part of the basin. Heat pulses have been responsible for the thermal doming and rifting, from the end of the Permian through Triassic times and for the obduction of the Hawasina nappes during the Campanian. Modelled present-day formation temperatures are confirmed by corrected borehole temperatures which vary from 140 to 160°C in the north-west, to 60–80°C in wells located close to the outcrop area, where burial did not exceed 2000 m. Representative burial graphs for the two areas show the different temperature histories (Fig. 4). Homogenization temperatures of primary fluid inclusions in late authigenic minerals (quartz, late Fe-dolomite), show values up to 15°C higher than present-day temperatures. These findings are consistent with the modelled burial history (Fig. 4), where maximum temperatures were reached during the

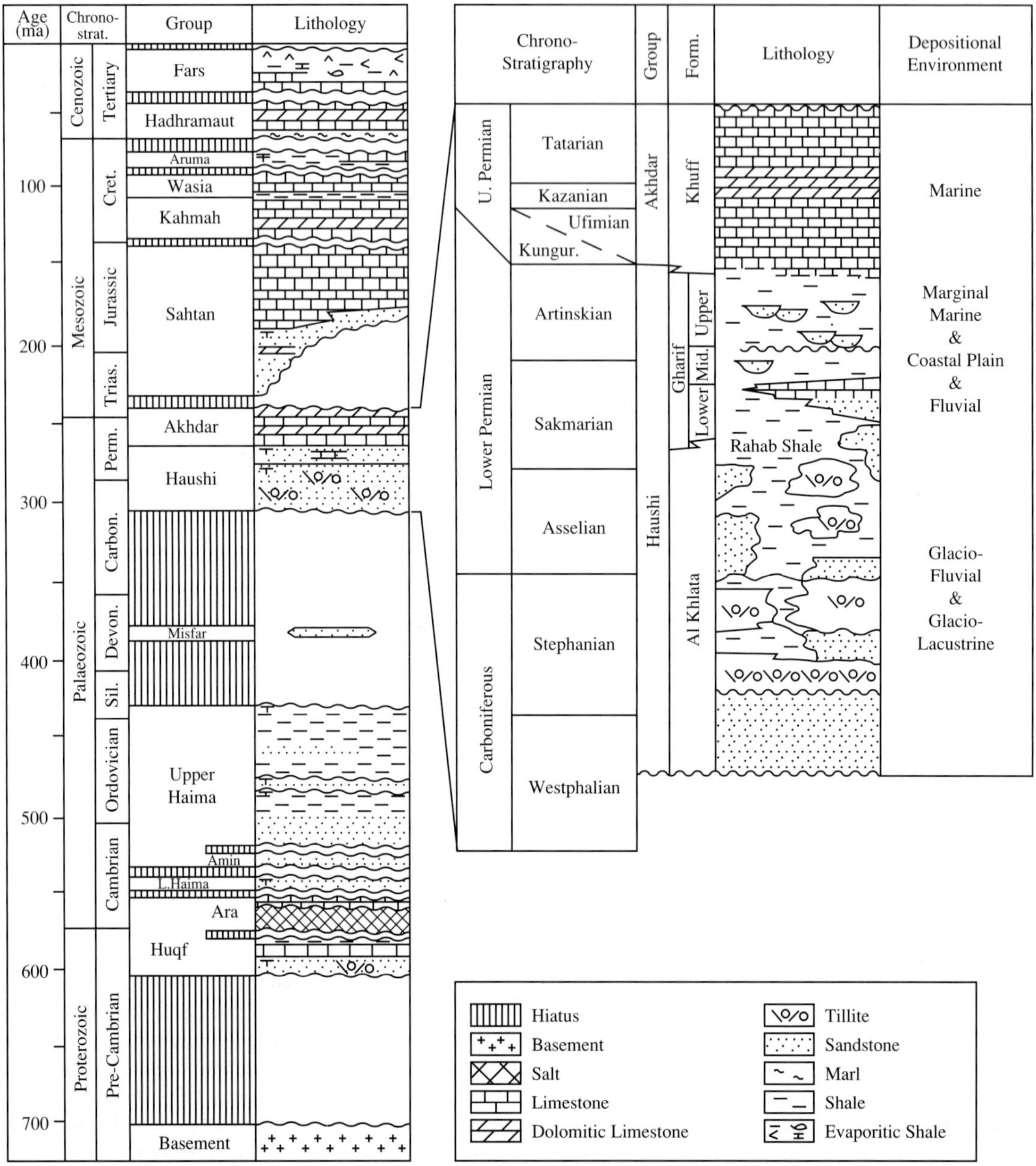


**Fig. 2.** Generalized stratigraphy of the Interior Oman Sedimentary Basin (modified after Loosveld *et al.*, 1996).

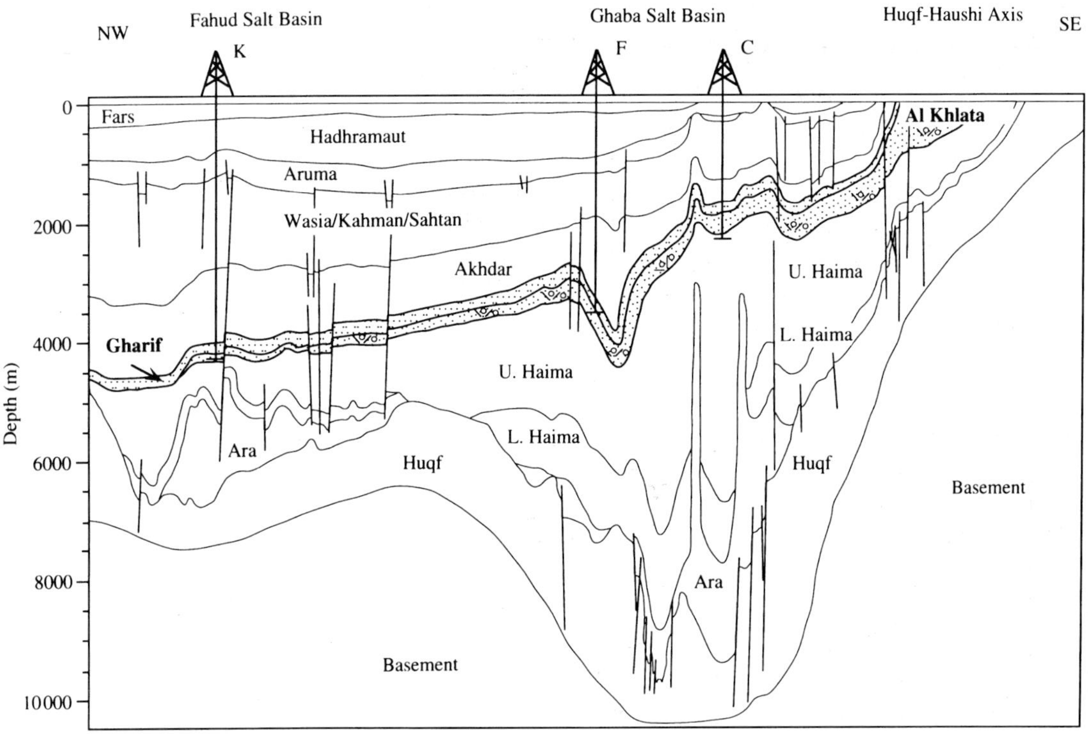


**Fig. 3.** Schematic geological cross-section through the study area based on seismic and well data (modified after Visser, 1991). For locations see Fig. 1.

emplacement of the Hawasina nappes during the Campanian (Hartmann, 1996).

## MATERIAL AND METHODS

### Sampled material

A total of 202 sandstone samples of the Gharif Formation and 80 sandstones from the Al Khlata Formation were taken from 26 deep wells situated in the central and northern part of the IOSB, from three shallow wells from the Huqf–Haushi area (location B on Fig. 1), and from the two outcrop locations in the Huqf–Haushi area (Fig. 1). The sample depths for deep core material range from 1280.8 m to 4861.0 m, whereas the shallow core samples were taken from 3.4 m to 16 m. Samples were taken from every 10–20 cm of the available cores for basic petrographic analysis. However, to reduce geochemical variables and to have a comparable data set between the Gharif and the Al Khlata samples, only matrix-free sandstones with a mean grain size between 0.6 mm and 0.15 mm were selected for this extensive quartz cementation study. In addition, all shaley and silty intervals were sampled for clay mineral analysis.

### Petrography

For petrographic analyses, samples were impregnated with blue-dyed epoxy resin. Polished thin sections were stained for carbonate identification using the method of Dickson (1966). All thin sections were point-counted (400 points) using an optical microscope. Additional quantitative analyses were performed by point counting 66 colour slides (400 points) taken under cathodoluminescence from representative areas of thin sections. A high sensitivity hot-cathode CL microscope was used (Ramseyer *et al.*, 1989). The applied beam current density was 0.3 μA/mm$^2$ at 30 keV electron energy. Pressure solution effects have been quantified through point counting (300 points) overlapping areas between detrital grains on cathodoluminescence images after the method of Houseknecht (1984). The amount of pressure solution is expressed as percentage of the rock volume present

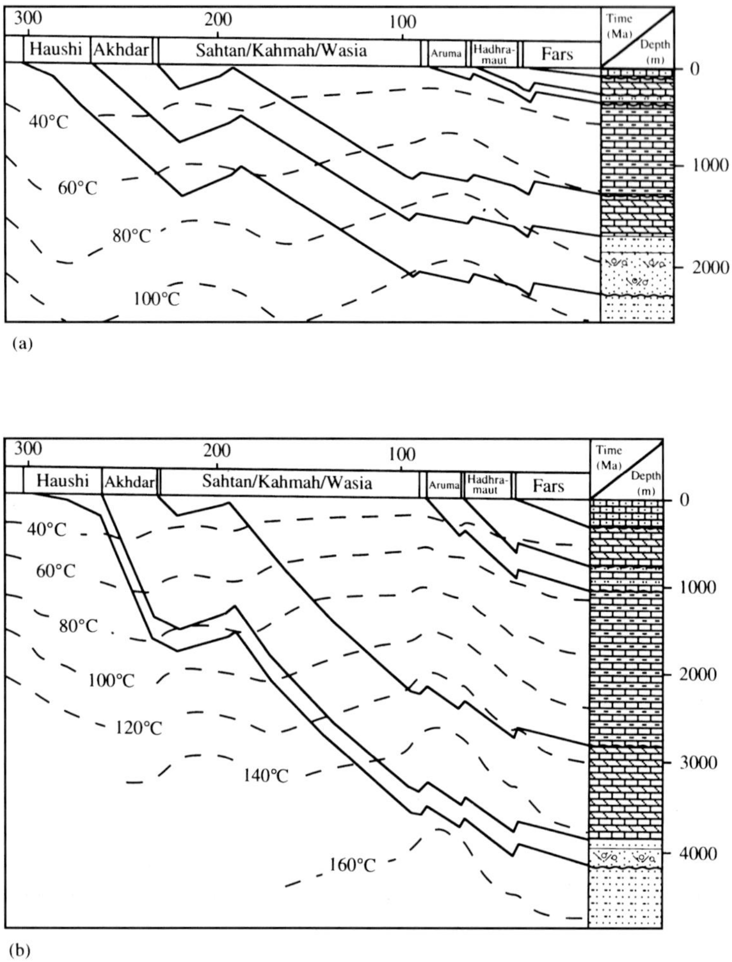


**Fig. 4.** Decompacted burial graphs and temperature history for representative wells in the northern part of the IOSB. (a) Well C located in the south-east. (b) Well K located in the north-west.

after pressure solution. The degree of confidence in estimating overlap quartz varies with the sphericity and roundness of the original detrital quartz grains (Houseknecht, 1988). The studied sandstones of the Al Khlata and Gharif formations have mainly spherical and rounded to well rounded quartz grains, therefore the amount of overlap quartz could be estimated with high confidence. Representative samples were examined on a CamScan S5 scanning electron microscope with a Tracor Northern 5400 EDS spectrometer. The operating conditions for EDS and BSI analyses were 15 kV accelerating voltage, 100 μA emission current and a 2 nA beam current.

## Clay-fraction XRD analysis

Clay mineral analyses from shales, siltstones and sandstones were performed on the < 2 μm fraction. Samples were run air-dried and after ethylene glycol saturation on a Philips PW-1710 diffractometer using Ni-filtered Cu–Kα radiation. Three regions, between 2–11 °2 θ, 14–20 °2 θ and 23–26 °2 θ Cu–Kα were numerically decomposed into 1–6 symmetric Gaussian or Lorenzian peaks with the help of 'DECOMPXR' (Lanson & Velde, 1992). Percentage illite in I–S mixed-layer clay minerals was determined after Moore & Reynolds (1989) from the difference in °2 θ of the numerically decomposed I–S (*001/002*) and I/S (*002/003*) peak positions in the ethylene glycol saturated samples. Quantitative analysis of different clay minerals was performed with the method described by Moore & Reynolds (1989). To differentiate the kaolin polytypes, random orientated samples from the < 6 μm size fraction of shales, siltstones and sandstones were studied.

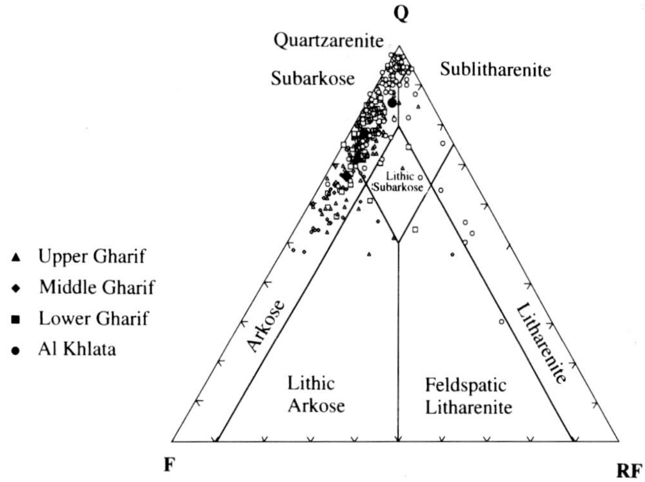


**Fig. 5.** Composition of sandstones of the Haushi Group. Filled symbols represent mean values.

### Isotope geochemistry

Authigenic quartz was separated from detrital grains using the method described by Lee & Savin (1985) and Brint *et al.* (1991). Through sieving, the separated material was concentrated into four size fractions: > 85 μm, 53–85 μm, 30–53 μm, and < 30 μm. Each size fraction was then checked by SEM to check that the overgrowths were free of detrital quartz. Oxygen stable isotope ratios in quartz were measured on a VG SIRA-10 mass spectrometer for all fractions. Results were corrected using standard procedures (Craig, 1957), and are quoted in per mil variation from the SMOW standard. Isotopic reproducibility of standard materials is better than 0.2‰ for $\delta^{18}O$.

### Fluid inclusion microthermometry

For fluid-inclusion microthermometric studies, samples were prepared as 100 μm thick wafers, polished on both sides. Measurements were performed using a Linkham™ semi-automatic, gas-flow freezing-heating stage, calibrated with synthetic fluid inclusions of known composition. The presence of fluorescent inclusions in various cement phases of the studied samples allows us to assume that the pore waters were saturated with respect to methane. Therefore, following the reasoning of Burley *et al.* (1989), no pressure correction was applied and the measured homogenization temperatures ($T_{hom}$) were considered as true trapping temperatures of the inclusions. The $T_{hom}$ measurements were reproducible within an accuracy of ±1°C, whereas for final ice melting temperatures ($Tm_{ice}$) the accuracy was ±0.25°C. Final ice melting temperature of all-liquid inclusions was measured by the method of partial decrepitation (Goldstein & Reynolds, 1994).

## RESULTS

### Detrital composition of sandstones

Monomineralic and polycrystalline quartz (48% to 98%), plagioclase and K-feldspar (2% to 49%), and various rock fragments (3% to 38%) are the main detrital components of the Haushi Group sandstones. The samples plot mainly in the quartz arenite, subarkose and arkose fields (Fig. 5) in a Q–F–RF diagram of McBride (1963). Samples with a higher percentage of rock fragments are exceptions; these are mainly Al Khlata outcrop samples and Gharif sandstones containing reworked caliche clasts. Samples from all three members of the Gharif and from the Al Khlata have comparable compositions, with Al Khlata sandstones being the most mineralogically mature samples. The only observed effect of grain size on mineralogy is a slightly higher percentage of rock fragments in the coarser fraction (Ramseyer, 1983; Hartmann, 1996). No depth-related grain size variation could be observed.

### Present-day pore-water composition

Chemical analyses of formation waters are scarce and only available for the Gharif aquifer from wells with a present-day burial depth between 1200 and 3000 m (Fig. 1). The waters are generally NaCl brines with various amounts of $Ca^{2+}$, $Mg^{2+}$, $SO_4^{2-}$ and $HCO_3^-$. A plume of

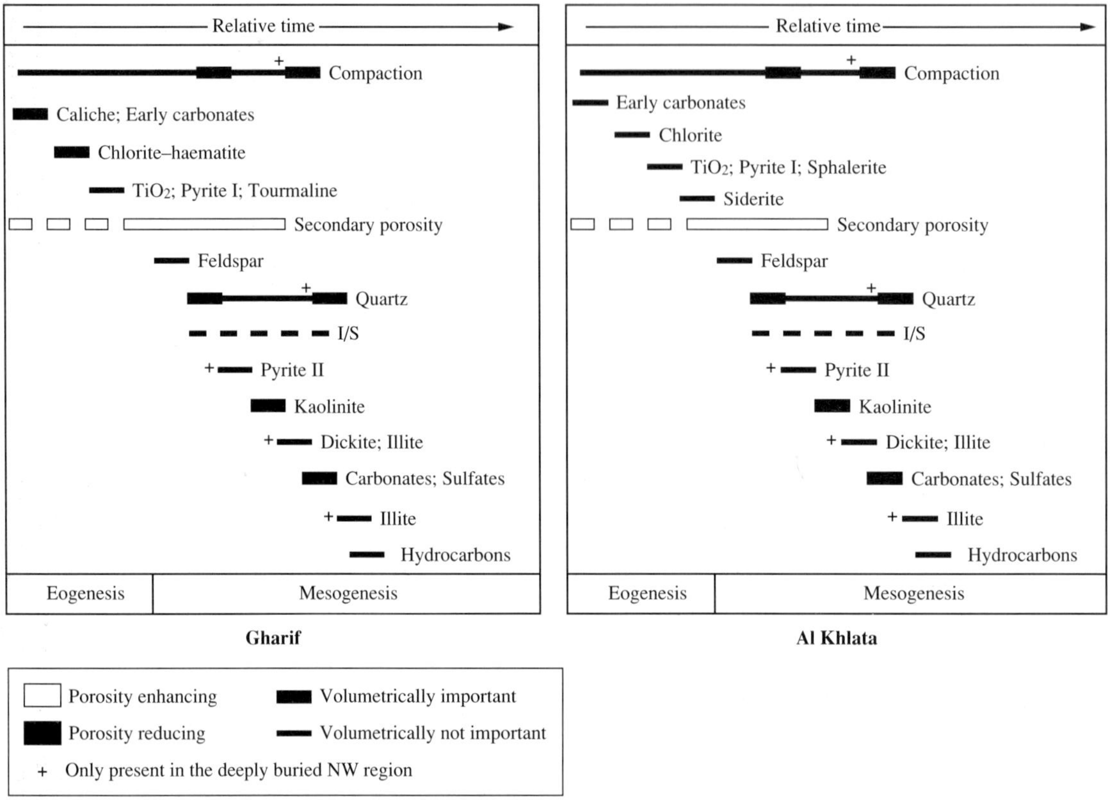


**Fig. 6.** Paragenetic sequences of sandstones of the Gharif and Al Khlata Formations. Note that the eogenetic alterations are different, whereas the mesogenetic evolutions are identical for the two formations.

low salinity waters extends from the recharge area in Dhofar, the Qara Arch (Fig. 1; Al Lamki & Terken, 1996), towards the north-east, reaching total dissolved solid values (TDS) of 38–41 g/L in well Y and 76–77 g/L in wells X and V (Fig. 1). Wells situated further to the north have TDS values between 218 and 263 g/L. No TDS values exist from the region of the Huqf–Haushi outcrop area, but API gravity data and lower subsurface temperatures suggest that biodegradation and thus also recharge was active along the Huqf–Haushi High during recent past pluvial periods, although there is no recharge today (Al Lamki & Terken, 1996).

## Paragenetic sequence

Diagenetic alterations have been previously studied in the Gharif Formation (Hartmann, 1996) and the Al Khlata Formation (Ramseyer, 1983; Juhász-Bodnár *et al.*, 1997). The different burial and thermal histories of the north-west and south-east (Fig. 4) are reflected in the paragenetic sequences of the Haushi Group sandstones by the presence or absence of quartz II, pyrite II, dickite and late illite. Figure 6 gives an overview of the authigenic phases and processes and their relative timing. For the purpose of this paper, of the diagenetic features recognized, only those important for quartz cementation are discussed and only quartz cement and penecontemporaneous silica-providing processes are described in detail.

Eogenetic phases reflect climatic conditions during deposition of the Haushi Group. Tangential grain-coating clay minerals, together with haematite crystallites in fluvial sandstones, dolocretes and calcretes indicate deposition under semiarid to arid red-bed diagenetic conditions in the Upper and Middle Gharif. These red-bed phases are absent in the marine Lower Gharif and glaciogenic Al Khlata.

Mesogenetic features vary between the north-west and south-east of the study area. The paragenetic sequence of wells D and E in the south-east (Figs 1 & 6) are taken as representative for mesogenetic alteration before the onset

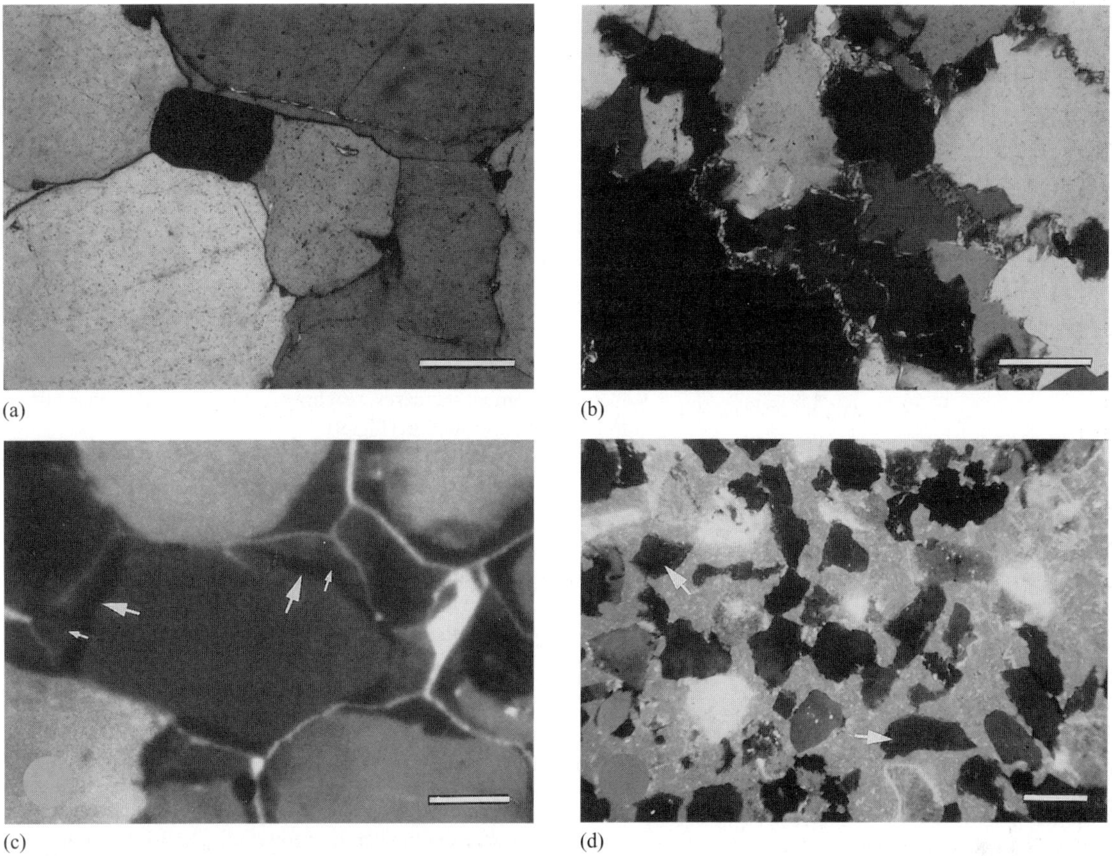


**Fig. 7.** (*Also reproduced in colour, see Plate 1, facing p. 254.*) (a) Sandstone (sample A, Figs 13 & 14) with 4% of overlapping grain boundaries and 20% quartz cement acting as silica importer (Al Khlata Formation, well I, depth 3592.5 m; scale bar is 200 μm). (b) Sandstone (sample B, Figs 13 & 14) with 1.5% quartz cement and 32% pressure solution, acting as silica exporter (Al Khlata Formation, well I, depth 3593.9 m; scale bar is 200 μm). (c) Early quartz I (large arrows) as non-luminescing overgrowth followed by later pore-filling, greenish-blue zoned quartz II (small arrows) (Gharif Formation, well F, depth 2738.8 m, CL image; scale bar is 200 μm). (d) Early diagenetic framework-stabilizing calcite II with floating grains (arrows) (Gharif Formation, well X, depth 1550.2 m, CL image; scale bar is 500 μm).

of intense burial, whereas well K in the north-west shows the effect of intense burial (Figs 1 & 6). Sandstones from the south-east are characterized by the presence of early calcite (Fig. 7d; see also Plate 1, facing p. 254) and abundant dissolution features. Framework-stabilizing calcite formed either before the onset of compaction, or as a replacement cement after dissolution of framework-stabilizing anhydrite or carbonate. Petrographic evidence for this early calcite is exclusively found in Gharif sandstones without eogenetic grain coating clay minerals. Dissolution and alteration of chemically unstable grains and cements such as feldspars, calcite and, to a lesser extent, rock fragments and detrital clay minerals, generated intragranular porosity in all sandstone samples from the surface to a depth of almost 2600 m. Dissolution provides a variety of ions and is closely connected to the precipitation of most mesogenetic phases such as authigenic quartz, feldspar, calcite, dolomite/ankerite, kaolinite and sulphates (Fig. 6).

The clay mineralogy of sandstones changes with exposure to higher burial temperatures. Chlorite shows an increase of the IIb $\beta = 97°$ polytype. The transformation of kaolinite into dickite also occurs, starting at temperatures around 100°C. A variety of expandability versus depth trends for I–S mixed-layer clay minerals is found in the different members of the Haushi Group (Fig. 8). These differences can be explained by the influence of distinct weathering conditions during sedimentation of the units, resulting in different initial compositions of the I–S mixed-layer clays (Hartmann *et al.*, 1995; Hartmann

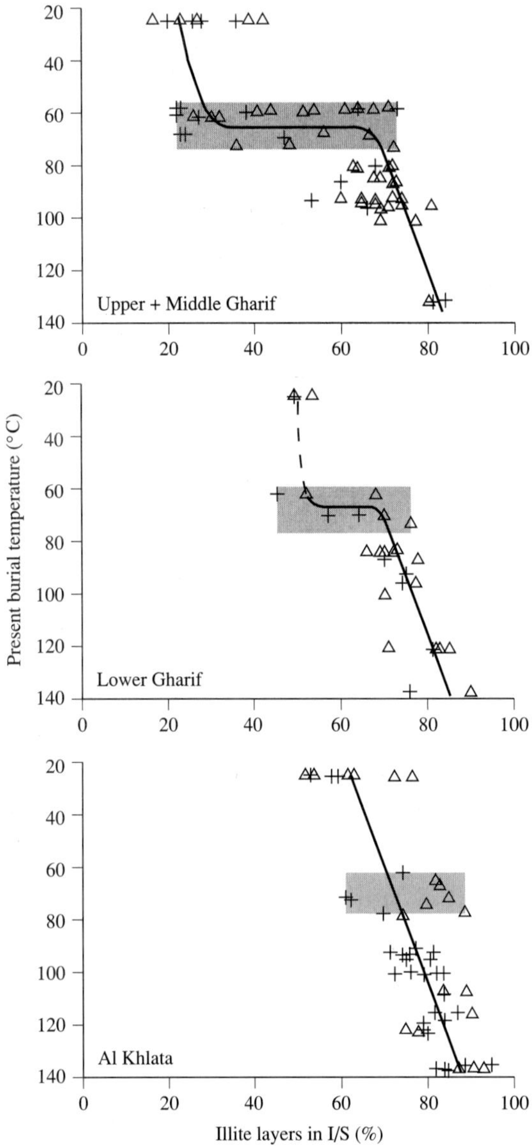


**Fig. 8.** Temperature–composition relations of I–S mixed-layer clay minerals from various stratigraphic units of the Haushi Group sandstones (triangles) and shales (crosses). Note that the starting composition and the depth and temperature trends of illite layers in I–S mixed-layer clay minerals depend on the stratigraphic unit. The shaded areas highlight the isothermal reaction zone where the smectite to illite transformation provides most of the silica. (From Hartmann *et al.*, 1999.)

*et al.*, 1999). Illite in the sandstones was formed by the reaction of earlier authigenic kaolinite/dickite minerals and detrital feldspar under elevated temperature. Sandstones from the north-west are characterized by intense compaction. Solid hydrocarbons appear in both areas as the final phase in the paragenetic sequence.

### *Quartz cement*

Quartz cement is present as syntaxial overgrowths, varying from 0.1 to 25.5 vol%. In samples from shallow cores and outcrops, quartz cement comprises only up to 4 vol%. With increasing present-day burial, the volume of quartz cement increases, reaching 25 vol% in sandstones buried below 3000 m (Fig. 9).

In sandstones buried to less than 2600 m, authigenic quartz (quartz I) has an average thickness of 20–30 μm, rarely exceeding 150 μm. The overgrowths post-date early framework-stabilizing calcite but pre-date late carbonates and sulphates (Figs 6 & 7d). In sandstones buried below 2600 m, quartz cement (quartz II), occludes most of the intergranular porosity. Quartz II post-dates or is synchronous with late sulphates and carbonates (Fig. 7a,c; see also Plate 1, facing p. 254). Cathodoluminescence petrography of deeply buried sandstones shows that quartz cement is of multiphase origin (Fig. 7c; Plate 1, facing p. 254). Quartz I is non-luminescent and is present in all samples, whereas quartz II shows a polyphased internal structure with alterating zones of greenish-blue luminescence and only occurs below 2600 m in the north-west.

The majority of fluid inclusions have been found in quartz II. Quartz I overgrowths are presumably too thin and precipitated too slowly to trap many measurable inclusions. Primary liquid–vapour fluid inclusions are trapped in quartz II, whereas quartz I contains assemblages of liquid–vapour inclusions together with all-liquid inclusions. The all-liquid inclusions are assumed to be metastable at room temperature as is common for inclusions trapped below approximately 80°C (Goldstein & Reynolds, 1994). Measured homogenization temperatures in quartz I are 76–78°C, and for quartz II range between 90 and 135°C. Final ice melting temperatures range between –4.7 and –24°C (Fig. 10). The measured eutectic temperatures range from –47 to –58°C which is typical for the eutecticum of a system including a $CaCl_2$-component (–52°C) or a system including additionally a $MgCl_2$-component (–57°C) (Goldstein & Reynolds, 1994).

Stable oxygen isotopic composition was measured in four size fractions of a sample from well S (Table 1). Results show an increase of the $\delta^{18}O$ values from 7.8 to 12.1‰ with decreasing size fraction.

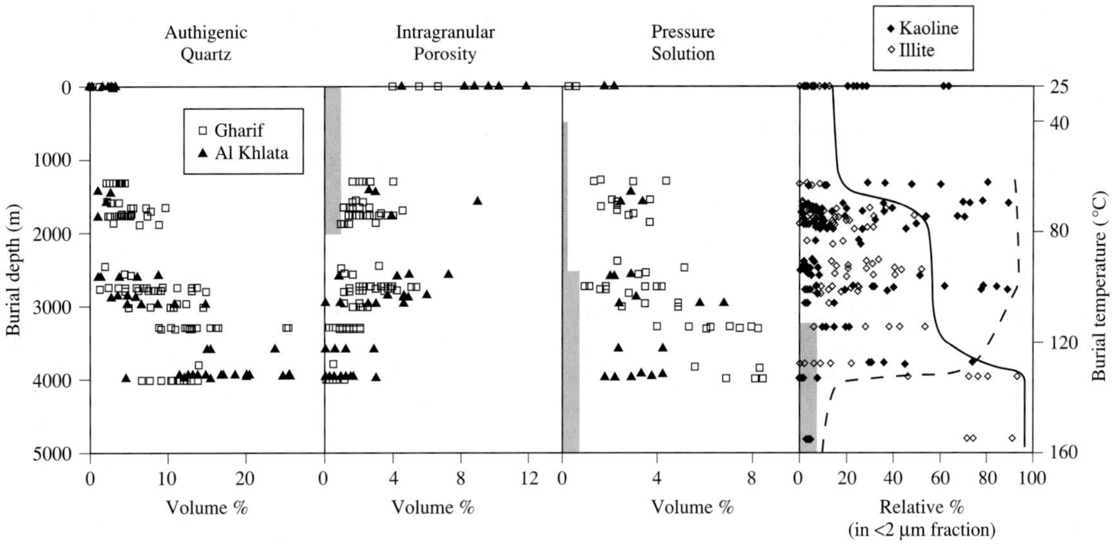


**Fig. 9.** Distribution of authigenic quartz, intragranular porosity, point-counted pressure solution, kaolinite and illite, versus present-day burial depth and temperature. Shaded areas represent the burial and temperature regime where diagenetic alterations involving the different phases provide silica for quartz cementation. The dashed and solid lines represent maximum values of kaolinite and illite, respectively.

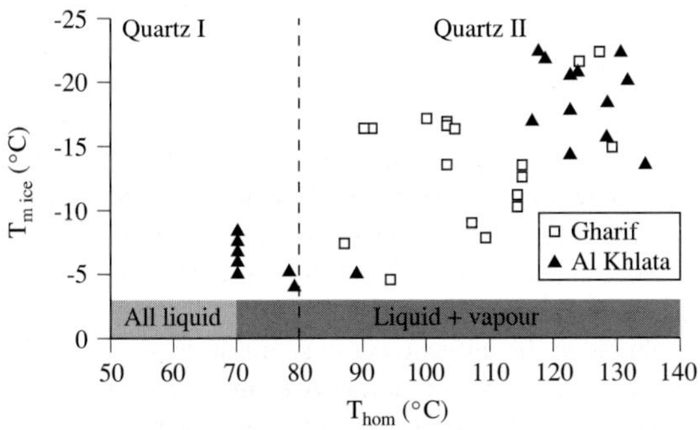


**Fig. 10.** Results of fluid inclusion microthermometry in quartz I and II of various samples from the Haushi Group.

**Table 1.** Oxygen isotopes of various size fractions from a Gharif sandstone

| Well | Depth (m) | $\delta^{18}O_{SMOW}$(‰) | | | |
|---|---|---|---|---|---|
| | | < 30 μm | 30–53 μm | 53–85 μm | > 85 μm |
| S | 2375.8 | 12.1 | 10.8 | 7.8 | 7.8 |

*Dissolution of aluminosilicates*

Evidence for dissolution of aluminosilicates is found in the porosity distribution of Haushi Group sandstones. While most of the porosity is intergranular, intragranular porosity in feldspars, and to a smaller extent in rock fragments, can be up to 12 vol% (Fig. 9). Dissolution locally generated 'ghost-structures', where clasts were completely dissolved, leaving only the clay coatings behind. Figure 9 gives the depth-related distribution of point-counted intragranular porosity.

*Smectite to illite transformation in mixed-layer clay minerals*

An increase in illite within I–S mixed-layer clay minerals during burial is observed (Fig. 8). Separating the samples into the different stratigraphic units of the Haushi Group reveals a different rate of increase in the percentage illite in I–S with depth for each unit (Fig. 8). In addition, a considerable degree of heterogeneity in the lowest temperature range exists (Fig. 8), probably reflecting a range of initial compositions of detrital I–S during and after deposition of the sediments (Pollastro, 1990). In the Upper + Middle and Lower Gharif units a drastic change in I–S composition occurs in the temperature interval between 60 and 80°C, which is almost isothermal.

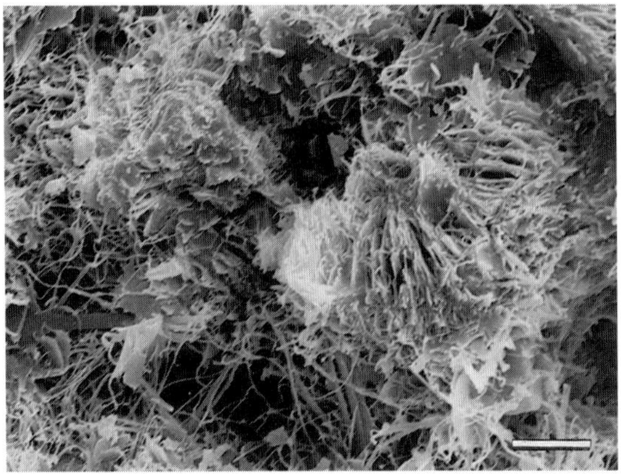

**Fig. 11.** SEM image of a vermicular kaolinite being altered to hairy illite (Al Khlata Formation, well I, depth 3592.5 m; scale bar is 10 μm).

*Pressure solution*

Concave–convex and sutured grain contacts and stylolitization are evidence of chemical compaction. Samples from the surface to 2600 m show mainly point and concave–convex grain contacts with minor pressure solution of up to 4 vol%. Within this depth range, no depth related trend is noticeable. Below 2600 m sutured grain contacts are more common and a marked increase in the point counted volume of pressure solution, from 4 to almost 9 vol%, is observable (Fig. 9). Stylolites occur below 3500 m in wells K, J, and I. Maximum observed pressure solution along grain contacts increases with deeper burial (Figs 9 & 7b).

*Illitization of kaolinite*

In deeply buried sandstones, intense illitization of kaolinite (Fig. 11), clay matrix and clay-rich clasts has taken place. The point counted volume of illite ranges from 0 to 2 vol% in samples below 2800 m. In well F, minor amounts of pore-bridging illite could be detected. Commonly, illite replaces feldspar grains, starting at cleavage boundaries and twinning planes. In some cases illite forms 'ghost structures', marking now completely dissolved feldspar clasts. Figure 9 shows this change from a kaolinite dominated clay fraction to an illite dominated one in samples from a depth of ~ 4000 m.

## DISCUSSION

The variation in the paragenetic sequences is reflected not only in the different temperature and pressure regimes in the shallow and deeply buried parts of the basin, but also in the pore-water evolution. Present-day pore-water composition (Fig. 1), API gravity data (Al Lamki & Terken, 1996) together with intense dissolution in sandstones buried < 2600 m, reveal the presence of a long-term meteoric influence, indicating a downdip flow towards the north-west away from the recharge areas in the Dhofar, the Huqf–Haushi area and the Oman Mountains (Al-Lamki & Terken, 1996). In contrast, in the deeply buried area of the north-west, salinity values from the present-day pore-waters and from fluid inclusions of late authigenic phases are considerably higher than expected for the connate waters of the Gharif. These saline brines are derived from the evaporites and dolomites of the Late Proterozoic Ara salt. Halokinesis brought the Ara sediments in close proximity to the Haushi Group, allowing the brines to invade the sandstones (Hartmann, 1996).

### Authigenic quartz precipitation

Primary fluid inclusions within quartz of the Haushi Group do not appear to have stretched or leaked, and thus can provide valuable information on the diagenetic evolution of quartz (Osborne & Haszeldine, 1993; Osborne & Haszeldine, 1995; Walderhaug, 1995). Final ice melting temperatures of primary fluids in authigenic quartz decrease with increasing homogenization temperatures (Fig. 10). Salinities deduced from these final ice melting temperatures increase with increasing precipitation temperature, showing generally lower values during quartz I crystallization than during quartz II formation. In sandstones from the south-east, this difference can be attributed to the influence of meteoric water during shallow burial, and to the increasing importance of highly saline brines derived from the Ara salt during deep burial in the north-west. This fluid evolution is also recorded in inclusions of other authigenic minerals of the Gharif Formation (Hartmann, 1996).

Cathodoluminescence petrography shows the presence of only quartz I in a sample from well S of the Gharif Formation, for which separation of authigenic quartz was partly successful and oxygen isotope data are available. Fortunately, the oxygen isotope evolution of Gharif pore-waters is well known from measurements of various carbonate phases and kaolinite (Hartmann, 1996). Together with the quantification of different quartz phases from thin sections and SEM images, this information allows an interpretation of the oxygen isotope data. The values obtained from well S are below the typical range of authigenic quartz, between 13 and 31‰ (Brint *et al.*, 1991). This indicates significant contamination by detrital

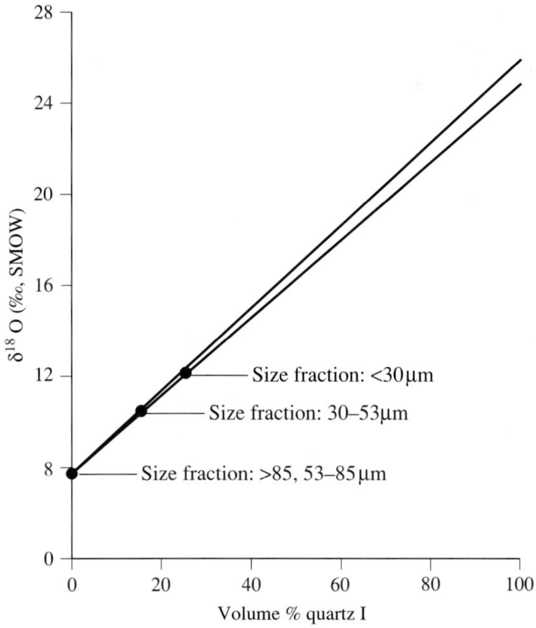


**Fig. 12.** Interpretation of the $\delta^{18}O$ data for a sample of the Gharif Formation (well S, depth 2378.5 m). The sample contains only quartz I and detrital quartz. Quantification of quartz I was achieved by interpreting SEM images of the different size fractions.

material in all size fractions (Lee & Savin, 1985). The SEM analysis shows that the two coarse sample fractions are almost exclusively detrital quartz. Thus, the average $\delta^{18}O$ value of 7.8‰ of these samples represents the oxygen isotope composition of detrital quartz. Based on SEM observations, the < 30 μm fraction contains approximately 25 vol% authigenic quartz and the 30–53 μm fraction has 15 vol% authigenic quartz. A linear interpolation of those values yields a $\delta^{18}O$ value for quartz I between 25 and 26‰ (Fig. 12). Using the fractionation equation from Friedman & O'Neil (1977) for authigenic quartz and assuming a temperature of precipitation around 70–80°C, the $\delta^{18}O$ values of quartz I translate into $\delta^{18}O$ values from −1 to +1‰ for the precipitating fluid. This estimation is supported by post-dating kaolinite, which has a $\delta^{18}O$ value of 16.5‰ (Hartmann, 1996) for samples of similar depth. Using the fractionation equation from Friedman & O'Neil (1977) for kaolinite and assuming a precipitation temperature slightly higher than for quartz (e.g. 80–90°C), then the $\delta^{18}O$ value of the pore-fluid (−1‰ to −2‰) is similar to that from which quartz I crystallized. The low $\delta^{18}O$ value of the pore-fluid supports the incursion of meteoric surface water during shallow burial.

## Silica-providing processes

Knowledge of the source and transport mechanism of silica is essential for a quantitative understanding of quartz cementation. To explain significant amounts of authigenic quartz, several authors (Haszeldine *et al.*, 1984; Land *et al.*, 1990; Gluyas & Coleman, 1992) have claimed that silica must have been introduced from some distance outside a sandstone by advective flow. Because of low silica solubility at pH < 9, large amounts of quartz precipitation from migrating pore-water is only possible if significant volumes of silica are transported as ionic complexes not yet identified (McBride, 1989), or if pore-water fluxes are very large. Based on the modelling of coupled carbonate–silicate reactions during convective pore-water flow, and on pre-Tertiary sandstone diagenesis and fluid evolution in North Sea reservoirs, Bjørlykke & Egeberg (1993) and Aplin *et al.* (1993) suggest that the source of silica precipitated as quartz in sandstones is predominantly local, transported by diffusion over a distance of few centimetres to metres.

Although there are many possible local silica-providing sources, the following four have been identified in the paragenetic sequence of the Haushi Group siliciclastics, as being contemporaneous with silica precipitation: dissolution of aluminosilicates, smectite to illite transformation in I–S mixed-layer clay minerals, pressure solution and illitization of kaolinite.

### *Dissolution of aluminosilicates*

Feldspar dissolution and kaolinitization are the primary diagenetic features produced during shallow burial of Haushi samples. They are caused by $CO_2$-charged meteoric water infiltration by gravitational force along the basin-margin into the subsurface. Low-salinity meteoric water presently affects the Gharif pore-fluids down to 2000 m (Al Lamki & Terken, 1996). Dissolution is thus an eogenetic to early mesogenetic feature.

Dissolution of feldspars and to a smaller extent rock fragments results in the creation of secondary porosity in the sandstones of the Haushi Group. Kaolinite minerals occur predominantly in secondary macropores, suggesting a genetic connection to the hydrolysis of feldspars according to the following reaction:

$$\underset{\text{K-feldspar}}{2KAlSi_3O_8} + 2H^+ + H_2O = \underset{\text{Kaolinite}}{Al_2Si_2O_5(OH)_4} + 2K^+ + 4SiO_2$$

| | K-feldspar | Kaolinite |
|---|---|---|
| Molar volume: (Robie & Hemingway, 1995) | 217.4 | 90.8 |
| Volume ratio: | 1.0 | 0.42 |

For every mole of K-feldspar altered to kaolinite, two moles of silica are released and made available for quartz cementation (Siever, 1957). Assuming that all the counted intragranular porosity has been created by K-feldspar dissolution and that all the dissolved K-feldspar has been altered to kaolinite, and all the silica released formed quartz cement, then 1–5 vol% intragranular porosity (Fig. 9) in K-feldspar is equal to 0.4–2.1 vol% quartz cement. Thus only a small fraction of the quartz cement observed in the samples may be sourced by this process. However, for shallow samples, with < 2600 m burial depth this source may be an important one and corresponds well with the occurrence of only quartz I in this depth range as is shown on Fig. 9.

### *Smectite to illite transformation in mixed-layer clay minerals*

Transformation of smectite to illite in I–S mixed-layer clay minerals, a common process in sandstones and shales of the Haushi Group, also provides silica during burial (Hartmann *et al.*, 1995; Hartmann *et al.*, 1999). Assuming K-feldspar as the most likely source of potassium, the transformation can be described by the following reaction (Abercrombie *et al.*, 1994):

$$\underset{\text{K-feldspar}}{KAlSi_3O_8} + \underset{\text{K-smectite}}{2K_{0.3}Al_{1.9}Si_4O_{10}(OH)_2}$$

$$= \underset{\text{Illite}}{2K_{0.8}Al_{1.9}(Al_{0.5}Si_{3.5})O_{10}(OH)_2} + 4SiO_{2(aq)}$$

Composition–temperature relationships for the Upper–Middle Gharif sandstones and shales reveal a considerable degree of heterogeneity in the lower temperature range, with a starting composition of 20–40% illite in the mixed-layer clay minerals (Fig. 8). An abrupt change in mixed-layer composition occurs at a depth of 1500 m (~60°C) over a small range of temperature (±20°C). Data for the Lower Gharif (Fig. 8) can be interpreted in a similar way, but indicate a starting composition of around 50% illite in the mixed-layer clays and a less abrupt change in the I–S composition at 1500 m depth. The Al Khlata data are considerably different from the Gharif. Mixed-layer clays in the Al Khlata with a starting composition of about 60% illite in the I/S mixed-layer minerals exhibit very little change in composition with increasing temperature (Fig. 8).

The abrupt transformation in the Gharif is characterized by an elevated reaction rate, providing larger volumes of silica at that depth. Since the rapid reaction is not observed in sandstones from the Al Khlata Formation, only Gharif sandstones should be affected by an increase in authigenic quartz at around 1500 m. In fact, the sandstones of the Gharif Formation show a higher amount of quartz cement at around 1500 m than the Al Khlata sandstones (Fig. 9).

### *Pressure solution*

Intergranular pressure solution is often thought to be the most important silica-providing process in quartzose sandstones (Robin, 1978; Bjørlykke *et al.*, 1986; Houseknecht, 1988; Bjørkum *et al.*, 1993). Porter & James (1986) evaluated the change of quartz solubility at grain contacts with increasing burial depth of quartzose sandstones. Influenced by pressure, temperature and pore-water salinity, the quartz solubility for near-neutral pH conditions is seven times greater at 4 km than at 1 km.

The change from mainly point and concave–convex to sutured grain contacts at around 2600 m is associated with a marked increase in the amount of pressure solution from 4 to almost 9 vol% (Fig. 9). In the same depth interval, quartz cementation also increases from 10 to over 20 vol%.

The silica budget (Houseknecht, 1988) in Fig. 13 shows that sandstones consisting of only quartz I are more or less balanced with respect to silica from pressure solution, whereas sandstones also containing quartz II cement are clearly silica importers with respect to silica created internally by pressure solution. This latter relationship has been described by numerous authors for sandstones of a wide age range from many sedimentary basins (Sibley & Blatt, 1976; Dutton, 1987; McBride, 1987). It indicates that there are other sources of silica and/or that the silica system is open on a thin-section scale. In this study we have evidence for both mechanisms, dissolution of aluminosilicates and illite to smectite transformation as additional sources, and the inhomogeneous distribution of pressure solution and quartz cement on a centimetre to metre scale indicating an open system. Figure 7(a,b) is representative of the wide variation in the amount of quartz cement and the degree of pressure solution within samples of the same well from almost identical depths. Figure 7(a) shows a sandstone buried 3592.5 m with 4% overlapping grain boundaries and 20% quartz cement (sample A). Sample B from 3593.9 m (Fig. 7b) is characterized by 1.5% quartz cement and 32% pressure solution, and must have acted as a silica exporter. Such large differences (Fig. 14) can be explained by differences in the clay content and the earlier diagenetic evolution of the two sandstones. The few compactional features of sample A are indicative of a framework-stabilizing early cement inhibiting compaction. In contrast, sample B shows an abundance of grain-coating illite. In many sandstones a strong negative

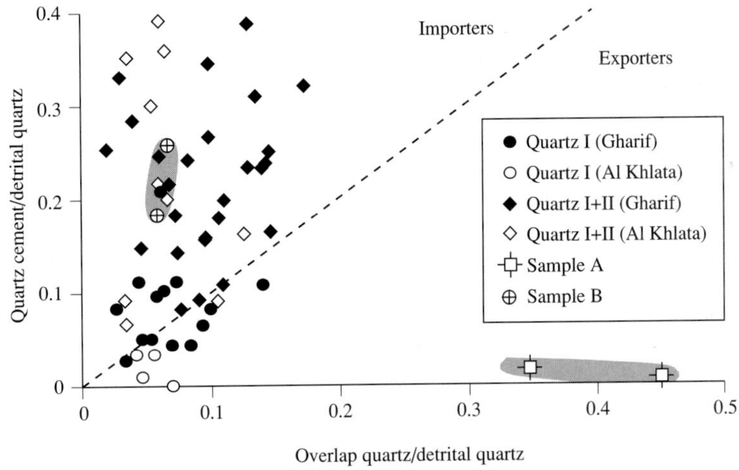


**Fig. 13.** Silica budget plot for Haushi Group sandstones. The dashed line represents mass-balance with respect to pressure solution. For a detailed explanation of the differences between samples A and B, refer to Fig. 14.

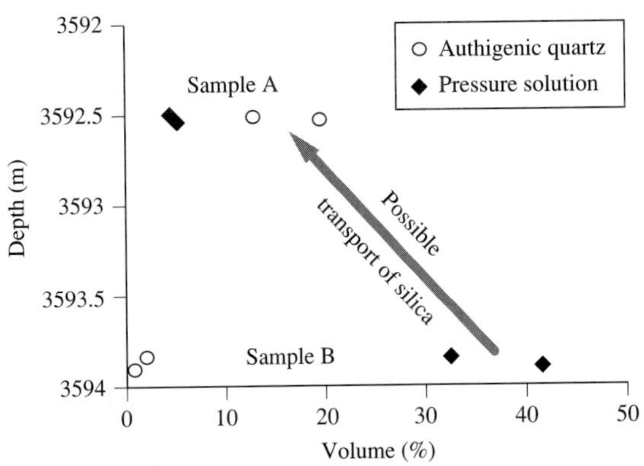


**Fig. 14.** Quantified pressure solution and quartz cement from two closely spaced sandstone samples A and B of the Al Khlata Formation from well I shown on Fig. 7. Note that sample A acted as a silica importer, whereas sample B is clearly exporting silica. This difference may be due to the larger content of grain-coating illite in sample B (see Fig. 7).

correlation between clay content and minus cement porosity can be observed (Tada & Siever, 1989). Several explanations have been proposed for the effect of clay on intergranular pressure solution. Clay on grain contacts enhances the rate of diffusion along the contacts, because a clay film consists of individual platelets with associated water films, thus providing numerous diffusion pathways (Weyl, 1959). Clay coatings also promote pressure solution by impeding quartz overgrowth formation in sandstones (Heald & Larese, 1974; Sibley & Blatt, 1976; Boles & Johnson, 1983). Thus, sandstone layers characterized by grain coating clay minerals are more likely to be severely affected by pressure solution and to act as a source for silica. The silica is exported to, and precipitates in, neighbouring clay-poor sandstones where framework-stabilizing early cement has preserved open pore space. Mullis (1991) concludes from the precipitation kinetics of silica that it is difficult to envisage diffusion of dissolved quartz over distances of more than a few metres. Since the two sandstone samples described above are only 1.4 m apart, diffusion is a plausible transport mechanism. Therefore, eogenetic features, e.g. tangential clay rims, play a crucial role in determining the effects of later diagenetic events such as compaction and late quartz cementation.

Grain-size differences can also influence pressure solution, with very fine sands having 3.3 times the pressure solution index of coarse sand (Porter & James, 1986). Thus, compactional processes in intraformational siltstones and shales may provide silica for quartz cementation in neighbouring sandstones (Shaw & Weaver, 1965; Füchtbauer, 1967; Evans, 1989). This mechanism has been described by Ramseyer (1983) for the Al Khlata Formation in one of the deepest wells in the western part of the IOSB.

### *Illitization of kaolinite*

Illitization of kaolinite may also supply silica for quartz precipitation at elevated temperature according to the equation (Bjørlykke, 1980):

$$\underset{\text{Kaolinite}}{Al_2Si_2O_5(OH)_4} + \underset{\text{K-feldspar}}{KAlSi_3O_8} = \underset{\text{Illite (muscovite)}}{KAl_3Si_3O_{10}(OH)_2} + 2SiO_{2(aq)} + H_2O$$

| | | |
|---|---|---|
| Molar volume: (Robie & Hemingway, 1995) | 140.7 | 45.4 |
| Volume ratio: | 1.0 | 0.32 |

Petrographic evidence from the Haushi Group shows that illitization of kaolinite (Fig. 11) begins at approximately 2800–3000 m at a temperature of approximately 100°C. This temperature estimate is in good agreement with observed illitization reactions in other siliciclastic sediments at temperatures above 100–140°C (Ehrenberg & Nadeau, 1989; Giles *et al.*, 1992; Lanson *et al.*, 1996). However, illitization of kaolinite is volumetrically insignificant as point count data shows only 0–2 vol% authigenic illite in the deeply buried samples of the Al Khlata Formation. Simple mass balance calculations suggest that the maximum amount of silica released by this mechanism would be only 0.7 vol%.

Pressure solution and illitization of kaolinite, occurring in the same temperature range, are coupled through the silica activity in the pore-water. Bjørkum *et al.* (1993) show that illitization can proceed at silica activities several times higher than quartz saturation. Dissolution of quartz along stylolitic grain boundaries on the other hand is very sensitive to silica supersaturation, thus partly explaining the high variability in quartz cementation of Haushi Group sandstones.

## CONCLUSIONS

**1** Dissolution of aluminosilicates, smectite to illite transformation, pressure solution, stylolitization and illitization of kaolinite have been identified as processes providing silica during burial in the Haushi Group sandstones.

**2** The dissolution of aluminosilicates during eogenesis and early mesogenesis contributes to quartz I precipitation.

**3** An increase in the rate of transformation of smectite to illite in mixed-layer clay minerals at a depth of ~ 1500 m releases additional silica. The more smectitic starting composition of the I–S clay minerals in the Gharif Formation relative to the Al Khlata Formation is reflected by the higher amount of quartz I in the Gharif Formation.

**4** Pressure solution and stylolitization are responsible for most quartz II cement. The marked increase in pressure solution at detrital grain boundaries at 2600 m corresponds to the onset of quartz II precipitation at the same depth.

**5** Illitization of kaolinite supplies only minor amounts of silica for quartz precipitation at burial temperatures above 100°C.

**6** Dissolution of aluminosilicates, pressure solution and stylolitization are the major silica-providing processes in the Haushi Group sandstones.

## ACKNOWLEDGEMENTS

This project was carried out with funding provided by the Swiss National Science Foundation (Grants nos. 20–43128.95 and 21–26579.89). Petroleum Development of Oman LLC is acknowledged for providing samples, various additional information and logistic support during field work. Many thanks to Joachim Amthor, Nawal Saoud Al-Kharusi, Han Boelens, Stuart D. Lake, Jos Terken and Kees van der Zwan (all PDO) for their support and discussions during various stays in Muscat. Yves Krüger and Larryn Diamond helped with the fluid inclusion study. Oxygen isotope analysis where carried out by Tony Fallick and Brian Davidson at the SURRC in Glasgow. We appreciate Stephen J. Burns's and Larryn Diamond's helpful comments on earlier drafts of the manuscript. Reviews by Richard Worden, James J. Hickey and an anonymous reviewer lead to significant improvements. We thank the Ministry of Petroleum and Minerals of the Sultanate of Oman and Petroleum Development of Oman LLC for permission to publish this paper.

## REFERENCES

Abercrombie, H.J., Hutcheon, I.E., Bloch, J.D. & de Caritat, P. (1994) Silica activity and the smectite–illite reaction. *Geology* **22**, 539–542.

Al Lamki, M.S.S. & Terken, J.J.M. (1996) The role of hydrogeology in Petroleum Development Oman. *GeoArabia* **1** (4), 495–510.

Al-Marjeby, A. & Nash, D. (1986) A summary of the geology and oil habitat of the Eastern Flank Hydrocarbon Province of South Oman. *Marine and Petroleum Geology* **3**, 306–314.

Aplin, A.C., Warren, E.A., Grant, S.M. & Robinson, A.G. (1993) Mechanism of quartz cementation in North Sea reservoir sandstones: constraints from fluid composition. In: *Diagenesis and Basin Development.* (eds Horbury, A.D. & Robinson, A.G.) pp. 7–22. American Association of Petroleum Geologists Studies in Geology 36, Tulsa.

Besems, R.E. & Schuurman, W.M.L. (1987) Palynostratigraphy of late Palaeozoic glacial deposits of the Arabian Peninsula with special reference to Oman. *Palynology* **11**, 37–53.

Bjørkum, P.A., Walderhaug, O. & Aase, N.E. (1993) A model for the effect of illitization on porosity and quartz cementation of sandstones. *Journal of Sedimentary Petrology* **63**, 1089–1091.

Bjørlykke, K. (1980) Clastic diagenesis and basin evolution. *Revista del Instituto Geológicas. Diputation Provincial, Universitad de Barcelona* **34**, 21–44.

Bjørlykke, K. & Egeberg, P.K. (1993) Quartz cementation in sedimentary basins. *American Association of Petroleum Geologists Bulletin* **77**, 1538–1548.

Bjørlykke, K., Aagaard, P., Dypvik, H., Hastings, D.S. & Harper, A.S. (1986) Diagenesis and reservoir properties of Jurassic sandstones from the Haltenbanken area, offshore mid-Norway. In: *Habitat of Hydrocarbons on the Norwegian Continental Shelf.* (ed. Spencer, A.M.) pp. 275–286. Graham & Trotman, London.

Boles, J.R. & Johnson, K.S. (1983) Influence of mica surfaces on pore-water pH. *Chemical Geology* **43**, 303–317.

Braakman, J.H., Levell, B.K., Martin, J.H., Potter, T.L. & van Vliet, A. (1982) Late Palaeozoic Gondwana glaciation in Oman. *Nature* **299**, 48–50.

Brint, J.F., Hamilton, P.J., Haszeldine, R.S., Fallick, A.E. & Brown, S. (1991) Oxygen isotopic analysis of diagenetic quartz overgrowth from the Brent Sands: a comparison of two preparation methods. *Journal of Sedimentary Petrology* **61**, 527–533.

Büker, C., Littke, R. & Welte, D.H. (1995) 2D-modelling of thermal evolution of Carboniferous and Devonian sedimentary rocks of the eastern Ruhr basin and northern Rhenish Massif, Germany. *Zeitscrift der Deutschen Geologischen Gesellschaft* **146**, 321–339.

Burley, S.D., Mullis, J. & Matter, A. (1989) Timing diagenesis in the Tartan Reservoir (UK North Sea): constraints from combined cathodoluminescence microscopy and fluid inclusion studies. *Marine and Petroleum Geology* **6**, 98–120.

Craig, H. (1957) Isotopic standards for carbon and oxygen and correction factor for mass spectrometric analysis of carbon dioxide. *Geochimica et Cosmochimica Acta* **12**, 133–149.

De la Grandville, B.F. (1982) Appraisal and development of a structural and stratigraphic trap oil field with reservoirs in glacial to periglacial clastics. In: *The Deliberate Search for the Subtle Trap.* (ed. Halbouty, M.T.) pp. 267–286. American Association of Petroleum Geologists Memoir 32.

Dickson, J.A.D. (1966) Carbonate identification and genesis as revealed by staining. *Journal of Sedimentary Petrols* **36**, 491–505.

Dutton, S.P. (1987) Diagenesis and burial history of the Lower Cretaceous Travis Peak Formation. *Texas Bureau of Economic Geology, Report of Investigations* **164**.

Ehrenberg, S.N. & Nadeau, P.H. (1989) Formation of diagenetic illite in sandstones of the Garn Formation, Haltenbanken Area, Mid-Norwegian continental shelf. *Clay Minerals* **24**, 233–253.

Evans, I.J. (1989) Geochemical fluxes during shale diagenesis, an example from the Ordovician of Marocco. In: *Water–Rock Interaction WRI-6.* (ed. Miles, D.L.) pp. 219–222. Balkema, Rotterdam.

Focke, J.W. & van Popta, J. (1989) Reservoir evaluation of the Permian Gharif Formation, Sultanate of Oman. *Society of Petroleum Engineers* **17978**, 517–528.

Friedman, I. & O'Neil, J.R. (1977) Compilation of stable isotope fractionation factors of geochemical interest. In: *Data of Geochemistry.* (ed. Fleischer, M.) pp. 1–12. United States Geological Survey Special Paper, 440-KK.

Füchtbauer, H. (1967) Influence of different types of diagenesis on sandstone porosity. *7th World Petroleum Congress* **2**, 353–369.

Giles, M.R., Stevenson, S., Martin, S.V. *et al.* (1992) The reservoir properties and diagenesis of the Brent Group: a regional perspective. In: *Geology of the Brent Group.* (eds Morton, A.C., Haseldine, R.S. Giles, M.R. & Brown, S.) pp. 289–327. Geological Society Special Publication 61, London,

Gluyas, J. & Coleman, M. (1992) Material flux and porosity changes during sediment diagenesis. *Nature* **315**, 52–54.

Goldstein, R.H. & Reynolds, T.J. (1994) *Systematics of Fluid Inclusions in Diagenetic Minerals.* Society of Economic Paleontologists and Mineralogists Short Course 31, Tulsa.

Guit, F.A., Al-Lawati, M.H. & Nederlof, P.J.R. (1994) Seeking new potential in the Early–Late Permian Gharif play. west Central Oman. *GeoArabia* **1**, 447–462.

Hartmann, B.H. (1996) Diagenesis and pore-water evolution of the Lower Permian Gharif Formation, Sultanate of Oman. Bern. PhD Thesis, University of Bern, Switzerland.

Hartmann, B.H., Hillier, S. & Ramseyer, K. (1995) Possible effects of palaeoclimatic conditions on smectite to illite transformation, Haushi Group, Sultanate of Oman. *1st SEPM Congress on Sedimentary Geology: Linked Earth Systems, St. Petersburg, FL.*

Hartmann, B.H., Juhász-Bodnár, K., Ramseyer, K. & Matter, A. (1999) Effect of Permo-Carboniferous climatic conditions on I/S mixed-layer clay composition, Haushi Group, Sultanate of Oman. *Clays and Clay Minerals* **47**, 131–143.

Haszeldine, R.S., Samson, I.M. & Cornford, C. (1984) Quartz diagenesis and convective fluid movement: Beatrice Oilfield, UK North Sea. *Clay Minerals* **19**, 391–402.

Heald, M.T. & Larese, R.E. (1974) Influence of coatings on quartz cementation. *Journal of Sedimentary Petrology* **44**, 1269–1274.

Hermanrud, C. (1993) Basin modelling techniques—an overview. In: *Basin Modelling: Advances and Applications.* (eds Doré, A.G.) pp. 1–34. NPF Special Publications 3. Elsevier, Amsterdam.

Houseknecht, D.W. (1984) Influence of grain size and temperature on intergranular pressure solution, quartz cementation, and porosity in a quartzose sandstone. *Journal of Sedimentary Petrology* **54**, 348–361.

Houseknecht, D.W. (1988) Intergranular pressure solution in four quartzose sandstones. *Journal of Sedimentary Petrology* **58**, 228–246.

Hughes Clarke, M.W. (1988) Stratigraphy and rock unit nomenclature in the oil-producing area of interior Oman. *Journal of Petroleum Geology* **11**, 5–60.

IES GmbH. (1994) PetroMod® *Manual. Vers. 2.4.* Jülich, 603pp.

Juhász-Bodnár, K., Büker, C., Ramseyer, K. & Matter, A. (1997) Changes in fluid composition during quartz cementation: Evidence from fluid inclusions, Upper Palaeozoic Haushi Group sandstones, Sultanate of Oman. In: *Gaea Heidelbergensis: Abstracts.* (eds Bechstädt, T., Bengtson, P., Greiling, R. & Schweizer, V.) 3, pp. 183–184.

Land, L.S., Mack, L.E., MacPherson, G.L. & Milliken, K.L. (1990) Open system diagenesis and the loss of provenance signatures: isotopic and elemental data from sandstones, Frio Formation, South Texas. *Association of Petroleum Geologists Bulletin* **74**, 699.

Lanson, B. & Velde, B. (1992) Decomposition of X-ray diffraction patterns: a convenient way to describe complex I/S diagenetic evolution. *Clays and Clay Minerals* **40**, 629–643.

Lanson, B., Beaufort, D., Berger, G., Baradat, J. & Lacharpagne, J.-P. (1996) Illitization of diagenetic kaolinite-to-dickite conversion series: late-stage diagenesis of the Lower Permian Rotliegend sandstone reservoir, offshore The Netherlands. *Journal of Sedimentary Research* **66**, 501–518.

Lee, M. & Savin, S.M. (1985) Isolation of diagenetic overgrowth on quartz sand grains for oxygen isotopic analysis. *Geochimica et Cosmochimica Acta* **49**, 497–501.

LEVELL, B.K., BRAAKMAN, J.H. & RUTTEN, K.W. (1988) Oil-bearing sediments of Gondwana glaciation in Oman. *American Association of Petroleum Geologists Bulletin* **72**, 775–796.

LOOSVELD, R., BELL, A. & TERKEN, J. (1996) The tectonic evolution of Interior Oman. *GeoArabia* **1**, 28–51.

MCBRIDE, E.F. (1963) A classification of common sandstones. *Journal of Sedimentary Petrology* **33**, 664–669.

MCBRIDE, E.F. (1987) Compaction of Norphlet sandstones, Rankin County, Mississippi. *Gulf Coast Association Geological Society Transactions* **37**, 98–107.

MCBRIDE, E.F. (1989) Quartz cementation in sandstones: a review. *Earth-Science Review* **26**, 69–112.

MERCADIER, C.G.L. & LIVERA, S.E. (1993) Applications of the formation micro-scanner to modelling of Palaeozoic reservoirs in Oman. In: *The Geological Modelling of Hydrocarbon Reservoirs and Outcrop Analogues.* (eds FLINT, S.S. & BRYANT, I.D.) pp. 125–142. International Association of Sedimentologists Special Publication 15.

MILLSON, J.A., MERCADIER, C.G.L., LIVERA, S.E. & PETERS, J.M. (1996) The Lower Palaeozoic of Oman and its context in the evolution of a Gondwana continental margin. *Journal of the Geological Society of London* **153**, 213–230.

MOORE, D.M. & REYNOLDS, J.R.C. (1989) *X-Ray Diffraction and the Identification and Analysis of Clay Minerals.* Oxford University Press, Oxford.

MULLIS, A.M. (1991) The role of silica precipitation kinetics in determining the rate of quartz pressure solution. *Journal of Geophysical Research* **96B**, 10007–10013.

OSBORNE, M. & HASZELDINE, S. (1993) Evidence for resetting of fluid inclusion temperatures from quart cements in oilfields. *Marine and Petroleum Geology* **10**, 271–278.

OSBORNE, M. & HASZELDINE, S.R. (1995) Evidence for resetting of fluid inclusion temperatures from quartz cements in oilfields. *Marine and Petroleum Geology* **12**, 570–574.

POLLASTRO, R.M. (1990) The illite/smectite geothermometer—concepts, methodology and application to basin history and hydrocarbon generation. In: *Applications of Thermal Maturity Studies to Energy Exploration.* (eds NUCCIO, V.F. & BARKER, C.E.) pp. 1–18. Society of Economic Paleontologists and Mineralogists Rocky Mountain Section, Tulsa.

PORTER, E.W. & JAMES, W.C. (1986) Influence of pressure, salinity, temperature and grain size on silica diagenesis in quartzose sandstones. *Chemical Geology* **57**, 359–369.

RAMSEYER, K. (1983) Bau eines Kathodenlumineszenz-Mikroskopes und Diagenese-Untersuchungen an permischen Sedimenten aus Oman. PhD Thesis, University of Bern, Switzerland.

RAMSEYER, K.J., FISCHER, J., MATTER, A., EBERHARDT, P. & GEISS, J. (1989) A cathodoluminescence microscope for low intensity luminescence. *Journal of Sedimentary Petrology* **59**, 619–622.

ROBIE, R.A. & HEMINGWAY, B.S. (1995) Thermodynamic properties of minerals and related substances at 298.15 K and 1 bar ($10^5$ pascals) pressure and at higher temperatures. *U.S. Geological Survey Bulletin* **2131**, 461.

ROBIN, P.-Y.F. (1978) Pressure solution at grain-to-grain contacts. *Geochimica et Cosmochimica Acta* **42**, 1383–1389.

SACHSENHOFER, R.F. (1994) Petroleum generation and migration in the Styrian Basin (Pannonian Basin system, Austria): an integrated geochemical and numerical modelling study. *Marine and Petroleum Geology* **11**, 684–701.

SHAW, D.B. & WEAVER, C.E. (1965) The mineral composition of shales. *Journal of Sedimentary Petrology* **35**, 213–222.

SIBLEY, D.F. & BLATT, H. (1976) Intergranular pressure solution and cementation of the Tuscarora orthoquartzite. *Journal of Sedimentary Petrology* **46**, 881–896.

SIEVER, R. (1957) Pennsylvanian sandstones of the Eastern Interior Coal Basin. *Journal of Sedimentary Petrology* **27**, 227–250.

TADA, R. & SIEVER, R. (1989) Pressure solution during diagenesis. *Annual Review of Earth and Planetary Science* **17**, 89–118.

TSCHOPP, R.H. (1967) The general geology of Oman. *7th World Petroleum Congress* **2**, 231–241.

VISSER, W. (1991) Burial and thermal history of Proterozoic source rocks in Oman. *Precambrian Research* **54**, 15–36.

WALDERHAUG, O. (1995) Evidence for resetting of fluid inclusion temperatures from quartz cements in oilfields-Discussion. *Marine and Petroleum Geology* **12**, 559–561.

WALKER, T.R. (1976) Diagenetic origin of continental red beds. In: *The Continental Permian in Central, West and South Europe.* (ed. FALKE, H.) pp. 240–282. Reidel, Dordrecht.

WELTE, D.H., HORSFIELD, B. & BAKER, D.R. (1997) *Petroleum and Basin Evolution: Insights From Petroleum Geochemistry. Geology and Basin Modeling.* Springer, Berlin.

WEYL, P.K. (1959) Pressure solution and the force of crystallisation—a phenomenological theory. *Journal of Geophysical Research* **64**, 2001–2025.

WILSON, M.D. & STANTON, P.T. (1994) Diagenetic mechanisms of porosity and permeability reduction and enhancement. In: *Reservoir Quality Assessment and Prediction in Clastic Rocks.* (ed. WILSON, M.D.) pp. 59–119. Society of Economic Paleontologists and Mineralogists Short Course 30, Tulsa.

WINTSCH, R.P. & KVALE, C.M. (1994) Differential mobility of elements in burial diagenesis of siliciclastic rocks. *Journal of Sedimentary* **A64**, 349–361.

*Spec. Publs int. Ass. Sediment.* (2000) **29**, 271–280

# The porosity-preserving effects of microcrystalline quartz coatings in arenitic sandstones: examples from the Norwegian continental shelf

J. JAHREN[1] *and* M. RAMM[2]

[1]*Department of Geology, University of Oslo, P. O. Box 1047, Blindern, N-0316 Oslo, Norway; and* [2]*Norsk Hydro Exploration, N-1321 Stabekk, Norway*

## ABSTRACT

Microcrystalline quartz coatings on detrital quartz grains are not uncommon in reservoir sandstones from the Norwegian continental shelf. The present burial depths for these sandstones range between 4150 and 4419 m. Effective pressures in the reservoirs range between 20 and 40 MPa. Scanning electron microscope investigation reveals various amounts and developments of macrocrystalline quartz overgrowths coexisting with the microcrystalline variety. The microcrystalline quartz crystals are between 0.5 and 5 μm in diameter with an average diameter around 1.5 μm. Authigenic quartz crystals larger than this are always in optical continuity with the detrital quartz. The presence of microcrystalline quartz in the system indicates that the porewater has been oversaturated with respect to macroquartz and microcrystalline quartz. The microcrystalline quartz coatings hindered quartz cementation and slowed down pressure dissolution. The reason for this is that pressure dissolution controlled quartz supersaturation, without effective reprecipitation or removal of silica from the system (system size, 10–100 m), will result in an elevated supersaturation of silica, balancing further pressure dissolution. In such situations porosity will be preserved to greater burial depths than expected.

## INTRODUCTION

Quartz cementation is the most important porosity reducing process in medium to deep burial diagenesis (2500–5000 m) in arenitic sandstones (Bjørlykke *et al.*, 1989; Ehrenberg, 1990; Nedkvitne *et al.*, 1993). Leder & Park (1986) reviewed the mechanisms of quartz cementation in sandstones, which include: (i) cooling of hot quartz-saturated solutions; (ii) lowering the pore pressure; and (iii) semimembrane osmosis through a clay rich shale. The source of silica includes dissolution of biogenic silica, dissolution of quartz in sandstones by convecting fluids, pressure dissolution, and supply of silica from quartz–muscovite grain contacts due to elevated pH. Silica derived from other diagenetic reactions such as illitization of smectite, illitization of kaolinite + K-feldspar, feldspar dissolution, is also important but the effect of mineral reactions within a closed system on net porosity is generally unimportant because the mineral volume is about the same before and after the reaction. Permeability, however, can be profoundly modified by isochemical mineral reactions due to differences in crystal morphology.

The mechanisms and importance of these diagenetic processes are still debated (Bjørlykke *et al.*, 1992; Giles *et al.*, 1992; Gluyas & Coleman, 1992; Grant & Oxtoby, 1992; Bjørkum *et al.*, 1993). Intergranular pressure dissolution is believed to be the most effective contributor of silica (Houseknecht, 1987; Bjørlykke *et al.*, 1992). Modelling and experimental studies (Tada & Siever, 1986; Dewers & Ortoleva, 1990) indicate that pressure dissolution is the main source of silica in medium to deep burial diagenesis.

Pressure dissolution (Weyl, 1959) is still an incompletely understood process. The rate of pressure dissolution is thought to be limited by diffusion along the grain-to-grain contact within a thin film of fluid (Rutter, 1983). An alternative explanation is that dissolution takes place at the periphery of the grain contact due to excess stress (Engelder, 1982; Tada *et al.*, 1987). The first mechanism is termed water film diffusion (WFD), and the second mechanism free face pressure solution (FFPS) (Tada *et al.*, 1987). Both mechanisms probably occur together (Tada *et al.*, 1987). Dewers & Ortoleva (1990) found that WFD operates in clay rich sandstones, while FFPS is favoured in clean sandstones. The reason for this is that WFD requires an aqueous intergranular film that can support shear stress and transport solutes at the same time (Rutter, 1983). In clay rich sandstones clay mineral coatings around quartz grains provide an additional diffusion path (Weyl, 1959) and enhance intergranular pressure dissolution. The formation of stylolites is one familiar feature which illustrates this process. On the other hand, the presence of detrital clays has been found to inhibit quartz overgrowth (Heald & Anderegg, 1960; Tillman & Almond, 1979; Houseknecht & Hathon, 1987).

In this paper silica diagenesis in two wells from the North Sea (15/3–3 and 2/1–6) is described as a base for a discussion of quartz diagenesis, pressure dissolution of quartz and of chemical and physical conditions of quartz cement growth.

Similar occurrences of microcrystalline quartz, as described herein have recently been published from the North Sea by Aase *et al.* (1996). Texturally similar microcrystalline quartz has been reported by Hendry & Trewin (1995) from a turbidite environment.

## MATERIALS AND METHODS

Seven samples from well 2/1–6 (Ramm & Forsberg, 1991), and four samples from well 15/3–3 (Ramm & Bang, 1991) were examined. The samples were taken from shallow marine, Upper Jurassic sandstone core intervals with pervasive microcrystalline quartz coatings on detrital quartz and little macroquartz cementation. The microcrystalline quartz coatings have been briefly described by Jahren (1993, 1995). Present burial depths for these sandstones range between 4150 and 4419 m and the effective pressures are between 25 and 40 MPa in well 2/1–6 and 20 and 30 MPa in well 15/3–3, respectively (Ramm & Bang, 1991; Ramm & Forsberg, 1991).

The samples were examined petrographically by a JEOL 6400 scanning electron microscope (SEM) equipped with a Link eXL Energy Dispersive Spectrometer (EDS). All images presented are from gold coated chips. The SEM was used to characterize quartz overgrowth, grain to grain contacts, microcrystalline quartz coatings, mechanical deformation and pressure dissolution features.

## RESULTS

Pervasive microcrystalline quartz coatings on detrital quartz grains occur in all samples. Various amounts and developments of macrocrystalline quartz overgrowth in optical continuity with the detrital quartz were also recorded (Fig. 1a,b). The microcrystalline quartz overgrowths display a typical random orientation of the individual crystals. The microcrystalline quartz crystals are between 0.5 and 5 μm in diameter (prism length) with an average diameter between 1 and 1.5 μm. Coarser quartz cement crystals (> 5 μm) have grown in crystallographic continuity with detrital quartz grains, as later overgrowths. Overgrowths found as a number of prismatic quartz crystals (Fig. 1c) growing on a detrital grain and overgrowing the earlier-formed microcrystalline quartz overgrowths are examples of this. In a close-up (Fig. 1d) the difficulty that macrocrystalline quartz must overcome in order to overgrow microcrystalline quartz is illustrated. Prismatic doubly terminated crystals are formed. Later

(a)

(b)

**Fig. 1.** (*Above and opposite.*) Quartz overgrowth features: (a) Pervasive microcrystalline quartz coating on a detrital quartz grain (4335 m burial depth in well 2/1–6). (b) Close-up of microcrystalline quartz crystals with characteristic chert morphology. A few fibrous illite laths can also be seen.

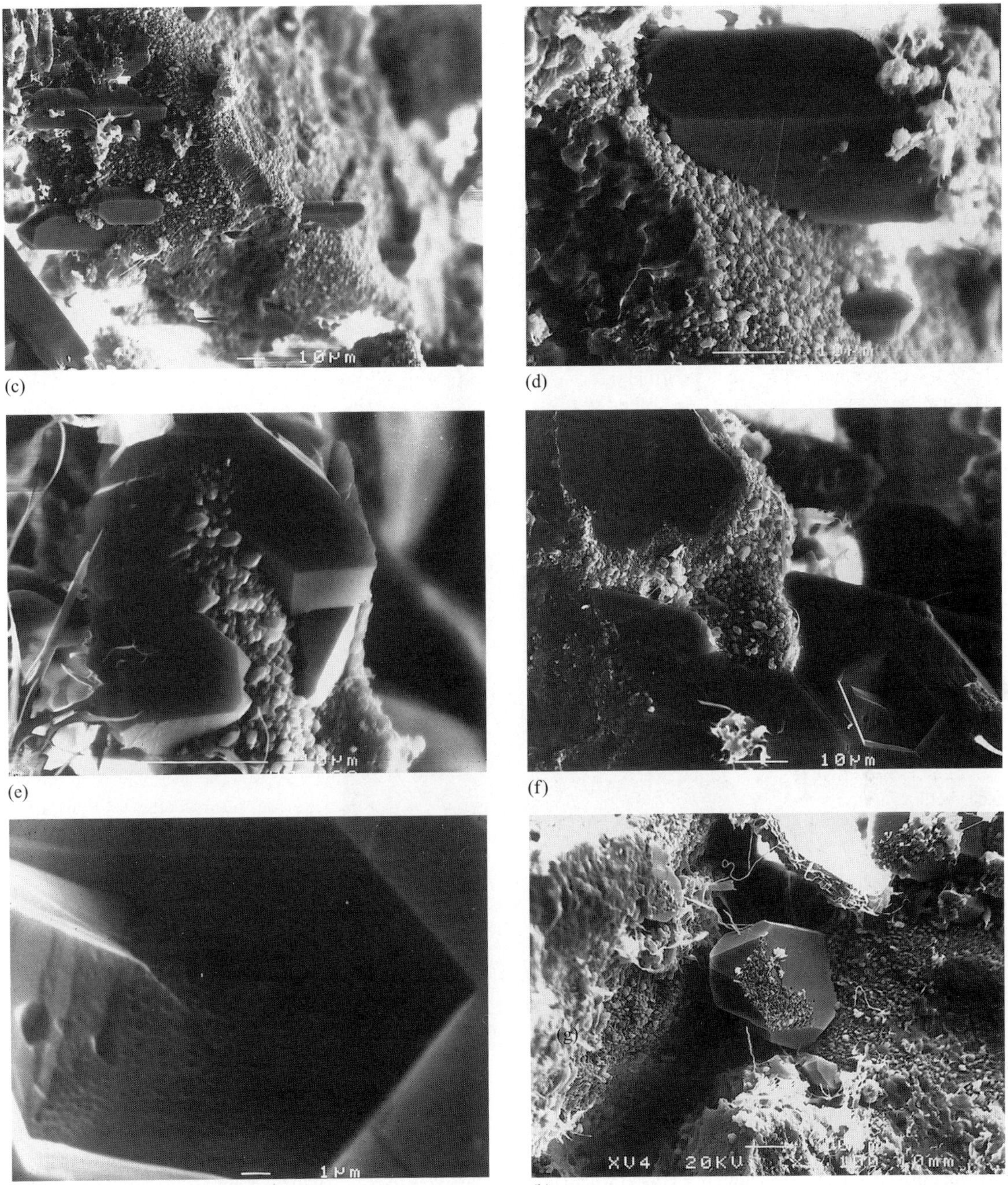


(c) (d) (e) (f) (g) (h)

**Fig. 1.** (*Contd.*) (c) Prismatic quartz crystals showing parallel growth probably in optical continuity with the detrital quartz grain hidden by microcrystalline quartz coating (4239 m burial depth from well 2/1–6). Note the tiny dissolution/fracture groove indicating very little compaction and pressure dissolution. (d) Detail from the same sample showing the prismatic macrocrystalline overgrowth. (e) Coalescence of prismatic quartz overgrowths covering most of the coating (same sample as [c]). (f) Quartz overgrowths with euhedral cavity from the same sample, formed because of growth problems. (g) Close-up of (f) showing pits and steps generated during cementation due to lattice mismatch between overgrowth and coating. (h) A macrocrystalline quartz overgrowth crystal which has passively grown on the microcrystalline coating and on to the detrital quartz grain opposite.

(a)

(b)

**Fig. 2.** Compaction/dissolution features: (a) Quartz dissolution grooves and ridges on a grain-to-grain contact (same sample as [a]). (b) Pressure dissolution feature with same kind of features as (a). Note the circular pressure dissolution feature which illustrates the pervasiveness of the coating.

more continuous quartz overgrowth are formed (Fig. 1e) probably by coalescence of several individual overgrowth crystals. Figure 1(f) shows one crystal where this process has almost gone to completion. The overgrowth covers most of the detrital quartz grain and the microquartz as well. On the bottom of the euhedral pit (Fig. 1g) texture spots separated by approximately the distance between the original microcrystalline quartz grains can be seen. Macrocrystalline quartz can even be pore-filling and simply overgrow the microcrystalline coating without a dissolution precipitation front to accommodate the difference in growth direction. Microcrystalline quartz is passively overgrown in this case.

Pressure dissolution features depicted in Fig. 2 (a) and (b) are similar to those observed by Pittman (1972) and inferred by Tada *et al.* (1987) to be a result of free face dissolution behaviour. The small distance between areas of dissolution and areas of microcrystalline coating with no apparent quartz precipitation shows that precipitation of quartz has been hindered.

## DISCUSSION

Redistribution of quartz in sandstones occurs when silica supersaturation reaches a level high enough to induce quartz precipitation. The supersaturation level is a function of the growth mechanism and presence of growth inhibitors. Growth mechanisms include spiral growth and two-dimensional growth (Sunagawa, 1981). Spiral growth, or screw dislocation growth, depends only on the availability of exposed screw dislocations on crystal faces for growth sites. Screw dislocations are ubiquitous in natural crystals so growth will take place if supersaturation is high enough (usually < 5% supersaturation is needed; Sunagawa, 1981) to overcome the activation energy related to the strain energy associated with the dislocation morphology. If physical or chemical barriers such as coatings or non-mineral forming atoms attached to growth sites, then higher supersaturation is required before growth can commence. Two-dimensional growth from discrete two-dimensional nucleii on flat crystal surfaces is less likely, because it requires a much higher supersaturation than spiral growth as each growth layer must be nucleated individually. The theoretical supersaturation needed for this mechanism in the quartz system would be around 35% (Ohara & Read, 1973). This supersaturation condition cannot be sourced by pressure dissolution (Oelkers *et al.*, 1996).

The presence of microquartz coatings indicates an amorphous silica precursor was present early in the system's diagenetic history (Maliva & Siever, 1988). This precursor could have been sponge spicules (Ramm & Forsberg, 1991; Aase *et al.*, 1996). The number of microquartz crystals present indicates that the dissolution rate of a microquartz precursor was fast enough to sustain silica saturation at the microquartz nucleation level, resulting in numerous crystals instead of a few large ones. This means that the dissolution rate of the precursor phase was higher than the precipitation rate of the product, indicating low-temperature growth (< 80°C) where growth rate of quartz is slow (Williams *et al.*, 1985). The resulting microquartz crystals are expected to grow according to Ostwald ripening related growth kinetics (Maliva & Siever, 1988) where growth is driven by differences in surface free energy (Lifshitz & Slyozov, 1961), i.e. crystals smaller than a critical size dissolve whereas crystals bigger than the critical size grow.

The relationship between quartz solubility and specific surface area can be expressed by the Freundlich–Ostwald equation (Williams *et al.*, 1985):

$$\mathrm{Log}\frac{S_A}{S_O} = \frac{\gamma M}{2{,}3RT}\frac{2A}{3}$$

where $S_A$ is quartz solubility of a crystal with surface area A; $S_O$ is quartz solubility of a crystal with infinite size; R is a gas constant; T is absolute temperature; M is molecular weight; A is specific surface area; $\gamma$ is surface free energy; then $\gamma$ can be estimated from known solubility data (Parks, 1984), for example the SUPBCRT 92 data base (Johnson *et al.*, 1992) and the size-solubility relation for quartz can be obtained. In sandstones $SiO_2$ occurs in various forms (amorphous silica, opal-CT, chalcedony, quartz, etc.), that have different solubilities in water. The solubility is a function of pressure, temperature, structure, disorder, surface area and size. In this paper it is assumed that the difference in free energy between macrocrystalline and microcrystalline quartz is mainly controlled by surface area. Experimental data (Blum *et al.*, 1990) indicate that dislocation density is unimportant in quartz dissolution. Internal disorder could be important during later growth and is believed to account for the dominance of macrocrystalline quartz growth over further growth of the microcrystalline quartz. The random geometrical distribution of the microcrystalline quartz will also prevent further growth of the individual crystals. Adsorbed alkali cations on quartz surfaces increase the reactivity of the surface (Dove & Crerar, 1990), but do not affect the stability of microquartz. Calculated solubility of macroscopic quartz at 150°C as a function of pressure (Fig. 3a) suggests a relatively linear relationship between pressure and solubility between 0.1 kbar and 1 kbar.

The temperature contribution to silica solubility (Fig. 3b) is the main kinetic factor for quartz dissolution/precipitation (Bjørlykke & Egeberg, 1993). The relationship between solubility and size for quartz shows that elevated quartz supersaturation is expected for quartz crystals < 5 μm in size (Fig. 4) (data from Parks, 1984; Williams *et al.*, 1985; Johnson *et al.*, 1992). The driving force for microquartz growth by Ostwald ripening is minimal in microcrystalline quartz populations with mean crystal size above a few micrometres, explaining the common occurrence of 0.5–5 μm size crystals (Maliva & Siever, 1988; Morse & Casey, 1988).

In sedimentary basins, pore fluids are often saturated or slightly supersaturated with respect to quartz in arenitic sandstones at burial depths of 2500 m (Egeberg & Aagaard, 1989). At shallower depths amorphous silica and chalcedony form, indicating that formation of quartz is slow. The most unstable silica polymorphs relative to quartz will dissolve and reprecipitate as quartz overgrowths below 2500 m and quartz equilibrium saturation will be approached. Pressure dissolution will induce quartz supersaturation. Depending on actual pressure, temperature, sorting, grain size and rounding, the potential quartz supersaturation from pressure dissolution will vary.

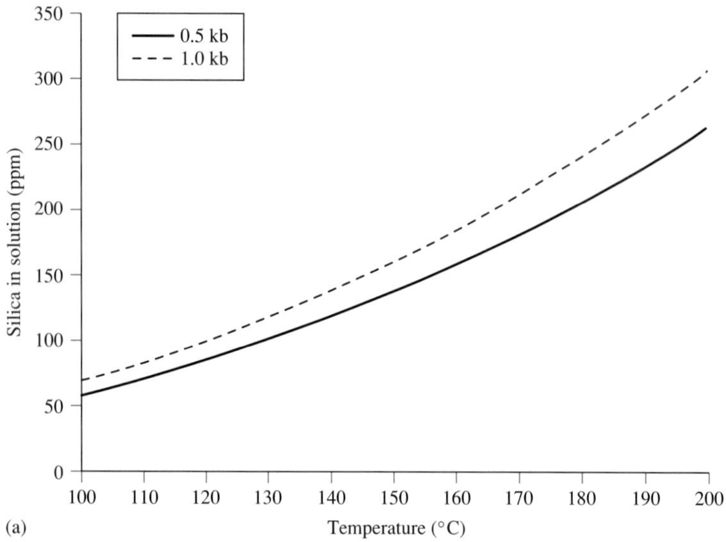


**Fig. 3.** Quartz solubility in formation waters; (a) silica in solution as a function of temperature; (b) the pressure dependence at 150°C corresponding to present-day formation conditions for the samples studied, at 0.5 kb and 1 kb, respectively. The calculations were performed using the SUPBCRT 92 package (Johnson *et al.*, 1992).

This excess silica in solution relative to quartz saturation must either be transported out of the system or precipitated within the system as quartz to allow further pressure dissolution to take place.

If silica transportation out of the system is slow (only diffusion) or if precipitation does not occur because the supersaturation reached is below the growth saturation level needed, further pressure dissolution will be halted. This is only possible if the supersaturation level needed for spiral growth is profoundly elevated by inhibitors like overgrowths (Fig. 5). Quartz solubility increases on average 3%, which is temperature independent in the region of interest, for every 10 MPa of pressure increase (Johnson *et al.*, 1992).

The effective pressure is the difference between the lithostatic and the fluid pressures. In sandstones, this is a function of the grain packing, grain contact area, compaction history, structural setting, etc. Brown & Hoek (1978) examined a number of published values of both vertical (lithostatic) and horizontal *in situ* stresses showing that the vertical stress is close to the weight of the rock column and that the horizontal stress can be described by a hyperbolic relation with limits $0.3 + 100/\text{depth} < k < 0.5 + 1500/\text{depth}$, where k is a constant relating vertical

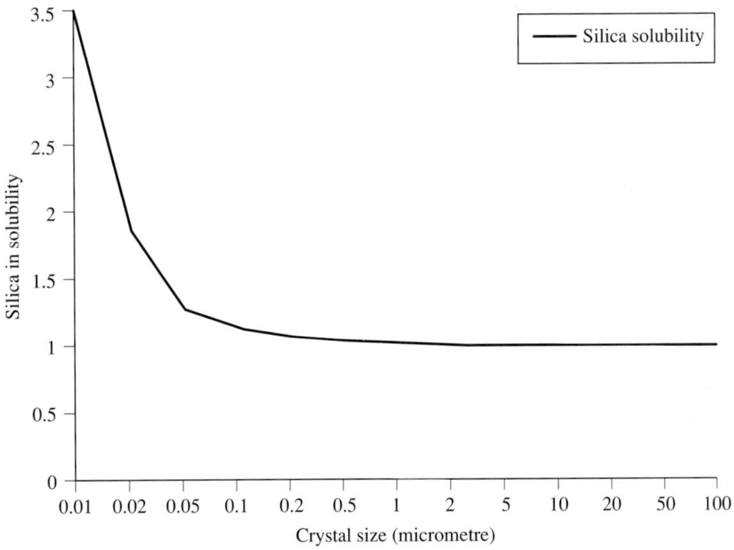


**Fig. 4.** Quartz solubility vs. crystal size. Note that the excess solubility related to surface area is negligible for crystals larger than a few micrometers. The calculation is based on the Ostwald–Freundlich relation (Williams *et al*., 1985) and data from Parks (1984).

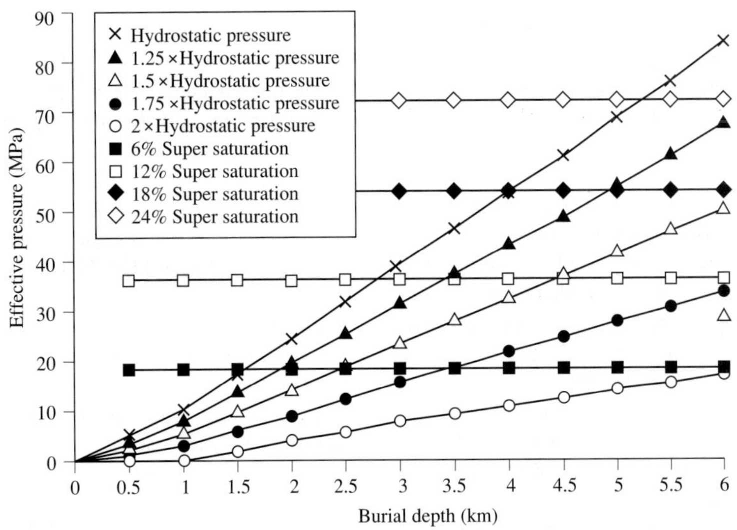


**Fig. 5.** Different supersaturation (horizontal lines) and overpressure (dipping lines) scenarios possible in burial diagenesis. Quartz cement growth is depicted as a function of effective pressure and burial depth only. Overpressure situations between hydrostatic pressure and a doubling of the fluid pressure relative to hydrostatic pressure are shown. The fluid pressure lines are based on a rock weight of 1.9 g/cm$^3$ at 0.5 km increasing to 2.5 g/cm$^3$ at 6 km burial depth. The supersaturation produced by pressure dissolution is based on SUPCRT92 (Johnson *et al*., 1992) calculations (Fig. 3b). The measured effective pressure in the samples varies between 20 and 40 MPa indicating that saturation levels above 15% are necessary to prevent cementation in hydrostatically pressured sandstones. Note that overpressure could play an important role in preserving porosity in coated sandstones if the cementation time limiting step is not the pressure dissolution process.

and horizontal stress. This horizontal stress relation indicates a mean value of about 0.8 relative to the vertical stress at 4000 m burial, which agrees well with values found in well summary data drawn from leak-off tests which supposedly measure the horizontal stress.

Pressure dissolution along stylolites contributes most of the silica during normal diagenesis. Stylolites are always subhorizontal in stable tectonic settings, but if folding or faulting takes plays dipping or vertical stylolites can be formed. This indicates that pressure dissolution is a function of the principal stress alone, which is vertical. Based on this, both mechanical and chemical compaction will be evenly distributed in sandstones and effective grain contact pressure equal to the difference between the lithostatic and the fluid pressures can be used (Fig. 5). The horizontal lines (Fig. 5) represent different silica supersaturation levels as a function of effective stress and the dipping lines represent different fluid overpressure situations (0% overpressure is equal to hydrostatic pressure). The role of overpressure on quartz cementation is to lower the pressure at grain to grain contacts and hence the supersaturation produced at the contact. At hydrostatic pressure, 2500 m of burial is needed to produce a high enough supersaturation to precipitate quartz if the supersaturation required for growth in the system is 10% (Fig. 5). In overpressured sequences the burial depth needed for quartz growth sourced from pressure dissolution would increase. For example, at a fluid pressure 50% above the hydrostatic pressure, more than 3500 m of burial would be needed to start quartz precipitation sourced by pressure dissolution if 10% supersaturation was needed. If the fluid pressure approaches the horizontal pressure (0.8 times the lithostatic pressure or about twice the hydrostatic pressure at 4000 m burial depth), pressure dissolution even in a sandstone free of grain coatings, would not start before around 4000 m burial if 3% supersaturation is assumed as a realistic cut-off value for quartz growth. On the other hand petrographic evidence presented by Bjørkum (1996) suggests that quartz cementation is precipitation controlled. If this is the case, processes at the growth surface are rate limiting and quartz cementation will be a simple function of the free quartz surface (surface free of mineral coatings) available for growth. If, on the other hand, the rate of quartz cementation is dissolution controlled, overpressure could be an important factor for preserving porosity below 4000 m. In any case, microcrystalline quartz coatings will enhance preservation of porosity.

## CONCLUSIONS

Petrographic observations and thermodynamic considerations suggest that microcrystalline quartz is stable together with macrocrystalline quartz in sandstones where the silica saturation level is at quartz saturation or above. Pressure dissolution is hindered by pervasive microcrystalline quartz coatings on detrital quartz surfaces. The inferred high supersaturation level needed for growth in these rocks, and the limited number of growth sites available between the microcrystalline quartz crystals would explain the apparent porosity preservation encountered in the reservoirs studied.

Microcrystalline quartz grains less than 5 μm are found in systems with a marked elevated silica saturation relative to macroquartz. Petrographic observations suggest a textural control favouring growth on detrital grains relative to growth on microcrystalline quartz grains. The late quartz overgrowths precipitate in optical continuity with the substrate and form initially as prisms which later coalesce as the microcrystalline quartz is completely overgrown.

The role played by overpressure on quartz cement sourced from pressure dissolution is not straightforward because the interplay between the timing of overpressure build up, eventual pressure release through hydrofracturing, the effect of grain coatings, solution chemistry, pattern of fluid movement, and actual effective grain contact pressure, during burial is so far poorly constrained.

The consequence of these findings to reservoir geology is that porosity will be preserved deeper than normal if pervasive microcrystalline quartz is present during deep burial diagenesis.

## ACKNOWLEDGEMENTS

Norsk Hydro is thanked for providing samples and financial support for this project. The SEM work was done at Department of Geology, University of Manchester. The manuscript was greatly improved by the reviews of Janos Matyas, Jeff Warner and Sadoon Morad.

## REFERENCES

Aase, N.E., Bjørkum, P.A. & Nadeau, P.H. (1996) The effect of grain-coating microquartz on preservation of reservoir porosity. *American Association of Petroleum Geologists Bulletin* **80**, 1654–1673.

Bjørkum, P.A. (1996) How important is pressure in causing dissolution of quartz in sandstones? *Journal of Sedimentary Research* **66**, 147–154.

Bjørkum, P.A., Walderhaug, O. & Aase, N.E. (1993) A model for the effect of illitization on porosity and quartz cementation of sandstones. *Journal of Sedimentary Petrology* **63**, 1089–1091.

Bjørlykke, K. & Egeberg, P.K. (1993) Quartz cementation in sedimentary basins. *American Association of petroleum Geologists Bulletin* **77**, 1538–1548.

Bjørlykke, K., Ramm, M. & Saigal, G.C. (1989) Sandstone diagenesis and porosity modification during basin evolution. *Geologishe Rundschau* **78**, 243–268.

Bjørlykke, K., Nedkvitne, T., Ramm, M. & Saigal, G.C. (1992) Diagenetic processes in the Brent Group (middle Jurassic) reservoirs of the North Sea; an overview. In: *Geology of the Brent Group*. (eds Morton., A.C., Hazeldine, R.S., Giles, M.R. & Brown, S.) pp. 263–287. Special Publications of the Geological Society of London 61.

Blum, A.E., Yund, R.A. & Lasaga, R.C. (1990) The effect of dislocation density on the dissolution of quartz. *Geochimica et Cosmochimica Acta* **54**, 283–297.

Brown, E.T. & Hoek, E. (1978) Trends in relationships between measured *in situ* stresses and depth. *International Journal of Rock Mechanics and Mineral Science* **15**, 211–215.

Dewers, T. & Ortoleva, P. (1990) A coupled reaction/transport/mechanical model for intergranular pressure solution, stylolites, and differential compaction and cementation in clean sandstones. *Geochimica et Cosmochimica Acta* **54**, 1609–1625.

Dove, P.M. & Crerar, D.A. (1990) Kinetics of quartz dissolution in electrolyte solutions using a hydrothermal mixed flow reactor. *Geochimica et Cosmochimica Acta* **54**, 955–969.

Egeberg, P.K. & Aagaard, P. (1989) Formation water chemistry in relation to the stability of detrital and authigenic minerals in clastic reservoirs from offshore Norway. In: *Water–Rock Interactions in the Diagenetic Environment; Deep Sea Sediments and Deeply Buried Chalks and Sandstones*. Unpublished Dr Sci. Thesis, University of Oslo.

Ehrenberg, S.N. (1990) Relationship between diagenesis and reservoir quality in sandstones of the Garn Formation, Haltenbanken, mid-Norwegian continental shelf. *American Association of Petroleum Geologists Bulletin* **74**, 1538–1558.

Engelder, T. (1982) A natural example of the simultaneous operation of free face dissolution and pressure solution. *Geochimica et Cosmochimica Acta* **46**, 69–74.

Giles, M.R., Stevenson, M.S.V., Cannon, S.J.C. *et al.* (1992) The reservoir properties and diagenesis of the Brent Group; a regional perspective. In: *Geology of the Brent Group.* (eds Morton, A.C., Hazeldine, R.S., Giles, M.R. & Brown, S.) pp. 289–337. Special Publication of the Geological Society of London 61.

Gluyas, J. & Coleman, M. (1992) Material flux and porosity change during sediment diagenesis. *Nature* **356**, 52–54.

Grant, S.M. & Oxtoby, N.H. (1992) The timing of quartz cementation in Mesozoic sandstones from Haltenbanken, offshore mid-Norway: fluid inclusion evidence. *Journal of the Geological Society of London* **149**, 479–482.

Heald, M.T. & Anderegg, R.C. (1960) Differential cementation in the Tuscarora sandstone. *Journal of Sedimentary Petrology* **76**, 568–577.

Hendry, J.P. & Trewin, N.H. (1995) Authigenic quartz microfabrics in Cretaceous turbidities: evidence for silica transformations in sandstones. *Journal of Sedimentary Research A* **65**, 380–392.

Houseknecht, D.W. (1987) Assessing the relative importance of compaction processes and cementation to reduction of porosity in sandstones. *American Association of Petroleum Geologists Bulletin* **71**, 633–642.

Houseknecht, D.W. & Hathon, L.A. (1987) Relationships among thermal maturity, sandstone diagenesis, and reservoir quality in Pennsylvanian strata of the Arkoma basin. *American Association of Petroleum Geologists Bulletin* **71**, 568–569.

Jahren, J.S. (1993) Microcrystalline quartz coatings in sandstones: a scanning electron microscopy study. In: *Extended Abstracts from the 45th Annual Meeting of the Scandinavian Society for Electron Microscopy* (ed. Karlson, G.) pp. 111–112.

Jahren, J.S. (1995) Relation between quartz coating on detrital quartz and pressure solution derived quartz cementation. Abstract. *V.M. Goldschmidt Conference, 24–26 May, Penn State University, USA*, p. 57.

Johnson, J.W., Oelkers, E.H. & Helgeson (1992) A SUPBCRT92: Software package for calculating the standard molal thermodynamic properties of minerals, gases, aqueous species, and reactions from 1 to 5000 bars and 0–1000°C. *Computers and Geosciences* **18**, 899–946.

Leder, F. & Park, W.C. (1986) Porosity reduction in sandstones by quartz overgrowth. *American Association of Petroleum Geologists Bulletin* **70**, 1713–1728.

Lifshitz, I.M. & Slyozov, V.V. (1961) The kinetics of precipitation from supersaturated solid solutions. *Physics and Chemistry of Solids* **19**, 35–50.

Maliva, R.G. & Siever, R. (1988) Pre-Cenozoic nodular cherts: Evidence for opal-CT precursors and direct quartz replacement. *American Journal of Science* **288**, 798–809.

Morse, J.W. & Casey, W.H. (1988) Ostwald processes and mineral paragenesis in sediments. *American Journal of Science* **288**, 537–560.

Nedkvitne, T., Karlsen, D.A., Bjørlykke, K. & Larter, S.R. (1993) Relationship between reservoir diagenetic evolution and petroleum emplacement in the Ula field, north North Sea. *Marine and Petroleum Geology* **10**, 255–270.

Oelkers, E.H., Bjørkum, P.A. & Murphy, N. (1996) A petrographic and computational investigation of quartz cementation and porosity reduction in North Sea sandstones. *American Journal of Science* **296**, 420–452.

Ohara, M. & Read, R.C. (1973) *Modelling Crystal Growth Rates from Solution*. Prentice Hall.

Parks, G.A. (1984) Surface and interfacial free energies of quartz. *Journal of Geophysical Research* **89**, 3997–4008.

Pittman, E.D. (1972) Diagenesis of quartz in sandstones as revealed by scanning electron microscopy. *Journal of Sedimentary Petrology* **42**, 507–519.

Porter, E.W. & James, W.C. (1986) Influence of pressure, salinity, temperature and grain size on silica diagenesis in quartzose sandstones. *Chemical Geology* **57**, 359–369.

Ramm, M. & Bang, N.A. (1991) Porosity distribution in middle Jurassic sandstones of the Southern Viking graben, North Sea. In: *Porosity Depth Trends in Reservoir Sandstones*. Unpublished Dr Sci. Thesis, University of Oslo.

Ramm, M. & Forsberg, A.W. (1991) Porosity versus depth trends in upper Jurassic sandstones from the Cod-Terrace area, Central North Sea. In: *Porosity Depth Trends in Reservoir Sandstones*. Unpublished Dr Sci. Thesis, University of Oslo.

Rutter, E.H. (1983) Pressure solution in nature, theory and experiment. *Journal of the Geological Society of London* **140**, 725–740.

Sunagawa, I. (1981) Characteristics of crystal growth in nature as seen from the morphology of mineral crystals. *Bulletin of Mineralogy* **104**, 81–87.

TADA, R. & SIEVER, R. (1986) Experimental knife-edge pressure solution of halite. *Geochimica et Cosmochimica Acta* **50**, 29–36.

TADA, R., MALIVA, R. & SIEVER, R. (1987) A new mechanism for pressure solution in porous quartzose sandstone. *Geochimica et Cosmochimica Acta* **51**, 2295–2301.

TILLMAN, R.W. & ALMOND, W.R. (1979) Diagenesis of the Frontier Formation offshore bar sandstones, Spearhead Ranch Field, Wyoming. In: *Aspects of Diagenesis*. (eds SCHOLLE, P.A. and SCHLUGER, P.R.) 337–378. Society of Economic Paleontologists and Mineralogists Special Publication 26. Special Publication.

WEYL, P.K. (1959) Pressure solution and the force of crystallization—A phenomenological theory. *Journal of Geophysical Research* **64**, 2001–2025.

WILLIAMS, L.A., PARKS, G.A. & CRERAR, D.A. (1985) Silica diagenesis, I. Solubility controls. *Journal of Sedimentary Petrology* **55**, 301–311.

*Spec. Publs int. Ass. Sediment.* (2000) **29**, 281–297

# High-temperature quartz cement and the role of stylolites in a deep gas reservoir, Spiro Sandstone, Arkoma Basin, USA

C. SPÖTL[1], D. W. HOUSEKNECHT[2] *and* L. R. RICIPUTI[3]

[1]*Institut für Geologie und Paläontologie, Universität Innsbruck, Innrain 52, 6020 Innsbruck, Austria;*

[2]*US Geological Survey, 956 National Center, Reston, VA 20192, USA; and*

[3]*Chemical and Analytical Sciences Division, Oak Ridge National Laboratory, Oak Ridge, TN 37831–6365, USA*

## ABSTRACT

The Spiro Sandstone, a natural gas play in the central Arkoma Basin and the frontal Ouachita Mountains preserves excellent porosity in chloritic channel-fill sandstones despite thermal maturity levels corresponding to incipient metamorphism. Some wells, however, show variable proportions of a late-stage, non-syntaxial quartz cement, which post-dated thermal cracking of liquid hydrocarbons to pyrobitumen plus methane. Temperatures well in excess of 150°C and possibly exceeding 200°C are also suggested by (i) fluid inclusions in associated minerals; (ii) the fact that quartz post-dated high-temperature chlorite polytype II*b*; (iii) vitrinite reflectance values of the Spiro that range laterally from 1.9 to ≥ 4%; and (iii) the occurrence of late dickite in these rocks. Oxygen isotope values of quartz cement range from 17.5 to 22.4‰ VSMOW (total range of individual *in situ* ion microprobe measurements) which are similar to those of quartz cement formed along high-amplitude stylolites (18.4–24.9‰). We favour a model whereby quartz precipitation was controlled primarily by the availability of silica via deep-burial stylolitization within the Spiro Sandstone. Burial-history modelling showed that the basin went from a geopressured to a normally pressured regime within about 10–15 Myr after it reached maximum burial depth. While geopressure and the presence of chlorite coats stabilized the grain framework and inhibited nucleation of secondary quartz, respectively, stylolites formed during the subsequent high-temperature, normal pressured regime and gave rise to high-temperature quartz precipitation. Authigenic quartz growing along stylolites underscores their role as a significant deep-burial silica source in this sandstone.

## INTRODUCTION

Quartz cementation in porous sandstones is frequently documented and reasonably well understood. Precipitation occurs predominantly as syntaxial overgrowths on detrital quartz grains at burial temperatures between ~ 80 and ~ 140°C (e.g. Bjørlykke & Egeberg, 1993; Aplin & Warren, 1994; Walderhaug, 1994a), although it is possible the current data set is biased toward shallow and more frequently drilled reservoirs. Because this temperature interval coincides largely with the liquid hydrocarbon 'window', understanding the controls of porosity reduction by quartz cementation is crucial for hydrocarbon exploration. Temperature is generally regarded as a major control on quartz precipitation, although other parameters are also known to affect the solubility and precipitation kinetics, including the composition of the quartz grain surfaces and the composition and availability of fluids (Walderhaug, 1994b, 1996; Bjørkum, 1996; Bolton *et al.*, 1996; Oelkers *et al.*, 1996).

Here we report on quartz cement in the gas-bearing Spiro Sandstone of the Arkoma Basin. Paragenetic information, in conjunction with vitrinite reflectance and sparse fluid inclusion data suggest that this cement probably formed at temperatures close to 200°C (Spötl *et al.*, 1996). Similar high-temperature quartz cement has been encountered in other deep-basin gas provinces, including the Norphlet Formation of the US Gulf Coast basin (McBride *et al.*, 1987; Dixon *et al.*, 1989; Thomas *et al.*, 1993) and the Rotliegende of the North German basin (R. Gaupp, personal communication, 1996). In this article we evaluate the origin of this cement using a variety of

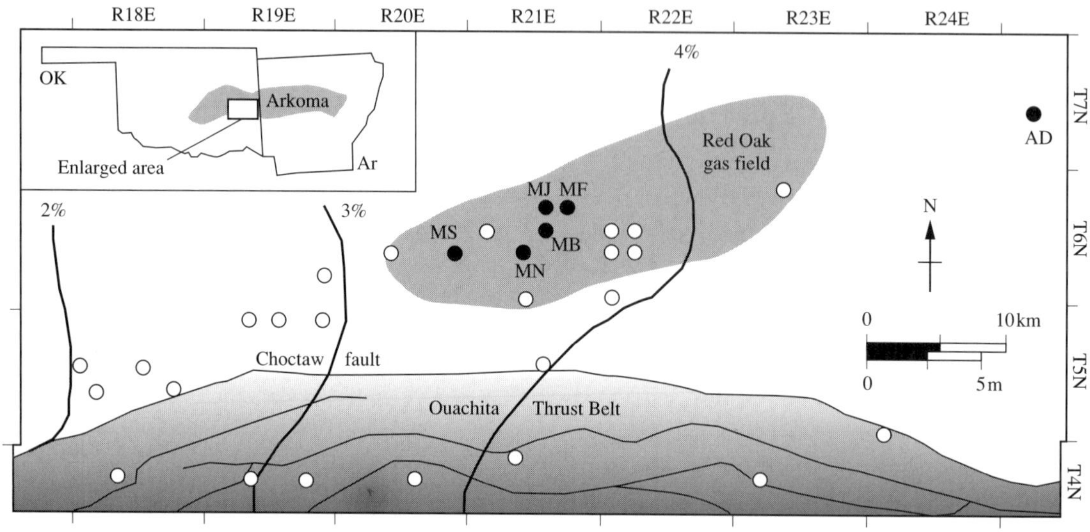


**Fig. 1.** Map of the central Arkoma Basin showing location of wells examined in this study (filled circles) and other wells previously studied for mineral diagenesis and organic maturity (open circles). Heavy lines are isocontours of rotational vitrinite reflectance at Spiro level. The reflectance of Spiro penetrated in well AD east of Red Oak gas field, used for quantitative burial history modelling, is 4.1%. Key to well abbreviations: MS: Midwest Sorrels; MN: Midwest Noah; MB: Midwest Booth; MJ: Midwest Jones; MF: Midwest Free; AD: Arkla Downes.

information, including new *in situ* ion microprobe oxygen isotope and fluid inclusion data and results from burial-history modelling.

## GEOLOGICAL SETTING

The Arkoma Basin is a foreland basin in south-east Oklahoma and west-central Arkansas that formed in response to the Late Palaeozoic Ouachita orogeny (Houseknecht, 1986; Arbenz, 1989). The basin and the adjacent Ouachita thrustbelt are currently one of the most actively explored deep-basin gas provinces in North America. The Spiro Sandstone, one of the major gas reservoirs in this basin and in the adjacent thrustbelt, is a ≤ 60 m thick blanket of mid-Pennsylvanian quartz arenite overlain by a several kilometre thick wedge of synorogenic sediments (Atoka Formation). Spiro deposition occurred in a near-shore environment ranging from fluvial and tidal environments in the north to a shallow-marine environment in the south, dominated by finer-grained, carbonaterich sediments (Lumsden *et al.*, 1971; Houseknecht & McGilvery, 1990; Gross *et al.*, 1995). Productive reservoirs are best developed in medium-grained sandstones deposited as dip-orientated channel-fills north of the thrust belt and as strike-orientated offshore bars within the thrust belt. Porosity and permeability in the reservoir range up to 25% and nearly 1 darcy, respectively, although values of 5–18% and 1–100 millidarcies are more typical.

An extensive vitrinite reflectance database that provides accurate constraints on the thermal maturity of the Spiro Sandstone is available (Houseknecht & Spötl, 1993). Although there has been some controversy regarding the levels of thermal maturity of the Spiro (e.g. Gross *et al.*, 1995), an extensive reflectance data set (Houseknecht & Weesner, 1997) together with palaeotemperature constraints published previously (Spötl *et al.*, 1993, 1994, 1996) provide a firm foundation for our inferences. Figure 1 shows that there is a slight lateral palaeothermal gradient across the study area (the Red Oak gas field). From west to east mean rotational vitrinite reflectance, which is essentially equivalent to $R_o$ measured in non-polarized light (Houseknecht & Weesner, 1997), increases from ≤ 3% to 4.5% (Fig. 1), i.e. the Spiro is clearly anchimetamorphic.

Core samples from 12 wells in the area of Red Oak field were examined (Spötl *et al.*, 1996), but this article is primarily based on new data from three wells (Fig. 1): Midwest Booth (15-6N-21E), Midwest Jones (10-6N-21E), and Midwest Free (11-6N-21E). These three wells are less than 3 km apart and vitrinite reflectance measurements indicate that they experienced a similar thermal history (3.5% reflectance at Spiro level).

## METHODS

Details of the petrographic and geochemical methods were given in our previous paper (Spötl *et al.*, 1996). In order to further elucidate the origin of quartz in this sandstone reservoir, additional data were obtained using a variety of methods, including *in situ* oxygen isotope analysis, hot-cathode cathodoluminescence (CL), and modelling of burial, temperature, and pressure history.

*In situ* oxygen isotope analysis was performed using a Cameca 4f ion microprobe located in the Chemical and Analytical Sciences Division at Oak Ridge National Laboratory. Oxygen isotope microanalysis of insulators such as quartz using the ion microprobe is a fairly novel technique that allows *in situ* measurements at a scale several orders of magnitude smaller than conventional laser-ablation mass spectrometric methods (e.g. Hervig *et al.*, 1995; Lyon *et al.*, 1995). The disadvantage of the ion microprobe is that the errors of individual measurements are 2–5 times higher than those obtained using conventional techniques. Oxygen isotope analyses were conducted using the extreme energy filtering technique (Hervig, 1992; Riciputi & Paterson, 1994). Samples were sputtered using a $^{133}Cs^{+}$ primary ion beam focused to a beam diameter of 20–25 μm. Negative secondary ions were analysed using an energy offset of 320 eV to suppress molecular interferences and provide stable analytical conditions; a normal-incident electron gun was used to neutralize sample charging (further details are available in Riciputi & Paterson, 1994). Each 25 minute analysis consisted of 200 individual $^{18}O/^{16}O$ ratio measurements. Based on counting statistics for each analysis the internal precision predicted was ~ 0.65‰, similar to the actual precision obtained (see Table 1). The instrumental fractionation was calibrated during each analytical session by repeated measurements ($n = 4–6$) on a quartz standard. These repeated measurements also allowed reproducibility to be estimated for the different analytical sessions, based on the standard deviation of the multiple analyses of the standard ($1\ \sigma = 0.24–0.70‰$, average of 0.4‰). Overall precision, including the error of each individual analysis and calibration errors are generally < 1‰.

Trace element analyses of Ca, K, Ti, Li, B, Na, Mg and Al were conducted using a $^{16}O^{-}$ primary beam (25–30 μm diameter), analysing positively charged secondary ions. A moderate energy offset (80 eV) was applied to suppress molecular interferences. Ion yields were quantified using NBS 610 glass, since we did not have a quartz standard available. Due to the difference in ion yields between the glass standard and quartz, absolute values are only semiquantitative. However, the relative magnitude of

**Table 1.** Ion microprobe analyses of $\delta^{18}O$ in Spiro quartz cements

| Sample | Description | $\delta^{18}O$ (‰VSMOW) | 1σ | Mean of sample |
|---|---|---|---|---|
| Quartz cement | | | | |
| MB 12115 | detrital quartz grain | 10.3 | 0.7 | |
| | area 4, near edge | 19.6 | 0.7 | |
| | area 4, near edge | 17.7 | 0.6 | |
| | area 4, near edge | 20.7 | 0.7 | 19.4 |
| | area 5, near edge | 18.7 | 0.7 | (± 1.1) |
| | area 5, core | 20.4 | 0.7 | |
| | area 6, core | 19.5 | 0.7 | |
| MF 11864 | area 4, core | 21.8 | 0.6 | |
| | area 4, near edge | 21.0 | 0.7 | |
| | area 4, core | 21.6 | 0.6 | |
| | area 4, near edge | 22.2 | 0.8 | 21.9 |
| | area 5, core | 22.4 | 0.7 | (± 0.4) |
| | area 5, edge | 22.1 | 0.7 | |
| | area 6, core | 22.1 | 0.7 | |
| | area 6, rim | 22.2 | 0.7 | |
| | area 6, toward rim | 22.0 | 0.7 | |
| MJ 11858 | detrital quartz grain | 10.3 | 0.8 | |
| | area 4, core | 18.1 | 0.7 | |
| | area 4, edge | 20.7 | 0.7 | |
| | area 5, edge | 19.2 | 0.7 | 19.1 |
| | area 6, edge | 17.5 | 0.7 | (± 1.3) |
| | area 6, core | 20.8 | 0.7 | |
| | area 6, edge | 19.0 | 0.7 | |
| | area 6, core | 18.1 | 0.6 | |
| MJ 11862 | area A, core | 20.6 | 0.7 | |
| | area A, rim | 19.4 | 0.7 | |
| | area A, core | 20.9 | 0.7 | |
| | area A, edge | 19.9 | 1.4 | 20.8 |
| | area B, edge | 22.0 | 0.7 | (± 0.9) |
| | area B, core | 21.6 | 0.7 | |
| | area B, edge | 21.3 | 0.6 | |
| Stylolitic quartz | | | | |
| MB 12113A | inclusion rich core | 18.9 | 0.6 | |
| | crystal A, rim | 20.3 | 0.7 | |
| | crystal A, rim | 22.9 | 0.8 | |
| | crystal A, inner area | 22.5 | 0.7 | |
| | crystal A, inner area | 22.9 | 0.7 | 21.8 |
| | crystal A, rim | 22.3 | 0.8 | (± 1.4) |
| | crystal B, inner area | 22.4 | 0.7 | |
| | crystal B, rim | 22.8 | 0.7 | |
| | crystal B, adj rim | 21.6 | 0.7 | |
| | inclusion rich core | 21.0 | 0.7 | |
| MB 12113B | crystal B, rim | 23.9 | 0.6 | |
| | crystal B, adj rim | 23.3 | 0.6 | |
| | crystal B, next rim | 22.6 | 0.8 | |
| | crystal B, core | 22.6 | 0.7 | 22.3 |
| | crystal B, core | 24.9 | 0.7 | (± 2.2) |
| | crystal A, near rim | 18.4 | 0.7 | |
| | crystal A, near rim | 20.5 | 0.7 | |
| MB 12113C | crystal A, rim | 23.1 | 0.7 | |
| (traverse rim | next | 24.0 | 0.7 | |
| to core) | next | 22.9 | 0.7 | |
| | next | 22.1 | 0.6 | 22.2 |
| | past core boundary | 23.6 | 0.6 | (± 1.7) |
| | next | 19.3 | 0.8 | |
| | next | 20.0 | 0.8 | |
| | crystal B, inner area | 22.9 | 0.7 | |

**Table 2.** Constraints used in modelling burial and thermal history of Spiro Sandstone

| Constraint | Summary/description | Reference |
|---|---|---|
| Stratigraphy | Strata penetrated by Arkla Downes (Oklahoma; section 20, T7N, R25E) | n/a |
| Thermal maturity | Profile of depth vs. mean rotational reflectance of vitrinite | Houseknecht & Weesner (1997) |
| Diagenetic temperatures | (a) Fluid inclusion data from deep burial quartz and calcite cements<br>(b) Temperatures inferred from C and O isotopic analysis of carbonate cements | Spötl *et al.* (1996)<br>Spötl *et al.* (1996) |
| Timing of thermal events | (a) Regional maximum temperatures based on radiometric dating of hydrothermal minerals<br>(b) Regional cooling to temperatures < 100°C | Summarized by Shelton *et al.* (1986)<br>Arne (1992) |
| Present geothermal gradient | Regional map of present geothermal gradients based on borehole temperature logs | Johnson (1986), Lee *et al.* (1996) |
| Palaeosurface temperatures | Palaeolatitude and present world surface temperatures | Habicht (1979)<br>Scotese & McKerrow (1990) |
| Structural deformation | (a) Early Atokan basement-rooted, syndepositional normal faults related to obduction of incipient Ouachita tectonic pile on to N. American crust<br>(b) Compressional structures affect youngest strata (Boggy Fm.) in Arkoma basin<br>(c) Regional isotopic age dates as young as middle Permian (257 Ma) | Houseknecht (1986)<br>Bradley & Kidd (1991)<br>Viele & Thomas (1989)<br>Viele & Thomas (1989) |

variations between different quartz analyses are quantitative, and allow us to determine whether there are differences between the different quartz types, or if zoning profiles are present.

The CL of quartz cement was studied using the hot-cathode luminoscope at the University of Bern (Ramseyer *et al.*, 1989) and a few samples were also investigated using a commercial cold-cathode machine (Technosyn Mk II) operating at 500–600 mA beam current and ~ 20 kV beam potential.

Modelling of burial, thermal, and pressure history was conducted using 1D BasinMod software (Platte River Associates). Input data included age, thickness, and lithology of stratigraphic units in the study area, estimates of additional section that has been eroded, and heat flow values typical of analogous geological settings. Modelling was conducted using heat flow and thermal conductivity (rather than geothermal gradient), transient heat flow equations, and compactional disequilibrium feedback options. We used the default parameters incorporated into BasinMod for the original porosity values for each lithology (e.g. 60% for shale). Porosity reduction was calculated using the Falvey & Middleton (1981) equation combined with burial history. Permeability was calculated using a modified Kozeny-Carman equation. Model output was compared with depth–vitrinite reflectance profiles, geothermometry estimates from fluid inclusion and isotopic studies, and other constraints listed in Table 2. Modelling iterations were conducted using adjusted values of input data until a close match to depth–vitrinite reflectance profiles and other model constraints was achieved.

## RESULTS

### Core observations

We examined cores from the three wells mentioned above and from two additional wells nearby (wells MN and MS – Fig. 1) in an attempt to characterize the type and abundance of stylolites in the Spiro. In general, most macroscopically recognizable stylolites are subhorizontally aligned relative to the bedding plane and show amplitudes between 0.5 and 2.5 cm with rare examples of up to 28 cm (Fig. 2b). High-amplitude stylolites classify as up-peak, rectangular-type and sutured-type stylolites according to Park & Schot (1968), whereas low-amplitude stylolites are more commonly of the primitive wave-like type. High-amplitude stylolites show up to 5 mm thick accumulations of carbonaceous material at the crests and troughs of stylolites (Fig. 2a,c). White, fibrous calcite is common in the pressure shadow of individual stylolite pillars and also occurs as thin layers along the flanks of such pillars (Fig. 2a–c).

Stylolites apparently formed both between sandstones of different composition (e.g. sandstone heavily cemented by ankerite and porous sandstone showing well

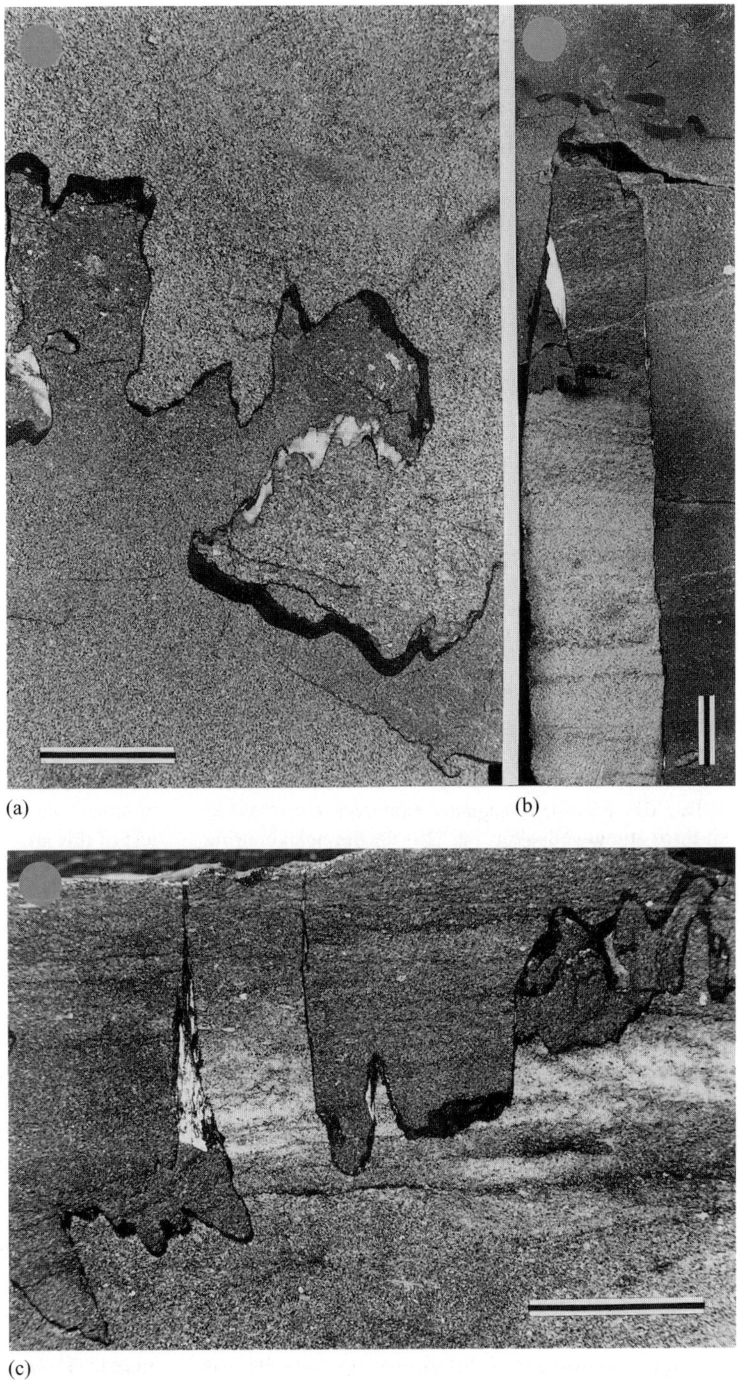

**Fig. 2.** High-amplitude stylolites in core slabs of Spiro Sandstone. Scale equals 2 cm on all photographs. (a) Complex digital stylolite showing thick accumulations of residual black organic rich matter and white fibrous calcite formed in pressure shadows. Sample MB 12125. (b) Large stylolite pillar formed at the interface between an ankerite cemented sandstone and a porous, chlorite-rich sandstone. Minor fibrous calcite formed in the upper part of the stylolite. Sample MB 12113. (c) High-amplitude stylolite showing evidence of stacked stylolites (right part of photograph). Note accumulations of black organic-rich residue in the troughs between the pillars. Sample MF 11853.

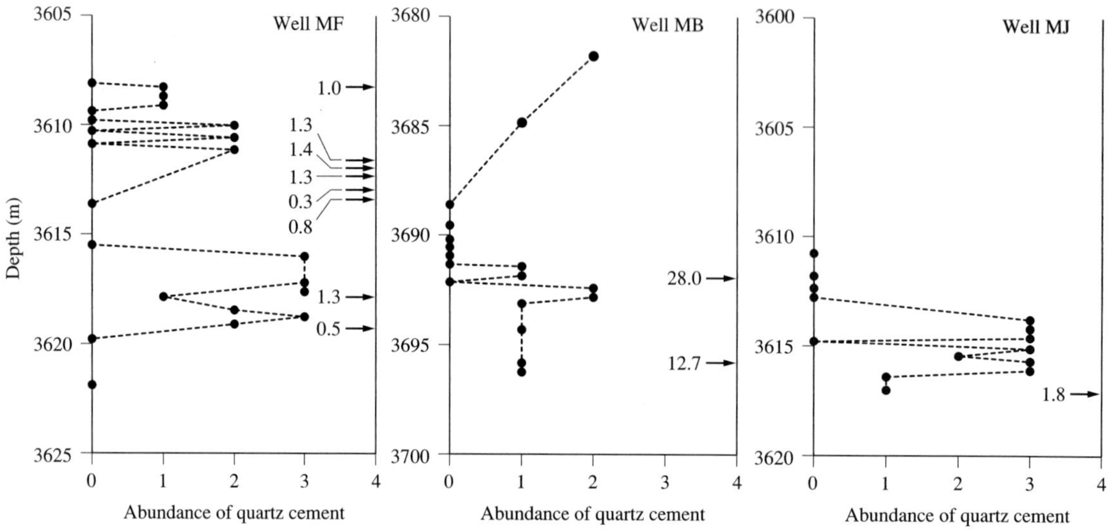


**Fig. 3.** Comparison of stylolite density and abundance of quartz cement in three Spiro cores. The abundance of quartz is based on visual thin-section estimates: 0 no quartz, 1 traces, 2 some, 3 abundant. Individual macroscopic stylolites recognized in core are indicated by arrows and the amplitude is given in centimetres.

developed chlorite coatings) and, less commonly, within more or less homogeneous sandstones (stylolites probably nucleated along pre-existing clay drapes).

Although stylolites are only abundant in one of the wells (MF, Fig. 3), spectacular examples are found in all three study wells (e.g. Fig. 2b). In the neighbouring Midwest Sorrells well (24-6N-20E) a total of 17 macroscopically visible stylolites showing amplitudes mostly between 0.5 and 3 cm were measured within a depth interval of 27 m. In yet another Spiro well with good core control, Midwest Noah (21-6N-2E), we noted seven stylolites (amplitudes 0.5–2.5 cm) within 18.5 m of core. Although these numbers may not be representative of the true stylolite density in the Spiro, they nevertheless provide some indication of stylolite abundance.

## Petrographic observations

The detrital components of Spiro channel-fill sandstones consist primarily of quartz (> 90%) and variable proportions of bioclastic debris (mostly crinoid fragments) and sand-size chloritic–phosphatic particles. Detrital feldspars are virtually absent. Cements include chlorite (as grain coatings and as diffuse matrix), ankerite and ferroan calcite, apatite, and quartz (Fig. 4). Chlorite (chamosite) grain coatings played a key role in the porosity evolution of these sandstones, inasmuch as these coatings both retarded intergranular pressure solution and largely inhibited nucleation and growth of authigenic quartz on the detrital grains (cf. Pittman & Lumsden, 1968; Houseknecht & McGilvery, 1990; Spötl *et al.*, 1994). Many of these chlorite-coated sandstones are still porous today and/or contain variable amounts of pyrobitumen (see below). Others, and these are the main subject of this article, show pore-fill quartz cement which is not in optical continuity with the adjacent quartz grains. This cement is unlike syntaxial quartz overgrowths and is characterized by extremely inclusion-poor, pseudopoikilotopic crystals occluding a fairly high intergranular volume, i.e. up to 15% (Fig. 5; see also Plate 1, facing p. 290). Crystals apparently nucleated on grain surfaces incompletely coated by chlorite ('outgrowths' rather than overgrowths according to Pittman & Lumsden, 1968). A characteristic feature of these samples is the very small extent of intergranular pressure solution prior to quartz precipitation (Fig. 5a–c; Plate 1, facing p. 290).

Closely spaced sampling of Spiro cores showed that quartz-cemented sandstones form layers of variable thickness (Fig. 3) intercalated with either porous sandstones showing thick chlorite rims, sandstones tightly cemented by ankerite or stained by pyrobitumen, or, least commonly, sandstone layers cemented by syntaxial quartz. The latter layers lack chlorite coatings and show significant pressure solution prior to quartz cementation. These intervals are regarded as lateral equivalents of the fine-grained interchannel sandstones (Houseknecht & McGilvery, 1990).

The presence of pyrobitumen provides important

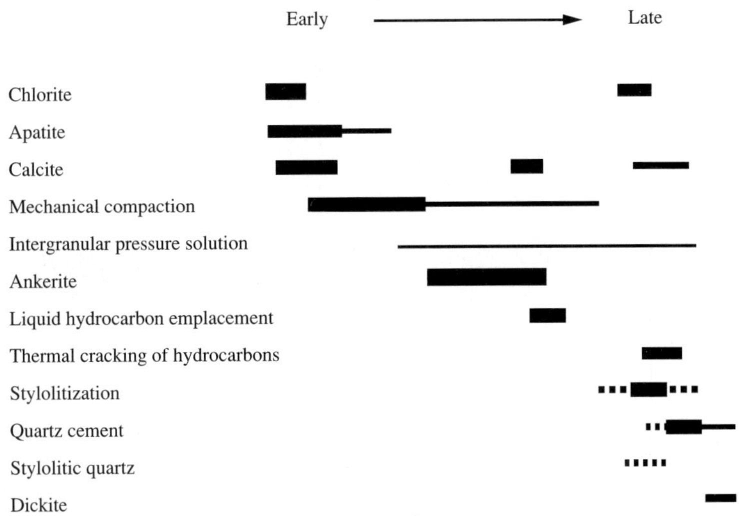


**Fig. 4.** Paragenetic sequence of diagenetic and anchimetamorphic events in the Spiro channel-fill sandstones.

paragenetic evidence regarding the diagenetic history of the Spiro Sandstone. Pyrobitumen was identified on the basis of petrographic and SEM observations (see Houseknecht & McGilvery, 1990) combined with the property of being insoluble in organic solvents. Pyrobitumen is observed as opaque material filling pre-existing porosity, moulded against earlier diagenetic products, and locally containing porous bubbles. Petrographic observations show that quartz precipitation post-dated pyrobitumen formation in many samples (Fig. 4), e.g. quartz filling shrinkage cracks in pyrobitumen (Houseknecht & McGilvery, 1990), although some textures suggest that both processes may have occurred contemporaneously (Fig. 4). Quartz cement also post-dated formation of chlorite polytype II*b* (Spötl *et al.*, 1994).

A second, volumetrically insignificant quartz variety locally formed along stylolites. This quartz occurs either as euhedral crystals 150–300 μm in diameter (locally ≤ 750 μm; Fig. 5d; Plate 1, facing p. 290) or, less commonly, as inclusion-rich crystals arranged in columnar aggregates parallel to the stylolite growth direction. A microcrystalline, chert-like quartz was observed locally as well. Textural observations show transitions between both types of stylolitic quartz, suggesting that they are genetically related (Fig. 4). Stylolitic quartz is intergrown with fibrous calcite, and rarely with sphalerite. In one Spiro core from the thrustbelt we found coarse crystalline chlorite within a high-amplitude stylolite.

Comparison of core and thin-section observations shows that there is no clear spatial relationship between sandstones containing late quartz cement and the occurrence of macroscopically visible stylolites (Fig. 3). In particular, intervals heavily cemented by late quartz commonly lack abundant high-amplitude stylolites (as in well MJ). Paragenetic relationships suggest that quartz cement within the host sandstone may have formed coevally with quartz precipitation along stylolites, but textural evidence is ambiguous.

## Cathodoluminescence

Quartz cement shows fairly intense greenish-blue CL emission when viewed in the hot-CL device, which is initially brighter than the emission of most detrital quartz grains (with the exception of minor red luminescent quartz). Unlike the short-lived CL described from other quartz samples (Ramseyer & Mullis, 1990; Perny *et al.*, 1992) emission intensity of quartz cement decays only slowly over a period of up to 2 min. In contrast, the same emission vanishes within a few seconds under the cold-CL which operates at a much higher beam power density. The CL images reveal that pore-fill quartz cement consists of a mosaic of individual crystals showing extensive development of sector and fine-scale concentric zoning (Fig. 5e,f; Plate 1, facing p. 290). Growth directions are entirely non-syntaxial and individual crystals extending over entire pores are not uncommon. Some quartz cement also apparently formed in microfractures within quartz framework grains. There is no evidence of an earlier (syntaxial) quartz generation in these samples (Fig. 5e,f; Plate 1, facing p. 290).

CL of stylolitic quartz is similar to that of quartz cement within the host sandstone. Initially bright emission is greenish-blue, which decays rapidly within

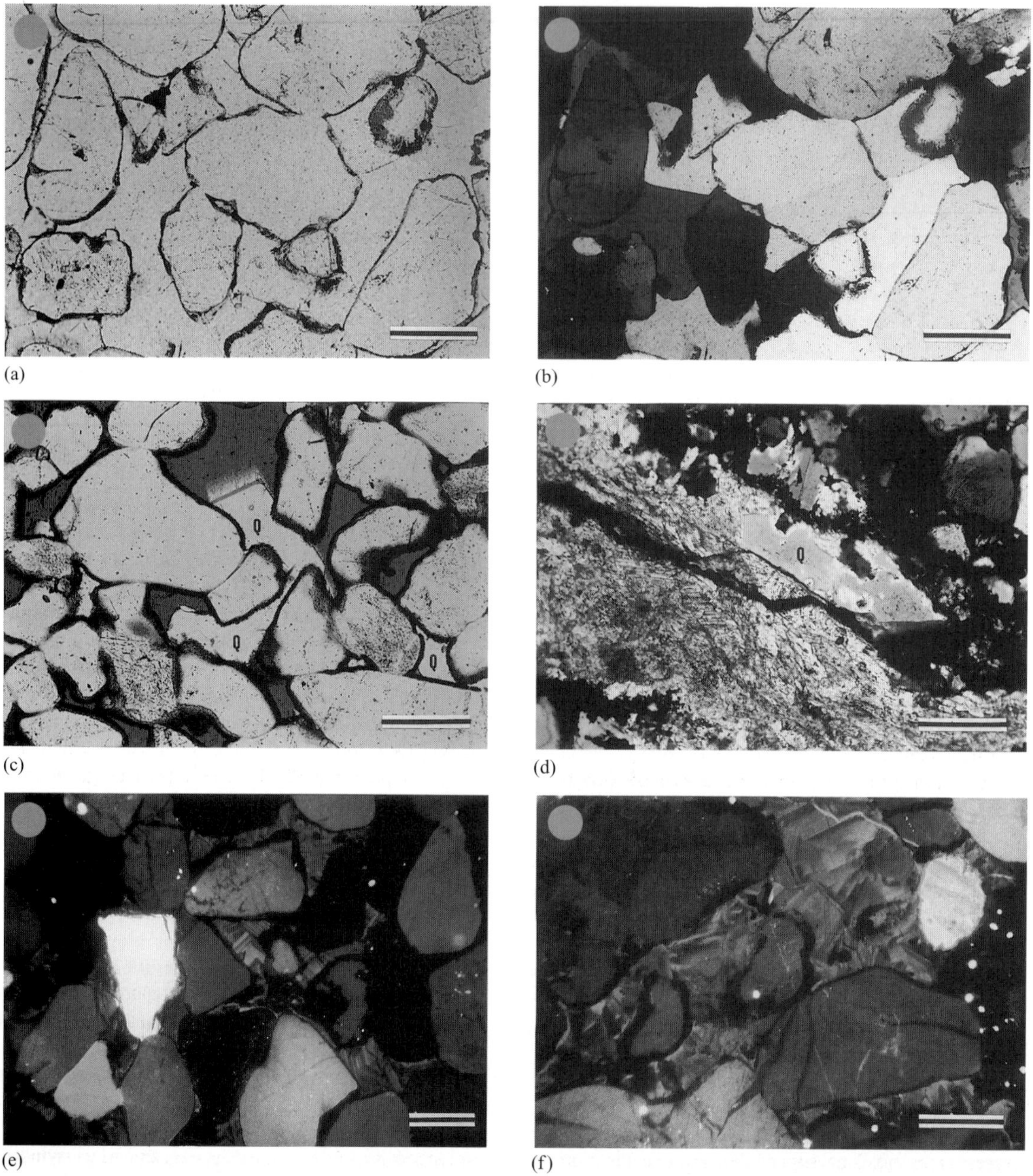


**Fig. 5.** (*Also reproduced in colour, see Plate 1, facing p. 290.*) Thin-section photomicrographs of deep-burial quartz cements in the Spiro reservoir. (a) Sandstone completely cemented by quartz. Note high intergranular volume and the presence of dark-brown coatings (chamosite) on detrital quartz grains. Sample MB 11858. Plain light photomicrograph. Scale: 200 μm. (b) Same, but crossed polars. Note the absence of syntaxial quartz rims. (c) Incipient growth of quartz cement (marked by Q) heavily coated by syndepositional chlorite coatings. Porosity is stained blue. Plain light photomicrograph. Sample MF 11872. Scale 200 μm. (d) Euhedral quartz (Q) growing within diagenetic calcite along a prominent stylolite surrounded by sandstone (upper right-hand and lower left-hand corners). Sample MB 12113. Crossed polars. Scale 200 μm. (e) Quartzose Spiro Sandstone heavily cemented by oscillatory and sector-zoned quartz. Non-luminescent rims of quartz framework grains are exclusively chlorite coatings. Bright yellow spots are apatite crystals. Overexposed detrital grain just left of picture centre is a fairly bright red luminescent quartz grain. Note absence of significant chemical compaction prior to late quartz precipitation. Sample MJ 11856.9. Hot-CL photomicrograph. Scale: 250 μm. (f) Complexly intergrown sector-zoned pore-fill quartz cement. Note the presence of minor authigenic quartz as fracture-fill in detrital quartz grains. Dark, non-luminescent rims around framework grains are chlorite coats. Sample MJ 11858. Hot-CL photomicrograph. Scale 200 μm.

seconds using the cold-cathode luminoscope, but shows similar slow decay in the hot-cathode device as quartz cement outside stylolites. Stylolitic quartz crystals commonly show inclusion-rich inner areas in normal transmitted light, but with CL microscopy we failed to detect evidence of detrital quartz cores in these crystals.

### Fluid inclusions

Quartz cement is exceptionally inclusion-poor and during our first study (Spötl *et al.*, 1996) we found no fluid inclusions in this cement. For the present study we examined an additional set of doubly polished thin-section wafers containing abundant quartz cement and located a total of four individual liquid–vapour inclusions in two samples. The inclusions were small (2–3 μm) and homogenization could not be recognized unambiguously. The four inclusions were apparently close to homogenization at 121–132°C (temperature at which a tiny vibrating bubble could barely be recognized), but actual homogenization probably occurred at somewhat higher temperatures. No freezing runs were performed because of the small size of the inclusions. The arrangement of these inclusions together with even smaller gaseous (?) and solid inclusions suggest that they may be secondary in origin (cf. Roedder, 1984).

Two-phase liquid–vapour inclusions are slightly more common in stylolitic megaquartz. In addition to the data presented in our previous publication (Spötl *et al.*, 1996) we measured homogenization temperatures of another 11 primary inclusions. The new values confirm the previously reported wide range in homogenization temperatures (from 150 to 219°C with a mean at 185 ±25°C). No salinity and gas measurements were performed on the new set of inclusions, but earlier measurements show evidence of clathrate formation (Spötl *et al.*, 1996), implying the presence of dissolved methane within the inclusions.

### Isotopic analyses

A total of 56 individual spot analyses were made on seven samples (Table 1; Fig 6). Isotopic analyses of quartz cement were performed on a few intergranular pores (called areas in Table 1) completely occluded by this cement. In stylolitic quartz samples individual euhedral crystals were probed. Isotopic variation within individual samples is larger in stylolitic quartz (between 4.0 and 6.5‰) than in quartz cement away from stylolites (between 2.2 and 3.3‰) and is also reflected in the larger standard deviations for the stylolitic quartz analyses (Table 1).

Mean oxygen isotope values of four quartz cement samples range from 19.1 to 21.9‰ (total range of individual spot analyses 17.5–22.4‰). No systematic difference between analyses of the interior and of the outer portion of pore-fill quartz crystals was detected (Table 1), although there are variations exceeding the analytical precision in three of the four samples (Table 1). Two analyses of detrital quartz grains gave identical $\delta^{18}O$ values of 10.3‰ (Table 1; Fig. 6). Three samples of authigenic quartz from a single, high-amplitude stylolite (Fig. 2b) yielded mean $\delta^{18}O$ values indistinguishable within the analytical precision (21.8–22.3‰; total range of individual spot analyses 18.4–24.9‰). No systematic core-to-rim isotopic variation was found nor is there evidence of detrital cores in the inner, commonly inclusion-rich areas of these crystals (Table 1).

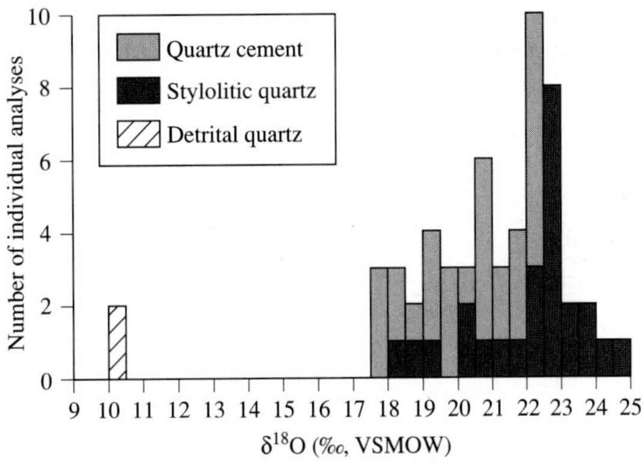


**Fig. 6.** Frequency distribution of individual *in situ* ion microprobe analyses of authigenic and detrital quartz in the Spiro Sandstone.

### Trace element analysis

Four quartz samples analysed isotopically were also studied for trace element concentrations using the ion microprobe. The present dataset comprises only 32 individual measurements and there are some uncertainties with regard to standardization and accuracy (see Methods). This reconnaissance study shows similar overall concentrations for Ca, K, Ti, Li, B, Na, Mg, and Al in stylolitic quartz and quartz cement. In contrast, detrital quartz yielded consistently higher Ti and Mg contents (mean values 55 and 13 p.p.m., respectively) than quartz cement (0.6 p.p.m. Ti and 6 p.p.m. Mg). Al concentrations are highly variable both within the detrital cores (8–4538 p.p.m.) and in secondary quartz (9–2513 p.p.m.). Li values are mostly below 1 p.p.m. in detrital and authigenic quartz, whereas B varies between 1 and 12 p.p.m. with no relation to quartz type. Stylolitic quartz crystals appear to be compositionally zoned with regard to most of these elements.

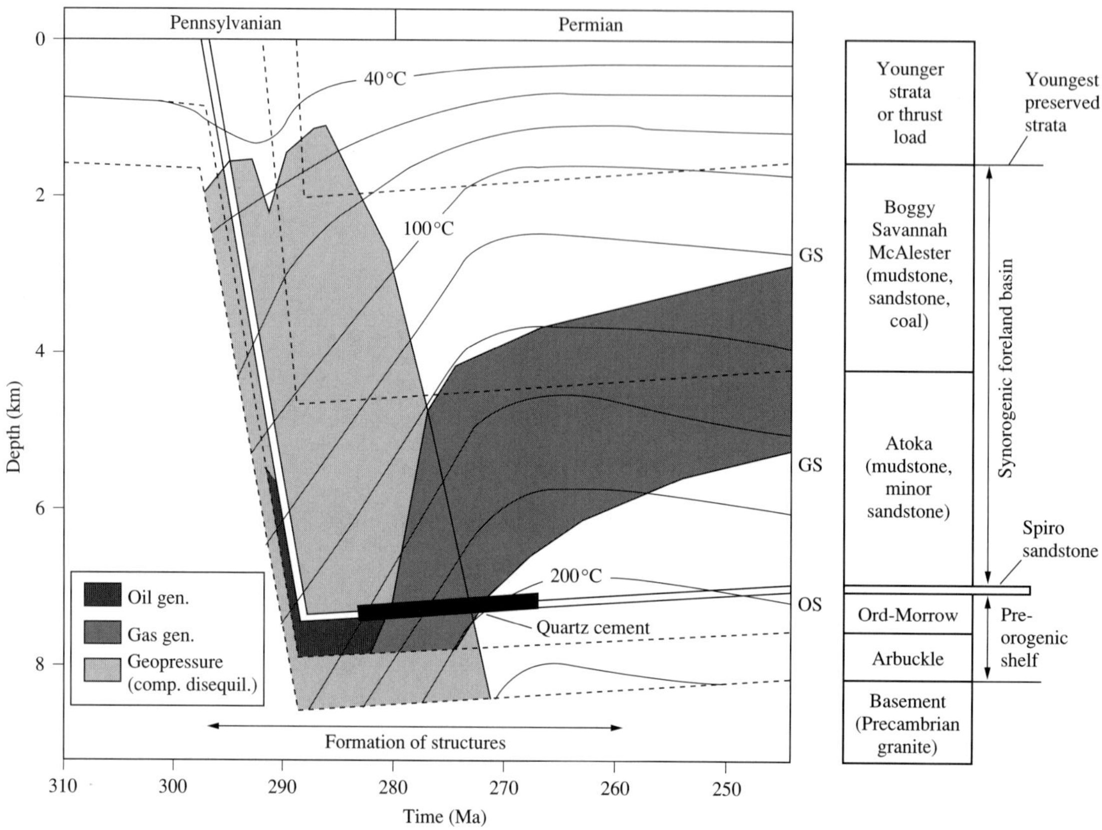


**Fig. 7.** Illustration of depth and inferred temperature history of Spiro Sandstone and associated strata during critical portion of burial history (history prior to 310 Ma and after 245 Ma deleted for purposes of emphasizing deep burial history). This model is based on the stratigraphy and thermal maturity profile observed in well AD (Fig. 1); the thermal maturity of the Spiro in this well is somewhat higher than in the wells sampled for petrographic and isotopic analysis. Therefore, the maximum temperatures shown in this model likely are slightly higher than those where quartz cement was analysed. Arrow representing timing of structural deformation includes basement rooted normal faulting and thrust faulting and folding associated with Ouachita orogenesis (see Table 2 for references). OS and GS represents oil and gas source rocks, respectively.

## Inferred burial and thermal history

Few case histories of high temperature quartz cement and preservation of good sandstone reservoir quality at thermal maturities as high as those documented for the Spiro Sandstone have been published (e.g. Dixon *et al.*, 1989; Gaupp *et al.*, 1993). Therefore, we have modelled the burial and thermal history of the Spiro Sandstone in the study area so that we can better interpret the deep diagenetic environment of this important gas reservoir. It is difficult to represent the burial and thermal history of the Spiro Sandstone with a single model because the deep Arkoma basin is characterized by abrupt lateral variations in stratigraphic burial and thermal maturity caused by growth faulting (Houseknecht, 1986; Houseknecht *et al.*, 1992). In addition, maximum burial is difficult to constrain because the existing sedimentary pile was undoubtedly buried by an unknown thickness of (i) stratigraphic section younger than any preserved in the basin and/or (ii) thrust sheets that once may have extended farther north from the Ouachita orogen. Moreover, the region is known to have been overprinted by advective heat (and mass) transport induced by tectonic compression and uplift associated with Ouachita orogenesis (Leach *et al.*, 1991; Ge & Garven, 1992; Cathles, 1993). Despite these complications, we believe the model presented here is representative of the deep diagenetic environment of the Spiro. The geological constraints used for modelling the burial and thermal history of the Spiro Sandstone in the study area are shown in Table 2.

The modelling results shown in Fig. 7 include several aspects that are important to consider in reconstructing

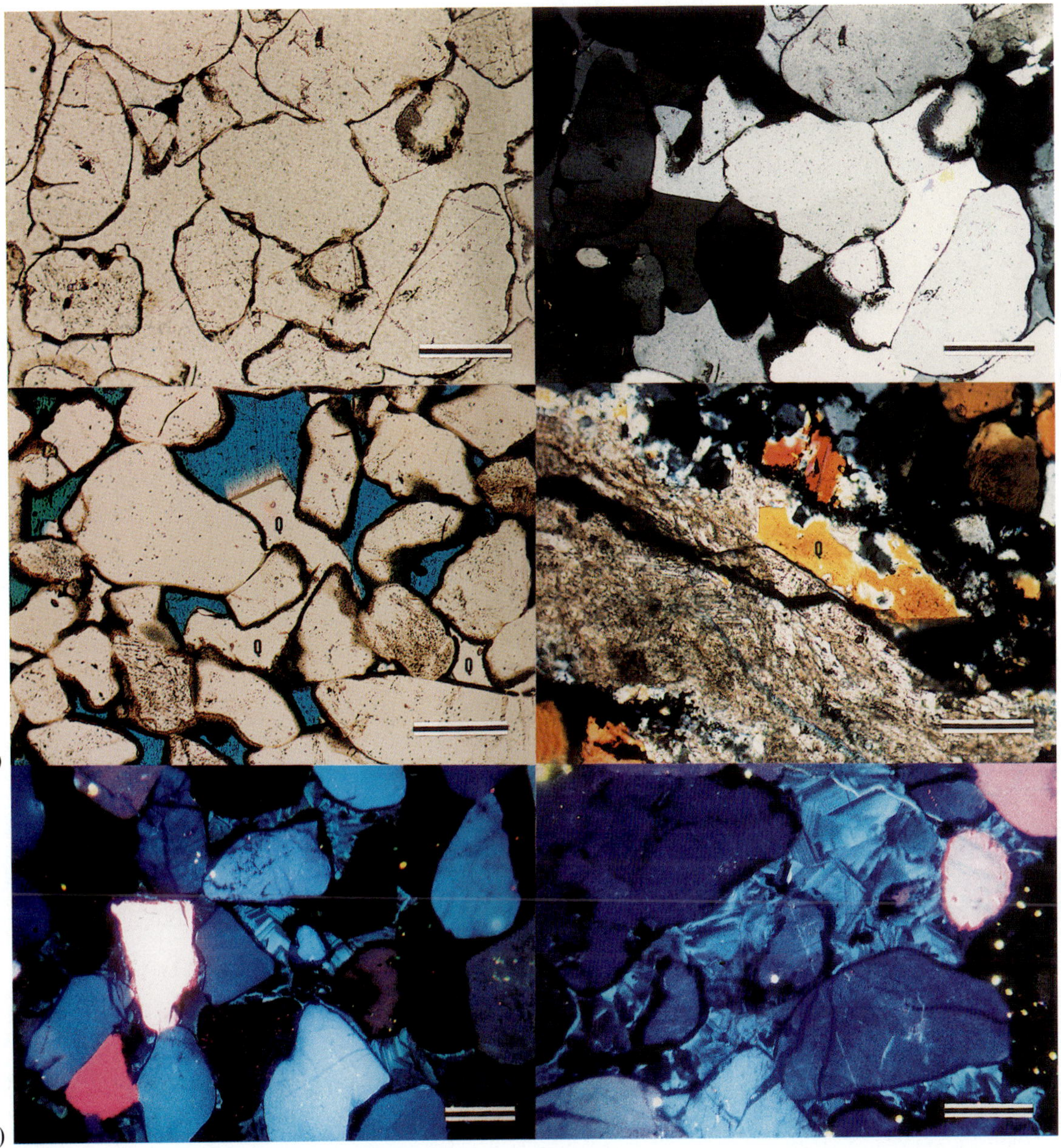

Plate 1. (a) (b) (c) (d) (e) (f)

**Plate 1.** Thin-section photomicrographs of deep-burial quartz cements in the Spiro reservoir. (a) Sandstone completely cemented by quartz. Note high intergranular volume and the presence of dark-brown coatings (chamosite) on detrital quartz grains. Sample MB 11858. Plain light photomicrograph. Scale: 200 μm (b) Same, but crossed polars. Note the absence of syntaxial quartz rims. (c) Incipient growth of quartz cement (marked by Q) heavily coated by syndepositional chlorite coatings. Porosity is stained blue. Plain light photomicrograph. Sample MF 11872. Scale 200 μm (d) Euhedral quartz (Q) growing within diagenetic calcite along a prominent stylolite surrounded by sandstone (upper right-hand and lower left-hand corners). Sample MB 12113. Crossed polars. Scale 200 μm (e) Quartzose Spiro sandstone heavily cemented by oscillatory and sector-zoned quartz. Non-luminescent rims of quartz framework grains are exclusively chlorite coatings. Bright yellow spots are apatite crystals. Overexposed detrital grain just left of picture centre is a fairly bright red luminescent quartz grain. Note absence of significant chemical compaction prior to late quartz precipitation. Sample MJ 11856.9. Hot-CL photomicrograph. Scale: 250 μm (f) Complexly intergrown sector-zoned pore-fill quartz cement. Note the presence of minor authigenic quartz as fracture-fill in detrital quartz grains. Dark, non-luminescent rims around framework grains are chlorite coats. Sample MJ 11858. Hot-CL photomicrograph. Scale 200 μm.

Spiro diagenesis in general and the origin of deep-burial cements in particular. First, sediment accumulation rates were extremely high for ~ 10 Myr following deposition of the Spiro. Second, geopressures were generated as the result of compactional disequilibrium induced by high sediment accumulation rates combined with the generally low permeability (i.e. mudstone-rich) character of post-Spiro strata. Third, during rapid sediment accumulation, temperatures within the sediment pile were depressed as the result of abnormally low thermal conductivities of the undercompacted, water-saturated sediments. Fourth, once rapid sediment accumulation ceased, geopressures began to dissipate, sediments began to dewater and compact, and temperatures within the sediment pile rose. In essence, heat that had been suppressed by the insulating effect of the undercompacted, geopressured sediments rose rapidly via conduction through the compacting sediment pile, resulting in a distinct upward bulge in the isotherms during the early to middle Permian (Fig. 7). This model suggests that the Spiro was exposed to maximum temperatures ~ 10–15 Myr after it reached maximum burial depth.

Inferred timing of hydrocarbon generation is also shown in Fig. 7. Generation of liquid hydrocarbons from oil-prone source rocks in pre-Spiro shelf strata apparently started during the latter part of rapid sediment accumulation (approximately at the end of Atoka Formation deposition) and continued until well past maximum burial (to ~ 280 Ma). Generation of gas, both by thermal cracking of previously formed oil and by primary generation from gas-prone source rocks in post-Spiro strata, started at ~ 280 Ma and continued for a long time (perhaps to the present), although the zone of active gas generation 'migrated' into younger strata through time. The top of the gas window rose rapidly through the stratigraphic section during ~ 280–275 Ma as a result of the rapid upward conductive heat flow that accompanied compaction of the sediment pile.

Significantly, oil and gas generation in source rocks stratigraphically near the Spiro mostly occurred while the Spiro was at or near maximum burial depth, but before the Spiro was exposed to maximum temperatures. This timing is consistent with the observation that the Spiro contains pyrobitumen, evidence that liquid hydrocarbons accumulated in the reservoir and then were cracked to methane as temperatures rose (Houseknecht & McGilvery, 1990; Spötl *et al.*, 1996). Moreover, the onset and peak of gas generation in the Spiro stratigraphic interval occurred while the section was still overpressured as a result of compactional disequilibrium (Fig. 7). It is likely that gas generation accentuated the overpressured conditions because of a significant pore fluid volume increase (e.g. Spencer, 1987; Barker, 1990; Luo & Vasseur, 1996; Osborne & Swarbrick, 1997), although we have not attempted to model the magnitude and duration of overpressures that resulted from gas generation.

Little is known about the post-Pennsylvanian burial history, because the youngest sediments in the basin are lowermost Permian in age. We therefore assumed linear uplift, an assumption that also fits available apatite fission-track cooling data for this region (Arne, 1992). Present-day fluid pressures in the Spiro reservoirs are generally hydrostatic, although a few cases of both slight overpressure (only in gas-bearing reservoirs; never in water-bearing formations) and underpressure have been encountered.

## DISCUSSION

The Spiro Sandstone clearly is a rather unusual example of an economic natural gas reservoir at thermal maturity levels commonly regarded as 'overmature'. In the following discussion we evaluate the processes that controlled, among other things, preservation of primary intergranular porosity as well as its local destruction by late quartz cement in this high-temperature sandstone reservoir.

### Constraints on late quartz formation

Quartz cement in the Spiro Sandstone is petrographically distinct from syntaxial quartz overgrowths, which are the dominant variety of intergranular quartz cement in sandstones from most sedimentary basins around the world studied to date. The paragenesis of this unusual quartz cement, pyrobitumen and high-temperature chlorite cement (Spötl *et al.*, 1994) suggests temperatures higher than those commonly reported from authigenic quartz in hydrocarbon reservoirs (i.e. > 140°C). Vitrinite reflectance values indicate maximum burial temperatures corresponding to anchimetamorphism and even approaching incipient greenschist metamorphism in some deep wells close to the thrustbelt.

Constraining the temperature of quartz cementation is difficult, but paragenetic information in conjunction with temperature estimates of pyrobitumen formation (Barker, 1990) and fluid-inclusion data from associated late-stage calcite cement (Spötl *et al.*, 1996) favour precipitation at temperatures close to 200°C. No reliable direct fluid inclusion evidence is available for authigenic quartz. Rare inclusions found in two samples are possibly secondary in origin and do not provide unequivocal temperature constraints. Fluid-inclusion data are available for stylolitic quartz cement, which is consistent with a

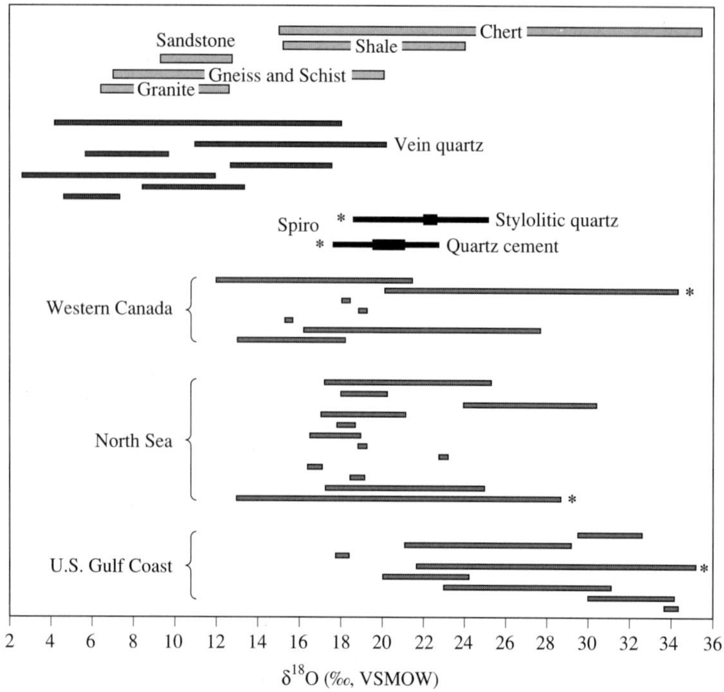


**Fig. 8.** Oxygen isotopic composition of Spiro quartz cements as compared with published data from authigenic quartz overgrowths in reservoir sandstones, vein quartz in metamorphic and hydrothermal settings, and quartz in various rock types. The bars indicate the total range of values. In the case of the Spiro, the thicker bars show the range of mean values for these samples. Data sets based on *in situ* microanalysis of oxygen isotope ratios are marked by an asterisk. Data sources: U.S. Gulf Coast basin (Aplin & Warren, 1994 and references therein; Land & Fisher, 1987; Williams *et al.* 1997a), North Sea basin (Aplin & Warren, 1994 and references therein; Brint *et al.*, 1991; Macaulay *et al.*, 1992; Hogg *et al.*, 1995; McLaughlin *et al.*, 1996; Williams *et al.*, 1997b), Western Canada Sedimentary Basin of Alberta (Aplin & Warren, 1994 and references therein; Longstaffe & Ayalon, 1991; Hervig *et al.*, 1995), vein quartz (Blattner, 1975; Nesbitt & Muehlenbachs, 1989; Conrad & Chamberlain, 1992; Kirschner *et al.*, 1993; Onasch & Vennemann, 1995), and rocks (Blatt, 1987).

high-temperature burial regime (185 ± 25°C). Although absolute proof is lacking, clathrate formation shows that small amounts of methane are present in these inclusions. As a consequence, pressure corrections were not employed, and the homogenization values are regarded as approximate fluid trapping temperatures.

These findings, in conjunction with results from burial history modelling, suggest quartz formation at, or subsequent to, peak burial when temperatures were rising (Fig. 7). During this time interval, constrained to ~ 285–265 Ma (at 7.3–7.1 km depth) by our burial history model, the reservoir temperatures rose to a maximum of approximately ~ 200°C, causing thermal cracking of previously reservoired oil to methane plus pyrobitumen residue.

Although paragenetic observations suggest that stylolitic quartz may have formed coevally with quartz cement, the isotope analyses do not support this strongly. The $\delta^{18}O$ values for stylolitic quartz (22.1 ± 1.7‰, $n = 25$) are systematically higher than those of quartz cement (20.4 ± 1.4‰, $n = 29$), although there is significant overlap (Fig. 6). Consequently, quartz cement formed either at different (slightly higher) temperatures, and hence time, than stylolitic quartz or both quartz types formed coevally from fluids of slightly different isotopic but similar trace elemental compositions. None of these interpretations can be ruled out definitely. The stylolite samples (MB 12113 A, B and C), which yielded mean values between 21.8 and 22.3‰, are only 60 cm upsection from sample MB 12115, whose quartz cement yielded a slightly lower mean $\delta^{18}O$ value of 19.4 ± 1.1‰. While this observation argues in favour of two fluids of slightly different isotopic composition and/or temperature (~ 35° temperature difference at constant fluid composition in this case), the data on quartz cement document that minor inhomogeneities in fluid composition and/or temperature existed as well (Table 1). These fluctuations in fluid composition and/or temperature possibly resulted in the complex compositional zoning of quartz cement as revealed by CL imaging.

Despite the documented intrasample inhomogeneity in Spiro quartz cements, these cements are isotopically fairly homogeneous when compared with data from syntaxial quartz overgrowths from other sedimentary basins. Figure 8 shows a compilation of published oxygen isotope data of syntaxial quartz overgrowths from three of the most extensively studied sedimentary basins, the US Gulf Coast basin, the North Sea basin and the Western Canada Sedimentary Basin. At first glance, data from the Spiro Sandstone are characterized by a fairly large range in values, in particular stylolitic quartz, whereas several other datasets (e.g. North Sea sandstones) have a narrower range of values. With the exception of a few studies (Hervig *et al.*, 1995; Williams *et al.*, 1997a,b), however, all data of quartz overgrowths were obtained by the conventional fluorination technique, a bulk-sample method masking intrasample and intracrystalline isotopic inhomogeneities despite superior analytical precision. It is thus not surprising that the ranges of *in-situ* $\delta^{18}O$ values in the Spiro Sandstone are somewhat large compared with bulk-sample datasets. The very large range of values reported from the ion microprobe studies of sandstones in Alberta (Hervig *et al.*, 1995; 20–34‰ in authigenic quartz and a 5–25‰ variation in detrital quartz grains), the Michigan Basin and the Wisconsin Arch (Graham *et al.*, 1996; 12.5–31.6‰ in authigenic quartz and a 5–13‰ variation in detrital quartz grains) and the North Sea (Williams *et al.*, 1997b; 13.0–28.4‰ in authigenic quartz and a 1.9–24.1‰ variation in detrital quartz grains) clearly underscores this point. In essence, despite the apparent range in individual spot analyses, the Spiro quartz cements are isotopically fairly homogeneous as compared with limited data available from ion microprobe studies of quartz overgrowths. We speculate that the high temperatures of quartz cementation in the Spiro and possibly also the associated fluid regime may have been responsible for this fact.

Figure 8 also shows that the values of Spiro quartz cements are similar to those of syntaxial quartz cements from other basins, overlapping with the lower $\delta^{18}O$ values of these datasets. Spiro quartz cement, however, is typically enriched in $^{18}O$ compared with vein quartz from metamorphic and hydrothermal settings (Fig. 8). In conjunction with temperature estimates, isotopic data allow more quantitative constraints to be placed on the palaeo-pore-fluid composition. Assuming isotopic equilibrium and using the fractionation equation of Friedman & O'Neil (1977) a mean $\delta^{18}O$ value of 22.1‰ for stylolitic quartz in conjunction with a mean fluid inclusion temperature of 185°C (± 25) of sample MB 12113 suggests an oxygen isotopic composition of the fluid of +8.7‰ (+1.6/−1.9) VSMOW (using the total range of values (18.4–24.9‰) a $\delta^{18}O$ value of the pore fluid of +5.0 to +11.5‰ can be calculated). Similar high $\delta^{18}O$ values were also calculated using oxygen isotope data from high-temperature chlorite cements (Spötl *et al.*, 1994). No such calculations can be made for quartz cement, because direct temperature constraints are not available, but the overlap of measured $\delta^{18}O$ values of both quartz varieties (Fig. 6) suggests similar temperature–fluid compositional relationships also for quartz cement. Two analyses of fibrous calcite growing along stylolite sample MB 12113 yielded $\delta^{18}O$ values of −11.1 and −11.2‰ VPDB, respectively. An equilibrium pore fluid $\delta^{18}O$ value of +9.3‰ VSMOW is obtained—if this calcite is formed at the same temperature as the coexisting stylolitic quartz—a value indistinguishable from that calculated for quartz (see above). The inferred fluid composition suggests that late quartz cement (and associated cements) formed from pore fluids that acquired their isotopically enriched signature by reactions with the surrounding rock, probably within a hydrologically restricted regime (low fluid-to-rock ratio).

### Silica sources

There are two principal internal sources of silica in the Spiro, intergranular pressure solution and dissolution along stylolites, and a possible external source, i.e. silica-rich fluids expelled from Atoka mudrocks.

Petrographic observations show that very little pressure solution has occurred in the Spiro channel-fill sandstones prior to late quartz precipitation. Evidence of moderate to strong pressure solution can be found in the interchannel sandstones, which lack chlorite coatings. These sandstones locally interfinger with the chloritic channel-fill facies, such as in well MB. It is unlikely, however, that these sandstones represent a significant silica source for the coarser-grained channel sandstones, because most of the silica dissolved along grain boundaries apparently was reprecipitated locally as syntaxial quartz rims within these units. The timing of this silica precipitation appears to be earlier than that of quartz cement, because pyrobitumen post-dates syntaxial quartz rims in these samples.

Stylolites are commonly invoked as significant sources of silica (and/or carbonate) in deep-burial settings (Füchtbauer, 1978; Bjørlykke *et al.*, 1989; Dutton & Diggs, 1990; Oelkers *et al.*, 1996) and are also a potential source in the Spiro. Macroscopic stylolites, although not particularly abundant in the cores that contain quartz cement, reach dimensions that are significantly larger than those reported from many other sandstones (cf. Tada & Siever, 1989, p. 93). We did not attempt to quantify the amount of silica dissolved along stylolites in the Spiro. We feel that there are inadequacies in all available techniques, including: (i) measuring stylolite amplitudes systematically (which gives only a minimum volume and tends to overlook incipient stylolites not visible macroscopically); (ii) multiplying the stylolite amplitude by some poorly constrained correction factor (Heald, 1955; Dutton & Diggs, 1990); (iii) comparing the concentration of heavy minerals within and outside the stylolite (results in large errors if the heavy mineral concentration was primarily low; some heavy minerals may dissolve along stylolites as previously pointed out by Heald [1955]); (iv) comparing the concentration of clay minerals within and outside the stylolite (can give erroneously high estimates because many stylolites nucleate along clay drapes, e.g. Stone & Siever, 1996); and (v) comparing the concentration of presumably immobile trace elements within and outside the stylolite (again, possibly large errors, if the abundance of the predominant carrier mineral (e.g. heavy minerals in the case of Ti) is primarily low; tedious and expensive if tracers such as the rare earth elements are to be analysed in small stylolites). Based on our current data of stylolite abundance and size we conclude, qualitatively, that stylolites are important in the Spiro. They apparently formed during deep burial and high temperatures (as indicated by fluid inclusion data and textural observations), and we regard them as a potential source of silica (and carbonate) released during deep burial. Our observations, however, argue against stylolites as the sole source of quartz cement in porous sandstones, unless mass transfer at the scale of up to a few metres was involved (cf. Figure 3). Such mass transport is beyond typical distances of ion diffusion in burial diagenetic regimes, in particular at temperatures above ~ 150°C, where the silica system is generally regarded as transport-controlled (Oelkers *et al.*, 1996), and can only be realized by fluid advection. Burial-history modelling suggests that the transition from an overpressured to a normal pressured regime during deep burial (Fig. 7) may have destabilized the grain framework, promoted stylolitization, and hence resulted in a local increase in the degree of silica concentration in the pore fluid. It is likely that pressure solution along stylolites was enhanced by tectonic stress associated with thrust faulting, folding, and thrust loading, which occurred during this critical time interval (Fig. 7). Finally, the transition from overpressure to normal pressure likely increased fluid mobility, providing an attractive mechanism for silica transport.

In addition to the two internal sources of silica in the Spiro there is also a possible external source that needs to be taken into account, i.e. mudrocks of the Atoka Formation. A thick succession of mudrocks directly overlies the Spiro and is locally juxtaposed along normal faults, providing possible pathways for shale-derived fluids into the Spiro reservoir. The question whether shale diagenesis is an isochemical process or involves significant mass transfer is still being debated (Evans, 1990; Bloch & Hutcheon, 1992; Awwiller, 1993; Wintsch & Kvale, 1994; Sutton & Land, 1996; Totten & Blatt, 1996). Analysis of clay mineralogy and crystallinity of Atoka mudstones suggests that most illite is detrital, which implies that only a limited volume of silica could have been released by illite/smectite-to-illite transformations within this mudstone section (Spötl *et al.*, 1993). The absence of high-temperature silicification in sandstones within the Atoka Formation (e.g. the Red Oak sandstone; Houseknecht & Ross, 1992) corroborates this assumption.

## CONCLUSIONS

Despite a strong effort to unravel the process of late-stage quartz cement in the hot Spiro Sandstone reservoir, its origin is still poorly understood. This dilemma is in part due to the apparent lack of primary fluid inclusions in this cement, which could provide critically needed information on the temperature and composition of these palaeofluids.

Our current data suggest quartz precipitation at temperatures in excess of ~ 150°C and possibly as high as ~ 200°C from fluids of heavy oxygen isotopic composition. High-amplitude stylolites are regarded as the major (single?) source of silica released during near-maximum burial. Intergranular pressure solution was not a significant silica source, because of the presence of chlorite coats (and possibly overpressure) stabilizing the grain fabric. Quartz growing along high-amplitude stylolites underscores their role as a silica source. The observed distribution of quartz cement relative to stylolites recognized in core suggests, however, that silica was transported from the stylolites into the porous sandstones not only by ionic diffusion, but also by fluid advection. The small differences in the oxygen isotopic composition between quartz cement and stylolitic quartz may be attributed to this process.

## ACKNOWLEDGEMENTS

CS acknowledges support from the Austrian Academy of Sciences (APART programme) during completion of this study and Karl Ramseyer for his generous help with the hot-CL. LR was partially supported by Basic Energy Sciences, Department of Energy, under contract DE-ACO5–96OR22464 to Oak Ridge National Laboratory, managed by Lockheed-Martin Energy Research Corporation. Portions of this work also were supported by the U.S. Geological Survey. The authors wish to thank Richard Worden, Stuart A. Barclay and an anonymous reviewer for their technical reviews.

## REFERENCES

Aplin, A.C. & Warren, E.A. (1994) Oxygen isotopic indications of the mechanisms of silica transport and quartz cementation in deeply buried sandstones. *Geology* **22**, 847–850.

ARBENZ, J.K. (1989) Ouachita thrust belt and Arkoma basin. In: *The Appalachian–Ouachita Orogen in the United States.* (eds HATCHER, R.D., THOMAS, W.A. & VIELE, G.W.) pp. 621–634. The Geology of North America, Vol. F-2. Geological Society of America, Boulder, Colorado.

ARNE, D.C. (1992) Evidence from apatite fission-track analysis for regional Cretaceous cooling in the Ouachita Mountain Fold Belt and Arkoma Basin of Arkansas. *American Association of Petroleum Geologists Bulletin* **76**, 392–402.

AWWILLER, D.N. (1993) Illite/smectite formation and potassium mass transfer during burial diagenesis of mudrocks: a study from the Texas Gulf Coast Paleocene–Eocene. *Journal of Sedimentary Petrology* **63**, 501–512.

BARKER, C. (1990) Calculated volume and pressure changes during thermal cracking of oil to gas in reservoirs. *American Association of Petroleum Geologists Bulletin* **74**, 1254–1261.

BJØRKUM, P.A. (1996) How important is pressure in causing dissolution of quartz in sandstones? *Journal of Sedimentary Research* **66**, 147–154.

BJØRLYKKE, K. & EGEBERG, P.K. (1993) Quartz cementation in sedimentary basins. *American Association of Petroleum Geologists Bulletin* **77**, 1538–1548.

BJØRLYKKE, K., RAMM, M. & SAIGAL, G.C. (1989) Sandstone diagenesis and porosity modification during basin evolution. *Geologische Rundschau* **78**, 243–268.

BLATT, H. (1987) Oxygen isotopes and the origin of quartz. *Journal of Sedimentary Petrology* **57**, 373–377.

BLATTNER, P. (1975) Oxygen isotopic composition of fissure-grown quartz, adularia, and calcite from Broadlands geothermal field, New Zealand. *American Journal of Science* **275**, 785–800.

BLOCH, J. & HUTCHEON, I.E. (1992) Silica mass transport during shale diagenesis: are shales a source or sink for silica?. In: *Water–Rock Interaction, Proceedings of the 7th International Symposium on Water–Rock Interaction. Park City, Utah.* (eds KHARAKA, Y.K. & MAEST, A.S.) Vol. 2, pp. 1157–1160. Balkema, Rotterdam.

BOLTON, E.W., LASAGA, A.C. & RYE, D.M. (1996) A model for the kinetic control of quartz dissolution and precipitation in porous media flow with spatially variable permeability: formulation and examples of thermal convection. *Journal of Geophysical Research, 101/B* **10**, 22157–22187.

BRADLEY, D.C. & KIDD, W.S.F. (1991) Flexural extension of the upper continental crust in collisional foredeeps. *Geological Society of the American Bulletin* **103**, 1416–1438.

BRINT, J.F., HAMILTON, P.J., HASZELDINE, R.S., FALLICK, A.E. & BROWN, S. (1991) Oxygen isotopic analysis of diagenetic quartz overgrowths from the Brent sands: a comparison of two preparation methods. *Journal of Sedimentary Petrology* **61**, 527–533.

CATHLES, L.M. (1993) A discussion of flow mechanisms responsible for alteration and mineralization in the Cambrian aquifers of the Ouachita–Arkoma basin–Ozark system. In: *Diagenesis and Basin Development.* (eds HORBURY, A.D. & ROBINSON, A.G.) pp. 99–112. American Association of Petroleum Geologists Studies in Geology 36.

CONRAD, M.E. & CHAMBERLAIN, C.P. (1992) Laser-based, *in situ* measurements of fine-scale variations in the $\delta^{18}O$ values of hydrothermal quartz. *Geology* **20**, 812–816.

DIXON, S.A., SUMMERS, D.M. & SURDAM, R.C. (1989) Diagenesis and preservation of porosity in Norphlet Formation (Upper Jurassic), southern Alabama. *American Association of Petroleum Geologists Bulletin* **73**, 707–728.

DUTTON, S.P. & DIGGS, T.N. (1990) History of quartz cementation in the Lower Cretaceous Travis Peak Formation, East Texas. *Journal of Sedimentary Petrology* **60**, 191–202.

EVANS, J. (1990) Quartz dissolution during shale diagenesis. Implications for quartz cementation in sandstones. In: *Second International Symposium Geochemistry of the Earth's Surface and of Mineral Formation.* (eds NOACK, Y. & NAHON, D.). *Chemical Geology* **84**, 239–240.

FALVEY, D.A. & MIDDLETON, M.F. (1981) Passive continental margins: evidence for a prebreakup deep crustal metamorphic subsidence mechanism. *Oceanologica Acta. Proceedings of the 26th Internal Geological Congress, Paris*, 1980. pp. 103–114.

FRIEDMAN, I. & O'NEIL, J.R. (1977) Compilation of stable isotope fractionation factors of geochemical interest. In: *Data of Geochemistry*, 6th edn (ed. FLEISCHER, M.) pp. 1–12. U. S. Geological Survey Professional Paper 440-KK.

FÜCHTBAUER, H. (1978) Zur Herkunft des Quarzzements. Abschätzung der Quarzauflösung in Silt- und Sandsteinen. *Geologische Rundschau* **67**, 991–1008.

GE, S. & GARVEN, G. (1992) Hydromechanical modeling of tectonically driven groundwater flow with application to the Arkoma foreland basin. *Journal of Geophysical Research* ***97/B* 6**, 9119–9144.

GRAHAM, C.M., VALLEY, J.W. & WINTER, B.L. (1996) Ion micropsobe analysis of $^{16}O/^{16}O$ in authigenic and detrital quartz in the St. Peter Sandstone, Michigan Basin and Wisconsin Arch, USA: Contrasting diagenetic histories. *Geochimica et cosmochimica Acta* **60**, 5101–5116.

GROSS, J.S., THOMPSON, S.A., CLAXTON, B.L. & CARR, M.B. (1995) Reservoir distribution and exploration potential of the Spiro Sandstone in the Choctaw trend, Arkoma basin, Oklahoma and Arkansas. *American Association of Petroleum Geologists Bulletin* **79**, 159–185.

HABICHT, J.K.A. (1979) Paleoclimate, paleomagnetism, and continental drift. *American Association of Petroleum Geologists Studies in Geology*, 9.

HEALD, M.T. (1955) Stylolites in sandstones. *Journal of Geology* **63**, 101–114.

HERVIG, R.L. (1992) Oxygen isotope analysis using extreme energy filtering. *Chemical Geology* **101**, 185–186.

HERVIG, R.L., WILLIAMS, L.B., KIRKLAND, I.K. & LONGSTAFFE, F.J. (1995) Oxygen isotope microanalyses of diagenetic quartz: possible low temperature occlusion of pores. *Geochimica et Cosmochimica Acta* **59**, 2537–2543.

HOGG, A.J.C., PEARSON, M.J., FALLICK, A.E. & HAMILTON, P.J. (1995) An integrated thermal and isotopic study of the diagenesis of the Brent Group, Alwyn South, U.K. North Sea. *Applied Geochemistry* **10**, 531–546.

HOUSEKNECHT, D.W. (1986) Evolution from passive margin to foreland basin: the Atoka Formation of the Arkoma basin, south-central U.S.A. In: *Foreland Basins.* (eds ALLEN, P.A. & HOMEWOOD, P.) pp. 327–345. Special Publications of the International Association of Sedimentology 8.

HOUSEKNECHT, D.W. & MCGILVERY, T.A. (1990) Red Oak field. In: *Structural Traps. II Traps Associated with Tectonic Faulting.* (eds BEAUMONT, E.A. & FOSTER, N.H.) pp. 201–225. Treatise of Petroleum Geology. Atlas of Oil and Gas Fields. American Association of Petroleum Geologists, Tulsa.

HOUSEKNECHT, D.W. & ROSS, L.M. (1992) Clay minerals in

Atokan deep-water sandstone facies, Arkoma basin: origins and influence on diagenesis and reservoir quality. In: *Origin, Diagenesis, and Petrophysics of Clay Minerals in Sandstones.* (eds HOUSEKNECHT, D.W. & PITTMAN, E.D.) pp. 227–240. Society of Economic Paleontologists and Mineralogists Special Publications 47.

HOUSEKNECHT, D.W. & SPÖTL, C. (1993) Empirical observations regarding methane deadlines in deep basins and thrust belts. In: *The Future of Energy Gas.* (ed. HOWELL, D.G.) pp. 217–231.U.S. Geological Survey Professional Paper 1570.

HOUSEKNECHT, D.W. & WEESNER, C.M.B. (1997) Rotational reflectance of dispersed vitrinite from the Arkoma Basin. *Organic Geochemistry* **26**, 191–206.

HOUSEKNECHT, D.W.L.A., HATHON & MCGILVERY, T.A. (1992) Thermal maturity of Paleozoic strata in the Arkoma basin. In: *Source Rocks in the Southern Midcontinent.* (eds JOHNSON, K.S. & CARDOTT, B.J.) pp. 122–132. Oklahoma Geological Survey Circular 93.

JOHNSON, C.R. (1986) Geology and pre-Demoinesian stratigraphy of the Arkoma basin, Oklahoma. In: *The Relationship Between Coal Rank and Present Geothermal Gradient in the Arkoma Basin, Oklahoma.* (eds CARDOTT, B.J., HEMISH, L.A., JOHNSON, C.R. & LUZA, K.V.) pp. 6–15. Oklahoma Geological Survey Special Publication 86–4.

KIRSCHNER, D.L., SHARP, Z.D. & TEYSSIER, C. (1993) Vein growth mechanisms and fluid sources revealed by oxygen isotope laser microprobe. *Geology* **21**, 85–88.

LAND, L.S. & FISHER, R.S. (1987) Wilcox sandstone diagensis, Texas Gulf Coast: a regional isotopic comparison with the Frio Formation. In: *Diagenesis of Sedimentary Sequences.* (ed. MARSHALL, J.D.) pp. 219–235. Geological Society of London Special Publication 36.

LEACH, D.L., PLUMLEE, G.S., HOFSTRA, A.H. *et al.* (1991) Origin of late dolomite cement by $CO_2$-saturated deep basin brines: evidence from the Ozark region, central United States. *Geology* **19**, 348–351.

LEE, Y., DEMING, D. & CHEN, K.F. (1996) Heat flow and heat production in the Arkoma Basin and Oklahoma Platform, southeastern Oklahoma. *Journal of Geophysical Research* **101/B11**, 25 387–25 401.

LONGSTAFFE, F.J. & AYALON, A. (1991) Mineralogical and O-isotope studies of diagenesis and porewater evolution in continental sandstones, Cretaceous Belly River Group, Alberta, Canada. *Applied Geochemistry* **6**, 291–303.

LUMSDEN, D.N., PITTMAN, E.D. & BUCHANAN, R.S. (1971) Sedimentation and petrology of Spiro and Foster Sands (Pennsylvanian), McAlester Basin, Oklahoma. *American Association of Petroleum Geologists Bulletin* **55**, 254–266.

LUO, X. & VASSEUR, G. (1996) Geopressuring mechanism of organic matter cracking: numerical modeling. *American Association of Petroleum Geologists Bulletin* **80**, 856–874.

LYON, I.C., SAXTON, J.M., MCKEEVER, P.J., CHATZITHEODORIDIS, E. & VAN LIERDE, P. (1995) Precision and reproducibility of *in situ* oxygen isotope ratio measurements on quartz obtained using an Isolab 54 ion microprobe. *International Journal of Mass Spectrometry Ion Processes* **151**, 1–16.

MACAULAY, C.I., HASZELDINE, R.S. & FALLICK, A.E. (1992) Diagenetic pore waters stratified for at least 35 million years: Magnus oil field, North Sea. *American Association of Petroleum Geologists Bulletin* **76**, 1625–1634.

MCBRIDE, E.F., LAND, L.S. & MACK, L.E. (1987) Diagenesis of eolian and fluvial feldspathic sandstones, Norphlet Formation (Upper Jurassic), Rankin County, Mississippi, and Moble County, Alabama. *American Association of Petroleum Geologists Bulletin* **71**, 1019–1034.

MCLAUGHLIN, O.M., HASZELDINE, R.S. & FALLICK, A.E. (1996) Quartz diagenesis in layered fluids in the south Brae oilfield, North Sea. In: *Siliciclastic Diagenesis and Fluid Flow: Concepts and Applications.* (eds CROSSEY, L.J., LOUCKS, R. & TOTTEN, M.W.) pp. 93–113. Society of Economic Paleontologists and Mineralogists Special Publication 55.

NESBITT, B.E. & MUEHLENBACHS, K. (1989) Origins and movements of fluids during deformation and metamorphism in the Canadian Cordillera. *Science* **245**, 733–736.

OELKERS, E.H., BJØRKUM, P.A. & MURPHY, W.M. (1996) A petrographic and computational investigation of quartz cementation and porosity reduction in North Sea sandstones. *American Journal of Science* **296**, 420–452.

ONASCH, C.M. & VENNEMANN, T.W. (1995) Disequilibrium partitioning of oxygen isotopes associated with sector zoning in quartz. *Geology* **23**, 1103–1106.

OSBORNE, M.J. & SWARBRICK, R.E. (1997) Mechanisms for generating overpressure in sedimentary basins: a reevaluation. *American Association of Petroleum Geologists Bulletin* **81**, 1023–1041.

PARK, W.C. & SCHOT, E.H. (1968) Stylolites: their nature and origin. *Journal of Sedimentary Petrology* **38**, 175–191.

PERNY, B., EBERHARDT, P., RAMSEYER, K., MULLIS, J. & PANKRATH, R. (1992) Microdistribution of Al, Li, and Na in alpha quartz: possible causes and correlation with short-lived cathodoluminescence. *American Mineralogists* **77**, 534–544.

PITTMAN, E.D. & LUMSDEN, D.N. (1968) Relationship between chlorite coatings on quartz grains and porosity, Spiro Sand, Oklahoma. *Journal of Sedimentary Petrology* **38**, 668–670.

RAMSEYER, K. & MULLIS, J. (1990) Factors influencing short-lived blue cathodoluminescence of alpha-quartz. *American Mineralogist* **75**, 791–800.

RAMSEYER, K., FISCHER, J., MATTER, A., EBERHARDT, P. & GEISS, J. (1989) A cathodoluminescence microscope for low intensity luminescence. *Journal of Sedimentary Petrology* **59**, 619–622.

RICIPUTI, L.R. & PATERSON, B.A. (1994) High-spatial-resolution microanalysis of O-isotope ratios in silicates and carbonates by ion microprobe. *American Mineralogy* **79**, 1227–1230.

ROEDDER, E. (1984) *Fluid Inclusions*. Reviews in Mineralogy 12. Mineralogical Society of America, Washington, D.C.

SCOTESE, C.R. & MCKERROW, W.S. (1990) Revised world maps and introduction. In: *Palaeozoic Palaeogeography and Biogeography.* (eds MCKERROW, W.S. & SCOTESE, C.R.) pp. 1–21. *Geological Society of London Memoir* 12.

SHELTON, K.L., READER, J.M., ROSS, L.M., VIELE, G.W. & SEIDEMANN, D.E. (1986) Ba-rich adularia from the Ouachita Mountains, Arkansas: implications for a postcollisional hydrothermal system. *American Mineralogist* **71**, 916–923.

SPENCER, C.W. (1987) Hydrocarbon generation as a mechanism for overpressuring in Rocky Mountain region. *American Association of Petroleum Geologists Bulletin* **71**, 368–388.

SPÖTL, C., HOUSEKNECHT, D.W. & JAQUES, R. (1993) Clay mineralogy and illite crystallinity of the Atoka Formation, Arkoma Basin, and frontal Ouachita Mountains. *Clays and Clay Minerals* **41**, 745–754.

SPÖTL, C., HOUSEKNECHT, D.W. & LONGSTAFFE, F.J. (1994) Authigenic chlorites in sandstones as indicators of high-temperature diagenesis, Arkoma foreland basin, USA. *Journal of Sedimentary Research* **A64**, 553–566.

SPÖTL, C., HOUSEKNECHT, D.W. & BURNS, S.J. (1996) Diagenesis of an 'overmature' gas reservoir: the Spiro sand of the Arkoma Basin, USA. *Marine and Petroleum Geology* **13**, 25–40.

STONE, W.N. & SIEVER, R. (1996) Quantifying compaction, pressure solution and quartz cementation in moderately- and deeply-buried quartzose sandstones from the greater Green River Basin, Wyoming. In: *Siliciclastic Diagenesis and Fluid Flow: Concepts and Applications.* (eds CROSSEY, L.J., LOUCKS, R. & TOTTEN, M.W.) pp. 129–150. Society of Economic Paleontologists and Mineralogists Special Publications 55.

SUTTON, S.J. & LAND, L.S. (1996) Postdepositional chemical alteration of Ouachita shales. *Geological Society of America Bulletin* **108**, 978–991.

TADA, R. & SIEVER, R. (1989) Pressure solution during diagenesis. *Annual Review of Earth Planetary Sciences* **17**, 89–118.

THOMAS, A.R., DAHL, W.M., HALL, C.M. & YORK, D. (1993) $^{40}Ar/^{39}Ar$ analyses of authigenic muscovite, timing of stylolitization, and implications for pressure solution mechanisms: Jurassic Norphlet Formation, offshore Alabama. *Clays and Clay Minerals* **41**, 269–279.

TOTTEN, M.W. & BLATT, H. (1996) Sources of silica from the illite to muscovite transformation during late-stage diagenesis of shales. In: *Siliciclastic Diagenesis and Fluid Flow: Concepts and Applications.* (eds CROSSEY, L.J., LOUCKS, R. & TOTTEN, M.W.) pp. 85–92. Society of Economic Paleontologists and Mineralogists Special Publications 55.

VIELE, G.W. & THOMAS, W.A. (1989) Tectonic synthesis of the Ouachita orogenic belt. In: *The Appalachian–Ouachita Orogen in the United States.* (eds HATCHER, R.D., THOMAS, W.A. & VIELE, G.W.) Vol. F-2, pp. 695–728. The Geology of North America. The Geological Society of North America, Boulder, Colorado.

WALDERHAUG, O. (1994a) Temperatures of quartz cementation in Jurassic sandstones from the Norwegian continental shelf—evidence from fluid inclusions. *Journal of Sedimentary Research,* **A64**, 311–323.

WALDERHAUG, O. (1994b) Precipitation rates for quartz cement in sandstones determined by fluid-inclusion microthermometry and temperature-history modeling. *Journal of Sedimentary Research* **A64**, 324–333.

WALDERHAUG, O. (1996) Kinetic modeling of quartz cementation and porosity loss in deeply buried sandstone reservoirs. *American Association of Petroleum Geologists Bulletin* **80**, 731–745.

WILLIAMS, L.B., HERVIG, R.L. & DUTTON, S.P. (1997a) Constraints on palaeofluid composition in the Travis Peak Formation, east Texas: evidence from microanalyses of oxygen isotopes in diagenetic quartz. In: *Basin-Wide Diagenetic Patterns: Integrated Petrologic, Geochemical, and Hydrological Considerations* (eds Montañez, I.P., Gregg, J.M. & Shelton, K.L.) pp. 269–280. Society for Sedimentary Geology Special Publication 57.

WILLIAMS, L.B., HERVIG, R.L. & BJØRLYKKE, K. (1997b) New evidence for the origin of quartz cements in hydrocarbon reservoirs revealed by oxygen isotope microanalysis. *Geochimica et Cosmochimica Acta* **61**, 2529–2538.

WINTSCH, R.P. & KVALE, C.M. (1994) Differential mobility of elements in burial diagenesis of siliciclastic rocks. *Journal of Sedimentary Research* **A64**, 349–361.

*Spec. Publs int. Ass. Sediment.* (2000) **29**, 299–316

# Oxygen isotope analysis of authigenic quartz in sandstones: a comparison of ion microprobe and conventional analytical techniques

I. C. LYON[1], S. D. BURLEY[2,3,5], P. J. McKEEVER[2]*, J. M. SAXTON[1] *and* C. MACAULAY[4]

[1]*Department of Earth Sciences, University of Manchester, Manchester M13 9PL, UK;*
[2]*Diagenesis Research Group, Department of Earth Sciences, University of Manchester, Manchester, M13 9PL, UK;*
[3]*BG Technology, Ashby Road, Loughborough, Leicestershire LE11 3GR, UK;*
[4]*Scottish Universities Research and Reactor Centre, East Kilbride, Glasgow G75 0QU, UK; and*
[5]*Basin Dynamics Group, School of Earth Sciences, University of Keele, Keele ST5 5BG, UK*

## ABSTRACT

This paper describes a methodological approach using an Isolab®54 ion microprobe to obtain accurate oxygen isotope ratio measurements at precise locations in authigenic quartz overgrowths. The analytical spot size may be as small as 10 μm or less and the position of analyses can be defined by previous examination of the sample using cathodoluminescence (CL) or scanning electron microscopy. The total volume analysed involves only picomoles of material, representing $10^{-7}$ of the sample size required by conventional fluorination oxygen isotope measurement. The data are compared with the methodology and results obtained by conventional fluorination techniques.

$^{18}O/^{16}O$ ratio measurements were obtained of quartz overgrowths from a sample of the Penrith Sandstone (Permian) from northern England and a Piper Formation sandstone (Jurassic) from the UK North Sea. Analyses yielded an average $\delta^{18}O$ value of $+28.8‰_{SMOW} \pm 0.9‰$ (1 SD) from nine analyses in one profile across an overgrowth in the Penrith Sandstone and $+27.3‰_{SMOW} \pm 1.0‰$ (1 SD) from a second profile of five spots. An average $\delta^{18}O$ value of $+22.7‰_{SMOW} \pm 1.8‰$ was obtained from six profiles across an overgrowth in the Piper Formation sandstone compared with an average value of $10.6‰_{SMOW} \pm 1.3‰$ (1 SD) from 17 spots on the detrital grain. These data are compared with $\delta^{18}O_{SMOW}$ measurements for overgrowths from the same sandstones using a technique involving point counting to determine the detrital quartz/overgrowth content of the sandstones followed by conventional fluorination analysis of sandstones without overgrowths and sandstones with abundant overgrowths. Agreement between $\delta^{18}O$ values obtained by both methods is high for the Piper Formation sandstone but more divergent for the Penrith Sandstone.

Although the analytical precision obtained using the ion microprobe is not as high as that obtained using conventional bulk fluorination methods, this is offset by the ability to measure isotopic ratios *in situ* and with high spatial resolution. Moreover, the inherent shortcomings of mineral separation necessary with conventional methods mean that the ion microprobe represents a major advance in the determination of the oxygen isotope composition of diagenetic silicate minerals.

## INTRODUCTION

Oxygen isotopes are commonly used to either constrain the temperature of authigenic mineral precipitation or define the oxygen isotope composition of aqueous fluids from which minerals precipitate (Longstaffe, 1989). The technique is commonly applied to authigenic quartz in sandstones because the oxygen isotope composition of quartz varies widely as a function of temperature under diagenetic conditions and may also reflect changes in mineral reactions that contribute dissolved silica to the diagenetic system.

Authigenic quartz usually occurs as syntaxial overgrowths to detrital quartz grains in sandstones (Waugh, 1970; Burley *et al*., 1985; McBride, 1989). A characteristic feature of quartz overgrowths is a complex internal growth zonation as revealed by cathodoluminescence

* Present address: Geological Survey of Northern Ireland, 20 College Gardens, Belfast BT9 6BS, Northern Ireland.

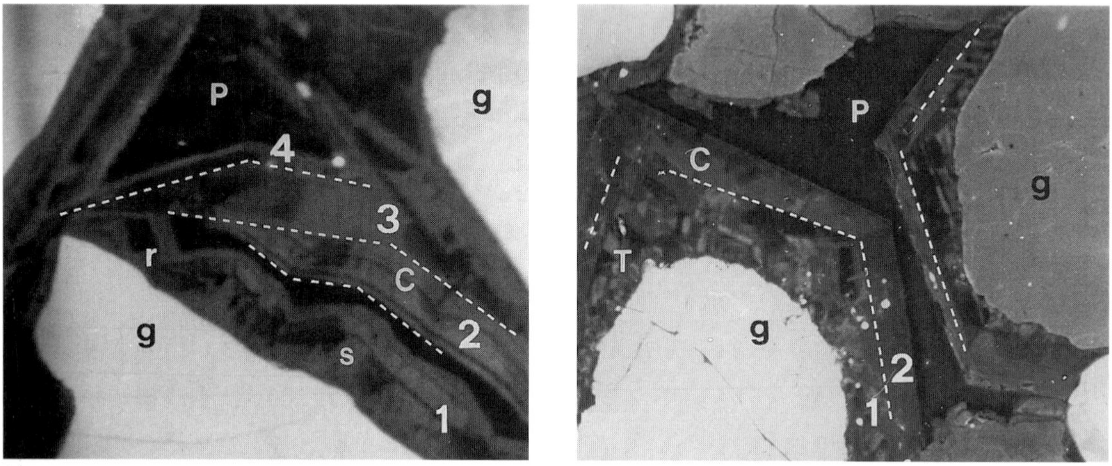


(a) (b)

**Fig. 1.** SEM–CL photomicrographs of growth zonation in quartz overgrowths. In both micrographs the detrital grains are annotated g and pore space is annotated P. (a) Piper Formation sandstone. Complex concentric growth zonation with four distinct growth generations (labelled 1–4, dashed white lines separating generations). The initial generation is dominated by rhombohedral growth faces (r) and exhibits sector zonation (s). This generation is overlain by three generations each showing more regular concentric zonation (c). Note the micron-scale concentric individual growth zones and the fact that individual growth zones are not necessarily continuous around the host detrital grains. Width of micrograph = 280 μm. (b) Penrith Sandstone. Two generations of large overgrowths (labelled 1 and 2, separated by a dashed white line). The inner overgrowth generation is dominated by a twinned growth fabric (T) whilst the outer growth zone displays a regular concentric fabric (c). Width of micrograph = 520 μm.

(CL) microscopy (Fig. 1a,b; Sippel, 1968; Burley *et al.*, 1989; Walker & Burley, 1991; Hogg *et al.*, 1992; Turner *et al.*, 1995). Growth zones are typically on the scale of tens of microns in width and probably reflect variations in the physico-chemical conditions during which the overgrowth precipitated (Perny *et al.*, 1992). The growth zonation observed in quartz overgrowths may be reflected in the oxygen isotope composition and could be a result of closed system diffusional processes, open system episodic growth or open system continuous growth processes (Fig. 2). Sector zonation may be superimposed on concentric growth fabrics and result in isotopic compositional variation related to growth rate (Onasch & Vennemann, 1995).

Oxygen isotope ratio measurements of authigenic quartz are, however, difficult to obtain because of the mineralogical similarity of the quartz overgrowth and its host detrital grain. The overgrowths are not easy to detach from detrital grain surfaces and the small size of the overgrowths, typically 20–100 μm in width, prohibits the use of microdrill techniques to obtain a pure overgrowth separate. Traditional methods of either physical or chemical separation are flawed because of the inherent difficulty in separation of the overgrowths from detrital grains which results in separates being mixtures of overgrowths and grains (Fig. 3), and potentially, other oxygen-bearing minerals. Moreover, the bulk nature of such sample separates excludes investigation of variation in the oxygen isotope composition within an individual overgrowth. This in turn prevents study of the precipitation process and the evolution of the palaeo-porefluids during quartz authigenesis.

This paper outlines a systematic methodological approach for studying the oxygen isotope composition of individual detrital quartz grains and their overgrowths. The method combines cathodoluminescence (CL) microscopy to reveal growth fabrics in quartz with a new, high resolution analytical technique for the *in situ* determination of oxygen isotopes that employs the Isolab®54 ion microprobe. Examples of oxygen isotope analysis of quartz overgrowths from two sandstones are presented to illustrate the application of the technique to the measurement of oxygen isotope composition across individual overgrowths. The results are compared with conventional oxygen isotope analyses on bulk samples.

## OXYGEN ISOTOPE ANALYSIS OF QUARTZ OVERGROWTHS IN SANDSTONES

### Conventional methods

Conventional oxygen isotope measurement techniques on authigenic quartz are essentially refinements of the

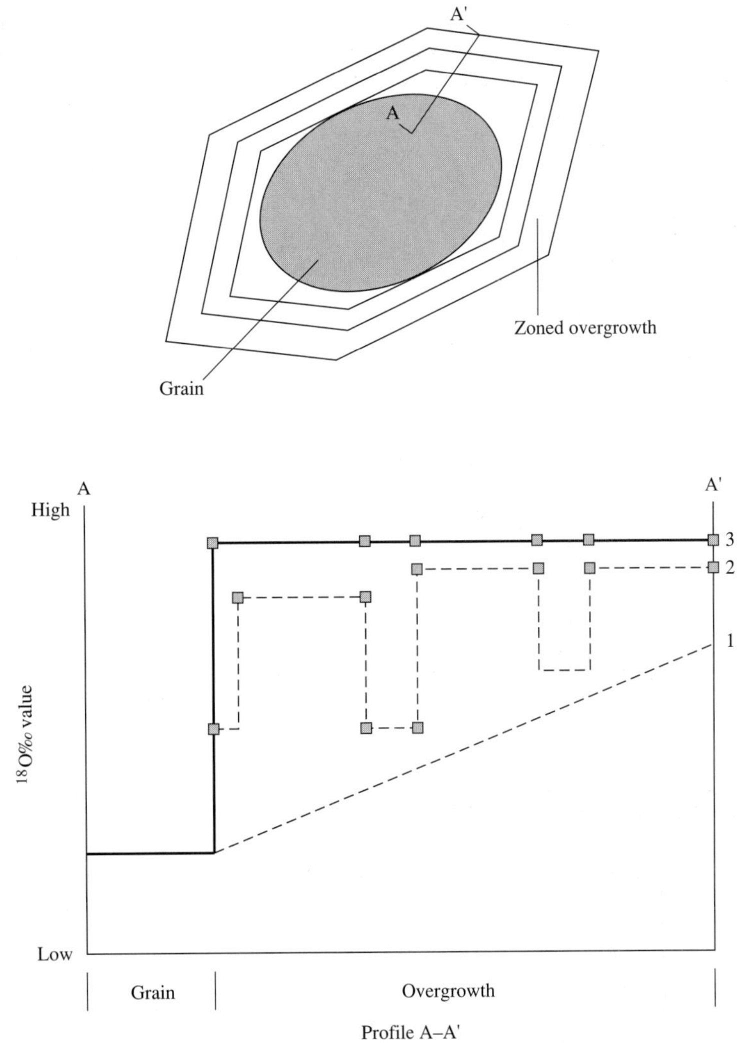


**Fig. 2.** Cartoon illustrating growth zonation in quartz overgrowths around a detrital grain and the probable end-member processes of quartz precipitation that may be reflected in the variation of oxygen isotope composition across a zoned quartz overgrowth: (1) closed system diffusional growth; (2) open system episodic (punctuated) growth; or (3) open system continuous growth.

fluorination technique first developed by Clayton & Mayeda (1963). The quartz overgrowths need to be separated from the detrital grains (e.g. see Fig. 3) and then the oxygen isotope composition of the overgrowth separate can be analysed using fluorination. Rotary and sonic microdrills, mounted to either normal or inverted optical microscopes, have been used to microsample authigenic minerals from their host sediments and in vein samples (Prezbindowski, 1980; Dettman & Lohmann, 1995). However, the small size of quartz overgrowths in most sandstones (< 100 μm) precludes microdrilling as a general technique for separation of overgrowths from their host detrital grains. There are two ways in which separation of the overgrowths from their detrital cores is normally achieved.

In the first method, the quartz overgrowths are physically removed from their detrital cores by a combination of etching in hydrofluoric acid (HF) followed by agitation in an ultrasonic bath to break off the overgrowths at their points of contact. Subsequent size fractionation by sieving concentrates the overgrowths into smaller grain size fractions for fluorination (see Fig. 3 and Lee & Savin, 1985). The second method derives the oxygen isotope composition of the overgrowths from the difference in the $\delta^{18}O$ composition between a bulk sandstone sample containing a known amount of overgrowths (determined from thin section modal analysis) and a sample in which either no overgrowths are present or from a sample in which the overgrowths have been removed by preferential dissolution in HF. The results are then extrapolated to a

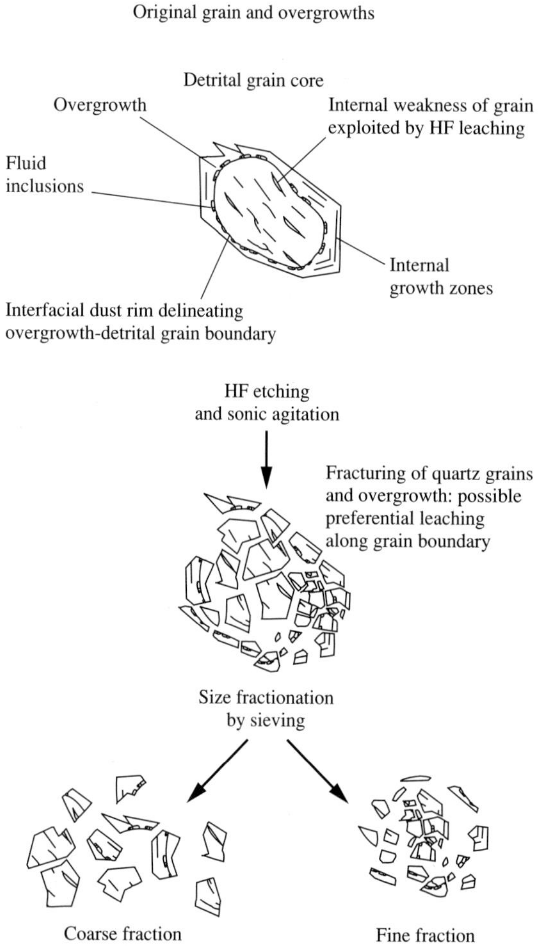


**Fig. 3.** Cartoon illustrating the inherent problems associated with physical and chemical methods for the separation of quartz overgrowths from their detrital grains. During HF leaching and ultrasonic agitation random fracturing of the detrital grains is likely to occur because of incomplete development of interfacial dust rims or irregular development of overgrowths. The resulting size fraction concentrates are probably mixtures of both detrital grains and overgrowths, and the overgrowth concentrate must also be a composite of all growth zones present in the sample. Modified after Brint *et al.* (1991).

pure overgrowth end-member value (e.g. Land & Dutton, 1978; Millikin *et al.*, 1981).

There are, however, shortcomings inherent in both of these methods which severely limit their usefulness (Brint *et al.*, 1991). Physical removal of the overgrowths employing the preferential HF etching technique relies on the presence of either solid or fluid inclusions at the interface between the overgrowth and the grain which create a weakness that can be preferentially attacked by the acid. In some sandstones, little interfacial material is present so the overgrowths are incompletely removed during acid etching. In fact, many other weaknesses can be present in the quartz grains (such as fractures or contacts between subgrains within the detrital grains) which are also preferentially attacked by the HF acid, causing disintegration of the quartz grain into small fragments. As a result, the various size fractions inevitably contain mixtures of detrital grain and overgrowth components. Similar problems apply to the successive dissolution method, particularly in cases where either the overgrowths are incompletely developed around the host grain or at points of grain contact. Neither separation techniques therefore can yield pure separates of the overgrowth material. Of equal concern in these techniques is the composite nature of a bulk sample separate. Not only do these methods assume that each overgrowth is isotopically homogeneous, but they also assume that individual overgrowths in a separate from the same sample will all have the same oxygen isotope compositions and all detrital grains will be similarly isotopically homogeneous. Such an approach can never address variation of oxygen isotope composition from within an individual overgrowth or between adjacent overgrowths from within a sample.

## Microbeam methods

Microbeam techniques employ either a focused laser or ion beam to provide high resolution, *in situ* analysis of oxygen isotopes in non-conducting material in a rock section. One approach tried in carbonates (Dickson *et al.*, 1990; Smalley *et al.*, 1992), sulphides (Crowe *et al.*, 1990; Fallick *et al.*, 1992) and also in silicates (Sharp, 1992; Mattey & Macpherson, 1993) has been to use a focused infra-red laser beam to heat the sample in a localized area. This approach has proved quite successful for carbonates where $CO_2$ is released directly from a small spot (down to 20–30 μm) and the $CO_2$ can be analysed directly by conventional mass spectrometers (Smalley *et al.*, 1992). However, in silicates, the laser provides a localized heat source to initiate a fluorination reaction between the silicate mineral and bromine pentafluoride or chlorine trifluoride and the reactions are much more complex. A large crater is produced which can limit spatial resolution to several hundred microns (Elsenheimer & Valley, 1992), and thermal gradients in the material surrounding the heated spot release oxygen from a large aureole (up to 1200 μm in diameter; Sharp, 1992). Isotopic fractionation effects, which are strongly dependent on matrix, and heating effects are also commonly reported. The heating effect of the laser is dependent on the colour of the sample but quartz overgrowths are colourless and therefore are heated inefficiently by the laser. Recent reports suggest that the formation of an alteration aureole can be reduced by the use of ultra-violet lasers which release oxygen through a photochemical reaction, but currently this technique still requires large sample sizes (> 0.1 mg) (Valley *et al.*, 1994; Weichart & Hoefs, 1994, 1995).

An alternative method to the laser probe is secondary ion mass spectrometry (SIMS) in which the sample is bombarded by a focused beam of heavy ions. Atoms and molecules are sputtered from the sample surface, a small fraction of which emerge as ions. These ions are accelerated and focused into a mass spectrometer where the isotopes are separated and measured by suitable ion detectors. There are, however, several factors which make the precise determination of isotope ratios difficult. Firstly, there is a large natural abundance difference between the oxygen isotopes ($^{18}O$ isotope abundance 0.2%; $^{16}O/^{18}O$ = 500) which results in a low intensity $^{18}O^-$ secondary ion beam. Secondly, reliable charge compensation of insulating mineral samples is required and instrumental fractionation effects occur (e.g. Shimizu & Hart, 1982; Lyon *et al.*, 1994) which make measured ratios vary, often by up to tens of permil. Two different approaches have been developed. One, known as extreme energy filtering (Hervig *et al.*, 1992), measures only high energy secondary ions (> 300 eV relative to the surface) to reduce variable fractionation caused by imperfect charge compensation. Molecular isobaric interferences are also usually eliminated, thus necessitating only low mass resolution. The disadvantage of this technique is that the high energy ions typically only constitute 0.01% of the secondary ion yield, thus drastically reducing the detection efficiency. Nevertheless, several workers have shown that the technique can yield reproducible isotope ratios on insulating materials and be used to measure oxygen isotope ratios in geological samples (e.g. Jamtveit & Hervig, 1994; Riciputi & Paterson, 1994; Hervig *et al.*, 1995; Graham *et al.*, 1996). The alternative method is to use low energy secondary ions (10–20 eV relative to the surface potential) which yields far higher count rates (e.g. Lyon *et al.*, 1995).

In the present study a Fisons Instruments Isolab®54 ion microprobe (Lyon & Turner, 1992; Saxton *et al.*, 1996) has been used to develop a technique for undertaking high spatial resolution *in situ* measurements of $^{18}O/^{16}O$ with an accuracy of around 1‰ (Lyon *et al.*, 1995). When combined with cathodoluminescence petrography, the technique provides the ability to rapidly profile the isotopic composition and isotopic variation across individual quartz overgrowths at precisely selected sample analysis sites.

## HIGH RESOLUTION OXYGEN ISOTOPE ANALYSIS

### Sample preparation

Samples for analysis consist of either polished thin sections (30 μm thick) or doubly polished wafers, up to approximately 80 μm thick. The ion probe sample changer accepts glass sections up to 3 cm by 1.5 cm. These are cut from the standard polished thin sections, or if fluid inclusion wafers are being examined in the ion probe, a fragment of the wafer is attached to an appropriately sized glass slide template with a conducting glue. As quartz is an electrical insulator, the sections and wafers are either gold or carbon coated to prevent charge build up on the sample surface. No other special preparation is required.

After a sample has been selected for study, the area chosen for oxygen isotope analysis has to be imaged at both low and high magnifications to allow easy re-location of the area chosen for analysis with the ion probe. Due to the petrographic similarity between quartz overgrowths and their host detrital grains in optical petrography and in the scanning electron microscope (SEM) in both secondary electron mode and back-scatter electron mode, cathodoluminescence (CL) microscopy in the SEM is required to distinguish the overgrowths and to reveal their internal growth fabrics.

Optical and SEM–CL photomicrographs are then used to identify the area of interest in the reflected light optical viewing system of the Isolab®54. The exact point at which the primary ion beam strikes the sample surface is previously calibrated in the optical imaging system of the ion probe and the sample manipulated so that a traverse may be made across the overgrowth. Scratch marks on the surface or previous ion beam sputter marks can be observed using secondary ion imaging in the ion probe. If the precise location of these features relative to the overgrowth is known then they also provide a useful guide to positioning the ion beam for the desired oxygen isotope traverse. Secondary ion imaging is achieved by rastering the primary ion beam across the surface of the sample and observing the resultant secondary ion signal in synchronism with the primary beam raster. $^{16}OH^-$ is particularly useful in this respect as water molecules tend to be ubiquitous and uniformly distributed over the surface. The $^{16}OH^-$ signal is usually enhanced or depleted in the sputtered spot relative to the surface so that it shows as a bright or dark spot. A traverse of several ion beam analyses is then made across a suitable area of the overgrowth and host grain. Solid and fluid inclusions imaged optically or in CL mode are avoided. Ion beam sputter marks

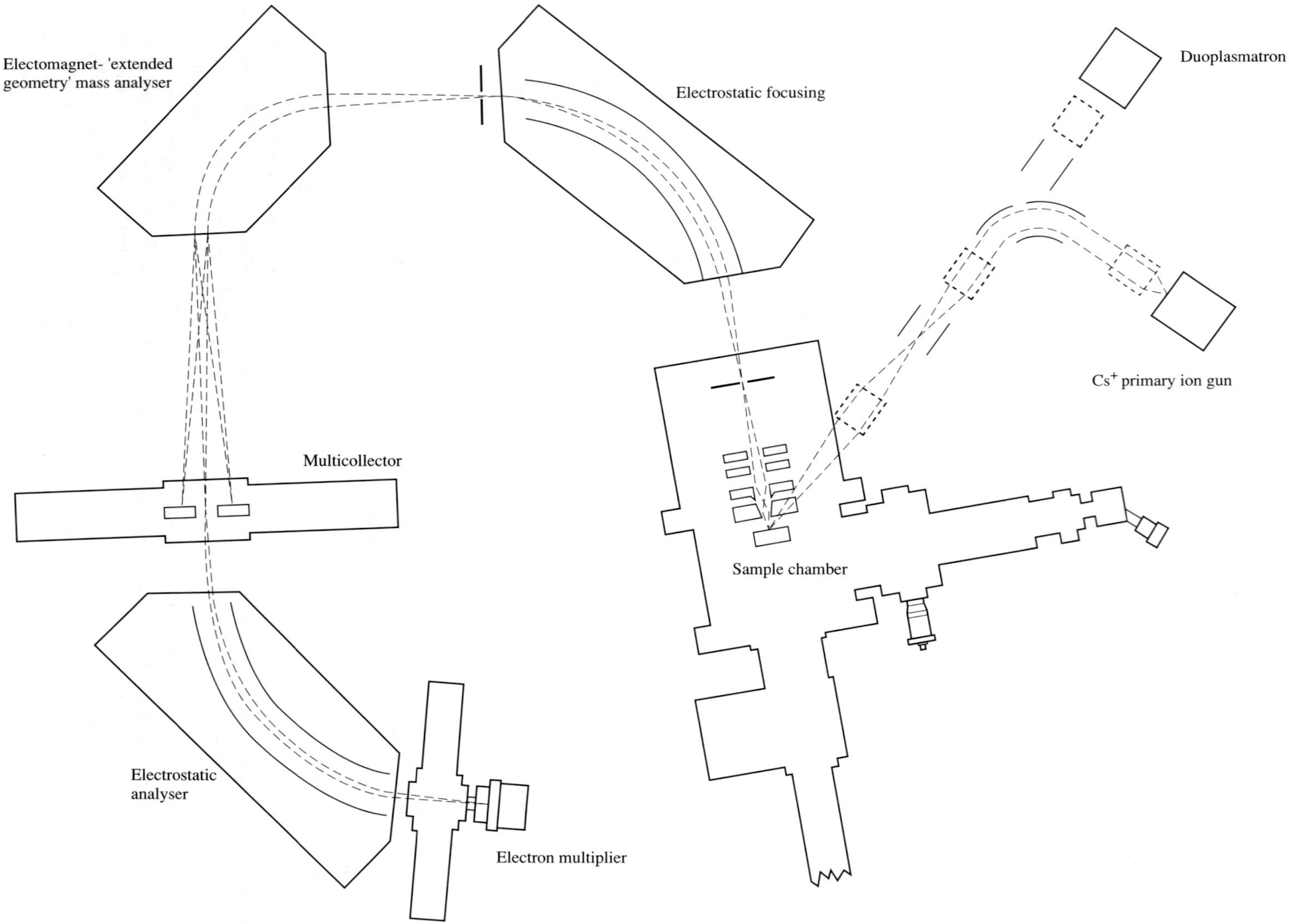


**Fig. 4.** A schematic representation of the Isolab®54 ion microprobe (simplified from Saxton *et al.*, 1996), see text for details.

are easily visible when later inspected using optical or electron microscopy as the conducting coating has been removed by sputtering revealing the exact points at which the analyses were made.

### Ion probe instrumentation

A schematic representation of the Isolab®54 ion microprobe showing the relevant components is illustrated in Fig. 4. The operation of the instrument is described in detail by Saxton *et al.* (1996). In essence, positive caesium primary ions ($Cs^+$) are produced from a caesium ion gun maintained at 10 kV and focused on to the sample by a series of lenses and deflector plates. The primary ion beam sputters the sample surface, and as the conducting coating is sputtered away, a focused electron beam (which impacts the surface with a net energy of 8 keV) is used to neutralize charge on the analysed spot, which would otherwise rapidly charge through gaining $Cs^+$ ions and losing secondary electrons and negative ions. The sample is held at −8 kV so that negative secondary ions from the surface are accelerated into a second system of optics which focus the secondary ions on to the source slit of a mass spectrometer. Multicollection is used to measure the oxygen isotopes simultaneously in different detectors; $^{16}O^-$ ions are measured by a Faraday cup and $^{18}O^-$ ions by a CDS detector. These technical details are expanded upon by Lyon *et al.* (1995) and Saxton *et al.* (1996).

### Standards, reproducibility and accuracy

The measurement of accurate isotope ratios by secondary ion mass spectrometry is hindered by variable and often large isotopic fractionation which appears as a natural consequence of the secondary ionization process. The fractionation appears in two ways. Firstly, fractionation results from a matrix effect in which the isotopic composition of the secondary ions is different to that of the bulk material (Havette & Slodzian, 1980; Lu & Mann, 1986) and is a complex function of the bulk matrix composition. Secondly, instrumental fractionation is due to wide angular and energy distributions of the secondary ions as they leave the sample surface (Shimizu & Hart, 1982; Lyon *et al.*, 1994). Matrix effects are still poorly understood and the current pragmatic solution is to analyse the sample in conjunction with a standard of essentially the same chemical composition. In the present study, both detrital grains and overgrowths are high purity quartz with only minor $Al^{3+}$ compensated substitution and thus matrix effects can be disregarded. Instrumental fractionation causes unequal focusing of the secondary ions on the source slit of the mass spectrometer leading to large and variable mass fractionation effects (Lyon *et al.*, 1994). This inherent instrumental fractionation can be controlled using a technique for integrating the secondary ion beams (Lyon *et al.*, 1994) which eliminates a major source of instrumental isotopic fractionation enabling the acquisition of reproducible ratios.

A thin section of an ultra-pure synthetic quartz crystal of known chemical composition (total < 1 ppm metal cation impurities, Al < 0.1 ppm) grown by GEC (Wembley) was selected for use as a quartz standard. This crystal was studied using CL microscopy and is unzoned except for the presence of the quartz seed plate on which the ultra-pure crystal nucleated. Repeated measurements collected on the synthetic quartz crystal standard show that reproducible $^{18}O/^{16}O$ ratios (SD = 1.4‰ over 4 days on measured ratios, SD ≈1‰ on any one day) are readily obtained (Lyon *et al.*, 1994) and SD = 1.3‰ over a period of 8 weeks on a different quartz standard (Saxton *et al.*, 1995). Although the number of $^{18}O$ counts obtained from the measurement of a spot analysis is usually greatly in excess of $10^6$, leading to an implied precision due to counting statistics alone of much less than 1‰, the confidence limits quoted in Lyon *et al.* (1994), Saxton *et al.* (1995), Lyon *et al.* (1995) and Saxton *et al.* (1996) and this present paper are based on the reproducibility of measurements made upon standards over periods of many days or weeks. The scatter in these measurements thus include all sources of error—instrumental fractionation, changes in gain of detectors, crater effects, standard inhomogeneity and incomplete charge compensation—and thus much better represent the accuracy (as opposed to precision) of isotopic ratios quoted in this paper.

## RESULTS

### Samples

Two sandstones were selected for study. These were (i) the Penrith Sandstone of the Vale of Eden in northern England, an aeolian sandstone of Permian age (Macchi, 1990) with well-developed quartz overgrowths in part of its outcrop; and (ii) the Piper Formation sandstone from the Outer Moray Firth, UK North Sea, an Upper Jurassic shallow marine sandstone (Turner *et al.*, 1984) with variable amounts of quartz cement. Aspects of the diagenesis of the Penrith Sandstone are given in Waugh (1970) and Turner *et al.* (1995) whilst details of Piper Formation sandstone diagenesis are given in Burley (1986), Burley *et al.* (1989), Guscott & Burley (1993) and McCants & Burley (1996).

## Methods

For each sandstone, samples with abundant quartz overgrowths (identified from initial petrographic study) were selected for oxygen isotope analysis. Polished thin sections were made of these samples and both optical and SEM–CL photomicrographs of individual overgrowths were taken. Oxygen isotope analysis traverses were made across several overgrowths and their host detrital grains. To compare the results of the ion microprobe analysis with conventional isotope analyses, samples of the Penrith and Piper Formation sandstones were chosen for analysis by conventional fluorination at the Scottish Universities Research and Reactor Centre. Modal analysis of each sandstone was undertaken by point counting to determine the amount of quartz cement present so that the oxygen isotope composition of the overgrowths could be determined according to the method of Land & Dutton (1978).

## Sputter marks and the sampling volume

Bombardment of the gold or carbon coated sections with the primary ion beam sputters away the semi-opaque coating and enables the area of primary beam interaction with the sample to be observed. A uniform ion density can be achieved by using an aperture in the primary ion column (a method of focusing the primary ions known as Köhler illumination, Benninghoven *et al.*, 1987). The resulting sputter mark is typically 10–12 μm in diameter and crudely circular or slightly elliptical in shape (Fig. 5a-d). Examination of the sputter marks in the SEM in secondary electron mode shows that they form shallow craters, 2–3 μm in depth (Fig. 5d) with shallow sides and a corrugated, basal surface (Fig. 5c,d). The typical volume of quartz volatilized in each sputter mark is thus approximately 3 picomoles, only $10^{-7}$ of the material required by conventional fluorination techniques. Moreover, only a few microns depth of the sample are sputtered so the thin section is not significantly penetrated and contamination by sampling underlying material is not a problem.

Optical micrographs of the ion beam sputter marks can give the impression that the sputter volumes are larger than this although careful observation of recoated sputter marks shows that the sputtered area consists of a central crater from which most material has been sputtered and an enclosing halo of variable diameter and shape. The halo is very shallow and is barely deeper than the conducting coat ($<<0.1$ μm) whereas the well defined core can be up to several microns in depth (Fig. 5c,d). The halo results from scattered primary ions and makes an insignificant contribution to the total measured oxygen signal. The measured volume is essentially therefore that of the central crater.

## Ion probe oxygen isotope measurements

### *Penrith sandstone*

Two $\delta^{18}O$ profiles were obtained from a large euhedral overgrowth in a polished thin section from the Penrith Sandstone (Fig. 6). Table 1 lists the oxygen isotope data reported here. The profiles were obtained by focusing the primary ion beam to a well resolved spot, approximately an $11 \times 8$ μm ellipse in the case of profile 1, and moving the sample in steps of between 10 and 15 μm to acquire a series of measurements. The profile was acquired by moving the ion beam from the pore space towards the overgrowth and was completed when the detrital grain was reached, although the data have been tabulated in a standardized format from grain to overgrowth (Table 1). Figure 6 illustrates the location of the analyses from profile 1, their size and shape, and shows the relationship of the sputter marks to the internal growth fabric of the overgrowth as revealed by CL. One analysis was taken on the detrital grain, one crosses the detrital-grain–overgrowth boundary and nine were taken across the overgrowth, including both growth zones. The profile was normalized to an absolute SMOW value by measuring the $^{18}O/^{16}O$ value of a standard (seed quartz, −4.79‰) before and after the profile. Profile 1 (Table 1) was obtained in August 1994 and reported in Saxton *et al.* (1995). Profile 2 (Table 1) was acquired in December 1995 in order to test the accuracy of Profile 1.

The $\delta^{18}O_{SMOW}$ values obtained from the detrital grain in the two profiles are +10.9 and +14.0‰$_{SMOW}$. For the overgrowth, the values range from +25.8 to +30.5‰$_{SMOW}$, averaging +28.8‰$_{SMOW}$. Figure 7 shows the data from profile 1 (Table 1) with position across the overgrowth. The error bars are 1 SD and include the uncertainty in the measurements on the standard. There is variation in $\delta^{18}O_{SMOW}$ values across the overgrowth, but the variation falls within a statistical scatter about a constant $\delta^{18}O$ value of +28.8‰$_{SMOW}$ with a 1-SD scatter of 0.9‰. The $\delta^{18}O_{SMOW}$ values obtained from profile 2 are in good agreement with profile 1, although the second profile averages a slightly lower value of +27.3 ± 1.0‰$_{SMOW}$. This difference is also within the normalization error and so is not statistically valid.

### *Piper Formation sandstone*

Six $\delta^{18}O$ profiles were obtained across a detrital quartz grain and its enclosing authigenic overgrowth from a

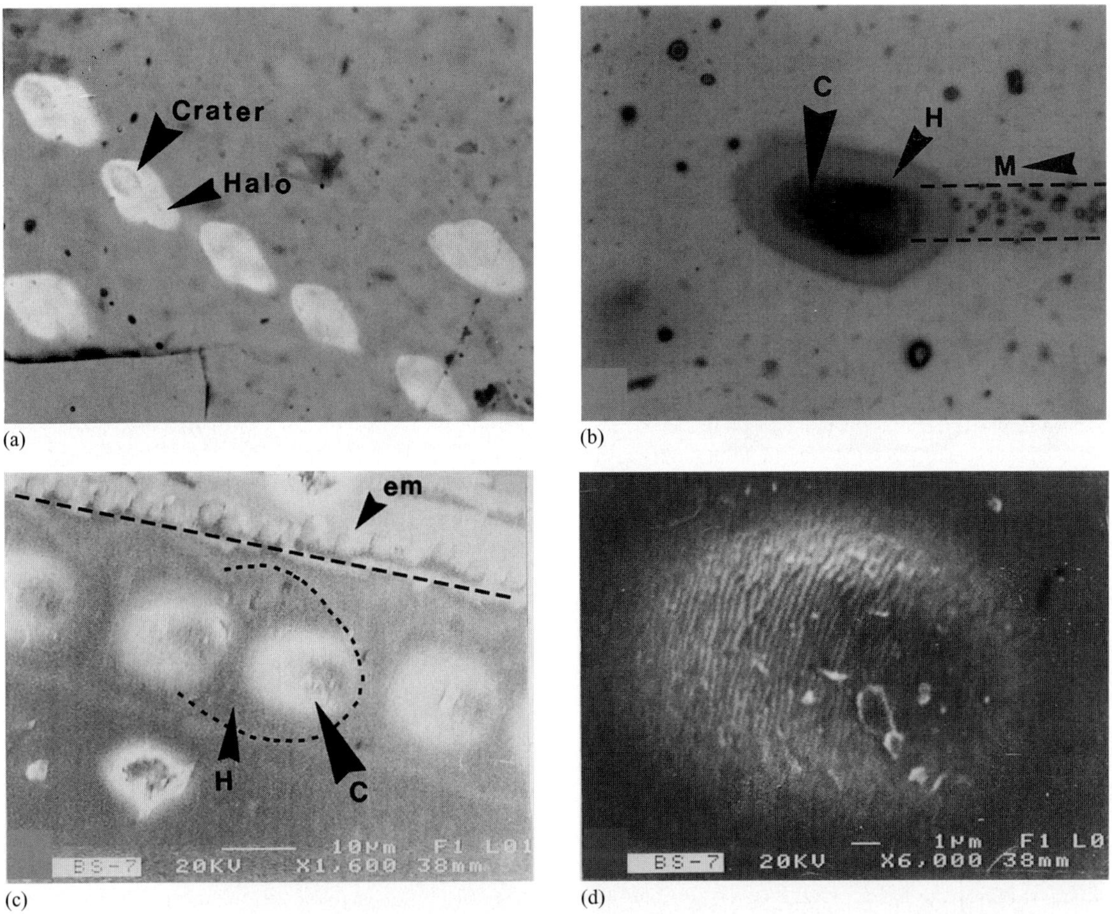


**Fig. 5.** Images of the ion probe sputter marks from the Penrith sample illustrating the size and nature of the sampling volume. (a) Optical photomicrograph in transmitted, plane polarized light showing a trail of elliptical sputter marks revealed by the absence of a carbon coating where the ion beam has sputtered away the coating and sample surface. Individual sputter marks average 20 μm in width, comprising a central, 10 μm-diameter sputter crater surrounded by a shallow, more elliptical sputter halo. Width of micrograph = 160 μm. (b) Detail of an individual sputter mark from an optical photomicrograph in transmitted, plane polarized light. This sample has been recoated after sputtering to emphasize the crater (c) and halo (h). The longest axis of the crater is 12 μm long. The direction of movement of the primary $Cs^+$ beam over the sample surface when moving the $Cs^+$ beam to the analysis spot is shown by the arrow (M) and the dotted lines that bracket the small sputter spots. Width of micrograph = 42 μm. (c) 'Side illumination' (illumination effect is from bottom right) of a sputter mark trail showing the asymmetric form of the sputter crater (example arrowed C). The enclosing sputter haloes are only poorly discernible in this image; one (arrowed H) has been highlighted with a dashed black line to show the extent of halo overlap. The trail arrowed em at the top of the micrograph is a line of 2 μm diameter electron microprobe analysis craters. Scanning electron backscatter micrograph. Width of micrograph = 66 μm. (d) Detail of a central sputter crater. Note the slightly elongate shape and depth profile of the crater, and the corrugated floor to the crater. From the size of the particles within the crater its depth is estimated to be 2–3 μm. Scanning electron backscatter micrograph. Width of micrograph = 16 μm.

sample of the Piper Formation sandstone. Cathodoluminescence microscopy reveals four main growth zones and the detrital-grain–overgrowth boundary is clearly identifiable (Fig. 8c). Table 2 lists the oxygen isotope data.

The sputter mark spots in all profiles were 10–18 μm in diameter. Error bars are 1 SD, which includes the error from the scatter of measurements on the seed quartz standard against which each profile was normalized to obtain an absolute $\delta^{18}O$ value. The six profiles were obtained at different times over a period of five months and the coincidence of measured values from all profiles on the detrital grain indicates that there are no large

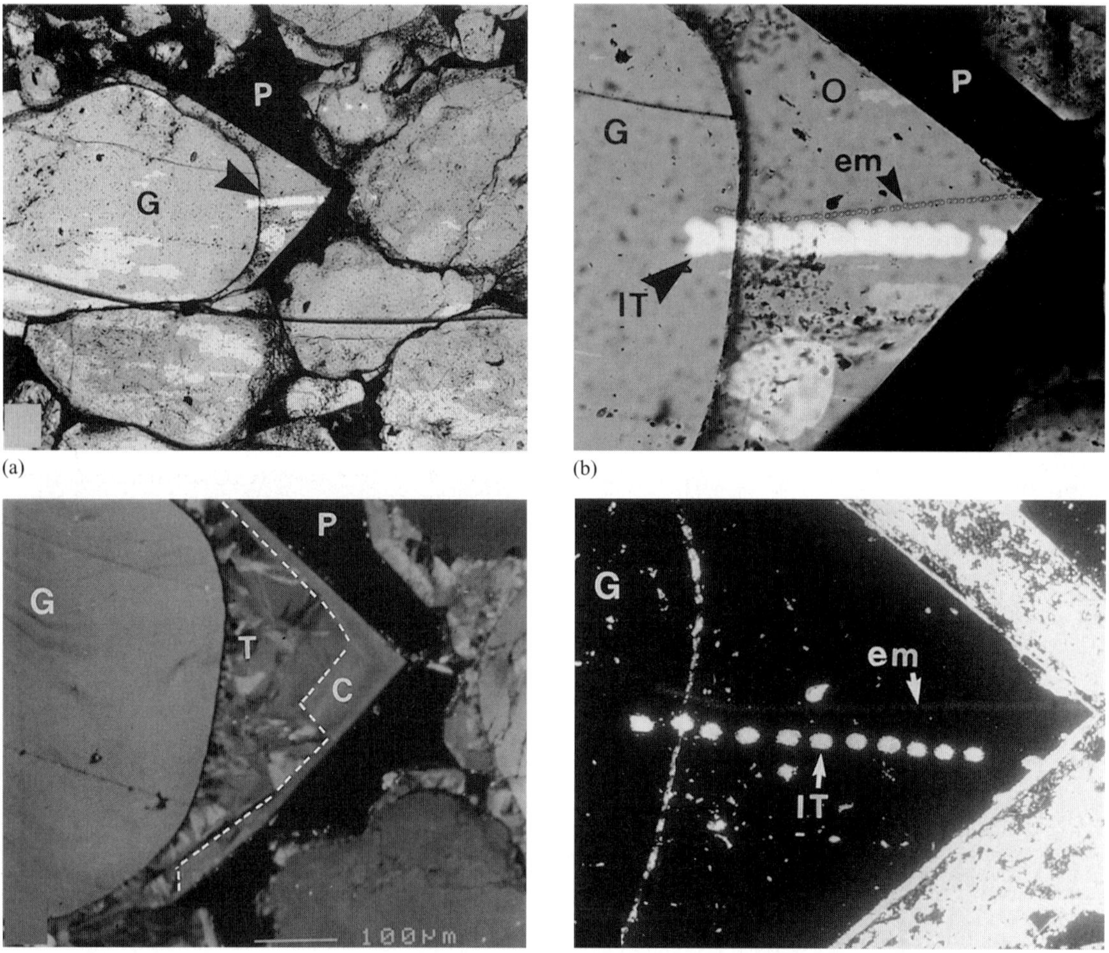


(a) (b) (c) (d)

**Fig. 6.** Transmitted light and SEM micrographs of the Penrith Sandstone showing ion probe analysis profile 1. Data are given in Table 1. (a) General view of the Penrith sandstone displaying well developed quartz overgrowths on rounded detrital grains. Note the clay and iron oxide pellicle coating the detrital grains. The ion probe analysis is arrowed IT. Optical photomicrograph, transmitted, cross polarized light. Width of micrograph = 1.7 mm; P: pore; G: grain. (b) Detail of the ion probe analysis profile (arrowed, IT). Quartz overgrowth annotated O. Electron microprobe analysis trail (with 2 μm diameter burn marks) is annotated em. Optical photomicrograph, transmitted, plane polarized light. Width of micrograph = 400 μm. (c). Internal growth fabric of the analysed overgrowth revealed by cathodoluminescence. An inner twinned growth zone (T) is enclosed by concentric growth layers (C). SEM–CL micrograph. Width of micrograph = 620 μm. (d) Scanning electron backscatter micrograph image of the overgrowth after ion probe sputtering (analysis trail IT). Individual sputter craters are slightly elongate and average 12 μm in diameter. The craters have been recoated with gold and the unsputtered surface carbon coated to show the size and shape of the ion beam craters. Width of micrograph = 280 μm.

systematic errors (> 1‰) affecting the absolute $\delta^{18}O$ values measured on the quartz overgrowth.

Oxygen isotope analyses yielded consistent $\delta^{18}O$ values on the detrital grain of +10.6‰ ± 1.3‰$_{SMOW}$. Six profiles across the overgrowth range from +20.6 to +25.9‰$_{SMOW}$ with the exception of two analysis spots in profile 4 which gave values of +17.4 and +16.6‰$_{SMOW}$. Statistically, this variation can be represented as a constant $\delta^{18}O$ value of +22.7‰$_{SMOW}$ with a 1-SD scatter of ±1.8‰. The two low values have been excluded from the data set as subsequent inspection of their craters showed them to have sampled a large fluid inclusion. Figure 9

**Table 1.** Summary listing of oxygen isotope ratios in the Penrith Sandstone. Profile 1 data were obtained on 4 August 1994 and absolute SMOW values were obtained by normalizing to five measured spots on a standard (seed quartz, –4.79‰$_{SMOW}$) obtained on 3, 4 and 5 August 1994. These yielded a measured mean $^{18}O/^{16}O$ ratio of $1.9845 \times 10^{-3}$ with a standard deviation of 1.1‰. Profile 2 data were obtained on 18 December 1995 and were normalized using 15 spot measurements on the same standard acquired between 6 December 1995 and 19 December 1995). These yielded $^{18}O/^{16}O = 1.996 \times 10^{-3}$ with a standard deviation of 1.2‰

| Sample | Analysis spot, microns along traverse | $^{18}O/^{16}O \times 10^6$ measured | $\delta^{18}O$‰ SMOW |
|---|---|---|---|
| | **Profile 1** | | |
| Detrital grain | 25 | 2022.1 | 14.0 |
| Grain boundary | 40 | 2033.9 | 19.9 |
| Overgrowth | 55 | 2051.6 | 28.8 |
| Overgrowth | 65 | 2053.0 | 29.5 |
| Overgrowth | 80 | 2048.0 | 27.5 |
| Overgrowth | 95 | 2050.5 | 28.4 |
| Overgrowth | 110 | 2050.6 | 28.4 |
| Overgrowth | 125 | 2051.1 | 28.6 |
| Overgrowth | 140 | 2052.7 | 29.4 |
| Overgrowth | 150 | 2051.1 | 28.6 |
| Overgrowth | 165 | 2054.9 | 30.5 |
| | **Profile 2** | | |
| Detrital grain | 20 | 2027.6 | +10.9 |
| Overgrowth | 35 | 2057.6 | +25.8 |
| Overgrowth | 50 | 2059.2 | +26.6 |
| Overgrowth | 65 | 2062.6 | +28.3 |
| Overgrowth | 80 | 2061.6 | +27.8 |
| Overgrowth | 95 | 2061.6 | +27.8 |
| | **Profile 1, 3 February 1994** | | |
| Detrital grain | 75 | 2002.4 | +11.5 |
| Overgrowth | 120 | 2025.2 | +23.0 |
| Overgrowth | 210 | 2024.6 | +22.7 |
| | **Profile 2, 10 February 1994** | | |
| Detrital grain | 50 | 2002.3 | +11.4 |
| Detrital grain | 90 | 2002.5 | +11.5 |
| Detrital grain | 130 | 1997.9 | +9.2 |
| Detrital grain | 180 | 1999.5 | +10.0 |
| Overgrowth | 240 | 2020.6 | +20.6 |
| Overgrowth | 280 | 2023.1 | +21.9 |
| Overgrowth | 340 | 2021.2 | +21.0 |
| | **Profile 3, 15 February 1994** | | |
| Detrital grain | 20 | 1997.3 | +8.9 |
| Detrital grain | 120 | 1998.4 | +9.4 |
| Overgrowth | 170 | 2022.4 | +21.6 |
| Overgrowth | 240 | 2023.2 | +22.0 |
| Overgrowth | 290 | 2021.2 | +21.0 |
| | **Profile 4, 15 February 1994** | | |
| Overgrowth | 50 | 2014.2 | +17.4 |
| Overgrowth | 80 | 2012.5 | +16.6 |
| Overgrowth | 120 | 2020.7 | +20.7 |
| Overgrowth | 150 | 2021.2 | +21.0 |

plots the data from profile 6, chosen because this profile samples all the overgrowth zones, with position across the overgrowth. The error bars in Fig. 9 are 1 SD and include the uncertainty in the measurements on the standard.

### Conventional oxygen isotope analyses

Six samples were chosen for conventional oxygen isotope analysis, three from each of the Penrith and Piper Formation sandstones to compare bulk sandstone isotopic analysis methods with the ion probe measurements. In each case, sandstone samples without visible overgrowths and with abundant quartz overgrowths, according to modal analysis were chosen. Results of petrographic modal analysis and conventional oxygen isotope analysis are reported in Table 3 and plotted in Fig. 10. Accuracy of the point count technique is ±2% of the component analysed in the area analysed (Pettijohn *et al.*, 1972). No attempt has been made to quantify the variability of quartz overgrowth abundance on a scale greater than that of the thin section analysed.

Bulk Penrith and Piper Formation sandstones without quartz overgrowths have similar oxygen isotope compositions of around 13‰$_{SMOW}$. This compares with measured $\delta^{18}O$ values of 14.0 ± 1.1‰$_{SMOW}$ for the detrital grains in the Penrith Sandstone and 10.6 ± 1.3‰$_{SMOW}$ for Piper Formation sandstone detrital grains. The Penrith Sandstone data form a three point regression line with an extrapolated overgrowth $\delta^{18}O$ composition of 34.6‰$_{SMOW}$. Piper Formation sandstone data form a two point regression line which extrapolates to an overgrowth $\delta^{18}O$ estimate of 22.9‰$_{SMOW}$.

## DISCUSSION

### Variations in $\delta^{18}O$ across overgrowths

*Penrith Sandstone overgrowth*

Zonation in the Penrith Sandstone overgrowth revealed by CL comprises concentric growth zonation developed on a scale of about 100 microns, sufficiently coarse to be easily resolved by the ion probe (Fig. 6c). There is no appreciable difference in the $\delta^{18}O$ composition of the inner twinned growth zone or the outer concentric growth zone (Fig. 7). The data in profile 1 across the grain and overgrowth suggest a possible $\delta^{18}O$ gradient, starting at +28.8‰$_{SMOW}$ close to the detrital-grain–overgrowth boundary, dropping to +27.5‰$_{SMOW}$ some 25 microns into the overgrowth and thence becoming heavier towards the outer edge of the overgrowth. However, this variation can be equally well represented as statistical scatter about a constant $\delta^{18}O$ value of +28.8‰$_{SMOW}$ with a 1-SD scatter of 0.9‰. Statistically therefore the overgrowth is isotopically homogeneous at the precision limit of the ion probe, conforming to profile 3 in Fig. 2. The second profile

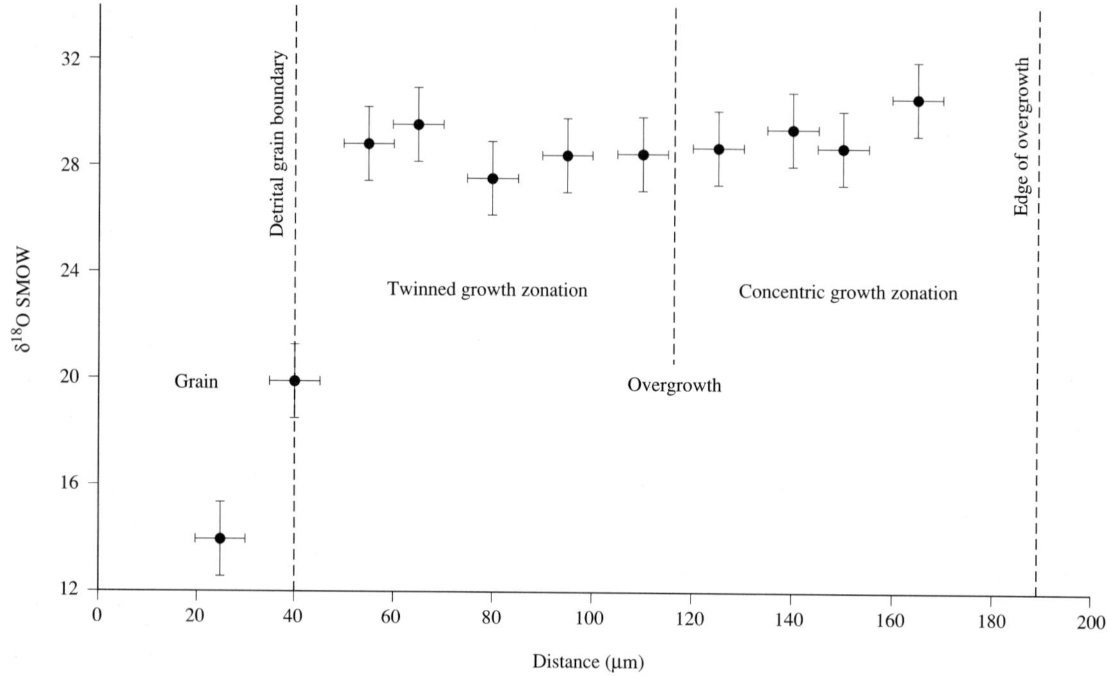


**Fig. 7.** Plot of the measured $\delta^{18}O$ data points across the detrital grain and overgrowth in profile 1 from the Penrith Sandstone. Data in Table 1. The error bars on each analysis point represent the width of the sputter mark and 1 SD in the $\delta^{18}O$ value. The mean $\delta^{18}O$ value of the overgrowth is +28.5‰$_{SMOW}$ (1 SD = ±1.1‰). Only profile 1 is shown as an illustration as the two profiles were taken across different thicknesses of the overgrowth and thus cannot be easily compared along the distance scale.

averages to a slightly lower value of +27.3 ± 1.0‰$_{SMOW}$ but the difference in these averages could be accounted for by a normalization error and so the difference is not statistically significant. Nevertheless, this second profile also suggests the presence of an isotopic gradient, becoming heavier towards the outer edge of the overgrowth, supporting the inference that there may be real variation in the $\delta^{18}O$ composition across the overgrowth.

If a gradient of increasing $\delta^{18}O$ is present across the overgrowth, then it indicates that either overgrowth formation occurred initially at higher temperatures and progressed during cooling or that the isotopic composition of the pore fluid from which the quartz precipitated became heavier with time.

### *Piper Formation sandstone overgrowth*

Growth zonation in the Piper Formation sandstone overgrowth is more complex than in the Penrith Sandstone. In the overgrowth examined, four major growth zones are recognized, typically 40–50 μm in width, but some of these zones also exhibit very fine, submicron scale growth banding (Fig. 8c). The ion beam is therefore capable of resolving the coarser scale growth zonation, but not the finer scale zonation.

A total of 23 analyses have been made on the overgrowth, although two from profile 4 are excluded from interpretation as explained previously. The standard deviation of the 21 measurements is 1.8‰ which is significantly larger than the scatter of results on either the detrital grain (22 spots, 1 SD = 1.3‰) or on the standard (1 SD = 1.3‰ in February 1994 and 1.1‰ in December 1994). This larger scatter implies that the overgrowth may not be isotopically homogeneous. Averaging the analysis values from spots which fall inside the growth zones 1, 2 and 3 gives mean values for each of these zones of 23.9 ± 0.7‰, 21.9 ± 0.4‰ and 22.2 ± 0.5‰, respectively. The errors given here are the standard error in the mean values ( $s\sqrt{n}$ ). The mean value for zone 1 is therefore different from the mean value of zone 2 at between the 95% and 99% (2–3 SD) level on the Student *t* test. The plot of $\delta^{18}O$ analyses in profile 6 across the overgrowth

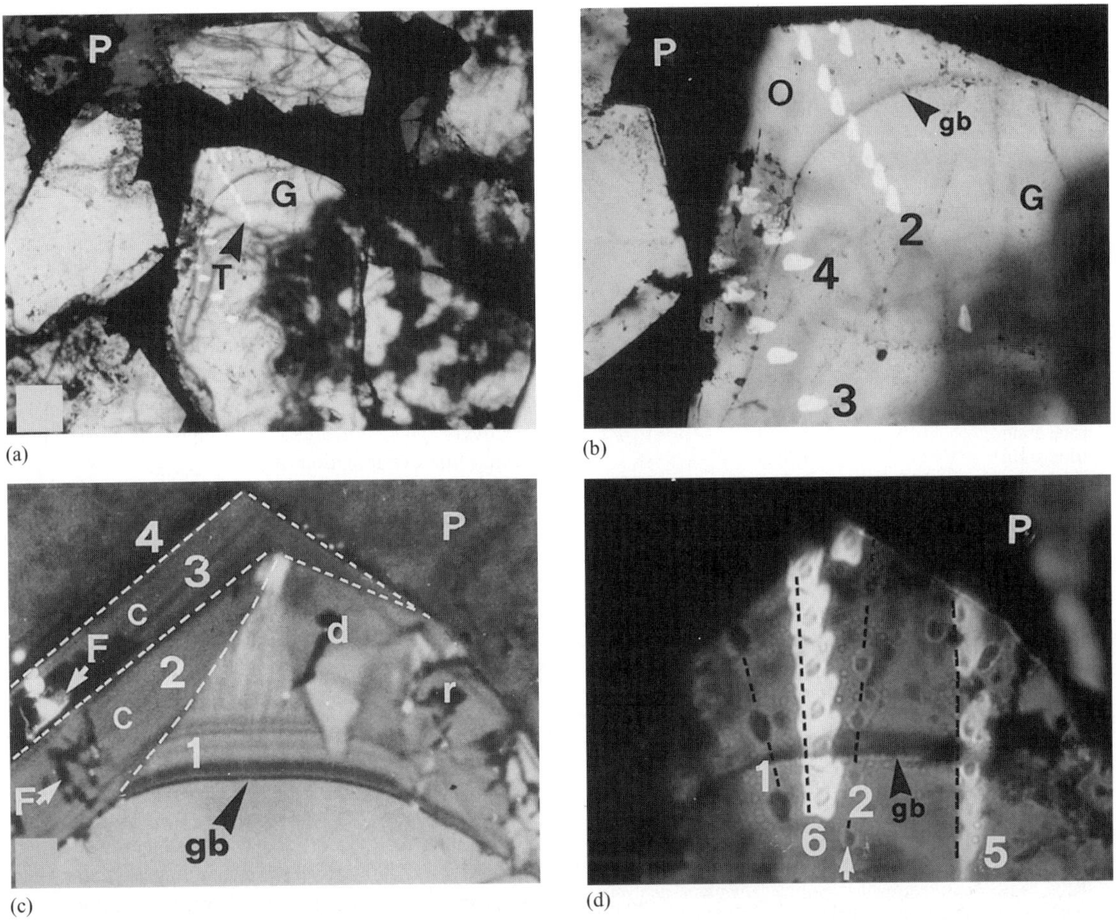


**Fig. 8.** Transmitted light and SEM micrographs of the Piper Formation sandstone showing ion probe analysis profiles 1, 2, 3, 4, 5 and 6. Data are given in Table 2. Detrital grains are annotated G and pore space is annotated P in all micrographs. Detrital grain boundary is arrowed GB. (a) General view of the coarse grained Piper Formation sandstone displaying quartz overgrowths around detrital grains. The ion probe analysis area is arrowed T. Optical photomicrograph, transmitted, plane polarized light. Width of micrograph = 2 mm. (b) Detail of ion probe analysis profiles 2, 3 and 4. Quartz overgrowth annotated O. Analysis profiles 3 and 4 both cross large fluid inclusions in the overgrowth (dark areas, also visible in Fig. 3c). Optical photomicrograph, transmitted, plane polarized light. Width of micrograph = 900 μm. (c) Internal growth fabric of the analysed overgrowth revealed by cathodoluminescence. Four growth zones (1–4) are visible. The inner growth zone displays rhombohedral face terminations (r), dislocation defects (d) as well as finely banded concentric growth structure. Growth zones 2, 3 and 4 exhibit more regular concentric zonation. Fluid inclusions are arrowed F. SEM–CL micrograph. Width of micrograph = 425 μm. (d) Detail of ion probe analysis profiles 5 and 6 (earlier analysis profiles, recoated, are also visible). The sputter mark haloes are asymmetric but the sputter craters are circular and ~ 12 μm in diameter. Analysis profile 5 crosses growth zone 1 whilst analysis profile 6 crosses all four growth zones. Optical photomicrograph, transmitted, plane polarized light. Width of micrograph = 520 μm.

(Fig. 9) supports the inference that the growth zones have different isotopic compositions in that, within any one growth zone, the isotopic composition is consistent to within < 1‰ but different from adjacent growth zones to > 1‰. There is also a trend of decreasing $\delta^{18}O$ across the overgrowth from $\approx +25‰_{SMOW}$ to $\approx +20‰_{SMOW}$. The number of data points in each zone is, however, small and it is possible that apparent trends in the data may be misleading. A bulk mean is the only valid conclusion which can be drawn.

**Table 2.** Summary listing of oxygen isotope ratios measured in the Piper Formation sandstone. Data were obtained between 3 February 1994 and 18 February 1994. Absolute SMOW values were obtained by normalizing to measured $^{18}O/^{16}O$ ratios on seed quartz, −4.79‰ SMOW. A total of 14 measurements between 1 February 1994 and 18 February 1994 yielded a measured mean $^{18}O/^{16}O$ ratio of $1.9702 \times 10^{-3}$ with a standard deviation of 1.3‰. Data from the 3 August 1994 were normalized as described in Table 1

| Sample | Analysis spot, microns along traverse | $^{18}O/^{16}O \times 10^6$ measured | $\delta^{18}O$‰ SMOW |
|---|---|---|---|
| | **Profile 5, 18 February 1994** | | |
| Detrital grain | 25 | 2001.4 | +11.0 |
| Detrital grain | 60 | 2000.4 | +10.5 |
| Detrital grain | 95 | 1996.0 | +8.2 |
| Detrital grain | 130 | 2001.3 | +10.9 |
| Detrital grain | 175 | 1995.8 | +8.1 |
| Overgrowth | 225 | 2025.6 | +23.2 |
| Overgrowth | 245 | 2024.6 | +22.7 |
| Overgrowth | 285 | 2031.1 | +25.9 |
| Overgrowth | 310 | 2030.2 | +25.5 |
| | **Profile 6, 3 August 1994** | | |
| Detrital grain | 30 | 2019.0 | +12.3 |
| Overgrowth | 72 | 2045.6 | +25.7 |
| Overgrowth | 96 | 2045.0 | +25.4 |
| Overgrowth | 145 | 2038.3 | +22.0 |
| Overgrowth | 165 | 2038.5 | +22.1 |
| Overgrowth | 185 | 2042.9 | +24.3 |
| Overgrowth | 230 | 2041.0 | +23.3 |
| Overgrowth | 245 | 2035.0 | +20.3 |

**Table 3.** Conventional fluorination O isotope data obtained at SURRC. % BSV of overgrowth determined my modal analysis of thin sections of sandstone samples from which bulk sandstone powders were prepared

| | Sample | % BSV overgrowth | $\delta^{18}O$‰ SMOW |
|---|---|---|---|
| Penrith Sandstone | CO1 | 0 | 12.87, 13.34 |
| | SR1 | 12 | 15.84 |
| | BW1 | 16 | 16.5 |
| Piper Formation sandstone | B726 | 0 | 13.73 |
| | B721 | 25.5 | 16.43 |
| | B708 | 24.9 | 15.62 |

## Comparison of bulk sample analysis with ion microprobe results

### *Detrital grains*

In the case of the Penrith Sandstone, the values determined by the two techniques for the detrital grain composition are similar; 13.1‰$_{SMOW}$ (average of +10.9 and +14.0‰$_{SMOW}$) for the conventional analysis compared with 12.5‰$_{SMOW}$ measured by the ion probe. There is not such good agreement between the detrital grain oxygen isotope compositions for the Piper Formation sandstone; 10.6 ± 1.3‰$_{SMOW}$ measured by the ion probe and 13.7‰$_{SMOW}$ according to the conventional analysis. However, variability in detrital grain isotopic composition may simply reflect variability in grain provenance.

### *Overgrowths*

The oxygen isotope composition of quartz overgrowths obtained with the Isolab®54 from the Penrith and Piper Formation sandstones are +28.3 ± 0.9‰ (weighted mean of two profiles) and +22.7 ± 1.8‰$_{SMOW}$ (six profiles), respectively. This compares with compositions based on a mixing line extrapolation of 34.6‰$_{SMOW}$ for the Penrith Sandstone and 22.9‰$_{SMOW}$ for the Piper Formation sandstone. In the case of the Piper Formation sandstone there is thus a good agreement between the two techniques. These data suggest that for the Piper Formation sandstone (i) the $\delta^{18}O$ composition of the quartz overgrowths on the scale of the samples used for isotope analysis is uniform; and (ii) that the point count modal analysis is sufficiently accurate for the conventional fluorination approach.

In the case of the Penrith Sandstone, the agreement between the ion probe and conventional $\delta^{18}O$ analyses is not so close. The extrapolated conventional $\delta^{18}O$ composition for the overgrowths of 34.6‰$_{SMOW}$ is significantly heavier than the 28.3‰$_{SMOW}$ determined using the ion probe. The dissimilarity between the conventional and ion probe $\delta^{18}O$ analysis in the Penrith Sandstone may result from several factors. These pertain to (i) inaccuracies in the quantification of the quartz overgrowth abundance in the Penrith Sandstone; (ii) variation in the internal $\delta^{18}O$ composition of the overgrowths; (iii) variation in the $\delta^{18}O$ composition of the overgrowths throughout the Penrith Sandstone; or (iv) variation in the bulk detrital grain isotope composition which will strongly affect estimates from conventional analytical techniques.

It is unlikely that the modal point count analysis is sufficiently inaccurate to severely skew the conventional isotope analysis. Quartz overgrowths in the Penrith Sandstone are separated from the host grains by a rim of iron oxide stained clays and other inclusions and are easy to identify in thin section. The statistical error associated with the modal analysis technique for mineral components in the 10–20% abundance range is low (< ±2%; Pettijohn *et al.*, 1972) and would not greatly affect the extrapolation. To bring the conventional and ion probe analyses into parity, the modal analysis would have to be in error by 56%. Moreover, if quartz overgrowths were

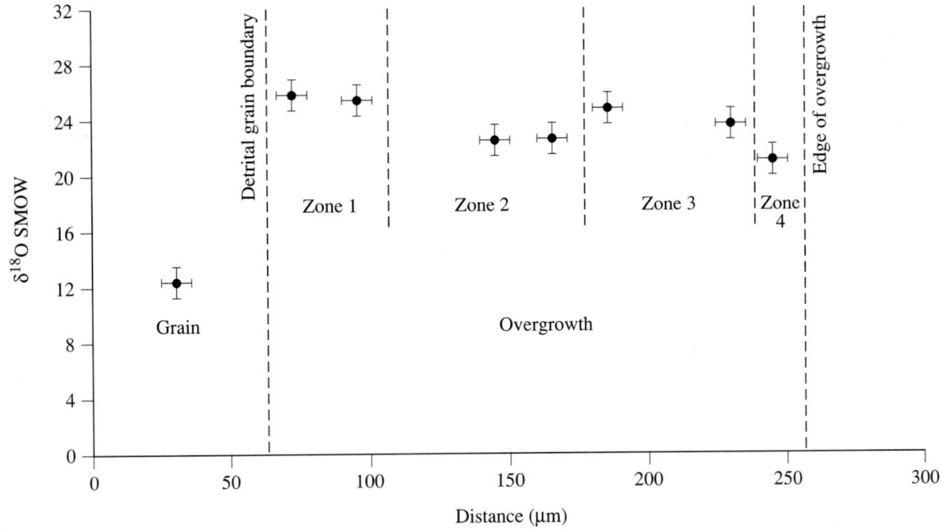


**Fig. 9.** Measured $\delta^{18}O$ data in profile 1, Piper Formation sandstone (Table 2). Error bars represent width of sputter mark and 1 SD in the $\delta^{18}O$ value. The mean $\delta^{18}O$ value of the overgrowth is $+22.7‰_{SMOW}$ (1 SD = ±1.8‰). Only 1 profile is plotted for illustration as the six profiles cross different thicknesses of the overgrowth and therefore cannot be easily compared on a simple distance scale.

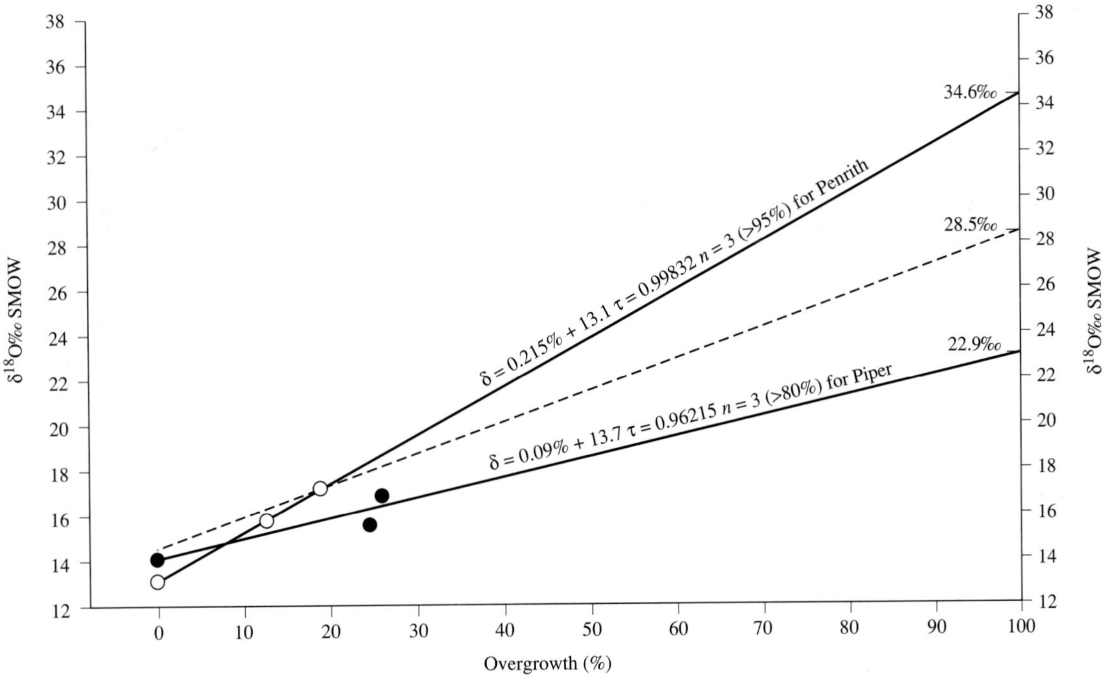


**Fig. 10.** Determination of the oxygen isotope composition of the quartz overgrowths from bulk samples of the Penrith and Piper Formation sandstones by regression of conventional fluorination oxygen isotope analyses. For the Penrith Sandstone, $\delta = 0.215\% + 13.1$ (where percentage = ‰ overgrowth), $n = 3$, $\tau = 0.99832$ (significant above the 95% confidence level); extrapolation to percentage = 100 gives a $\delta^{18}O$ composition of $+34.6‰_{SMOW}$ for the overgrowths. For the Piper Formation sandstone, $\delta = 0.092\% + 13.7$, $n = 3$, $\tau = 0.96215$ (significant above the 80% confidence level); extrapolation to percentage = 100 gives a $\delta^{18}O$ composition of $+22.9‰_{SMOW}$ for the overgrowths. The dashed line is the extrapolation to 0‰ overgrowths given the ion probe composition of $+28.5‰_{SMOW}$ for the Penrith Sandstone overgrowths and illustrates the minor shift required in the bulk sandstone $\delta^{18}O$ composition required to reconcile the ion probe and conventional measurements of the Penrith Sandstone overgrowths.

actually present in the 0% overgrowth sandstone sample then this would make the extrapolation to the overgrowth $\delta^{18}O$ composition even heavier. It is possible that the modal analysis on a thin section scale is not representative of the abundance of quartz overgrowths in the volume of Penrith Sandstone crushed for bulk rock conventional fluorination analysis. However, this sample volume was in all cases taken from the slices left over after preparation of the thin section, and petrographic observations do not suggest that the abundance of quartz cements varies greatly on a centimetre scale in the Penrith Sandstone.

There is no statistically valid evidence in the ion probe traverses presented here that the $\delta^{18}O$ composition varies significantly across individual Penrith Sandstone overgrowths. However, the number of analyses available to date is not exhaustive, and ±2.3‰ variation in the $\delta^{18}O$ composition of quartz overgrowths in the Penrith Sandstone is a possibility in the available data that could be tested by further ion probe analyses. However, to account for the observed variation, some parts of the overgrowths would have to have $\delta^{18}O$ compositions of $> +40‰_{SMOW}$ which seems geologically unreasonable. Alternatively, the $\delta^{18}O$ composition of the quartz overgrowths in the Penrith Sandstone could be extremely variable, either between grains, or on a sample to sample basis. Such an interpretation is supported by the $22.1 \pm 0.7‰_{SMOW}$ estimate of $\delta^{18}O$ composition of quartz overgrowths in the Penrith Sandstone from Nunnery Walks reported by Sullivan (1991). In this case the overgrowths were physically separated according to the method of Lee & Savin (1985). However, the likelihood of significant pore fluid compositional changes is low in the Penrith Sandstone given the uniformity of growth fabrics revealed by CL throughout the Penrith Sandstone (Fowles & Burley, 1993) and the uniformity of the salinity of the palaeofluid from which the overgrowths precipitated (Turner *et al.*, 1995).

It thus seems more likely that the dissimilarity between the quartz overgrowth conventional and ion probe analyses is either due to problems with the overgrowth separation technique (in the case of Sullivan, 1991) or with the characterization and analysis of the bulk detrital grain composition. Notice that in the extrapolation of the Penrith Sandstone overgrowth isotope composition, only a very slight variation in the $\delta^{18}O$ composition of the bulk sandstone sample without overgrowths results in a large variation in the end-member quartz overgrowth composition. For example, a $\delta^{18}O$ composition of $14.2‰_{SMOW}$ for a Penrith Sandstone sample with 0% overgrowth would give an estimate of $28.3‰_{SMOW}$ for the overgrowths, directly comparable with the ion probe measurements.

## CONCLUSIONS

A technique to measure the $^{18}O/^{16}O$ ratios of silicate minerals such as quartz in a cylindrical volume of 10 μm in diameter and a few microns in depth to an accuracy of approximately 1‰ has been successfully developed using the Isolab®54 ion microprobe. This volume of quartz contains only a few picomoles of material, a reduction of seven orders of magnitude in sample size over conventional fluorination techniques for oxygen isotope ratio determination. Although this precision is an order of magnitude poorer than conventional fluorination, the enormous gain in spatial resolution and the fact that the analyses are obtained *in situ* from conventional polished thin sections or wafers permits direct measurement of growth zones within single quartz overgrowths which have been characterized by SEM–CL petrography.

This technique has been applied to the study of quartz overgrowths from two sandstones where oxygen isotope profiles across individual overgrowths and their host detrital grains were obtained. The $\delta^{18}O$ values obtained were $+22.7‰ \pm 1.8‰_{SMOW}$ in the Piper Formation sandstone overgrowth and $+28.3‰ \pm 1.1‰_{SMOW}$ in the Penrith Sandstone overgrowth. No conclusive proof of isotopic variation in different growth fabrics in these samples was demonstrated and a model of uniform isotopic composition throughout the overgrowths remains as a valid fit to the present data. Although these $\delta^{18}O$ values are similar to $\delta^{18}O$ values obtained using conventional fluorination methods, the microprobe technique allows very high resolution, *in situ* analysis of several spots within an individual overgrowth and therefore offers the capability of investigating isotopic evolution across overgrowths. This represents a major improvement in analytical technique over conventional approaches to oxygen isotope analysis.

## ACKNOWLEDGEMENTS

This work forms part of a programme of development and application of the Isolab®54 supported by the Science and Engineering Research Council (now the Particle Physics and Astronomy Research Council) who also provided a research fellowship for one of us (JMS). Stuart Burley is the BG Geology Senior Honorary Lecturer in Manchester and thanks BG E & P for supporting research on the applications of the Isolab®54 ion probe to petroleum geology. Patrick McKeever undertook some of this research whilst on the tenure of a Mobil North Sea Ltd postdoctoral Research Fellowship and acknowledges their unconstrained support. BG and Mobil North

Sea Ltd. generously allowed publication of this work. Professor A.E. Fallick is thanked for facilitating the conventional oxygen isotope analyses at SURRC for standard testing and comparison of the ion probe analyses with conventional analyses. Mr D. Croxall of GEC Wembley kindly provided samples of ultrapure and standard industrial quartz crystals. Dr Simon Guscott provided the sample of the Piper Formation sandstone. We also thank Drs S. Kelley and C. Smalley for constructive reviews.

None of this work would have been possible without the support and assistance given by Professor Grenville Turner and we gratefully acknowledge his contributions. As always, we are indebted to Dave Blagburn and Bevist Clementson without whose help the ion probe would cease to function.

## REFERENCES

BENNINGHOVEN, A., RUDENAUER, F.G. & WERNER, H.W. (1987) Secondary ion mass spectrometry: Basic concepts, instrumental aspects, applications and trends. In: *Chemical Analysis, 86.* (eds ELVING, P.J. & WINEFORDNER, J.D.) pp. 411–419. John Wiley, Chichester.

BRINT, J.F., HAMILTON, P.J., FALLICK, A.E., HASZELDINE, S. & BROWN, S. (1991) Oxygen isotopic analysis of diagenetic quartz overgrowths from the Brent Sands: a comparison of two preparation methods. *Journal of Sedimentary Petrology* **61**, 527–533.

BURLEY, S.D. (1986) The development and destruction of porosity within Upper Jurassic reservoir sandstones of the Piper and Tartan oilfields, Outer Moray Firth, North Sea. *Clay Mineralogy* **21**, 649–694.

BURLEY, S.D., KANTOROWICZ, J.D. & WAUGH, B. (1985) Clastic diagenesis. In: *Sedimentology: Recent and Applied Aspects.* (eds BRENCHLEY, P.J. & WILLIAMS, B.P.J.) pp. 189–226. Special Publications of the Geological Society of London 18.

BURLEY, S.D., MULLIS, J. & MATTER, A. (1989) Timing diagenesis in the Tartan reservoir (UK North Sea): constraints from combined cathodoluminescence microscopy and fluid inclusion studies. *Marine and Petroleum Geology* **6**, 21–42.

CLAYTON, R.N. & MAYEDA, T.K. (1963) The use of bromine pentafluoride in the extraction of oxygen from oxides and silicates for isotopic analysis. *Geochimica et Cosmochimica Acta* **27**, 43–52.

CROWE, D.E., VALLEY, J.W. & BAKER, K.L. (1990) Microanalysis of sulphur isotope ratios and zonation by laser microprobe. *Geochimistry et Cosmochimica Acta* **54**, 2075–2092.

DETTMAN, D.L. & LOHMANN, K.C. (1995) Microsampling carbonates for stable isotope and minor element analysis: physical separation of samples in a 20 micrometer scale. *Journal of Sedimentary Research* **65**, 566–569.

DICKSON, J.A.D., SMALLEY, P.C., RAHEIM, A. & STIJFHOORN, D.E. (1990) Intracrystalline carbon and oxygen isotope variations in calcite revealed by laser microsampling. *Geology* **18**, 809–811.

ELSENHEIMER, D. & VALLEY, J.W. (1992) *In situ* oxygen isotope analysis of feldspar and quartz by Nd: Yag laser microprobe. *Chemical Geology* **101**, 21–42.

FALLICK, A.E., MCCONVILLE, P., BOYCE, A.J., BURGESS, R. & KELLEY, S.P. (1992) Laser microprobe stable isotope measurements on geological materials: some experimental considerations (with special reference to $\delta^{34}S$ in sulphides). *Chemical Geology* **101**, 53–61.

FOWLES, J. & BURLEY, S.D. (1993) Textural and permeability characteristics of faulted, high porosity sandstones. *Marine and Petroleum Geology* **11**, 608–623.

GRAHAM, C.M., VALLEY, J.W. & WINTER, B.L. (1996) Ion microprobe analysis of $^{18}O/^{16}O$ in authigenic and detrital quartz in the St. Peter Sandstone, Michigan Basin and Wisconsin Arch, USA: Contrasting diagenetic histories. *Geochimica et Cosmochimica Acta* **60**, 5101–5116.

GUSCOTT, S.C. & BURLEY, S.D. (1993) A systematic approach to reconstructing palaeofluid evolution from fluid inclusions in authigenic quartz overgrowths. In: *Geofluids '93 Extended Abstracts.* (ed. PARNELL, J.) pp. 323–328.

HAVETTE, A. & SLODZIAN, G. (1980) Matrix effects in secondary ion emission: quantitative analysis of silicates. *Journal de Physique–Lettres* **41**, 247–250.

HERVIG, R.L., WILLIAMS, P., THOMAS, R.M., SCHAUER, S.N. & STEELE, I.M. (1992) Microanalysis of oxygen isotopes in insulators by secondary ion mass spectrometry. *International Journal of Mass Spectroscopy and Ion Process* **120**, 45–63.

HERVIG, R.L., WILLIAMS, L.B., KIRKLAND, I.K. & LONGSTAFFE, F.J. (1995) Oxygen isotope microanalysis of diagenetic quartz: possible low temperature occlusion of pores. *Geochimica et Cosmochimica Acta* **59**, 2537–2543.

HOGG, A.J.C., SELLIER, E. & JOURDAN, A.J. (1992) Cathodoluminescence of quartz cements in Brent Group Sandstones, Alwyn South, UK North Sea. In: *Geology of the Brent Group.* (eds MORTON, A.C., HASELDINE, R.S., GILES, M.R. and BROWN, S.) pp. 421–440. Special Publications of the Geological Society of London 61.

JAMTVEIT, J. & HERVIG, R.L. (1994) Constraints on transport and kinetics in hydrothermal systems from zoned garnet crystals. *Science* **263**, 505–508.

LAND, L.S. & DUTTON, S.P. (1978) Cementation of a Pennsylvanian sandstone: isotopic data. *Journal of Sedimentary Petrology* **48**, 1167–1176.

LEE, M. & SAVIN, S.M. (1985) Isolation of diagenetic overgrowths on sand grains for oxygen isotope analysis. *Geochemistry et Cosmochemica Acta* **49**, 497–501.

LONGSTAFFE, F.J. (1989) Stable isotopes as tracers in clastic diagenesis. In: *Burial Diagenesis Mineralogy Association of Canada Short Course Handbook.* (ed. HUTCHEON, I.C.) 15, pp. 201–277.

LU, M.L. & MANN, K. (1986) Bond breaking and the ionization of sputtered atoms. *Physical Review Letters* **57**, 1476–1479.

LYON, I.C. & TURNER, G. (1992) The Isolab 54 ion microprobe. *Chemical Geology* **101**, 197–199.

LYON, I.C., SAXTON, J.M. & TURNER, G. (1994) Isotopic fractionation in secondary ionisation mass spectrometry. *Rapid Communications in Mass Spectrometry* **8**, 837–843.

LYON, I.C., SAXTON, J.M., MCKEEVER, P.J., CHATZITHEODORIDIS, E. & VAN LIERDE, P. (1995) Precision and reproducibility of *in situ* oxygen isotope ratio measurements on quartz obtained using an Isolab 54 ion microprobe. *International Journal of Mass Spectrometry and Ion Process* **151**, 1–16.

Macchi, L. (1990) *A Field Guide to the Continental Permo-Triassic Rocks of Cumbria.* Liverpool Geological Society. Excursion Guide, 88pp.

Mattey, D. & Macpherson, C. (1993) High-precision oxygen isotope microanalysis of ferromagnesian minerals by laser-fluorination. *Chemical Geology* **105**, 305–318.

McBride, E.F. (1989) Quartz cement in sandstones: a review. *Earth Science Reviews* **26**, 69–112.

McCants, C.Y. & Burley, S.D. (1996) Reservoir architecture and porosity evolution in deep, downthrown fault block plays: the Lowlander Prospect of block 14/20b, Witch Ground Graben, Outer Moray Firth. In: *Geology of the Humber Group: Central Graben and Moray Firth*, UKCS, 251–285.

Millikin, K.L., Land, L.S. & Loucks, R.G. (1981) History of burial diagenesis determined from isotopic geochemistry, Frio Formation, Brazoria County, Texas. *American Association of Petroleum Geologists Bulletin* **65**, 1397–1413.

Onasch, C.M. & Vennemann, T.W. (1995) Disequilibrium partitioning of oxygen isotopes associated with sector zoning in quartz. *Geology* **23**, 113–1106.

Perny, B., Eberhardt, P., Ramseyer, K., Mullis, J. & Pankrath, R. (1992) Microdistribution of Al, Li and Na in α-quartz: possible causes and correlation with short lived cathodoluminescence. *American Mineralogy* **77**, 534–544.

Pettijohn, E.J., Potter, P.E. & Siever, R. (1972). *Sand and Sandstone*, 2nd edn. Springer-Verlag, Heidelberg, 553pp.

Prezbindowski, D. (1980) Microsampling technique for stable isotope analyses of carbonates. *Journal of Sedimentary Petrology* **50**, 643–644.

Riciputi, L.R. & Paterson, B.A. (1994) *In-situ* analysis of oxygen isotope ratios in silicates and carbonates by ion microprobe. *Mineral Magazine* **58A**, 770–771.

Saxton, J.M., Lyon, I.C. & Turner, G. (1995) High precision, *in-situ* oxygen isotope ratio measurements obtained from geological and extra-terrestrial material using an Isolab 54 ion microprobe. *Analyst* **120**, 1321–1326.

Saxton, J.M., Lyon, I.C., Chatzitheodoridis, E., Perera, I.K., Van Lierde, P., Freedman, P. & Turner, G. (1996). The Manchester Isolab 54 Ion Microprobe. *International Journal of Mass Spectrometry and Ion Process* **154**, 98–128.

Sharp, Z.D. (1992) Laser microprobe techniques for stable isotope analysis. *Chemical Geology* **101**, 3–19.

Shimizu, N. & Hart, S.R. (1982) Isotope fractionation in secondary ion mass spectrometry. *Journal of Applied Physics* **53**, 1303–1311.

Sippel, R.F. (1968) Sandstone petrology, evidence from luminescence petrography. *Journal of Sedimentary Petrology* **38**, 530–554.

Smalley, P.C., Maile, C.N., Coleman, M.L. & Rouse, J.E. (1992) LASSIE (laser ablation sampler for stable isotope extraction) applied to carbonate minerals. *Chemical Geology* **101**, 43–52.

Sullivan, M.D. (1991) *Diagenetic Study of the Lower Permian Leman Sandstone, Leman Field, Southern North Sea.* PhD Thesis, University of Glasgow.

Turner, C.C., Richards, P.C., Shallow, J.L. & Grimshaw, S.P. (1984) Upper Jurassic stratigraphy and sedimentary facies in the central Outer Moray Firth Basin, North sea. *Journal of Marine Petroleum Geology* **1**, 105–117.

Turner, P., Burley, S.D., Rey, D. & Prosser, J. (1995) Burial history of the Penrith Sandstone (Lower Permian) deduced from the combined study of fluid inclusions and palaeomagnetic data. In: *Palaeomagnetic Applications in Hydrocarbon Exploration and Production.* Special Publications of the Geological Society of London 98, pp. 43–78.

Valley, J.W., Eiler, J.M., Kohn, M.J. *et al.* (1994) Contrasting styles of oxygen isotope exchange. *Mineral Magazine* **58A**, 924–925.

Walker, G. & Burley, S.D. (1991) Luminescence petrography and spectroscopic studies of diagenetic minerals. In: *Luminescence Microscopy: Quantitative and Qualitative Applications* (eds Barker, C.E. & Kopp, O.C.) pp. 83–92. Special Publications, Society of Economic Paleontologists and Mineralogists 25.

Waugh, B. (1970) Petrology, provenance and silica diagenesis of the (Penrith Sandstone Lower Permian) of north-west England. *Journal of Sedimentary Petrology* **40**, 1226–1240.

Weichart, U. & Hoefs, J. (1994) Initial results with a UV laser based microanalytical preparation technique for the in-situ determination of oxygen isotope ratios in silicates and oxides. *Mineral Magazine* **58A**, 973–974.

Weichart, U. & Hoefs, J. (1995) An excimer laser-based microanalytical preparation technique for in-situ oxygen isotope analysis of silicate and oxide minerals. *Geochimica et Cosmochimica Acta* **59**, 4093–4101.

*Spec. Publs int. Ass. Sediment.* (2000) **29**, 317–331

# Significance of trace element composition of quartz cement as a key to reveal the origin of silica in sandstones: an example from the Cretaceous of the Barrow Sub-basin, Western Australia

G. M. KRAISHAN[1], M. R. REZAEE[1]* *and* R. H. WORDEN[2]

[1]*NCPGG, The Barton Campus, The University of Adelaide, Adelaide, SA, 5005, Australia; and*
[2]*School of Geosciences, The Queen's University of Belfast, BT7 1NN, UK*

## ABSTRACT

Trace elements have been used to constrain the origin of silica in quartz cement: one of the most abundant authigenic minerals in marine sandstones of the Lower Cretaceous Barrow Group, Barrow Sub-basin, Australia. Petrographic, fluid inclusion, electron microprobe and cathodoluminescence data from quartz cements in these sandstones indicate multiple stages of cementation at different temperatures. Three quartz cementation phases were detected by using the cathodoluminescence imaging. Homogenization temperature calculations from fluid inclusions indicate precipitation temperature in a range of 60–140°C; phase I between 60 and 90°C; phase II between 90 and 105°C; phase III between 105 and 140°C. Oil stains and bitumen coat the second phase of quartz overgrowths and were trapped between the middle and the late phases, implying that the late phase formed after oil migration commenced. Trace element concentrations and mass balance calculations suggest two predominant sources for silica: feldspar alteration and pressure dissolution. Aluminium varies consistently between each cement phase with an average 245 p.p.m. for phase I, 805 p.p.m. for phase II and 89 p.p.m. for phase III. Textural analysis and depth-related patterns suggest that silica in the two first phases was probably a by-product of feldspar alteration reactions whilst the last phase was probably a result of pressure dissolution. The lack of correlation between aluminium content and temperature implies that the primary source of the silica is the main control on the aluminium content of quartz cement.

## INTRODUCTION

Numerous mechanisms have been suggested in the literature as possible sources for silica in quartz cement. The most important sources, summarized by Leder & Park (1986), include: intergranular pressure dissolution, silica dissolved in advecting pore fluids by flow over quartz grains in sandstones, silica liberated from mudstones including the dissolution of amorphous opaline silica, and silica liberated from alteration of silicate minerals. Some of the main factors that control the amount of quartz cement in sandstones are thought to include: framework grain composition, residence time in the 'silica mobility window', fluid composition, and flow volume and pathways (McBride, 1989).

The main objectives of this study are: to document the trace elements and their correspondence with the cathodoluminescence colours of the quartz cements, to determine the timing of quartz cementation, and to constrain silica sources from trace elements using a wide range of techniques including quartz cement chemical analysis, petrography, fluid inclusion studies and thermal history modelling.

## GEOLOGICAL SETTING

The Barrow Sub-basin, situated offshore in the northern part of the Carnarvon Basin on the North-west Shelf of Australia contains numerous commercial oil and gas accumulations. More than 40 oil and gas accumulations have been discovered to date and more than 870 million

* Present address: Geology Department, Faculty of Science, Tehran University, Tehran, Iran.

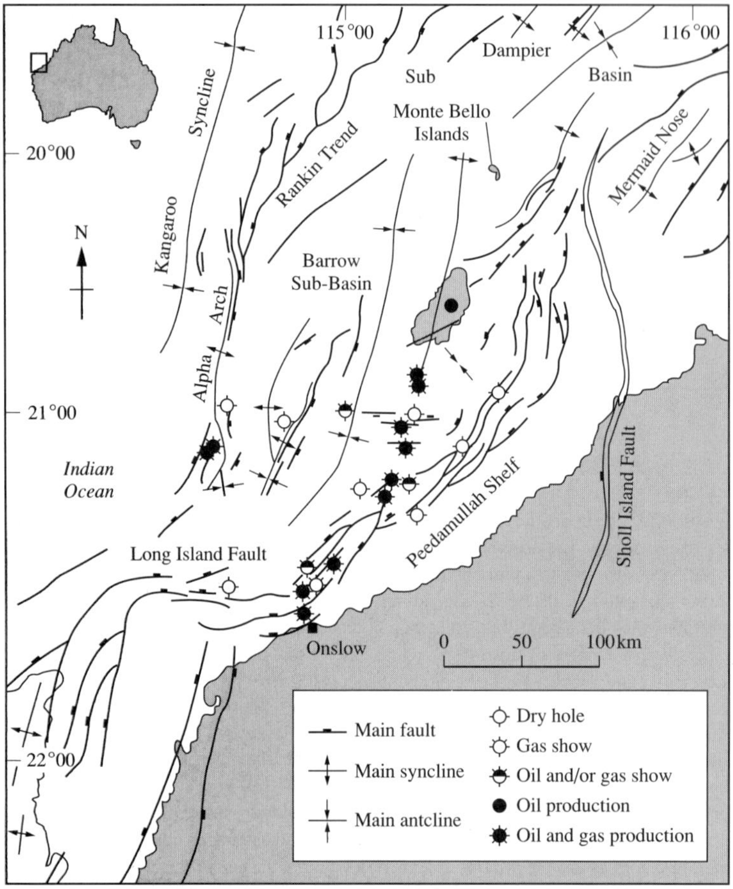


**Fig. 1.** Location map showing the main structural features of the Barrow Sub-basin and wells studied.

barrels (MMBO) of crude oil and condensate and more than 14 trillion cubic feet (TCF) of gas have thus far been produced (Kopsen, 1994).

The sub-basin occupies an area about 20 000 km$^2$ that covers the offshore region south of the Monte Bello Islands in the north to Onslow town in the south (Fig. 1). The sub-basin currently has an asymmetric profile with a deep synclinal trough parallel to Rankin Platform and Alpha Arch along its western margin and a series of shallow shelves that onlap the basement rocks (Pilbara Block) along the eastern margin.

The Basin was the site of various depositional environments during the Palaeozoic to Cenozoic eras, with greater than 4000 m of sediments deposited along the eastern margin and more than 15 000 m deposited within the depocentre. The break-up of Gondwana during the mid-Jurassic led to the formation of the sub-basin and controlled its fill. The break-up of the Indo-Australian plate has affected the geological setting of the sub-basin and the surrounding areas.

The Palaeozoic sediments have only been intersected by wells along the southern part of the Carnarvon Basin (Hocking, 1994). The Barrow Sub-basin is bound to the north by the Dampier Sub-basin with the two sub-basins arbitrarily separated by the change of the structural orientation near the Deepdale–Mermaid Nose (Kopsen & McGann, 1985). The southern edge of the Barrow Sub-basin is defined by the Long Island Fault System, the western edge by the Rankin Platform and its extension, the Alpha Arch, and the eastern margin by Peedamullah Shelf (Fig. 1).

The Lower Cretaceous Barrow Group is generally considered to be the result of a progradational delta (Kopsen & McGann, 1985). This delta prograded north-eastwards along the length of the Barrow Sub-basin. Tait (1985) subdivided the Barrow Group into the Malouet and Flacourt Formations. McGilvery (1996) defined the Barrow Group as slope/basin to shore face and progradational slope complexes that can be grouped into three main associations; mounded turbidite lobes, slope apron and marine-

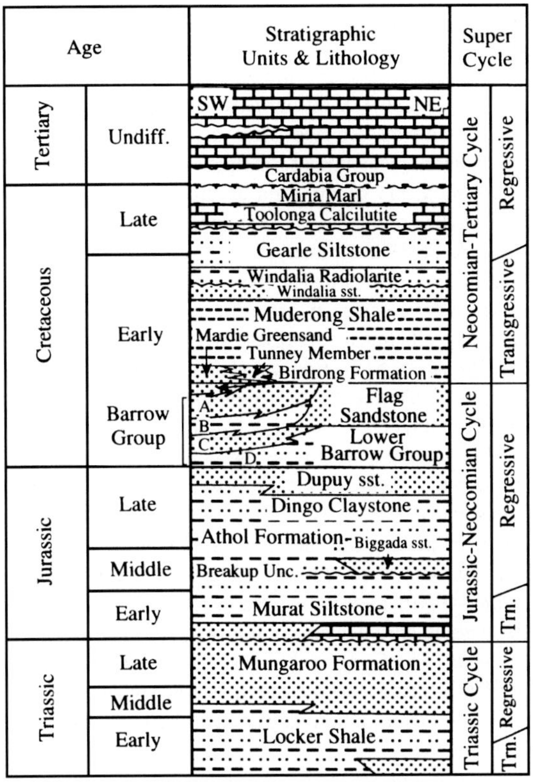


Fig. 2. Generalized stratigraphic column of the Barrow Sub-basin.

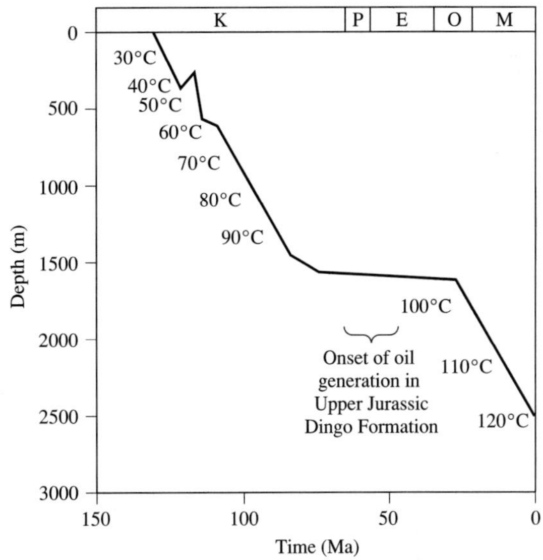


Fig. 3. Simple burial history diagram for the top of the Barrow Group with the probable time of the main phase of oil generation (from the underlying Upper Jurassic Dingo Claystone) indicated.

reworked strandplain sediments. The upper part (Flacourt Formation) is the principal reservoir in the Barrow Sub-basin. It typically consists of several-metre-thick clean sandstone units, interbedded with sandy bioturbated sandstones. The Flacourt Formation has been interpreted to represent marine-reworked strandplain and shoreface deposits. The lower part (Malouet Formation), however, is composed predominantly of slope apron deposits, consisting of tight, matrix-rich, non-productive sandstones. Barrow Group sediments were subdivided by Williams & Poynton (1985) into a series of seismic intervals (A, B, C, and D; Fig. 2). Intervals A and C are of most interest since they contain most of the discovered hydrocarbons in the area. The Barrow Group unconformably overlies the Dingo Claystone and underlies the Winning Group (Fig. 2).

Most of the hydrocarbons in the Barrow Sub-basin are thought to have been generated from the Upper Jurassic Dingo Claystone (Alexander *et al.*, 1980; 1983; Volkman *et al.*, 1983). The optimum time for generation and migration was the early Tertiary (Volkman *et al.*, 1983). Peak oil generation occurred at about 130°C requiring about 2800 m of overburden rocks at the anticipated heat flow at that time (Fig. 3).

Burial at the sampled sites occurred progressively through time with only a very minor phase of uplift in the early part of the basin history (Kaiko, personal communication 1997, Fig. 3). Most of the wells in the depocentre and the western margins of the sub-basin have had similar burial history. In the eastern margins, several wells have experienced elevated palaeo-temperatures (Kaiko & Tingate, 1996). Samples for this study were collected predominantly from the depocentre and the western margins (Fig. 1). The temperature of the Barrow Group in these areas has steadily increased to reach maximum temperatures of about 120–140°C (depending on the exact well location).

## METHODOLOGY

Core samples were collected from the Barrow Group from subsurface depths between 900 and 3500 m from 24 exploration wells located in and around the Barrow Sub-basin (Fig. 1). The majority of the samples consist of relatively clean, medium to coarse-grained sandstones which were representative of the different lithofacies recognized during core description. Several samples were collected near oil–water contacts, around thin shale beds, and from different concretions, nodules and cemented zones in order to sample non-reservoir but diagenetically important lithologies. In addition, most samples were collected adjacent to where a core plug had been taken.

Both polished-sections and thin-sections were prepared for all the samples. They were cut perpendicular to the bedding plane after impregnation with blue epoxy resin to facilitate porosity identification. The modal composition for 96 samples was determined by counting 400–1000 points per thin section using standard techniques (Zuffa, 1985; Pettijohn *et al.*, 1987). The mean grain size and degree of sorting were estimated for all samples by measuring the length of 100 detrital grains per thin section.

X-ray diffraction (XRD) analyses were carried out on all samples. Whole-rock samples were X-rayed for their bulk mineralogy. Samples were crushed manually to fine powder in ethanol using an agate mortar and pestle and then oven-dried at less than 60°C to prevent collapse of clay structures. All samples were prepared as randomly orientated powders using a front-mounting press in an aluminium holder. The clay fraction (< 2 μm) of 16 representative samples of Barrow Group sandstones was separated by gravity settling and smeared on glass slides to assess the transformation of authigenic clays as a possible source for silica. Three X-ray runs were made for each sample: (i) air-dried at room temperature; (ii) after exposing the samples to ethylene glycol vapour for at least 16 h at 60°C; and (iii) after heating to 550°C for 2 h. The XRD runs for glycol-saturated fractions were prepared on ceramic plates held under vacuum. X-ray analyses for the whole-rock and the clay fraction samples were carried out on a Philips PW 1050 X-ray diffractometer system using monochromatic $CoK_\alpha$ radiation, at the following settings: 50 kV, 35 mA, 1° divergent slit, 0.5° receiving slit and a scan speed of 2°/min and 1°/min. Whole-rock samples were scanned between 3° and 75°2θ with a step size of 0.05°, whilst the orientated clay samples were scanned between 3° and 35°2θ with a step size of 0.02°. Mineral identification was checked by comparison with Joint Committee on Powder Diffraction Standards (JCPDS) files using XPLOT software developed by CSIRO Division of Soils.

A Technosyn Model 8200 Mk II cathodoluminoscope and a Leitz Orthomat E automatic microscope camera were used to examine selected polished sections bonded with thermo-stable epoxy resin. The CL photomicrographs were taken at 1–4 minute exposure times in integral metering mode with Kodak Ektapress 1600 ASA film. To generate suitable quartz luminescence, an electron gun voltage of 12–20 kV and a beam current of 200–300 μA were used. A correction factor of 4 was used to correct for reciprocity failure and a light metering factor of 80% (dark field) was used to obtain suitable film exposure.

A scanning electron microscope equipped with an energy-dispersive X-ray system (SEM–EDS) was used on 32 samples selected after detailed petrographical and mineralogical studies. Samples were examined both as polished thin section and as broken chips. Selected polished thin sections were examined using back scattered electron (BSE) image analysis.

Quantitative chemical analyses of quartz cements were obtained by electron microprobe of four polished thin sections. Chemical analyses were performed on a CAMECA SX 51 instrument at standard operating conditions (15 kV acceleration potential, 20 nA and beam current, 2 μm beam diameter). Counting times of 60 s for Al, Cu, Mn, Fe and Ti were used and all elements were measured on both peaks and background wavelength dispersively. Standards used were garnet for Al and Fe; copper metal for Cu; rudonite for Mn and rutile for Ti. The raw data were reduced using the Pouchou–Pichoir method (PAP). Detection limits (99% probability) for the elements analysed were as follows: Al = 40 ppm, Fe = 150 ppm, Ti = 75 ppm, Mn = 125 ppm, Cu = 250 ppm. Microprobe data were measured on both single points and traverses, covering both detrital quartz grains and quartz overgrowths.

Four doubly thick (~ 140 μm) polished thin sections were prepared from Barrow Group sandstones for microthermometric measurements. Medium to very coarse-grained, moderately to well-sorted and matrix-free sandstones were selected for fluid inclusion studies. The samples were examined by a Leitz standard polarizing microscope with a 100 × objective lens and 12.5 × eyepieces, giving a total magnification of 1250 ×. Homogenization temperatures were performed using a Reynolds (Fluid Inc.), USGS type gas-flow heating–cooling stage (Reynolds, 1978). The heating stage was calibrated with synthetic fluid inclusion standards (SYN FLINC™) with melting points of 20°C and 375°C. Only primary fluid inclusions of both aqueous and hydrocarbons (mostly aqueous) were selected from quartz cement to determine their minimum precipitation temperatures (Roedder, 1979). The primary fluid inclusions were heated incrementally, using a heating rate of 5°C/min until homogenization. Homogenization temperatures were defined as the mean of the lowest temperature where the gas bubble disappeared (Walderhaug, 1990). After homogenization had occurred, temperature was lowered slightly (~ 10°C) to observe the reappearance of the gas bubble. This step was performed at least three time in order to allow precise determination of the homogenization temperatures. Reproducibility of measurements was approximately ±1°C.

## RESULTS

### Detrital grain composition

Barrow Group sandstones are generally well sorted and consist of fine to coarse-grained sandstones. Sands are quartz-rich with minor quantities of mica, feldspar and detrital clay matrix, and trace amounts of rock fragments. Feldspars are mainly potassic (orthoclase and microcline) with small quantities of plagioclase. Rock fragments mainly have a sedimentary origin (chert and siltstone), or are plutonic (granite) and metamorphic (mica schist and quartzite). Glauconite is abundant in some marine sandstone and occurs mainly as faecal pellets (Kraishan, 1997).

For classification purposes glauconite was excluded from the main rock components. The Barrow Group sandstones are classified as quartz arenites, subarkose (subfeldsparenites) and sublitharenites according to Folk's classification (Fig. 4; after Folk, 1974). The Flacourt Formation sandstones are generally richer in feldspar and contain less detrital quartz and rock fragments than those of the Malouet Formation. Conversely the Malouet Formation is richer in mica and detrital clay matrix than Flacourt sandstones. The remnant detrital feldspar content decreases with depth in Barrow Group sandstones (Fig. 5). Rock fragments in both formations represent less than 2% by volume and consist mainly of low-grade metamorphic quartzite, mica schists and sedimentary chert, shale and silt clasts. Mica occurs in trace amounts in both formations. Apart from the presence of matrix, there are no significant differences in the detrital framework grain composition of the different depositional environments.

The original feldspar content was calculated from the total existing potassium feldspar and plagioclase volumes plus partially dissolved skeletal feldspar masses, kaolinized and carbonate-replaced feldspar grains. The pre-diagenesis average framework composition of each formation was relatively feldspathic and averaged $Q_{86}F_{11}R_3$ and $Q_{83}F_{15}R_2$ for the Malouet and Flacourt Formations, respectively (as witnessed in more shallow parts of the basin; Kraishan, 1997). The average framework compositions of the Malouet and Flacourt Formation sandstones presently are $Q_{92}F_5R_3$ and $Q_{89}F_9R_2$ representing an approximate decrease in feldspar content of 6% for both formations

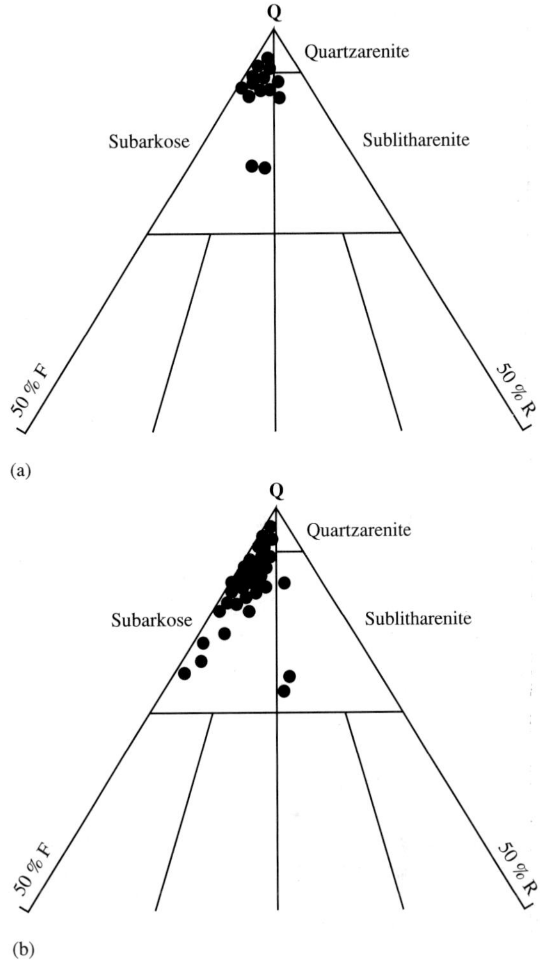


**Fig. 4.** QFR ternary sandstone diagram showing the present-day framework grain component compositions of (a) Flacourt Formation and (b) Malouet Formation.

### Diagenetic mineralogy and textures

Fifteen authigenic minerals have been identified although only quartz, kaolinite and authigenic carbonate minerals are volumetrically significant (Table 1). Diagenetic glauconite, pyrite, K-feldspar, albite, clay minerals (other than kaolinite and illite) and barite are present in minor to trace amounts. The authigenic minerals of the Barrow Group sandstones appear to be homogeneously distributed on a regional scale. Framboidal pyrite and micritic siderite are the earliest diagenetic mineral cements. Partial dissolution of feldspars (Plate 1a, facing p. 326) and the local growth of minor feldspar cement followed. Poikilotopic calcite encloses the earlier authigenic minerals (Plate 1b, facing p. 326) and was itself followed by extensive quartz cement and siderite cement (sparry, in contrast to the early diagenetic phase). Later in the diagenetic history, potassium feldspar underwent widespread (though not total) dissolution, ankerite cement was precipitated,

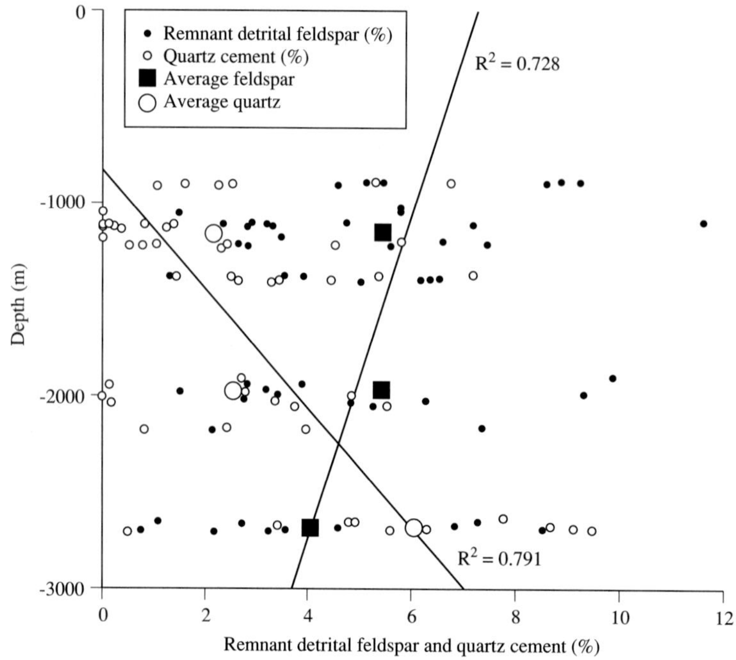


**Fig. 5.** Variation of detrital K-feldspar and diagenetic quartz cements with depth. There is a significant decrease in the average amount of detrital K-feldspar with increasing depth. There is also a significant increase in the amount of quartz cement with increasing depth.

**Table 1.** Point count data from the Barrow Group sandstones

| | Malouet Formation | | | | Flacourt Formation | | | |
|---|---|---|---|---|---|---|---|---|
| | Av. | Min. | Max. | St. dev. | Av. | Min. | Max. | St. dev. |
| Quartz (mono) | 65.1 | 29.9 | 84.1 | 10.0 | 58.2 | 33.7 | 79.6 | 8.0 |
| Quartz (poly) | 0.9 | 0.0 | 3.4 | 0.9 | 1.2 | 0.0 | 9.2 | 1.5 |
| K-feldspar | 4.6 | 0.0 | 14.5 | 4.3 | 5.2 | 0.0 | 16.6 | 3.4 |
| Plagioclase | 0.2 | 0.0 | 1.6 | 0.4 | 0.1 | 0.0 | 1.3 | 0.3 |
| SRF | 0.8 | 0.0 | 4.0 | 0.9 | 0.7 | 0.0 | 6.5 | 1.1 |
| VRF | 0.1 | 0.0 | 0.8 | 0.2 | 0.1 | 0.0 | 0.9 | 0.2 |
| MRF | 0.8 | 0.0 | 4.5 | 1.1 | 0.5 | 0.0 | 5.5 | 1.1 |
| Mica | 0.8 | 0.0 | 9.9 | 1.7 | 0.5 | 0.0 | 3.4 | 0.8 |
| Glauconite | 1.0 | 0.0 | 11.7 | 2.4 | 0.2 | 0.0 | 2.2 | 0.5 |
| Calcite | 1.8 | 0.0 | 21.1 | 5.1 | 0.9 | 0.0 | 28.7 | 4.7 |
| Dolomite | 2.4 | 0.0 | 20.4 | 4.4 | 2.2 | 0.0 | 25.6 | 5.7 |
| Siderite | 2.8 | 0.0 | 59.8 | 10.8 | 1.0 | 0.0 | 25.3 | 4.0 |
| Pyrite | 1.1 | 0.0 | 6.5 | 1.5 | 2.0 | 0.0 | 32.8 | 4.8 |
| Kaolin | 1.9 | 0.0 | 14.8 | 3.6 | 3.9 | 0.0 | 12.8 | 3.6 |
| Quartz cement | 1.6 | 0.0 | 10.4 | 2.1 | 3.5 | 0.0 | 11.6 | 3.2 |
| Feldspar cement | 0.0 | 0.0 | 0.0 | 0.0 | 0.3 | 0.0 | 7.5 | 1.2 |
| Organic matter | 0.4 | 0.0 | 4.2 | 0.8 | 0.8 | 0.0 | 16.1 | 2.3 |
| Heavy minerals | 0.0 | 0.0 | 0.3 | 0.1 | 0.0 | 0.0 | 0.4 | 0.1 |
| Matrix | 7.8 | 0.0 | 30.5 | 10.0 | 1.9 | 0.0 | 19.6 | 4.0 |
| Other | 0.0 | 0.0 | 0.4 | 0.1 | 0.2 | 0.0 | 5.1 | 0.8 |
| Primary porosity | 4.1 | 0.0 | 17.3 | 4.5 | 13.6 | 0.3 | 25.6 | 6.7 |
| Secondary porosity | 1.4 | 0.0 | 4.6 | 1.5 | 2.8 | 0.2 | 8.1 | 2.0 |
| IGV | 19.6 | 3.6 | 63.9 | 12.8 | 28.0 | 11.1 | 54.0 | 9.9 |
| Grain size (mm) | 0.23 | 0.12 | 0.62 | 0.12 | 0.35 | 0.10 | 0.86 | 0.22 |
| Sorting (ϕ) | 0.9 | 0.6 | 2.7 | 0.5 | 0.7 | 0.4 | 1.3 | 0.2 |

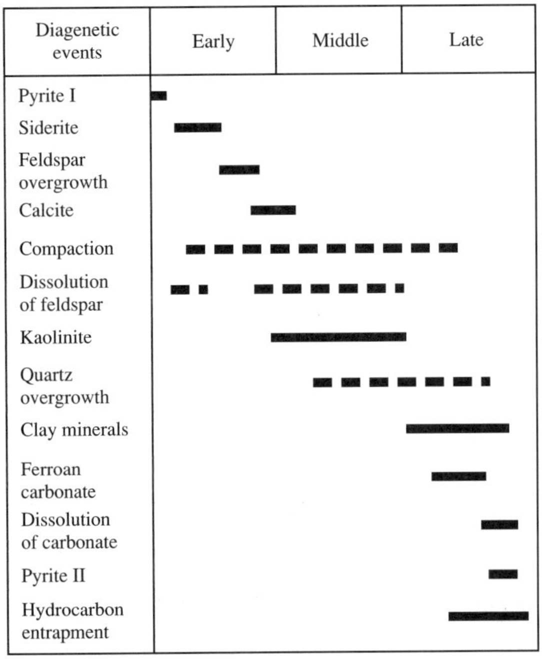


**Fig. 6.** Paragenetic sequence of diagenetic features in sandstones of the Barrow Group.

followed by the last diagenetic mineral: cubic pyrite typically associated with trapped oil.

Quartz cement appears to have grown at various stages throughout the burial history; more details of this will be given subsequently. Quartz cement volumes vary considerably with an average of about 3% (exact value = 2.8%) for the entire volume of sandstone and a local maximum value of more than 11%. Quartz cement contents tend to increase with depth of burial. A summary of the diagenetic history of the reservoir sandstones of the Barrow Group is given in Fig. 6.

## Quartz cement textures and cathodoluminescence

Authigenic quartz in the Barrow Group sandstones is present as euhedral syntaxial overgrowths and is best developed in quartz-rich, matrix-free, medium to coarse-grained sandstones, where it surrounds intergranular voids (Plate 1c, facing p. 326). 'Dust' rims, euhedral faces and relative lack of inclusions between the authigenic cements and the detrital quartz grains were used to distinguish detrital and authigenic quartz, aided by CL microscopy. The 'dust' rims are composed of yellowish brown cryptocrystalline detrital clay (Plate 1c, facing p. 326).

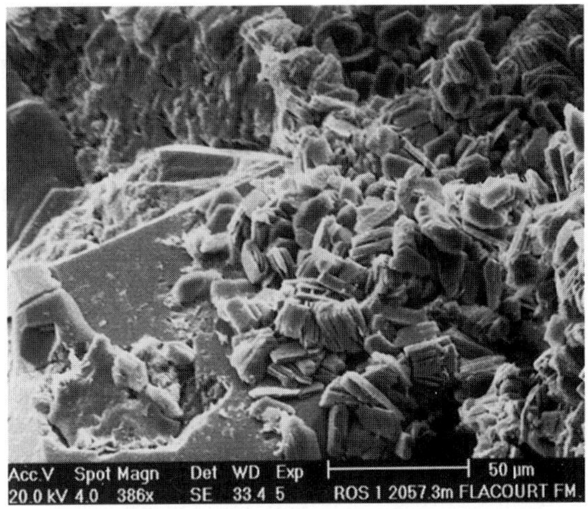


**Fig. 7.** SEM micrograph illustrating the kaolinite booklets that have been engulfed by quartz overgrowths, suggesting at least simultaneous formation of both authigenic minerals, Flacourt Formation, 2060.3 m, Rosily 1 A.

Inherited quartz cements were recognized using the criteria described by Sanderson (1984).

Cathodoluminescence microscopy allowed the discrimination of three quartz cement phases (Plate 1d, facing p. 326) in Barrow Group sandstones. Each phase has typically dull-luminescent colours (dark brown or non-luminescence to purple to dark blue to light brown) but are texturally and/or chromatographically discrete (Plate 1d, facing p. 326).

The earliest phase (I) is characterized by brown luminescence and is relatively thin (< 20 µm in thickness; Plate 1d, facing p. 326). This early phase is observed filling and healing fractures in detrital quartz grains, suggesting that the early quartz cement postdated the onset of compaction.

The second phase (II) represents the main phase of quartz cementation (in terms of volume) and is locally intergrown with kaolinite booklets (Fig. 7) The second phase is characterized by colours varying from dark brown or non-luminescence to purple to dark blue luminescence. Despite the great volume of phase II, it is unevenly distributed and has variable thickness (Plate 1d, facing p. 326). The boundaries between the second phase of quartz cement and kaolinite booklets are irregular, suggesting the synchronous precipitation of these two authigenic minerals. The last phase (III) is characterized by similar luminescence colours to the early phase (brown luminescence) and irregular thickness.

Quartz cement is most abundant in the deepest samples where the quartz cements may locally exceed 11% of the

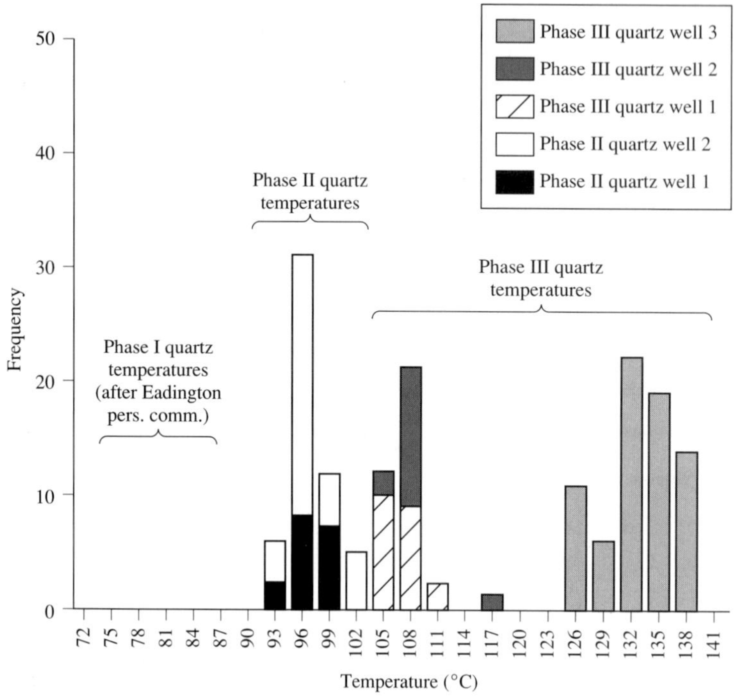


**Fig. 8.** Measured aqueous homogenization temperatures of quartz cement phases II and III with the span of temperatures reported from Eadington for phase I indicated.

whole-rock volume (Fig. 5). The quartz cement becomes thicker (> 100 μm) and almost completely surrounds the detrital quartz grains in the deepest samples. Quartz cements in the deeper samples commonly have sutured contacts.

## Fluid inclusion analysis of quartz cements

Fluid inclusions in quartz cements are typically found trapped at the grain–cement interface but are also isolated within intergranular quartz cement. They range in diameter from 1 to 10 μm but with an average of less than 5 μm and are generally spherical or subspherical. Inclusions are exclusively two phase (liquid and vapour) with the liquid consistently being the dominant phase. Both colourless aqueous and fluorescent hydrocarbon fluid inclusions were observed. Homogenization temperatures for quartz cement were measured for both aqueous and hydrocarbon fluid inclusions from four samples and are illustrated in Fig. 8. Homogenization temperatures from phase II and III show a wide range from 80 to 140°C. Homogenization temperatures are consistent in different samples in different wells for each phase of quartz cement.

The first phase of quartz cement seems to occur at temperature lower than 70°C (as suggested by Lisk, CSIRO, personal communication, 1996) but may occur in a range of 60–90°C. The second phase (II) of quartz cement is the main phase by volume and thus most of the measurements come from that phase. The homogenization temperatures range from 90 to 105°C. The last phase (III) of quartz cement has the highest homogenization temperatures (range between 105 and 140°C) and approaches the present-day calculated formation temperature (Kaiko & Tingate, 1996).

Hydrocarbon fluid inclusions were found mainly in the second phase of quartz cement indicating that quartz cementation (phase II) occurred when some petroleum was in the sandstone. Also oil and oil stains have been observed coating the second stage of quartz overgrowth (Plate 2a,b, facing p. 326). Similarly, indirect evidence from the temperature and timing of source rock maturation indicates that hydrocarbon migration occurred at the same time as the growth of the second phase of quartz cement (see de Boer & Collins, 1988).

## Chemical composition of quartz

The electron microprobe analysis of Barrow Group quartz cements reveals different chemical compositions for each quartz cement phase. Five elements (Ti, Al, Mn,

**Table 2.** Chemical data from quartz cements from the Barrow Group (Flacourt Formation) sandstones.

| Sample | Depth (m) | Formation | Phase | CL colour | Ti (ppm) | Al (ppm) | Mn (ppm) | Fe (ppm) | Cu (ppm) |
|---|---|---|---|---|---|---|---|---|---|
| GR1-4 | 2707.2 | Flacourt | Early | Brown | 247 | 161 | 165 | bdl | bdl |
| GR1-4 | 2707.2 | Flacourt | Early | Brown | bdl | 112 | bdl | bdl | bdl |
| GR1-4 | 2707.2 | Flacourt | Early | Brown | 185 | 103 | bdl | bdl | bdl |
| GR1-4 | 2707.2 | Flacourt | Early | Brown | 147 | 456 | bdl | bdl | bdl |
| GR1-2 | 2714.7 | Flacourt | Middle | Blue | 78 | 290 | bdl | bdl | 516 |
| GR1-2 | 2714.7 | Flacourt | Middle | Blue | bdl | 448 | 172 | bdl | 298 |
| GR1-2 | 2714.7 | Flacourt | Middle | Blue | 78 | 1090 | bdl | bdl | bdl |
| GR1-2 | 2714.7 | Flacourt | Middle | Blue | 195 | 266 | 103 | 152 | 216 |
| GR1-2 | 2714.7 | Flacourt | Middle | Blue | 117 | 1066 | bdl | bdl | bdl |
| GR1-4 | 2707.2 | Flacourt | Middle | Blue | bdl | 1249 | bdl | bdl | bdl |
| GR1-4 | 2707.2 | Flacourt | Middle | Blue | 101 | 818 | 152 | bdl | bdl |
| GR1-4 | 2707.2 | Flacourt | Middle | Blue | 141 | 1298 | bdl | bdl | bdl |
| GR1-4 | 2707.2 | Flacourt | Middle | Blue | 122 | 1226 | bdl | 184 | bdl |
| GR1-2 | 2714.7 | Flacourt | Late | Brown | 122 | bdl | 407 | 116 | bdl |
| GR1-2 | 2714.7 | Flacourt | Late | Brown | bdl | 106 | 133 | bdl | 302 |
| GR1-2 | 2714.7 | Flacourt | Late | Brown | 151 | bdl | bdl | bdl | bdl |
| GR1-2 | 2714.7 | Flacourt | Late | Brown | bdl | bdl | 103 | 236 | bdl |
| GR1-2 | 2714.7 | Flacourt | Late | Brown | 130 | bdl | 121 | 121 | bdl |
| GR1-2 | 2714.7 | Flacourt | Late | Brown | 136 | 239 | bdl | 100 | bdl |
| GR1-2 | 2714.7 | Flacourt | Late | Brown | 346 | 112 | 121 | 225 | bdl |
| GR1-2 | 2714.7 | Flacourt | Late | Brown | 109 | bdl | 152 | bdl | bdl |
| GR1-2 | 2714.7 | Flacourt | Late | Brown | 193 | 94 | 200 | 246 | bdl |
| GR1-2 | 2714.7 | Flacourt | Late | Brown | 90 | bdl | 200 | 236 | bdl |
| GR1-2 | 2714.7 | Flacourt | Late | Brown | 101 | 186 | 200 | 178 | bdl |
| GR1-2 | 2714.7 | Flacourt | Late | Brown | 124 | 110 | bdl | bdl | 403 |
| GR1-2 | 2714.7 | Flacourt | Late | Brown | bdl | 126 | bdl | bdl | bdl |
| GR1-2 | 2714.7 | Flacourt | Late | Brown | bdl | 198 | bdl | bdl | bdl |
| GR1-2 | 2714.7 | Flacourt | Late | Brown | bdl | 172 | bdl | bdl | bdl |
| GR1-2 | 2714.7 | Flacourt | Late | Brown | bdl | 141 | bdl | 147 | bdl |
| GR1-4 | 2707.2 | Flacourt | Late | Brown | 88 | bdl | bdl | bdl | bdl |
| GR1-4 | 2707.2 | Flacourt | Late | Brown | 168 | bdl | bdl | 126 | 209 |
| GR1-4 | 2707.2 | Flacourt | Late | Brown | bdl | bdl | bdl | 196 | bdl |
| GR1-4 | 2707.2 | Flacourt | Grain | Brown | 130 | bdl | bdl | bdl | bdl |
| GR1-4 | 2707.2 | Flacourt | Grain | Blue | 185 | 73 | 128 | bdl | bdl |
| GR1-4 | 2707.2 | Flacourt | Grain | Red | bdl | 212 | bdl | bdl | 440 |

bdl: below detection limit.

Fe and Cu) were analysed as they are typically enhanced by luminescence (Sprunt, 1981; Matter & Ramseyer, 1985). The substitution of these elements for Si leaves unsatisfied bonds and charge imbalance. This charge imbalance can be re-balanced by the uptake of some charge balancing monovalent cations (e.g. $H^+$, $Li^+$, $Na^+$ and $K^+$). Ramseyer & Mullis (1990) and Perny *et al.* (1992) have attributed the changing luminescent colours in quartz cements to the influence of these cations.

Chemical analyses were carried out away from the kaolinite cement to avoid the contamination of Al from the intergrown kaolinite. The summary of the chemical analyses is shown in Table 2 and Fig. 9. Aluminium is the only element that shows significant and systematic variation between the three quartz cement phases. The early phase (I) has an average of 245 p.p.m. of aluminium. The middle, and most volumetric, phase (II) has the highest aluminium concentration with a mean value of 805 ppm. The late phase (III) contains the lowest average aluminium content with an average of 89 ppm assuming that the detection limit was 40 ppm (or an average of 118 ppm if all analyses that were below detection for aluminium are ignored). The chemistry of the detrital quartz grains does not show any correlation with cathodoluminescence colours.

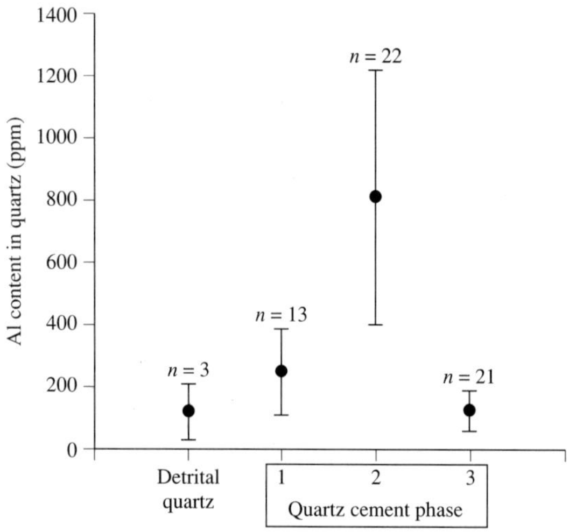


**Fig. 9.** Summary diagram to illustrate the quantities of different trace elements in the different phases of quartz cement, the size of the sample sets and the aluminium content of detrital quartz for reference.

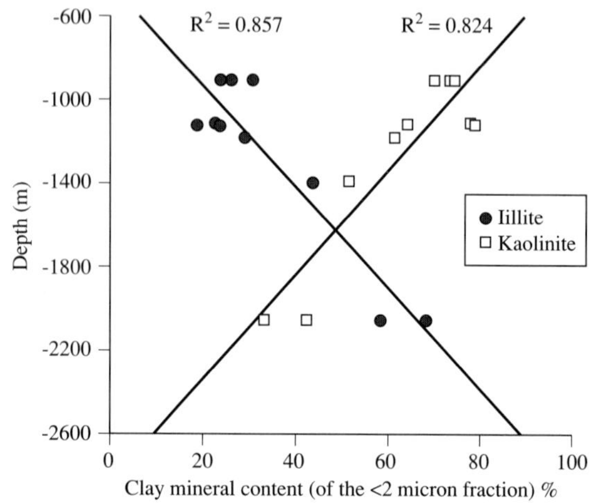


**Fig. 10.** Abundance of XRD-defined kaolinite and illite as a function of depth in Barrow Group sandstones.

**Table 3.** Clay mineral abundances from the Barrow Group sandstones from the less than 2 μm fraction

| Sample | Depth (m) | Ch (%) | I (%) | I–S (%) | K (%) |
|---|---|---|---|---|---|
| ROL-01 | 902.1 | 0 | 25 | 1 | 74 |
| ROL-02 | 903.0 | 0 | 23 | 4 | 73 |
| ROL-04 | 905.7 | 0 | 30 | 0 | 70 |
| ROS-01 | 2057.3 | 0 | 58 | 0 | 42 |
| ROS-02 | 2060.3 | 0 | 68 | 0 | 32 |
| THE-01 | 1176.8 | 11 | 28 | 0 | 61 |
| WPP-02 | 1393.0 | 0 | 43 | 6 | 51 |
| YAM-04 | 1116.8 | 0 | 22 | 0 | 78 |
| YAM-05 | 1120.7 | 0 | 18 | 3 | 79 |
| YAM-07 | 1123.0 | 13 | 23 | 0 | 64 |

Ch: chlorite; I: illite; I–S: illite–smectite mixed layer; and K: kaolinite.

### Clay mineralogy

The XRD patterns of the < 2 μm size fraction in the Barrow Group sandstones show that the clays are dominated mainly by kaolinite and illite (Table 3) with minor illite–smectite of low expandability and very minor chlorite (Kraishan, 1997; Kraishan *et al.*, 1997). Note that kaolinite tends to decrease in significance with depth whilst illite increases in significance (Fig. 10). The illite–smectite is typically only slightly expandable and is dominated by illite (Kraishan, 1997). The relative abundance of authigenic clay minerals is summarized in Table 3 and was determined semiquantitatively using basal plane peak heights compared with standard mixtures.

## DISCUSSION

### Temperature and sequence of quartz cement growth

Fluid inclusion homogenization temperatures must be considered carefully before they are interpreted as indicators of growth conditions. The two key factors are (i) the nature of the fluid phase and (ii) the integrity of inclusion shape and volume.

A controversy has recently been raging about the integrity of fluid inclusions within quartz in rocks undergoing progressive heating (e.g. Haszeldine & Osborne, 1993; Osborne & Haszeldine, 1993; Walderhaug, 1995; Worden *et al.*, 1995). Despite the fact that homogenization temperatures are concentrated close to the present-day formation temperatures, they do not show evidence of resetting (re-equilibration) during burial. The evidence against re-equilibration of quartz cements in the Barrow Group sandstones include:

**1** The discrete range of homogenization temperatures obtained from each sample for each phase of quartz overgrowths (Fig. 8).

**2** The distribution of homogenization temperatures does not show a high temperature 'tail' or a wide spread (Fig. 8).

**3** The inclusions in the Barrow Group are small; these have more resistance to re-equilibration (Bodnar & Bethke, 1984; Worden *et al.*, 1995).

**Plate 1.** Thin section micrographs. (a) Partially dissolved detrital K-feldspar with abundant secondary porosity, Flacourt Formation, 2700.9 m, Griffin-1. (b) Poikilotopic calcite cement surrounding thick feldspar overgrowths. Note that quartz cement is absent, Flacourt Formatio, 904.5 m, Roller-1. (c) Euhedral quartz overgrowth with grain boundary demarked by a prominent dust-rim, Flacourt Formation, 2714.5 m, Grffin-1. (d) CL micrograph of quartz cement phases showing the inner dark brown/non-luminescent phase I, purple to dark blue luminescent phase II and the light brown luminescence of phase III, Flacourt Formation, 2652.4 m, Griffin-1.

(a)

(b)

**Plate 2.** Thin section micrographs: (a) Oil stains coat the second stage of quartz overgrowth as evidenced by fluid inclusions and CL petrography, Flacourt Formation, 2714.7 m, Griffin-1. (b) Oil or hydrocarbon meterials trapped between detrital and quartz overgrowths, Flacourt Formation, 2653.4 m, Griffin-1.

**Plate 3.** Thin section micrograph of an interconnecting network stylolites. Stylolites of this type are only abundant at depth greater than 2.0 km, Flacourt Formation, 2714.4 m, Griffin-1.

The homogenization temperature of pure water trapped in minerals during burial must be corrected for the effect of pressure to reveal the mineral growth temperature. Isochores (contours of equal density) have a steep but positive slope in P–T space and must be used to track up from surface pressure to the estimated pressure of growth. However, the addition of methane to the water system changes the situation. The two phase (gas–water) envelope reaches considerable temperatures and thus minimizes the effect of the pressure correction (Hanor, 1980). There are numerous source rocks in the section (Fig. 2) and there is abundant petroleum in the permeable part of the system. It is thus reasonable to assume that the formation water from which quartz precipitated was saturated with methane. The homogenization temperatures for fluid inclusions in quartz are thus approximately equal to the trapping temperatures.

Phase I quartz cement grew at between about 60 and 90°C. Note also that calcite cement grew before phase I quartz at about 40–50°C (Kraishan *et al.*, 1997), underlining the lower temperature limit for quartz cement phase I. Phase II quartz cement grew at between 90 and 105°C and phase III quartz cement grew at between 105 and 140°C. Kaolinite, intergrown with phase II quartz cement must have grown at between 90 and 105°C. Ferroan carbonates commenced growth during the formation of phase III quartz and thus grew at temperatures greater than about 105°C.

### Sources of quartz cement

Silicate reactions in sandstones (often involving feldspars) typically liberate silica and lead to quartz cement (e.g. Walker, 1960; Hower *et al.*, 1976; Boles & Franks, 1979; Leder & Park, 1986; Morad & Al Dahan, 1987). The volume of silica that was potentially liberated from the alteration of the detrital feldspar to produce quartz and a clay mineral (e.g. kaolinite or illite) can be estimated from the point count data. The silica released from the dissolution of feldspar minerals can be as high as 0.43 $cm^3$ per each $cm^3$ of the dissolved feldspar (Bjørlykke, 1984, 1988; Leder & Park, 1986; Morad & Al Dahan, 1987) depending on the original composition and the type of feldspar minerals. Detrital feldspar comprised 11 and 15% of the framework grains of the Malouet and Flacourt Formations sandstones, respectively, at the time of deposition, which accounts for as much as twice of the present-day content (5 and 9%, respectively). Thus for both the Malouet and Flacourt Formations the feldspar content fell by 6% leading to a possible increase in quartz content by approximately 2.5% (i.e. $0.43 \times 6$). Feldspar decay reactions may thus account for about 2.5% by volume of the quartz cement total. This leaves a small volume of quartz cement to account for since the average amount of quartz cement is approximately 3%. Whilst the simple reaction of K-feldspar to a clay mineral plus quartz cement may account for some of the textures and mineral abundance patterns, it is also possible that there has been reaction between the early diagenetic kaolinite and detrital K-feldspar to produce illite and quartz. This is suggested by the clay mineral abundance patterns with depth (Fig. 10).

Pressure solution can occur at grain contacts during advanced burial of siliciclastic rocks (e.g. Waldschmidt, 1941; Heald, 1955; Dutton & Diggs, 1990). Based on petrographic and core observations, pressure solution and stylolites were found mostly in deeper samples (below 2000 m, Plate 3, facing p. 326); they increase in abundance with increasing depth. The amount of silica released by this process has been estimated by applying the method proposed by Dutton & Diggs (1990). They used the present values of minus-cement porosity to determine the amount of silica that derived from intergranular pressure solution. Minus-cement porosity is the summation of primary intergranular porosity plus all cements that occupy intergranular spaces (Heald, 1956). Well-sorted sandstones have at least 40% porosity at the time of deposition (Pryor, 1973). Porosity can be reduced to 26% (closest packing) in well-rounded, matrix-free, ductile-free sandstones by mechanical compaction (Houseknecht, 1987). Dutton and Diggs assumed that mechanical compaction caused the reduction in the minus-cement porosity to 26%, while greater losses in minus-cement porosity were caused by intergranular pressure solution (chemical compaction). They were thus able to estimate the volume of silica derived by intragranular pressure solution by assuming that all clean sandstones with minus-cement porosity less than 26% have undergone dissolution of quartz grains at point contacts. Dutton & Diggs (1990) concluded that this method overestimates the amount of silica released by intergranular pressure solution as the overlap quartz and porosity relationship is not strictly linear (Houseknecht & Hathon, 1987).

We have adopted the Dutton and Diggs approach to estimate the amount of pressure-solution-derived quartz in the Malouet and Flacourt Formations in the Barrow Sub-basin. Each thin section is 30 μm thick, the exact average amount of quartz cement is 2.8% and there are 96 thin sections. The absolute volume of silica in quartz cement in the rock section studied represents about 8.06 mm of the total thickness of the thin sections ($96 \times 2.8 \times 0.03$). Minus-cement porosity values of less than 26% were observed in 24 thin sections of clean sandstones with an average of 20.6%. For these samples, they have lost an average of 5.4% volume due to chemical compaction. The amount of silica lost by intergranular pressure solution in these samples is thus estimated to be equivalent to 3.88 mm (24 × 5.4 × 0.03). The maximum amount of silica derived from pressure dissolution in these sandstones is thus 48% of the total quartz cement in all samples (3.88/8.06). Therefore, the absolute maximum amount of silica derived by intergranular pressure solution in the Barrow Group sandstones must be less than 48% of the total quartz cement.

### Controls on the Al content in quartz cement

Quartz is a very pure mineral not typically known for its trace element content. However, Al can be found in measurable amounts in quartz cement. The principal factors thought to influence the Al incorporation in quartz crystals include:

**1** growth direction (Brown & Thomas, 1960; Cohen, 1960; Pavlishin *et al.*, 1978);

**2** crystallization conditions (Pavlishin *et al.*, 1978);

**3** solution chemistry (Brown & Thomas, 1960; Pavlishin *et al.*, 1978; Ramseyer & Mullis, 1990; Perny *et al.*, 1992);

**4** temperature and pressure (Dennen *et al.*, 1970; Pavlishin *et al.*, 1978);

**5** alkalinity (Pavlishin *et al.*, 1978; Marino *et al.*, 1989).

Growth rate and temperature seem unlikely to be controls on aluminium in quartz in this case. Phase I to III quartz cement grew at progressively increasing temperature (as shown by the fluid inclusion data) but there is no pattern of correlation between temperature and the aluminium content of quartz. The same argument probably holds for pressure. All other factors being equal, quartz will grow at faster rates as temperature increases (due to increasing rates of the various potential rate-controlling steps; Murphy *et al.*, 1989). Thus quartz I, should have grown at the slowest rate and quartz III should have grown most quickly. Again, there is no evidence of aluminium being controlled by growth rate.

Marino *et al.* (1989) suggested that a silicate growing from an aqueous solution should preferentially incorporate tetrahedrally co-ordinated aluminium ($Al_{tet}$, $[Al(OH)_4]^-$) rather than the octahedral form ($Al_{oct}$, $[Al(H_2O)_6]^{3+}$). The transition from octahedral to tetrahedral aluminium is pH dependent at a given temperature. The octahedral form is dominant at low pH values. The transition takes place between pH 5.5 and 6.5 at 25°C and between pH 4 and 5 at 100°C. Consequently there is likely to be a pH control on the Al content in quartz. At the temperatures at which quartz cement grew (> 60, < 140°C), to incorporate substantial aluminium in quartz, the pH must have been greater than about 5. Perny *et al.* (1992) attribute trace element variations in zones which had grown in the same directions to changes in the pore-water chemistry rather than the effect of temperature or growth rate.

In the Barrow Group sandstones, diagenetic kaolinite is coeval with quartz cement phase II. The formation of both minerals occurred in the middle stage of the diagenetic history and was accompanied by the dissolution of detrital K-feldspars. Thus at the time of growth of the most Al-rich phase of quartz cement (II), there was an abundant source of aluminium in the system. However, the pH of the overall water–rock system is constrained by the need to have $Al_{tet}$ as the stable form of dissolved aluminium whilst being appropriate for converting K-feldspar into kaolinite plus quartz. This reaction is controlled by the potassium ion ($K^+_{aq}$), and aqueous silica concentrations as much as pH but must have been considerably constrained to have the appropriate pH for $Al_{tet}$ to be stable.

From the very low Al content, it may be anticipated that phase III quartz cement grew at lower pH values than the first two phases (less than pH 4 using the Marino *et al.*, 1989 model). However, this is not necessarily the control for this phase of quartz. The final phase of quartz was not sourced from K feldspar, rather it was most likely sourced from stylolites and dissolution seams. These do not necessarily involve the introduction of aluminium into the diagenetic system. Thus, the low aluminium content in phase III quartz may equally be due to supply as well as pH. Indeed present-day formation water has an average pH of 6.0 (after geochemical recalculation to correct for subsurface temperature and gas geochemistry; data from Kraishan, 1997). If this water is similar to the water that mediated the growth of quartz phase III, then it follows that pH is not the control on aluminium in the final stages of quartz diagenesis.

Given the factors that affect the incorporation of Al in quartz, it is most likely that Al incorporation into quartz cements is controlled by (i) the source of the silica and the attendant fluctuations in aluminium concentrations of the pore-water during clay–feldspar diagenesis; (ii) the pH of the formation water (possibly as a secondary, limiting control). The other factors (temperature, pressure, growth rate, etc.) seem unlikely to be able to explain the fact that Al concentrations vary between the three phases of quartz cement.

From this analysis, it follows that it may be possible ultimately to relate the aluminium content of quartz cement in sandstones to the source of quartz cement if the pH can be independently assessed (e.g. by defining the relevant mineral paragenesis).

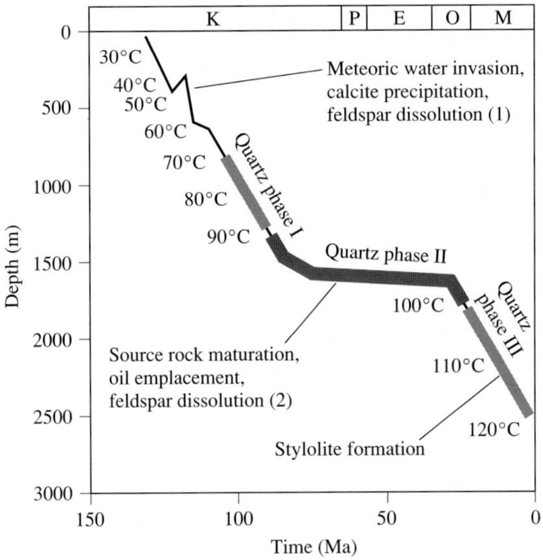


**Fig. 11.** Burial/thermal history diagram for the top of the Barrow Group with the likely times of the quartz cement phases superimposed (from fluid inclusion temperature data).

## Diagenetic synthesis

Hydrocarbons in the Barrow Sub-basin originated mostly from the Upper Jurassic Dingo Claystone (Alexander *et al.*, 1980, 1983; Volkman *et al.*, 1983). Generation occurred at about 130°C when about 2800 m of overburden rocks were required to produce this degree of thermal maturity. It is believed that the early phase of the quartz cement (I) occurred relatively early in the diagenetic history during or after the early episode of feldspar dissolution and during or soon after the early uplift of the sub-basin (Fig. 11). This is supported by the fluid inclusion data for the early phase of quartz cement (I) which is in the range of 60–90°C. The possible source of silica for the first quartz cement phase was the dissolution of the feldspar due to early phases of meteoric water invasion during basin margin uplifting (Kraishan, 1997; Kraishan *et al.*, 1997). Based on isotopic studies reported by Kraishan (1997) poikilotopic calcite precipitated early between 40 and 50°C from meteoric waters. This meteoric water also partially dissolved detrital feldspar.

Quartz cement (II) precipitated from waters which must have been rich in tetrahedral aluminium in a temperature range of 90–105°C (Fig. 11). The homogenization temperature range of this phase is in accordance with the thermal maturation of organic matter and the secondary migration of hydrocarbons from the underlying Upper Jurassic Dingo Claystone (de Boer & Collins, 1988; Kraishan, 1997; Kraishan *et al.*, 1997). $CO_2$ and organic acids are typically released at the earliest stages of maturation; such acids are thought to cause the dissolution of feldspars (Huang & Keller, 1970; Surdam *et al.*, 1984). Feldspar dissolution could transiently enrich local pore waters with aluminium. Considering the association of the second phase of quartz cement (II) and kaolinite, it is reasonable to deduce that the source of both aluminium and silica was the dissolution of detrital feldspar. It is thus possible to conclude that both quartz cement phase (II) and kaolinite formed prior to or soon after the start of the main phase of oil generation and migration into the sandstone (during the Early Tertiary) from fluid with a pH of greater than 4.5–5.5 that was rich in aluminium due to feldspar dissolution (Kraishan, 1997).

The high precipitation temperatures (up to 140°C but averaging 120°C in the case of the well selected for 1-dimensional modelling in Fig. 11) and the low aluminium content of quartz cement (III) suggest a source other than feldspar dissolution. Quartz (III) is most common in lithologies rich in evidence of intergranular pressure dissolution and stylolites. The third phase occurred late in the diagenetic history when the sediments were deeply buried during the Late Miocene through to the present (Fig. 11). The source for the silica in the quartz cement (III) was probably the dissolution of detrital quartz grains during deep burial and stylolitization, common in deeper samples. The low aluminium content is in agreement with this source of silica.

## CONCLUSIONS

**1** Quartz cement, a common authigenic mineral in the Cretaceous marine sandstones of the Barrow Group, Barrow Sub-basin, Australia, precipitated in a temperature range of 60–140°C.

**2** Three phases of quartz cement were identified using cathodoluminescence and microprobe analysis: phase (I), low temperature with a moderate aluminium content; phase (II), moderate temperature with a high aluminium content and phase (III), high temperature with a low aluminium content.

**3** The main sources of silica were feldspar dissolution and pressure dissolution.

**4** The early and middle phases were a consequence of feldspar dissolution during (i) meteoric water invasion; and (ii) early source rock maturation leading to the dissolution of Al-rich feldspar resulting in a high aluminium content within quartz cement.

**5** The late phase was sourced from the pressure dissolution of Al-poor detrital quartz resulting in a low aluminium content in quartz cement derived from that source.

**6** We suggest that measurement of the aluminium content of quartz cement may be a helpful tool in determining the origin of the silica.

## ACKNOWLEDGEMENTS

The authors would like to thank Barrow Project Sponsors; Australian Petroleum Co-operative Research centre (APCRC), Mobil Exploration and Producing Australia, Western Mining Corporation and Boral Energy Resources. The earlier version of the manuscript has benefited from the constructive comments by Sadoon Morad, Attila Juhász and Mogens Ramm and has sharpened the final product. G.M.K gratefully acknowledges the financial support during his PhD study in NCPGG by Overseas Postgraduate Research Scholarship (OPRS), APCRC and Adelaide University Scholarships. Finally, we would like to thank ACS Laboratories Pty Ltd for their financial support during the final production stages of this paper.

## REFERENCES

ALEXANDER, R., KAGI, R. & WOODHOUSE, G. (1980) Origin of the Windalia oil, Barrow Island, Western Australia. *Australian Petroleum Exploration Association Journal* **20**, 250–256.

ALEXANDER, R., KAGI, R.I., WOODHOUSE, G.W. & VOLKMAN, J.K. (1983) The geochemistry of some biodegraded Australian oils. *Australian Petroleum Exploration Association Journal* **23**, 53–63.

BJØRLYKKE, K. (1984) Formation of secondary porosity: How important is it? In: *Clastic Diagenesis.* (eds MCDONALD, D.M. & SURDAM, R.C.) pp. 277–286. American Association of Petroleum Geologists Memoir 37.

BJØRLYKKE, K. (1988) Sandstone diagenesis in relation to preservation, destruction and creation of porosity. In: *Diagenesis, I.* (eds CHILINGARIAN, G.V. & WOLF, K.H.) pp. 531–565. Developments in Sedimentology 41. Elsevier, New York.

BODNAR, R.J. & BETHKE, P.M. (1984) Systematics of stretching in fluid inclusions—I: fluorite and sphalerite at 1 atmosphere confining pressure. *Economic Geology* **79**, 141–161.

DE BOER, R. & COLLINS, L.B. (1988) Petrology and diagenesis of the Flag Sandstone, Harriet field, Barrow Sub-basin. In: *The North West Shelf, Australia.* (eds PURCELL, P.G. & R.R.) pp. 225–235. Proceedings of Petroleum Exploration Society of Australia Symposium, Perth.

BOLES, J.R. & FRANKS, S.G. (1979) Clay diagenesis in Wilcox sandstones of southwest Texas: implications of smectite diagenesis on sandstones cementation. *Journal of Sedimentary Petrology* **49**, 55–70.

BROWN, C.S. & THOMAS, L.A. (1960) The effect of impurities on the growth of synthetic quartz. *Physics and Chemistry of Solids* **13**, 337–343.

COHEN, A.J. (1960) Substitutional and interstitial aluminum impurity in Quartz, structure and color center interrelationships. *Physics and Chemistry of Solids* **13**, 321–325.

DENNEN, W.H., BLACKBURN, W.H. & QUESADA, A. (1970) Aluminium in quartz as a geothermometer. *Contributions in Mineralogy and Petrology* **27**, 332–342.

DUTTON, S.P. & DIGGS, T.N. (1990) History of quartz cementation in the Lower Cretaceous Travis Peak Formation, East Texas. *Journal of Sedimentary Petrology* **60**, 191–202.

FOLK, R.L. (1974). *Petrology of Sedimentary Rocks.* Hemphill, Austin, Texas.

HANOR, J.S. (1980) Dissolved methane in sedimentary brines: potential effect on the PVT properties of fluid inclusions. *Economic Geology* **75**, 603–609.

HASZELDINE, R.S. & OSBORNE, M. (1993) Fluid inclusion temperatures in diagenetic quartz reset by burial: implications for oil field cementation. In: *Diagenesis and Basin Development.* (eds HORBURY, A.D. & ROBINSON, A.G.) pp. 35–46. American Association of Petroleum Geologists Studies in Geology, 36.

HEALD, M.T. (1955) Stylolites in sandstones. *Journal of Geology* **63**, 101–114.

HEALD, M.T. (1956) Cementation of Simpson and St. Peter sandstones in parts Oklahoma, Arkansas, and Missouri. *Journal of Geology* **64**, 16–30.

HOCKING, R.M. (1994) *Subdivisions of Western Australian Neoproterozoic and Phanerozoic Sedimentary Basins.* Western Australian Geological Survey, Record 1994/4.

HOUSEKNECHT, D.W. (1987) Assessing the relative importance of compactional processes and cementation to the reduction of porosity in sandstones. *American Association of Petroleum Geologists Bulletin* **71**, 633–246.

HOUSEKNECHT, D.W. & HATHON, L.A. (1987) Petrographic constraints on methods of intergranular pressure solution in quartzose sandstones. *Applied Geochemistry* **2**, 507–521.

HOWER, J., ESLINGER, E.V., HOWER, M.E. & PERRY, E.A. (1976) Mechanism of burial metamorphism of argillaceous sediment: I. Mineralogical and chemical evidence. *Geological Society of the America Bulletin* **87**, 725–737.

HUANG, W.H. & KELLER, W.D. (1970) Dissolution of rock-forming silicate minerals in organic acids: simulated first stage weathering of fresh mineral surfaces. *American Mineralogist* **55**, 2076–2094.

KAIKO, A. & TINGATE, P. (1996) Suppressed vitrinite reflectance and its effect on thermal history modeling in the Barrow and Dampier Sub-basins. *Australian Petroleum Production Exploration Association Journal* **36**, 428–444.

KOPSEN, E. (1994) Northern Carnarvon Basin hydrocarbon distribution and future petroleum potential. In: *The Sedimentary Basins of Western Australia.* (ed. PURCELL, P.G. & R.R.) pp. 127–139. Proceedings of the West Australia Symposium, Perth, Western Australia.

KOPSEN, E. & MCGANN, G. (1985) A review of the Hydrocarbon Habitat of the Eastern and central Barrow–Dampier Sub-basin, Western Australia. *Australian Petroleum Exploration Association Journal* **25**, 154–176.

KRAISHAN, G.M. (1997) *Controls on Sandstone Reservoir Quality: Permian to Early Cretaceous Sequences, Barrow Sub-basin North West Shelf, Western Australia.* Unpublished PhD Thesis, NCPGG, The University of Adelaide.

KRAISHAN, G.M., LEMON, N.M. & TINGATE, P.R. (1997) Depositional and diagenetic controls on the quality of Barrow

Group sandstone reservoirs, Barrow Sub-basin, North West Shelf, Western Australia. *Australian Petroleum Production Exploration Association Journal* **37**, 214–231.

LEDER, F. & PARK, W. (1986) Porosity reduction in sandstone by quartz overgrowth. *American Association Petroleum Geologists Bulletin* **70**, 1713–1728.

MARINO, E., HARVEY, C. & MURRAY, H.A. (1989) Aqueous-chemical control of the tetrahedral-aluminum content of quartz, halloysite, and other low-temperature silicates. *Clay and Clay Minerals* **37**, 135–142.

MATTER, A. & RAMSEYER, K. (1985) Cathodoluminescence microscopy as a tool for provenance studies of sandstones. In: *Provenance of Arenites.* (ed. ZUFFA, G.G.) pp. 191–211. North Atlantic Treaty Organization (NATO), Advanced Study Institute Series 148. V.C.D. Reidel, Dordrecht.

MCBRIDE, E.F. (1989) Quartz cement in sandstones: a review. *Earth Science Reviews* **26**, 69–112.

MCGILVERY, T. (1996) *Stratigraphic Framework, Facies Architecture, and Depositional History of the Barrow Group (Early Cretaceous) in the Barrow Sub-basin, North West Shelf, Australia.* Unpublished PhD Thesis, The University of Texas at Austin.

MORAD, S. & AL DAHAN, A.A. (1987) A SEM study of diagenetic kaolinitization and illitization of detrital feldspar in sandstones. *Clay Minerals* **22**, 237–243.

MURPHY, W.M., OELKERS, E.H. & LICHTNER, P.C. (1989) Surface reaction versus diffusion control of mineral dissolution and growth rates in geochemical processes. *Chemical Geology* **78**, 357–380.

OSBORNE, M. & HASZELDINE, S. (1993) Evidence for resetting of fluid inclusion temperatures from quartz cements in oilfields. *Marine and Petroleum Geology* **10**, 271–278.

PAVLISHIN, V.I., MAZYKIN, V.V., MATYASH, I.V. & VOZNYAK, D.K. (1978) Variations in the proportion of substitutional aluminium during growth of a quartz crystal. *Geochemistry International* **15**, 158–165.

PERNY, B., EBERHARDT, P., RAMSEYER, K., MULLIS, J. & PANKRATH, R. (1992) Micro distribution of Al, Li, and Na in a quartz: possible causes and correlation with short-lived cathodoluminescence. *American Mineralogist* **77**, 534–544.

PETTIJOHN, F.J., POTTER, P.E. & SIEVER, R. (1987). *Sand and Sandstone*. Springer-Verlag, New York.

PRYOR, W.A. (1973) Permeability–porosity patterns and variations in some Holocene sand bodies. *American Association of Petroleum Geologists Bulletin* **57**, 162–189.

RAMSEYER, K. & MULLIS, J. (1990) Factors influencing short-lived blue cathodoluminescence of $\alpha$-quartz. *American Mineralogist* **75**, 791–800.

REYNOLDS, T.J. (1978) *Fluid Inclusions Adapted U.S.G.S. Gas Heating/Freezing Instruction Manual.* Fluid Incorporated, Denver, Colorado.

ROEDDER, E. (1979) Fluid inclusion evidence of the environments of sedimentary diagenesis, a review. In: *Aspects of Diagenesis* (eds SCHOLE, P.A. & SCHLUGER, P.R.), Society of Economic Palaeontologists and Mineralogists Special Publication **26**, 89–107.

SANDERSON, I.D. (1984) Recognition and significance of inherited quartz overgrowths in quartz arenites. *Journal of Sedimentary Petrology* **54**, 473–486.

SPRUNT, E.S. (1981) Causes of quartz luminescence colours. *Scanning Electron Microscopy* **1**, 525–535.

SURDAM, R.C., BOLES, S.W. & CROSSEY, L.J. (1984) The chemistry of secondary porosity. In: *Clastic Diagenesis.* (eds MCDONALD, D.A. & SURDAM, R.C.) pp. 127–150. *American Association of Petroleum Geologists Memoir 37.*

TAIT, A.M. (1985) A depositional model for the Dupuy Member and the Barrow Group in the Barrow Sub-basin, Western Australia. *Australian Petroleum Exploration Association Journal* **25**, 282–290.

VOLKMAN, J.K., ALEXANDER, R., KAGI, R.I., NOBLE, R.A. & WOODHOUSE, G.W. (1983) A geochemical reconstruction of oil generation in the Barrow Sub-basin of Western Australia. *Geochimica et Cosmochimica Acta* **47**, 2091–2105.

WALDERHAUG, O. (1990) A fluid inclusion study of quartz-cemented sandstones from offshore Mid-Norway—possible evidence for continued quartz cementation during oil emplacement. *Journal of Sedimentary Petrology* **60**, 203–210.

WALDERHAUG, O. (1995) Discussion of 'Evidence for resetting fluid inclusion temperatures from quartz cements in oilfields' by Osborne and Haszeldine (1993). *Marine and Petroleum Geology* **12**, 559–561.

WALDSCHMIDT, W.A. (1941) Cementation materials in sandstones and their influence on the migration of oil. *American Association of Petroleum Geologists Bulletin* **25**, 1839–1879.

WALKER, T.R. (1960) Carbonate replacement of detrital crystalline silicate minerals as a source of authigenic silica in sedimentary rocks. *Geological Society of America Bulletin* **71**, 145–152.

WILLIAMS, A.F. & POYNTON, D.J. (1985) The geology and evolution of the South Pepper hydrocarbon accumulations. *Australian Petroleum Exploration Association Journal* **25**, 235–247.

WORDEN, R.H., WARREN, E.A., SMALLEY, P.C., PRIMMER, T.J. & OXTOBY, N.H. (1995) Discussion of 'Evidence for resetting fluid inclusion temperatures from quartz cements in oilfields' by Osborne and Haszeldine (1993). *Marine and Petroleum Geology* **12**, 566–570.

ZUFFA, G.G. (1985) Optical analysis of arenites: Influence of methodology on compositional results. In: *Provenance of Arenites.* (eds ZUFFA, G.G.) pp. 165–189. North Atlantic Treaty Organization (NATO), Advanced Study Institute Series 148. V.C.D. Reidel, Dordrecht.

*Spec. Publs int. Ass. Sediment.* (2000) **29**, 1–20

# Quartz cementation in oil field sandstones: a review of the key controversies

R. H. WORDEN[1] *and* S. MORAD[2]

[1]*School of Geosciences, The Queen's University, Belfast, BT7 1NN, UK; and*
[2]*Department of Earth Sciences, Uppsala University, S-752 36 Uppsala, Sweden*

## ABSTRACT

Despite quartz cement being the most important pore-occluding mineral in deeply buried (> 2500 m) sandstones, its origin and the controls on its distribution are still subject to disagreement and debate. Depending on the primary mineralogy and conditions of pressure and temperature, internal $SiO_2$ sources include pressure dissolution of quartz grains, feldspar-alteration reactions, clay-mineral transformations and dissolution of amorphous silica. However, sources external to a given sandbody are still proposed, despite the lack of direct evidence and problems of the low solubility of silica and the vast quantities of water required to accomplish quartz cementation. Redistribution of silica between facies in sandstone sequences and a systematic bias of sampling only reservoir facies may be partially responsible for apparent advective import.

Quartz cementation is strongly affected by temperature, quartz only being a major cement in rocks that have been heated to above about 80°C. However, quartz cementation is likely to be a kinetically controlled process such that small amounts may develop slowly even at low temperatures. It is not yet proven whether quartz cement forms continuously at slow rates or rapidly over short periods. Pressure (effective stress) is potentially an important control on quartz cementation, since quartz cement tends to be less abundant in overpressured sandstones than in normally pressured sandstones. Primary lithofacies exerts a strong control on quartz cementation because quartz needs clean substrates to form: quartz cementation is most likely to be inhibited by the presence of grain coating clay, infiltrated clays and microquartz.

The emplacement of petroleum in a sandstone probably inhibits quartz cementation although the degree of inhibition would depend on the remaining water saturation, wettability and fluid pressure in the sandstone. Import of silica through advection will be entirely halted in oil fields because the two-phase relative permeability of a reservoir to water becomes extremely low at high oil saturations. Direct empirical evidence of the effects of petroleum on quartz cementation are as yet lacking due to incomplete data collection.

Fundamental rate data for many of the component processes involved in quartz cementation are known to different degrees. Although no predictive models yet incorporate all the sources and controls discussed in this paper, it should be possible, at some time in the future, to bring all the controls into a unified model. Realistic quantitative modelling of quartz cementation based upon all the potential sources of silica may ultimately be possible for a sandstone unit if silica sources, thermal history, transport mechanism, pore pressure history, local variations in lithology and petroleum-filling history for the reservoir are known or predictable.

## INTRODUCTION: WHY WORRY ABOUT QUARTZ CEMENT?

Economic evaluation of a petroleum accumulation demands a knowledge of the distribution of porosity and permeability in sedimentary rocks. Porosity and permeability are key properties of a petroleum reservoir because they have profound effects upon the reserves (volume of petroleum) and production rate. Reservoir quality in a sandstone buried to any degree is controlled by three factors: (i) the depositional porosity and permeability which are strongly influenced by sorting, grain size, grain morphology and sand/mud matrix ratio (ii) the degree of mechanical and chemical compaction; and (iii) the amount and type of pore-filling cement. The creation of new, additional secondary porosity due to grain and cement dissolution is probably of minor significance in many, though not all reservoirs. Permeability, in turn, is controlled by porosity as well as the nature and connectedness of the pore network. Compaction during burial to depths commensurate with common quartz cementation (2000–3000 m) typically decreases the depositional porosity of a quartzose sandstone (35–45%) by up to 20% leaving post-compactional porosities of 15–25% at these depths. It is often the presence, absence or precise quantity of cement that dictates whether a sandstone-hosted petroleum reservoir will prove economic.

Quartz is probably the most volumetrically important mineral in sandstones for the simple reason that quartz grains act both as substrate for quartz precipitation and as a source of silica. Quartz cement is most common in sandstones buried to depths that are commensurate with heating to ~ 79 70–80°C. The ability to predict the distribution pattern of quartz cement, be it deterministically or stochastically, or the reservoir scale, will bring rich rewards in terms of better pay analysis, prospect evaluation, and reservoir management and also possibly faster petroleum recovery.

The most important and controversial questions needing to be addressed considering quartz cementation in sandstones are:

**1** Is silica in quartz cement derived internally to the cemented sandstone or externally?

**2** Which of supply, transport or precipitation is the dominant process controlling the rate of quartz cementation?

**3** Which transport process (diffusion or advection) is dominant?

**4** Is quartz cementation a long, drawn-out affair or does it happen in short bursts?

**5** What physico-chemical conditions lead to quartz cementation?

**6** Is quartz cementation stopped by the emplacement of oil in sandstones?

In this paper we review some of these key controversies and address the terms of the debate about the origin of quartz cement in sandstones. We have not tackled silcretes as these are controlled by entirely different processes than burial cements. We are not yet at the solution to the problems of quartz cement prediction, but this paper, and the volume as a whole, should help to refocus the attention of the end-users of technology, researchers, and data collectors, on the critical issues.

## NATURE OF QUARTZ CEMENT IN SANDSTONES

Quartz cement grows via both homogeneous and heterogeneous nucleation mechanisms. Heterogeneous nucleation is the most common process evidenced by the domination of syntaxial quartz overgrowths relative to crystallization of pore-filling, discrete quartz crystals. Syntaxial overgrowths typically have the same crystallographic orientation as the detrital quartz grain substrates. Some quartz cement occurs as an equal thickness overgrowth. However, some rocks have quartz outgrowth rather than overgrowths (McBride, 1989). Outgrowths are pronounced localized projections into the pore rather than rims around the majority of the detrital grain. Outgrowths are syntaxial, too, but they tend to occupy the adjacent pore and thus cause more damage to permeability than overgrowths. Quartz cement also grows in fractures in mechanically broken detrital quartz grains (Fig. 1a). Luminescence and zonation of quartz cement have been attributed to the presence of up to a few thousand p.p.m. $Al_2O_3$ and a few hundred p.p.m. transition metal oxides (Kraishan *et al.*, this volume, pp. 317–331). Cathodoluminescence (CL) imaging has also revealed that zones of 'disordered quartz' are often present in quartz cements. These zones have no clear or systematic CL zonation but do have localized variations in CL intensity. This is usually found closest to the detrital grain suggesting that multiple nucleation sites may have occurred followed by competition for space during subsequent growth and dominance of one particular nuclei as the overgrowth developed (Hendry & Trewin, 1995; Hogg *et al.*, 1995).

Microcrystalline quartz has been reported from several oil fields. This often occurs as polynuclear coatings on the surfaces of detrital quartz grains but can also occur as discrete crystals or aggregations of crystals in the pore. The crystal size of quartz has been assigned to the rate of quartz growth, the degree of silica oversaturation and the lack of a suitable substrate for the homogeneous nucleation of macrocrystalline quartz overgrowths. The occurrence of microcrystalline quartz has also been associated with depressed rates of macrocrystalline quartz growth due to reduced chemical potential gradients from sites of pressure dissolution to the pore (Aase *et al.*, 1996; Jahren & Ramm, this volume, pp. 271–279).

## GEOCHEMICAL CONTROLS ON QUARTZ CEMENTATION

The occurrence of quartz cement in sandstones depends on two key controls. For quartz cementation to occur at all, there must be a thermodynamic driving force, i.e. an overall reduction in the free energy of the system. Such a driving force may be due to chemical equilibria, such as smectite breakdown to illite and quartz (Abercrombie *et al.*, 1994) or to textural instability such as Ostwald ripening during which high surface area fine grains dissolve and reprecipitate on the surface of coarser grains (Steefel & Van Cappellen, 1990). However, for a thermodynamically favourable process to occur, the rate of that process must also be favourable (Fig. 2). The overall rate

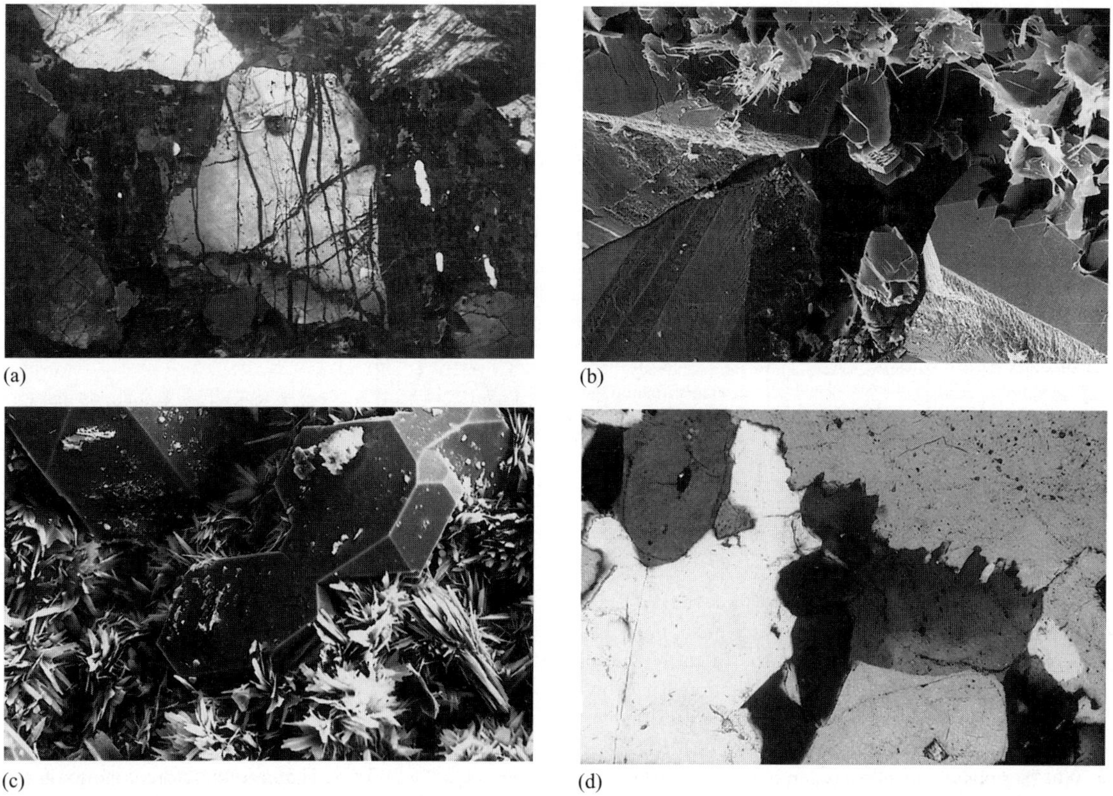


**Fig. 1.** Photomicrographs of textures typically associated with quartz cementation. (a) BSE/CL image of a quartz grain that contains microfractures healed by diagenetic quartz. Scale: 1.0 cm = 90 μm. (b) SEM micrograph of a sandstone that is pervasively cemented by quartz and in which kaolinite cement is replaced by illite and dickite. Scale: 1.0 cm = 10 μm. (c) SEM micrograph of a sandstone showing illite grain coating partially enveloped by quartz overgrowth. Scale: 1.0 cm = 10 μm. (d) Optical micrograph (x-nichols) showing a sandstone that displays pressure dissolution of quartz grains along a stylolitic surface. Note the presence of abundant quartz overgrowths around quartz grains. Scale: 1.3 cm = 90 μm.

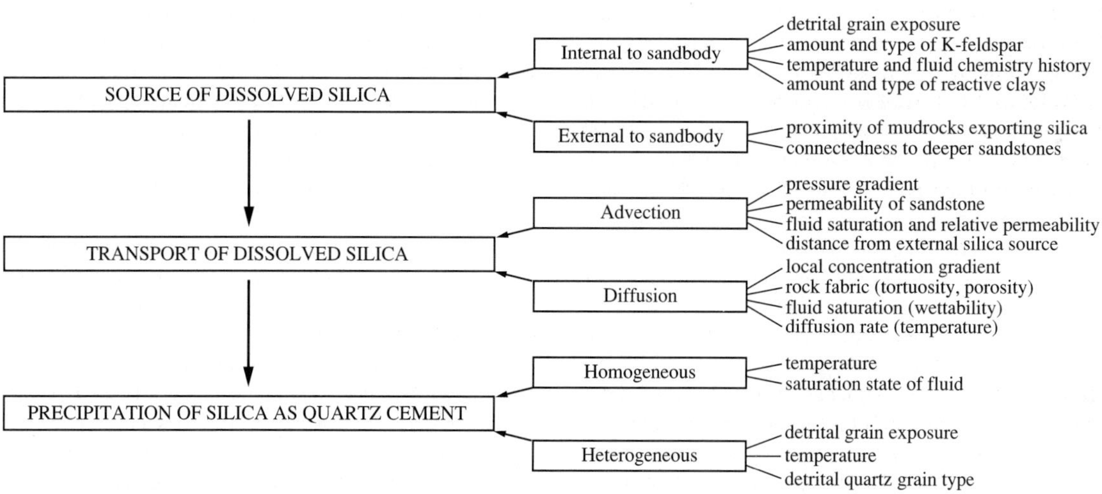


**Fig. 2.** Schematic diagram for the geochemical controls on quartz cementation. The three fundamental controls are the rates of supply to aqueous solution, transport and precipitation from aqueous solution. These have been subdivided into the key secondary controls and the main influences on the rate of these key secondary controls are listed on the right of the diagram.

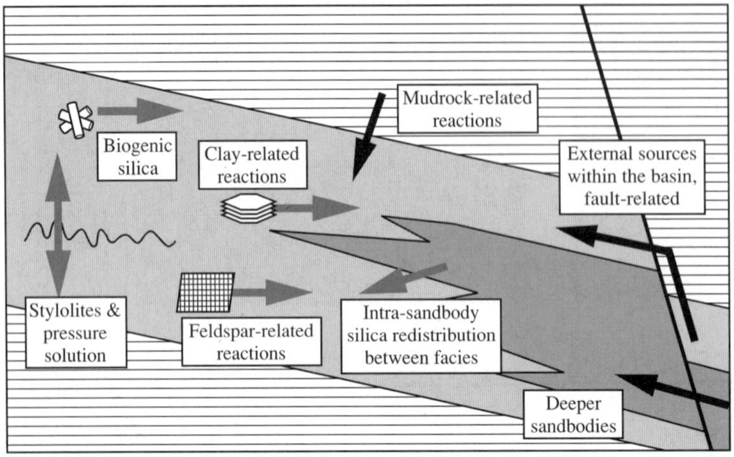


**Fig. 3.** Schematic diagram for the general sources of quartz cementation. External sources are indicated by a black arrow; internal sources by a grey arrow.

of quartz cementation is related to three linked steps which must occur in series: silica source or supply, silica transportation from the site of the source to the site of cementation, and quartz cement precipitation. The slowest step will be rate-limiting, dictating the overall rate of quartz cementation (Lasaga & Kirkpatrick, 1981). Even if it is thermodynamically favourable, significant quartz cementation will not occur if the rate of one of the component processes is too slow to occur over the available geological time.

## ORIGINS OF QUARTZ CEMENT IN SANDSTONES

There are several potential sources of silica for quartz cementation (Fig. 3). None should be automatically discounted although some are likely to be volumetrically more significant than others. In this section and other sections of this paper, we use the terms internal and external. By these we are referring to sources within the sandbody in question (internal) or sources from outside the sandbody in question (external). Strictly, these terms are scale-independent but external sources typically require greater transport distances than internal sources. The latter probably involve material movement on the scale of 1–10 m or much less whilst the former could also occur on that scale (i.e. from mudrocks encasing sandbodies) or could occur on a much larger scale (many tens of metres to kilometres). The scale of diffusion depends on the concentration gradient and the exact rate-controlling step (Oelkers *et al*., 1992). The scale of advection depends on the size of the pressure 'cell', the pressure gradient and the dominance of transport rate over surface controlled processes.

### Internal sources of quartz cement

#### *Detrital feldspar-related sources*

Detrital feldspars in sandstones are mainly derived from igneous and metamorphic rocks formed at elevated temperatures ($\gg 300°C$). They are therefore commonly out of equilibrium with diagenetic pressure, temperature and pore-fluid chemistry, and are thus subject to a variety of diagenetic alteration processes in order to reach a new equilibrium state. Feldspar minerals are a common component of many sandstones when they are deposited. However, their abundance often decreases with increasing depth of burial below a critical depth. K-feldspar, especially, decreases in abundance in a fairly regular pattern in many basins such as the U.S. Gulf Coast (e.g. Hower *et al*., 1976) and Northern North Sea (Giles *et al*., 1992; Glassman, 1992). Albite, on the other hand, is a common, diagenetically late mineral suggesting that the geochemical system is undergoing redistribution of material rather than whole-sale loss or conversion. Most important among the processes of feldspar alteration are dissolution, replacement by clay minerals and albitization. Dissolution, which may create secondary mouldic porosity (albeit typically with local precipitation of material) and liberate ions (Ca, K and Na) to the formation waters, may ultimately result in the formation of diagenetic quartz arenite (e.g. Harris, 1989).

The clay minerals that replace feldspars are typically controlled by the chemistry of formation waters and temperature. For instance, the percolation of meteoric waters in shallow-buried sandstones can cause alteration of feldspars to kaolinite. The main reason that feldspars are sources of silica is that they have higher Si/Al ratios than the clay minerals that replace them. Alkali feldspars

(K–Na series) have an Si/Al ratio of 3 whereas illite and kaolinite typically have Si/Al ratios of about 1. All reactions that involve the growth of clay minerals (such as kaolinite and illite) at the expense of feldspar lead to silica-release and thus the potential for quartz cementation.

The stoichiometric replacement of K-feldspar by kaolinite or illite is sensitive to the pH and the activity of aqueous potassium:

$$\underset{\text{K-feldspar}}{2KAlSi_3O_8} + 2H^+ + H_2O = \underset{\text{kaolinite}}{Al_2Si_2O_5(OH)_4} + \underset{\text{quartz}}{4SiO_2} + 2K^+ \quad (1a)$$

$$\underset{\text{K-feldspar}}{3KAlSi_3O_8} + 2H^+ = \underset{\text{Illite}}{KAl_3Si_3O_{10}(OH)_2} + \underset{\text{quartz}}{6SiO_2} + 2K^+ \quad (1b)$$

In many sandstones, the formation of quartz and kaolinite or illite is accompanied by the coprecipitation of carbonate cements such as calcite. The precipitation of calcite is initially encouraged by reactions 1 because the water loaded with calcium and bicarbonate has a low pH. Subsequently, the precipitation of each mole of calcite produces one proton that can in turn be used in further dissolution of feldspar, as follows:

$$Ca^{2+} + HCO_3^- = CaCO_3 + H^+ \quad (2)$$

Thus silicate diagenesis is intimately linked to the carbonate diagenetic system through their mutual, yet opposite, sensitivity to pH.

Reaction 1b is typically effective at the lower end of the range of temperatures associated with quartz cementation (see later). At elevated temperatures (~ 125°C), remaining K-feldspars and diagenetic kaolinite often react pervasively to form illite and quartz, as follows (Fig. 3a):

$$\underset{\text{K-feldspar}}{KAlSi_3O_8} + \underset{\text{kaolinite}}{Al_2Si_2O_5(OH)_4} = \underset{\text{illite}}{KAl_3Si_3O_{10}(OH)_2} + \underset{\text{quartz}}{2SiO_2} + H_2O \quad (3)$$

This overall reaction, which also results in the precipitation of quartz cement, is controlled largely by temperature; the reaction is neutral for both pH and formation water cation geochemistry. The reaction results from the movement of the K-feldspar–illite–kaolinite stability fields in temperature–water-chemistry space (Fig. 4). A consequence of this reaction is the common association of diagenetic illite and quartz (Fig. 1b).

Formation waters from many oil fields from the North Sea and offshore Norway fall within the stability fields of illite and K-feldspar and not kaolinite, meaning that kaolinite is thermodynamically unstable even at temperatures as low as 60°C. However, its common presence in reservoirs with temperatures < 90–100°C indicates that illitization (reaction 2 above) is kinetically controlled. Kaolinite in K-feldspar poor sandstones is replaced by the $Al_2Si_2O_5(OH)_4$ polymorph: dickite. Dickite is thermodynamically less unstable than kaolinite and will thus be more resistant to illitization (Fig. 1c; and see Morad *et al.*, 1994).

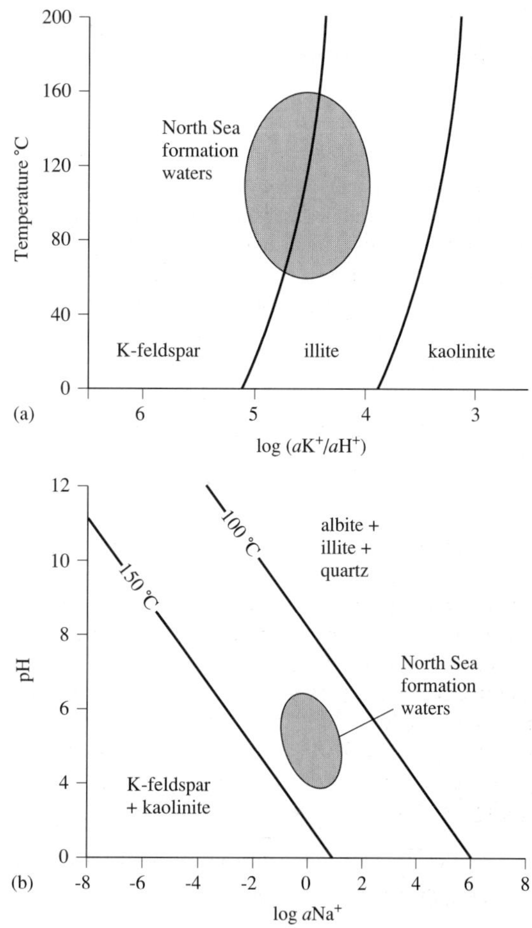


**Fig. 4.** Activity diagram indicating the effects of fluid geochemistry and temperature on the reaction of K-feldspar with water and hydrogen ions leading to the growth of clay minerals and quartz (modified from Morad *et al.* (1990) and Egeberg & Aagaard (1989)).

The commonly reported decrease in the amounts of detrital K-feldspar with increasing depth does not always mean a simple decrease in the total volume of framework feldspar. This situation may be achieved by pseudomorphic albitization (i.e. the shape of detrital feldspar is preserved; Morad, 1986) rather than dissolution of detrital K-feldspars. This can be written as follows (Fig. 4b):

$$2KAlSi_3O_8 + 2.5Al_2Si_2O_5(OH)_4 + Na^+ = NaAlSi_3O_8 \quad \text{(K-feldspar, kaolinite, albite)}$$

$$+ 2KAl_3Si_3O_{10}(OH)_2 + 2SiO_2 + 2.5H_2O + H^+ \quad \text{(illite, quartz)} \qquad (4)$$

Feldspar albitization is a common form of burial diagenetic alteration in oil field sandstones, such as those in the North Sea (e.g. Morad *et al.*, 1990). The onset of feldspar albitization may be at temperatures as low as ≈ 65°C, but wholesale albitization occurs at temperatures of about ⩾ 100–130°C. Reaction 4 is strongly controlled by temperature and formation water chemistry ($a\mathrm{Na}^+/a\mathrm{H}^+$) with the assemblage albite–illite–quartz becoming more stable at elevated temperatures. Albitization will thus be sensitive to changes in the bulk salinity and pH of the formation water (Fig. 4b).

Reaction of one unit volume of K-feldspar by reaction 1 (including reactions 1a and 1b) produces 0.43 unit volumes of quartz. Reaction 3 is a less efficient way of producing quartz cement: each unit volume of reactants produces about 0.10 unit moles of quartz. Albitization (reaction 4) results in the least quartz cement (for the volume of the reactants) of the three listed silicate reactions. For the albitization reaction as written, the molar volumes of reactants and products, for each unit volume of reactants, only 0.05 unit volumes of quartz cement are produced. Albitization of plagioclase, which is accomplished prior to K-feldspars, may either consume silica (Boles, 1982) or contributes little silica for quartz cementation (Morad *et al.*, 1990).

### *Illitization and chloritization of smectite*

Smectite in sandstones occurs as infiltrated clays, cement and mud intraclasts, and is also common in mudstones. Smectite varies greatly in chemical composition, but has a relatively high Si/Al ratio and thus is a potential source of silica for quartz cementation (Boles & Franks, 1979). Smectite in mudrocks is subject to kinetically controlled illitization, related to rate of $K^+$ flux and thermal conditions. The illite/smectite ratio is often used as a thermal maturity indicator. Thus smectite inside sandbodies and in neighbouring mudstones may be a source of quartz cement. A model reaction for the conversion of one particular type of smectite to illite was described by Abercrombie *et al.* (1994):

$$KAlSi_3O_8 + 2K_{0.3}Al_{1.9}Si_4O_{10}(OH)_2 \quad \text{(K-feldspar, K-smectite)}$$

$$= 2K_{0.8}Al_{2.4}Si_{3.5}O_{10}(OH)_2 + 4SiO_2 \quad \text{(illite, quartz)} \qquad (5)$$

This is analogous to reaction 3. This reaction could occur within sandstones or within adjacent mudrocks. Silica from mudrock versions of reaction 5 may diffuse into adjacent sandstones and precipitate as quartz overgrowths (Gluyas *et al.*, this volume, pp. 199–217).

Chloritization of smectite, which occurs at similar temperature range as illitization, may cause the release of silica to pore waters. Chloritization requires an elevated $a\mathrm{Fe}^{2+}/(a\mathrm{H}^+)^2$ ratio and low $a\mathrm{K}^+/a\mathrm{H}^+$ ratio in the pore waters, and thus occurs in sediments enriched in detrital Fe-silicates, Fe-oxides and Fe–Ti oxides, whereas illitization occurs in sediments enriched in K-feldspars. The original composition of smectite may also have a substantial influence on the transformation pathway. Mg–Fe-rich smectites tend to be chloritized (Chang *et al.*, 1986), whereas K-, Ca- and Na-rich smectites tend to be illitized.

### *Pressure dissolution and stylolites within sandstones*

It has widely been recognized that some silica in quartz cement may be sourced by pressure dissolution: quartz grain dissolution with local reprecipitation. Net movement occurs due to diffusion under a local concentration gradient. The anomalously elevated silica activity at the site of dissolution leads to diffusion toward zones of lower silica concentration (i.e. the sites of silica precipitation). Evidence for this comes from interpenetrating quartz grains and stylolite seams in deeply buried sandstones (Fig. 1d). The main controversy surrounding pressure dissolution is whether it is initiated by increasing lithostatic pressure, increasing temperature, or the presence of clay minerals along grain interfaces (or a combination of these).

Pressure-induced changes in silica solubility were widely thought to be the key control on chemical compaction (Rutter, 1976). Diagenetic systems were initially thought to operate under the same series of rules and mechanisms that lead to the precipitation of quartz in regions of relatively low stress in metamorphic rocks. These processes in such high-temperature rocks lead to the formation of quartz concentrations in pressure shadows (e.g. near to equant 'pressure-favouring' minerals like garnet) and at low stress regions of crenulated rocks (e.g. in the interstices between layers on a fold hinge). However, the critical role of temperature and the negligible role of pressure have recently been emphasized for silica dissolution in diagenetic rocks (Bjørkum, 1996). A consensus is beginning to form around the idea that temperature-controlled (i) silica solubility and (ii) rates of dissolution, transport and precipitation may be the key influences (Bjørkum, 1996; Oelkers *et al.*, 1996). The pressure regime between grains will be important, though, at least in controlling the thickness of the aqueous diffusive medium on impinging grain surfaces (Renard *et al.*, this volume, pp. 67–77) and possibly in forming a subsidiary control on relative silica solubility between the grain contact and the pore.

Illite coatings around detrital quartz grains are thought to enhance quartz pressure dissolution (Fisher *et al.*, this volume, pp. 129–146). This is likely to be especially true for illite that occurs as infiltrated clay forming thin (< 10 μm) platelets lying parallel to the surfaces of quartz grains. Infiltrated clays are most efficiently emplaced under semiarid climatic conditions in depositional environments that are characterized by high amounts of suspended mud, fluctuating water level and negligible sediment reworking (point bar, alluvial channels, crevasse splay and delta plains). In contrast to infiltrated clays, deep-burial diagenetic illite (typically with a hair-like habit that lines the adjacent pore space) does not seem to enhance pressure dissolution of quartz. Similar to infiltrated, grain-coating illite, detrital muscovite and biotite grains enhance the pressure dissolution of quartz at areas of their intergranular contacts. It is unclear whether or not clays other than illite have a catalytic effect on intergranular pressure dissolution. However, it has recently been argued (Oelkers *et al.*, 1996) that pressure dissolution of quartz grains occurs exclusively at mica and illitic clay interfaces with quartz due to poorly defined, temperature-dependent, chemically catalysing processes and that 'pressure-solution' is in fact a misnomer (temperature, and not pressure, being seen as the key). Indeed, dissolution of quartz grains that are totally embedded in micaceous matrices is common during low-grade metamorphism (Passchier, 1996). The key conclusion from this is that pressure dissolution and the processes causing stylolite formation may be the same.

The scale of dissolved silica migration from the zones of dissolution to sites of precipitation may in some cases be difficult to constrain. The amount of quartz cement will be greatest close to the stylolitic surfaces if the process is transport-controlled thus permitting a local assessment of sources. However, Oelkers *et al.* (1996) found that the distribution pattern of quartz cement about stylolitic surfaces depends on the burial temperatures and on the interstylolite distances. In cores taken from 3.6 km depth, the distribution of quartz cement was independent of distance from stylolite. Sandstones buried at depths of about 5 km showed greater concentration of quartz cement close to the stylolitic surfaces only if these surfaces were not closely spaced ($\geqslant \approx 50$ cm).

Sandstones showing evidence of both (i) intergranular pressure dissolution and/or stylolite formation; and (ii) precipitation of quartz overgrowths on thin-section scale are fairly common, but not ubiquitous. According to Oelkers *et al.* (1996), the distance between stylolite surfaces decreases continuously as pressure dissolution leads to chemical compaction of the entire section.

A high degree of silica saturation, a condition that may inhibit pressure dissolution of quartz, is obtained when: (i) dissolved silica is prevented from precipitating due, for example, to the presence of thick infiltrated clay, authigenic chlorite, or microcrystalline quartz coatings around the framework grains; and (ii) much silica is supplied locally from processes such as the dissolution of feldspar and rock fragments and the illitization of kaolinite. Conversely, low silica concentrations, and hence low quartz saturation, are achieved when the precipitation of quartz occurs around clean quartz-grain surfaces, such as in aeolian sandstones.

Fine- to medium-grained sandstones are usually more susceptible to pressure dissolution than coarse-grained sandstones. This is because of relative thermodynamic instability of mineral surfaces relative to mineral interiors. Ostwald ripening leads to the preferential dissolution of small grains and reprecipitation of the dissolved silica as a contribution to larger grains. Large-scale pressure dissolution of silica due to stylolite formation is most common in siltstones and fine-grained sandstones that are rich in clay-mineral matrix and detrital mica.

The enrichment of sandstones in ductile rock fragments (mudstones, mud intraclasts and schist or other mica/clay-rich rock fragments) retards the pressure dissolution of quartz grains due to the formation of abundant pseudomatrix. Such sandstones undergo a rapid loss of porosity due to mechanical, rather than chemical, compaction and are relatively poor in late cements, including quartz overgrowths (Worden *et al.*, 1997). Quartz cementation is inhibited in pseudomatrix-rich sandstones because pressure dissolution is minimal and because the detrital quartz grain surfaces are covered with clay. The same effect is also true of grain-coating clay minerals that have achieved almost total detrital grain coverage (e.g. Thompson, 1971).

Pressure dissolution/stylolite formation is probably controlled by temperature and the presence of mica and illite minerals on quartz–quartz interfaces. However, the precise role of pressure is as yet uncertain.

### *Dissolution of amorphous silica*

Amorphous silica in sediments occurs mainly as biogenic and volcanic grains. Siliceous, opaline fossils are especially common in Upper Jurassic and Lower Cretaceous sandstones but have existed since the Cambrian through to the present. The spines (or sponge spicules) of *Rhaxella perforata* are enriched in the sand fraction of some Upper Jurassic and Lower Cretaceous sediments and are a relatively reactive form of silica because they are composed of opaline silica. Volcanic glass is also composed of amorphous silica. These silica forms (opal, glass, etc.) are thermodynamically unstable forms of hydrous silica that are significantly more soluble than quartz. The quartz cement that results from the dissolution and reprecipitation of this type of detrital silica has a number of forms including chalcedonic pore fillings, microcrystalline grain coatings and mesocrystalline quartz overgrowths (Vagle *et al.*, 1994; Hendry & Trewin, 1995). Quartz cement resulting from the dissolution of unstable silica polymorphs (primary opal but even secondary chalcedony) and reprecipitation of the more stable form (quartz) may occur at relatively lower temperatures than quartz sourced from feldspars or pressure dissolution. Vagle *et al.* (1994) found up to 40% microcystalline quartz cement by volume in rocks that had been buried to depths resulting in temperatures of no more than 60°C. This will be a major source of silica in rocks that were deposited with a significant amount of siliceous debris, and are thus mainly limited to certain shallow marine sediments.

### External sources of quartz cement

It is pertinent to examine some possible external sources of silica and to contemplate possible transport processes and geological pathways for fluid movement. Silica is only sparingly soluble in water at diagenetic temperatures and to advectively transport significant volumes of silica in aqueous dissolution would require enormous fluxes of water (Bjørlykke, 1994). There have been many cases where an external source of quartz cement has been proposed. External sources include rocks adjacent to sandbodies (i.e. mudrocks encasing sandbodies) but also includes sources, and thus transport, on a much larger scale (many tens of metres to kilometres including other more deeply buried sandstones, deeply buried mudrocks, granite intrusions, basement rocks, etc.).

Despite geochemical arguments about limited silica solubility, Gluyas & Coleman (1992) demonstrated that silica seems to have been added to porous sandstones relative to the non-porous and impermeable early carbonate cemented nodules, suggesting import of silica. The scale of silica transportation was not discussed.

Improbably high fluxes of water ($10^8$ $cm^3$ water passing through each $cm^2$ of rock) would be required to cause advection-related quartz cementation (Bjørlykke, 1994). Although there is ample evidence of fluid flow in the subsurface, the limited potential for the enormous fluid volumes required to accomplish quartz cementation in sandstones in deep-burial regimes, indicates that local sources of silica must be favoured over large scale advection-controlled quartz cementation. Indeed, little or no quartz cementation occurs along or adjacent to the fault planes that have been proposed (e.g. Burley, 1993) to be the most likely route for the movement of large volumes of aqueous silica in the subsurface (Fisher *et al.*, this volume, pp. 129–146).

The solution to this paradox may be the scale of the system under examination. In addition to the local presence of both pressure dissolution and quartz precipitation sites on a thin-section scale, differences in behaviour may occur on a mesoscopic scale between sandstone beds within the same sequence. Some beds may display predominant evidence of pressure dissolution (Si-exporters; Houseknecht, 1984) whilst others may have evidence for quartz cementation (Si-importers). Whether a sandstone becomes an exporter or importer of silica probably depends on the local degree of silica saturation. Trewin & Fallick (this volume, pp. 219–229) have demonstrated that silica has been transported from one depositional facies to another within the same formation. Had only one part of the system, one lithofacies, been examined, the conclusion would have been bulk export or bulk import. However, an examination of a whole system revealed that the issue is one of redistribution on the sand accumulation-scale rather than a mass movement scenario.

Localized diffusion- or advection-controlled silica transportation into sandstones from adjacent mudrocks has been suggested by the occurrence of local enrichment of quartz cement in portions of sandstones that are in contact with mudstones (Sullivan & McBride, 1991). The particular source of silica from the mudrock could have been any of the feldspar or clay breakdown reactions or stylolite-pressure dissolution (discussed earlier).

One possible way for making an externally sourced, advective silica-supply more geologically reasonable is that our understanding of the controls on silica solubility may be faulty. Several studies seem to have demonstrated that silica achieves higher concentrations when it is complexed by organic species (e.g. Bennet & Siegel, 1987). Thus if silica concentrations in fluids could be increased by several orders of magnitude, the volume of advecting fluid could presumably be reduced by several orders of magnitude. However, recent work on realistic formation water compositions and a careful review of all the experimental and empirical evidence suggests that silica solubility is in no way enhanced in the presence of oil and that silica is not complexed by natural petroleum fluids or organic species in natural formation waters (Fein, this volume, pp. 119–127). Thus enhanced silica concentrations in fluids is unlikely to be achieved by any form of organo-silica complexing.

## TIMING AND DURATION OF QUARTZ CEMENTATION

Quartz cementation has been viewed as both (i) a prolonged process that occurs slowly over a very long period of time; and (ii) as an episodic process in which short bursts of activity are punctuated by long periods of quiescence. Both scenarios may occur depending upon the details of the lithology and burial-, temperature- and fluid-evolution histories. Both models have been used to describe quartz cementation and advocates of the two models are still in profound disagreement.

### Episodic rapid versus continuous slow scenario

Continuous quartz cementation is when the growth of quartz occurs at all temperatures greater than a critical (kinetically defined) minimum temperature. Episodic quartz cementation is when quartz grows for short periods above the minimum temperature. The exact duration of each period is dependent on the burial history and age of the basin. For Mesozoic basins, continuous cementation may occur over many tens of millions of years whereas episodic cementation might occur over a few million years.

Ironically, fluid-inclusion data have been used to support both the slow-continuous and rapid-episodic models. Walderhaug (1990) used fluid inclusion temperatures to characterize the minimum temperature of quartz cementation and concluded that it commences at about 70–80°C. These data were further developed to reveal the rate of quartz cementation. This was achieved by subtracting the minimum fluid inclusion temperature from the present day temperature and converting the temperature interval into a time interval (by reference to the thermal history, e.g. Walderhaug, 1994). The time interval was subsequently used to convert the known amount of quartz cement into a precipitation rate per unit surface area per unit time using sample-specific textural information. This approach implicitly employed the assumption that quartz cementation is a continuous process from the minimum temperature of inclusion trapping through to the present. The approach also used the assumptions that (i) the quartz cement was apparently sourced only from stylolites within the reservoir; and (ii) that rate of quartz cementation was controlled by the precipitation rate of quartz. According to this methodology, at 80°C quartz cementation occurs at a rate of about

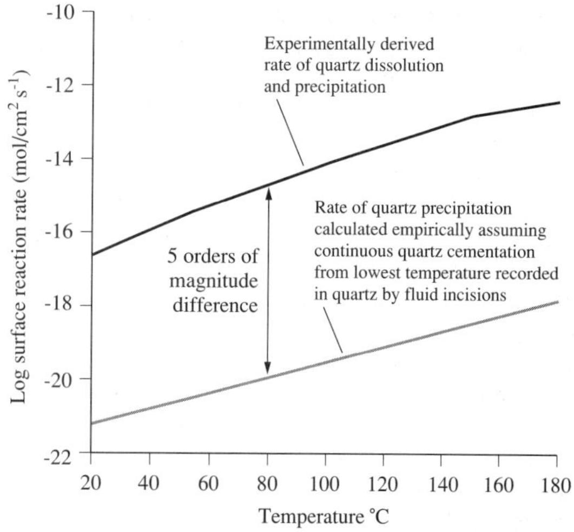


**Fig. 5.** Comparison of the overall rates of quartz cementation suggested by Walderhaug (1994) with experimental data from Crerar *et al.* (1988). The experimental rates are many orders of magnitude faster than the derived empirical rates. Note that the empirical rates were calculated assuming *continuous* quartz cementation—i.e. assuming that the growth of quartz occurred over a very long period of time.

$1 \times 10^{-20}$ moles/cm$^2$ s$^{-1}$ whilst at 140°C quartz cementation occurs at a rate of about $5 \times 10^{-19}$ moles/cm$^2$ s$^{-1}$ (Fig. 5). It must be appreciated that these rates are only realistic if quartz cementation was a continuous process. Also, the rates of quartz cementation revealed by this approach cannot be used if (i) quartz has a source other than internal stylolites; (ii) the rate of precipitation is not the rate controlling step; or (iii) quartz cementation is an episodic rather than a continuous process.

In contrast to this approach, Grant and Oxtoby (1992) and Gluyas *et al.* (1993) examined fluid-inclusion temperatures from sandstone reservoirs and observed discrete fluid inclusion homogenization temperature populations. They made the assumption that the time interval over which quartz cementation occurred was fully represented in the population and that discrete temperature populations could be converted to discrete time intervals for cementation by reference to a thermal history plot. This approach, which was not followed through to produce kinetic parameters for quartz cementation, would lead to much faster rates (by several orders of magnitude) of quartz cementation than that revealed by Walderhaug (1994). Robinson and Gluyas (1992) concluded that quartz cementation occurred during periods of rapid subsidence when the fluid dynamics were disturbed leading to periods of intense fluid circulation within pressure compartments in the basin.

Rates of some of the component process involved in quartz cementation have been experimentally investigated (e.g. Crerar *et al.*, 1988; Brady & Walther, 1990). The rates of quartz dissolution and precipitation revealed by this approach are many orders of magnitude faster than the rate interpreted by Walderhaug (1994). The overall rate of the component processes of quartz cementation sourced from stylolites has been modelled as lying between $10^{-16}$ moles/cm$^2$ s$^{-1}$ at about 80°C and at a rate of about $10^{-14}$ moles/cm$^2$ s$^{-1}$ at 140°C. However, the integrated rates must depend hugely on the geometry of the system as well as other complications, such as the effect of grain coating clays and microquartz, the source and distance of silica transportation and the timing of oil emplacement and the degree of silica supersaturation in the formation water.

The rates modelled by Walderhaug *et al.* (this volume, pp. 39–49) and Murphy *et al.* (1989) are for silica sourced at stylolites only. If quartz cement was supplied by diagenetic processes, such as the dissolution of K-feldspar reaction, smectite to illite reactions and K-feldspar plus kaolinite reactions, then the rate of quartz cementation may be controlled by the rates of those supply processes if they are slower than the rates of silica diffusion and quartz precipitation. The conclusion is that different models for quartz cementation are needed for different geological circumstances. There is no single model that can be used to predict quartz cementation in all sandstones.

The examination of textures or fabrics within minerals has often been used to assess the mechanism of mineral growth. Cathodoluminescence (CL) patterns are probably the best (and certainly the easiest) way to examine quartz. The CL zonation that occurs in quartz overgrowths in sandstone sequences from various basins is often attributed to variations in trace contents of aluminium which replaces silica atoms, but may also be related to local charge imbalances (or vacancies within the mineral—typically associated with impurities; Kraishan *et al.*, this volume, pp. 317–331). The common occurrence of CL zonation in quartz cement suggests that (i) quartz cementation is an episodic, non-steady state process; and (ii) changes in the chemical composition of pore waters may have occurred during growth. However, it is not immediately clear what causes changes in formation-water composition during quartz cementation. Variations in the concentrations of dissolved aluminium content may be related to changes in the extent of dissolution of feldspar or other silicate (Kraishan *et al.*, this volume, pp. 317–331). The key conclusion to this is that textural (CL) evidence seems to suggest that quartz cementation is *not* a steady process.

## FACTORS INFLUENCING QUARTZ CEMENTATION

### Temperature

Temperature may influence quartz cementation in two main ways. It can affect the stability of diagenetic assemblages causing geochemical processes such as reactions 1, 3, 4 and 5 above and thus releasing silica. Additionally, the rates of silica dissolution, diffusion and precipitation are all strongly controlled by temperature. Temperature thus affects both the equilibrium thermo-dynamics and the kinetics of geochemical processes that cause quartz cementation.

Based on maximum burial depths of sandstones and fluid-inclusion microthermometry, it is apparent that the precipitation of quartz overgrowths in most sedimentary basins occurs during deep-burial diagenesis (> 2.5 km), at elevated temperatures, typically 90–130°C (Fig. 5; and Giles *et al.*, 1992; Gluyas *et al.*, 1993). The temperature range of quartz cementation deduced from fluid-inclusion thermometry depends on the subsidence history of the basin, and hence the residence time of the sequence within a certain temperature interval. In slowly subsiding, cratonic basins, considerable quartz cementation may occur during long residence times (tens of millions of years) at relatively low temperatures (< 100°C) (Morad *et al.*, 1994). Conversely, cementation at higher temperatures may take place over a shorter time span.

Small amounts of quartz overgrowths may precipitate during shallow-burial which may be in sufficient quantities to support the sandstone framework and inhibit compactional loss of intergranular porosity. These pores may later be occluded by further precipitation of quartz overgrowths or other cements, such as carbonates. Initiation of precipitation of quartz overgrowths at temperatures as low as ≈ 40–60°C has been reported in sandstones (Pagel, 1975; Haszeldine *et al.*, 1984; Burley *et al.*, 1989). Within a single sandstone sequence, quartz cementation may then continue over a temperature range of 40°C or more (Pagel, 1975; Haszeldine *et al.*, 1984; Burley *et al.*, 1989; Glasmann *et al.*, 1989; Grant & Oxtoby, 1992; Nedkvitne *et al.*, 1993), and hence over depth ranges greater than 1.5 km. However, in other sequences, cementation may occur during rapid burial over a narrower temperature range of 25°C or less (Ehrenberg, 1990; Walderhaug, 1990).

The increase in the amounts of quartz overgrowths with depth in actively subsiding basins of the US Gulf Coast and North Sea, and the occurrence of homogenization temperatures that are close to present-day *in situ* temperatures, has been interpreted to indicate ongoing quartz cementation (cf. Walderhaug, 1994). There has been considerable debate about the significance of fluid inclusion requilibration in quartz cement. Exponents of this argument cite the concordance of maximum fluid inclusion temperatures equivalent to present day temperatures as a key line of evidence (e.g. Osborne & Haszeldine, 1993). There are counter-arguments to this (e.g. Walderhaug, 1995; Worden *et al.*, 1995) and the debate remains open. The considerable range of burial depths and temperatures over which quartz cementation occurs, coincides with the onset of several important temperature-dependent, silica releasing diagenetic processes, such as intergranular pressure dissolution of detrital quartz, stylolite formation in siltstones and sandstones, dissolution, albitization and alteration into clay minerals of detrital feldspars, and illitization of kaolinite and smectite. Accordingly, quartz cementation should be studied within the context of the overall burial-diagenetic evolution of the host sandstones and not simply as a function of the temperature history.

### Pressure

Pressure at any depth in sedimentary basins can be divided between the pressure of the fluid phase (hydrostatic pressure) and the pressure exerted by the rock columns (lithostatic pressure). Because the fluid phase supports some of the weight of the rock, the effective stress at grain contacts is equivalent to the difference between the hydrostatic and lithostatic pressures and is known as the effective stress. The effective stress increases with increasing depth as the (normally pressured) hydrostatic and lithostatic pressure gradients diverge with increasing depth. If the fluid pressure is elevated above the hydrostatic pressure gradient, the fluid is said to be overpressured. The main consequence of overpressuring is that the effective stress at grain contacts decreases. Pressure was formerly thought to be all-important in controlling 'pressure solution' (Tada & Siever, 1989) although the effective stress was the actual control being considered. However, recent work has cast doubt on this, alternatively suggesting that temperature in combination with the presence of illitic clay coatings are critical to the localized dissolution of quartz at grain contacts (e.g. Bjørkum, 1996). Pressure was thought to be critical in increasing the local solubility of silica thus permitting preferential dissolution at grain contacts, diffusion down concentration gradients and reprecipitation on open pore surfaces.

Chemical compaction of North Sea sandstones has been effectively modelled to incorporate the pressure variations on the relative local solubility of silica at grain–grain contacts and in adjacent pore-spaces (Renard *et al.*, this volume, pp. 67–78). The effective-stress inclusive model seems to fit real data suggesting that this factor may be important in quartz cementation. Since fluid pressure (effective stress) and temperature tend to increase together during burial it is difficult to separate the two controls from empirical data sets. However, anomalously high fluid pressures (overpressure), leading to reduced effective stresses between grains should, in principle, inhibit pressure dissolution—if pressure is the key control. Reservoirs at 5000 m and more with anomalously high fluid pressures are proving to be valid exploration prospects in the North Sea. These reservoirs are commonly at > 150°C and should have experienced rapid pressure dissolution and quartz cementation if temperature was the main control on this process. However, these deep, overpressured reservoirs commonly have 20–30% porosity and much less quartz cement than expected for their depth and temperature. This suggests that pressure (or more accurately, effective stress) is at least a significant factor in the overall process of pressure dissolution (Ramm & Bjørlykke, 1994). It is probably significant that overpressured sandstones (that consequently have a relatively low effective stress between mineral grains) tend to have less quartz cement than non-overpressured sandstones (Swarbrick, 1994; Osborne & Swarbrick, 1999). It seems that the fluid pressure of the system, in conjunction with the lithostatic pressure, has a significant influence upon patterns of quartz cementation.

### Fluid composition

The formation of quartz depends upon many processes and both the origin and rate of quartz dissolution/precipitation can be affected by water chemistry. Feldspar-alteration reactions (e.g. reactions 1, 3, 4 and 5, above) are strongly influenced by water chemistry. The main variables that influence feldspar stability are the pH and the activities of potassium and sodium cations. A supply of hydrogen ions (e.g. from reaction 2 above) can lead to feldspar-decay reactions by moving the formation water $aK^+/aH^+$ ratio into the illite or kaolinite stability fields (and see Fig. 3). Note that reaction 2 requires a steady supply of $CO_2$ and of Ca (or Ca and Mg if dolomite precipitation occurred) for this reaction to lead to volumetrically substantial feldspar replacement by quartz (and other minerals).

The $aK^+/aH^+$ ratio can also be reduced by altering the chemical activity of aqueous potassium. This could occur if the concentration of potassium decreased (e.g. through an influx of fresh water). Conversely, an influx of high ionic strength waters (e.g. after localized halite dissolution) could decrease the activity of potassium because activity is the product of concentration and activity coefficient. The activity coefficient is a complex function,

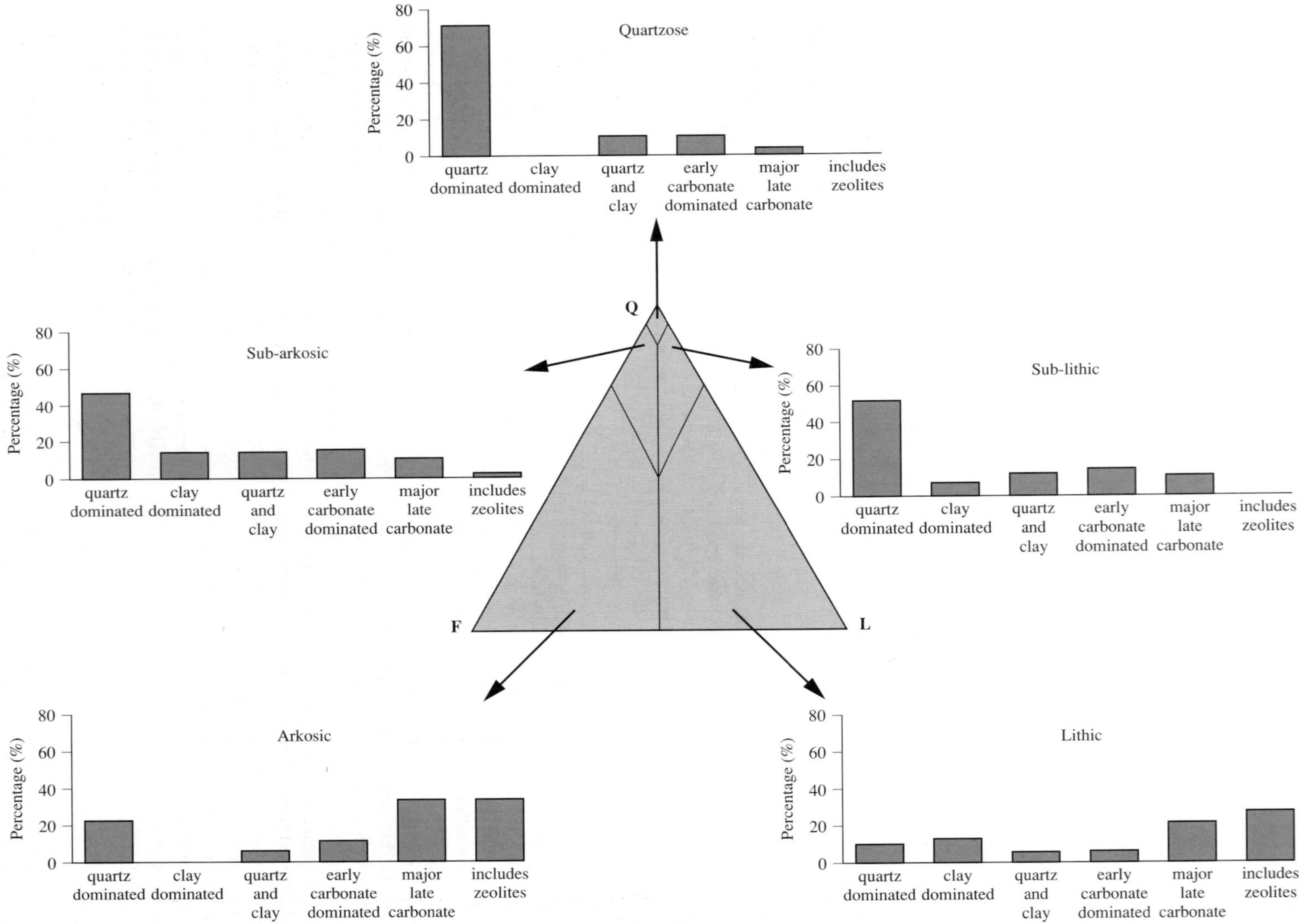


**Fig. 6.** Diagram of the variation of the relative dominance of quartz cement in different sandstone lithofacies (adapted from Primmer *et al.*, 1997). The primary rock type seems to have a very important control on the relative amount of quartz cement.

# Index

Please note: page numbers in *italic* refer to figures, and those in **bold** to tables.